Applied Numerical Analysis

FIFTH EDITION

Curtis F. Gerald
Patrick O. Wheatley

California Polytechnic State University

ADDISON-WESLEY PUBLISHING COMPANY
Reading, Massachusetts • Menlo Park, California • New York
Don Mills, Ontario • Wokingham, England • Amsterdam • Bonn
Sydney • Singapore • Tokyo • Madrid • San Juan • Milan • Paris

Sponsoring Editor: Laurie Rosatone
Production Supervisor: Karen Wernholm
Marketing Manager: Andrew Fisher
Text Designer: Mark Ong
Cover Designer: Leslie Haimes
Manufacturing Manager: Roy Logan

Library of Congress Cataloging-in-Publication Data

Gerald, Curtis, F., 1915–
    Applied numerical analysis / Curtis F. Gerald, Patrick O. Wheatley.
— 5th ed.
      p. cm.
    Includes bibliographical references and index.
    ISBN 0-201-56553-6
    1. Numerical analysis. I. Wheatley, Patrick O. II. Title.
QA297.G47 1994   93-1996
519.4—dc20   CIP

Cover: 3D wireframe model by Mark Button using Autocad and LANDCADD.

DERIVE is a registered trademark of Soft Warehouse, Inc. *Mathematica* is a registered trademark of Wolfram Research, Inc. *Mathematica* is not associated with Mathematica, Inc., Mathematica Policy Research, Inc., or MathTech, Inc. Maple is a registered trademark of Waterloo Maple Software.

3 4 5 6 7 8 9 10-DOC-9695

Preface

This fifth edition of *Applied Numerical Analysis* retains many of the features of the fourth edition. At the same time, there are significant changes; the details of these changes are presented later.

Applied Numerical Analysis is written for sophomores and juniors in engineering, science, mathematics, and computer science. It is also valuable as a sourcebook for practicing engineers and scientists who need to use numerical procedures to solve problems. We have been gratified to find that many who purchased this book as students keep it as a permanent part of their reference library because its readability and breadth allow them to expand their knowledge of the subject by self-study.

While it is assumed that the reader has a good knowledge of calculus, appropriate topics are reviewed in the context of their use, and an appendix gives a summary of the most important items that are used to develop and analyze numerical procedures. The mathematical notation is purposely kept simple for clarity.

Purpose

The purpose of this fifth edition is the same as in previous editions: to give a broad coverage of the field of numerical analysis, emphasizing its practical applications rather than theory. At the same time, methods are compared, errors are analyzed, and relationships to the fundamental mathematical basis for the procedures are presented so that a true understanding of the subject is attained. Clarity of exposition, development through examples, and logical arrangement of topics aid the student to become more and more adept at applying the methods.

Content Features

Applied Numerical Analysis has enjoyed significant success because of several outstanding features. These are retained and amplified in this fifth edition:

- The unusually large number of exercises allows the instructor to select those that are appropriate for the background and interests of the students. The exercises are keyed to the corresponding section of the chapter to assist in this. When the reader is using the book for self-study, the many exercises are an important supplement to the text. In addition to the practice exercises, each chapter (except 0) has "Applied Problems and Projects," which are challenging and illustrate many fields of application of the various numerical procedures.

- By solving the same problem with several different methods, the relative efficiency and effectiveness of the methods become clear.

- A short summary of its contents is given at the beginning of each chapter to provide a preview of what is coming.

- Most chapters are introduced with an easily understood example that applies the material of the chapter. This motivates the student and shows that the material is of real utility.

- Each chapter ends with a summary that reminds the student of what has been covered and suggests that appropriate review will ensure that nothing has been overlooked or not learned.

- The coverage of partial-differential equations in an easily understood manner is unusual in a book at this level.

- Postponing the treatment of computer arithmetic and errors until after the student has been exposed to a numerical method puts this important topic into proper context and helps in the appreciation of its significance as one factor in the accuracy of the computed result.

- Computer programs in several languages are given at the conclusion of most chapters. They implement the more important algorithms and serve as easily understood examples of how the computer can be used to carry out the computations. They do not pretend to be at a professional level of programming, because their purpose is illustrative and clarity would otherwise be sacrificed.

Features of This Edition

There are several important changes in this edition of the book. Some of these changes were suggested by our reviewers, for which we thank them. The changes are not only in content but in organization.

A new chapter, Chapter 0, has been added. (It seemed appropriate to begin the numbering at zero for a text so closely related to computers.) This chapter contains a rewrite of material that was formerly in Chapter 1 of earlier editions and shows how computers and numerical analysis are related. It introduces two items that are new: (1) A description of computer algebra systems (also known as symbolic algebra programs), which offer an alternative to writing a program to solve a problem, and (2) a discussion of parallel processing and how this can be adapted to numerical procedures.

Organizational changes. One organizational change that should help the reader to better relate practice and theory has been to move much of the theoretical discussions to a separate section, near the end of each chapter. This change will also permit the instructor to minimize theoretical discussions if the emphasis of the course is on using the methods. At the same time, those items of theory that are essential background for a method are presented along with its development. This change occurs in each chapter.

We moved the discussion of least squares from Chapter 10 to Chapter 3 because most instructors want to cover this topic early in the course.

In Chapter 4, Romberg integration is included in the section describing the trapezoidal rule. Adaptive integration is covered earlier.

Content changes. There are significant changes in content. As already mentioned, each chapter discusses the ways that parallel processing can speed up the various numerical methods. These discussions are not in depth because full discussion would require a separate book. Another addition to nearly every chapter is a section on how computer algebra systems, specifically DERIVE, MAPLE, and *Mathematica,* can assist with the topics of the chapter.

Important changes have been made to the programs. First, we reduced the number of programs, to provide more opportunities for the student to write his or her own program rather than to copy from the book. Further, instead of using only FORTRAN, we have included programs in several other languages—BASIC, Pascal, C, Ada—to emphasize the wide choice of languages for programming numerical procedures. We have also included two examples of how a spreadsheet program can carry out numerical algorithms.

Other significant changes or additions include the following:

- Material on actual computer number systems in Chapter 0.

- New motivational examples in Chapters 1, 2, and 5.

- A new section on the difficulties with multiple roots, in Chapter 1. The material on Graeffe's method is expanded and a discussion of Laguerre's method is added.

- In Chapter 2, an introduction to the characteristic polynomial and the eigenvalues of a matrix.

- A description of Neville's method, an addition to the section on Lagrangian polynomials in Chapter 3; and a better comparison of divided differences and ordinary differences. Interpolating polynomials, other than the Newton–Gregory forward polynomial, are given only brief mention and inverse interpolation is omitted.

- In Chapter 4, a rewrite of the sections on derivatives from difference tables and adaptive integration. The sections on Lozenge diagrams and improper integrals were deleted.

- In Chapter 5, a new section on multivalued methods.

- In Chapter 6, the addition of collocation and Galerkin's method to the

Rayleigh–Ritz section and a new section that covers finite elements in some detail. The discussion of the *QR* method is revised.

- A rewrite of the section on finite elements in Chapter 7, and a new section, the Theta Method, in Chapter 8.

- A new section—Fourier Series—in Chapter 10 that gives necessary background for the fast Fourier transform. This is also related more closely to getting coefficients numerically.

- The updating of several of the appendices.

- A rewrite of many of the exercises.

Pedagogical Features

We recognize that the student is the most important part of the teaching/learning system and have tried to facilitate his or her understanding by including several pedagogical tools:

- Sections that preview the material lead off each chapter.

- A chapter summary at the end of each chapter reminds the student of what should have been learned.

- The second color makes it easier to recognize the more important equations and algorithms. Color in the illustrations adds interest and clarifies their message.

Solved examples are used extensively to illustrate the various methods presented. Most of the computations were done in single precision. If the student duplicates these, there may be minor differences in the values, due to the way that the calculations are rounded off or if double precision is used.

Supplements

A number of supplements are available to assist the instructor:

- The *Instructor's Solutions Manual* gives the answers to nearly all the exercises and to some of the applied problems and projects. Hints for other problems or projects are also provided. The *Solutions Manual* gives suggestions for selecting from the text when the instructor does not have time to cover it entirely. Since our coverage of topics in numerical analysis is unusually broad, such selection is frequently necessary. In addition, suggestions on how to incorporate computer algebra systems in a numerical analysis course are offered.

- Instructors who adopt this book may obtain electronic copies of the programs in ASCII form from the publisher. This disk is in IBM PC-compatible form on 3.5-inch disks.

- These same programs are also available on Internet via electronic mail. Interested professors may select some or all of these programs. Detailed instructions on accessing these programs via FTP are located on the following page.

Acknowledgments

We especially want to thank the multitude of our own students whose feedback has helped us to improve over previous editions. Our wives have been supportive during this revision and have helped us with proofreading. Many instructors have given valuable suggestions and constructive criticism. In particular we thank our colleagues Ed Garner and Cinda Heeren. The latter helped significantly with the material on parallel processing.

In addition, we mention those whose thorough reviews have helped make this edition better:

Charles Ashley, Villanova University

Thomas R. Chase, University of Minnesota

Ralph E. Ekstrom, University of Nebraska

Steven L. Hodge, Hampton University

Brenton LeMesurier, University of Arizona

Richard H. Pletcher, Iowa State University

Clyde Scandrett, Naval Postgraduate School

Charles Waters, Mankato State University

Peter J. Welcher, U.S. Naval Academy

Abdou Youssef, George Washington University

Hong Zhang-Sun, Clemson University

We also want to express our thanks to those at Addison-Wesley who have worked extensively with us to ensure the publication of another quality edition. Our former editor, Michael Payne, our current editor, Laurie Rosatone, and our production supervisor, Karen Wernholm, have guided and encouraged us during this revision.

C.F.G.
P.O.W.

San Luis Obispo, California

FTP Access to Programs in ASCII

The sample programs presented in this book are available on line in ASCII form for your personal, noncommercial use. The program files are available on the Internet on host aw.com via anonymous FTP. To access the files from an Internet host, enter the following commands:

% ftp aw.com
Name: anonymous
Password: <your e-mail address>
cd aw.mathematics
cd gerald5
ls (to display files to select programs)

A 3.5 inch program disk is also available for professors who adopt this text. To order the disk, call Addison-Wesley at 617-944-3700, extension 2280.

Contents

4   Numerical Differentiation and Numerical Integration   308

9   Hyperbolic Partial-Differential Equations   663

10   Approximation of Functions   696

Appendixes

0

Numerical Computing and Computers

Contents of This Chapter

We begin each chapter of this book with a preview of its contents. This chapter introduces you to numerical analysis by discussing several important concepts about the subject.

0.1 Introduction
Describes how numerical analysis differs from analytical analysis and shows where each has special advantages. It gives a brief listing of topics that will be covered in later chapters.

0.2 Using a Computer to Do Numerical Analysis
Explains why computers and numerical analysis are intimately related and tells about three methods for employing a computer to perform the steps of a numerical analysis.

0.3 A Typical Example
Tells how a typical problem is solved, introducing you to a widely used method for solving a nonlinear equation that is usually called bisection but sometimes is known as interval halving. A symbolic algebra program is used to draw a graph.

0.4 Implementing Bisection
Develops a computer program to carry out the bisection method for the example problem of Section 0.3.

0.5 Computer Arithmetic and Errors
Examines the important topic of accuracy of computation and the different types of errors that may occur. Those errors, due to the way that computers store numbers and do arithmetic, are discussed in some detail.

0.6   Theoretical Matters

Points out the theoretical aspects of the numerical methods of the chapter. In the present chapter, only one method, interval halving, is included, but later chapters will cover several related methods.

0.7   Parallel Processing

Discusses how doing numerical analysis can be speeded up by using a computer system that has many separate processing units operating in parallel on the same problem, and when one might need this potential increase in speed. Some of the special problems that are involved in parallel processing are mentioned.

Chapter Summary

Gives you a checklist against which you can measure your understanding of the topics of this chapter. Obviously, you should restudy those sections where your comprehension is not complete.

0.1 Introduction

Numerical analysis is a way to do higher mathematics problems on a computer, a technique widely used by scientists and engineers to solve their problems. A major advantage for numerical analysis is that a numerical answer can be obtained even when a problem has no "analytical" solution. For example, the following integral, which gives the length of one arch of the curve of $y = \sin(x)$, has no closed form solution:

$$\int_0^\pi \sqrt{1 + \cos^2(x)}\, dx. \qquad (0.1)$$

Numerical analysis can compute the length of this curve by standard methods that apply to essentially any integrand; there is never a need to make special substitutions or to do integration by parts to get the result. Further, the only mathematical operations required are addition, subtraction, multiplication, and division plus the making of comparisons. Since these simple operations are the only functions that computers can perform, computers and numerical analysis make a perfect combination.

It is important to realize that a numerical analysis solution is always numerical. Analytical methods usually give a result in terms of mathematical functions that can then be evaluated for specific instances. There is thus an advantage to the analytical result, in that the behavior and properties of the function are often apparent; this is not the case for purely numerical results. However, numerical results can be plotted to show some of the behavior of the solution.

Another important distinction is that the result from numerical analysis is an approximation, but results can be made as accurate as desired. (There are limitations to the achievable level of accuracy, because of the way that computers do arithmetic;

we will explain these limitations later.) To achieve high accuracy, many, many separate operations must be carried out, but computers do them so rapidly without ever making mistakes that this is no significant problem. Actually, evaluating an analytical result to get the numerical answer for a specific application is subject to the same errors.

The analysis of computer errors and the other sources of error in numerical methods is a critically important part of the study of numerical analysis. This subject will occur often throughout this book.

Here are some of the operations that numerical analysis can do and that are described in this book:

Solve for the roots of a nonlinear equation.

Solve large systems of linear equations.

Get the solutions of a set of nonlinear equations.

Interpolate to find intermediate values within a table of data.

Find derivatives of any order for functions even when the function is known only as a table of values.

Integrate any function even when known only as a table of values. Multiple integrals can also be obtained.

Solve ordinary differential equations when given initial values for the variables. These can be of any order and complexity.

Solve boundary-value problems and determine eigenvalues and eigenvectors.

Obtain numerical solutions to all types of partial differential equations.

Fit curves to data by a variety of methods.

Find efficient and effective approximations of functions.

0.2 Using a Computer to Do Numerical Analysis

Numerical methods require such tedious and repetitive arithmetic operations that only when we have a computer to carry out these many separate operations is it practical to solve problems in this way. A human would make so many mistakes that there would be little confidence in the result. Besides, the manpower cost would be more than could normally be afforded. (Once upon a time military firing tables were computed by hand using desk calculators, but that was a special case of national emergency before computers were available.)

Of course, a computer is essentially dumb and must be given detailed and complete instructions for every single step it is to perform. In other words, a computer program must be written so the computer can do numerical analysis. As you study this book you will learn enough about the many numerical methods available that you will be able to write programs to implement them. The computer language is not very important; programs can be written in BASIC (many dialects), FORTRAN, Pascal, C, and even assembly language. Most of the methods will be described fully through pseudocode in such a form that translating this code into a program is relatively straightforward. A number of example programs are given to illustrate the process. These programs have been designed to be easy to read rather than to be examples of professional caliber.

Actually, writing programs is not always necessary. Numerical analysis is so important that extensive commercial software packages are available. The IMSL (International Mathematical and Statistical Library) MATH/LIBRARY has hundreds of routines, of efficient and of proven performance, written in FORTRAN and C that carry out the methods. Recently, LAPACK (Linear Algebra Package) has been made available at nominal cost. This package of FORTRAN programs incorporates the subroutines that were contained in the earlier packages of LINPACK and EISPACK. Appendix C of this book gives information on these and other programs. The bimonthly newsletter of the Society for Industrial and Applied Mathematics (*SIAM News*) contains discussions and advertisements on some of the latest packages. A set of books, *Numerical Recipes*, lists and discusses numerical analysis programs in a variety of languages: FORTRAN, Pascal, and C.

One important trend in computer operations is the use of several processors working in parallel to carry out procedures with greater speed than can be obtained with a single processor. Some numerical analysis procedures can be carried out this way. Special programming techniques are needed to utilize these fast computer systems.

An alternative to using a program written in one of the higher-level languages is to use a kind of software sometimes called a *symbolic algebra program*. (This name is not very standardized and not too descriptive.*) These programs mimic the way that humans solve mathematical problems. Such a program is designed to recognize the type of function (polynomial, transcendental, etc.) presented and then to carry out requested mathematical operations on the function or expression. It does so by looking up in tables the new expressions that result from doing the operation or by using a set of built-in rules. For example, a program can use the ordinary rules for finding derivatives, employ tables of integrals to do integrations, and factor a polynomial or expand a set of factors. These are only a few of the capabilities. If an analytical answer cannot be given, most of these programs allow the user to derive an answer by numerical methods.

In connection with numerical analysis, an important feature of many symbolic algebra programs is the ability to write utility files that are essentially macros: A sequence of the built-in operations are defined to perform a desired larger task or one not inherent in the program. A succession of operations, each of which uses the results of the previous one—a procedure called *iteration*—is also possible. Many numerical analysis procedures are iterative.

Many such symbolic algebra programs are available, including *Mathematica*,

---

* Such programs are also called *computer algebra systems* (CAS).

DERIVE, MAPLE, MathCad, MATLAB, and MacSyma. In this chapter, we will explain how one of these, the DERIVE program, can be very useful in doing numerical analysis as well as some of the more analytical steps that are preliminary to the numerical method. We will shorten the name "the DERIVE program" to the single word "DERIVE" even though the word DERIVE really should be used as an adjective. In later chapters we will explain the use of two other computer algebra systems.

One special feature of most of these programs is their ability to carry out many operations with exact arithmetic. An interesting example is to see π displayed to 100 decimal places. Ordinarily, we must be satisfied with a limited number of digits of precision when a normal computer program is employed.

Of particular importance with programs like DERIVE is that the plotting of functions, even functions of two independent variables (which require a three-dimensional plot), is built in. In *Mathematica* this graphical capability is especially well developed, but the program is so big that lower-level personal computers are inadequate (at least at the present time).

Another alternative to writing a computer program to do numerical analysis is to employ a spreadsheet program. In two places later in this book, we will show how such widely available software can carry out numerical procedures.

0.3 A Typical Example

We will introduce the subject of numerical analysis by showing a typical problem solved numerically. If you worked for a mining company, Example 0.1 might be a problem you would be asked to solve.

EXAMPLE 0.1   *The Ladder in the Mine.*   There are two intersecting mine shafts that meet at an angle of 123°, as shown in Fig. 0.1(a). The straight shaft has a width of 7 ft, and the entrance shaft is 9 ft wide. What is the longest ladder that can negotiate the turn at the intersection of the two shafts? Neglect the thickness of the ladder members, and

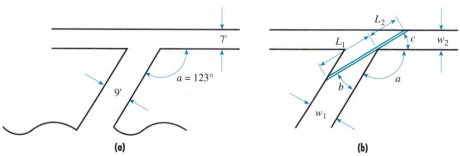

(a)   (b)

Figure 0.1

assume it is not tipped as it is maneuvered around the corner. Provide for the general case in which the angle a is a variable as well as for the widths of the shafts.

Whenever a scientific or engineering problem is to be solved, there are four general steps to be followed:

1. State the problem clearly, including any simplifying assumptions.
2. Develop a mathematical statement of the problem in a form that can be solved for a numerical answer. This process may involve, as in the present case, the use of calculus. In other situations, other mathematical procedures may be employed. When this statement is a differential equation, appropriate initial conditions and/or boundary conditions must be specified.
3. Solve the equation(s) that result from step 2. Sometimes the method will be algebraic, but frequently more advanced methods will be needed. This text may provide the method that is needed. The result of this step is a numerical answer or set of answers.
4. Interpret the numerical result to arrive at a decision. This will require experience and an understanding of the situation in which the problem is embedded. This interpretation is the hardest part of solving problems and must be learned on the job. This book will emphasize step 3 and will deal to some extent with steps 1 and 2, but step 4 cannot be meaningfully treated in the classroom.

The above description of the problem has taken care of step 1. Now for step 2.

Here is one way to analyze our ladder problem. Visualize the ladder in successive locations as we carry it around the corner; there will be a critical position in which the two ends of the ladder touch the walls while a point along the ladder touches the corner where the two shafts intersect (see Fig. 0.1b). Let c be the angle between the ladder and the wall when in this critical position. It is usually preferable to solve problems in general terms, so we work with variables a, b, c, w_1, and w_2.

Consider a series of lines drawn in this critical position—their lengths vary with the angle c, and the following relations hold (angles are expressed in radian measure):

$$L_1 = \frac{w_1}{\sin(b)}, \qquad L_2 = \frac{w_2}{\sin(c)}$$

$$b = \pi - a - c, \qquad (0.2)$$

$$L = L_1 + L_2 = \frac{w_1}{\sin(\pi - a - c)} + \frac{w_2}{\sin(c)}.$$

The maximum length of ladder that can negotiate the turn is the minimum of L as a function of the angle c. We hence set $dL/dc = 0$.

This is not a difficult derivative to do by hand, but DERIVE can do it for us. Let us see how this is done.

DERIVE is a menu-driven program. When the program is initiated, a screen appears, giving a copyright notice and a command menu, at the bottom. We invoke a command from the menu by pressing the space bar until the command is highlighted and then pressing ⟨ENTER⟩, or just by pressing the capitalized letter in the command. We first declare w_1 and w_2 variables through the Declare/Variable command. (The Declare command brings up a submenu from which we invoke Variable. We will specify such combinations in this manner.) Then we use the Author command to enter the expression for L. When we have done this, the display shows the expression with builtup fractions, as seen in the third line of Fig. 0.2. We then invoke the succession of commands Calculus/Differentiate and specify that the differentiation variable is c, getting the fourth line of Fig. 0.2.

Actually, all that line 4 of Fig. 0.2 does is to exhibit the differentiation operation. To get the result, we invoke Simplify, getting line 5. Observe that DERIVE performs some trigonometric simplifications in the process. Notice the use of $SIN(c)^2$ to represent the square of $SIN(c)$, although mathematicians write it as $\sin^2 c$. The equation that we need to solve to get the minimum for c is the expression in line 5 set to zero. The Manage/Substitute command lets us enter values for some of the variables, as shown in line 6.

After using DERIVE to develop the equation, we must now solve it. There are several numerical methods that we can use. We plan to use one of the most ancient and surely the simplest to understand. It is the *bisection method*, also known as *halving the interval*. To solve the expression $f(x) = 0$, we first must know an interval in which a root lies. (A root is a value of x that satisfies the expression. We also call it a zero of $f(x)$.) In our example, c is the variable, so we want to solve the equation $f(c) = 0$.

From the nature of this problem, we know that there must be a root, and Fig. 0.1(b) suggests that it might be of the order of $c = 30°$ (about 0.5 radian).

1:    w1 : =

2:    w2 : =

3:    $$\frac{w1}{SIN\,(\pi - a - c)} + \frac{w2}{SIN\,(c)}$$

4:    $$\frac{d}{dc}\left[\frac{w1}{SIN\,(\pi - a - c)} + \frac{w2}{SIN\,(c)}\right]$$

5:    $$-\frac{w1\,COS\,(a + c)}{SIN\,(a + c)^2} - \frac{w2\,COS\,(c)}{SIN\,(c)^2}$$

6:    $$-\frac{9\,COS\,(2.147 + c)}{SIN\,(2.147 + c)^2} - \frac{7\,COS\,(c)}{SIN\,(c)^2}$$

Figure 0.2

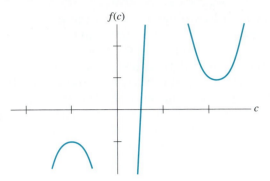

Figure 0.3

In more general terms, the bisection method solves for a root of $f(x) = 0$ by starting with two values that bracket the root. We can be sure that there is at least one root in the interval between these values if $f(x)$ is continuous and changes sign. So it is a good idea to begin by graphing $f(x)$ to find a good interval to begin with. Suppose we use DERIVE to give us this graph.

Our first step is to tell DERIVE the expression we want to plot. It is already present in line 6. It is highlighted to indicate that it is the "current line," the one on which action will be taken. (We can move the highlight with the arrow keys to select a different current line.) All we need to do to get a graph of the expression is to invoke the Plot command. (In getting the plot, we need to set some parameters for the plot, but we omit these details.)

The result is shown in Fig. 0.3. The tiny tick marks on the axes are spaced 1.0 apart in the c-direction and 10 apart vertically. These values are indicated by scale values that appear below the graph. Two-dimensional plots in DERIVE have a small cross that can be moved with the arrow keys. After the plot was created, the cross was moved to coincide closely with the crossing of the c-axis. It is hard to see this cross in the figure, but its location is exhibited as x (same as c) = 0.4666, y (the vertical coordinate) = 0. This tells us that the zero of our function is about $c = 0.4666$.

0.4 Implementing Bisection

The last section mentioned the bisection method. We now demonstrate this technique in an example. Figure 0.4 shows graphically how successive values converge on a root of $f(x)$ when we begin with a pair of values that bracket the root. x_3 is halfway between x_1 and x_2. x_4 is halfway between x_2 and x_3. We always take the next x-value as the midpoint of the last pair that bracket the root: These values bracket the root when there is sign change of $f(x)$ at the two points. As the process is continued, it is clear that we will converge to the root.

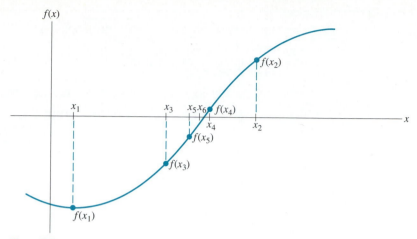

Figure 0.4

We intend to write a computer program to carry out this algorithm. (*Algorithm* means a rule or prescription for carrying out a certain computation.) Before we do so, we will express the algorithm more specifically in *pseudocode*, a way to express the steps of a program in generic form; it describes the program logic in language-independent terms.

An Algorithm for Halving the Interval (Bisection)

To determine a root of $f(x) = 0$ that is accurate within a specified tolerance value, given values x_1 and x_2 such that $f(x_1) * f(x_2) < 0$,

> REPEAT
>> Set $x_3 = (x_1 + x_2)/2$.
>> IF $f(x_3) * f(x_1) < 0$:
>>> SET $x_2 = x_3$.
>> ELSE Set $x_1 = x_3$.
>> ENDIF.
> UNTIL $(|x_1 - x_2| <$ tolerance value$)$ or $f(x_3) = 0$.

The final value of x_3 approximates the root; it is in error by not more than $(1/2)|x_1 - x_2|$.

Note. The method may give a false root if $f(x)$ is discontinuous on $[x_1, x_2]$.

It is not difficult to write a computer program when we have the algorithm expressed in pseudocode. Figure 0.5 exhibits a BASIC program that does bisection for the ladder in the mine problem. (The version of BASIC is QuickBasic 4.5. This

```
' USES BISECTION METHOD TO FIND A ROOT OF F(X) = 0
' GETS X1, X2, TOLERANCE VALUE FROM USER
' CHANGE DEF FNX() TO MATCH REQUIRED F(X)
DEF FNX (C) = 9 * COS(A + C) / SIN(A + C) ^ 2 + 7 * COS(C) / SIN(C) ^ 2
A = 123 * 3.14159 / 180
LIMIT = 50
INPUT "ENTER VALUES FOR X1, X2, TOLERANCE ", X1, X2, TOL
' DO A HEADING
PRINT "ITER NO", "X1", "X2", "X3", "F(X3)": PRINT
ITER = 1
DO
  F1 = FNX(X1)
  F2 = FNX(X2)
  IF F1 * F2 > 0 THEN
    PRINT "VALUES DO NOT BRACKET A ROOT": EXIT DO
  END IF
  X3 = (X1 + X2) / 2
  F3 = FNX(X3)
  PRINT ITER, X1, X2, X3, F3
  IF F3 * F1 < 0 THEN X2 = X3 ELSE X1 = X3
  ITER = ITER + 1
LOOP UNTIL ABS(X1 - X2) / 2 < TOL OR F3 = 0 OR ITER > LIMIT
```

Figure 0.5   A program for bisection

software is compatible with Microsoft's QBASIC.) Figure 0.6 shows the output when the program is run. Observe that the root, 0.4677734, is near 0.4666, which we found graphically.

Knowing that the critical angle is 0.4678 radian (26.8°), we can compute the maximum length of ladder from Eq. 0.2. Using this relation, we find that $L = 33.42$ ft. It would appear that ladders up to this length will be adequate for use in the mine.

In the example program of Fig. 0.5, a function statement is used to define $f(x)$. It might be preferable to define $f(x)$ in a separate procedure and have the code for bisection call the procedure to compute values of $f(x)$. This step would make the code a general-purpose program; only the function defining procedure would have to be changed.

The program adds a test, although the pseudocode does not include it, to terminate the computations if more than a certain number of iterations have been done.

ITER NO	X1	X2	X3	F(X3)
1	.4	.5	.45	4.661504
2	.45	.5	.475	-1.893359
3	.45	.475	.4625	1.364828
4	.4625	.475	.46875	-.2675005
5	.4625	.46875	.465625	.5476606
6	.465625	.46875	.4671875	.1398538
7	.4671875	.46875	.4679688	-6.388076E-02
8	.4671875	.4679688	.4675781	3.797662E-02
9	.4675781	.4679688	.4677734	-1.295546E-02

Figure 0.6   Program output for $x_1 = 0.4$, $x_2 = 0.5$, tolerance = 1E-4

It is always best to include this test to avoid an endless loop. (This program will be in an endless loop if the tolerance is set too small, an effect of precision in the computer, a topic that we will discuss later.)

As an alternative to writing a program in a computer language like BASIC or FORTRAN or C, the ITERATES command in DERIVE could be used to carry out the successive steps. This command involves the IF function to perform the various tests. We will exhibit such a program in Chapter 1.

0.5 Computer Arithmetic and Errors

We have mentioned the likelihood of errors due to the way that computers perform arithmetic computations. For example, the bisection program can go into an endless loop if the tolerance values are set very small. It is time that we discuss this most important subject of errors in numerical analysis. When we do numerical analysis there are several possible sources of error in addition to those due to the inexact arithmetic of the computer.

Truncation Error

The term *truncation error* refers to those errors caused by the method itself (the term originates from the fact that numerical methods can usually be compared to a truncated Taylor series). For instance, we may approximate e^x by the cubic

$$p_3(x) = 1 + \frac{x}{1!} + \frac{x^2}{2!} + \frac{x^3}{3!}.$$

However, we know that to compute e^x really requires an infinitely long series:

$$e^x = p_3(x) + \sum_{n=4}^{\infty} \frac{x^n}{n!}.$$

We see that approximating e^x with the cubic gives an inexact answer. The error is due to truncating the series and has nothing to do with the computer or calculator. For iterative methods, this error can usually be reduced by repeated iterations, but since life is finite and computer time is costly, we must be satisfied with an approximation to the exact analytical answer.

Round-Off Error

All computing devices represent numbers, except for integers, with some imprecision. (DERIVE and similar programs work with integers and rational fractions to achieve results of higher precision.) Digital computers will nearly always use floating-point numbers of fixed word length; the true values are not expressed exactly by such representations. We call the error due to this computer imperfection the *round-off* error. When numbers are rounded when stored as floating-point numbers, the round-off error is less than if the trailing digits were simply chopped off. We discuss this in more detail later in the chapter.

Error in Original Data

Real-world problems, in which an existing or proposed physical situation is modeled by a mathematical equation, will nearly always have coefficients that are imperfectly known. The reason is that the problems often depend on measurements of doubtful accuracy. Further, the model itself may not reflect the behavior of the situation perfectly. We can do nothing to overcome such errors by any choice of method, but we need to be aware of such uncertainties; in particular, we may need to perform tests to see how sensitive the results are to changes in the input information. Since the reason for performing the computation is to reach some decision with validity in the real world, sensitivity analysis is of extreme importance. As Hamming says, "the purpose of computing is insight, not numbers."

Blunders

You will likely always use a computer or at least a programmable calculator in your professional use of numerical analysis. You will probably also use such computing tools extensively while learning the topics covered in this text. Such machines make mistakes very infrequently, but since humans are involved in programming, operation, input preparation, and output interpretation, blunders or gross errors do occur more frequently than we like to admit. The solution here is care, coupled with a careful examination of the results for reasonableness. Sometimes a test run with known results is worthwhile, but it is no guarantee of freedom from foolish error. When hand computation was more common, check sums were usually computed—they were designed to reveal the mistake and permit its correction.

Propagated Error

Propagated error is more subtle than the other errors. By *propagated error* we mean an error in the succeeding steps of a process due to an occurrence of an earlier error—such error is in addition to the local errors. It is somewhat analogous to errors in the initial conditions. Some root-finding methods find additional zeros by changing the function to remove the first root; this technique is called *reducing* or *deflating the equation*. Here the reduced equations reflect the errors in the previous stages. The solution, of course, is to confirm the later results with the original equation.

In examples of numerical methods treated in later chapters, propagated error is of critical importance. If errors are magnified continuously as the method continues, eventually they will overshadow the true value, destroying its validity; we call such a method *unstable*. For a *stable* method—the desirable kind—errors made at early points die out as the method continues. This issue will be covered more thoroughly in later chapters.

Each of these types of error, while interacting to a degree, may occur even in the absence of the other kinds. For example, round-off error can occur even if truncation error is absent, as in an analytical method. Likewise, truncation errors can cause inaccuracies even if we can attain perfect precision in the calculation. The usual error analysis of a numerical method treats the truncation error as though such perfect precision did exist.

Floating-Point Arithmetic

To examine round-off error in detail, we need to understand how numeric quantities are represented in computers. In nearly all cases, numbers are stored as floating-point quantities, which are very much like scientific notation.* For example, the fixed-point number 13.524 is the same as the floating-point number $.13524 * 10^2$, which is often displayed as .13524E2. Another example: -0.0442 is the same as $-.442E-1$.

Different computers use slightly different techniques, but the general procedure is similar. In computers, floating-point numbers have three parts: the *sign* (which requires one bit); the *fraction part*—often called the *mantissa* but better characterized by the name *significand*; and the *exponent part*—often called the *characteristic*. The three parts of numbers have a fixed total length that is often 32 or 64 bits (sometimes even more). The fraction part uses most of these bits, perhaps 23 to as many as 52 bits, and that number determines the precision of the representation. The exponent part uses 7 to as many as 11 bits, and this number determines the range of the values.

Computers represent their floating-point numbers in the general form

$$\pm .d_1 d_2 d_3 \ldots d_p * B^e,$$

where the d_i's are digits or bits with values from zero to $B - 1$ and

$B = $ the number base that is used, usually 2, 16, or 10.

$p = $ the number of significand bits (digits), that is, the precision.

$e = $ an integer exponent, ranging from $E_{\min}$ to $E_{\max}$, with the values going from negative $E_{\min}$ to positive $E_{\max}$.

The significand bits (digits) constitute the fractional part of the number. In almost all cases, numbers are normalized, meaning that the fraction digits are shifted and the exponent adjusted so that d_1 is nonzero. Zero is a special case; it usually has a

---

* Another name often used for floating-point numbers is *real numbers*, but here we reserve the term *real* for the continuous (and infinite) set of numbers on the "number line." When printed as a number with a decimal point, it is called *fixed point*. The essential concept is that these numbers are in contrast to integers.

fraction part with all zeros and a zero exponent. This kind of zero is not normalized and never can be.

In hand calculators, the base B is usually 10; in computers the base is often 2, but sometimes a base of 16 is used. Most computers permit two or even three types of numbers: *single precision*, which is equivalent to 6 to 7 significant decimal digits; *double precision*, equivalent to 13 to 14 significant decimal digits; and *extended precision*, which may be equivalent to 19 to 20 decimal digits.

Examples of Numbers as Represented in a Computer

Working with binary or hexadecimal digits is awkward, so we will begin with the more familiar base 10 in some examples. Suppose $B = 10$ and $p = 4$. Then these numbers would be represented as

$$27.39 \quad \rightarrow +.2739 * 10^2;$$
$$-0.00124 \rightarrow -.1240 * 10^{-2};$$
$$37000 \quad \rightarrow +.3700 * 10^5.$$

Observe that we have *normalized* the fractions—the first fraction digit is nonzero. If the base is 2, that means that the first fraction bit is always 1. Some systems take advantage of that fact and do not store the first bit, thus gaining one bit of precision. When this first bit is suppressed, it is referred to as a *hidden bit*.

Because the number of bits used for a floating-point number is fixed, there is a finite number of distinct values in the computer's number system—in great contrast to the real number system. Thus there are gaps within the computer's number system. To illustrate, let us use a greatly simplified case where $B = 2$, $p = 2$, and $-2 \le e \le 3$. For this system, all the normalized numbers would be of either of these forms:

$$\pm.10_2 * 2^e \quad \text{or} \quad \pm.11_2 * 2^e, \qquad -2 \le e \le 3.$$

Since the binary fractions $.10_2 = \frac{1}{2}$ and $.11_2 = \frac{1}{2} + \frac{1}{4} = \frac{3}{4}$, these numbers range from $-.11_2 * 2^3 = -6$ to $+.11_2 * 2^3 = +6$. Here is a list of all the positive numbers:

$$.10_2 * 2^{-2} = \frac{1}{8}, \qquad\qquad .11_2 * 2^{-2} = \frac{3}{16},$$

$$.10_2 * 2^{-1} = \frac{1}{4}, \qquad\qquad .11_2 * 2^{-1} = \frac{3}{8},$$

$$.10_2 * 2^0 = \frac{1}{2}, \qquad\qquad .11_2 * 2^0 = \frac{3}{4},$$

$$.10_2 * 2^1 = 1, \qquad\qquad .11_2 * 2^1 = \frac{3}{2},$$

$$.10_2 * 2^2 = 2, \qquad\qquad .11_2 * 2^2 = 3,$$

$$.10_2 * 2^3 = 4, \qquad\qquad .11_2 * 2^3 = 6.$$

The following diagram shows the distribution of these positive values. The negative values will be similarly distributed. Gaps are present that are uneven in size.

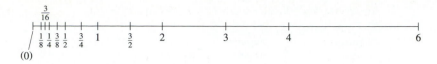

These interior gaps are in addition to the underflow and overflow gaps that we will describe.

Observe that zero is not present. To represent zero, we define a special case: Zero is a value with all zeros in the fraction (it is not normalized) with a zero exponent. This corresponds to the standard in almost all computer systems.

Pay particular attention to the important consequence of these gaps in all computer number systems. In the tiny system illustrated in the preceding diagram, the value 2.3 will be stored as 2; to the computer, the two values 2.2 and 2.4 are precisely identical—this is true for all values greater than 2 and up to 3. This explains why the bisection program will loop indefinitely if the tolerance value is too small. (In one instance, after 21 iterations, $x_1 = 0.4677237$, $x_2 = 0.4677238$, and $x_3 = 0.4677238$; the computer could not distinguish between x_2 and x_3, so there can be no further approach to the true value of the root.)

Actual Computer Number Systems

In actual computer systems of floating-point numbers, the format is similar but not as easy to comprehend because the base is not 10. Table 0.1 compares the composition of three different systems. The fact that these systems are not identical means

Table 0.1 Comparative methods for storing floating-point numbers

Method	Total length	Bits in fraction	Bits in exponent	Bias value[††]	Base	Max. exponent	Min. exponent	Largest number	Smallest number	Approx. precision, no. decimal digits
IEEE										
single	32	23*	8	127	2	127	−126	1.701E38	1.755E-38	7
double	64	52*	11	1023	2	1023	−1022	8.988E307	2.225E-308	16
extended	80	64	15	16383	2	16383	−16382	6E4931	3E-4931	19
VAX										
single	32	23	8	127	2	127	−127	1.701E38	5.877E-39	7
double-1	64	55	8	127	2	127	−127	1.701E38	5.877E-39	16
double-2	64	52	11	1023	2	1023	−1023	8.988E307	1.123E-308	15
extended	128	112	15	16383	2	16383	−16383	6E4931	1E-4931	33
IBM										
single	32	24	7	63	16	63	−64	7.237E75	8.636E-78	7
double	64	56	7	63	16	63	−64	7.237E75	8.636E-78	16
extended	128[†]	112	7	63	16	63	−64	7.237E75	8.636E-78	33

* Plus 1 "hidden bit."   †† The bias value is added to the exponent so that unsigned numbers represent the entire range.
† 8 bits not used.

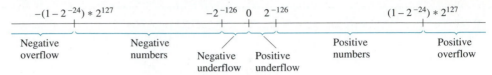

Figure 0.7   Numbers in the IEEE standard format

that different computers can give different results from the same set of computations, thus making programs not always readily transportable. Figure 0.7 illustrates the ranges of numbers representable in the IEEE standard, shown in Table 0.1.

Arithmetic Accuracy in Computers

We again avoid the confusion of thinking in number bases other than 10, by discussing accuracy of floating-point operations with examples using normalized base-10 numbers. The bases used in computers behave analogously. To keep it simple, we assume only three digits in the fraction part and one decimal digit for the exponent; $B = 10$, $p = 3$, $-9 \le e \le 9$. We compare rounding and cropping of the results.

When two floating-point numbers are added or subtracted, the digits in the number with the smaller exponent must be shifted to align the decimal points (normalization is forgotten in the adder). This shifting can lose some of the significant digits of one of the values. The result may also need to be shifted and the exponent adjusted to normalize it. Some computers automatically round the final answer, others just chop off extra digits beyond the precision of the system.

When multiplied (divided), the fractions are just multiplied (divided) and the exponents added (subtracted). The result is normalized.

Some examples follow.

EXAMPLE 0.2   Compute $1.37 + .0269 = .137 * 10^1 + .269 * 10^{-1}$.

$$\left.\begin{array}{r} .137 \ \ * 10^1 \\ + \ .00269 * 10^1 \end{array}\right\} \quad \text{Align decimal points.}$$
$$\overline{.13969 * 10^1}$$

$$\begin{array}{ll} \text{Chop} \rightarrow & .139 * 10^1. \\ \text{Round} \rightarrow & .140 * 10^1. \end{array}$$

Rounding involves an extra operation, which may be done either in hardware or through a software routine.

EXAMPLE 0.3   Compute $4850 - 4820 = .485 * 10^4 - .482 * 10^4$.

$$.485 * 10^4$$
$$- .482 * 10^4$$
$$\overline{.003 * 10^4}$$
$$.300 * 10^2 \quad \text{Normalize.}$$

$$\text{Chop} \rightarrow \quad .300 * 10^2.$$
$$\text{Round} \rightarrow \quad .300 * 10^2.$$

In Example 0.3 observe that there is really only one digit of accuracy in the result even though the difference is represented as though the trailing zeros were significant. This loss of accuracy when two nearly equal numbers are subtracted is a major source of error in floating-point operations. In this case, rounding and chopping reach the same final answer.

EXAMPLE 0.4   Compute $3780 - .321 = .378 * 10^4 - .321 * 10^0$.

$$\left.\begin{array}{r} .378 * 10^4 \\ - .0000321 * 10^4 \end{array}\right\} \quad \text{Align decimal points.}$$
$$\overline{.3779679 * 10^4}$$

$$\text{Chop} \rightarrow \quad .377 * 10^4.$$
$$\text{Round} \rightarrow \quad .378 * 10^4.$$

In Example 0.4 shifting to align the decimal points has completely lost the significant digits of the subtrahend!

EXAMPLE 0.5   Compute $403000 * .0197 = .403 * 10^6 * .197 * 10^{-1}$.

$$\left.\begin{array}{r} .403 \\ *.197 \end{array}\right\} \text{Multiply fractions.} \qquad \left.\begin{array}{r} 6 \\ -1 \end{array}\right\} \text{Add exponents.}$$
$$\overline{.079391} \qquad\qquad\qquad \overline{5}$$

$$.079391 * 10^5 \quad \text{Combine fraction and exponent.}$$
$$.79391 * 10^4 \quad \text{Normalize.}$$

$$\text{Chop} \rightarrow \quad .793 * 10^4.$$
$$\text{Round} \rightarrow \quad .794 * 10^4.$$

In multiplying two n-digit numbers, a product with $2n$ digits results. Internally, double-length registers are used (but this is often accomplished by joining two single-length registers). The final result is truncated to the length of a single register.

EXAMPLE 0.6   Compute $.0356/1560 = .356 * 10^{-1}/.156 * 10^4$.

$$\left.\begin{array}{r} .356 \\ \div \ \underline{.156} \\ 2.28205 \end{array}\right\} \text{ Divide fractions.} \qquad \left.\begin{array}{r} -1 \\ \underline{-4} \\ -5 \end{array}\right\} \text{ Subtract exponents.}$$

$$2.28205 * 10^{-5} \quad \text{Combine fraction and exponent.}$$
$$.228205 * 10^{-4} \quad \text{Normalize.}$$

$$\text{Chop} \rightarrow \quad .228 * 10^{-4}$$
$$\text{Round} \rightarrow \quad .228 * 10^{-4}$$

Double-length registers are also involved in doing division.

▲

The time required to perform the different arithmetic operations will vary. Multiplication may be several times (2.5 to 10 times) slower than addition or subtraction. (Subtraction is equivalent to addition because it is done by adding two's-complements.) Floating-point division may be the slowest of all (4 to 25 times that for addition). These statements apply to computers of the 1980s. In some of these early computers, multiplication and division were performed through software routines—the timing differences were even greater in these machines.

On today's personal computers equipped with math coprocessors, the differences in computing time are much less—some experiments with an Intel 386 CPU with coprocessor chip showed that multiplication took only about 3 percent longer than addition, and division about 9 percent longer. (These figures are not precise because the timing was affected by the intervals between clock ticks.) An Intel 286 processor without math coprocessor indicated a time for multiplication about 11 percent greater and for division about 35 percent greater than for addition.

Errors in Converting Values

The numbers that are input to a computer are ordinarily base-10 values. Thus the input must be converted to the computer's internal number base, normally base 2. This conversion itself causes some errors. Terminating decimal fractions can be nonterminating in base 2: $(0.6)_{10} = (0.100110011001\dots)_2$. Because of the gaps in the computer's number system, we have to map the infinite set of real numbers into a finite set of computer numbers. For Examples 0.2 to 0.6, which have three decimal-digit fractions, there are only 900 different fractional values—all the mathematical numbers between 0.1 and 1.0 must be translated to one of these 900 values. In each decade, as represented by a constant value of the exponent, there are also only 900 values. The spacing between values in the different decades is therefore different, as we have said.

The representation of numbers smaller than $\frac{1}{8}$ in magnitude in the first base-2 example system is impossible. Such an input results in *exponent underflow*. This same phenomenon occurs with all computer number systems (see Fig. 0.7). The range of underflow values depends on the number of fraction bits and the number base. Some computers distinguish between *positive underflow* and *negative underflow*. In some systems, when this occurs, either from an attempt to input a value or from an intermediate computation, the value is replaced by zero (and, we hope, an error message is generated).

Similar things happen with number values larger than those that can be represented. Our simplified system has *exponent overflow* when values larger in magnitude than 6 are encountered. With actual computer number systems, much larger values can be accommodated, but there is always a limit. Some systems replace the number with the largest possible value when overflow occurs.

Machine eps

One important measure in computer arithmetic is how small a difference between two values the computer can recognize. This quantity is termed the *computer eps*, where "eps" is a shortening of the Greek letter epsilon. This measure of *machine accuracy* is standardized by finding the smallest floating-point number that, when added to floating-point 1.000, produces a result different from 1.000. (Numbers smaller than eps are effectively zero in the computer.) In the exercises, you are asked to determine the size of the eps of machines available to you. This determination may even depend on the computer language in which a program is written, since some languages use only double precision for their floating-point numbers.

Peculiar things happen in floating-point arithmetic. For example, adding 0.001 one thousand times may not equal 1.0 exactly. In some instances, multiplying a number by unity does not reproduce the number. In many computations, changing the order of the calculations will produce different results.

Absolute Versus Relative Error, Significant Digits

The accuracy of any computation is always of great importance. There are two common ways to express the size of the error in a computed result: *absolute error* and *relative error*. The first is defined as

$$\text{absolute error} = \text{true value} - \text{approximate value}.$$

If we know the absolute error, we can get the correct (exact) value by adding the absolute error to the approximation. (The catch is that we rarely know the absolute error.)

A given size of error is usually more serious when the magnitude of the true value is small. For example 1036.52 ± 0.010 is accurate to five significant digits and may be of adequate precision, whereas 0.005 ± 0.010 is a clear disaster.

Using relative error is a way to compensate for this problem. Relative error is defined as

$$\text{relative error} = \frac{\text{true value} - \text{approximate value}}{\text{true value}}.$$

The relative error is more independent of the scale of the value, a desirable attribute. When the true value is zero, the relative error is undefined. It follows that the round-off error due to a finite length of the fractional part of floating-point numbers is more nearly constant when expressed as relative error than when expressed as absolute error. Some people define these errors in terms of magnitudes, in which case the error is always a positive quantity.

Another term that is commonly used to express accuracy is *significant digits*, that is, how many digits in the number have meaning. Extra digits that show up when numbers are shifted to normalize them are meaningless; this is a real problem when there are trailing zeros in a number. We may not know whether they are really zero or just fillers.

A more formal definition of significant digits follows.

1. Let the true value have digits $d_1 d_2 \ldots d_n d_{n+1} \ldots d_p$.
2. Let the approximate value have $d_1 d_2 \ldots d_n e_{n+1} \ldots e_p$.

where $d_1 \neq 0$ and with the first difference in the digits occurring at the $(n + 1)$st digit. We then say that (1) and (2) agree to n significant digits if $|d_{n+1} - e_{n+1}| < 5$. Otherwise, we say they agree to $n - 1$ significant digits.

EXAMPLE 0.7   Let the true value = 10/3 and the approximate value = 3.333.

The absolute error is $0.000333\ldots = 1/3000$.

The relative error is $(1/3000)/(10/3) = 1/10000$.

The number of significant digits is 4.

0.6 Theoretical Matters

Any user of a mathematical procedure should be concerned with its theoretical underpinnings because they explain the limitations of the procedure and the conditions that must be true for the procedure to produce reliable results. This book, while mentioning theory in the body of chapters, will reserve a fuller discussion of theoretical matters for special sections like this one. In this chapter, only one method has been described—bisection (interval halving)—so this section is short. However, it introduces the kind of questions about theory that generally should be asked when a numerical method is employed. In addition, this section will discuss some general points on how theory will be presented.

When theory is to be discussed, two questions arise:

1. Where do we start—what background is assumed? Does every definition and postulate have to be stated before the pertinent theorems are developed? In this book we assume that the reader has a background in calculus and knows the definitions of that body of mathematics.
2. How are theorems presented? Is it better to use the rather condensed notations of mathematicians and their special symbols or to use language and style that is more accessible to the average person? We have opted for the latter, feeling that it appeals to the average student who is using this book as a text or reference.

Most of the methods of numerical analysis are iterative: approximate answers are obtained through a sequence of improved estimates. There are four questions we always ask about an iterative method:

1. Under what conditions does the method apply? For what kinds of functions does the method work, and how can we know that the conditions are satisfied?
2. Does the method converge? Do the successive approximations reach the true answer to a given accuracy?
3. What bounds can be placed on the error of each estimate? Can we know in advance the maximum size of the error after a certain number of iterations?
4. How rapidly do the errors of the successive estimates decrease? For example, do errors decrease proportionally to the number of iterations or is the accuracy improved more rapidly than linearly, a most desirable situation? How accurately do we know the error (or a bound to the error)?

When finding a root of $f(x)$ by interval halving, we require that $f(x)$ be continuous and of opposite sign at different values of x. This requirement guarantees that there is at least one root between the two x-values. The proof of this statement comes from visualizing the graph of $f(x)$—the function must cut the x-axis at some point(s) between the two values.

If we let x be the midpoint of the interval, $f(x)$ must either be zero at this new point (and the method is finished) or be nonzero, meaning that $f(x)$ at the new x-value is of sign opposite to one of the original points. We then select the two points where $f(x)$ is of opposite sign as a new starting pair and subdivide again.

Since the interval is cut in half at each step and a root lies in this last interval, the method obviously converges to the true value of a root. If we say that the midpoint of this last interval is our best estimate of the root, the error is certainly no greater than half the last interval. We see that the magnitude of this error estimate is precisely

$$\text{error} = \left(\frac{1}{2^n}\right)\left|b - a\right|,$$

where n is the number of steps and a and b are the initial pair of x-values. Bisection or interval halving is one of the few methods of numerical analysis where we know in advance how many steps we must take to reduce the error to a prescribed size.

Be sure to recognize that such error analysis does not take machine errors into account. The theory of numerical analysis almost always assumes perfect computational precision. The effect of the machine is normally handled separately (and, more often than not, entirely neglected because so little can be said about it quantitatively).

As we consider the theory, it is instructive to ask the effect if the assumptions are violated. For example, why must $f(x)$ be continuous? Certainly, if there is jump discontinuity so that $f(x)$ changes sign without cutting the x-axis, there is no guarantee of a root in the starting interval. What if there is a single-jump discontinuity but the jump is not across the x-axis? The exercises will pose questions like this last one.

0.7   Parallel Processing

A computer normally runs its instructions sequentially—one after another—but another trend is emerging. Throughout the historical development of computers, faster and faster machines have been built, but today we have about reached the limits of speed improvements. The fastest machines now can operate with clock times of about 1 to 3 nanoseconds, giving of the order of 10^9 floating-point operations per second ("flops"). Machines with such high performance are today's very expensive supercomputers. But for really large scale problems, such as short-term weather forecasting, simulation to predict aerodynamics performance, image processing, and artificial intelligence, these speeds, while almost mind-boggling, are inadequate. Many of these applications involve the solution of very large sets of simultaneous equations by numerical methods.

One of the first techniques to increase the operating speed of a computer was "pipelining"—that is, performing a second instruction within the CPU before the previous instruction is completed. This technique takes advantage of the fact that doing a single "instruction" actually involves several microcoded steps and that the initial microsteps can be applied to an additional instruction even though the first sequence of microsteps has not yet finished. Pipelining permits a speedup by a factor of two or more.

Another technique has been to build vector processing operations into the CPU. Since the individual steps required to solve sets of equations involve many multiplications of a vector times another vector, these machines offer significant speed improvements but only by a factor of 5 or 10—not by the factor of 10,000 that is really desired. Further, this feature increases the cost of mainframes considerably.

The current trend is to use parallel processing, that is, to put several machines to work on a single problem, dividing the steps of the solution process into many steps that can be performed simultaneously. Not all problems permit such parallel operations, but many important problems of applied mathematics can be so structured. Obtaining many or even several supercomputers is outrageously costly, however. An alternative is to employ a massive number of low-cost microprocessors, say, of the order of a thousand (1024 is a practical number). Although the individual speed of a microprocessor is not equal to that of a supercomputer, the difference in speed is made up for by the larger number of machines that are combined.

Classification of Computers

The most familiar grouping of computers has been into microcomputers, workstations, minicomputers, mainframes, and supercomputers. This breakdown ranks systems on the basis of cost, size, and speed; the peripherals connected to each group also vary similarly. The groupings are not precise. Microcomputers, better known as personal computers, have increased so much in power but with so little increase in price as to push minicomputers almost out of the market.

Mainframes and minis are nearly always used now as host machines for several to many terminals or as file servers. Often a number of interconnected personal computers use their local memories for much of their work, accessing files on the host for shared information or for copies of application programs. Supercomputers are needed when very large programs must be run or when rapid access to huge data bases is required.

Another classification, a taxonomy suggested in 1966 for multiprocessing machines, subdivides computers on the basis of how instructions and data are processed. A sequential-instruction, single-data-stream machine (SISD) is the typical von Neumann computer that processes data through a single CPU, executing one instruction after another. Such a machine may involve pipelining.

A single-instruction, multiple-data-stream machine (SIMD) has several processing units, all supervised by a single control unit. Each processor receives the same instruction at any cycle but utilizes different data coming in separate data streams. A vector processor (also called an array processor) is of this type.

The development of a multiple-instruction, single-data-stream machine (MISD), is possible but such a machine seems to be impractical, for no real computers of this type have ever been built.

Multiple-instruction, multiple-data-stream machines (MIMDs) are the most promising of assemblies for parallel processing. MIMDs use many processing units, each executing instructions and utilizing data that are independent of the instructions and data used by the other CPUs.

Special Problems in Parallel Computing

If parallel computing is to be used to solve a large problem rapidly, several new aspects come into play. Is the data stream provided from a single shared memory, or do the separate units have individual memories? If the memory is distributed, how is communication between the units accomplished? What type of bus provides the data channels to the separate units, and can separate units read and write data at the same time? What sort of intercommunication is there between the individual processors, and can they exchange data without going through memory?

Other questions remain. Do the units operate synchronously, with all controlled by a single clock, or do they run asynchronously? If operation is asynchronous, how does one unit know when to accept data from a prior operation of a different unit, or do all units operate "chaotically"? How can the loads for the separate processors be balanced—will some units sit idle while others are running at capacity? (It would be preferable for all units to run at full loading.) What about the programs for parallel processing? Does the programmer have to be concerned with synchronization and intercommunications? Is the code portable to other machines?

The questions about programming a parallel system are not yet settled. If it were possible to have the compiler recognize parallelism within a conventional program written for sequential operations and have it develop the changed code to be run on the parallel system, the task of programming would be much easier. On the other hand, writing code that specifically takes advantage of the parallel CPUs should be more optimum, but this task is tricky and complicated. It requires a skill that few programmers currently have. This mode would involve knowing exactly how the hardware is organized and what communications problems are involved. Further, it is likely that the best algorithm (solution procedure) for a parallel machine will not always be the optimum one for sequential processing.

We do not intend to explain all these many aspects of parallel processing in this book. We must be content to show where parallelism exists for the various kinds of problems that we attack numerically. For example, here is a simple classical problem that exhibits the advantage of parallel processing. Suppose we are to add together n values. We can show the successive steps by a "directed acyclic graph" (dag), as shown in Fig. 0.8. Now, imagine that we have many separate processors that can be applied to the job. Figure 0.9 shows that the number of time steps can be decreased from seven to three.

The term *speedup* is used to describe the increased performance of a parallel system compared to a single processor. It is the ratio of the execution time for the original sequential process, using a single processor, to the time for the same job using parallel processoring. In the preceding simple example, the speedup is $\frac{7}{3} =$ 2.333. In computing the speedup, we use the time for the optimum sequential procedure (or for the best-known procedure if the actual optimum procedure is not known) and for the best parallel algorithm. Another term, the *efficiency,* is based

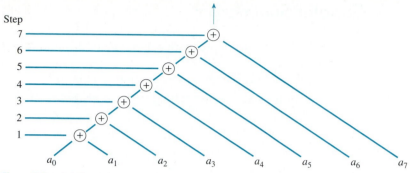

Figure 0.8   Adding eight numbers sequentially

on how the speedup compares to the number of processors used. Theoretically, if we have n processors, we should be able to do the job n times as fast. In our example, the efficiency is $2.333/4 = 0.583$. We have less than an efficiency of 1.00 because some of the processors are idle after the first step. Sometimes the speedup and efficiency are reduced because the size of the problem does not fit to the number of processors. For example, if we were to add only seven numbers in this example, we would still require four processors to get the sum in three steps, but now the speedup would be only $\frac{6}{3} = 2.000$ and the efficiency would drop to 0.5. On the other hand, what if we were limited to four processors but had 15 numbers to add? We would subdivide the problem and, while trivial in this example, it would have a best solution that is not obvious.

You can see why a complete exposition of how parallel processing and numerical analysis interact will not be attempted in this book, whose major emphasis is on the numerical solution to applied problems.

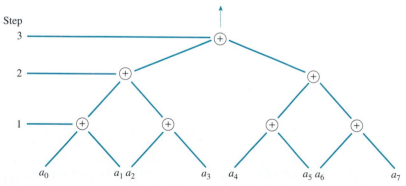

Figure 0.9   Adding eight numbers with parallel processors

Chapter Summary

We will conclude each chapter with a summary of the important items of the chapter, as in this section.

After studying this chapter, you should be able to

1. Tell how a numerical solution differs from an analytical one, and know the advantages of each.

2. Describe the kinds of software that can be used to instruct a computer to carry out the procedures of numerical analysis.

3. Outline the four steps that should be followed in solving any scientific or engineering problem.

4. Show a fellow student the basic principles of bisection to find a root of a function, and explain how to get initial values that bracket the root.

5. Employ the program listed in the chapter to solve a nonlinear equation. Perhaps you can rewrite it in a different computer language and modify it so that the function evaluation is performed in a subprogram.

6. Distinguish among five types of errors and explain how each can be minimized.

7. Discuss the relative advantages of using relative error versus absolute error as a measure of accuracy.

8. Explain how floating-point numbers are stored in computers and tell what factors affect their accuracy and range. Be able to compute the eps value from the number of bits used to store the different parts of a floating-point number.

9. Tell why parallel processing is currently of great interest to numerical analysts, and describe some of the problems associated with this technique. Explain how a directed acyclic graph (dag) can show the difference in times required for sequential and parallel processing and how a dag can be used to compute the speedup and efficiency values.

Exercises*

Section 0.3

1. Use the DERIVE program (or another symbolic algebra program) to evaluate the following derivatives.
   a. $d/dx[x \sin(x^3/2)]$
   b. $d^2/dy^2[(x + y)/(y - x)]$
   c. $d/dx[e^{-(x+a)}\{x \sin(bx)\}]$
   d. $d/dy[d/dx(x + y/x^2)]$

2. Use DERIVE (or some other computer plotting routine) to reproduce Fig. 0.3, except scale it so the portion between [0, 1] occupies essentially the entire width of the screen. Can you now find a better estimate of the root? If so, what is it? Why is such "zooming" not really a good idea?

3. Suppose the height of the inlet and straight shafts of Example 0.1 is 6 ft. If you are permitted to tip the ladder as the corner is negotiated, what maximum length of ladder can now be taken into the mine?

---

* Answers are given at the end of the text for exercises marked by ▶.

▶**4.** A circular well is 5.6 ft in diameter and is 14.3 ft deep, with a flat bottom. A ladder that is 17 in. wide (outside measurement) has side rails that measure 1 in. by 3 in. What is the longest length of ladder that can be placed in the well if its top is to be exactly even with the surface of the ground? Is a computer essential for solving this problem?

Section 0.4

5. Enter the program of Fig. 0.5 into your computer and run it. Do the results duplicate those of Fig. 0.6? If they do not, explain the difference. (You may have to modify the program if you use a different dialect of BASIC.)

6. Translate the program of Fig. 0.5 into some other computer language and run the program. If your results do not duplicate Fig. 0.6, explain the difference.

▶**7.** Modify the program of Fig. 0.5 to eliminate the test for maximum number of iterations. At what tolerance level does the program now enter an endless loop? How will you stop the program when this loop occurs?

8. Modify the program of Fig. 0.5 to print the result after the iterations terminate as:

    THE LENGTH OF THE LADDER IS
    XX.XXXX +/- .YYYY FT

where .YYYY is the calculated error in the length.

9. Modify the program of Fig. 0.5 to obtain roots of these equations.
a. $e^x - 3x^2 = 0$ within $[-2, 0]$
▶**b.** $x^3 - 2x^2 + 1.1 = 0$ within $[0, 1.5]$
c. $x \cos[x/(x - 2)] = 0$ within $[-3, 1]$

10. Repeat Exercise 7, but now compare double-precision to single-precision arithmetic.

Section 0.5

11. Express the quantities below as floating-point numbers in the form

$$0.xxx\ldots xx \text{ E } yy$$

where the first digit after the decimal point is nonzero and E stands for "times 10 to the power" and yy is an integer.
a. 1234.5678
b. −0.001020304
c. 1234567890
d. 0.000,000,001

12. Parallel the simplified examples of floating-point numbers in Section 0.5 except with $B = 16$, $p = 4$, and $-4 \le e \le 5$.
a. How many distinct numbers are there in this system?
b. What is the largest positive number?
c. What is the smallest (most negative) number?
d. What is the positive number of least magnitude?
e. What is the negative number of least magnitude?

▶**13.** Repeat Exercise 12 except for $B = 10$.

14. For some computer system available for your use, find the values of B, p, and E_{max} and E_{min}. Are these language-dependent (meaning does the answer vary with the computer language)? Answer the questions of Exercise 12 for this system. Your study should include the various precision levels that are provided (single, double, extended precision).

▶**15.** Using the simplified number system of Examples 0.2–0.6, determine the results with both rounding and chopping for each of the following.
a. 12.3 + 0.0234
b. −0.0321 + 0.000136
c. 12.3 − 0.0234
d. −321 + 32.1
e. 132 ∗ 0.987
f. −2.14/0.000137
g. (−.111 + .222) ∗ .00111/999 (in left to right order)

16. Write a computer program to determine experimentally the relative speeds of addition, subtraction, multiplication, and division on some computer system available to you. Be sure to do enough repetitions to overcome the effect of the clock ticks of your system and be sure to compensate for the loop overheads. Are these results language-dependent? Does the system have special hardware for doing arithmetic operations?

17. For some computer system available to you, find out what the machine does when there is exponent overflow and when there is underflow. Are the results language-dependent?

18. Write a computer program that determines the machine eps for some system available for your use. Is this experimentally determined value equal to what you would expect from the machine's number system?

19. Compute the absolute and relative errors of each result of Exercise 15.

▶**20.** Evaluate the following polynomial for $x = 1.07$, using both chopping and rounding to three digits, proceeding through the polynomial term by term from left to right. What are the absolute and relative errors of your results?

$$2.75x^3 - 2.95x^2 + 3.16x - 4.67$$

21. Repeat Exercise 20, except this time proceed from right to left, term by term.

22. Evaluating a polynomial in "nested form" is more efficient. The nested form of the polynomial in Exercise 20 is

$$((2.75x - 2.95)x + 3.16)x - 4.67.$$

Repeat Exercise 20, but this time do it with the nested form.

23. Compute the absolute and relative errors for each result in Exercises 20, 21, and 22.

24. Write a computer program that computes the sums below. Print out results that should equal 0.1, 0.2, 0.3, ... What are the absolute and relative errors of each final result?
a. 0.001 added 1000 times.
b. 0.0001 added 10,000 times.
c. 0.00001 added 100,000 times.

▶**25.** The infinite series $1 + \frac{1}{2} + \frac{1}{3} + \frac{1}{4} + \cdots$ is divergent. Write a computer program to evaluate this sum. Is the computer series divergent? If not, why not?

26. Repeat Exercise 18 (that determines the machine eps for some system available to you), but do this with double and extended precision if your machine and language permit.

Section 0.6

27. Why is it necessary for $f(x)$ to be continuous on $[a, b]$ for bisection (interval halving) to be successful?

28. Find an $f(x)$ that is discontinuous on $[a, b]$ but for which interval halving does find the root.

▶**29.** Suppose that $f(x)$ has five roots within $[a, b]$. Which of the roots will bisection find if it begins with $x = a$ and $x = b$?

30. In Section 0.5, the following expression is given for the error after n iterations, suggesting that the errors decrease steadily as n increases. Show that this is not always true.

$$\text{error} = (1/2^n)|b - a|$$

Section 0.7

31. Do any of the computer systems in your organization do pipelining? Do any of them have multiple CPUs?

32. Can bisection be speeded up by the use of parallel processing? If not, why not?

▶**33.** How can one recognize when parallel processing can speed up an iterative numerical procedure?

1

Solving Nonlinear Equations

1.2 Interval Halving (Bisection) Revisited

Reviews the method described in the previous chapter with greater emphasis on its error analysis. Its advantages and disadvantages are discussed, and times when it cannot be used to find a root are pointed out.

1.3 Linear Interpolation Methods

Describes two methods, the secant method and the method of false position, both of which are based on approximating $f(x)$ by a straight line in the vicinity of the root.

1.4 Newton's Method

Explains a method that is perhaps the most widely used technique for finding roots of functions. It is more rapidly convergent than almost all other methods, but there are pitfalls that you should know about. Complex valued roots can be obtained by Newton's method if complex arithmetic is used in the computations.

1.5 Muller's Method

Shows that approximating $f(x)$ near a root with a quadratic polynomial (a parabola) significantly improves the rate of convergence over linear interpolation methods, but it still does not equal Newton's method in this regard.

1.6 Fixed-Point Iteration: $x = g(x)$ Method

Departs from the approach used in the previous methods by finding a rearrangement of the form $x = g(x)$ that may converge to a root from a starting value x_0 if this is substituted into $g(x)$ and the resulting x is used as the argument of $g(x)$ and the process repeated. This method not only is feasible for root finding, but it is used to establish important theoretical results.

1.7 Newton's Method for Polynomials

Tells about some special techniques that are useful when $f(x)$ is a polynomial. The section also illustrates how parallel processing can speed up certain operations.

1.8 Bairstow's Method for Quadratic Factors

Discusses a specialized method for finding complex-valued roots of polynomials without using complex arithmetic.

1.9 Other Methods for Polynomials

Describes four other methods that can be used to find the roots of polynomials. Three of these methods have the advantage of not needing starting values. The fourth is extremely rapid in convergence.

1.10 Multiple Roots

Shows some ways to cope with the problems that arise when $f(x)$ has multiple roots. This situation often gives the standard methods great difficulty.

1.11 Theoretical Matters

Examines the important subject of errors in the estimates of roots when the methods of this chapter are used. It develops the proofs for some statements about errors and convergence rates that are merely stated in earlier sections.

1.12 Using DERIVE

Gives information on this valuable "mathematical assistant" in the area of root finding.

Chapter Summary

Gives you a checklist against which you can measure your understanding of the topics of this chapter. Obviously, you should restudy those sections where your comprehension is not complete.

Computer Programs

Illustrates the process of writing programs to find roots, and shows that the language used to implement them is not very important.

1.1 Improved Ideal Gas Laws

If you took a chemistry course, you almost certainly learned that the pressure, p, volume, v, and temperature, T, are interrelated. Increasing the temperature of a confined quantity of gas causes the pressure to increase in proportion to the absolute temperature. (The absolute temperature is measured above absolute zero, which is $-273°C$, or $-460°F$.) When expressed as degrees Centigrade above $-273°$, the temperature scale is known as degrees kelvin, written as °K. This relation is known as Charles's law.

On the other hand, if the volume is decreased—say, by pushing a sliding piston to compress the gas—the pressure increases in proportion to the decrease in volume, provided that the temperature is held constant. This is known as Boyle's law.

The combination of these laws is expressed by the ideal gas law,

$$pv = RT,$$

when the quantity of gas is a weight equal to the molecular weight in grams. R is called the universal gas constant and is equal to 1.98 when pressures are measured in atmospheres (the standard atmosphere is 14.7 pounds per square inch), the volume in liters, and the temperature in °K.

But gases are not "ideal." Particularly at very high pressures and/or at very low temperatures, there is a significant departure from the ideal gas law. One widely known improved gas law is the van der Waals equation:

$$\left(p + \frac{a}{v}\right)(v - b) = RT, \quad \text{or} \quad p = \frac{RT}{(v - b)} - \frac{a}{v},$$

where a and b are constants for the particular gas. (The constant b is supposed to account for the volume occupied by the molecules of the gas, and a is related to the molecular motion.) A more modern improved gas law, one that even more closely shows the relationship, is the Redlich–Kwong equation:

$$p = \frac{RT}{(v - b)} - \frac{A(t)}{v(v + b)},$$

where $A(t)$ is a constant that depends on the temperature and b is another constant.

In one set of experiments, these values were measured: $p = 87.3$, $T = 486.9$, and $v = 12.005$. It is known the $A(t) = 0.0837$ under these conditions. (R is 1.98, the universal gas constant.) We need to know the value of b that satisfies the equation.

Finding roots of $f(x) = 0$ is a common and important problem in applied mathematics. Several methods have been developed to solve this problem, and one of them, interval halving (better known as bisection), has been discussed in Chapter 0. So one way to find the value of b in the preceding example is to transfer p to the right-hand side to form $f(b) = 0$ and then use bisection. Another way would be to plot the $f(b)$ versus b and see where the graph crosses the axis.

We will explore several other root-finding methods in this chapter. Some of these find the root more quickly than the bisection method.

But wait a minute! Actually, this example is just an algebra problem because the equation is really just a quadratic polynomial in b [multiply by $(v^2 - b^2)$]. If we do so, then the quadratic formula can be used to solve for b. (And DERIVE can carry out these steps for us.)

Sometimes an equation can be rearranged to exhibit the unknown as a function of the other parameters, and all we need do is plug in the known values. (DERIVE can do this rearrangement for us, so we don't even need to perform the algebra!) Every linear equation in one unknown can be so rearranged; a root-finding routine is required only for nonlinear relations. Be sure not to jump into a sophisticated root-finding operation when there is simpler way to solve the problem!

If we were not able to find a rearrangement that shows b as a function of the other parameters, many other ways are available to us to find the value of b that satisfies the equation. We could use trial and error, trying different values of b until the two sides of the equation agree. A systematic search would be even better, trying values of b in a succession of values to find a match. We can think of interval halving as a kind of search method.

1.2 Interval Halving (Bisection) Revisited

This ancient but effective method for finding a zero of $f(x)$ was discussed in Chapter 0. Its strategy is to begin with two values of x—a and b—that bracket a root of $f(x) = 0$. It determines that the values $x = a$ and $x = b$ do bracket a root by finding that $f(a) * f(b) < 0$ (because they are of opposite signs). The method then successively divides the interval in half and replaces one endpoint with the midpoint so that again the root is bracketed. One knows in advance that the error in the estimate of the root is less than $|(b - a) * (\frac{1}{2^n})|$, where n is the number of iterations performed. It is required that $f(x)$ be continuous in the interval.

This chapter will describe several different methods for solving $f(x) = 0$. To compare them, we will solve the same equation by each method. This standard equation is

$$f(x) = x^3 + x^2 - 3x - 3 = 0.$$

One can almost see by inspection that a root is $\sqrt{3}$. We will then be able to see how quickly successive iterates converge on the value 1.732150808. Using the algorithm whose pseudocode was given in the previous chapter, the results of Table 1.1 are obtained.

Table 1.1   Finding a root of $f(x) = x^3 + x^2 - 3x - 3$ starting with $a = 1$, $b = 2$, and tolerance of $1E - 4$ by interval halving

Iteration	x_1	x_2	x_3	$F(x_3)$	Maximum error	Actual error
1	1.000000	2.000000	1.500000	−1.875000	0.500000	−0.232051
2	1.500000	2.000000	1.750000	0.171875	0.250000	0.017949
3	1.500000	1.750000	1.625000	−0.943359	0.125000	−0.107051
4	1.625000	1.750000	1.687500	−0.409424	0.062500	−0.044551
5	1.687500	1.750000	1.718750	−0.124786	0.031250	−0.013301
6	1.718750	1.750000	1.734375	0.022030	0.015625	0.002324
7	1.718750	1.734375	1.726563	−0.051756	0.007813	−0.005488
8	1.726563	1.734375	1.730469	−0.014957	0.003906	−0.001582
9	1.730469	1.734375	1.732422	0.003512	0.001953	0.000371
10	1.730469	1.732422	1.731445	−0.005728	0.000977	−0.000605
11	1.731445	1.732422	1.731934	−0.001109	0.000488	−0.000117
12	1.731934	1.732422	1.732178	0.001202	0.000244	0.000127
13	1.731934	1.732178	1.732056	0.000046	0.000122	0.000005

Tolerance met

The main advantage of interval halving is that it is guaranteed to work if $f(x)$ is continuous in $[a, b]$ and if the values $x = a$ and $x = b$ actually bracket a root. Another important advantage that few other root-finding methods share is that the number of iterations to achieve a specified accuracy is known in advance. Since the interval $[a, b]$ is halved each time, the last value of x_3 differs from the true root by less than $\frac{1}{2}$ the last interval. So we can say with surety that

$$
\text{error after } n \text{ iterations} < \left| \frac{(b-a)}{2^n} \right|.
$$

The major objection of interval halving has been that it is slow to converge. Other methods require fewer iterations to achieve the same accuracy (but then we do not always know a bound on the accuracy).

Observe in Table 1.1 that the estimate of the root may be better at an earlier iteration than at later ones. (The second iterate is closer to the true root than are the next two; we are closer at iteration 6 than at iteration 7.) Of course, in this example we have the advantage of knowing the answer, which is never the case. However, the values of $f(x_3)$ themselves show that these better estimates are closer to the root. (This is not an absolute criterion—some functions may be nearly zero at points not so near the root, but, for smooth functions, a small value of the function is a good indicator that we are near to the root. This is especially true when we are quite close to the root.) The methods we consider in later sections use the values of $f(x)$ to find the root more rapidly.

With speedy computers so prevalent today the slowness of the bisection method is of less concern. When the values of Table 1.1 were computed from a program, the results were seen in less than a second.

When the roots of functions must be computed many, many times (this may be a requirement of some other program that does engineering analysis), the efficiency of interval halving may be inadequate. This will be particularly true if $f(x)$ is not given explicitly but, instead, is developed internally within the other program. In that case, finding values of x that bracket the root may also be a problem.

In spite of arguments that other methods find roots with fewer iterations, interval halving is an important tool in the applied mathematician's arsenal. Bisection is generally recommended for finding an approximate value for the root, and then this value is refined by more efficient methods. The reason is that most other root-finding methods require a starting value near to a root—lacking this, they may fail completely.

Do not overlook other techniques that may seem mundane for getting a first approximation to the root. Graphing the function is always helpful in showing where roots occur, and with programs like DERIVE available that do plots so handily, getting the graph before beginning a root-finding routine is a good practice. Searching methods should also be considered as a preliminary step. Stepping through the interval $[-1, 1]$ and testing whether $f(x)$ changes sign will show whether there are roots in that interval. Roots of larger magnitude can be found by stepping through that same interval with x replaced by $1/y$, because the roots of this modified function are the reciprocals of the roots of the original function. Experience with the particular types of problems that are being solved may also suggest approximate values of roots. Even intuition can be a factor. Acton (1970) gives an especially interesting and illuminating discussion.

When there are multiple roots, interval halving may not be applicable, because the function may not change sign at points on either side of the roots. Here a graph will be most important to reveal the situation. In this case, we may be able to find the roots by working with $f'(x)$, which will be zero at a multiple root.

1.3 Linear Interpolation Methods

Although the bisection (interval halving) method is easy to compute and has simple error analysis, it is not very efficient. For most functions, we can improve the speed at which the root is approached through a different scheme. Almost every function can be approximated by a straight line over a small interval. We begin from a value—say, x_0—that is near to a root, r. (We would get x_0 from a graph or from a few applications of bisection.)

The Secant Method

Suppose we assume that $f(x)$ is linear in the vicinity of the root r. Now we choose another point, x_1, which is near to x_0 and also near to r (which we don't know yet), and we draw a straight line through the two points. Figure 1.1 illustrates the situation with the distances along the x-axis exaggerated. (Since we don't know the value of the root yet, the two points could be on opposite sides of the root or to the left of the root, rather than as shown.)

If $f(x)$ were truly linear, the straight line would intersect the x-axis at the root. But $f(x)$ will never be exactly linear because we would never use a root-finding method on a linear function! That means that the intersection of the line with the

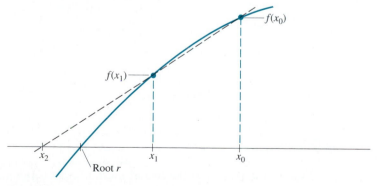

Figure 1.1

x-axis is not at $x = r$ but that it should be close to it. From the obvious similar triangles we can write

$$\frac{(x_0 - x_2)}{f(x_0)} = \frac{(x_0 - x_1)}{f(x_0) - f(x_1)}$$

and from this solve for x_2:

$$x_2 = x_0 - f(x_0) * \frac{(x_0 - x_1)}{f(x_0) - f(x_1)}.$$

Since $f(x)$ is not exactly linear, x_2 is not equal to r, but it should be closer than either of the two points we began with. We can continue to get better estimates of the root if we do this repeatedly, always choosing the two x-values nearest to r for drawing the straight line. Since each newly computed value should be nearer to the root, we can do this easily after the second iterate has been computed, by always using the last two computed points. But after the *first* iteration there aren't "two last computed points." So we make sure to start with x_1 closer to the root than x_0 by testing $f(x_0)$ and $f(x_1)$ and swapping if the first function value is smaller. The net effect of this rule is to set $x_0 = x_1$ and $x_1 = x_2$ after each iteration. The exceptions to this rule are pathological cases, which we consider next.

The technique we have described is known as the secant method because the line through two points on the curve is called the secant line. Here is pseudocode for the secant method algorithm:

An Algorithm for the Secant Method

To determine a root of $f(x) = 0$, given two values, x_0 and x_1, that are near the root,

    IF $|f(x_0)| < |f(x_1)|$:
        Swap x_0 with x_1.
    REPEAT
        Set $x_2 = x_0 - f(x_0) * (x_0 - x_1)/[f(x_0) - f(x_1)]$.
        Set $x_0 = x_1$.
        Set $x_1 = x_2$.
    UNTIL $|f(x_2)| <$ tolerance value.

Note: If $f(x)$ is not continuous, the method may fail.

An alternative stopping criterion for the secant method is when the pair of points being used are sufficiently close together.

An Example

Table 1.2 shows the results of the secant method on the function $f(x) = x^3 + x^2 - 3x - 3 = 0$ with starting values of $x = 1$ and $x = 2$. Values not very close to the root (which is at $x = 1.73205$) were deliberately chosen. Table 1.3 shows the results on a transcendental function $f(x) = 3x + \sin(x) - e^x$, starting from $x = 0$ and $x = 1$. This latter function has a zero at $x = 0.360421703$.

Table 1.2 Secant method on $f(x) = x^3 + x^2 - 3x - 3$

0	1	2	1.571429	-1.364432
1	2	1.571429	1.705411	-0.2477449
2	1.571429	1.705411	1.735136	2.925562E-02
3	1.705411	1.735136	1.731996	-5.147391E-04
4	1.735136	1.731996	1.732051	-1.422422E-06

At $x = 1.732051$, tolerance of .00001 met!

Table 1.3 Secant method on $f(x) = 3x + \sin(x) - e^x$

Iteration	x_0	x_1	x_2	$F(x_2)$
0	1	0	0.4709896	0.2651588
1	0	0.4709896	0.3722771	2.953367E-02
2	0.4709896	0.3722771	0.3599043	-1.294787E-03
3	0.3722771	0.3599043	0.3604239	5.552969E-06
4	0.3599043	0.3604239	0.3604217	3.554221E-08

At $x = .3604217$, tolerance of .0000001 met!

An objection is sometimes raised about the secant method. If the function is far from linear near the root, the successive iterates can fly off to points far from the root, as seen in Fig. 1.2. Another pathological case with a result similar to Fig. 1.2 arises when an iterate duplicates a previous one, resulting in an endless loop that never reaches the true value for the root.

If the method is being carried out by a program that displays the successive iterates, the user can interrupt the program should such improvident behavior be observed. Also, if the function was plotted before starting the method, it is unlikely that the problem will be encountered, because a better starting value would be used. There are times when this remedy is not possible: when the routine is being used within another program that needs to find a root before it can proceed.

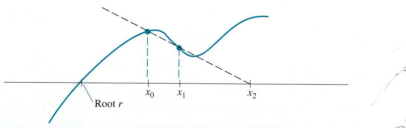

Figure 1.2 A pathological case for the secant method

Linear Interpolation (False Position)

A way to avoid such pathology is to ensure that the root is bracketed between the two starting values and remains between the successive pairs. When this is done, the method is known as *linear interpolation*, or, more often, as the *method of false position* (in Latin, *regula falsi*). This technique is similar to bisection except the next iterate is taken at the intersection of a line between the pair of x-values and the x-axis rather than at the midpoint. Doing so gives faster convergence than does bisection, but at the expense of a more complicated algorithm.

Here is the pseudocode for *regula falsi* (method of false position).

An Algorithm for the Method of False Position (*regula falsi*)

To determine a root of $f(x) = 0$, given values of x_0 and x_1 that bracket a root, that is, $f(x_0)$ and $f(x_1)$ are of opposite sign,

> REPEAT
> > Set $x_2 = x_0 - f(x_0) * (x_0 - x_1)/(f(x_0) - f(x_1))$.
> > IF $f(x_2)$ of opposite sign to $f(x_0)$:
> > > Set $x_1 = x_2$.
> > ELSE Set $x_0 = x_2$.
> > ENDIF.
> UNTIL $|f(x_2)| <$ tolerance value.

Note: The method may fail if $f(x)$ is discontinuous on the interval.

Table 1.4 compares the result of three methods—interval halving (bisection), linear interpolation, and the secant method—on $f(x) = 3x + \sin(x) - e^x = 0$. Observe that the speed of convergence is best for the secant method, poorest for interval

Table 1.4 Comparison of methods, $f(x) = 3x + \sin(x) - e^x = 0$

	Interval halving		False position		Secant method	
Iteration	x	$f(x)$	x	$f(x)$	x	$f(x)$
1	0.5	0.330704	0.470990	0.265160	0.470990	0.265160
2	0.25	−0.286621	0.372277	0.029533	0.372277	0.029533
3	0.375	0.036281	0.361598	$2.94 * 10^{-3}$	0.359904	$-1.29 * 10^{-3}$
4	0.3125	−0.121899	0.360538	$2.90 * 10^{-4}$	0.360424	$5.53 * 10^{-6}$
5	0.34375	−0.041956	0.360433	$2.93 * 10^{-5}$	0.360422	$2.13 * 10^{-7}$
Error after 5 iterations	0.01667		$-1.17 * 10^{-5}$		$<-1 * 10^{-7}$	

(Exact value of root is 0.360421703.)

halving, and intermediate for false position. Notice that false position converges to the root from only one side, slowing it down, especially if that end of the interval is farther from the root. There is a way to avoid this result, called modified linear interpolation. We omit the details of this method, but they can be found in earlier editions of this book.

1.4 Newton's Method

One of the most widely used methods of solving equations is Newton's method.* Like the previous ones, this method is also based on a linear approximation of the function, but does so using a tangent to the curve. Figure 1.3 gives a graphical description. Starting from a single initial estimate, x_0, that is not too far from a root, we move along the tangent to its intersection with the x-axis, and take that as the next approximation. This is continued until either the successive x-values are sufficiently close or the value of the function is sufficiently near zero.†

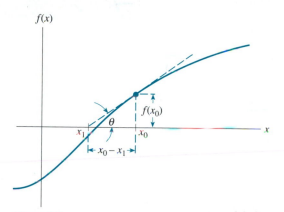

Figure 1.3

The calculation scheme follows immediately from the right triangle shown in Fig. 1.3, which has the angle of inclination of the tangent line to the curve at $x = x_0$ as one of its acute angles:

$$\tan \theta = f'(x_0) = \frac{f(x_0)}{x_0 - x_1}, \qquad x_1 = x_0 - \frac{f(x_0)}{f'(x_0)}.$$

---

*Newton did not publish an extensive discussion of this method, but he solved a cubic polynomial in *Principia* (1687). The version given here is considerably improved over his original example.
† Which criterion should be used often depends on the particular physical problem to which the equation applies. Customarily, agreement of successive x-values to a specified tolerance is required.

We continue the calculation scheme by computing

$$x_2 = x_1 - \frac{f(x_1)}{f'(x_1)},$$

or, in more general terms,

$$x_{n+1} = x_n - \frac{f(x_n)}{f'(x_n)}, \qquad n = 0, 1, 2, \ldots$$

Newton's algorithm is widely used because, at least in the near neighborhood of a root, it is more rapidly convergent than any of the methods discussed so far. We show in a later section that the method is quadratically convergent, by which we mean that the error of each step approaches a constant K times the square of the error of the previous step. The net result of this is that the number of decimal places of accuracy nearly doubles at each iteration. However, offsetting this is the need for two function evaluations at each step, $f(x_n)$ and $f'(x_n)$.*

When Newton's method is applied to $f(x) = 3x + \sin x - e^x = 0$, we have the following calculations:

$$f(x) = 3x + \sin x - e^x,$$
$$f'(x) = 3 + \cos x - e^x.$$

If we begin with $x_0 = 0.0$, we have

$$x_1 = x_0 - \frac{f(x_0)}{f'(x_0)} = 0.0 - \frac{-1.0}{3.0} = 0.33333;$$

$$x_2 = x_1 - \frac{f(x_1)}{f'(x_1)} = 0.33333 - \frac{-0.068418}{2.54934} = 0.36017;$$

$$x_3 = x_2 - \frac{f(x_2)}{f'(x_2)} = 0.36017 - \frac{-6.279 \times 10^{-4}}{2.50226} = 0.3604217.$$

After three iterations, the root is correct to seven significant digits. Comparing this with the results in Table 1.3, we see that Newton's method converges considerably more rapidly than the previous methods. In comparing numerical methods, however, we usually count the number of times functions must be evaluated. Because Newton's method requires two function evaluations per step, the comparison is not as one-sided in favor of Newton's method as at first appears; the three iterations with Newton's method required six function evaluations. Five iterations with the previous methods required seven evaluations.

---

* Another problem with Newton's method is that finding $f'(x)$ may be difficult. Computer algebra systems can be a real help.

A more formal statement of the algorithm for Newton's method, suitable for implementation in a computer program, is shown here.

Newton's Method

To determine a root of $f(x) = 0$, given a value x_0 reasonably close to the root,

> Compute $f(x_0), f'(x_0)$.
> Set $x_1 = x_0$.
> IF $(f(x_0) \neq 0)$ AND $(f'(x_0) \neq 0)$
>   REPEAT
>   Set $x_0 = x_1$
>   Set $x_1 = x_0 - f(x_0)/f'(x_0)$.
>   UNTIL $(|x_0 - x_1| < $ tolerance value 1) OR
> $(|f(x_1)| < $ tolerance value 2).

Note: The method may converge to a root different from the expected one or diverge if the starting value is not close enough to the root.

When Newton's method is applied to polynomial functions, special techniques facilitate such application. We consider these in a later section of this chapter.

In some cases Newton's method will not converge. Figure 1.4 illustrates this situation. Starting with x_0, one never reaches the root r because $x_6 = x_1$ and we are in an endless loop. Observe also that if we should ever reach the minimum or maximum of the curve, we will fly off to infinity. We will develop the analytical condition for this in a later section and show that Newton's method is quadratically convergent in most cases.

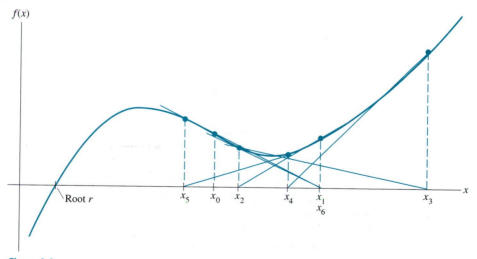

Figure 1.4

Relating Newton's Method to Other Methods

It is of interest to notice that the previous interpolation methods are closely related to Newton's method. For linear interpolation, whose algorithm we can write as

$$x_{n+1} = x_n - \frac{f(x_n)}{\dfrac{f(x_n) - f(x_{n-1})}{x_n - x_{n-1}}},$$

we see that the denominator of the fractional term is exactly the definition of the derivative except not taken to the limit as the two x-values approach each other. This *difference quotient* is an approximation to the derivative, as we will explain in detail in a later chapter. Since the denominator of the fractional term is an approximation to the derivative of f, we see the close resemblance to Newton's method.

The secant method has exactly this same resemblance to Newton's method because it is just linear interpolation without the requirement that the two x-values bracket the root. Since these two values usually are closer together than for linear interpolation, the approximation to the derivative is even better.

From this we see that there is an alternative way to get the derivative for Newton's method. If we compute $f(x)$ at two closely spaced values for x and divide the difference in the function values by the difference in x-values, we have the derivative (nearly) without having to differentiate. While this sounds like spending an extra function evaluation, we avoid having to evaluate the derivative function and so it breaks even. (Convergence will not usually be as fast, however.)

Complex Roots

Newton's method works with complex roots if we give it a complex value for the starting value. Here is an example.

EXAMPLE 1.1   Use Newton's method on $f(x) = x^3 + 2x^2 - x + 5$.

Figure 1.5 shows the graph of $f(x)$. It has a real root at about $x = -3$, while the other two roots are complex because the x-axis is not crossed again.

If we begin Newton's method with $x_0 = 1 + i$ (we used this in the lack of knowledge about the complex root), we get these successive iterates:

1. $0.486238 + 1.04587i$
2. $0.448139 + 1.23665i$
3. $0.462720 + 1.22242i$
4. $0.462925 + 1.22253i$
5. $0.462925 + 1.22253i$

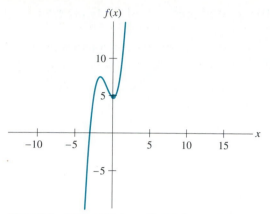

Figure 1.5 Plot of $f(x) = x^3 + 2x^2 - x + 5$

Since the fourth and fifth iterates agree to six significant figures, we are sure that we have an estimate good to at least that many figures. The second complex root is the conjugate of this: $0.462925 - 1.22253i$. If we begin with $x_0 = 1 - i$, the method converges to the conjugate. DERIVE does complex arithmetic; we do need to define x as complex valued, but that is all there is to it. We got the preceding results from DERIVE.

If we begin with a real starting value—say, $x_0 = -3$—we get convergence to the root at $x = -2.92585$.

1.5 Muller's Method

Most of the root-finding methods that we have considered so far have approximated the function in the neighborhood of the root by a straight line. Obviously, this is never true; if the function were linear, finding the root would take practically no effort. Muller's method is based on approximating the function in the neighborhood of the root by a quadratic polynomial. This gives a much closer match to the actual curve.

A second-degree polynomial is made to fit three points near a root, $[x_0, f(x_0)]$, $[x_1, f(x_1)]$, $[x_2, f(x_2)]$, and the proper zero of this quadratic, using the quadratic formula, is used as the improved estimate of the root. The process is then repeated using the set of three points nearest the root being evaluated.

The procedure for Muller's method is developed by writing a quadratic equation that fits through three points in the vicinity of a root, in the form $av^2 + bv + c$. (See Fig. 1.6.) The development is simplified if we transform axes to pass through the middle point, by letting $v = x - x_0$.

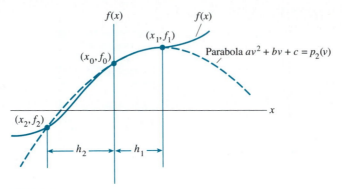

Figure 1.6

Let $h_1 = x_1 - x_0$ and $h_2 = x_0 - x_2$. We evaluate the coefficients by evaluating $p_2(v)$ at the three points:

$$v = 0: \quad a(0)^2 + b(0) + c = f_0;$$
$$v = h_1: \quad ah_1^2 + bh_1 + c = f_1;$$
$$v = -h_2: \quad ah_2^2 - bh_2 + c = f_2.$$

From the first equation, $c = f_0$. Letting $h_2/h_1 = \gamma$, we can solve the other two equations for a and b:

$$a = \frac{\gamma f_1 - f_0(1 + \gamma) + f_2}{\gamma h_1^2(1 + \gamma)}, \qquad b = \frac{f_1 - f_0 - ah_1^2}{h_1}.$$

After computing a, b, and c, we solve for the root of $av^2 + bv + c = 0$ by the quadratic formula, choosing the root nearest to the middle point x_0. This value is

$$\text{root} = x_0 - \frac{2c}{b \pm \sqrt{b^2 - 4ac}},$$

with the sign in the denominator taken to give the largest absolute value of the denominator (that is, if $b > 0$, choose plus; if $b < 0$, choose minus; if $b = 0$, choose either).

We take the root of the polynomial as one of a set of three points for the next approximation, taking the three points that are most closely spaced (that is, if the root is to the right of x_0, take x_0, x_1, and the root; if to the left, take x_0, x_2, and the root). We always reset the subscripts to make x_0 be the middle of the three values.

An algorithm for Muller's method is

> **Muller's Method**
>
> Given the points x_2, x_0, x_1 in increasing value,
>
> 1. Evaluate the corresponding function values f_2, f_0, f_1.
> 2. Find the coefficients of the parabola determined by the three points.
> 3. Compute the two roots of the parabolic equation.
> 4. Choose the root closest to x_0 and label it x_r.
> 5. IF $x_r > x_0$ THEN rearrange x_0, x_r, x_1 into x_2, x_0, x_1
>   ELSE rearrange x_2, x_r, x_0 into x_2, x_0, x_1.
> 6. IF $|f(x_r)| <$ FTOL, THEN return (x_r)
>   ELSE go to 1.

Muller's method, like Newton's, will find a complex root if given complex starting values. Of course, the computations must use complex arithmetic.

EXAMPLE 1.2   Find a root between 0 and 1 of the same transcendental function as before: $f(x) = 3x + \sin(x) - e^x$. Let

$$x_0 = 0.5, \quad f(x_0) = 0.330704 \qquad h_1 = 0.5,$$
$$x_1 = 1.0, \quad f(x_1) = 1.123189 \qquad h_2 = 0.5,$$
$$x_2 = 0.0, \quad f(x_2) = -1 \qquad\qquad \gamma = 1.0.$$

Then

$$a = \frac{(1.0)(1.123189) - 0.330704(2.0) + (-1)}{1.0(0.5)^2(2.0)} = -1.07644,$$

$$b = \frac{1.123189 - 0.330704 - (-1.07644)(0.5)^2}{0.5} = 2.12319,$$

$$c = 0.330704,$$

and

$$\text{root} = 0.5 - \frac{2(0.330704)}{2.12319 - \sqrt{(2.12319)^2 - 4(-1.07644)(0.330704)}}$$
$$= 0.354914.$$

For the next iteration, we have

$$x_0 = 0.354914, \quad f(x_0) = -0.0138066 \qquad h_1 = 0.145086,$$
$$x_1 = 0.5, \qquad\quad f(x_1) = 0.330704 \qquad h_2 = 0.354914,$$
$$x_2 = 0, \qquad\qquad f(x_2) = -1 \qquad\qquad\quad \gamma = 2.44623.$$

Then

$$a = \frac{(2.44623)(0.330704) - (-0.0138066)(3.44623) + (-1)}{2.44623(0.145086)^2(3.44623)} = -0.808314,$$

$$b = \frac{0.330704 - (-0.0138066) - (-0.808314)(0.145086)^2}{0.145086} = 2.49180,$$

$$c = -0.0138066,$$

$$\text{root} = 0.354914 - \frac{2(-0.0138066)}{2.49180 - \sqrt{(2.49180)^2 - 4(-0.808314)(-0.0138066)}}$$

$$= 0.360465.$$

After a third iteration, we get 0.3604217 as the value for the root, which is identical to that from Newton's method after three iterations.

▲

Experience shows that Muller's method converges at a rate that is similar to that for Newton's method.[*] It does not require the evaluation of derivatives, however, and (after we have obtained the starting values) needs only one function evaluation per iteration. There is an initial penalty in that one must evaluate the function three times, but this is frequently overcome by the time the required precision is attained.

1.6 Fixed-Point Iteration: $x = g(x)$ Method

The method known as *fixed-point iteration* (we also call it the $x = g(x)$ method) is a very useful way to get a root of $f(x) = 0$. This method is also the basis for some important theory. To use the method, we rearrange $f(x)$ into an equivalent form $x = g(x)$, which usually can be done in several ways. Observe that if $f(r) = 0$, where r is a root of $f(x)$, it follows that $r = g(r)$. Whenever we have $r = g(r)$, r is said to be a fixed point for the function g.

Under suitable conditions that we explain later, the iterative form

$$x_{n+1} = g(x_n) \qquad n = 0, 1, 2, 3, \ldots,$$

converges to the fixed point r, a root of $f(x)$.

Here is a simple example:

$$f(x) = x^2 - 2x - 3 = 0.$$

$f(x)$ is easy to factor to show roots at $x = -1$ and $x = 3$. (We pretend that we don't know this.)

Suppose we rearrange to give this equivalent form:

$$x = g_1(x) = \sqrt{2x + 3}.$$

_____

[*] Atkinson (1978) shows that each error is about proportional to the previous error to the 1.85th power.

If we start with $x = 4$ and iterate with the fixed-point algorithm, successive values of x are

$$x_0 = 4,$$
$$x_1 = \sqrt{11} = 3.31662,$$
$$x_2 = \sqrt{9.63325} = 3.10375,$$
$$x_3 = \sqrt{9.20750} = 3.03439,$$
$$x_4 = \sqrt{9.06877} = 3.01144,$$

and it appears that the values are converging on the root at $x = 3$.

Other Rearrangements

Another rearrangement of $f(x)$ is

$$x = g_2(x) = \frac{3}{(x - 2)}.$$

Let us start the iterations again with $x_0 = 4$. Successive values then are

$$x_0 = 4,$$
$$x_1 = 1.5,$$
$$x_2 = -6,$$
$$x_3 = -0.375,$$
$$x_4 = -1.263158,$$
$$x_5 = -0.919355,$$
$$x_6 = -1.02762,$$
$$x_7 = -0.990876,$$
$$x_8 = -1.00305,$$

and it seems that we now converge to the other root, at $x = -1$. We also see that the convergence is oscillatory rather than monotonic, as in the first case.

Consider a third rearrangement:

$$x = g_2(x) = \frac{(x^2 - 3)}{2}.$$

Starting again with $x_0 = 4$, we get

$$x_0 = 4,$$
$$x_1 = 6.5,$$
$$x_2 = 19.625,$$
$$x_3 = 191.070,$$

and the iterates are obviously diverging.

This difference in behavior of the three rearrangements is interesting and worth further study. First, though, let us look at the graphs of the three cases. The fixed point of $x = g(x)$ is the intersection of the line $y = x$ and the curve $y = g(x)$ plotted against x. Figure 1.7 shows the three cases.

Observe that we always get the successive iterates by this construction: Start on the x-axis at the initial x_0, go vertically to the curve, then horizontally to the line $y = x$, then vertically to the curve, and again horizontally to the line. Repeat this process until the points on the curve converge to a fixed point or else diverge. It appears that the different behaviors depend on whether the slope of the curve is greater, less, or of opposite sign to the slope of the line (which equals $+1$).

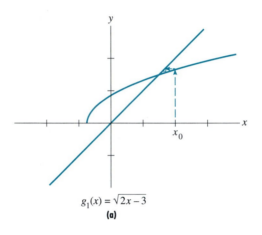

$$g_1(x) = \sqrt{2x - 3}$$

(a)

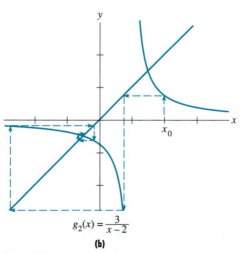

$$g_2(x) = \frac{3}{x - 2}$$

(b)

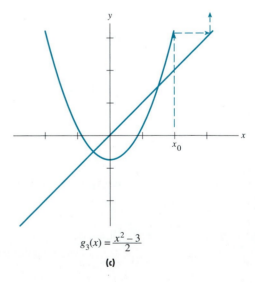

$$g_3(x) = \frac{x^2 - 3}{2}$$

(c)

Figure 1.7

Here is pseudocode for the fixed-point ($x = g(x)$) method:

Iteration with the Form $x = g(x)$

To determine a root of $f(x) = 0$, given a value x_1 reasonably close to the root,

Rearrange the equation to an equivalent form $x = g(x)$.
Set $x_2 = x_1$.
REPEAT
    Set $x_1 = x_2$.
    Set $x_2 = g(x_1)$.
UNTIL $|x_1 - x_2| <$ tolerance value.

Note: The method may converge to a root different from the expected one, or it may diverge. Different rearrangements will converge at different rates.

We will give the details of the proof of the following statement in a later section, but the statement is true.

If $g(x)$ and $g'(x)$ are continuous on an interval about a root r of the equation $x = g(x)$, and if $|g'(x)| < 1$ for all x in the interval, then $x_{n+1} = g(x_n)$, $n = 0, 1, 2, \ldots$, will converge to the root $x = r$, provided that x_1 is chosen in the interval. Note that this is a sufficient condition only, since for some equations, convergence is secured even though not all the conditions hold.*

Examine again the three graphs of Fig. 1.7. Observe that in part (a) the slope of the curve $[g'(x)]$ is positive but less than 1. There will be monotonic convergence. In part (b), the slope of the curve is negative and less than 1 in value; there is oscillatory convergence. In part (c) the slope of the curve is greater than 1, so the iterates diverge. It is easy to see from similar graphs that convergence is faster if the slope of the curve is nearer to zero.

The preceding conditions for convergence are based on the fact that the error at each step of the iterations is

$$|e_{n+1}| = |g(\xi_n)| * |e_n|,$$

where ξ_n is a value between x_n and r, the fixed point, and e_i is the error of the ith iterate.

---

* The analytical test that $|g'(x)| < 1$ is often awkward to apply. A constructive test is merely to observe whether the successive x_i values converge. In a computer program it is worthwhile to determine whether $|x_3 - x_2| < |x_2 - x_1|$.

When we are near to r, the derivative of $g(x)$ is essentially constant. That is, each successive error is a fraction of the preceding one, or, in other words, the error at any step is approximately proportional to the previous one.

The fixed-point method is particularly easy to program. In DERIVE, the ITER-ATES function implements the algorithm exactly. We will discuss the use of DERIVE later in this chapter in a section that explains DERIVE's applicability to this and the other methods. Rather than incorporate a test in the program for divergence, it is easiest just to exhibit the successive iterates and manually stop the program if the values don't converge.

The major difficulty one finds with fixed-point iteration is determining a suitable $g(x)$ that converges to the desired root.

Accelerating Convergence

Even though the statement that each error is proportional to the previous one is only an approximation, assuming it to be true allows us to accelerate the convergence. This is known as *Aitken acceleration*.

Assume that

$$e_n = x_n - r = K^{n-1}e_1$$

or

$$x_n = r + K^{n-1}e_1.$$

Similarly, if

$$e_{n+1} = x_{n+1} - r = K^n e_1$$

and

$$e_{n+2} = x_{n+2} - r = K^{n+1}e_1,$$

then

$$x_{n+1} = r + K^n e_1$$

and

$$x_{n+2} = r + K^{n+1}e_1.$$

Substitute these expressions into

$$\frac{x_n x_{n+2} - x_{n+1}^2}{x_{n+2} - 2x_{n+1} + x_n}.$$

It is found that

$$\frac{x_n x_{n+2} - x_{n+1}^2}{x_{n+2} - 2x_{n+1} + x_n} = \frac{(r + K^{n-1}e_1)(r + K^{n+1}e_1) - (r + K^n e_1)^2}{(r + K^{n+1}e_1) - 2(r + Ke_1) + (r + K^{n-1}e_1)}$$

$$= \frac{r(K^{n+1} - 2K^n + K^{n-1})e_1}{(K^{n+1} - 2K^n + K^{n-1})e_1} = r.$$

From three successive estimates of the root, x_1, x_2, and x_3, we extrapolate to an improved estimate. Since the assumption of constant ratio between successive errors is not normally true, our extrapolated value is not exact, but it is usually improved. One uses this technique by calculating two new values, extrapolating again, and so on.

A different form is useful to avoid the round-off problem that occurs in subtracting large numbers of nearly the same magnitude. Define

$$\Delta x_i = x_{i+1} - x_i,$$
$$\Delta^2 x_i = \Delta(\Delta x_i) = \Delta(x_{i+1} - x_i) = x_{i+2} - 2x_{i+1} + x_i.$$

Our acceleration scheme becomes

$$r = x_n - \frac{(\Delta x_n)^2}{\Delta^2 x_n} = \frac{x_n x_{n+2} - x_{n+1}^2}{x_{n+2} - 2x_{n+1} + x_n}.$$

The differences are most readily computed in a table. We illustrate with the iterates from the first example in this section:

$$f(x) = x^2 - 2x - 3 = 0,$$
$$x_{n+1} = \sqrt{2x_n + 3}, \qquad x_1 = 4,$$

x	Δx	$\Delta^2 x$
$x_1 = 4.000$		
	0.684	
$x_2 = 3.316$		0.472
	0.212	
$x_3 = 3.104$		

The accelerated estimate is

$$r = 4.000 - \frac{(0.684)^2}{0.472} = 3.009.$$

We have jumped ahead about two iterations. If $g(x)$ is expensive to compute, we have gained. Often Aitken acceleration gives bigger jumps than this. In fact, there is an excellent measure to determine when to use Aitken acceleration.

Suppose that for some n we have $x_n, x_{n+1}, x_{n+2}, x_{n+3}$. Then evaluate

$$C = \frac{\sum x_i x_{i+1} - \frac{1}{3}\sum x_i \sum x_{i+1}}{\sqrt{\left[\sum x_i^2 - \frac{1}{3}\left(\sum x_i\right)^2\right]\left[\sum x_{i+1}^2 - \frac{1}{3}\left(\sum x_{i+1}\right)^2\right]}},$$

where the sums are from $i = n$ to $i = n + 2$. If C is close to ± 1, then Aitken acceleration is most effective. In the present example we have the values

$$x_0 = 4.000,$$
$$x_1 = 3.316,$$
$$x_2 = 3.104,$$
$$x_3 = 3.034.$$

We find that, with $n = 0$ in the formula,

$$C = \frac{32.974 - \frac{1}{3}(10.42)(9.454)}{\sqrt{(36.631 - 36.1921)(29.8358 - 29.7927)}}$$
$$= 0.99992.$$

See Jones (1982) for this and further extensions for improving the acceleration method. Aitken acceleration can be applied in any iterative process where the errors can be assumed to decrease proportionately, not just to this $x = g(x)$ method.

1.7   Newton's Method for Polynomials

Polynomial functions are of special importance. We will see throughout the remainder of this book that many valuable numerical procedures are based on polynomials. This important role of polynomial functions is due to their "nice" behavior: They are everywhere continuous, they are smooth, their derivatives are also continuous and smooth, and they are readily evaluated. Descartes' rule of signs (see Appendix A) lets us predict the number of positive roots. Polynomials are particularly well adapted to computers because the only mathematical operations they require for evaluation are addition, subtraction, and multiplication, all of which are speedy operations on computers.

Because of this special importance of polynomials, we now consider how our root-finding methods can be applied to them. For most of the methods previously discussed there is nothing new to say, but for Newton's method there are significant new ideas to consider. We begin on an historical note, with a procedure that saves time in hand computations. However, we will see that this same procedure is also the basis for computer calculations as well. We assume in this section that we are finding a simple root of the polynomial. Functions with multiple roots are discussed in another section.

In applying Newton's method to polynomials in a hand computation, it is most efficient to evaluate $f(x_n)$ and $f'(x_n)$ by use of synthetic division.* We illustrate this

---

* The mechanics of synthetic division, whereby we divide a polynomial by the factor $x - x_i$, are explained in most algebra books. In the example, when $x^3 + x^2 - 3x - 3$ is divided by $(x - 2)$, the result is $x^2 + 3x + 3$, with a remainder of 3. In brief, as shown in the example, the coefficients of the polynomial are first written in a row. Below a line, the first coefficient is copied down. This is multiplied by x_0 and added to the second coefficient. The result is written below the line. This is then multiplied by x_0 and added to the next coefficient, and the process is repeated. The last result is equal to $f(x_0)$.

by the same cubic polynomial that we used before, $x^3 + x^2 - 3x - 3 = 0$, which has a root at $x = \sqrt{3}$. We begin with the value $x = 2$. We utilize the remainder theorem to evaluate $f(2)$, and evaluate $f'(2)$ as the remainder when the reduced polynomial (of degree 2 here) is divided by $(x - 2)$:

$$
\begin{array}{r|rrrr}
x_0 = 2 & 1 & 1 & -3 & -3 \\
        &   & 2 & 6 & 6 \\
\hline
        & 1 & 3 & 3 & ③ \leftarrow \quad \text{Remainder} = f(2). \\
        &   & 2 & 10 & \\
\hline
        & 1 & 5 & ⑬ \leftarrow\!\!\!-\!\!\!- \quad \text{Second remainder} = f'(2). \\
\end{array}
$$

We use the values from the synthetic division—$f(2) = 3$ and $f'(2) = 13$—to get an improved estimate of the root by Newton's method:

$$x_1 = 2 - \frac{3}{13} = 1.76923\ldots.$$

Continuing,

$$
\begin{array}{r|rrrr}
x_1 = 1.76923 & 1 & 1 & -3 & -3 \\
              &   & 1.76923 & 4.89940 & 3.36048 \\
\hline
              & 1 & 2.76923 & 1.89940 & 0.36048 \\
              &   & 1.76923 & 8.02957 & \\
\hline
              & 1 & 4.53846 & 9.92897 & \\
\end{array}
$$

$$x_2 = 1.76923 - \frac{0.36048}{9.92897} = 1.73292.$$

Similarly,

$$x_3 = 1.73292 - \frac{0.00823}{9.47487} = 1.73205.$$

The value of x_3 is correct to five decimals. To observe the improvement in accuracy, consider the successive errors:

	Error	**Number of correct figures**
$x_0 = 2$	0.26895	1
$x_1 = 1.76923$	0.03718	2
$x_2 = 1.73292$	0.00087	4
$x_3 = 1.73205$	0.00000	6+

To compute with five decimal places, as in this example, we used a calculator. (If you have access to a calculator with storage for two or more values, you will find it especially well adapted to this method.)

The initial value at which Newton's method is begun can make a considerable difference. For example, if this problem is started with $x = 1$, the following values result:

x	$f(x)$	$f'(x)$
$x_0 = 1$	-4	2
$x_1 = 3$	24	30
$x_2 = 2.2$	5.888	15.92
$x_3 = 1.83015$		

From here on, the convergence is rapid, for we are using iterates just about as near the root as in the previous example.

After a first root is found (as shown by a remainder that is very small), one normally proceeds to determine additional roots from the reduced polynomial (whose coefficients are in the third row of the synthetic-division tableau). This technique makes the computations somewhat shorter. In the example, the reduced equation is a quadratic, so the quadratic formula would be used, but if a higher-degree polynomial were being solved, Newton's method employing synthetic division would be employed to improve an initial estimate of a second root. The process is then repeated until the reduced equation is of second degree.

This technique of working with the reduced function can be used even if the function is not a polynomial. After a root r of $f(x) = 0$ has been found, the new function $F(x) = f(x)/(x - r)$ will have all the roots of $f(x)$ except the root r. This procedure is called *deflating the function*. One must remember that a discontinuity has been introduced at $x = r$, however. Figure 1.8 shows the example $P_3(x)$ and the deflated polynomial, $P_2(x)$. We suggest that you explore how deflation works on nonpolynomial functions by graphing $f(x) = (x - 1)(e^x - \cos x)$, then $g(x) =$

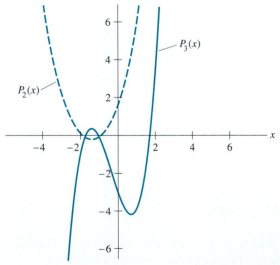

Figure 1.8   A polynomial and its deflated descendant

$f(x)/x$, $h(x) = f(x)/(x - 1)$, and comparing the graphs. You may also wish to compare to $h(x)/x$ and to functions derived by deflating $f(x)$ with roots other than $x = 0$ and $x = 1$.

It should be observed that using deflated functions can result in unexpected errors. If the first root is determined only approximately, the coefficients of the reduced equation are themselves not exact and the succeeding roots are subject not only to round-off errors and the errors that occur when iterations are terminated too soon, but also to inherited errors residing in the nonexact coefficients. This is an example of propagated error. Some functions are extremely sensitive, in that small changes in the value of the coefficients cause large differences in the roots.* Removing roots in order of increasing magnitude is said to minimize the difficulty, and the use of double-precision arithmetic will further help preserve accuracy.

It is of interest to develop the synthetic-division algorithm and to establish the remainder theorems. The scheme is also the most efficient way to evaluate polynomials and their derivatives in a computer program.

Write the nth-degree polynomial as

$$P_n(x) = a_0 x^n + a_1 x^{n-1} + \cdots + a_{n-1} x + a_n.$$

We wish to divide this by the factor $(x - x_1)$, giving a reduced polynomial $Q_{n-1}(x)$ of degree $n - 1$, and a remainder, b_n, which is a constant:

$$\frac{P_n(x)}{x - x_1} = Q_{n-1}(x) + \frac{b_n}{x - x_1}.$$

Rearranging yields

$$P_n(x) = (x - x_1)Q_{n-1}(x) + b_n.$$

Note that at $x = x_1$,

$$P_n(x_1) = (0)[Q_{n-1}(x_1)] + b_n,$$

which is the remainder theorem: The remainder on division by $(x - x_1)$ is the value of the polynomial at $x = x_1$, $P_n(x_1)$.

If we differentiate $P_n(x)$, we get

$$P_n'(x) = (x - x_1)Q_{n-1}'(x) + (1)Q_{n-1}(x) + 0.$$

Letting $x = x_1$, we have

$$P_n'(x_1) = Q_{n-1}(x_1).$$

We evaluate the Q-polynomial at x_1 by a second division whose remainder equals $Q_{n-1}(x_1)$. This verifies that the second remainder from synthetic division yields the value for the derivative of the polynomial.

---

* A classic example has been given by Wilkinson (1959). The 20th-degree polynomial with roots $-1, -2, \ldots, -20$ begins as $x^{20} + 210x^{19} + \cdots + 20! = 0$. If 2^{-23} is added to the coefficient of x^{19}, the roots change noticeably; five of them become complex with real parts that differ from the original roots by as much as 0.73.

We now develop the synthetic-division algorithm, writing $Q_{n-1}(x)$ in form similar to $P_n(x)$:

$$P_n(x) = a_0 x^n + a_1 x^{n-1} + \cdots + a_{n-1} x + a_n$$
$$= (x - x_1) Q_{n-1}(x) + b_n$$
$$= (x - x_1)(b_0 x^{n-1} + b_1 x^{n-2} + \cdots + b_{n-2} x + b_{n-1}) + b_n.$$

Multiplying out and equating coefficients of like terms in x, we get

$$\begin{array}{lll}
\text{Coef. of } x^n: a_0 = b_0 & & b_0 = a_0 \\
x^{n-1}: a_1 = b_1 - b_0 x_1 & & b_1 = a_1 + b_0 x_1 \\
x^{n-2}: a_2 = b_2 - b_1 x_1 & \text{or} & b_2 = a_2 + b_1 x_1 \\
x: a_{n-1} = b_{n-1} - b_{n-2} x_1 & & b_{n-1} = a_{n-1} + b_{n-2} x_1 \\
\text{Const.}: a_n = b_n - b_{n-1} x_1 & & b_n = a_n + b_{n-1} x_1.
\end{array}$$

The general form is $b_i = a_i + b_{i-1} x_1$, by which all the b's except b_0 may be calculated. If this is compared to the preceding synthetic divisions, it is seen to be identical, except that we now have a vertical array. The horizontal layout is easier for hand computation. For evaluation of the derivative, a set of c-values is computed from the b's in the same way in which the b's are computed from the a's.

Synthetic division is also known as the *nested multiplication method* of evaluating polynomials. Consider the fifth-degree polynomial, evaluated at $x = x_1$:

$$a_0 x_1^5 + a_1 x_1^4 + a_2 x_1^3 + a_3 x_1^2 + a_4 x_1 + a_5.$$

We can rewrite this as

$$(((((a_0 x_1 + a_1) x_1 + a_2) x_1 + a_3) x_1 + a_4) x_1 + a_5.$$

In the original form, $5 + 4 + 3 + 2 + 1 = 15$ multiplications are required, plus five additions. In the nested form, only five multiplications are required, plus five additions; it is obviously the more efficient method.

Comparing this with the equations $b_1 = a_1 + b_0 x_1$ and $b_i = a_i + b_{i-1} x_1$ for synthetic division, we see that the successive terms are formed in exactly the same way, so that synthetic division and nested multiplication are two names for the same thing.

Horner's Method

The historic name for synthetic division is Horner's method. It is not hard to show that for an nth-degree polynomial, the reduction in multiplies is from $n(n + 1)/2$ multiplies for the standard form to only n for Horner's method. There are n adds in either case.

We can write the algorithm for nested multiplication (Horner's method) in pseudocode as follows:

Algorithm for Horner's Method

To divide a given polynomial, $P_n(x)$, of degree n by $(x - x_1)$, where

$$P_n(x) = a_0 x^n + a_1 x^{n-1} + a_2 x^{n-2} + \cdots + a_{n-1} x + a_n,$$

SET $b_0 = a_0$.
REPEAT
$$b_i = a_i + b_{i-1} x_1$$
FOR $i = 1, 2, \ldots, n$.

The remainder after the division, b_n, is $P(x_1)$.

DERIVE can divide a polynomial by $(x - x_1)$ very easily. If we enter $(x^\wedge 3 + x^\wedge 2 - 3x - 3)/(x - 2)$ with the Author command and then Expand this, we get

$$\frac{3}{x - 2} + x^2 + 3x + 3.$$

(DERIVE writes the remainder first, then the reduced polynomial.)

Parallel Processing

Horner's method for evaluating a polynomial is one of the classic examples where we can speed up a computation by using parallel processors. The directed acyclic graphs (dag's) for the sequential and parallel algorithms are shown in Fig. 1.9. While we have more operations (five multiplies and three adds) with the parallel scheme (compared to three multiplies and three adds), the time required to produce the result is reduced from six steps to four steps. The time savings comes from doing some operations in parallel rather than in succession, of course. Observe that the most efficient method for sequential processing (Horner's method) is not used in parallel processing.

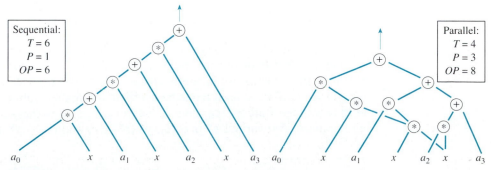

Figure 1.9   dag's for evaluating a polynomial of degree 3

1.8 Bairstow's Method for Quadratic Factors

The methods considered so far are difficult to use to find a complex root of a polynomial. It is true that Newton's and Muller's methods work satisfactorily, provided that we begin with initial estimates that are complex-valued; however, in a hand computation, performing the multiplications and divisions of complex numbers is awkward. There is no problem in a computer program if complex arithmetic capabilities exist, but the execution is slower.

For polynomials, the complex roots occur in conjugate pairs if the coefficients are all real-valued. For this case, if we extract the quadratic factors that are the products of the pairs of complex roots, we can avoid complex arithmetic because such quadratic factors have real coefficients. We first develop the algorithm for synthetic division by a trial quadratic, $x^2 - rx - s$, which, hopefully, is near to the desired factor of the polynomial:

$$
\begin{aligned}
P_n(x) &= a_0 x^n + a_1 x^{n-1} + \cdots + a_n \\
&= (x^2 - rx - s)Q_{n-2}(x) + \text{remainder} \\
&= (x^2 - rx - s)(b_0 x^{n-2} + b_1 x^{n-3} + \cdots + b_{n-3}x + b_{n-2}) \\
&\quad + b_{n-1}(x - r) + b_n.
\end{aligned}
$$

(The remainder is the linear term $b_{n-1}(x - r) + b_n$, written in this form to provide later simplicity. If $x^2 - rx - s$ is an exact divisor of $P_n(x)$, then b_{n-1} and b_n will both be zero.) The negative signs in the factor are also for later simplification.

On multiplying out and equating coefficients of like powers of x, we get

$$
\begin{array}{lll}
a_0 = b_0 & & b_0 = a_0 \\
a_1 = b_1 - rb_0 & & b_1 = a_1 + rb_0 \\
a_2 = b_2 - rb_1 - sb_0 & & b_2 = a_2 + rb_1 + sb_0 \\
a_3 = b_3 - rb_2 - sb_1 & \text{or} & b_3 = a_3 + rb_2 + sb_1 \qquad \textbf{(1.1)} \\
\quad \vdots & & \quad \vdots \\
a_{n-1} = b_{n-1} - rb_{n-2} - sb_{n-3} & & b_{n-1} = a_{n-1} + rb_{n-2} + sb_{n-3} \\
a_n = b_n - rb_{n-1} - sb_{n-2} & & b_n = a_n + rb_{n-1} + sb_{n-2}
\end{array}
$$

We would like both b_{n-1} and b_n to be zero, for that would show $x^2 - rx - s$ to be a quadratic factor of the polynomial (because the remainder is zero). This will normally not be so; if we properly change the values of r and s, we can make the remainder zero, or at least make its coefficients smaller. Obviously b_{n-1} and b_n are both functions of the two parameters r and s. Expanding these as a Taylor series for a function of two variables* in terms of $(r^* - r)$ and $(s^* - s)$, where $(r^* - r)$ and

---

* Appendix A reviews this.

$(s^* - s)$ are presumed small so that terms of higher order than the first are negligible, we obtain

$$b_{n-1}(r^*, s^*) = b_{n-1}(r, s) + \frac{\partial b_{n-1}}{\partial r} (r^* - r) + \frac{\partial b_{n-1}}{\partial s} (s^* - s) + \cdots,$$

$$b_n(r^*, s^*) = b_n(r, s) + \frac{\partial b_n}{\partial r} (r^* - r) + \frac{\partial b_n}{\partial s} (s^* - s) + \cdots.$$

Let us take (r^*, s^*) as the point at which the remainder is zero, and

$$r^* - r = \Delta r, \qquad s^* - s = \Delta s.$$

(Δr and Δs are increments to add to the original r and s to get the new values r^* and s^* for which the remainder is zero.) Then

$$b_{n-1}(r^*, s^*) = 0 \simeq b_{n-1} + \frac{\partial b_{n-1}}{\partial r} \Delta r + \frac{\partial b_{n-1}}{\partial s} \Delta s,$$

$$b_n(r^*, s^*) = 0 \simeq b_n + \frac{\partial b_n}{\partial r} \Delta r + \frac{\partial b_n}{\partial s} \Delta s.$$

All the terms on the right are to be evaluated at (r, s). We wish to solve these two equations simultaneously for the unknown Δr and Δs, so we need to evaluate the partial derivatives.

Bairstow showed that the required partial derivatives can be obtained from the b's by a second synthetic division by the factor $x^2 - rs - s$ in just the same way that the b's are obtained from the a's. Define a set of c's by the following relations shown at the left, and compare these to the partial derivatives in the right columns:

$$c_0 = b_0 \qquad \frac{\partial b_0}{\partial r} = \frac{\partial a_0}{\partial r} = 0 \qquad \frac{\partial b_0}{\partial s} = \frac{\partial a_0}{\partial s} = 0$$

$$c_1 = b_1 + rc_0 \qquad \frac{\partial b_1}{\partial r} = r\frac{\partial b_0}{\partial r} + b_0 = b_0 = c_0 \qquad \frac{\partial b_1}{\partial s} = \frac{\partial a_1}{\partial s} + r\frac{\partial b_0}{\partial s} = 0$$

$$c_2 = b_2 + rc_1 + sc_0 \qquad \frac{\partial b_2}{\partial r} = r\frac{\partial b_1}{\partial r} + b_1 = c_1 \qquad \frac{\partial b_2}{\partial s} = r\frac{\partial b_1}{\partial s} + s\frac{\partial b_0}{\partial s} + b_0$$

$$= b_0 = c_0$$

$$c_3 = b_3 + rc_2 + sc_1 \qquad \frac{\partial b_3}{\partial r} = r\frac{\partial b_2}{\partial r} + b_2 + s\frac{\partial b_1}{\partial r} \qquad \frac{\partial b_3}{\partial s} = r\frac{\partial b_2}{\partial s} + s\frac{\partial b}{\partial s} + b_1$$

$$= b_2 + rc_1 + sc_0 = c_2 \qquad = b_1 + rc_0 = c_1$$

$$\vdots \qquad\qquad \vdots \qquad\qquad \vdots$$

$$c_{n-1} = b_{n-1} + rc_{n-2} + sc_{n-3} \qquad \frac{\partial b_{n-1}}{\partial r} = r\frac{\partial b_{n-2}}{\partial r} + b_{n-2} + s\frac{\partial b_{n-3}}{\partial r} \qquad \frac{\partial b_{n-1}}{\partial s} = r\frac{\partial b_{n-2}}{\partial s} + s\frac{\partial b_{n-3}}{\partial s} + b_{n-3}$$

$$= b_{n-2} + rc_{n-3} + sc_{n-4} \qquad = b_{n-3} + rc_{n-4} + sc_{n-5}$$

$$= c_{n-2} \qquad\qquad = c_{n-3}$$

Hence the partial derivatives that we need are equal to the properly corresponding c's. Our simultaneous equations become, where Δr and Δs are unknowns to be solved for,

$$-b_{n-1} = c_{n-2}\,\Delta r + c_{n-3}\,\Delta s,$$
$$-b_n = c_{n-1}\,\Delta r + c_{n-2}\,\Delta s.$$

We express the solution as ratios of determinants:

$$\Delta r = \frac{\begin{vmatrix} -b_{n-1} & c_{n-3} \\ -b_n & c_{n-2} \end{vmatrix}}{\begin{vmatrix} c_{n-2} & c_{n-3} \\ c_{n-1} & c_{n-2} \end{vmatrix}},$$

$$\Delta s = \frac{\begin{vmatrix} c_{n-2} & -b_{n-1} \\ c_{n-1} & -b_n \end{vmatrix}}{\begin{vmatrix} c_{n-2} & c_{n-3} \\ c_{n-1} & c_{n-2} \end{vmatrix}}.$$

As an exercise, the student should write the algorithm for this method.

EXAMPLE 1.3 Find the quadratic factors of

$$x^4 - 1.1x^3 + 2.3x^2 + 0.5x + 3.3 = 0.$$

Use $x^2 + x + 1$ as starting factor ($r = -1, s = -1$). (Frequently $r = s = 0$ are used as starting values if no information as to an approximate factor is known.) Equations (1.1) lead to a double synthetic-division scheme as follows:

Note that the equations for b_1 and c_1 have no term involving s. The dashes in the preceding tableau represent these missing factors. Then

$$\Delta r = \frac{\begin{vmatrix} 0.8 & -3.1 \\ -0.7 & 5.5 \end{vmatrix}}{\begin{vmatrix} 5.5 & -3.1 \\ -3.2 & 5.5 \end{vmatrix}} = \frac{2.23}{20.33} = 0.11, \qquad r^* = -1 + 0.11 = -0.89,$$

$$\Delta s = \frac{\begin{vmatrix} 5.5 & 0.8 \\ -3.2 & -0.7 \end{vmatrix}}{20.33} = \frac{-1.29}{20.33} = -0.06, \qquad s^* = -1 - 0.06 = -1.06.$$

The second trial yields

	1	−1.1	2.3	0.5	3.3	
−0.89		−0.89	1.77	−2.68	0.06	
−1.06		—	−1.06	2.11	−3.17	
	1	−1.99	3.01	−0.07	0.17	← *b*'s
		−0.89	2.56	−4.01		
		—	−1.06	3.05		
		−2.88	4.51	−1.03		← *c*'s

$$\Delta r = \frac{\begin{vmatrix} 0.07 & -2.88 \\ -0.17 & 4.51 \end{vmatrix}}{\begin{vmatrix} 4.51 & -2.88 \\ -1.03 & 4.51 \end{vmatrix}} = \frac{-0.175}{17.374} = -0.010, \qquad r^* = -0.89 - 0.010 = -0.900,$$

$$\Delta s = \frac{\begin{vmatrix} 4.51 & 0.07 \\ -1.03 & -0.17 \end{vmatrix}}{17.374} = \frac{-0.694}{17.374} = -0.040, \qquad s^* = -1.06 - 0.040 = -1.100.$$

The exact factors are $(x^2 + 0.9x + 1.1)(x^2 - 2x + 3)$.

Observe that the other factor has its coefficients in the row of *b*'s.

1.9 Other Methods for Polynomials*

Of the many other methods for finding the roots of polynomials, we discuss four in this section. Three methods—the QD algorithm, Graeffe's root-squaring method, and Lehmer's method—do not require that we start with a value near to a root. The fourth, Laguerre's method, does need a starting value, but it is remarkably efficient.

---

* The methods of this section are not widely used. You may therefore want to skip this section.

The QD Algorithm

The QD, or quotient-difference method, is quite efficient. We present it without elaboration.*

For the nth-degree polynomial

$$P_n(x) = a_0 x^n + a_1 x^{n-1} + \cdots + a_{n-1} x + a_n,$$

we form an array of q and e terms, starting the tableau by calculating a first row of q's and a second row of e's:

$$q^{(0)} = -\frac{a_1}{a_0}, \quad \text{All other } q\text{'s are zero.}$$

$$e^{(i)} = \frac{a_{i+1}}{a_i} \quad i = 1, 2, \ldots, n-1,$$

$$e^{(0)} = e^{(n)} = 0.$$

The start of the array is

$e^{(0)}$	$q^{(0)}$	$e^{(1)}$	$q^{(1)}$	$e^{(2)}$	$q^{(2)}$	$\cdots$	$e^{(n-1)}$	$q^{(n-1)}$	$e^{(n)}$
	$\dfrac{-a_1}{a_0}$		0		0	$\cdots$		0	
0		$\dfrac{a_2}{a_1}$		$\dfrac{a_3}{a_2}$		$\cdots$	$\dfrac{a_n}{a_{n-1}}$		0

A new row of q's is computed by the equation

$$\text{New } q^{(i)} = e^{(i+1)} - e^{(i)} + q^{(i)},$$

using terms from the e and q rows just above. Note that this algorithm is "e to right minus e to left plus q above."

A new row of e's is now computed by the equation

$$\text{New } e^{(i)} = \left(\frac{q^{(i)}}{q^{(i-1)}} \right) e^{(i)};$$

"q to right over q to left times e above." The example in Table 1.5 isolates the roots of the quartic

$$P_4(x) = 128x^4 - 256x^3 + 160x^2 - 32x + 1$$

by continuing to compute rows of q's and then e's until all the e-values approach zero. When this occurs, the q-values assume the values of the roots. Since the method is slow to converge, it is generally used only to get approximate values, which are then improved by Newton's method.

---

* Henrici (1964) discusses the method in some detail.

Table 1.5   Example of QD method for $P(x) = 128x^4 - 256x^3 + 160x^2 - 32x + 1$

$e^{(0)}$	$q^{(0)}$	$e^{(1)}$	$q^{(1)}$	$e^{(2)}$	$q^{(2)}$	$e^{(3)}$	$q^{(3)}$	$e^{(4)}$
	2.000		0		0		0	
0		−0.625		−0.200		−0.031		0
	1.375		0.425		0.169		0.031	
0		−0.193		−0.079		−0.006		0
	1.182		0.539		0.242		0.037	
0		−0.088		−0.036		−0.001		0
	1.094		0.591		0.277		0.038	
0		−0.048		−0.017		−0.000		0
	1.046		0.622		0.294		0.038	
0		−0.028		−0.008		−0.000		0
	1.018		0.642		0.302		0.038	
0		−0.018		−0.004		−0.000		0
	1.000		0.656		0.304		0.038	
0		−0.012		−0.002		−0.000		0
	0.988		0.666		0.306		0.038	
0		−0.008		−0.001		−0.000		0
	0.980		0.673		0.307		0.038	
0		−0.005		−0.001		−0.000		0
	0.975		0.677		0.308		0.038*	

*The true values of the roots are 0.96194, 0.69134, 0.30866, and 0.03806.

What if there are complex roots? If the polynomial has a pair of conjugate complex roots, one of the e's will not approach zero but will fluctuate in value. The sum of the two q-values on either side of this e will approach r, and the product of the q above and to the left times the q below and to the right approaches $-s$ in the factor $x^2 - rx - s$. Two equal roots behave similarly.

Table 1.6 shows the result of the method for the polynomial

$$(x - 1)(x - 4)(x^2 - x + 3) = x^4 - 6x^3 + 12x^2 - 19x + 12.$$

Table 1.6   QD method with complex roots, for $P(x) = x^4 - 6x^3 + 12x^2 - 19x + 12$

$e^{(0)}$	$q^{(0)}$	$e^{(1)}$	$q^{(1)}$	$e^{(2)}$	$q^{(2)}$	$e^{(3)}$	$q^{(3)}$	$e^{(4)}$
	6.000		0		0		0	
0		−2.000		−1.583		−0.632		0
	4.000		0.417		0.951		0.632	
0		−0.208		−3.610		−0.420		0
	3.792		−2.985		4.141		1.052	
0		0.164		5.008		−0.107		0
	3.956		1.859		−0.974		1.159	
0		0.077		−2.624		0.127		0
	4.033		−0.842		1.777		1.032	
0		−0.016		5.538		0.074		0
	4.017		4.712		−3.687		0.958	
0		−0.019		−4.333		−0.019		0
	3.998		0.398		0.627		0.977	
0		−0.002		−6.826		−0.030		0
	4.000		−6.426		7.423		1.007	
0		0.003		7.885		−0.004		0
	4.003		1.456		−0.466		1.010	

For our example, the factors are $(x - 4)(x - 1)(x^2 - x + 3)$. We have:

$$q^{(0)} \text{ converging to } 4,$$
$$q^{(3)} \text{ converging to } 1.$$

Since $e^{(2)}$ does not approach zero, $q^{(1)}$ and $q^{(2)}$ represent a quadratic factor. We compute

$$r = q^{(1)} + q^{(2)} = 1.456 - 0.466 = 0.990;$$
$$s = -(-6.426)(-0.466) = -2.995.$$

This quadratic factor is $x^2 - rx - s = x^2 - 0.990x - (-2.995)$.

What if some a's are zero? Note that we cannot compute the first q and e rows if one of the coefficients in the polynomial is zero, for division by zero is undefined. In such a case, we change the variable to $y = x - 1$. (Subtracting 1 from the roots of the equation is an arbitrary choice, but this facilitates the reverse change of variable to get the roots of the original equation after the roots of the new equation in y have been found.)

For example, if $f(x) = x^4 - 2x^2 + x - 1 = 0$, we let $y = x - 1$ and use repeated synthetic division to determine the coefficients of $f(y) = 0$. The successive remainders on dividing by $x - 1$ are the coefficients of $f(y)$:

```
1│   1        0       -2        1       -1
             1        1       -1        0
     ─────────────────────────────────────
     1        1       -1        0      ⊖1
              1        2        1
     ──────────────────────────
     1        2        1       ①
              1        3
     ─────────────────
     1        3       ④
              1
     ──────────
     1       ④
     ①
```

Therefore,

$$f(y) = y^4 + 4y^3 + 4y^2 + y - 1.$$

We proceed to find the roots of $f(y) = 0$, and then get the roots of $f(x) = 0$ by adding 1.

Graeffe's Root-Squaring Method

Graeffe's method transforms $P_n(x)$ into another polynomial of the same degree but whose roots are the squares of the roots of the original polynomial. Thus if the roots of $P_n(x)$ are all real and distinct, the roots of the new polynomial are spread more widely apart than in the old polynomial. This is particularly so for roots greater than 1 in absolute value. Repeating this process until the roots are really far apart, we can compute the roots directly from the coefficients.

This simple example shows how to get the new polynomial when the initial polynomial is of the third degree. Let

$$P(x) = (x - 1)(x + 2)(x - 3).$$

Then

$$P(-x) = (-x - 1)(-x + 2)(-x - 3)$$
$$= (-1)^3(x + 1)(x - 2)(x + 3).$$

Multiplying,

$$P(x) * P(-x) = (-1)^3(x^2 - 1^2)(x^2 - 2^2)(x^2 - 3^2)$$
$$= (-1)^3(x^2 - 1)(x^2 - 4)(x^2 - 9)$$
$$= (-1)^3(z - 1)(z - 4)(z - 9),$$

and we have a polynomial in $x^2 = z$ whose roots are indeed the squares of the roots of $P(x)$. [The power on the (-1) factor is always n, the degree of the polynomial.] Of course, we never have $P(x)$ in factored form, but the result is the same. Observe that the squaring process loses the signs of the roots so that "distinct" applies to the magnitudes of the roots.

Suppose we have squared the original k times. We then can estimate each of the roots with the 2^kth root of

$$\left| \frac{a_j}{a_{j-1}} \right| \quad j = 1, \ldots, n,$$

n being the degree of the polynomial.

EXAMPLE 1.4   Use root squaring on $x^3 - 3x^2 - 6x + 8$. The first three new polynomials (with x replacing x^2) are

$$k = 1: \quad x^3 - 21x^2 + 84x - 64,$$
$$k = 2: \quad x^3 - 273x^2 + 4368x - 4096,$$
$$k = 3: \quad x^3 - 65793x^2 + 16843008x - 16777216.$$

(DERIVE did the squaring easily.) Estimates of the roots from the first new polynomial are

$$\sqrt{\frac{64}{84}} = 0.8729, \qquad \sqrt{\frac{84}{21}} = 2, \qquad \sqrt{\frac{21}{1}} = 4.5826.$$

From the second, we get

$$\sqrt[4]{\frac{4096}{4368}} = 0.9841, \qquad \sqrt[4]{\frac{4368}{273}} = 2, \qquad \sqrt[4]{\frac{273}{1}} = 4.0648.$$

From the third,

$$\sqrt[8]{\frac{16777216}{16843008}} = 0.9995, \qquad \sqrt[8]{\frac{16843008}{65793}} = 2, \qquad \sqrt[8]{\frac{65793}{1}} = 4.0020.$$

The exact values are $1, -2, 4$. We must determine the signs of the roots from the estimates by substituting them into the original polynomial. If the sign of the root is positive, the result from the substitution is nearly zero; otherwise, the root is negative.

▲

The method fails when there are roots of equal magnitude (and this will always be true for imaginary roots). Ralston (1965) shows how to overcome this difficulty.

Lehmer's Method

This method, also known as the Lehmer–Schur method, can be applied to find complex roots as well as real roots, and works with polynomials with complex coefficients. It does not require an initial approximation to a root. We will only outline the procedure. For details, Ralston (1965) is a good source.

We require that the polynomial not have a root at $x = 0$. (This will be obvious because there will then be no constant term in $P(x)$, so we reduce the polynomial until there is none.) We begin with a unit circle in the complex plane. Through a fairly complicated test procedure, we can determine whether the polynomial has any root(s) inside this circle. If it does, we test a circle half as large. If this circle also contains a root, we decrease the radius again and repeat until we find a circle that contains no roots. By this means we locate an annulus whose inner radius is one-half the outer radius that contains one or more roots. Now we draw a circle that covers at least $\frac{1}{6}$ of the annulus and test again for the presence of roots. We continue to test such smaller circles that move around the annulus by an amount equal to $\pi/3$ until, in at most six such trials, we locate a small circle that contains one or more roots. We apply the test on circles that shrink inside that circle to define another annulus, test still smaller circles that cover the annulus, and so on, until a root is located as precisely as desired. Of course, we continue to test within each annulus until it is entirely covered.

If we find no roots within the initial unit circle, instead of testing smaller circles, we enlarge the radius by a factor of two. If no roots are there, we enlarge again until we find the presence of a root. We now have a (larger) annulus that contains root(s) whereby we proceed analogously.

Although this method has wide applicability, it is less efficient than other root-finding procedures, even if the finding of initial values for, say, Newton's method, is by searching. After all, Newton's method works for complex roots if we are willing to do complex arithmetic (or we can use Bairstow if the polynomial has real coefficients).

Laguerre's Method

Suppose that $P_n(x) = (x - x_1)(x - x_2)(x - x_3) \cdots (x - x_n)$. By computing the derivative of $\ln|P_n|$, it is easy to show that

$$A = \frac{P_n'}{P_n} = \sum_{i=1}^{n} \frac{1}{(x - x_i)}.$$

By computing the second derivative of $\ln|P_n|$, we find that

$$B = \frac{-a^2(\ln|P_n|)}{ax^2} = \left(\frac{P_n'}{P_n}\right)^2 - \frac{P_n''}{P_n}$$

$$= \sum_{i=1}^{n} \frac{1}{(x - x_i)^2}.$$

We see that B is always positive.

Let x_1 be the root we want to determine. Assume that all other roots are distant from x_1 and are bunched closely together at some point $x = X$. (A pretty rash assumption! But it works!!) Define $a = x - x_1$ and $b = x - X$. We can then rewrite the equations for A and B as

$$A = \frac{1}{a} + \frac{(n - 1)}{b},$$

$$B = \frac{1}{a^2} + \frac{(n - 1)}{b^2}.$$

From these two equations, eliminate b to get a:

$$a = \frac{n}{A \pm \sqrt{(n - 1)(nB - A^2)}},$$

where we use plus if A is positive and minus if A is negative. We begin with a value for x_0 that is near to the desired root. Using this value, we compute A, B, and a. The next iterate is $x_1 = x_0 - a$. We repeat this until a is sufficiently small.

EXAMPLE 1.5
$$P(x) = x^3 - 8.6x^2 + 22.41x - 16.236$$

Let us start with $x_0 = 1.0$. From the formulas for A and B, we compute A and B and then a, getting $a = -0.199973$. So $x_1 = 1.19973$. Continuing, we find the next $a = -2.800E\text{-}5$, so $x_2 = 1.200000$, which is exactly the smallest root of $P(x)$.

If we start with $x_0 = 5.0$, we get $a = 0.870960$; $x_1 = 4.12904$. Repeating, we get $a = 0.029025$ and $x_2 = 4.10000$, which is exactly the largest root. Since the sum of the roots is 8.6, the third root is at $x = 3.3$.

1.10 Multiple Roots

A function can have more than one root of the same value. When that is true, the graph will resemble Fig. 1.10. The curve on the left has a triple root at $x = -1$ [the function is $f(x) = (x + 1)^3$], while there is a double root at $x = 1$ for the curve on the right: $f(x) = (x - 1)^2$. If there were more than two or three roots, somewhat similar curves would result, but they would be flatter near the x-axis and rise more steeply as the curve departs from the x-axis.

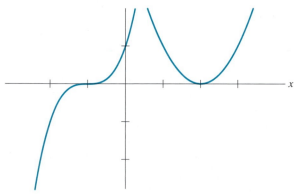

Figure 1.10   Functions with multiple roots

The methods we have studied do not work well with multiple roots. For example, Newton's method converges only linearly to a double root, as will be shown in the next section, while the method converges quadratically to a single (simple) root. (*Quadratic convergence* means that each error is proportional to the square of the preceding error. The net effect of quadratic convergence is that the number of significant figures in the estimates approximately doubles at each iteration.)

Still, our methods work to get a double root, except for bisection and linear iteration (because the function may not change sign at the root). Table 1.7 compares the convergence of Newton's method, the secant method, and Muller's method on $f(x) = (x - 1) * (e^{(x-1)} - 1)$, which has a double root at $x = 1$.

Table 1.7   Getting a double root, for $f(x) = (x - 1) * (e^{(x-1)} - 1)$

	Secant method	Newton's method	Muller's method
Estimate after 9 iterations	1.00331	1.00126	1.00058
Start value(s)	1.2, 1.5	2.0	0, 1.2, 1.5

From Table 1.7, we see that Newton's method is faster than the secant method, even though the latter starts with values nearer the root. Muller's method gets a more accurate estimate in nine iterations. (Comparing the closeness of starting values is not readily done.) Often Muller's method seems to converge more rapidly—but be careful! Some sets of starting values cause the method to fail, because the square root of a negative number is needed. For example, using starting values of 0, 0.5, and 1.5 with the function in Table 1.7 creates a parabola that does not cross the x-axis, so the parabola has complex roots.

Table 1.8 shows that convergence is linear with Newton's method, since each error is almost exactly half of the preceding error, especially as we get near the root. For a triple root, convergence will be even slower although all of our methods work, even those that require the function to change sign.

Table 1.8 Successive errors with Newton's method, for $f(x) = (x - 1) * (e^{(x-1)} - 1)$

Iteration	Error	Iteration	Error
0	1.0	6	0.0199
1	0.5	7	0.0100
2	0.2798	8	0.0050
3	0.1494	9	0.0025
4	0.0775	10	0.00125
5	0.0395	11	0.000625

Table 1.9 compares results from the bisection, Newton's, and Muller's methods on $f(x) = (x + 1)^3 = 0$, with a triple root at $x = -1$. Because this function is symmetrical about the root value, bisection will give immediate success if starting values of -2 and 0 are used. Remarkably, Muller's method converges immediately with many sets of starting values, but, again, a set such as $-1.5, -1.1, 0$ runs into a complex-valued estimate. Table 1.10 shows that Newton's method is again linearly convergent, since each error is $\frac{2}{3}$ times the previous one, and that convergence is slower than in the double-root example.

Table 1.9 Getting a triple root, for $f(x) = (x + 1)^3 = 0$

	Bisection method	Newton's method	Muller's method
Estimate after 9 iterations	−0.99902	−1.01301	−1.00000*
Start value(s)	−1.5, 0	−1.5	−1.5, −0.5, 0

*Muller's method gave exact answer on first iteration!

In addition to a slow convergence, there is another disadvantage to using these methods to find multiple roots: imprecision. Since the curve is "flat" in the neighborhood of the root—$f'(x)$ will always be zero at a multiple root, as is apparent from Fig. 1.10—there is a "neighborhood of uncertainty" around the root where values of $f(x)$ are very small. Thus the imprecise arithmetic of almost all computational devices will find $f(x)$ "equal" to zero throughout this neighborhood; that is, the

Table 1.10 Successive errors with Newton's method, for $f(x) = (x + 1)^3 = 0$

Iteration	Error	Iteration	Error
0	0.5	6	0.0439
1	0.3333	7	0.0293
2	0.2222	8	0.0195
3	0.1482	9	0.0130
4	0.0988	10	0.00867
5	0.0658		

program cannot distinguish which x-value is really the root. Using double precision will decrease the neighborhood of uncertainty. In fact, DERIVE in exact mode can give as much precision as desired, even to 100 significant figures, so this "neighborhood" can be very small.

Remedies for Multiple Roots

Section 1.11 will examine the rate of convergence of Newton's method. We will see that, if there is a root of multiplicity k (there are k roots at $x = r$), we can restore quadratic convergence by modifying the algorithm to

$$x_{n+1} = x_n - k * \frac{f(x_n)}{f'(x_n)}.$$

Using this method to get the root of $f(x) = (x - 1) * (e^{(x-1)} - 1)$, we find that the third iterate is $x = 1.00088$ with $f(x) = 0.00000$. We also find that $e_{n+1} = 0.24 * e_n^2$, confirming quadratic convergence.

This algorithm would seem to solve the problem of multiple roots using Newton's method, but we don't know the multiplicity of the root in advance! (This objection is a little academic as the following argument shows.)

We might guess at the value for k and see whether we get quadratic convergence, or we could try several values and see what happens. Better yet, we could compare a graph of $f(x)$ with the plots of $(x - r)^k$, using an approximate value for r and various values for k. The "flatness" of the curves will be the same for $f(x)$ and the plot of equivalent multiplicity. But we wonder whether all such effort is justified—why not just live with the linear convergence? We will find the root with sufficient accuracy from that operation long before we complete the alternative explorations.

Another solution to multiple roots is tempting to consider. We can divide $f(x)$ by $(x - r)$ and deflate the function, reducing the multiplicity by one. The problem here is that we don't know r. But dividing by $(x - s)$, where s is an approximation of r, does almost the same thing. We suggest that you might want to explore this idea.

Be warned that the division creates an indeterminate form at $x = r$ and a strong discontinuity at $x = s$.

Using the Derivative Function

Since $f'(x) = 0$ at a multiple root, we can converge quadratically by applying Newton's method to $f'(x)$ to get a double root of $f(x)$. We need to apply this method to $f''(x)$ to get quadratic convergence to a triple root. The agony we sometimes experience in differentiating complicated functions is largely relieved if we let DERIVE do the work.

A most tempting scheme is to compute the roots of $f(x)/f'(x)$. It is easy to show that, when $f(x)$ has a root of multiplicity n at $x = r$, $f'(x)$ has a root of multiplicity $n - 1$ at that point. Let

$$f(x) = (x - r)^n F(x).$$

Then

$$f'(x) = n(x - r)^{n-1}F(x) + (x - r)^n F'(x)$$
$$= (x - r)^{n-1}[nF(x) + (x - r)F''(x)].$$

Hence, if we divide $f(x)$ by $f'(x)$, we effectively deflate the function $(n - 1)$ times and we now work with a new function that has only a single root at $x = r$.

Some warnings are in order, however. This technique works fine with polynomials, but when $f(x)$ involves transcendentals, there may be difficulties. The deflated function will have infinite discontinuities at the maxima and minima of $f(x)$, which can distort the reduced function and make it difficult to converge to some single roots of $f(x)$. Figure 1.11 shows the plots of $f(x) = (x - 1)^3 \sin(x)$ and of $f(x)/f'(x)$. Observe the discontinuities at points where $f'(x) = 0$ and the distortions at the zeros of the sine function.

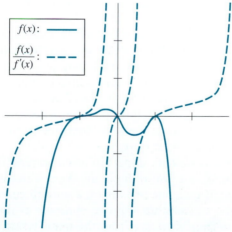

Figure 1.11 Plots of $f(x)/f'(x)$ when $f(x) = (x - 1)^3 \sin(x)$

Nearly Multiple Roots

A problem related to multiple roots is a function that has two or more roots very close together. If these roots are all within the region of uncertainty (which is a function of the arithmetic precision we are using), they are effectively multiple roots because for all of them $f(x)$ is computationally equal to zero.

Newton's method is again essentially linearly convergent when we have nearly equal roots, provided that we start outside the interval that holds the roots. Unfortunately, modifying the method by considering them to be multiple roots doesn't work; often an infinite loop occurs. If we are so unlucky as to start between two almost equal roots, Newton's method can fly off to "outer space," as we previously observed.

Whenever we want to find roots that are near $f'(x) = 0$, we are in trouble. We strongly recommend that you graph the function, before jumping into a root-finding routine, to see in advance whether such problems will arise. With programs like DERIVE available, this graphing is no great burden.

1.11 Theoretical Matters

While we mentioned rates of convergence earlier in this chapter, it is essential that we substantiate these statements. We also need to be more precise in how we express convergence rates.

We define the *order of convergence* for a method by the following:

> If $|e_{n+1}|$ approaches $K * |e_n|^p$ as n becomes infinite,

we say that the method is of order p. (It is almost always true that the errors do not decrease this fast until we are near the root.) How do we know a method is of a certain order; why can we say, for example, "Newton's method is of order 2—its convergence is quadratic"?

One way to find the order of a method is to perform experiments with the method and see how the errors decrease. (We must be sure not to be too hasty in reaching the conclusion, because the order is attained only as a limit as the root is approached.) Figure 1.12 is a plot of $\ln|e_{n+1}|$ versus $\ln|e_n|$ for Newton's method on the cubic $x^3 - 2x - 3$ with a starting value of 5. A line of slope 2 is shown, and the points for Newton's method do seem to be about parallel. The figure also shows results when fixed-point iteration is used on the same function $[g(x) = \sqrt{2 + 3/x}]$. These points seem to parallel a line whose slope is 1.

Plotting the logarithms as we have done will give a straight line of slope p if the relation is true everywhere, not just in the limit. We see from Fig. 1.12 that while the points do not lie exactly on a line or even on a smooth curve, the slopes appear to match the orders that we claim for the two methods as the iterations proceed (points at the left end of the curves are from the later iterations).

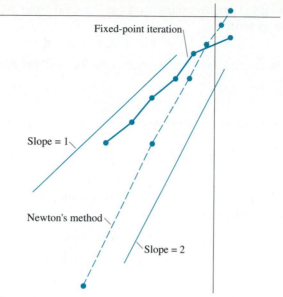

Slope = 1

Newton's method

Slope = 2

Fixed-point iteration

Figure 1.12 Plots of $\ln |e_{n+1}|$ versus $\ln |e_n|$ for Newton's method and for fixed-point iteration

The trouble with this experimental approach is that it is not general enough; we cannot conclude the order of a method from experiments with a few functions no matter how cleverly we chose them. That is why theoretical studies are so important; they give conclusions that are true for all functions. We must emphasize, though, that our analysis of errors in this theoretical discussion assumes perfect arithmetic. The errors are errors of the *method*, not errors resulting from imperfections of the computing device.

Fixed-Point Iteration

We begin with an analysis of fixed-point iteration (which we also call the $x = g(x)$ method). Recall that we iterate with

$$x_{n+1} = g(x_n).$$

If we let $x = r$ be a solution of $f(x) = 0$, then $f(r) = 0$ and $r = g(r)$. Subtracting, and then multiplying and dividing by $(x_n - r)$, we have

$$x_{n+1} - r = g(x_n) - g(r) = \frac{g(x_n) - g(r)}{(x_n - r)}(x_n - r).$$

If $g(x)$ and $g'(x)$ are continuous on the interval from r to x_n, the mean-value theorem* lets us write

$$x_{n+1} - r = g'(\xi_n) * (x_n - r),$$

where ξ_n lies between x_n and r. If we define the error of the ith iterate as $e_i = r - x_i$, we can write

$$e_{i+1} = g'(\xi_i) * e_i.$$

Taking absolute values (because the successive iterates may oscillate around the root), we get

$$|e_{i+1}| = |g'(\xi_i)| * |e_i|.$$

Now suppose that $|g'(x)| < K < 1$ for all values of x in an interval of radius h about r. If x_0 is chosen in this interval, all succeeding iterates will be in this interval and the method will converge, since

$$|e_{n+1}| < K * |e_n| < K^2 * |e_{n-1}| < K^3 * |e_{n-1}| < \cdots < K^n * |e_0|.$$

We therefore conclude that fixed-point iteration is of order 1.

In summary, if $g(x)$ and $g'(x)$ are continuous on an interval about a root r of the equation $x = g(x)$, and if $|g(x)| < 1$ for all x in the interval, then $x_{n+1} = g(x_n), n = 0, 1, 2, \ldots$, will converge to the root r, provided that x_0 is chosen in the interval. Note that this is a sufficient condition only, since for some equations, convergence is secured, even though not all the conditions hold.†

Ralston (1965) points out that measuring the worth of methods by their order is not enough. We should be concerned with efficiency by accounting for the "cost" of the iterations. We probably should determine the cost factor by totaling the number of operations required and the time for each; often, just counting the number of function evaluations is enough, since they occupy most of the computer's time.

Order of Convergence for Newton's Method

Newton's method uses the iterative scheme

$$x_{n+1} = x_n - \frac{f(x_n)}{f'(x_n)} = g(x_n),$$

from which we see that it is of the same form as fixed-point iteration. Hence, for

---

* Appendix A reviews certain principles of calculus, including this theorem.

† The analytical test that $|g'(x)| < 1$ is often awkward to apply. A constructive test is merely to observe whether the successive x_i values converge. In a computer program it is worthwhile to determine whether $|x_{n+1} - x_n| < |x_n - x_{n-1}|$ or, alternatively, whether the iterates "jump" far away.

this scheme we can use the preceding theory: Successive iterates converge if $|g'(x)| < 1$. We assume that the root at $x = r$ is simple. Differentiating $g(x)$ gives

$$g'(x) = 1 - \frac{f'(x)*f'(x) - f(x)*f''(x)}{[f'(x)]^2} = \frac{f(x)*f''(x)}{[f'(x)]^2}.$$

Hence, if

$$\frac{f(x)*f''(x)}{[f'(x)]^2} < 1$$

on an interval about the root r, the method will converge for any initial value x_0 in the interval. The condition is sufficient only; it requires the usual conditions of continuity and existence for $f(x)$ and its derivatives. Note that $f'(x)$ must not be zero within the interval as it is for the pathological situation of Fig. 1.4. More specifically, $f'(r)$ must not be zero as it is for a multiple root at $x = r$. (See the next subsection).

Now we show that Newton's method is quadratically convergent. Since r is a root of $f(x) = 0$, $r = g(r)$. Since $x_{n+1} = g(x_n)$, we can write

$$x_{n+1} - r = g(x_n) - g(r).$$

Let us expand $g(x_n)$ as a Taylor series in terms of $(x_n - r)$, with the second derivative term as the remainder:

$$g(x_n) = g(r) + g'(r)*(x_n - r) + \frac{g''(\xi)}{2}(x_n - r)^2, \qquad (1.2)$$

where ξ lies in the interval from x_n to r.

Since

$$g'(r) = \frac{f(r)*f''(r)}{[f'(r)]^2} = 0 \qquad (1.3)$$

because $f(r) = 0$ (r is a root), we have

$$g(x_n) = g(r) + \frac{g''(\xi)}{2}(x_n - r)^2.$$

Then, if we let $(x_n - r) = e_n$, we have

$$e_{n+1} = x_{n+1} - r = g(x_n) - g(r) = \frac{g''(\xi)}{2}e_n^2. \qquad (1.4)$$

As the iterates approach r, so does ξ. Thus each error is (in the limit) proportional to the second power of the previous error, and Newton's method has convergence of order 2—it is quadratically convergent.

Newton's Method with Multiple Roots

If $f(x) = 0$ has k zeros at $x = r$, we say that r is a root of multiplicity k. Thus we can factor out $(x - r)$ k times to give

$$f(x) = (x - r)^k Q(x), \qquad (1.5)$$

where $Q(x)$ has no root at $x = r$ [it is deflated from $f(x)$]. That means that $Q'(r)$ is not zero. However, each of $f'(r), f''(r), \ldots, f^{(k-1)}(r)$ is zero, as is readily found by differentiating $f(x)$ repeatedly.

We see that Eq. (1.3) has a denominator equal to zero if the root at $x = r$ is multiple. Although the ratio is indeterminant, we cannot say that $g'(r) = 0$ and that Eq. (1.4) therefore does not apply. Since the $g'(r)$ term does not drop out of the Taylor series expansion, it must be true that Newton's method is linearly convergent if r is a multiple root.

We have stated that if the multiplicity k of the root is known, we can regain quadratic convergence by modifying Newton's method to read

$$x_{n+1} = x_n - k * \frac{f(x_n)}{f'(x_n)}. \tag{1.6}$$

We need to demonstrate that this modification does restore quadratic convergence. As before, since $f(r) = 0$, $g(r) = r$. Here we have

$$g(x) = x - k * \frac{f(x_n)}{f'(x_n)}.$$

If we substitute $f(x) = (r - x)^k Q(x)$ and differentiate, we get

$$g'(x) = \frac{(r - x)\{k(r - x)QQ'' + Q'[2kQ - (k - 1)(r - x)Q']\}}{[(r - x)Q' + kQ]^2}$$

(DERIVE really helped with this differentiation!), and we see that $g'(r) = 0$. From the preceding argument, then, the modified Newton's method now converges quadratically at a multiple root. (It also does so at a simple root with $k = 1$, of course.)

The only problem with this technique is that we may not know the proper value for k, although we gave some suggestions in Section 1.10.

Acton (1970) gives another technique by which we may obtain a multiple root with quadratic convergence. If $f(x)$ has a root of multiplicity k at $x = r$, we have $f(x) = (r - x)^k * Q(x)$. Let $S(x)$ be $f(x)/f'(x)$, so that

$$S(x) = \frac{(x - r)^k Q(x)}{k(x - r)^{k-1}Q(x) + (x - r)^k Q'(x)} = \frac{(x - r)Q(x)}{kQ(x) + (x - r)Q'(x)},$$

which has a simple root at $x = r$. When $S(x)$ is used in the Newton formula, we get

$$x_{n+1} = x_n - \frac{S(x)}{S'(x)} = x_n - \frac{f(x) * f'(x)}{[f'(x)]^2 - f(x) * f''(x)},$$

and we see that we need to evaluate three functions at each iteration: $f(x), f'(x)$, and $f''(x)$. Acton also points out that there are nearly equal quantities being subtracted in the denominator, a source of arithmetic error in addition to the errors we are discussing here.

Convergence Order for Other Methods

Bisection (Interval Halving). For this method, at each iteration we divide the original interval in half, confining the root in the new interval. If we take the midpoints of the successive intervals to be estimates of the root, obviously one-half the current interval is an upper bound to the error. While we cannot say that $|e_{n+1}| = 0.5 * |e_n|$, if we are willing to call the upper bounds the estimates of the errors, we may say that bisection is linearly convergent. This does not mean, as we have seen, that each error is smaller than the previous one.

Linear Interpolation (False Position) and the Secant Method. Both of these methods use iteration of the form

$$x_{n+1} = x_n - \frac{f(x_n)}{f(x_n) - f(x_{n-1})} (x_n - x_{n-1}) \tag{1.7}$$

to estimate r, a root of $f(x) = 0$. At each iteration, we replace one of the previous x-values with the new estimate. With false position, the replacement is such that the root is bracketed. It is usually found that one end of the interval does not change when this is done, so we can think of x_{n-1} as a constant in the formula. Thus Eq. (1.7) is of the form $x = g(x)$ and we can use the preceding development directly. We need to know whether $g'(r)$ is zero. We first write out $g(x)$:

$$g(x) = x - \frac{f(x)}{f(x) - F} (x - X),$$

where we use F and X to emphasize that they are fixed quantities. On differentiating we get

$$g'(x) = \frac{f'(x) * F * (x - X) - F * [f(x) - F]}{[f(x) - F]^2},$$

and we see that $g'(r)$ is not zero. Hence, since this is exactly the same as for fixed-point iteration, linear interpolation (false position) is linearly convergent.

The secant method differs, in that neither of the two previous x-values is fixed. We must therefore consider the iterations to be

$$x_{n+1} = g(x_n, x_{n-1}),$$

so that g is a function of two independent variables. We can parallel the development for Newton's method, but now we must remember that

$$g'(x_n, x_{n-1}) = g_{x_n} + g_{x_{n-1}},$$

where the subscripts on g indicate partial derivatives.

The partial derivatives are

$$g_{x_n} = \frac{f(x_{n-1})[(x_n - x_{n-1})f'(x_n) - f(x_n) + f(x_{n-1})]}{[f(x_n) - f(x_{n-1})]^2},$$

$$g_{x_{n-1}} = \frac{f(x_n)[(x_n - x_{n-1})f'(x_{n-1}) - f(x_n) + f(x_{n-1})]}{[f(x_n) - f(x_{n-1})]^2},$$

and we observe that $g'(r,r)$ is zero because both partials are zero. This technique is similar to Newton's method, so we look at the next term in the Taylor series expansion of g. This process involves both variables and the second partial derivatives. We omit the details (the derivatives are very messy), but the partials are not zero at $x = r$. Although the secant method is similar to Newton's method, we cannot conclude that the secant method is quadratically convergent, because it turns out that

$$e_{n+1} = \frac{g''(\xi_1, \xi_2)}{2}(e_n)(e_{n-1}).$$

Since the error is proportional not to the square of a previous error but to the product of the two previous errors, we conclude that the secant method is better than the linear method but poorer than the quadratic method. Pizer (1975) shows that through solving an approximate difference equation, the order of the secant method is

$$\frac{1 + \sqrt{5}}{2} = 1.62.$$

Muller's Method. Muller's method is more difficult to analyze for its rate of convergence. (The iterating function g is a function of three variables.) Ralston (1965) outlines a technique that can be used to show that Muller's method has a rate of convergence of order 1.84.

Aitken Acceleration. If Aitken acceleration is applied after every third iteration of fixed-point iteration, and the improved estimate is used to begin each set of iterates, the accelerated iterates converge quadratically. This is known as Steffenson's method. Henrici (1964) should be consulted for the proof.

Special Methods for Polynomials

We omit the details, but it has been shown that Laguerre's method for polynomials is cubically convergent (Householder, 1970). The quotient-difference method, Lehmer's method, and Graeffe's method do not lend themselves to this kind of error analysis, since they are not iterative in the same sense.

1.12 Using DERIVE

The DERIVE program, whose subtitle is "A Mathematical Assistant for Your Personal Computer," can be a real assistant in many ways in the root-finding operation. We already pointed out that DERIVE can divide polynomials. For example, let $P(x) = x^4 - 2x^3 + x^2 + 5x - 3$. One way to evaluate $P(x)$ at $x = a$ is to divide by $(x - a)$ and find the remainder that equals $P(a)$. (This is a fairly trivial task for integer values of a, but if we want $P(1.23456)$, it isn't so easy. Doing it with a calculator would best use nested multiplication.)

DERIVE can do the calculation for us in two ways. We can use Author to enter the expression $(x^4 - 2x^3 + x^2 + 5x - 3)/(x - a)$. DERIVE then displays the expression as a builtup fraction:

$$\frac{x^4 - 2x^3 + x^2 + 5x - 3}{x - a}.$$

The Manage/Substitute command will permit us to use any value of a we desire. When we invoke it we are first asked to enter a value for x (we just press ENTER to show we don't want to substitute for x). We are then asked for a value for a. If we respond with 1.2, DERIVE then redisplays the ratio but with 1.2 substituted for a. If we now request to Expand, we get

$$\frac{1910}{125(5x - 6)} + x^3 - \frac{4x^2}{5} + \frac{x}{25} + \frac{631}{125},$$

because we have DERIVE in Exact arithmetic mode. For this reason, DERIVE can give us accuracy to almost as many places as we desire; it always works with integers. If we preferred all values expressed as decimal fractions, we would reset the mode to Approximate through Options/Precision. Alternatively, we could leave it in Exact and ask to Approximate the previous expression. Doing so gives

$$\frac{15.288}{5x - 6} + x^3 - 0.8x^2 + 0.04x + 5.048.$$

Actually, we could have used Manage/Substitute on $P(x)$ itself, requesting that x take on the value 1.2. The result is $15.288/5 = 3.0576$, just as expected.

Bairstow's method divides $P(x)$ by a quadratic. DERIVE can perform the arithmetic for us. If we enter $P(x)/(x^2 - ax - b)$, then substitute values for a and b, then Expand, we get the same result as when done by hand (or through a computer program). For example, if we use $a = 1$ and $b = 1$, the result is

$$\frac{5x}{x^2 - x - 1} - \frac{2}{x^2 - x - 1} + x^2 - x + 1,$$

which shows the two remainders and the reduced polynomial.

DERIVE, of course, can find the roots of a quadratic directly. To get them, we use the soLve command. For example, the reduced polynomial from our example, $x^2 - x + 1$, has roots of $0.5 \pm 0.866025i$. The professional version of DERIVE even knows the complicated formulas to solve cubics and quartics! (The student version can find the roots but does so by a numerical method.)

The ITERATES Function

Of special interest in this chapter on root finding is DERIVE's ITERATES function. Its format is

$$\text{ITERATES}(g, x, a, n).$$

Its action is to begin with $x_0 = a$, then continue with the iterations $x_n = g(x_{n-1})$, doing it n times. This is exactly the fixed-point $[x = g(x)]$ method!

It is convenient to define our own function to do fixed-point iterations. Using Author, we enter

```
FIX_PT(g, x, a, n) := ITERATES(g, x, a, n)
```

Once we have defined our own FIX_PT function, we can use it to solve for roots. For example, $f(x) = x^3 - e^x - 1$ has a root near $x = 2$. (Doing a plot in DERIVE tells us this.) Several rearrangements are possible; one is $g(x) = (e^x + 1)/x^2$. If we now enter (through Author) FIX_PT$((e^x + 1)/x^2, x, 2, 4)$ to request four iterations with this $g(x)$, we get the vector

```
[2,   2.09726,   2.07885,   2.08145,   2.08106]
```

(As explained more fully in the next chapter, vectors are multivalued quantities; DERIVE encloses the elements of a vector in square brackets.) The successive iterates are displayed as the elements of a vector as seen already. As explained, the first element is the starting value, $x = 2$, and the next four values are the iterates that seem to be converging.

A variation on the ITERATES function is ITERATE. Its format is identical; it also performs fixed-point iterations but now only the final value is displayed. If we enter

```
ITERATE((e^x + 1)/x², x, 2, 20)
```

we get the single value (obtained by DERIVE in 20 iterations) of 2.08111. We see that we were almost at convergence in four iterations.

Since Newton's method uses iterations of

$$x_{n+1} = x_n - \frac{f(x_n)}{f'(x_n)},$$

it obviously fits the fixed-point formulation. We can define

```
NEWTON(u, x, a, n) := ITERATES(x - u/DIF(u, x), x, a, n)
```

to perform n iterations of Newton's method on the function $u(x) = 0$. "DIF(u, x)" is the way we indicate to DERIVE that we want the derivative of u with respect to x. (It is displayed in an imitation of the standard mathematical style.) The result of NEWTON$(x^3 - e^x - 1, x, 2, 3)$ produces the vector

```
[2,   2,08437,   2.08112,   2.08111],
```

and we see that Newton's method reaches the value correct to six significant figures in three iterations. DERIVE also told us that it took 0.3 second to do it. DERIVE supplies a utility file, with Newton's method already defined, that can be applied to a set of functions. We will discuss solving sets of simultaneous nonlinear functions in the next chapter.

Bisection

While the interval halving method does not fit the standard iteration formula, we can build a function that does. We will have to work with a pair of values that are the endpoints of the current interval rather than a single-valued quantity, in other words, a vector. We will use $[a, b]$ to represent the vector whose two components are a and b.

While we could enter a formula for bisection in one long ITERATES statement, it is easier to follow the logic if we break it into subcommands. We assume that we are to find a root between $x = a$ and $x = b$ for a function $u(x)$. Our strategy is this:

1. See whether there is a sign change between $x = a$ and $x = c$, where c is the midpoint of $[a, b]$.
2. If there is a sign change, change the interval to $[a, c]$; if not, change it to $[c, b]$.
3. Repeat this process as many times as desired.

We first define some variables (we can use Declare/Variable):

```
1: V1(v) := ELEMENT(v,1)
2: V2(v) := ELEMENT(v,2)
3: VC(v) := (ELEMENT(v,1) + ELEMENT(v,2))/2
```

The first two just extract the first or second element of vector v. The third finds the midpoint.

An auxiliary function is helpful:

```
4: B_A(u,x,a,b,c) := IF(lim(u,x,a)*lim(u,x,c) < 0,[a,c],[c,b])
```

The IF statement in DERIVE tests a condition (its first argument, here whether there is a sign change between $x = a$ and $x = c$, the midpoint), returning the second argument if true or the third argument if false. The $\lim(u, x, a)$ computes the limit of $u(x)$ as $x \to a$, which is really just $u(a)$ for most functions. The second limit is really $u(c)$, so we have a sign change if the product is negative. We now put these together in an ITERATES statement:

```
5: BISECT(u,x,a,b,n) :=
ITERATES(B_A(u,x,V1(v),V2(v),VC(v)), v, [a,b], n)
```

From the previous discussion of ITERATES, we see that this function will perform n iterations on the vector v, starting with $v = [a, b]$, then changing it to $[a, c]$ or $[c, b]$, depending on where there is a sign change. If we enter this statement:

```
BISECT(x^2 − 2, x, 1, 2, 4)
```

and Approximate, we get the array:

```
 1          2
 1          1.5
 1.25       1.5
 1.375      1.5
 1.375      1.4375
```

which shows convergence to $\sqrt{2}$.

DERIVE can do more of our root-finding operations, but this is enough to show its power.

All of the things that we have done with DERIVE can be done with most other computer algebra systems. The names of the functions may be different, as well as the technique of accessing the program, but all computer algebra programs provide similar features. They normally have programming capabilities, so a user can combine built-in functions to perform other functions that are not available. In subsequent chapters, we will show how two other computer algebra systems can carry out numerical procedures.

Chapter Summary

If you have understood this chapter on solving nonlinear equations, you are now able to

1. Recognize that a problem, even if it is disguised, may be one that does not require the powerful methods of this chapter.

2. Use these methods to find solutions to $f(x) = 0$:

   Bisection (also called interval halving)

   Linear interpolation (also called false position)

   Secant method

   Newton's method

   Muller's method

   Fixed-point iteration [also called $x = g(x)$ method].

3. Apply special methods to find roots of polynomials:

   Newton's method with synthetic division

   Bairstow's method

   QD algorithm

   Graeffe's root-squaring method

   Laguerre's method.

4. Explain how to use Aitken acceleration to speed the convergence of a method that is linearly convergent.

5. Show how Horner's method (synthetic division, nested multiplication) reduces the number of operations required to evaluate a polynomial.

6. Draw a directed acyclic graph (dag) to indicate the successive operations for a sequential procedure and a corresponding equivalent parallel processing procedure; show that the latter may run faster even though requiring more operations.

7. Realize that multiple roots cause problems in root finding, and know several techniques to overcome these difficulties.

8. Write the relation that indicates that the iterative method is of order p, and outline the proofs that substantiate these statements:

> Fixed-point iteration is linearly convergent.
>
> Newton's method converges with order 2 at a simple root.
>
> Newton's method is linearly convergent at a multiple root.
>
> The method of false position converges linearly.

9. For one or more of the methods of this chapter, make a plot that indicates the order of the method and explain why this is not the best way to determine the order.

10. Use the DERIVE program to evaluate $f(x)$ at $x = a$, to get the derivative of a function, to perform fixed-point iterations, and to carry out Newton's method.

11. Use a program given at the end of the chapter to solve a nonlinear equation.

12. Write a program to implement one or more of the algorithms of the chapter that is not shown by any of the listed programs.

Computer Programs

Beginning here, we conclude each chapter with a few example programs that implement some of the algorithms discussed in the chapter. We do not attempt to include code for every algorithm, so that the student can gain experience in writing his/her own programs. That skill is essential to the budding numerical analyst, even though prewritten packages may supply most of the computer program needs of a professional in this field. While most of the more useful algorithms may be found in these commercial packages, one cannot read the programs nor decide which is best suited for the task at hand without a background of creating at least simple programs.

We also use a variety of computer languages in these programs to emphasize that the choice of language is not very important. In this chapter, we exhibit three programs using the FORTRAN and Pascal languages. One of the "programs" is actually implemented through a spreadsheet program. We do so to alert the reader to the fact that spreadsheets can carry out most of the procedures of numerical analysis. (A good reference that shows a wide variety of such spreadsheet implementations is Orvis, 1987. See also Etter, 1993).

It is always good practice to incorporate comments within a computer program that aid in understanding what the program does and how it does it. We have commented liberally throughout our programs to make the logic easy to follow.

Program 1.1 Muller's Method

Figure 1.13 is a listing of a FORTRAN program that uses Muller's method. A subroutine MULLR is used. It has parameters FCN, XR, H, XTOL, FTOL, NLIM, and I:

FCN   The function whose zero is to be found. It is defined in a separate function subprogram.

XR   The value of the root that is returned. It is also used to pass an initial estimate of the root to the routine.

H   The size of the displacements from the initial estimate that are used to construct the first quadratic approximation to the function.

XTOL   A stopping criterion based on x-values.

FTOL   A stopping criterion based on $f(x)$-values.

NLIM   A limit to the maximum number of iterations that are to be used in the procedure.

I   A signal that shows how the routine ended:
  I = 1, XTOL is met
  I = 2, FTOL is met
  I = −1, NLIM is exceeded.
 I is also used to send a message to the routine when it is called:
  I = 0, print the successive values
  I ≠ 0, do not print values.

The function subprogram FCN receives an x-value through its single parameter X and returns (in FCN) the function value, $f(x)$. To obtain the root of a different function, the definition of the function in this module should be changed.

Figure 1.13 shows a typical calling program that gets a root through the subroutine MULLR. (Such a calling program is often called a "driver" program.) Observe that a DATA statement is used to define the values for XR, H, XTOL, FTOL, NLIM, and I. These values also should be changed appropriately. FCN must be declared as EXTERNAL in the driver program because its name is to be passed as a parameter.

At the end of the program listing in Figure 1.13 can be found the output printed by the program (it was called with I = 0).

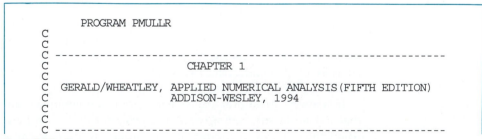

```
      PROGRAM PMULLR
C
C
C  -------------------------------------------------------------
C                          CHAPTER 1
C
C  GERALD/WHEATLEY, APPLIED NUMERICAL ANALYSIS (FIFTH EDITION)
C                    ADDISON-WESLEY, 1994
C
C
C  -------------------------------------------------------------
```

Figure 1.13 Program 1.1

Figure 1.13   *(continued)*

```
C
C  THIS DRIVER CALLS SUBROUTINE MULLR
C
C
C ------------------------------------------------------------
C
      REAL FCN,XR,H,XTOL,FTOL
      INTEGER I,NLIM
      EXTERNAL FCN
      DATA XR,H,XTOL,FTOL,NLIM,I/0.5,0.2,0.0001,0.00001,50,0/
C
      CALL MULLR(FCN,XR,H,XTOL,FTOL,NLIM,I)
C
C  IF I = 1 OR 2, THE ROOT IS EQUAL TO  XR
C
      STOP
      END
C
C
C ------------------------------------------------------------
C
      REAL FUNCTION FCN(X)
C
      REAL X
      FCN = 3*X + SIN(X) - EXP(X)
      RETURN
      END
C
C
C ------------------------------------------------------------
C
      SUBROUTINE MULLR(FCN,XR,H,XTOL,FTOL,NLIM,I)
C
C
C ------------------------------------------------------------
C
C
C     SUBROUTINE MULLR :
C     THIS SUBROUTINE FINDS THE ROOT OF F(X) = 0 BY
C     QUADRATIC INTERPOLATION ON THREE POINTS - MULLER'S METHOD.
C
C
C ------------------------------------------------------------
C
C
C     PARAMETERS ARE :
C
C     FCN   -FUNCTION THAT COMPUTES VALUES FOR F(X). MUST BE
C             DECLARED EXTERNAL IN CALLING PROGRAM. IT HAS
C             ONE ARGUMENT, X.
C     XR    -INITIAL APPROXIMATION TO THE ROOT. USED TO BEGIN
C             ITERATIONS. ALSO RETURNS THE VALUE OF THE ROOT.
C     H     -DISPLACEMENT FROM X USED TO BEGIN CALULATIONS.
C             THE FIRST QUADRATIC IS FITTED AT F(X),
C             F(X+H), F(X-H).
C     XTOL,
C     FTOL -TOLERANCE VALUES FOR X, F(X) TO TERMINATE ITERATIONS.
C     I          -A SIGNAL FOR HOW ROUTINE TERMINATED.
C      I = 1     MEETS TOLERANCE FOR X VALUES.
C      I = 2     MEETS TOLERANCE FOR F(X).
C      I = -1    NLIM EXCEEDED.
C
C  WHEN THE SUBROUTINE IS CALLED, THE VALUE OF I
C  INDICATES WHETHER TO PRINT EACH VALUE OR NOT.
C  I=0 MEANS PRINT THEM, I.NE.0 MEANS DON'T.
C
C ------------------------------------------------------------
C
```

Figure 1.13 *(continued)*

```
         REAL FCN,XR,H,XTOL,FTOL
         INTEGER NLIM,I,J
         REAL Y(3),F1,F2,F3,H1,H2,G,A,B,C,DISC,FR,DELX
C
C
C    ------------------------------------------------------------
C
C  SET INITIAL VALUES INTO Y ARRAY AND EVALUATE F AT THESE POINTS
C
         Y(1) = XR - H
         Y(2) = XR
         Y(3) = XR + H
         F1   = FCN( Y(1) )
         F2   = FCN( Y(2) )
         F3   = FCN( Y(3) )
C
C    ------------------------------------------------------------
C
C
C  BEGIN ITERATIONS
C
         DO 20 J=1,NLIM
         H1 = Y(2) - Y(1)
         H2 = Y(3) - Y(2)
         G = H1 / H2
         A = ( F3*G - F2*(1.0 + G) + F1 ) / ( H1*(H1 + H2))
         B = ( F3 - F2 - A*H2*H2 ) / H2
         C = F2
         DISC = SQRT( B*B - 4.0*A*C )
         IF ( B .LT. 0.0 ) DISC = -DISC
C
C    ------------------------------------------------------------
C
C  FIND ROOT OF QUADRATIC :  A * V**2  + B * V  +  C  =  0
C
         DELX = -2.0 * C / ( B + DISC )
C
C  UPDATE XR
C
         XR = Y(2) + DELX
         FR = FCN(XR)
         IF ( I .EQ. 0 ) PRINT 199, J,XR,FR
C
C    ------------------------------------------------------------
C
C  CHECK STOPPING CRITERIA
C
         IF ( ABS(DELX) .LE. XTOL ) THEN
           I = 1
           PRINT 202, J,XR,FR
           RETURN
         END IF
C
         IF ( ABS(FR) .LE. FTOL ) THEN
           I = 2
           PRINT 203, J,XR,FR
           RETURN
         END IF
C
C    ------------------------------------------------------------
C
C  SELECT THE THREE POINTS FOR THE NEXT ITERATION.
C   WHEN DELX .GT. 0, CHOOSE Y(2), Y(3), & XR,
C   BUT WHEN DELX .LT. 0 CHOOSE Y(1), Y(2), & XR.
C
C  ENTER THE PROPER SET INTO Y ARRAY SO THEY ARE
C              IN ASCENDING ORDER.
C
```

Figure 1.13   *(continued)*

```
          IF ( DELX .GE. 0 ) THEN
            Y(1) = Y(2)
            F1   = F2
            IF ( DELX .GT. H2 ) THEN
              Y(2) = Y(3)
              F2   = F3
              Y(3) = XR
              F3   = FR
            ELSE
              Y(2) = XR
              F2   = FR
            END IF
          ELSE
            Y(3) = Y(2)
            F3   = F2
            IF ( ABS(DELX) .GT. H1 ) THEN
              Y(2) = Y(1)
              F2   = F1
              Y(1) = XR
              F1   = FR
            ELSE
              Y(2) = XR
              F2   = FR
            END IF
C
          END IF
C
20        CONTINUE
C
C    ------------------------------------------------------
C
C  WHEN LOOP IS NORMALLY TERMINATED, NLIM IS EXCEEDED
C
          I = -1
          PRINT 200, NLIM,XR,FR
          RETURN
C
C    ------------------------------------------------------
C
199       FORMAT(' AT ITERATION',I3,3X,' X =',E12.5,4X,
     +         'F(X) = ',E12.5)
200       FORMAT(/' TOLERANCE NOT MET AFTER ',I4,' ITERATIONS  X = ',
     +         E12.5,' F(X) = ',E12.5)
201       FORMAT(/' FUNCTION HAS SAME SIGN AT X1 & X2 ')
202       FORMAT(/' TOLERANCE MET IN ',I2,' ITERATIONS   X = ',E10.5,
     +         '  F(X) = ',E10.5)
203       FORMAT(/' F TOLERANCE MET IN ',I4,
     +         ' ITERATIONS   X =',E10.5,
     +         '  F(X) = ',E10.5)
C
          END


 *************************************************************

                      OUTPUT FOR PMULLR.F


 AT ITERATION  1    X =  .35995E+00    F(X) =  -.11837E-02
 AT ITERATION  2    X =  .36042E+00    F(X) =   .15497E-05

 F TOLERANCE MET IN  2 ITERATIONS
             X = .36042E+00   F(X) = .15497E-05
```

Program 1.2   Fixed-Point Iteration: $x = g(x)$ Method

Figure 1.14(a) is an extremely simple implementation of the fixed-point $[x = g(x)]$ method with a spreadsheet. For those of you not familiar with spreadsheet programs, they display a form that simulates an accountant's page, with rows designated by numbers and columns designated by letters. Each junction of a row and column is a "cell," in which the user can input either text, a numerical value, or a formula.

Formulas can use any of the ordinary symbols for mathematical operations. Numeric constants, as well as references to the value in another cell, can be included in the formulas. Such cell references are comprised of the column–row designation—for example, B4 indicates the cell intersecting column B and row 4.

Spreadsheet formulas can also incorporate tests of the IF . . THEN . . ELSE variety, but we do not show this construct in our simple example. A great variety of functions can be used, especially financial computations such as present worth and values of annuities.

Figure 1.14(a) is a listing of the entries in cells A1 through B19 that find a root through fixed-point iteration. The first six entries are just a heading. Cell A6 contains a starting value $(x_0 = 4)$ for solving $f(x) = x^2 - 2x - 3 = 0$ through the rearrangement $x = \sqrt{2x + 3}$ (the example of Section 1.6). The right-hand side of this expression is put into cell B6 as a formula, referencing the value in cell A6 as its argument.

The procedure is very simple: We start at cell B6 and compute a new x (this will be x_1) by inserting the value in cell A6 into the formula. This new value is then copied into cell A7, and another iterate is computed from it by the formula in cell B7. The process is continued in rows 8 through 19 (providing for 14 repetitions of the method). (A terminating test would be a more sophisticated approach.)

Entering the repeated formulas into the spreadsheet is facilitated by the ability to copy cell contents to another cell or even a block of cells. Unless specially designated, any cell reference that is copied is treated as a relative reference rather than as an absolute cell address, so each formula uses as its argument the value in the cell to its left.

Figure 1.14(b) is how the display looks when all of the entries of Fig. 1.14(a) have been entered. (Normally, computations are automatically carried out.) Observe that convergence to a root at $x = 3.0000$ is shown; it turns out that we did not need as many iterations as was called for. To use the spreadsheet for another problem, changes must be made to the entries in cells A6 and B6 (as well as to the heading), and the new formula in B6 must be copied to the cells below it.

```
Implementation of fixed-point iteration using a spreadsheet.

A1  TEXT  "Solving f(x) = x^2 - 2x - 3 = 0"
A2  TEXT  "   through x = g(x) = SQRT(2x + 3)"
A4  TEXT  "        x "
B4  TEXT  "        f(x)"
```

Figure 1.14(a)   Program 1.2

Figure 1.14(a)   *(continued)*

```
A5  TEXT "------------"
B5  TEXT "------------"
A6  FORMULA = 4
B6  FORMULA =sqrt(2*A6+3)
A7  FORMULA = B6
B7  FORMULA =sqrt(2*A7+3)
A8  FORMULA = B7
B8  FORMULA =sqrt(2*A8+3)
A9  FORMULA = B8
B9  FORMULA =sqrt(2*A9+3)
A10 FORMULA = B9
B10 FORMULA =sqrt(2*A10+3)
A11 FORMULA = B10
B11 FORMULA =sqrt(2*A11+3)
A12 FORMULA = B11
B12 FORMULA =sqrt(2*A12+3)
A13 FORMULA = B12
B13 FORMULA =sqrt(2*A13+3)
A14 FORMULA = B13
B14 FORMULA =sqrt(2*A14+3)
A15 FORMULA = B14
B15 FORMULA =sqrt(2*A15+3)
A16 FORMULA = B15
B16 FORMULA =sqrt(2*A16+3)
A17 FORMULA = B16
B17 FORMULA =sqrt(2*A17+3)
A18 FORMULA = B17
B18 FORMULA =sqrt(2*A18+3)
A19 FORMULA = B18
B19 FORMULA =sqrt(2*A19+3)
```

```
Display after entries of Fig. 1.14(a) have been entered.

      Solving f(x) = x^2 - 2x - 3 = 0
      through x = g(x) = SQRT(2x + 3)

          x                f(x)
      ----------------------------
       4.0000            3.3166
       3.3166            3.1037
       3.1037            3.0344
       3.0344            3.0114
       3.0114            3.0038
       3.0038            3.0013
       3.0013            3.0004
       3.0004            3.0001
       3.0001            3.0000
       3.0000            3.0000
       3.0000            3.0000
       3.0000            3.0000
       3.0000            3.0000
       3.0000            3.0000
```

Figure 1.14(b)

Program 1.3 Bairstow's Method

The third program of this chapter uses Pascal to carry out Bairstow's method to resolve a polynomial into quadratic factors. Pascal programs are a sequence of procedures, the last of which is the driver program. In this program there are ten procedures (one is a function) that precede the main module. Figure 1.15 is a listing.

One feature of Pascal that reduces programming errors is the need to declare all variables before they are used. The global variables for PROGRAM BRSTOW are summarized here:

a	An array that holds the coefficients of the original polynomial. Space is provided for up to 12 coefficients.
b	An array that holds the coefficients when the polynomial is divided by a trial quadratic factor.
c	An array that holds the partial derivatives as explained in Section 1.8.
r, s	Parameters in the trial factor: $x^2 - rx - s$.
delr, dels	Adjustments to improve r and s.
tol	A stopping criterion.
denom	A symbol used in computing delr and dels.
n	The degree of the polynomial.
nPLUS2, i, j, k	Loop control variables.
continue	A Boolean used for loop control.

Here is a list of the procedures in BRSTOW:

initializeDATA: Defines the coefficients of the polynomial (array a), and sets starting values for r and s. The program does this for the following fifth-degree polynomial:

$$x^5 - 17.8x^4 + 99.41x^3 - 261.218x^2 + 352.611x - 134.105.$$

This module also sets a value for tol and zeros the b and c arrays.

printHEADING: Writes a heading and displays the original polynomial.

computeBandCarrays: Carries out the computation of the b's and c's.

findRandS: Computes delr and dels and updates r and s.

tolMET: Sees whether the accuracy criterion has been met.

changeRandSandK: Reinitializes r, s, and k to get the next factor.

printLINEARequation: Displays a linear factor.

printQUADRATICequation: Displays a quadratic factor.

reducePOLYNOMIAL: Resets the a array from the b's to get another factor.

checkTOLMET: Displays a factor when the process has converged, and repeats the adjustment of r and s if not.

This Pascal program clearly shows how structured programming is done and its advantages. With all the working procedures defined, the main program consists of a sequence of calls to them. (There are other calls between the working modules as well.)

The output from the example program is shown at the end of the listing.

To use program BRSTOW for a different polynomial, only the initializeDATA procedure has to be changed.

```
PROGRAM BRSTOW(INPUT,OUTPUT);
(*
                            Chapter 1

      Gerald/Wheatley, APPLIED NUMERICAL ANALYSIS(fifth edition)
                      Addison-Wesley, 1994

      ----------------------------------------------------------------
      This program uses Bairstow's Method of extracting the
      quadratic factors of a polynomial.
*)
VAR
      a,                      (* The original polynomial *)
      b, c:                   (* Working polynomials *)
            Array[0..12] OF Real;
      r,s,                    (* Estimates for quadratic factor *)
      delr, dels,             (* Changes in r and s from previous *)
      tol,                        (* Stopping criterion *)
      denom:          Real;

      n,                      (* Degree of the given Polynomial, a *)
      nPLUS2, i,j,k:          Integer;

      continue:               BOOLEAN;

PROCEDURE initializeDATA;
(*
      ----------------------------------------------------------
      Here we enter the given matrix in descending order as
      as well as initializing the working variables.
*)
VAR
      i :     Integer;
BEGIN
      n := 5;
      nPLUS2 := n + 2;
      a[2] := 1.0; a[3] := -17.8; a[4] := 99.41; a[5] := -261.218;
      a[6] :=  352.611;  a[7] := -134.105;

      r := 0.0; s := 0.0;
(*
      Set up B and C arrays
*)
      FOR i := 0 TO 12 DO BEGIN
          b[i] := 0.0; c[i] := 0.0 END;
          tol := 0.001;
END;  (* initializeDATA *)

PROCEDURE printHEADING;
(*
      ---------------------------------------------------
      Print out a heading and the coefficients of the
      given polynomial.
*)
```

Figure 1.15 Program 1.3

Figure 1.15   *(continued)*

```
        i : Integer;
BEGIN
    WriteLn; WriteLn(' QUADRATIC FACTORS BY BAIRSTOW METHOD.');
    WriteLn; WriteLn;
    WriteLn(' THE ORIGINAL POLYNOMIAL IS: ');
    WriteLn; WriteLn(' POWER OF X ', '':5, ' COEFFICIENT');

    FOR i := 2 TO nPLUS2 DO
        WriteLn( (nPLUS2-i):6, '':10, a[i]:10:3);
    WriteLn; WriteLn;
    WriteLn('*****THE FACTORS OF THE POLYNOMIAL ARE:*****');
    WriteLn
END;    (* printHEADING *)

PROCEDURE computeBandCarrays;
(*
    ------------------------------------------------------------
    We compute the B and C arrays which are needed to
    find the quadratic factor of the polynomial stored
    in the vector, A.
*)
VAR
    j : Integer;
BEGIN
    FOR j := 2 TO nPLUS2 DO BEGIN
        b[j] := a[j] + r*b[j-1] + s*b[j-2];
        c[j] := b[j] + r*c[j-1] + s*c[j-2]    END
END;  (* computeBandCarrays *)

PROCEDURE findRandS;
(*
    ------------------------------------------------------------
    R and S are the coefficients of the quadratic equation we
    are trying to factor from the given polynomial.
*)
BEGIN
    delR := (-b[n+1]*c[n] + b[n+2]*c[n-1])/denom;
    delS := (-c[n]*b[n+2] + c[n+1]*b[n+1])/denom;
    r := r + delR;
    s := s + delS
END;     (* findRandS *)

FUNCTION tolMET: BOOLEAN;
(*
    ------------------------------------------------------------
    Check whether the accuracy condition is met.
*)
BEGIN
    tolMET := (ABS(delR) + ABS(delS)) < tol;
END;  (* tolMET *)

PROCEDURE changeRandSandK;
(*
    ------------------------------------------------------------
    Change R and S values and start over when the denominator
    is zero.
*)
BEGIN
    r := r + 1.0; s := s + 1.0; k := 1
END;     (* changeRandSandK *)

PROCEDURE printLINEARequation;
(*
    ------------------------------------------------------------
    Print a linear equation.
```

Figure 1.15   *(continued)*

```
*)
BEGIN
    WriteLn(b[n+1]:11:5, ' X + ', b[n+2]:10:5);
    continue := FALSE
END;    (* printLINEARequation *)

PROCEDURE printQUADRATICequation;
(*
    ------------------------------------------------------------
    Print a quadratic equation;
*)
BEGIN
    WriteLn(b[n]:11:5,' X**2 + ', b[n+1]:10:5,' + ',b[n+2]:10:5);
    continue := FALSE
END;    (* printQUADRATICequation *)

PROCEDURE reducePOLYNOMIAL;
(*
    ------------------------------------------------------------
    Get new polynomial after the quadratic equation
    has been factored out.
*)
VAR
    i : Integer;
BEGIN
    nPLUS2 := n + 2; k := 1;
    FOR i := 2 TO nPLUS2 DO a[i] := b[i]
END;    (* reducePOLYNOMIAL *)

PROCEDURE checkTOLMET;
(*
    ------------------------------------------------------------
    If the tolerance condition is met we can factor out
    the quadratic equation and check on the degree of the
    remaining polynomial.
*)
BEGIN
    IF tolMET THEN                                          BEGIN
        WriteLn('   X**2 - ', r:10:5, ' X - ', s:10:5);
        n := n-2;

        CASE n OF                         (* Reduced Equation is of *)
        1     : printLINEARequation;   (*      Degree ONE         *)
        2     : printQUADRATICequation; (*      Degree TWO         *)
        3..12 : reducePOLYNOMIAL        (*    Degree > TWO         *)
                                          END  (* CASE *)   END
    ELSE   (* not tolMET *)
        k := k + 1
END;   (* checkTOLMET *)

BEGIN (* MAIN *)
    initializeDATA;
    printHEADING; k := 1; continue := (k<20);
    WHILE continue DO                   BEGIN  (* WHILE *)
        computeBandCarrays;
        denom := c[n]*c[n] - c[n+1]*c[n-1];
        IF (denom <> 0.0) THEN                    BEGIN (* denom IF *)
            findRandS;
            checkTOLMET                           END (* denom <> 0 *)

        ELSE   (* denom = 0 *)
            changeRandSandK;                      (* denom IF *)

    continue := (k < 20 ) AND continue END    (* WHILE *)
END.
```

Figure 1.15 *(continued)*

```
**********************************************************************

        OUTPUT FOR BAIRSTOW.PAS


    QUADRATIC FACTORS BY BAIRSTOW METHOD.

    THE ORIGINAL POLYNOMIAL IS:

    POWER OF X          COEFFICIENT
         5                  1.000
         4                -17.800
         3                 99.410
         2               -261.218
         1                352.611
         0               -134.105


    *****THE FACTORS OF THE POLYNOMIAL ARE:*****

    X**2 -      4.20000 X -     -2.09998
    X**2 -      3.30002 X -     -6.20009
         1.00000 X +   -10.30000
```

Exercises

Section 1.2

1. $e^x - 4x^2$ has a zero at $x = 0.714805912$. Beginning with $[0, 1]$, use eight iterations of interval halving to approximate the zero. Tabulate the error after each iteration, and tabulate the estimates of maximum error. Is the actual error always less than the maximum error estimate? Do the actual errors continually decrease?

2. The quadratic $f(x) = (x - 0.4)(x - 0.6) = x^2 - x + 0.24$ has zeros at $x = 0.4$ and $x = 0.6$, of course. Observe that the endpoints of $[0, 1]$ are not satisfactory to begin bisection. Graph the function, and from this deduce the boundaries of intervals that will converge to each of the roots. If you start with $[0.5, 1]$, what is error bound after six iterations? What is the actual error after six applications of the method?

▶**3.** Find where the graphs of $y = x - 3$ and $y = \ln(x)$ intersect with bisection. Get the intersection value correct to four decimals.

4. Use interval halving to find the smallest positive root of each of the following equations with a relative accuracy of 0.5%. Find good starting values from graphs of the equations.

a. $e^{-x} = \cos(x)$
b. $x^3 - 2x + 1 = 0$
c. $\sin(2x) - e^{x-1} = 0$
d. $\ln(x - 1) = x^2$

5. The function $f(x) = x \cos(x/(x - 2))$ has many zeros, especially near $x = 2$, where $f(x)$ is undefined. Graph the function. Determine the first four roots for $x > 0$ by bisection. Get each root correct to four significant digits.

Section 1.3

6. Repeat Exercise 4, this time using the secant method. Compare the number of iterations required with the number used in bisection.

7. Repeat Exercise 4, this time using the method of false position. Compare the number of iterations required with the number used in bisection and with the number used with the secant method (Exercise 6).

▶**8.** Find where the cubic $y = x^3 - x + 1$ intersects the parabola $y = 2x^2$. From a graph of the two functions, locate the intersection approximately, and then use both *regula falsi* and the secant method to refine the approximate root until you are sure you have it correct to five significant digits. How many iterations are needed in each case?

9. Explain why the secant method usually converges to the root with a specified precision more rapidly than either bisection or the method of linear interpolation.

10. Implement the algorithms given in Section 1.3 with computer programs using any computer language that appeals to you.

Section 1.4

11. Find the root near 0.7 of $f(x) = e^x - 4x^2$ by Newton's method, beginning with $x_0 = 1$. How accurate is the estimate after four iterations? How many iterations with Newton does it take to match the accuracy achieved after eight iterations of bisection (Exercise 1)? Tabulate the number of correct digits at each iteration with Newton, and determine whether these double each time.

12. Repeat Exercise 4, but this time use Newton's method. Compare the number of iterations required with bisection (Exercise 4), the secant method (Exercise 6), and *regula falsi* (Exercise 7).

▶13. Use Newton's method on the equation $x^2 = N$ to derive the algorithm for obtaining the square root of N:

$$x_{i+1} = \frac{1}{2}\left(x_i + \frac{N}{x_i}\right).$$

14. Repeat Exercise 13, this time deriving algorithms for the third and fourth roots of N.

15. If the algorithm of Exercise 13 is applied twice, show that

$$N^{1/2} \simeq \frac{(A + B)}{4} + \frac{N}{(A + B)},$$

where $N = A * B$.

16. Show that the error of the algorithm in Exercise 15 is approximately

$$\frac{1}{8}\left(\frac{A - B}{A + B}\right)^4.$$

17. Expand $f(x)$ about the point $x = a$ in a Taylor series, and from this derive Newton's method. (See Appendix A if you have forgotten Taylor series.)

▶18. $f(x) = (x - 1)^3(x - 2)$ has roots at $x = 1$ and $x = 2$. The first is a triple root, and the second is a simple root. Using starting values that differ from

the roots by 0.1, compare the number of iterations taken by Newton's method to reach each of the roots. (The problem that Newton's method has with multiple roots is discussed in Sections 1.10 and 1.11.)

19. Beginning with the interval $[0.9, 1.1]$, use the secant method on the root at $x = 1$ in Exercise 18. Can you explain why the secant method works better than Newton's method in this case?

▶20. What starting values cause Newton's method to fail totally in Exercise 18?

21. Repeat Exercise 20 for each of the equations in Exercise 4 (if invalid starting values exist).

22. Each of the following has complex roots. Apply Newton's method to find them:
   a. $x^2 = -2$
   b. $x^3 - x^2 - 1 = 0$
   c. $x^4 - 3x^2 + x + 1 = 0$
   d. $x^2 = (e^{-2x} - 1)/x$

Section 1.5

23. Use Muller's method to solve the following problems. Apply four iterations, and determine how the successive errors are related to each other. Use starting values that differ by 0.2 from one another.
   a. $2x^3 + 4x^2 - 2x - 5 = 0$, root near 1.0 (exact value is 1.07816259)
   b. $e^x - 3x^2 = 0$, root near 4.0 (exact value is 3.73307903)
   c. $e^x - 3x^2 = 0$, root near 1.0
   d. $\tan(x) - x - 1 = 0$, root near 1.1

▶24. Muller's method is sometimes used in a "self-starting" form. Instead of specifying values near a root, the algorithm automatically begins with the values 0, 0.5, −0.5. Use these starting values on the equations of Exercise 4. We are supposed to get the root nearest the origin by this technique. Will this always be true?

25. An extension to the self-starting principle is the deflation of the function after finding a root. *Deflating* refers to the removal of a root by dividing by $(x - r)$, resulting in a new function that lacks the root $x = r$. (Of course, a discontinuity is introduced by the division, at $x = r$). Compare the graphs of

$f_1(x) = x(x - 3)(x - 1)$, $f_2(x) = x(x - 3)$,
$f_3(x) = x(x - 1)$,   and $f_4(x) = (x - 1)(x - 3)$

to see that this is true. After beginning as in Exercise 24, continue with deflation to get all the roots of the following:

a. $x^3 - 2x^2 - 1 = 0$
b. $x^3 - 2x^2 - 1.185 = 0$
c. $\exp(x^2 - 1) + 10 \sin(2x) - 5 = 0$

▶**26.** A problem with Muller's method is that the parabola that is created may not intersect the x-axis. When will you know that this occurs? What can you do to recover if this happens?

27. Since Muller's method subtracts function values that may be almost equal, there is a possibility of large relative errors due to the imprecise arithmetic of the computer. Investigate this problem by using starting values that are close together (but not near the root) on the equations of Exercise 23.

28. Muller's method finds complex roots, but we must use complex arithmetic. Find the complex roots of the equations of Exercise 22 by Muller's method.

Section 1.6

29. $f(x) = e^x - 3x^2 = 0$ has three real roots. An obvious rearrangement is

$$x = \pm \sqrt{\frac{e^x}{3}}.$$

Show that convergence is to the root near -0.5 if we begin with $x_0 = 0$ and use the negative value. Show also that convergence to a second root near 1.0 is obtained if $x_0 = 0$ and the positive value is used. Show, however, that this form does not converge to the third root near 4.0 even though a starting value very close to the root is used. Find a different rearrangement that will converge to the root near 4.0.

▶**30.** One root of the quadratic $x^2 + x - 1 = 0$ is at $x = 0.6180$. The form

$$x = \frac{1}{(x + 1)}$$

converges to this root for $x_0 = 1$. How many iterations of the fixed-point method are required to get the root correct to four digits? If Aitken acceleration is used after three estimates are available, how many iterations are needed?

31. For what range of starting values does the fixed-point operation of Exercise 30 converge to the root near 0.6? (Looking at a graph may help.) Do not stop with a division by zero, that is, for $x_0 = -1$, $x_1 = 1/0$. Use $x_3 = \lim_{x_2 \to \infty} (1/(x_2+1)) = 0$.

▶**32.** The cubic $2x^3 + 3x^2 - 3x - 5 = 0$ has a root near

$x = 1.25$. Find at least three rearrangements that will converge to this root.

33. Use the criterion for the effectiveness of Aitken acceleration that is given in Section 1.6 to show that the acceleration in Exercise 30 will be quite rapid.

Section 1.7

34. Do this exercise by hand computation: A fourth-degree polynomial is

$$P(x) = x^4 + 4.6x^3 + 6.6x^2 - 11x - 14.$$

Using synthetic division, find $P(-1)$. From the result, write $P(x)$ as the product of two polynomials, one of degree 3. Find the one positive root of the cubic polynomial by Newton's method, again using synthetic division. Finally, use the quadratic formula to get the last two roots.

▶**35.** Do this exercise by hand computation: Using a calculator, find the two real negative roots of $P(x)$ correct to seven decimal places. Then determine the other roots from the deflated quadratic, again to seven places.

$$P(x) = 2x^4 + 7x^3 - 4x^2 + 29x + 14$$

36. Write a computer program in any language that implements the algorithm for Horner's method.

37. Working with deflated polynomials to get additional roots is a way to save computational effort, but any error in the first roots is propagated and causes the successive roots to be incorrect. Investigate this characteristic with the polynomial

$$P(x) = x^3 - 4x^2 + 3x + 1$$

by deliberately getting a first root imperfectly (say, correct to only 1% relative accuracy), and see how this affects the accuracy of the other roots.

The roots of $P(x)$ are

$$x_1 = 1.44504, \quad x_2 = 2.80193, \quad x_3 = -0.246979.$$

Work this procedure three times, getting each individual root imperfectly, to see its effect on the others obtained from the quadratic formula. You may want to see how the amount of error in the first root affects the accuracy of the other roots.

▶**38.** This exercise simulates the effect of inaccuracies such as those from imprecise experimental measurements. Suppose the coefficients of the polynomial in Exercise 35 are not perfectly known. Determine the influence of such imperfections by finding how much the roots change when any one of the five coefficients is varied by 1%, observing the effect when each coefficient is varied independently. Which coefficient causes the greatest change in the roots due to such variation?

39. Sometimes one or more of the coefficients of a polynomial are zero. How does this affect the speedup from parallel processing? In answering this, sketch the directed acyclic graphs for sequential and for parallel processing of a fourth-degree polynomial that has
a. One missing term,
b. Two missing terms.

Does it make a difference which term(s) are missing? Your graphs for parallel processing should maximize the speedup.

40. Where do the graphs of $y = x^2 - 3$ and $xy + 2y^2 = 4$ intersect? Find all of the intersections by Newton's method.

Section 1.8

41. Using hand computation, and beginning with the trial factor $x^2 - 2x + 4$, use Bairstow's method to obtain the two quadratic factors of

$$x^4 + 2.1x^3 - 3.42x^2 + 8.19x + 7.13.$$

From these, get the zeros of the polynomial.

▶**42.** Use Bairstow's method to resolve into quadratic factors:

$$x^4 - 5.7x^3 + 26.7x^2 - 42.21x + 69.23.$$

You should find that all four zeros are complex.

43. Solve the equation of Exercise 34 using the Bairstow method.

▶**44.** Use Bairstow's method to get the zeros of the polynomial in Exercise 35.

45. When the modulus of one pair of complex roots is the same as for another pair, Bairstow's method is slow to converge. Observe that this is true by using hand computation to get the quadratic factors of

$$x^4 - x^3 + x^2 - x + 1,$$

which factors into

$$(x^2 + 0.61803x + 1)(x^2 - 1.618034x + 1).$$

What is the modulus of the roots?

46. No formal algorithm is given in Section 1.8 for Bairstow's method, although the operations are shown clearly. Translate these operations into an algorithm of the nature used in previous sections of this chapter. Your algorithm should include a test to see whether additional repetitions of the procedure are required to get a quadratic factor with a specified degree of accuracy.

47. Write a computer program that implements your algorithm of Exercise 46. In the program, use $r = 0$, $s = 0$ as the starting values rather than have the user input them.

Section 1.9

48. Use the QD algorithm to get the roots of the following:
▶**a.** $x^3 - x^2 - 2x + 1 = 0$
▶**b.** $x^4 + 2.1x^3 - 3.42x^2 + 8.19x + 7.13 = 0$
c. $x^4 - 5.7x^3 + 26.7x^2 - 42.21x + 69.23 = 0$

In parts (b) and (c), compare your result to those from Exercises 41 and 42.

49. The QD algorithm also has difficulty with polynomials such as those in Exercise 45. Try your patience on that polynomial.

50. Write the algorithm for the QD method, and use this algorithm to write a computer program that carries it out.

51. Repeat Exercise 41, but this time use Graeffe's method. What happens when there are roots of equal magnitude?

▶**52.** Repeat Exercise 41, this time using Laguerre's method. Is there a difficulty with roots of equal magnitude?

Section 1.10

53. This polynomial has a double root at $x = 1$ and a triple root at $x = 3$:

$$(x - 1)^2(x - 3)^3$$
$$= x^5 - 11x^4 + 46x^3 - 90x^2 + 81x - 27.$$

Can you get all of the roots by bisection? Get the root at $x = 3$ starting with the interval $[2, 4]$.

54. Using the polynomial in Exercise 53, can you get all of the roots by the secant method? Which root do you get starting with the interval $[0, 4]$? Try to predict this root in advance.

55. Using the polynomial in Exercise 53, start Newton's method at $x = 2$. Does it converge? If so, to which root? Can you predict this in advance? Tabulate the error of the successive iterates when you start with $x = 2.9$. Does the number of accurate digits about double each time?

▶**56.** Using the polynomial in Exercise 53, use Bairstow's algorithm on the polynomial to get all the roots.

57. Using the polynomial in Exercise 53, try to determine the roots with Muller's method. Is it successful in finding both roots? If not, explain. Are there certain starting values that are unsuccessful?

58. Use the polynomial in Exercise 53, except now use the modified Newton's method:

$$x_{n+1} = x_n - k * \frac{f(x_n)}{f'(x_n)},$$

where k is the multiplicity of the root. Show from the successive iterates that convergence is now quadratic.

▶**59.** Solve for the roots of the polynomial of Exercise 53 by applying Newton's method to $P(x)/P'(x)$. Show that quadratic convergence is secured for both roots.

60. This quadratic has nearly equal roots:

$$(x - 1.99)(x - 2.01) = x^2 - 4x + 3.9999.$$

a. Use Newton's method with a starting value of 2.1. Is there quadratic convergence?
b. Use Newton's method with a starting value of 1.9. Is there quadratic convergence?
c. Use Newton's method with a starting value of 2.0. What happens?
d. Repeat part (c), but use a quadratic that has roots at 2.01 and 2.03, starting with $x = 2.02$. If the results are different, explain.

Section 1.11

61. $f(x) = e^x \sin(x) - x^2$ has a zero near $x = 2.6$. By comparing the errors of successive iterates with Newton's method (beginning at $x = 3$), determine the order of convergence.

62. Repeat Exercise 61, but this time start at $x = 2.1$. Why is the first iterate so far from the zero of $f(x)$?

▶**63.** Repeat Exercise 61, but this time use these starting values:
a. $x = 2.05$
b. $x = 2.00$

64. Repeat Exercise 61, this time using:
a. The *regula falsi* method (begin with $x = 2, x = 3$).
b. The secant method (begin with $x = 2, x = 3$).
c. Muller's method (begin with $x = 2, 2.5, 3$).

65. Make a sketch similar to Fig. 1.7 for a $g(x)$ where fixed-point iteration does converge even though not all the conditions given in Section 1.11 hold.

66. Show that Newton's method is not quadratically convergent by comparing errors for:
a. $e^{-x}(x^2 - 2x + 1) = 0$ (root at 1).
b. $1 - \cos(x) = 0$ (root at 2π).
▶**c.** $(e^x - 1)(x)[\sin(x)]$ (root at zero).

67. Repeat Exercise 66, but this time use the modifications to Newton's method given in Section 1.10.

68. Do an experiment similar to that of Exercise 61, this time for Laguerre's method on some quadratic polynomial.

Section 1.12

▶**69.** Use DERIVE to get the smallest positive root of $\sin(x) = x * \cos(x)$.

70. Modify the DERIVE program of Section 1.12 to have it use the secant method.

71. Modify the DERIVE program of Section 1.12 to have it use the *regula falsi* method. Incorporate a test to make sure that the initial values are of opposite sign.

Applied Problems and Projects

Beginning with this chapter and in each subsequent chapter, we present some problems that either are more involved or apply to the "real world."

72. Given

$$x'' + x + 2y' + y = f(t),$$
$$x'' - x + y = g(t), \qquad x(0) = x'(0) = y(0) = 0.$$

In solving this pair of simultaneous second-order differential equations by the Laplace transform method, it becomes necessary to factor the expression

$$(S^2 + 1)(S) - (2S + 1)(S^2 - 1) = -S^3 - S^2 + 3S + 1,$$

so that partial fractions can be used in getting the inverse transform. What are the factors?

73. DeSantis (1976) has derived a relationship for the compressibility factor of real gases of the form

$$z = \frac{1 + y + y^2 - y^3}{(1 - y)^3},$$

where $y = b/(4v)$, b being the van der Waals correction and v the molar volume. If $z = 0.892$, what is the value of y?

74. In studies of solar-energy collection by focusing a field of plane mirrors on a central collector, Vant-Hull (1976) derives an equation for the geometrical concentration factor C:

$$C = \frac{\pi(h/\cos A)^2 F}{0.5\pi D^2(1 + \sin A - 0.5 \cos A)},$$

where A is the rim angle of the field, F is the fractional coverage of the field with mirrors, D is the diameter of the collector, and h is the height of the collector. Find A if $h = 300$, $C = 1200$, $F = 0.8$, and $D = 14$.

75. Lee and Duffy (1976) relate the friction factor for flow of a suspension of fibrous particles to the Reynolds number by this empirical equation:

$$\frac{1}{\sqrt{f}} = \left(\frac{1}{k}\right) \ln(\mathrm{RE}\sqrt{f}) + \left(14 - \frac{5.6}{k}\right).$$

In their relation, f is the friction factor, RE is the Reynolds number, and k is a constant determined by the concentration of the suspension. For a suspension with 0.08% concentration, $k = 0.28$. What is the value of f if RE = 3750?

76. Based on the work of Frank–Kamenetski in 1955, temperatures in the interior of a material with embedded heat sources can be determined if we solve this equation:

$$e^{-(1/2)t}\cosh^{-1}(e^{(1/2)t}) = \sqrt{\tfrac{1}{2}L_{\mathrm{cr}}}.$$

Given that $L_{\mathrm{cr}} = 0.088$, find t.

77. Suppose we have the 555 Timer Circuit

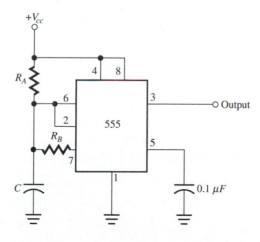

whose output waveform is

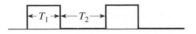

where

$$T_1 + T_2 = \frac{1}{f}$$

$$f = \text{frequency}$$

$$\text{Duty cycle} = \frac{T_1}{T_1 + T_2} \times 100\%.$$

It can be shown that

$$T_1 = R_A C \ln(2)$$

$$T_2 = \frac{R_A R_B C}{R_A + R_B} * \ln\left(\left|\frac{R_A - 2R_B}{2R_A - R_B}\right|\right).$$

Given that $R_A = 8670$, $C = 0.01 \times 10^{-6}$, $T_2 = 1.4 \times 10^{-4}$,
a. Find T_1, f, and the duty cycle,
b. Find R_B using any program you have written.
c. Select an f and duty cycle, then find T_1 and T_2.

78. The solution of boundary-value problems by an analytical (Fourier series) method often involves finding the roots of transcendental equations to evaluate the coefficients. For example,

$$y'' + \lambda y = 0, \qquad y(0) = 0, \qquad y(1) = y'(1),$$

involves solving $\tan z = z$. Find three values of z other than $z = 0$.

79. Find all max/min points of the function

$$f(x) = [\sin(x)]^6 * e^{20x} * \tan(1 - x)$$

on the interval $[0, 1]$. Compare your own root-finding program with the IMSL subroutine ZBRENT. [Note the disadvantage in trying to solve $f'(x) = 0$ using Newton's method.]

80. In Chapter 4, a particularly efficient method for numerical integration of a function, called *Gaussian quadrature*, is discussed. In the development of formulas for this method, it is necessary to evaluate the zeros of Legendre polynomials. Find the zeros of the Legendre polynomial of sixth order:

$$P_6(x) = \frac{1}{48}(693x^6 - 945x^4 + 315x^2 - 15).$$

(*Note:* All the zeros of the Legendre polynomials are less than one in magnitude and, for polynomials of even order, are symmetrical about the origin.)

81. The Legendre polynomials of Problem 80 are one set of a class of polynomials known as *orthogonal* polynomials. Another set are the *Laguerre* polynomials. Find the zeros of the following:
a. $L_3(x) = x^3 - 9x^2 + 18x - 6$
b. $L_4(x) = x^4 - 16x^3 + 72x^2 - 96x + 24$

82. Still another set of orthogonal polynomials are the *Chebyshev* polynomials. (We will use these in Chapter 10.) Find the roots of

$$T_6(x) = 32x^6 - 48x^4 + 18x^2 - 1 = 0.$$

(Note the symmetry of this function. All the roots of Chebyshev polynomials are also less than one in magnitude.)

83. A sphere of density d and radius r weighs $\frac{4}{3}\pi r^3 d$. The volume of a spherical segment is $\frac{1}{3}\pi(3rh^2 - h^3)$. Find the depth to which a sphere of density 0.6 sinks in water as a fraction of its radius. (See the accompanying figure.)

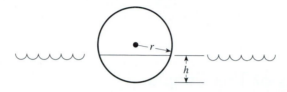

84. For several functions that have multiple roots, investigate whether Aitken acceleration improves the rate of convergence. Do this for several methods.

85. Make experimental comparisons of the rates of conversion for Newton's method, for Newton's method with the derivative estimated numerically, and for the secant method. Make a table that shows how the errors decrease for each method, then make a log plot of the errors.

86. Muller's method is said to converge with an order of convergence equal to 1.85. Verify this experimentally. Is it true if there is a multiple root?

2

Solving Sets of Equations

Contents of This Chapter

This chapter covers the extremely important topic of how to solve a large system of equations. We emphasize sets of simultaneous linear equations, although we also treat less thoroughly the more difficult area of solving systems of nonlinear equations.

Linear systems are perhaps the most widely applied numerical procedures when real-world situations are to be simulated. Linear systems are used in statistical analysis in many applications. Methods for numerically solving ordinary and partial-differential equations depend on them.

2.1 Computing the Forces in a Planar Truss
Shows how a system of linear equations can model an engineering structure known as a pin-jointed truss.

2.2 Matrix Notation
Reviews the elementary concepts of matrices and vectors that are used extensively in this chapter.

2.3 The Elimination Method
Describes the classic elimination method that changes a system of linear equations so that back-substitution can find the solution.

2.4 Gauss and Gauss–Jordan Methods
Explains the key LU method, which is the basis for solving systems of equations. The use of partial pivoting minimizes the effects of round-off error and may be required to avoid a division by zero. The "Big O" notation is introduced as a way to compare different methods.

2.5 Other Direct Methods
Explains additional ways to carry out Gaussian elimination. The LU decomposition of the coefficient matrix is applied to solve a system with a series of right-hand sides.

2.15 Parallel Processing

Tells how parallel computing can be applied to the solution of linear systems. An algorithm is developed in several stages that allows a many-fold reduction in processing time.

Chapter Summary

Lists the topics that you should understand after studying this chapter.

Computer Programs

Presents examples of computer programs that implement some of the algorithms of this chapter.

2.1 Computing the Forces in a Planar Truss

A *truss* is a structure used for supporting loads such as in bridges or buildings. A truss is composed entirely of straight members arranged to form one or more triangles, giving the structure stability. Usually these assumptions are made when a truss is analyzed: (1) All joints are *pin connections* (a pin joint is a frictionless connection that is free to rotate and hence cannot apply a bending moment to any member that is connected to it); (2) all loads are applied only at the joints, including the support loads (called *reactions*); and (3) the weight of the members of the truss is neglected. With these assumptions, it is clear that the truss must be supported only at some of the joints. Supports may be fixed or may allow a translation in one direction. Figure 2.1 shows a typical truss.

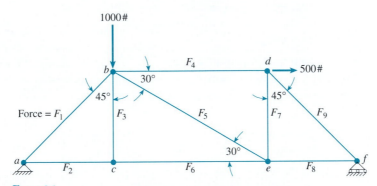

Figure 2.1

In the example truss, there are six pins (joints) and nine members. Each joint represents two degrees of freedom (signifying that two independent forces act there). We resolve the force acting at each of the pins into x- and y-components.

The forces that act on the truss, in view of the assumptions, are pointed along each of the members. It is conventional to consider the forces to act toward the center of the member, away from the joint. Since there are nine members, we have nine member-forces, $F_i, i = 1, \ldots, 9$. These forces can be determined by setting the sum of all forces acting on the pins horizontally or vertically to zero, since the system is at equilibrium (the structure itself does not move).

For instance, at joints b and d, we have

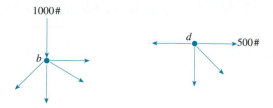

If we start from pin b and progress to each other pin, we can write nine equations:

$$
\begin{aligned}
\cos(45°) * F_1 - F_4 - \cos(30°) * F_5 &= 0 \\
\sin(45°) * F_1 + F_3 + \sin(30°) * F_5 &= -1000 \\
F_2 - F_6 &= 0 \\
- F_3 &= 0 \\
F_4 - \sin(45°) * F_9 &= 500 \\
F_7 + \cos(45°) * F_9 &= 0 \\
\cos(30°) * F_5 + F_6 - F_8 &= 0 \\
\sin(30°) * F_5 + F_7 &= 500 \\
F_8 + \cos(45°) * F_9 &= 0
\end{aligned}
\tag{2.1}
$$

These nine equations are exactly enough to solve for the nine unknown member-forces. Such a truss is called *statically determinant*. If we should add another member—say, a cross brace from pin c to pin d—we would not have enough equations to solve the system. That case is called *statically indeterminant*. (A solution can be found for this case, but additional properties must be considered; we do not discuss this situation, but see Applied Problem 86.)

This chapter describes methods for solving a set of equations such as those of Eq. (2.1). Our methods are adapted to a computer because, without such a device, getting the solution is most tedious and prone to error. The methods we will discuss can be used for solving several hundred simultaneous linear equations.

2.2 Matrix Notation

Our discussion of methods to solve sets of linear equations will be facilitated by some of the concepts and notation of matrix algebra. Only the more elementary ideas will be needed.

A *matrix* is a rectangular array of numbers in which not only the value of the number is important but also its position in the array. The size of the matrix is described by the number of its rows and columns. A matrix of n rows and m columns is said to be $n \times m$. The elements of the matrix are generally enclosed in brackets, and double-subscripting is the common way of indexing the elements. The first subscript always denotes the row, and the second denotes the column in which the element occurs. Capital letters are used to refer to matrices. For example,

$$A = \begin{bmatrix} a_{11} & a_{12} \ldots a_{1m} \\ a_{21} & a_{22} \ldots a_{2m} \\ \vdots & \\ a_{n1} & a_{n2} \ldots a_{nm} \end{bmatrix} = [a_{ij}], \qquad i = 1, 2, \ldots, n, \qquad j = 1, 2, \ldots, m.$$

Enclosing the general element a_{ij} in brackets is another way of representing matrix A, as just shown. Sometimes we will enclose the name of the matrix in brackets, $[A]$, to emphasize that A is a matrix.

Two matrices of the same size may be added or subtracted. The sum of

$$A = [a_{ij}] \qquad \text{and} \qquad B = [b_{ij}]$$

is the matrix whose elements are the sum of the corresponding elements of A and B:

$$C = A + B = [a_{ij} + b_{ij}] = [c_{ij}].$$

Similarly we get the *difference* of two equal-sized matrices by subtracting corresponding elements. If two matrices are not equal in size, they cannot be added or subtracted. Two matrices are equal if and only if each element of one is the same as the corresponding element of the other. Obviously, equal matrices must be of the same size. Some examples will help make this clear.

If

$$A = \begin{bmatrix} 4 & 7 & -5 \\ -4 & 2 & 12 \end{bmatrix} \qquad \text{and} \qquad B = \begin{bmatrix} 1 & 5 & 4 \\ 2 & -6 & 3 \end{bmatrix},$$

we say that A is 2×3 because it has two rows and three columns. B is also 2×3. Their sum C is also 2×3:

$$C = A + B = \begin{bmatrix} 5 & 12 & -1 \\ -2 & -4 & 15 \end{bmatrix}.$$

The difference D of A and B is

$$D = A - B = \begin{bmatrix} 3 & 2 & -9 \\ -6 & 8 & 9 \end{bmatrix}.$$

Multiplication of two matrices is defined as follows, when A is $n \times m$ and B is $m \times r$:

$$[a_{ij}][b_{ij}] = [c_{ij}] = \begin{bmatrix} (a_{11}b_{11} + a_{12}b_{21} + \cdots + a_{1m}b_{m1}) \ldots (a_{11}b_{1r} + \cdots + a_{1m}b_{mr}) \\ (a_{21}b_{11} + a_{22}b_{21} + \cdots + a_{2m}b_{m1}) \ldots (a_{21}b_{1r} + \cdots + a_{2m}b_{mr}) \\ \vdots \\ (a_{n1}b_{11} + a_{n2}b_{21} + \cdots + a_{nm}b_{m1}) \ldots (a_{n1}b_{1r} + \cdots + a_{nm}b_{mr}) \end{bmatrix},$$

$$c_{ij} = \sum_{k=1}^{m} a_{ik}b_{kj}, \qquad i = 1, 2, \ldots, n, \qquad j = 1, 2, \ldots, r.$$

It is simplest to select the proper elements if we count across the rows of A with the left hand while counting down the columns of B with the right. Unless the number of columns of A equals the number of rows of B (so the counting comes out even), the matrices cannot be multiplied. Hence if A is $n \times m$, B must have m rows or else they are said to be "nonconformable for multiplication" and their product is undefined. In general $AB \neq BA$, so the order of factors must be preserved in matrix multiplication.

If a matrix is multiplied by a scalar (a pure number), the product is a matrix, each element of which is the scalar times the original element. We can write

$$\text{If } kA = C, \qquad c_{ij} = ka_{ij}.$$

A matrix with only one column, $n \times 1$ in size, is termed a *column vector*, and one of only one row, $1 \times m$ in size, is called a *row vector*. When the unqualified term *vector* is used, it nearly always means a *column* vector. Frequently the elements of vectors are only singly subscripted.

Some examples of matrix multiplication follow.

$$\text{Suppose } A = \begin{bmatrix} 3 & 7 & 1 \\ -2 & 1 & -3 \end{bmatrix}, \qquad B = \begin{bmatrix} 5 & -2 \\ 0 & 3 \\ 1 & -1 \end{bmatrix}, \qquad x = \begin{bmatrix} -3 \\ 1 \\ 4 \end{bmatrix}, \qquad y = \begin{bmatrix} y_1 \\ y_2 \\ y_3 \end{bmatrix}.$$

$$A * B = \begin{bmatrix} 16 & 14 \\ -13 & 10 \end{bmatrix}; \qquad B * A = \begin{bmatrix} 19 & 33 & 11 \\ -6 & 3 & -9 \\ 5 & 6 & 4 \end{bmatrix};$$

$$A * x = \begin{bmatrix} 2 \\ -5 \end{bmatrix}; \qquad A * y = \begin{bmatrix} 3y_1 + 7y_2 + y_3 \\ -2y_1 + y_2 - 3y_3 \end{bmatrix}.$$

Since A is 2×3 and B is 3×2, they are conformable for multiplication and their product is 2×2. When we form the product of $B * A$, it is 3×3. Observe that not only is $AB \neq BA$; AB and BA are not even the same size. The product of A and the vector x (a 3×1 matrix) is another vector, one with two components. Similarly, Ay has two components. We cannot multiply B times x or B times y; they are nonconformable.

The product of the scalar number 2 and A is

$$2A = \begin{bmatrix} 6 & 14 & 2 \\ -4 & 2 & -6 \end{bmatrix}.$$

Since a vector is just a special case of a matrix, a column vector can be multiplied by a matrix, as long as they are conformable in that the number of columns of the matrix equals the number of elements (rows) in the vector. The product in this case will be another column vector. The size of a product of two matrices, the first $m \times n$ and the second $n \times r$, is $m \times r$. An $m \times n$ matrix times an $n \times 1$ vector gives an $m \times 1$ product.

The general relation for $Ax = b$ is

$$b_i = \sum_{k=1}^{\substack{\text{No. of cols.}}} a_{ik} x_k, \qquad i = 1, 2, \ldots, \text{No. of rows.}$$

Two vectors, each with the same number of components, may be added or subtracted. Two vectors are equal if each component of one equals the corresponding component of the other.

This definition of matrix multiplication permits us to write the set of linear equations

$$a_{11} x_1 + a_{12} x_2 + \cdots + a_{1n} x_n = b_1,$$
$$a_{21} x_1 + a_{22} x_2 + \cdots + a_{2n} x_n = b_2,$$
$$\vdots$$
$$a_{n1} x_1 + a_{n2} x_2 + \cdots + a_{nn} x_n = b_n,$$

much more simply in matrix notation, as $\boxed{Ax = b,}$ where

$$A = \begin{bmatrix} a_{11} & a_{12} \ldots a_{1n} \\ a_{21} & a_{22} \ldots a_{2n} \\ \vdots & \\ a_{n1} & a_{n2} \ldots a_{nn} \end{bmatrix}, \qquad x = \begin{bmatrix} x_1 \\ x_2 \\ \vdots \\ x_n \end{bmatrix}, \qquad b = \begin{bmatrix} b_1 \\ b_2 \\ \vdots \\ b_n \end{bmatrix}.$$

For example,

$$
\begin{bmatrix} 3 & 2 & 4 \\ 1 & -2 & 0 \\ -1 & 3 & 2 \end{bmatrix} * x = \begin{bmatrix} 14 \\ -7 \\ 2 \end{bmatrix}
$$

is the same as the set of equations

$$
\begin{aligned}
3x_1 + 2x_2 + 4x_3 &= 14, \\
x_1 - 2x_2 &= -7, \\
-x_1 + 3x_2 + 2x_3 &= 2.
\end{aligned}
$$

A very important special case is the multiplication of two vectors. The first must be a row vector if the second is a column vector, and each must have the same number of components. For example,

$$
\begin{bmatrix} 1 & 3 & -2 \end{bmatrix} * \begin{bmatrix} 4 \\ -1 \\ 3 \end{bmatrix} = \begin{bmatrix} -5 \end{bmatrix}
$$

gives a "matrix" of one row and one column. The result is a pure number, a scalar. This product is called the *scalar product* of the vectors, also called the *inner product*.

If we reverse the order of multiplication of these two vectors, we obtain

$$
\begin{bmatrix} 4 \\ -1 \\ 3 \end{bmatrix} * \begin{bmatrix} 1 & 3 & -2 \end{bmatrix} = \begin{bmatrix} 4 & 12 & -8 \\ -1 & -3 & 2 \\ 3 & 9 & -6 \end{bmatrix}.
$$

This product is called the *outer product*. Although not as well known as the inner product, the outer product is very important in nonlinear optimization problems.*

Certain square matrices have special properties. The diagonal elements are the line of elements a_{ii} from upper left to lower right of the matrix. If only the diagonal terms are nonzero, the matrix is called a *diagonal matrix*. When the diagonal elements are each equal to unity while all off-diagonal elements are zero, the matrix is said to be the *identity matrix of order n*. The usual symbol for such a matrix is I_n, and it has properties similar to unity. For example, the order-4 identity matrix is

$$
\begin{bmatrix} 1 & 0 & 0 & 0 \\ 0 & 1 & 0 & 0 \\ 0 & 0 & 1 & 0 \\ 0 & 0 & 0 & 1 \end{bmatrix} = I_4.
$$

The subscript is omitted when the order is clear from the context.

---

* Another important product of three-component vectors is the *vector product*, also known as the *cross product*.

A vector whose length is one is called a *unit vector*. A vector that has all its elements equal to zero except one element, which has a value of unity, is called a *unit basis vector*. There are three distinct unit basis vectors for order-3 vectors:

$$\begin{bmatrix} 1 \\ 0 \\ 0 \end{bmatrix}, \quad \begin{bmatrix} 0 \\ 1 \\ 0 \end{bmatrix}, \quad \text{and} \quad \begin{bmatrix} 0 \\ 0 \\ 1 \end{bmatrix}.$$

If all the elements above the diagonal are zero, a matrix is called *lower-triangular*; it is called *upper-triangular* when all the elements below the diagonal are zero. For example, these order-3 matrices are lower- and upper-triangular:

$$L = \begin{bmatrix} 1 & 0 & 0 \\ 4 & 6 & 0 \\ -2 & 1 & -4 \end{bmatrix}, \quad U = \begin{bmatrix} 1 & -3 & 3 \\ 0 & -1 & 0 \\ 0 & 0 & 1 \end{bmatrix}.$$

Tridiagonal matrices are those that have nonzero elements only on the diagonal and in the positions adjacent to the diagonal; they will be of special importance in certain partial-differential equations. An example of a tridiagonal matrix is

$$\begin{bmatrix} -4 & 2 & 0 & 0 & 0 \\ 1 & -4 & 1 & 0 & 0 \\ 0 & 1 & -4 & 1 & 0 \\ 0 & 0 & 1 & -4 & 1 \\ 0 & 0 & 0 & 2 & -4 \end{bmatrix}.$$

The *transpose* of a matrix is the matrix that results when the rows are written as columns (or, alternatively, when the columns are written as rows). The symbol A^T is used for the *transpose of A*.

EXAMPLE 2.1

$$A = \begin{bmatrix} 3 & -1 & 4 \\ 0 & 2 & -3 \\ 1 & 1 & 2 \end{bmatrix}; \quad A^T = \begin{bmatrix} 3 & 0 & 1 \\ -1 & 2 & 1 \\ 4 & -3 & 2 \end{bmatrix}.$$

It should be obvious that the transpose of A^T is just A itself.

When a matrix is square, a quantity called its *trace* is defined. The trace of a square matrix is the sum of the elements on its main diagonal. For example, the traces of the above matrices are

$$\text{tr}(A) = 3 + 2 + 2 = 7; \qquad \text{tr}(A^T) = 3 + 2 + 2 = 7.$$

It should be obvious that the trace remains the same if a square matrix is transposed.

We present here some additional examples of arithmetic operations with matrices.

$$(3) * \begin{bmatrix} 1 & 2 \\ 3 & 4 \end{bmatrix} = \begin{bmatrix} 3 & 6 \\ 9 & 12 \end{bmatrix}.$$

$$\begin{bmatrix} 1 & 3 & 2 \\ -1 & 0 & 4 \end{bmatrix} + \begin{bmatrix} -1 & 0 & 2 \\ 4 & 1 & -3 \end{bmatrix} = \begin{bmatrix} 0 & 3 & 4 \\ 3 & 1 & 1 \end{bmatrix};$$

$$\begin{bmatrix} 2 & 1 \\ 0 & -4 \\ 7 & 2 \end{bmatrix} - \begin{bmatrix} 3 & -2 \\ 4 & 1 \\ 0 & -2 \end{bmatrix} = \begin{bmatrix} -1 & 3 \\ -4 & -5 \\ 7 & 4 \end{bmatrix}.$$

$$\begin{bmatrix} 2 & 0 & -1 \\ 3 & 2 & 6 \end{bmatrix} * \begin{bmatrix} -1 \\ 2 \\ 1 \end{bmatrix} = \begin{bmatrix} -3 \\ 7 \end{bmatrix} \quad \text{but} \quad \begin{bmatrix} 6 & 1 \\ 3 & -2 \end{bmatrix} * \begin{bmatrix} -1 \\ 2 \\ 1 \end{bmatrix} \text{ is not defined.}$$

$$\begin{bmatrix} 1 & 3 \\ 2 & -1 \end{bmatrix} * \begin{bmatrix} 0 & 3 \\ -1 & 1 \end{bmatrix} = \begin{bmatrix} -3 & 6 \\ 1 & 5 \end{bmatrix} \quad \text{but} \quad \begin{bmatrix} 0 & 3 \\ -1 & 1 \end{bmatrix} * \begin{bmatrix} 1 & 3 \\ 2 & -1 \end{bmatrix} = \begin{bmatrix} 6 & -3 \\ 1 & -4 \end{bmatrix}.$$

Division of a matrix by another matrix is not defined, but we will discuss the *inverse* of a matrix later in this chapter.

The *determinant* of a square matrix is a number. For a 2×2 matrix, the determinant is computed by subtracting the product of the elements on the minor diagonal (from upper right to lower left) from the product of terms on the major diagonal. For example,

$$A = \begin{bmatrix} -5 & 3 \\ 7 & 2 \end{bmatrix}, \qquad \det(A) = (-5)(2) - (7)(3) = -31;$$

$\det(A)$ is the usual notation for the determinant of A. Sometimes the determinant is symbolized by writing the elements of the matrix between vertical lines (similar to representing the absolute value of a number).

For a 3×3 matrix, you may have learned a crisscross way of forming products of terms (we call it the "spaghetti rule") that probably should be forgotten, for it applies only to the special case of a 3×3 matrix; it won't work for larger systems.

The general rule that applies in all cases is to expand in terms of the minors of some row or column. The *minor* of any term is the matrix of lower order formed by striking out the row and column in which the term is found. The determinant is found by adding the product of each term in any row or column by the determinant of its minor, with signs alternating + and −. We expand each of the determinants of the minor until we reach 2×2 matrices. For example,

$$\text{Given } A = \begin{bmatrix} 3 & 0 & -1 & 2 \\ 4 & 1 & 3 & -2 \\ 0 & 2 & -1 & 3 \\ 1 & 0 & 1 & 4 \end{bmatrix},$$

$$\det(A) = 3\begin{vmatrix} 1 & 3 & -2 \\ 2 & -1 & 3 \\ 0 & 1 & 4 \end{vmatrix} - 0\begin{vmatrix} 4 & 3 & -2 \\ 0 & -1 & 3 \\ 1 & 1 & 4 \end{vmatrix} + (-1)\begin{vmatrix} 4 & 1 & -2 \\ 0 & 2 & 3 \\ 1 & 0 & 4 \end{vmatrix} - 2\begin{vmatrix} 4 & 1 & 3 \\ 0 & 2 & -1 \\ 1 & 0 & 1 \end{vmatrix}$$

$$= 3\left\{(1)\begin{vmatrix} -1 & 3 \\ 1 & 4 \end{vmatrix} - (3)\begin{vmatrix} 2 & 3 \\ 0 & 4 \end{vmatrix} + (-2)\begin{vmatrix} 2 & -1 \\ 0 & 1 \end{vmatrix}\right\}$$

$$+ (-1)\left\{(4)\begin{vmatrix} 2 & 3 \\ 0 & 4 \end{vmatrix} - (1)\begin{vmatrix} 0 & 3 \\ 1 & 4 \end{vmatrix} + (-2)\begin{vmatrix} 0 & 2 \\ 1 & 0 \end{vmatrix}\right\}$$

$$- 2\left\{(4)\begin{vmatrix} 2 & -1 \\ 0 & 1 \end{vmatrix} - (1)\begin{vmatrix} 0 & -1 \\ 1 & 1 \end{vmatrix} + (3)\begin{vmatrix} 0 & 2 \\ 1 & 0 \end{vmatrix}\right\}$$

$$= 3\{(1)(-7) - (3)(8) + (-2)(2)\} + (-1)\{(4)(8) - (1)(-3) + (-2)(-2)\}$$
$$- 2\{(4)(2) - (1)(1) + (3)(-2)\}$$

$$= 3(-7 - 24 - 4) + (-1)(32 + 3 + 4) - 2(8 - 1 - 6)$$

$$= 3(-35) + (-1)(39) - 2(1) = -146.$$

In computing the determinant, the expansion can be about the elements of any row or column. To get the signs, give the first term a plus sign if the sum of its column number and row number is even; give it a minus if the sum is odd, with alternating signs thereafter. (For example, in expanding about the elements of the third row we begin with a plus; the first element a_{31} has $3 + 1 = 4$, an even number.) Judicious selection of rows and columns with many zeros can hasten the process, but this method of calculating determinants is a lot of work if the matrix is of large size. Methods that triangularize a matrix, as described in Sections 2.4 and 2.5, are much better ways to get the determinant.

One special application of determinants is in the computation of the *characteristic polynomial* and the *eigenvalues* of a matrix. The eigenvalues are most important in applied mathematics. For a square matrix A, we define its characteristic polynomial as $p_A(\lambda) = |A - \lambda I| = \det(A - \lambda I)$.

For example, if

$$A = \begin{bmatrix} 1 & 3 \\ 4 & 5 \end{bmatrix},$$

then

$$p_A(\lambda) = |A - \lambda I| = \det \begin{bmatrix} (1-\lambda) & 3 \\ 4 & (5-\lambda) \end{bmatrix}$$
$$= (1-\lambda)(5-\lambda) - 12$$
$$= \lambda^2 - 6\lambda - 7.$$

(The characteristic polynomial is always of degree n if A is $n \times n$.) If we set the characteristic polynomial to zero and solve for the roots, we get the eigenvalues of A. For this example, these are $\lambda_1 = 7, \lambda_2 = -1$, or, in more symbolic mathematical notation,

$$\Lambda(A) = \{7, \quad -1\}.$$

We also mention the notion of an *eigenvector* corresponding to an eigenvalue. The eigenvector is a nonzero vector **w** such that

$$Aw = \lambda w, \quad \text{i.e.,} \quad (A - \lambda I)w = 0. \tag{2.2}$$

In the current example, the eigenvectors are

$$w_1 = \begin{bmatrix} 1 \\ 2 \end{bmatrix}, \qquad w_2 = \begin{bmatrix} -3 \\ 2 \end{bmatrix}.$$

We leave it as an exercise to show that these eigenvectors satisfy Eq. (2.2).

Observe that the trace of A is equal to the sum of the eigenvalues: $\text{tr}(A) = 1 + 5 = \lambda_1 + \lambda_2 = 7 + (-1) = 6$. This is true for any matrix: The sum of its eigenvalues equals its trace.

For now, we limit the finding of eigenvalues and eigenvectors to small matrices because getting these through the characteristic polynomial is difficult. In Chapter 6, we examine other more efficient ways to get these important quantities.

2.3 The Elimination Method

The first method we will study for the solution of a set of equations is just an enlargement of the familiar method of eliminating one unknown between a pair of simultaneous equations. Generally called *Gaussian elimination*, this method is the basic pattern of a large number of methods that are classified as *direct methods*. (This classification distinguishes them from indirect, or iterative, methods, which we will discuss later.)

Consider the simple example of three equations:

$$4x_1 - 2x_2 + x_3 = 15,$$
$$-3x_1 - x_2 + 4x_3 = 8,$$
$$x_1 - x_2 + 3x_3 = 13.$$

Multiplying the first equation by 3 and the second by 4 and adding these will eliminate x_1 from the second equation. Similarly, multiplying the first by -1 and the third by 4 and adding eliminates x_1 from the third equation. (We prefer to multiply by the negatives and add, to avoid making mistakes when subtracting quantities of unlike sign.) The result is

$$4x_1 - 2x_2 + x_3 = 15,$$
$$-10x_2 + 19x_3 = 77,$$
$$-2x_2 + 11x_3 = 37.$$

We now eliminate x_2 from the third equation by multiplying the second by 2 and the third by -10 and adding to get

$$4x_1 - 2x_2 + x_3 = 15,$$
$$-10x_2 + 19x_3 = 77,$$
$$-72x_3 = -216.$$

Obviously, $x_3 = 3$ from the third equation. Back-substitution into the second equation gives $x_2 = -2$, and substituting these into the first gives $x_1 = 2$.

We now present the same problem, solved in exactly the same way, but this time using matrix notation:

$$\begin{bmatrix} 4 & -2 & 1 \\ -3 & -1 & 4 \\ 1 & -1 & 3 \end{bmatrix} \begin{bmatrix} x_1 \\ x_2 \\ x_3 \end{bmatrix} = \begin{bmatrix} 15 \\ 8 \\ 13 \end{bmatrix}.$$

The arithmetic operations that we have performed affect only the coefficients and the constant terms, so we work with the matrix of coefficients *augmented* with the right-hand side vector:

$$A \,|\, b = \begin{bmatrix} 4 & -2 & 1 & \vdots & 15 \\ -3 & -1 & 4 & \vdots & 8 \\ 1 & -1 & 3 & \vdots & 13 \end{bmatrix}$$

(The dashed line is usually omitted.) We now perform *elementary row transformations** to convert A to *upper-triangular form*:

$$\begin{bmatrix} 4 & -2 & 1 & 15 \\ -3 & -1 & 4 & 8 \\ 1 & -1 & 3 & 13 \end{bmatrix}, \quad \begin{matrix} \\ 3R_1 + 4R_2 \rightarrow \\ (-1)R_1 + 4R_3 \rightarrow \end{matrix} \begin{bmatrix} 4 & -2 & 1 & 15 \\ 0 & -10 & 19 & 77 \\ 0 & -2 & 11 & 37 \end{bmatrix},$$

$$\begin{matrix} \\ \\ 2R_2 - 10R_3 \rightarrow \end{matrix} \begin{bmatrix} 4 & -2 & 1 & 15 \\ 0 & -10 & 19 & 77 \\ 0 & 0 & -72 & -216 \end{bmatrix}. \tag{2.3}$$

---

*Elementary row operations are arithmetic operations that are obviously valid rearrangements of a set of equations: (1) Any equation can be multiplied by a constant; (2) the order of the equations can be changed; (3) any equation can be replaced by its sum with another of the equations.

The array in Eq. (2.3) represents the equations

$$4x_1 - 2x_2 + x_3 = 15,$$
$$-10x_2 + 19x_3 = 77,$$
$$-72x_3 = -216.$$

(2.4)

The *back-substitution* step can be performed quite mechanically by solving the equations of (2.4) in reverse order:

$$x_3 = \frac{-216}{(-72)} = 3,$$

$$x_2 = \frac{(77 - 19(3))}{(-10)} = -2,$$

$$x_1 = \frac{(15 - 1(3) - (-2)(-2))}{4} = 2.$$

Thinking of this procedure in terms of matrix operations, we transform the augmented coefficient matrix by elementary row transformations until an upper-triangular matrix is created on the left. We then are ready to back-substitute using the elements in the rightmost column.

Note that there exists the possibility that the set of equations has no solution, or that the prior procedure will fail to find it. During the triangularization step, if a zero is encountered on the diagonal, we cannot use that row to eliminate coefficients below that zero element. However, in that case, we can continue by interchanging rows and eventually achieve an upper-triangular matrix of coefficients. The real stumbling block is finding a zero on the diagonal after we have triangularized. If that occurs, the back-substitution fails, for we cannot divide by zero. It also means that the determinant is zero: There is no solution.

It is worthwhile to explain in more detail what we mean by the *elementary row operations* that we have used here, and to see why they can be used in solving a linear system. There are three of these operations:

1. We may multiply any row of the augmented coefficient matrix by a constant.
2. We can add a multiple of one row to a multiple of any other row.
3. We can interchange the order of any two rows (this was not used earlier).

The validity of these row operations is intuitively obvious if we think of them applied to a set of linear equations. Certainly, multiplying one equation through by a constant does not change the truth of the equality. Adding equal quantities to both sides of an equality results in an equality, and this is the equivalent of the second transformation. Obviously, the order of the set is arbitrary, so rule 3 is valid.

These operations, which do not change the relationships represented by a set of equations, can be applied to an augmented matrix, because this is only a different notation for the equations. (We need to add one proviso: Since round-off error is related to the magnitude of the values when we express them in fixed-word-length computer representations, some of our previous operations may have an effect on the accuracy of the computed solution.)

We should also observe that the "back-substitution" phase, when it is done by making the coefficients above the diagonal zero and then reducing the coefficient matrix to the identity matrix, is exactly the same as back-substitution in the more explicit sense. The order of operation is changed, but each of the steps is identical.

MAPLE V, the computer algebra system that we discuss in a later section, solves the preceding example by the command

```
Solve ({4x − 2y + z = 15, −3x − y + 4z = 8, x − y + 3z = 13},{x, y, z});
```

which returns the solution as

$$\{x = 2, y = -2, z = 3\}.$$

2.4 Gauss and Gauss–Jordan Methods

While the procedure of the previous section is satisfactory for hand calculations on small systems, there are several objections that we should eliminate before we write a computer program to perform Gaussian elimination. In a large set of equations, and that is the situation we must prepare for, the multiplications will give very large and unwieldy numbers that may overflow the computer's registers. We will therefore eliminate the first coefficient in the ith row by subtracting a_{i1}/a_{11} times the first equation from the ith equation. (This is equivalent to making the leading coefficient 1 in the equation that retains that leading term.) We use similar ratios of coefficients in eliminating coefficients in the other columns.

We must also guard against dividing by zero. Observe that zeros may be created in the diagonal positions even if they are not present in the original matrix of coefficients. A useful strategy to avoid (if possible) such zero divisors is to rearrange the equations so as to put the coefficient of largest magnitude on the diagonal at each step. This is called *pivoting*. Complete pivoting may require both row and column interchanges. This is not frequently done. Partial pivoting, which places a coefficient of larger magnitude on the diagonal by row interchanges only, will guarantee a nonzero divisor if there is a solution to the set of equations, and will have the added advantage of giving improved arithmetic precision. The diagonal elements that result are called *pivot elements*. (When there are large differences in magnitude of coefficients in one equation compared to the other equations, we may need to *scale* the values; we consider this later.)

We repeat the example of the previous section, incorporating these ideas and carrying four significant digits in our work. We begin with the augmented matrix.

$$\begin{bmatrix} 4 & -2 & 1 & | & 15 \\ -3 & -1 & 4 & | & 8 \\ 1 & -1 & 3 & | & 13 \end{bmatrix} \tag{2.5}$$

$$\begin{array}{c} \\ (3/4)R_1 + R_2 \rightarrow \\ -(1/4)R_1 + R_3 \rightarrow \end{array} \begin{bmatrix} 4 & -2 & 1 & | & 15 \\ 0 & -2.5 & 4.75 & | & 19.25 \\ 0 & -0.5 & 2.75 & | & 9.25 \end{bmatrix}$$

$$\begin{array}{c} \\ \\ -(-0.5/-2.5)R_2 + R_3 \rightarrow \end{array} \begin{bmatrix} 4 & -2 & 1 & | & 15 \\ 0 & -2.5 & 4.75 & | & 19.25 \\ 0 & 0.0 & 1.80 & | & 5.40 \end{bmatrix} \tag{2.6}$$

The method we have just illustrated is called *Gaussian elimination*. (In this example, no pivoting was required to make the largest coefficients be on the diagonal.) Back-substitution, as presented in Eqs. (2.4), gives us, as before, $x_3 = 3, x_2 = -2, x_1 = 2$. We have come up with the exact answer to this problem. Often it will turn out that we shall obtain answers that are just close approximations to the exact answer because of round-off error. When there are many equations, the effects of round-off (the term is applied to the error due to chopping as well as when rounding is used) may cause large effects. In certain cases, the coefficients are such that the results are particularly sensitive to round-off; such systems are called *ill-conditioned*.

In the example just presented, the zeros below the main diagonal show that we have reduced the problem (Eq. (2.3)) to solving an upper-triangular system of equations as in Eqs. (2.4). However, at each stage, if we had stored the ratio of coefficients in place of zero (we show them in parentheses), our final form would have been

$$\begin{bmatrix} 4 & -2 & 1 & | & 15 \\ (-0.75) & -2.5 & 4.75 & | & 19.25 \\ (0.25) & (0.20) & 1.80 & | & 5.4 \end{bmatrix}.$$

Then, in addition to solving the problem as we have done, we find that the original matrix

$$A = \begin{bmatrix} 4 & -2 & 1 \\ -3 & -1 & 4 \\ 1 & -1 & 3 \end{bmatrix}$$

can be written as the product:

$$\underbrace{\begin{bmatrix} 1 & 0 & 0 \\ -0.75 & 1 & 0 \\ 0.25 & 0.20 & 1 \end{bmatrix}}_{L} * \underbrace{\begin{bmatrix} 4 & -2 & 1 \\ 0 & -2.5 & 4.75 \\ 0 & 0 & 1.80 \end{bmatrix}}_{U} \tag{2.7}$$

This procedure is called an *LU decomposition* of A. In this case,

$$A = L * U,$$

where L is *lower-triangular* and U is *upper-triangular*. As we shall see in the next example, usually $L * U = A'$, where A' is just a permutation of the rows of A due to row interchange from pivoting.

Finally, since the determinant of two matrices, $B * C$, is the product of each of the determinants, for this example we have

$$\det(L * U) = \det(L) * \det(U) = \det(U).$$

Thus for the example given in Eqs. (2.7), we have

$$\det(A) = \det(U) = (4) * (-2.5) * (1.8) = -18.$$

From this example, we see that Gaussian elimination does the following:

1. It solves the system of equations.
2. It computes the determinant of a matrix very efficiently.
3. It can provide us with the *LU* decomposition of the matrix of coefficients, in the sense that the product of the two matrices, $L * U$, may give us a permutation of the rows of the original matrix.

With regard to item (2), when there are row interchanges,

$$\det(A) = (-1)^m * u_{11} * \cdots u_{nn}, \tag{2.8}$$

where the superscript m represents the number of row interchanges.

Let us summarize the operations of Gaussian elimination in a form that will facilitate the writing of a computer program. Note that in the actual implementation of the algorithm, the L and U matrices are actually stored in the space of the original matrix A.

Gaussian Elimination

To solve a system of n linear equations: $Ax = b$.

FOR j = 1 TO $(n - 1)$

   pvt = $|a[j,j]|$
   pivot[j] = j

   FOR i = $j + 1$ TO n (Find pivot row)

     IF $|a[i,j]|$ > pvt
      pvt = $|a[i,j]|$
      ipvt_store = i END IF   END FOR i

```
        (Switch rows if necessary)
        IF pivot[j] <>ipvt_store                    THEN
            pivot[j] = ipvt_store
            pivot[ipvt_store] = j
            FOR k = 1 TO n
                temp = a[j,k]
                a[j,k] = a[pivot(j),k]
                a[pivot(j),k] = temp                END FOR k

            temp = b[j]
            b[j] = b[pivot(j)]
            b[pivot(j)] = temp                      END IF

    FOR i = j + 1 TO n (Store multipliers)
        a[i,j] = a[i,j]/a[j,j]      END FOR i

    FOR i = j + 1 TO n (*Create zeros below main diagonal*)
        FOR k = j + 1 TO n
            a[i,k] = a[i,k] − a[i,j] * a[j,k]       END FOR k

        b[i] = b[i] − a[i,j] * b[j]                 END FOR i

                                                    END FOR j

(Back-substitution part)

x[n] = b[n]/a[n,n]
FOR j = n − 1 DOWN TO 1
    x[j] = b[j]

    FOR k = n DOWN TO j + 1
        x[j] = x[j] − x[k] * a[j,k]                 END FOR k
        x[j] = x[j]/a[j,j]                          END FOR j
```

Some computer programs do not actually move rows to interchange all the elements of the rows when pivoting. In these programs, one keeps track of the order in which the rows are to be used in a vector whose elements represent row order. (In the next example, we shall make use of such a vector.) When an interchange is indicated, only the elements of this ordering vector are changed. These numbers are then used to locate the positions of the elements in the matrix of coefficients that are to be operated on, during both the reduction step and the back-substitution. This method can reduce the computer time for large systems, but since it adds to the complexity of the program, we shall make the actual row interchanges.

The algorithm for Gaussian elimination will be clarified by an additional numerical example. Solve the following system of equations using Gaussian elimination.

In addition, from the method used, compute the determinant of the coefficient matrix and the *LU* decomposition of this matrix.

Given the system of equations, solve

$$
\begin{aligned}
2x_2 \quad\quad\quad + x_4 &= 0, \\
2x_1 + 2x_2 + 3x_3 + 2x_4 &= -2, \\
4x_1 - 3x_2 \quad\quad\quad + x_4 &= -7, \\
6x_1 + x_2 - 6x_3 - 5x_4 &= 6.
\end{aligned}
\tag{2.9}
$$

The augmented coefficient matrix is

$$
\begin{bmatrix}
0 & 2 & 0 & 1 & 0 \\
2 & 2 & 3 & 2 & -2 \\
4 & -3 & 0 & 1 & -7 \\
6 & 1 & -6 & -5 & 6
\end{bmatrix}.
\tag{2.10}
$$

We cannot permit a zero in the a_{11} position because that element is the pivot in reducing the first column. We could interchange the first row with any of the other rows to avoid a zero divisor, but interchanging the first and fourth rows is our best choice. This gives

$$
\begin{bmatrix}
6 & 1 & -6 & -5 & 6 \\
2 & 2 & 3 & 2 & -2 \\
4 & -3 & 0 & 1 & -7 \\
0 & 2 & 0 & 1 & 0
\end{bmatrix}.
\tag{2.11}
$$

We make all the elements in the first column zero by subtracting the appropriate multiple of row one:

$$
\begin{bmatrix}
6 & 1 & -6 & -5 & 6 \\
0 & 1.6667 & 5 & 3.6667 & -4 \\
0 & -3.6667 & 4 & 4.3333 & -11 \\
0 & 2 & 0 & 1 & 0
\end{bmatrix}.
\tag{2.11a}
$$

We again interchange before reducing the second column, not because we have a zero divisor, but because we want to preserve accuracy.* Interchanging the second and third rows puts the element of largest magnitude on the diagonal. (We could also interchange the fourth column with the second, giving an even larger diagonal element, but we do not do this.) After the interchange, we have

$$
\begin{bmatrix}
6 & 1 & -6 & -5 & 6 \\
0 & -3.6667 & 4 & 4.3333 & -11 \\
0 & 1.6667 & 5 & 3.6667 & -4 \\
0 & 2 & 0 & 1 & 0
\end{bmatrix}.
\tag{2.12}
$$

---

* A numerical example that demonstrates the improved accuracy when partial pivoting is used will be found in Section 2.9.

Now we reduce in the second column

$$\begin{bmatrix} 6 & 1 & -6 & -5 & 6 \\ 0 & -3.6667 & 4 & 4.3333 & -11 \\ 0 & 0 & 6.8182 & 5.6364 & -9.0001 \\ 0 & 0 & 2.1818 & 3.3636 & -5.9999 \end{bmatrix}.$$ **(2.13)**

No interchange is indicated in the third column. Reducing, we get

$$\begin{bmatrix} 6 & 1 & -6 & -5 & 6 \\ 0 & -3.6667 & 4 & 4.3333 & -11 \\ 0 & 0 & 6.8182 & 5.6364 & -9.0001 \\ 0 & 0 & 0 & 1.5600 & -3.1199 \end{bmatrix}.$$ **(2.14)**

Back-substitution gives

$$x_4 = \frac{-3.1199}{1.5600} = -1.9999,$$

$$x_3 = \frac{-9.0001 - 5.6364(-1.9999)}{6.8182} = 0.33325,$$

$$x_2 = \frac{-11 - 4.3333(-1.9999) - 4(0.33325)}{-3.6667} = 1.0000,$$

$$x_1 = \frac{6 - (-5)(-1.9999) - (-6)(0.33325) - (1)(1.0000)}{6} = -0.50000.$$

The correct answers are $-2, \frac{1}{3}, 1$, and $-\frac{1}{2}$ for x_4, x_3, x_2, and x_1. In this calculation we have carried five significant figures and rounded each calculation. Even so, we do not have five-digit accuracy in the answers. The discrepancy is due to round-off. The question of the accuracy of the computed solution to a set of equations is a most important one, and at several points in the following discussion we will discuss how to minimize the effects of round-off and avoid conditions that can cause round-off errors to be magnified.

In this example, if one had replaced the zeros below the main diagonal with the ratio of coefficients at each step, the resulting augmented matrix would be

$$\begin{bmatrix} 6 & 1 & -6 & -5 & 6 \\ (0.66667) & -3.6667 & 4 & 4.3333 & -11 \\ (0.33333) & (-0.45454) & 6.8182 & 5.6364 & -9.0001 \\ (0.0) & (-0.54545) & (0.32) & 1.5600 & -3.1199 \end{bmatrix}.$$ **(2.15)**

This gives the *LU* decomposition as

$$
\begin{bmatrix}
1 & 0 & 0 & 0 \\
0.66667 & 1 & 0 & 0 \\
0.33333 & -0.45454 & 1 & 0 \\
0.0 & -0.54545 & 0.32 & 1
\end{bmatrix}
*
\begin{bmatrix}
6 & 1 & -6 & -5 \\
0 & -3.6667 & 4 & 4.3333 \\
0 & 0 & 6.8182 & 5.6364 \\
0 & 0 & 0 & 1.5600
\end{bmatrix}
\tag{2.16}
$$

It should be noted that the product of the matrices in (2.16) produces a permutation of the original matrix, call it A', where

$$
A' =
\begin{bmatrix}
6 & 1 & -6 & -5 \\
4 & -3 & 0 & 1 \\
2 & 2 & 3 & 2 \\
0 & 2 & 0 & 1
\end{bmatrix},
$$

since rows 1 and 4 were interchanged in (2.11) and rows 2 and 3 in (2.12). The determinant of the original matrix of coefficients—the first four columns of (2.10)—can be easily computed from (2.14) or (2.15) according to the formula

$$
\det(A) = (-1)^2 * (6) * (-3.6667) * (6.8182) * (1.5600) = -234.0028,
$$

which is close to the exact solution: -234. The exponent 2 is required, since there were two row interchanges in solving this system. To summarize, you should note that the Gaussian elimination method applied to (2.9) produces the following:

1. The solution to the four equations.
2. The determinant of the coefficient matrix

$$
\begin{bmatrix}
0 & 2 & 0 & 1 \\
2 & 2 & 3 & 2 \\
4 & -3 & 0 & 1 \\
6 & 1 & -6 & -5
\end{bmatrix}.
$$

3. The *LU* decomposition of the matrix, A', which is just the original matrix, A, after we have interchanged its rows in the process.

There are many variants to the Gaussian elimination scheme. The back-substitution step can be performed by eliminating the elements above the diagonal after the triangularization has been finished, using elementary row operations and proceeding upward from the last row. This technique is equivalent to the procedures described in the following example. The diagonal elements may all be made ones as a first step before creating zeros in their column; this does the divisions of the back-substitution phase at an earlier time.

One variant that is sometimes used is the *Gauss–Jordan* scheme. In it, the elements above the diagonal are made zero at the *same time* that zeros are created below the diagonal. Usually the diagonal elements are made ones at the same time that the reduction is performed; this transforms the coefficient matrix into the identity matrix. When this has been accomplished, the column of right-hand sides has been transformed into the solution vector. Pivoting is normally employed to preserve arithmetic accuracy.

The previous example, solved by the Gauss–Jordan method, gives this succession of calculations. The original augmented matrix is

$$\begin{bmatrix} 0 & 2 & 0 & 1 & 0 \\ 2 & 2 & 3 & 2 & -2 \\ 4 & -3 & 0 & 1 & -7 \\ 6 & 1 & -6 & -5 & 6 \end{bmatrix}.$$

Interchanging rows one and four, dividing the first row by 6, and reducing the first column gives

$$\begin{bmatrix} 1 & 0.16667 & -1 & -0.83335 & 1 \\ 0 & 1.6667 & 5 & 3.6667 & -4 \\ 0 & -3.6667 & 4 & 4.3334 & -11 \\ 0 & 2 & 0 & 1 & 0 \end{bmatrix}.$$

Interchanging rows two and three, dividing the second row by -3.6667, and reducing the second column (operating above the diagonal as well as below) gives

$$\begin{bmatrix} 1 & 0 & -1.5000 & -1.2000 & 1.4000 \\ 0 & 1 & 2.9999 & 2.2000 & -2.4000 \\ 0 & 0 & 15.000 & 12.400 & -19.800 \\ 0 & 0 & -5.9998 & -3.4000 & 4.8000 \end{bmatrix}.$$

No interchanges are required for the next step. We divide the third row by 15.000 and make the other elements in the third column into zeros:

$$\begin{bmatrix} 1 & 0 & 0 & 0.04000 & -0.58000 \\ 0 & 1 & 0 & -0.27993 & 1.5599 \\ 0 & 0 & 1 & 0.82667 & -1.3200 \\ 0 & 0 & 0 & 1.5599 & -3.1197 \end{bmatrix}.$$

We now divide the fourth row by 1.5599 and create zeros above the diagonal in the fourth column:

$$\begin{bmatrix} 1 & 0 & 0 & 0 & -0.49999 \\ 0 & 1 & 0 & 0 & 1.0001 \\ 0 & 0 & 1 & 0 & 0.33326 \\ 0 & 0 & 0 & 1 & -1.9999 \end{bmatrix}.$$

The solution is essentially the same as with the usual Gaussian method; round-off errors have created inaccuracies in a slightly different way than did the previous computation.

Although it might seem that the work to accomplish the Gauss–Jordan method is similar to that for Gauss elimination, this assumption is not true. In fact, the Gauss–Jordan technique requires almost 50% more arithmetic operations (Exercise 19). It can be shown that the total number of multiplications/divisions to solve an $n \times n$ system through Gaussian elimination equals $n^3/3 + n^2 - n/3$, whereas Gauss–Jordan takes $n^3/2 + n^2 - 7n/2 + 2$ multiplications/divisions. (We usually count only multiplications/divisions because these operations are generally much slower than additions/subtractions.)

Observe that for large n (where operation counts are significant), the sums that give the count of operations are dominated by the first terms. It is customary to consider only the dominant term in such cases, using a special terminology: We say that it takes "of the order of n^3 operations." This expression is often written as $O(n^3)$ and is called "the Big O notation." If we consider only the first terms in the operation counts, it is clear that the Gauss–Jordan method takes about 1.5 times as many operations as does the Gauss method.

A frequently occurring situation is the need to solve a set of equations with the same coefficient matrix but with a number of different right-hand sides. For example, in the design of a truss (see Section 2.1), one usually wishes to determine the stresses under a variety of external loads—this causes only the right-hand-side terms to vary.

If all of the different right-hand-side vectors are known in advance, the multiple solutions of the system can be obtained simultaneously using our Gaussian elimination method. One augments the coefficient matrix with all of the right-hand-side vectors, and treats each augmentation column in the same way as a single added column. In the back-substitution phase, each of the columns is employed to give a solution vector. At the end of this chapter, a subroutine is exhibited that does this.

EXAMPLE 2.2 Solve the system $Ax = b$, with multiple values of b, by Gaussian elimination:

$$A = \begin{bmatrix} 3 & 2 & -1 & 2 \\ 1 & 4 & 0 & 2 \\ 2 & 1 & 2 & -1 \\ 1 & 1 & -1 & 3 \end{bmatrix}, \quad b^{(1)} = \begin{bmatrix} 0 \\ 0 \\ 1 \\ 0 \end{bmatrix}, \quad b^{(2)} = \begin{bmatrix} -2 \\ 1 \\ 3 \\ 4 \end{bmatrix}, \quad b^{(3)} = \begin{bmatrix} 2 \\ 2 \\ 0 \\ 0 \end{bmatrix}.$$

We augment A with all of the b's, then triangularize:

$$\begin{bmatrix} 3 & 2 & -1 & 2 & 0 & -2 & 2 \\ 1 & 4 & 0 & 2 & 0 & 1 & 2 \\ 2 & 1 & 2 & -1 & 1 & 3 & 0 \\ 1 & 1 & -1 & 3 & 0 & 4 & 0 \end{bmatrix} \rightarrow$$

$$
\begin{bmatrix}
3 & 2 & -1 & 2 & 0 & -2 & 2 \\
(0.333) & 3.333 & 0.333 & 1.333 & 0 & 1.667 & 1.333 \\
(0.667) & -0.333 & 2.667 & -2.333 & 1 & 4.333 & -1.333 \\
(0.333) & 0.333 & -0.667 & 2.333 & 0 & 4.667 & -0.667
\end{bmatrix} \rightarrow
$$

$$
\begin{bmatrix}
3 & 2 & -1 & 2 & 0 & -2 & 2 \\
(0.333) & 3.333 & 0.333 & 1.333 & 0 & 1.667 & 1.333 \\
(0.667) & (-0.100) & 2.700 & -2.200 & 1 & 4.500 & -1.200 \\
(0.333) & (0.100) & -0.700 & 2.200 & 0 & 4.500 & -0.800
\end{bmatrix} \rightarrow
$$

$$
\begin{bmatrix}
3 & 2 & -1 & 2 & 0 & -2 & 2 \\
(0.333) & 3.333 & 0.333 & 1.333 & 0 & 1.667 & 1.333 \\
(0.667) & (-0.100) & 2.700 & -2.200 & 1 & 4.500 & -1.200 \\
(0.333) & (0.100) & (0.259) & 1.630 & 0.259 & 5.667 & -1.111
\end{bmatrix} \cdot
$$

$$
\underset{c^{(1)}}{\uparrow} \qquad \underset{c^{(2)}}{\uparrow} \qquad \underset{c^{(3)}}{\uparrow}
$$

We obtain the three solution vectors by back-substitution, employing the proper b'-vector. (These are indicated above as $c^{(i)}$.)

$$
x^{(1)} = \begin{bmatrix} 0.137 \\ -0.114 \\ 0.500 \\ 0.159 \end{bmatrix}, \qquad
x^{(2)} = \begin{bmatrix} -0.591 \\ -1.340 \\ 4.500 \\ 3.477 \end{bmatrix}, \qquad
x^{(3)} = \begin{bmatrix} 0.273 \\ 0.773 \\ -1.000 \\ -0.682 \end{bmatrix}.
$$

At the end of Section 2.5, we shall consider an example of the power of the LU method in Gaussian elimination for the case of solving a system of equations where we have several right-hand sides, but they are introduced sequentially rather than simultaneously, as in this example.

Scaling

We have mentioned that the rows of the augmented coefficient matrix may need to be scaled before a proper choice of pivot element can be made. *Scaling* is the operation of adjusting the coefficients of a set of equations so that they are all of the same order of magnitude. In some instances, a set of equations may involve relationships between quantities measured in widely different units (microvolts versus kilovolts, for example, or nanoseconds versus years). This may result in some of the equations having very large numbers and others very small. If we select the pivot elements without scaling, pivoting may put numbers on the diagonal that are not large in comparison to others in their row; this can actually create the round-off errors that pivoting was supposed to avoid. An example will clarify the concept.

$$\text{Given} \begin{bmatrix} 3 & 2 & 100 \\ -1 & 3 & 100 \\ 1 & 2 & -1 \end{bmatrix} x = \begin{bmatrix} 105 \\ 102 \\ 2 \end{bmatrix}.$$

Carrying only three digits to emphasize round-off, and using partial pivoting, we find that the triangularized system is

$$\begin{bmatrix} 3 & 2 & 100 & 105 \\ 0 & 3.66 & 133 & 137 \\ 0 & 0 & -82.6 & -82.7 \end{bmatrix},$$

from which $x_3 = 1.00, x_2 = 1.09, x_1 = 0.94$; the exact solution vector should be $[1.00, 1.00, 1.00]$.

If we scale the values before reduction by dividing each row by the magnitude of the largest coefficient, so that the system is

$$\begin{bmatrix} 0.03 & 0.02 & 1.00 \\ -0.01 & 0.03 & 1.00 \\ 0.50 & 1.00 & -0.50 \end{bmatrix} x = \begin{bmatrix} 1.05 \\ 1.02 \\ 1.00 \end{bmatrix},$$

we get, with the same arithmetic precision, the triangularized system

$$\begin{bmatrix} 0.50 & 1.00 & -0.50 & 1.00 \\ 0 & 0.05 & 0.99 & 1.04 \\ 0 & 0 & 1.82 & 1.82 \end{bmatrix},$$

from which $x_3 = 1.00, x_2 = 1.00, x_1 = 1.00$. The reason for the improvement is that rows are interchanged after scaling has been done. No interchanges are indicated in the unscaled equations.

Whenever the coefficients in one column are widely different from those in another column, scaling is beneficial. When all values are about the same order of magnitude, scaling should be avoided, for the additional round-off error incurred during the scaling operation itself may adversely affect the accuracy. The usual way to scale is as we have done here, by dividing each row by the magnitude of the largest term. Some authorities recommend scaling so that the sum of the magnitudes of the coefficients in each row is the same. This is probably very slightly more economical to do in a computer program.

2.5 Other Direct Methods

The term *direct method* indicates a method that solves a set of equations by techniques such as the Gaussian elimination method of Section 2.4. These methods are in contrast to *iterative methods* that we will describe in Section 2.10.

Although Gaussian elimination is the best known of the direct LU decomposition methods, another direct method is widely used. We will examine this method and then look at how the Gaussian LU method can be used efficiently to solve problems involving multiple right-hand sides.

Crout reduction (also called Cholesky reduction) transforms the coefficient matrix, A, into the product of two matrices, L and U, where U has ones on its main diagonal. (This technique differs from the method of the previous section in which L has the ones on its diagonal. Another name for that technique is *Doolittle's method*.) We now develop the procedure for Crout reduction.

We have previously seen that a matrix that has been triangularized and combined with the lower-triangular matrix formed from the ratios used in its reduction forms an LU pair. But LU pairs take many other forms. In fact, any matrix that has all diagonal elements nonzero can be written as a product of a lower-triangular and an upper-triangular matrix in an infinity of ways. For example,

$$\underset{A}{\begin{bmatrix} 2 & -1 & -1 \\ 0 & -4 & 2 \\ 6 & -3 & 1 \end{bmatrix}} = \underset{L_1}{\begin{bmatrix} 2 & 0 & 0 \\ 0 & -4 & 0 \\ 6 & 0 & 4 \end{bmatrix}} \underset{U_1}{\begin{bmatrix} 1 & -\frac{1}{2} & -\frac{1}{2} \\ 0 & 1 & -\frac{1}{2} \\ 0 & 0 & 1 \end{bmatrix}}$$

$$= \underset{L_2}{\begin{bmatrix} 1 & 0 & 0 \\ 0 & 1 & 0 \\ 3 & 0 & 1 \end{bmatrix}} \underset{U_2}{\begin{bmatrix} 2 & -1 & -1 \\ 0 & -4 & 2 \\ 0 & 0 & 4 \end{bmatrix}}$$

$$= \underset{L_3}{\begin{bmatrix} 1 & 0 & 0 \\ 0 & 2 & 0 \\ 3 & 0 & 1 \end{bmatrix}} \underset{U_3}{\begin{bmatrix} 2 & -1 & -1 \\ 0 & -2 & 1 \\ 0 & 0 & 4 \end{bmatrix}}$$

= and so on.

Of the entire set of LUs whose product equals matrix A, in the Crout method we choose the pair in which U has only ones on its diagonal, as in the first pair above. We get the rules for such an LU decomposition from the relationship that $LU = A$. In the case of a 4×4 matrix:

$$\begin{bmatrix} \ell_{11} & 0 & 0 & 0 \\ \ell_{21} & \ell_{22} & 0 & 0 \\ \ell_{31} & \ell_{32} & \ell_{33} & 0 \\ \ell_{41} & \ell_{42} & \ell_{43} & \ell_{44} \end{bmatrix} \begin{bmatrix} 1 & u_{12} & u_{13} & u_{14} \\ 0 & 1 & u_{23} & u_{24} \\ 0 & 0 & 1 & u_{34} \\ 0 & 0 & 0 & 1 \end{bmatrix} = \begin{bmatrix} a_{11} & a_{12} & a_{13} & a_{14} \\ a_{21} & a_{22} & a_{23} & a_{24} \\ a_{31} & a_{32} & a_{33} & a_{34} \\ a_{41} & a_{42} & a_{43} & a_{44} \end{bmatrix}.$$

Multiplying the rows of L by the first column of U, we get $\ell_{11} = a_{11}$, $\ell_{21} = a_{21}$, $\ell_{31} = a_{31}$, $\ell_{41} = a_{41}$; the first column of L is the same as the first column of A.

We now multiply the first row of L by the columns of U:

$$\ell_{11} u_{12} = a_{12}, \qquad \ell_{11} u_{13} = a_{13}, \qquad \ell_{11} u_{14} = a_{14}, \tag{2.17}$$

from which

$$u_{12} = \frac{a_{12}}{\ell_{11}}, \qquad u_{13} = \frac{a_{13}}{\ell_{11}}, \qquad u_{14} = \frac{a_{14}}{\ell_{11}}. \tag{2.18}$$

Thus the first row of U is determined.

In this method we alternate between getting a column of L and a row of U, so we next get the equations for the second column of L by multiplying the rows of L by the second column of U:

$$\ell_{21} u_{12} + \ell_{22} = a_{22},$$
$$\ell_{31} u_{12} + \ell_{32} = a_{32}, \tag{2.19}$$
$$\ell_{41} u_{12} + \ell_{42} = a_{42},$$

which gives

$$\ell_{22} = a_{22} - \ell_{21} u_{12},$$
$$\ell_{32} = a_{32} - \ell_{31} u_{12}, \tag{2.20}$$
$$\ell_{42} = a_{42} - \ell_{41} u_{12}.$$

Proceeding in the same fashion, the equations we need are

$$u_{23} = \frac{a_{23} - \ell_{21} u_{13}}{\ell_{22}}, \qquad u_{24} = \frac{a_{24} - \ell_{21} u_{14}}{\ell_{22}},$$

$$\ell_{23} = a_{33} - \ell_{31} u_{13} - \ell_{32} u_{23}, \qquad \ell_{43} = a_{43} - \ell_{41} u_{13} - \ell_{42} u_{23},$$

$$u_{34} = \frac{a_{34} - \ell_{31} u_{14} - \ell_{32} u_{24}}{\ell_{33}},$$

$$\ell_{44} = a_{44} - \ell_{41} u_{14} - \ell_{42} u_{24} - \ell_{43} u_{34}.$$

The general formula for getting elements of L and U corresponding to the coefficient matrix for n simultaneous equations can be written

$$\ell_{ij} = a_{ij} - \sum_{k=1}^{j-1} \ell_{ik} u_{kj}, \qquad j \leq i, \qquad i = 1, 2, \ldots, n, \tag{2.21}$$

$$u_{ij} = \frac{a_{ij} - \sum\limits_{k=1}^{i-1} \ell_{ik} u_{kj}}{\ell_{ii}}, \qquad i \leq j, \qquad j = 2, 3, \ldots, n. \tag{2.22}$$

$\Bigg($ For $j = 1$, the rule for ℓ reduces to

$$\ell_{i1} = a_{i1}.$$

For $i = 1$, the rule for u reduces to

$$u_{1j} = \frac{a_{1j}}{\ell_{11}} = \frac{a_{1j}}{a_{11}}.$$

The reason this method is popular in programs is that storage space may be economized. There is no need to store the zeros in either L or U, and the ones on the diagonal of U can also be omitted. (Since these values are always the same and are always known, it is redundant to record them.) One can then store the essential elements of U where the zeros appear in the L array. Examination of Eqs. (2.17) through (2.22) shows that, after any element of A, a_{ij}, is once used, it never again appears in the equations. Hence its place in the original $n \times n$ array A can be used to store an element of either L or U. In other words, the A array can be transformed by the above equations and becomes

$$
\begin{bmatrix}
a_{11} & a_{12} & a_{13} & a_{14} \\
a_{21} & a_{22} & a_{23} & a_{24} \\
a_{31} & a_{32} & a_{33} & a_{34} \\
a_{41} & a_{42} & a_{43} & a_{44}
\end{bmatrix}
\rightarrow
\begin{bmatrix}
\ell_{11} & u_{12} & u_{13} & u_{14} \\
\ell_{21} & \ell_{22} & u_{23} & u_{24} \\
\ell_{31} & \ell_{32} & \ell_{33} & u_{34} \\
\ell_{41} & \ell_{42} & \ell_{43} & \ell_{44}
\end{bmatrix}.
$$

Because we can condense the L and U matrices into one array and store their elements in the space of A, this method is often called a *compact scheme*.

EXAMPLE 2.3   Consider the matrix A:

$$
A = \begin{bmatrix}
3 & -1 & 2 \\
1 & 2 & 3 \\
2 & -2 & -1
\end{bmatrix}.
$$

Applying the equations for the ℓ's and u's, we obtain

$$\ell_{11} = 3, \quad \ell_{21} = 1, \quad \ell_{31} = 2; \quad u_{12} = -\tfrac{1}{3}, \quad u_{13} = \tfrac{2}{3}.$$

$$\ell_{22} = 2 - (1)(-\tfrac{1}{3}) = \tfrac{7}{3}, \quad \ell_{32} = -2 - (2)(-\tfrac{1}{3}) = -\tfrac{4}{3}.$$

$$u_{23} = \frac{3 - (1)(\tfrac{2}{3})}{\tfrac{7}{3}} = 1, \quad \ell_{33} = -1 - (2)(\tfrac{2}{3}) - (-\tfrac{4}{3})(1) = -1.$$

$$
L = \begin{bmatrix}
3 & 0 & 0 \\
1 & \tfrac{7}{3} & 0 \\
2 & -\tfrac{4}{3} & -1
\end{bmatrix}, \quad
U = \begin{bmatrix}
1 & -\tfrac{1}{3} & \tfrac{2}{3} \\
0 & 1 & 1 \\
0 & 0 & 1
\end{bmatrix}.
$$

If the quantities are written in the compact form as they are computed, we have

$$
LU = \begin{bmatrix} 3 & -\frac{1}{3} & \frac{2}{3} \\ 1 & \frac{7}{3} & 1 \\ 2 & -\frac{4}{3} & -1 \end{bmatrix}. \quad \begin{matrix} \leftarrow ② \\ \leftarrow ④ \\ \ \end{matrix}
$$

$$
\begin{matrix} \uparrow & \uparrow & \uparrow \\ ① & ③ & ⑤ \end{matrix}
$$

The circled numbers show the order in which columns and rows of the new matrix are obtained.

Here is an algorithm for LU decomposition. It does not compute the L and U matrices in place, but sets them up as separate matrices.

LU Decomposition

Part 1. To transform an $n \times n$ matrix A into $L * U$:

```
FOR i = 1 TO n
  L[i, 1] = a[i, 1]            END FOR i

FOR j = 1 TO n
  U[1, j] = a[1, j]/L[1, 1]   END FOR j

FOR j = 2 TO n
  FOR i = j TO n
    sum = 0.0
    FOR k = 1 TO j − 1
      sum = sum + L[i, k] * U[k, j]        END FOR k
    U[i, j] = a[i, j] − sum                 END FOR i
  U[j, j] = 1

  FOR i = j + 1 TO n
    sum = 0.0
    FOR k = 1 TO j − 1
      sum = sum + L[j, k] * U[k, i] END FOR k
    U[j, i] = (A[j, i] − sum)/L[j, j]        END FOR i
                                              END FOR j
```

The solution of the set of equations $Ax = b$ is readily obtained with the L and U matrices. Once the coefficient matrix has been converted to its LU equivalent, we are prepared to find the solution to the set of equations that corresponds to any given right-hand-side vector b. The L matrix is really a record of the operations required to make the coefficient matrix A into the upper-triangular matrix U. We apply these same transformations to the right-hand-side vector b, converting it to a new vector b'. If we augment b' to U and back-substitute, the solution appears.

The general equation for the reduction of b (it is exactly the same as the rule for forming the elements of U) is

$$b_1' = \frac{b_1}{\ell_{11}},$$

$$b_i' = \frac{b_i - \sum\limits_{k=1}^{i-1} \ell_{ik} b_k'}{\ell_{ii}}, \qquad i = 2, 3, \ldots, n.$$

The equations for the back-substitution are

$$x_n = b_n',$$

$$x_j = b_j' - \sum\limits_{k=j+1}^{n} u_{jk} x_k, \qquad j = n - 1, n - 2, \ldots, 1.$$

For example, if

$$A = \begin{bmatrix} 3 & -1 & 2 \\ 1 & 2 & 3 \\ 2 & -2 & -1 \end{bmatrix},$$

we get

$$L = \begin{bmatrix} 3 & 0 & 0 \\ 1 & \frac{7}{3} & 0 \\ 2 & -\frac{4}{3} & -1 \end{bmatrix} \quad \text{and} \quad U = \begin{bmatrix} 1 & -\frac{1}{3} & \frac{2}{3} \\ 0 & 1 & 1 \\ 0 & 0 & 1 \end{bmatrix}.$$

For $b = \begin{bmatrix} 12 \\ 11 \\ 2 \end{bmatrix}$,

$$b' = \begin{bmatrix} 4 \\ 3 \\ 2 \end{bmatrix},$$

because

$$b_1' = \frac{12}{3} = 4,$$

$$b_2' = \frac{11 - (1)(4)}{\frac{7}{3}} = 3,$$

$$b_3' = \frac{2 - (2)(4) - (-\frac{4}{3})(3)}{-1} = 2.$$

Augmenting b' to U and back-substituting

$$\begin{bmatrix} 1 & -\frac{1}{3} & \frac{2}{3} & | & 4 \\ 0 & 1 & 1 & | & 3 \\ 0 & 0 & 1 & | & 2 \end{bmatrix},$$

we get

$$x_3 = 2,$$

$$x_2 = 3 - 1(2) = 1,$$

$$x_1 = 4 - \frac{2}{3}(2) - (-\frac{1}{3})(1) = 3.$$

Examination of these operations reveals that we can get b' by augmenting L with b and solving that triangular system (a kind of "forward" substitution):

$$L \mathbin{|} b = \begin{bmatrix} 3 & 0 & 0 & | & 12 \\ 1 & \frac{7}{3} & 0 & | & 11 \\ 2 & -\frac{4}{3} & -1 & | & 2 \end{bmatrix}$$

and

$$b_1' = \frac{12}{3} = 4,$$

$$b_2' = \frac{11 - (1)(4)}{\frac{7}{3}} = 3,$$

$$b_3' = \frac{2 - (2)(4) - (-\frac{4}{3})(3)}{-1} = 2.$$

We do not write the algorithm for reducing the b-vector and back-substituting. This is left as an exercise for the student.

A special advantage of these LU methods is that we can accumulate the sums in double precision. This gives us greater accuracy by just using one or two double-precision variables. This is not easily done with the Gaussian elimination method of Section 2.4. Moreover, the LU method can be easily adapted to solve a system of new right-hand-side vectors with great economy of effort.* The number of arithmetic

---

* A numerical example that illustrates using the LU method with multiple right-hand sides will be found in Section 2.7.

operations to get the solution corresponding to each b turns out to be exactly the same as to multiply an $n \times n$ matrix by an n-component vector.

Pivoting with the LU method is somewhat more complicated than with Gaussian elimination because we do not usually handle the right-hand-side vector simultaneously with our reduction of the A matrix. This means we must keep a record of any row interchanges made during the formation of L and U so that the elements of the right-hand-side vector can be similarly interchanged. We do the interchanges immediately after computing each column of L, choosing the value to appear on the diagonal so as to have the one of largest magnitude. We illustrate with an example:

$$\text{Given } A = \begin{bmatrix} 0 & 2 & 1 \\ 1 & 0 & 0 \\ 3 & 0 & 1 \end{bmatrix}.$$

We will keep a record of row order in a vector: $O = [1, 2, 3]$, representing the original ordering.

The first column of L is

$$\begin{bmatrix} 0 \\ 1 \\ 3 \end{bmatrix};$$

we need to interchange rows 3 and 1. To keep track of this, we interchange the first and third elements of O, so O becomes $[3, 2, 1]$.

Interchange the rows of A and compute the first row of the U matrix. (We use the compact scheme.)

$$\begin{bmatrix} 3 & 0 & \frac{1}{3} \\ 1 & 0 & 0 \\ 0 & 2 & 1 \end{bmatrix}, \qquad \text{with } O = [3, 2, 1].$$

Now compute the second column of L; it is $\begin{bmatrix} 0 \\ 0 \\ 2 \end{bmatrix}$.

We must interchange again, the second row with the third, and O becomes $[3, 1, 2]$. Making the interchange of rows and computing the second row of U gives

$$\begin{bmatrix} 3 & 0 & \frac{1}{3} \\ 0 & 2 & \frac{1}{2} \\ 1 & 0 & 0 \end{bmatrix}, \qquad \text{with } O = [3, 1, 2].$$

Completing the reduction, we find $\ell_{33} = 0 - (1)(\frac{1}{3}) = -\frac{1}{3}$, giving

$$LU = \begin{bmatrix} 3 & 0 & \frac{1}{3} \\ 0 & 2 & \frac{1}{2} \\ 1 & 0 & -\frac{1}{3} \end{bmatrix}, \qquad \text{with } O = [3, 1, 2].$$

To solve the problem $Ax = b$, with $b^T = [5, -1, -2]$, we rearrange the elements of b in the order given by O and compute b':

$$L \vert b = \begin{bmatrix} 3 & 0 & 0 & \vert & -2 \\ 0 & 2 & 0 & \vert & 5 \\ 1 & 0 & -\frac{1}{3} & \vert & -1 \end{bmatrix}, \qquad b' = \begin{bmatrix} -\frac{2}{3} \\ \frac{5}{2} \\ 1 \end{bmatrix},$$

so

$$U \vert b' = \begin{bmatrix} 1 & 0 & \frac{1}{3} & \vert & -\frac{2}{3} \\ 0 & 1 & \frac{1}{2} & \vert & \frac{5}{2} \\ 0 & 0 & 1 & \vert & 1 \end{bmatrix}, \qquad \text{giving } x = \begin{bmatrix} -1 \\ 2 \\ 1 \end{bmatrix}.$$

The Gaussian elimination method of Section 2.4 can be used in the same way to get the solution if we have multiple right-hand sides. Suppose we want to solve $Ax = b^{(1)}$, then $Ax = b^{(2)}$, where the matrices A are the same but the right-hand sides are different.

As an example, we will solve this system by Gaussian elimination:

$$\begin{aligned} 4x - 3y \quad\;\;\; &= -7, \\ 2x + 2y + 3z &= -2, \\ 6x + \;\; y - 6z &= \;\;\; 6. \end{aligned} \qquad (2.23)$$

In augmented matrix form, we get

$$\begin{bmatrix} 4 & -3 & 0 & -7 \\ 2 & 2 & 3 & -2 \\ 6 & 1 & -6 & 6 \end{bmatrix}, \qquad \text{Row order vector } O = [1, 2, 3].$$

After using the Gaussian elimination method of the previous section, the system is

$$\begin{bmatrix} 6 & 1 & -6 & 6 \\ (0.6667) & -3.667 & 4 & -11 \\ (0.3333) & (-0.4545) & 6.8182 & -9 \end{bmatrix}, \qquad \text{Row order vector } O = [3, 1, 2].$$

After performing back-substitution, we find that the solution vector for this problem is

$$[x = -0.5800, \quad y = 1.5600, \quad z = -1.3200],$$

and the LU decomposition is

$$\begin{bmatrix} 1 & 0 & 0 \\ 0.6667 & 1 & 0 \\ 0.3333 & -0.4545 & 1 \end{bmatrix} * \begin{bmatrix} 6 & 1 & -6 \\ 0 & -3.6667 & 4 \\ 0 & 0 & 6.8182 \end{bmatrix},$$

which gives, for A',

$$\begin{bmatrix} 6 & 1 & -6 \\ 4 & -3 & 0 \\ 2 & 2 & 3 \end{bmatrix} = A',$$

agreeing with the order given in the vector: $O = [3, 1, 2]$.

In solving a system of n equations, the number of multiplications/divisions used is of the order $O(n^3/3)$. Suppose we wish to solve the same system (2.23) with a different right-hand side:

$$\begin{aligned} 4x - 3y \qquad &= 14, \\ 2x + 2y + 3z &= 9, \\ 6x + y - 6z &= -8. \end{aligned} \qquad (2.24)$$

We first rearrange this system of equations to fit the above form of $A'x = b'$, using the order vector. We now have the system

$$\begin{aligned} 6x + y - 6z &= -8, \\ 4x - 3y \qquad &= 14, \\ 2x + 2y + 3z &= 9, \end{aligned} \qquad (2.25)$$

which we can express as: $LUx = b'$ or $L(Ux) = b'$. Our solution method will be the following:

1. Solve $Ly = b'$, where the vector $y = Ux$ (which can be done with $O(n^2/2)$ multiplications/divisions).
2. Now solve $Ux = y$, where y is the solution from step 1 (this also takes $O(n^2/2)$ multiplications/divisions).

In our current example, we have the lower-triangular system

$$L * y = \begin{bmatrix} 1 & 0 & 0 \\ 0.6667 & 1 & 0 \\ 0.3333 & -0.4545 & 1 \end{bmatrix} * \begin{bmatrix} y_1 \\ y_2 \\ y_3 \end{bmatrix} = \begin{bmatrix} -8 \\ 14 \\ 9 \end{bmatrix}, \qquad (2.26)$$

which gives us the intermediate solution: $y_1 = -8, y_2 = 19.3333, y_3 = 20.4545$. We now solve the upper-triangular system:

$$U * x = \begin{bmatrix} 6 & 1 & -6 \\ 0 & -3.6667 & 4 \\ 0 & 0 & 6.9182 \end{bmatrix} * \begin{bmatrix} x_1 \\ x_2 \\ x_3 \end{bmatrix} = \begin{bmatrix} -8 \\ 19.3333 \\ 20.4545 \end{bmatrix}. \qquad (2.27)$$

Solving this system gives us the solution to the original problem in (2.24): $x_1 = 2$, $x_2 = -2$, and $x_3 = 3$. Although these two intermediate steps seem rather awkward, the actual total number of multiplications/divisions for the LU system that we now have is $O(n^2)$, in comparison to the $O(n^3/3)$ operations for the original system when we did not know the LU decomposition of the matrix A. This shows the power and usefulness of the Gaussian elimination for solving a variety of linear systems, evaluating the determinant, and obtaining the LU decomposition of a square matrix. Finally, in Section 2.7, we shall find that we can use Gaussian elimination in evaluating the inverse of a square matrix.

2.6 Pathology in Linear Systems—Singular Matrices

When a real physical situation is modeled by a set of linear equations, we can anticipate that the set of equations will have a solution that matches the values of the quantities in the physical problem, at least as far as the equations truly do represent it.* Because of round-off errors, the solution vector that is calculated may imperfectly predict the physical quantity, but there is assurance that a solution exists, at least in principle. Consequently, it must always be theoretically possible to avoid divisions by zero when the set of equations has a solution.

An arbitrary set of equations may not have such a guaranteed solution, however. There are several such possible situations, which we term "pathological." In each case, there is *no unique solution* to the set of equations.

First, if the number of equations relating the variables is less than the number of *unknowns*, we certainly cannot solve for unique values of the unknown variables. It turns out, in this case, that there is an infinite set of solutions, for we may arrange the n equations with all but n of the variables on the right-hand sides, grouped with the constant terms. We may assign almost any desired values to these segregated variables (combining them with the constant terms) and then solve for the n remaining variables.† Assigning new values to the variables on the right-hand sides gives another set of values for the unknowns, and so on. For example,

$$\begin{cases} x_1 - 2x_2 + x_3 = 4, \\ x_1 - x_2 + x_3 = 5. \end{cases}$$

Rewrite as

$$x_1 - 2x_2 = 4 - x_3,$$
$$x_1 - x_2 = 5 - x_3.$$

If $x_3 = 0$,

$$x_1 = 6, \qquad x_2 = 1.$$

If $x_3 = 1$,

$$x_1 = 5, \qquad x_2 = 1.$$

---

* There are certain problems for which values of interest are determined from a set of equations that do not have a unique solution; these are called *eigenvalue* problems and are discussed in Chapter 6.
† An important instance of this situation is the solution of a linear programming problem by the simplex method. In this method, the segregated variables are all assigned the value of zero.

If $x_3 = -1$,

$$x_1 = 7, \qquad x_2 = 1, \qquad \text{and so on.}$$

A second situation where we might not expect a set of equations to have a solution is that in which the number of equations is greater than the number of unknowns. If there are n unknowns, we can normally find a subset of the equations that can be solved for the unknowns. There are two subcases to consider.

If the remaining equations are satisfied by the values of the unknowns we have just determined (we would say these equations are *consistent* with the others), there exists a unique solution to the set of equations. Really, of course, there are not truly more equations than unknowns in this case; the extra equations are *redundant*.

The other subcase is a pathological one. If the solution to the first n equations does not satisfy the remaining ones, the set is clearly *inconsistent*, and *no* solution exists that satisfies the system.

Realizing that an equation may be redundant in the prior situation makes us reexamine our more standard case of n equations in n unknowns. What if there is redundancy there? How can we recognize redundancy when it is present? In this example it is obvious:

$$x + y = 3, \qquad 2x + 2y = 6.$$

The second equation is clearly redundant and contains no information not already given by the first. This system will then have an *infinity* of values for x and y; it is an example of fewer equations than unknowns.

Inconsistency may also be present:

$$x + y = 3, \qquad 2x + 2y = 7.$$

In this case there is *no* solution.

If $n \times n$ systems do not have a unique solution, they have a (square) coefficient matrix that is called *singular*. If the coefficient matrix can be triangularized without having zeros on the diagonal (hence the set of equations has a solution), the matrix is said to be *nonsingular*.

Larger systems may have redundancy or inconsistency even though it is not obvious at a glance. Even in a 3×3 system, it is not easy to tell:

$$
\begin{aligned}
x_1 - 2x_2 + 3x_3 &= 5, \\
2x_1 + 4x_2 - x_3 &= 7, \\
-x_1 - 14x_2 + 11x_3 &= 2.
\end{aligned}
$$

Are these inconsistent or redundant? (In other words, is the coefficient matrix singular?) Or do they have a unique solution? (That is, is the matrix nonsingular?) Is there a rule that we can apply, especially one that works for large systems? The answer is yes, there is a rule, or rather, there are several tests we can apply. The standard response from mathematics is to determine the *rank* of the coefficient matrix. If this value is less than n, the number of equations, then no unique solution exists; the equations are either inconsistent or one or more are redundant, depending on the right-hand-side values.

But how does one determine the rank? One practical method is to triangularize by Gaussian elimination: If no zeros show up on the diagonal of the final triangularized coefficient matrix, the rank is equal to n (the matrix is said to be of *full rank*) and a unique solution exists. If, in spite of pivoting, one or more zeros occur on the final diagonal, there is no unique solution. The set will be consistent (and have redundancy) if back-substitution gives (0/0) indeterminate forms. When we would need to divide a nonzero term by zero in back-substituting, inconsistency occurs. Let us apply this to our earlier 3×3 example:

$$\begin{bmatrix} 1 & -2 & 3 & \vline & 5 \\ 2 & 4 & -1 & \vline & 7 \\ -1 & -14 & 11 & \vline & 2 \end{bmatrix}.$$

On reduction we get

$$\begin{bmatrix} 1 & -2 & 3 & \vline & 5 \\ 0 & 8 & -7 & \vline & -3 \\ 0 & 0 & 0 & \vline & 1 \end{bmatrix}.$$

We see that there is no solution and that the equations are inconsistent. If the constant term in the third equation were 1 rather than 2, reduction would give

$$\begin{bmatrix} 1 & -2 & 3 & \vline & 5 \\ 0 & 8 & -7 & \vline & -3 \\ 0 & 0 & 0 & \vline & 0 \end{bmatrix}.$$

This system is redundant.

Another way to find whether a set of equations has a unique solution is to test the rows or columns of the coefficient matrix for *linear dependency*. Vectors are called linearly dependent if a linear combination of them can be found that equals the zero vector (one with all components equal to zero). (Of course the linear combination $a\bar{x} + b\bar{y} + c\bar{z}$ always equals zero if all the coefficients are zero; we rule out this possibility in our test for linear dependency. When $a = b = c = 0$, we say we have the *trivial case*.)

If the vectors are linearly independent, the only way a weighted sum of them can equal the zero vector is to weight each of them with a zero coefficient.

Our singular 3×3 system has columns that form vectors that are linearly dependent:

$$(-10)\begin{bmatrix} 1 \\ 2 \\ -1 \end{bmatrix} + (7)\begin{bmatrix} -2 \\ 4 \\ -14 \end{bmatrix} + (8)\begin{bmatrix} 3 \\ -1 \\ 11 \end{bmatrix} = \begin{bmatrix} 0 \\ 0 \\ 0 \end{bmatrix}.$$

Similarly, the rows form linearly dependent vectors:

$$(-3)[1 \quad -2 \quad 3] + (2)[2 \quad 4 \quad -1] + (1)[-1 \quad -14 \quad 11] = [0 \quad 0 \quad 0].$$

In the general case, we say that vectors $\bar{x}_1, \bar{x}_2, \bar{x}_3, \ldots, \bar{x}_n$ are *linearly dependent* if we can find scalar coefficients, $a_1, a_2, \ldots, a_n$ (with not all the a_i simultaneously zero), for which

$$\sum_{i=1}^{n} a_i \bar{x}_i = \bar{0}. \tag{2.28}$$

If the only linear combination of the $\bar{x}_i$ that equals the zero vector requires that all the a_i be zero, the set of vectors is called linearly independent. It follows that, if a set of vectors is linearly dependent, at least one of the vectors can be written as a linear combination of the others. If the set is linearly independent, none of the vectors can be written as a linear combination of the others. As a practical matter, we do not usually test the columns (or rows) for linear dependency to determine whether a matrix is singular.

If we are interested in determining the coefficients $a_1, a_2, \ldots, a_n$ that appear in the linear combination of Eq. (2.28), it turns out that we have to solve a set of linear equations to obtain them.

It is worthwhile to summarize the concepts and terminology of this section. The following lists of terms are all equivalent expressions. If a square matrix can be shown to have one property, it has all the others.

Equivalent Properties of Singular or Nonsingular Matrices

The matrix is singular.	The matrix is nonsingular.
A set of equations with these coefficients has no unique solution.	A set of equations with these coefficients has a unique solution.
Gaussian elimination cannot avoid a zero on the diagonal.	Gaussian elimination proceeds without a zero on the diagonal.
The rank of the matrix is less than n.	The rank of the matrix equals n.
The rows form linearly dependent vectors.	The rows form linearly independent vectors.
The columns form linearly dependent vectors.	The columns form linearly independent vectors.

In the next section we will consider two other properties of the matrix: its determinant and its inverse. This adds two more attributes to our lists: A singular matrix has a zero determinant and a nonsingular matrix has a nonzero determinant. A singular matrix has no inverse and a nonsingular matrix does have an inverse.

2.7 Determinants and Matrix Inversion

You have perhaps wondered why there has been no reference so far in this chapter to the solution of linear equations by determinants (Cramer's rule). The reason is that, except for systems of only two or three equations, the determinant method is too inefficient. For example, for a set of 10 simultaneous equations, about 70,000,000 multiplications and divisions are required if the usual method of expansion in terms of minors is used. A more efficient method of evaluating the determinants can reduce this to about 3000 multiplications, but even this is inefficient compared to Gaussian elimination, which would require about 380.

In fact, the evaluation of a determinant can perhaps best be done by adapting the Gaussian elimination procedure. Its utility derives from the fact that the determinant of a triangular matrix (either upper- or lower-triangular) is just the product of its diagonal elements. This is easily seen, in the case of an upper-triangular matrix, by expansion in terms of minors of the first column at each step. For example,

$$
\begin{vmatrix}
a_{11} & a_{12} & a_{13} & a_{14} \\
0 & a_{22} & a_{23} & a_{24} \\
0 & 0 & a_{33} & a_{34} \\
0 & 0 & 0 & a_{44}
\end{vmatrix}
= a_{11}
\begin{vmatrix}
a_{22} & a_{23} & a_{24} \\
0 & a_{33} & a_{34} \\
0 & 0 & a_{44}
\end{vmatrix}
- 0 + 0 - 0
$$

$$
= a_{11} \left(a_{22}
\begin{vmatrix}
a_{33} & a_{34} \\
0 & a_{44}
\end{vmatrix}
- 0 + 0 \right)
$$

$$
= a_{11} a_{22} (a_{33} a_{44} - 0) = a_{11} a_{22} a_{33} a_{44}.
$$

Adding a multiple of one row to another row of a matrix does not change the value of its determinant. The other row transformations change the value in predictable ways: Interchanging two rows changes its sign, and multiplying a row by a constant multiplies the value of the determinant by the same constant. If these changes are allowed for, using the procedure of Gaussian elimination to convert to upper-triangular is a simple way to evaluate the determinant.

EXAMPLE 2.4 Find the value of the determinant by using elementary row transformations to make it upper-triangular:

$$
\begin{vmatrix}
1 & 4 & -2 & 3 \\
2 & 2 & 0 & 4 \\
3 & 0 & -1 & 2 \\
1 & 2 & 2 & -3
\end{vmatrix}
=
\begin{vmatrix}
1 & 4 & -2 & 3 \\
0 & -6 & 4 & -2 \\
0 & -12 & 5 & -7 \\
0 & -2 & 4 & -6
\end{vmatrix}
=
\begin{vmatrix}
1 & 4 & -2 & 3 \\
0 & -6 & 4 & -2 \\
0 & 0 & -3 & -3 \\
0 & 0 & \frac{8}{3} & -\frac{16}{3}
\end{vmatrix}
$$

$$
=
\begin{vmatrix}
1 & 4 & -2 & 3 \\
0 & -6 & 4 & -2 \\
0 & 0 & -3 & -3 \\
0 & 0 & 0 & -8
\end{vmatrix}
= (1)(-6)(-3)(-8) = -144.
$$

For programming, an easy and very efficient method for computing the determinant is to use the algorithm in Section 2.4. Then the determinant of the matrix is just the product of the diagonal elements, with a reversed sign if there were an odd number of row interchanges: $\pm a_{11} * a_{22} * \cdots * a_{nn}$, where $+$ is used if there were 0 or an even number of row interchanges (otherwise we use $-$).

Applying this algorithm to the example, but using row interchanges, we see

$$
\begin{bmatrix}
1 & 4 & -2 & 3 \\
2 & 2 & 0 & 4 \\
3 & 0 & -1 & 2 \\
1 & 2 & 2 & -3
\end{bmatrix}
\rightarrow
\begin{bmatrix}
3 & 0 & -1 & 2 \\
(0.667) & 4 & -1.677 & 2.333 \\
(0.333) & (0.5) & 3.167 & -4.833 \\
(0.333) & (0.5) & (0.474) & 3.789
\end{bmatrix}.
$$

Since $3 * 4 * 3.167 * 3.789 = 144$ and there were 3 row interchanges in the process, we have the determinant $= -144$.

▲

While division of matrices is not defined, the matrix inverse gives the equivalent result. If the product of two square matrices is the identity matrix, the matrices are said to be inverses. If $AB = I$, we write $B = A^{-1}$; also $A = B^{-1}$. Inverses commute on multiplication, which is not true for matrices in general: $AB = BA = I$. Not all square matrices have an inverse. Singular matrices do not have an inverse, and these are of extreme importance in connection with the coefficient matrix of a set of equations, as discussed earlier.

The inverse of a matrix can be defined in terms of the matrix of the minors of its determinant, but this is not a useful way to find an inverse. The Gauss–Jordan technique can be adapted to provide a practical way to invert a matrix. The procedure is to augment the given matrix with the identity matrix of the same order. One then reduces the original matrix to the identity matrix by elementary row transformations, performing the same operations on the augmentation columns. When the identity matrix stands as the left half of the augmented matrix, the inverse of the original stands as the right half. It should be apparent that this is equivalent to solving a set of equations with n different right-hand sides; each of the right-hand sides is a *unit basis vector*, in which the position of the element whose value is unity changes from row 1 to row 2 to row 3 ... to row n.

EXAMPLE 2.5 Find the inverse of

$$
A = \begin{bmatrix}
1 & -1 & 2 \\
3 & 0 & 1 \\
1 & 0 & 2
\end{bmatrix}.
$$

Augment A with the identity matrix and then reduce:

$$
\begin{bmatrix}
1 & -1 & 2 & 1 & 0 & 0 \\
3 & 0 & 1 & 0 & 1 & 0 \\
1 & 0 & 2 & 0 & 0 & 1
\end{bmatrix}
\rightarrow
\begin{bmatrix}
1 & -1 & 2 & 1 & 0 & 0 \\
0 & 3 & -5 & -3 & 1 & 0 \\
0 & 1 & 0 & -1 & 0 & 1
\end{bmatrix}
$$

$$
\overset{(1)}{\rightarrow}
\begin{bmatrix}
1 & -1 & 2 & 1 & 0 & 0 \\
0 & 1 & 0 & -1 & 0 & 1 \\
0 & 0 & -5 & 0 & 1 & -3
\end{bmatrix}
\overset{(2)}{\rightarrow}
\begin{bmatrix}
1 & -1 & 0 & 1 & \frac{2}{5} & -\frac{6}{5} \\
0 & 1 & 0 & -1 & 0 & 1 \\
0 & 0 & 1 & 0 & -\frac{1}{5} & \frac{3}{5}
\end{bmatrix}
$$

$$
\rightarrow
\begin{bmatrix}
1 & 0 & 0 & 0 & \frac{2}{5} & -\frac{1}{5} \\
0 & 1 & 0 & -1 & 0 & 1 \\
0 & 0 & 1 & 0 & -\frac{1}{5} & \frac{3}{5}
\end{bmatrix}.
$$

[1] Interchange the third and second rows before eliminating from the third row.
[2] Divide the third row by -5 before eliminating from the first row.

We confirm the fact that we have found the inverse by multiplication:

$$
\begin{bmatrix}
1 & -1 & 2 \\
3 & 0 & 1 \\
1 & 0 & 2
\end{bmatrix}
\begin{bmatrix}
0 & \frac{2}{5} & -\frac{1}{5} \\
-1 & 0 & 1 \\
0 & -\frac{1}{5} & \frac{3}{5}
\end{bmatrix}
=
\begin{bmatrix}
1 & 0 & 0 \\
0 & 1 & 0 \\
0 & 0 & 1
\end{bmatrix}.
$$

▲

However, it is more efficient to use the Gaussian elimination algorithm of Section 2.4 by adding additional unit vectors to the augmented matrix.

Doing steps 2 through 4 gives us

$$
\begin{bmatrix}
1 & -1 & 2 & 1 & 0 & 0 \\
3 & 0 & 1 & 0 & 1 & 0 \\
1 & 0 & 2 & 0 & 0 & 1
\end{bmatrix}
\rightarrow
\begin{bmatrix}
3 & 0 & 1 & 0 & 1 & 0 \\
(0.333) & -1 & 1.667 & 1 & -0.333 & 0 \\
(0.333) & (0) & 1.667 & 0 & -0.333 & 1
\end{bmatrix}.
$$

Now applying back-substitution on the last three columns, we get

$$
\begin{bmatrix}
3 & 0 & 1 & 0 & 0.4 & -0.2 \\
(0.333) & -1 & 1.667 & -1 & 0 & 1 \\
(0.333) & (0) & 1.667 & 0 & -0.2 & 0.6
\end{bmatrix},
$$

where the last three columns store the inverse matrix. This method is actually more efficient than the Gauss–Jordan method, which takes about $3n^3/2$ versus $4n^3/3$ multiplications and/or divisions to compute the inverse of a nonsingular matrix. Moreover, in the Gaussian elimination method, we have also found the LU matrix, which we can use to solve the system for other right-hand sides.

The inverse of the coefficient matrix provides a way of solving the set of equations $Ax = b$ because, when we multiply both sides of the relation by A^{-1}, we get

$$A^{-1}Ax = A^{-1}b,$$
$$x = A^{-1}b.$$

The second equation follows because the product $A^{-1}A = I$, the identity matrix, and $Ix = x$. If we know the inverse of A, we can solve the system for any right-hand-side b simply by multiplying the b-vector by A^{-1}. This would seem like a good way to solve systems of equations, and one finds frequent references to it.

If we care about the efficiency of our method of solving the equations, however, this is not the preferred method, because solving the system with the LU decomposition of A, and doing the equivalent of two back-substitutions, requires exactly the same effort as multiplying b by the matrix. We compare the efficiency of the two schemes, then, by comparing the work needed to get the inverse and that to get the LU equivalent. Getting the inverse is more work, because it is the equivalent of solving the system with n right-hand sides, while getting the LU is the equivalent of doing only the reduction to triangular form.

Even though the inverse is not the most efficient way to solve a set of simultaneous equations, the inverse is very important for theoretical reasons and is essential to the understanding of many situations in applied mathematics. The use of the inverse concept and notation often simplifies the development of some fundamental relationships. We illustrate this by again considering the LU decomposition.

Find a pair of matrices such that $LU = A$.

Then $Ax = b$ can be written as

$$LUx = b.$$

Multiply both sides by L^{-1},

$$(L^{-1}L)Ux = L^{-1}b,$$

so $Ux = L^{-1}b$ because $L^{-1}L = I$.

We see that Ux (which is a vector because the product of a matrix times a vector always yields a vector) is equal to the vector formed by $L^{-1}b$. Call this vector b'.

Then

$$Ux = b', \qquad b' = L^{-1}b, \qquad \text{or} \qquad Lb' = b.$$

We can get the vector b' by solving the system $Lb' = b$. This is particularly easy to do because L is triangular; all we need to do is the back-substitution phase (actually a forward-substitution because L is lower-triangular).

Once we have b', we can solve for x from the system $Ux = b'$. This is also easy because U is triangular.

Observe how using the concept and notation of inverses helps to clarify and prove the validity of the LU method.

2.8 Norms

When we discuss multicomponent entities like matrices and vectors, we frequently need a way to express their magnitude—some measure of "bigness" or "smallness." For ordinary numbers, the absolute value tells us how large the number is, but for a matrix there are many components, each of which may be large or small in magnitude. (We are not talking about the *size* of a matrix, meaning the number of elements it contains.)

Any good measure of the magnitude of a matrix (the technical term is *norm*) must have four properties that are intuitively essential:

1. The norm must always have a value greater than or equal to zero, and must be zero only when the matrix is the zero matrix (one with all elements equal to zero).
2. The norm must be multiplied by k if the matrix is multiplied by the scalar k.
3. The norm of the sum of two matrices must not exceed the sum of the norms.
4. The norm of the product of two matrices must not exceed the product of the norms.

More formally, we can state these conditions, using $\|A\|$ to represent the *norm of matrix* A:

$$
\begin{aligned}
&1. \ \|A\| \geq 0 \text{ and } \|A\| = 0 \text{ if and only if } A = 0. \\
&2. \ \|kA\| = |k|\,\|A\|. \\
&3. \ \|A + B\| \leq \|A\| + \|B\|. \\
&4. \ \|AB\| \leq \|A\|\,\|B\|.
\end{aligned}
\tag{2.29}
$$

The third relationship is called the *triangle inequality*. The fourth is important when we deal with the product of matrices.

For the special kind of matrices that we call vectors, our past experience can help us. For vectors in two- or three-space, the length satisfies all four requirements and is a good value to use for the <u>norm of a vector</u>. This norm is called the *Euclidean norm*, and is computed by $\sqrt{x_1^2 + x_2^2 + x_3^2}$.

We compute the Euclidean norm of vectors with more than three components by generalizing:

$$\|x\|_e = \sqrt{x_1^2 + x_2^2 + \cdots + x_n^2} = \left(\sum_{i=1}^n x_i^2\right)^{1/2}.$$

This is not the only way to compute a vector norm, however. The sum of the absolute values of the x_i can be used as a norm; the maximum value of the magnitudes of the x_i will also serve. These three norms can be interrelated by defining the *p*-norm as

$$\|x\|_p = \left(\sum_{i=1}^n |x_i|^p\right)^{1/p}.$$

From this it is readily seen that

$$\|x\|_1 = \sum_{i=1}^n |x_i| = \text{sum of magnitudes;}$$

$$\|x\|_2 = \left(\sum_{i=1}^n x_i^2\right)^{1/2} = \text{Euclidean norm;}$$

$$\|x\|_\infty = \max_{1 \le i \le n} |x_i| = \text{maximum-magnitude norm.}$$

Which of these vector norms is best to use may depend on the problem. In most cases, satisfactory results are obtained with any of these measures of the "size" of a vector.

EXAMPLE 2.6   Compute the 1-, 2-, and ∞-norms of the vector x, if $x = (1.25, 0.02, -5.15, 0)$.

$$\|x\|_1 = |1.25| + |0.02| + |-5.15| + |0| = 6.42$$
$$\|x\|_2 = [(1.25)^2 + (0.02)^2 + (-5.15)^2 + (0)^2]^{1/2} = 5.2996$$
$$\|x\|_\infty = |-5.15| = 5.15$$

The norms of a matrix are developed by a correspondence to vector norms. Matrix norms that correspond to the above, for matrix A, can be shown to be

$$\|A\|_1 = \max_{1 \leq j \leq n} \sum_{i=1}^{n} |a_{ij}| = \text{maximum column sum};$$

$$\|A\|_\infty = \max_{1 \leq i \leq n} \sum_{j=1}^{n} |a_{ij}| = \text{maximum row sum}.$$

The matrix norm $\|A\|_2$ that corresponds to the 2-norm of a vector is not readily computed. It is related to the eigenvalues of the matrix. It sometimes has special utility because no other norm is smaller than this one. It therefore provides the "tightest" measure of the magnitude of a matrix, but is also the most difficult to compute. This norm is also called the *spectral norm*.

For an $m \times n$ matrix, we can paraphrase the Euclidean (also called Frobenius) norm* and write

$$\|A\|_e = \left(\sum_{i=1}^{m} \sum_{j=1}^{n} a_{ij}^2 \right)^{1/2}.$$

EXAMPLE 2.7 Compute the Euclidean norms of A, B, and C, and the ∞-norms, given that

$$A = \begin{bmatrix} 5 & 9 \\ -2 & 1 \end{bmatrix}; \quad B = \begin{bmatrix} 0.1 & 0 \\ 0.2 & -0.1 \end{bmatrix}; \quad \text{and} \quad C = \begin{bmatrix} 0.2 & 0.1 \\ 0.1 & 0 \end{bmatrix}.$$

$\|A\|_e = \sqrt{25 + 81 + 4 + 1} = \sqrt{111} = 10.54;$ $\qquad \|A\|_\infty = 14.$

$\|B\|_e = \sqrt{0.01 + 0 + 0.04 + 0.01} = \sqrt{0.06} = 0.2449;$ $\qquad \|B\|_\infty = 0.3.$

$\|C\|_e = \sqrt{0.04 + 0.01 + 0.01 + 0} = \sqrt{0.06} = 0.2449;$ $\qquad \|C\|_\infty = 0.3.$

The results of our examples look quite reasonable; certainly A is "larger" than B or C. While $B \neq C$, both are equally "small." The Euclidean norm is a good measure of the magnitude of a matrix.

We see then that there are a number of ways that the norm of a matrix can be expressed. Which way is preferred? There are certainly differences in their cost; for

---

*Be alert to a situation that may be confusing: For a vector, the 2-norm is the same as the Euclidean norm, but for a matrix, the 2-norm is not the same as the Euclidean norm.

example, some will require more extensive arithmetic than others. The spectral norm is usually the most "expensive." Which norm is best? The answer to this question depends in part on the use for the norm. In most instances, we want the norm that puts the smallest upper bound on the magnitude of the matrix. In this sense, the spectral norm is "best." We observe, in the next example, that not all the norms give the same value for the magnitude of a matrix.

EXAMPLE 2.8

$$A = \begin{bmatrix} 5 & -5 & -7 \\ -4 & 2 & -4 \\ -7 & -4 & 5 \end{bmatrix}.$$

$\|A\|_e = \text{Euclidean norm} = 15;$

$\|A\|_\infty = 17;$

$\|A\|_1 = 16;$

$\|A\|_2 = \text{spectral norm} = 12.$

If a matrix is a diagonal matrix, all p-norms have the same value, however.

▲

Why are norms important? For one thing, they let us express the accuracy of the solution to a set of equations in quantitative terms by stating the norm of the error vector (the true solution minus the approximate solution vector). Norms are also used to study quantitatively the convergence of iterative methods of solving linear systems (which we will cover in a later section).

2.9 Condition Numbers and Errors in Solutions

When we solve a set of linear equations, $Ax = b$, we hope that the calculated vector $\bar{x}$ is a close representation of the true solution vector x. In the previous examples, we have seen how round-off can make the computed solution differ from the exact solution, and several times the use of pivoting has been recommended as a way to minimize the effect.

Pivoting

The next example will demonstrate how much the accuracy can be improved by pivoting. We will exaggerate the effect by using only four-digit arithmetic (rounding to four digits after each operation). The same effects are observed with more precise arithmetic in large systems.

Assume $Ax = b$ with

$$A = \begin{bmatrix} -0.002 & 4.000 & 4.000 \\ -2.000 & 2.906 & -5.387 \\ 3.000 & -4.031 & -3.112 \end{bmatrix}$$

and

$$b = (7.998, -4.481, -4.143)^T.$$

After Gaussian elimination without pivoting, the triangularized matrix (augmented with the b-vector) is

$$\begin{bmatrix} -0.002 & 4.000 & 4.000 & 7.998 \\ 0 & -3.997 & -4.005 & -8.002 \\ 0 & 0 & -10.00 & 0.000 \end{bmatrix},$$

which gives the computed solution

$$\bar{x} = (-1496, 2.000, 0.000)^T.$$

The exact solution is $x = (1.000, 1.000, 1.000)^T$, and we see that round-off errors (particularly obvious in the last row of the triangularized system) have given a completely incorrect result. In this example the very small value of a_{11} has been the source of difficulty.

If the equations are reordered to give

$$A = \begin{bmatrix} 3.000 & -4.031 & -3.112 \\ -0.002 & 4.000 & 4.000 \\ -2.000 & 2.906 & -5.387 \end{bmatrix}, \quad \text{and} \quad b = \begin{bmatrix} -4.413 \\ 7.998 \\ -4.481 \end{bmatrix},$$

and we repeat the four-digit computations, the triangularized system becomes

$$\begin{bmatrix} 3.000 & -4.031 & -3.112 & -4.143 \\ 0 & 3.997 & 3.998 & 7.995 \\ 0 & 0 & -7.680 & -7.680 \end{bmatrix},$$

which gives $\bar{x} = (1.000, 1.000, 1.000)^T$, whose error is nil.

We conclude from this example that the algorithm we employ can have a very significant effect on the accuracy.

Ill-Conditioned Systems

Unfortunately, the error due to round-off is sometimes large even with the best available algorithm, because the problem itself may be very sensitive to the effects of small errors. Consider this example:

Assume $Ax = b$, with

$$A = \begin{bmatrix} 3.02 & -1.05 & 2.53 \\ 4.33 & 0.56 & -1.78 \\ -0.83 & -0.54 & 1.47 \end{bmatrix} \quad \text{and} \quad b = \begin{bmatrix} -1.61 \\ 7.23 \\ -3.38 \end{bmatrix}.$$

If we solve the system by Gaussian elimination, using pivoting and carrying three digits rounded, the triangularized system is

$$\begin{bmatrix} 4.33 & 0.56 & -1.78 & 7.23 \\ 0 & -1.44 & 3.77 & -6.65 \\ 0 & -0 & -0.00362 & 0.00962 \end{bmatrix}.$$

The computed solution is $\bar{x} = (0.875, -2.33, -2.66)^T$ while the true solution vector is $x = (1, 2, -1)^T$. The very small number on the diagonal in the third row is a sign of such inherent sensitivity to round-off. Such a system is said to be *ill-conditioned*. One strategy to use with ill-conditioned systems is greater precision in the arithmetic operations. For the preceding example, when six-digit computations are used, we get marked improvement:

$$x = (0.9998, 1.9995, -1.002)^T;$$

but note that there is still an appreciable error. A large ill-conditioned system would require even more digits of accuracy than six if we wished to compute a solution anywhere near the exact answer.

There are some important problems involving linear systems in which the coefficient matrix is always nearly singular, and hence ill-conditioned. One example is fitting a polynomial of relatively high degree to a set of points by the least-squares method. This is covered in Chapter 3.

Another way to look at ill-conditioned systems is to examine the effect of small changes in the coefficients. In the preceding system, changing the a_{11} coefficient from 3.02 to 3.00 gives a solution (exact) of

$$x_3 = 0.73326, \qquad x_2 = 6.52208, \qquad x_1 = 1.12767.$$

This property of ill-conditioned systems—that their solution is extremely sensitive to small changes in the coefficients—explains why they are also so sensitive to round-off error. The small inaccuracies in values as the computation proceeds, caused by round-off errors, are equivalent to numbers we would encounter in using exact arithmetic on a problem with slightly altered coefficients.

A more vivid understanding of ill-conditioning is obtained by examining a system of two equations with two unknowns. Consider the system $Ax = b$:

$$\begin{bmatrix} 1.01 & 0.99 \\ 0.99 & 1.01 \end{bmatrix} \begin{bmatrix} x \\ y \end{bmatrix} = \begin{bmatrix} 2.00 \\ 2.00 \end{bmatrix}. \tag{2.30}$$

It is obvious that the solution is $x = 1, y = 1$. Suppose we modify b, the right-hand side, just slightly to have

$$\begin{bmatrix} 1.01 & 0.99 \\ 0.99 & 1.01 \end{bmatrix} \begin{bmatrix} x \\ y \end{bmatrix} = \begin{bmatrix} 2.02 \\ 1.98 \end{bmatrix}. \tag{2.31}$$

Now the obvious solution is $x = 2, y = 0$. Finally, if there were another slightly different right-hand-side b-vector,

$$\begin{bmatrix} 1.01 & 0.99 \\ 0.99 & 1.01 \end{bmatrix} \begin{bmatrix} x \\ y \end{bmatrix} = \begin{bmatrix} 1.98 \\ 2.02 \end{bmatrix}, \tag{2.32}$$

we would have $x = 0, y = 2$!

It will be helpful to think of the system, $Ax = b$, as a *linear system solver machine*. In this view, the preceding example can be considered as having *inputs* to the machine of the three right-hand sides, the b's. The *outputs* from the machine are the solutions, the x's, for a fixed set of coefficients, A.

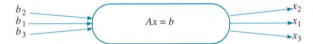

Even though the three inputs are "close together"—$b_1 = (2, 2)^T$, $b_2 = (2.02, 1.98)^T$, and $b_3 = (1.98, 2.02)^T$—we get very "distant" outputs—$x_1 = (1, 1)^T, x_2 = (2, 0)^T$, $x_3 = (0, 2)^T$. This modest example shows the basic idea of an ill-conditioned system: For *small changes in input*, *we get large changes in the output*.

In some situations, one can combat ill-conditioning by transforming the problem into an equivalent set of equations that are not ill-conditioned. The efficiency of this scheme is related to the relative amount of computation required for the transformation, compared to the cost of doing the calculations in higher precision.*

An interesting phenomenon of an ill-conditioned system is that we cannot test for the accuracy of the computed solution merely by substituting it into the equations to see whether the right-hand sides are reproduced. Consider again the ill-conditioned example we have previously examined:

$$A = \begin{bmatrix} 3.02 & -1.05 & 2.53 \\ 4.33 & 0.56 & -1.78 \\ -0.83 & -0.54 & 1.47 \end{bmatrix}, \qquad b = \begin{bmatrix} -1.61 \\ 7.23 \\ -3.38 \end{bmatrix}.$$

If we compute the vector Ax, using the exact solution $x = (1, 2, -1)^T$, we of course get

$$Ax = (-1.61, 7.23, -3.38)^T = b.$$

---

* Double precision is not required throughout the computations. When the system is solved through LU decomposition, the accumulation of inner products in double precision is sufficient.

But if we substitute a clearly erroneous vector

$$\bar{x} = (0.880, -2.34, -2.66)^T,$$

we get $A\bar{x} = (-1.6047, 7.2348, -3.3716)^T$, which is very close to b.

We define the residual of a solution vector as the difference between b and $A\bar{x}$, where $\bar{x}$ is the computed solution:

$$r = b - A\bar{x}.$$

Our example shows that the norm of r is not a good measure of the norm of the error vector ($e = x - \bar{x}$) for an ill-conditioned system.

Condition Numbers

Because the degree of ill-condition of the coefficient matrix is so important in determining the magnitude of round-off effects, it is valuable to have a quantitative measure. The *condition number* is normally defined as the product of two matrix norms:

$$\text{Condition}(A) = \|A\| \|A^{-1}\|.$$

Unfortunately, this is not an inexpensive quantity to compute, for it requires us to invert A. Because inverting a matrix amounts to solving a linear system (solving it with n different right-hand sides, actually), and the computed solution for an ill-conditioned system may be inexact, we will not compute A^{-1} very accurately. This suggests that the condition number will not be computed very exactly either. Ordinarily this causes no great difficulty; if the condition number is large, we know we are in serious trouble. Observe that condition numbers will always be at least unity, which corresponds to the condition number of the identity matrix.

For our previous example, we have

$$A = \begin{bmatrix} 3.02 & -1.05 & 2.53 \\ 4.33 & 0.56 & -1.78 \\ -0.83 & -0.54 & 1.47 \end{bmatrix}, \quad A^{-1} = \begin{bmatrix} 5.661 & -7.273 & -18.55 \\ 200.5 & -268.3 & -669.9 \\ 76.85 & -102.6 & -255.9 \end{bmatrix}.$$

Using matrix ∞-norms, we find that the condition number is

$$\|A\| \|A^{-1}\| = (6.67)(1138.7) = 7595.$$

The element of A^{-1} will be large relative to the elements of A when A is ill-conditioned. However, this can also be true when the elements of A are small, even in the absence of ill-conditioning. Multiplying the two norms has a normalizing effect, so the condition number is large only for an ill-conditioned system.

The condition number lets us relate the magnitude of the error in the computed solution to the magnitude of the residual. We use norms to express the magnitude of the vectors.

Let $e = x - \bar{x}$, where x is the exact solution to $Ax = b$ and $\bar{x}$ is an approximate solution. Let $r = b - A\bar{x}$, the residual. Since $Ax = b$, we have

$$r = b - A\bar{x} = Ax - A\bar{x} = A(x - \bar{x}) = Ae. \tag{2.33}$$

Hence,

$$e = A^{-1}r.$$

Taking norms and recalling Eq. (2.29), line 4, for a product, we write

$$\|e\| \leq \|A^{-1}\| \|r\|. \tag{2.34}$$

From $r = Ae$, we also have $\|r\| \leq \|A\| \|e\|$, which combines with Eq. (2.34) to give

$$\frac{\|r\|}{\|A\|} \leq \|e\| \leq \|A^{-1}\| \|r\|. \tag{2.35}$$

Applying the same reasoning to $Ax = b$ and $x = A^{-1}b$, we get

$$\frac{\|b\|}{\|A\|} \leq \|x\| \leq \|A^{-1}\| \|b\|. \tag{2.36}$$

Taking Eqs. (2.35) and (2.36) together, we reach a most important relationship:

$$\frac{1}{\|A\| \|A^{-1}\|} \frac{\|r\|}{\|b\|} \leq \frac{\|e\|}{\|x\|} \leq \|A\| \|A^{-1}\| \frac{\|r\|}{\|b\|},$$

or

$$\frac{1}{(\text{Condition no.})} \frac{\|r\|}{\|b\|} \leq \frac{\|e\|}{\|x\|} \leq (\text{Condition no.}) \frac{\|r\|}{\|b\|}. \tag{2.37}$$

Equation (2.37) shows that the relative error in the computed solution vector $\bar{x}$ can be as great as the relative residual multiplied by the condition number. Of course it can also be as small as the relative residual divided by the condition number. Therefore, when the condition number is large, the residual gives little information about the accuracy of $\bar{x}$. Conversely, when the condition number is near unity, the relative residual is a good measure of the relative error of $\bar{x}$.

When we solve a linear system, we are normally doing so to determine values for a physical system for which the set of equations is a model. We use the measured values of the parameters of the physical system to evaluate the coefficients of the equations, so we expect these coefficients to be known only as precisely as the measurements. When these are in error, the solution of the equations will reflect these errors. We have already seen that an ill-conditioned system is extremely sensitive to small changes in the coefficients. The condition number lets us relate the change in the solution vector to such errors in the coefficients of the set of equations $Ax = b$.

Assume that the errors in measuring the parameters cause errors in the coefficients of A so that the actual set of equations being solved is $(A + E)\bar{x} = b$, where $\bar{x}$ represents the solution of the perturbed system and A represents the true (but unknown) coefficients. We let $\bar{A} = A + E$ represent the perturbed coefficient matrix. We desire to know how large $x - \bar{x}$ is.

Using $Ax = b$ and $\bar{A}\bar{x} = b$, we can write

$$x = A^{-1}b = A^{-1}(\bar{A}\bar{x}) = A^{-1}(A + \bar{A} - A)\bar{x}$$
$$= [I + A^{-1}(\bar{A} - A)]\bar{x}$$
$$= \bar{x} + A^{-1}(\bar{A} - A)\bar{x}.$$

Since $\bar{A} - A = E$, we have

$$x - \bar{x} = A^{-1}E\bar{x}.$$

Taking norms, we get

$$\|x - \bar{x}\| \le \|A^{-1}\|\,\|E\|\,\|\bar{x}\| = \|A^{-1}\|\,\|A\|\frac{\|E\|}{\|A\|}\|\bar{x}\|,$$

so that

$$\frac{\|x - \bar{x}\|}{\|\bar{x}\|} \le (\text{Condition no.})\frac{\|E\|}{\|A\|}. \tag{2.38}$$

This says that the error of the solution relative to the norm of the computed solution can be as large as the relative error in the coefficients of A multiplied by the condition number. The net effect is that, if the coefficients of A are known to only four-digit precision and the condition number is 1000, the computed vector x may have only one digit of accuracy.

When the solution to the system $Ax = b$ has been computed, and, because of round-off error, we obtain the approximate solution vector $\bar{x}$, it is possible to apply iterative improvement to correct $\bar{x}$ so that it more closely agrees with x. Define $e = x - \bar{x}$. Define $r = b - A\bar{x}$. As already shown (Eq. (2.33)),

$$Ae = r. \tag{2.39}$$

If we could solve this equation for e, we could apply this as a correction to $\bar{x}$.

Furthermore, if $\|e\|/\|\bar{x}\|$ is small, it means that $\bar{x}$ should be close to x. In fact, if the value of $\|e\|/\|\bar{x}\|$ is 10^{-p}, we know that $\bar{x}$ is probably correct to p digits.

The process of iterative improvement is based on solving Eq. (2.39). Of course this is also subject to the same round-off error as the original solution of the system for $\bar{x}$, so we actually get $\bar{e}$, an approximation to the true error vector. Even so, unless the system is so ill-conditioned that $\bar{e}$ is not a reasonable approximation to e, we will get an improved estimate of x from $\bar{x} + \bar{e}$. One special caution is important to observe: The computation of the residual vector r must be as precise as possible. One always uses double-precision arithmetic; otherwise, iterative improvement will not be successful. An example will make this clear.

We are given

$$A = \begin{bmatrix} 4.23 & -1.06 & 2.11 \\ -2.53 & 6.77 & 0.98 \\ 1.85 & -2.11 & -2.32 \end{bmatrix}, \qquad b = \begin{bmatrix} 5.28 \\ 5.22 \\ -2.58 \end{bmatrix},$$

whose true solution is

$$x = \begin{bmatrix} 1.000 \\ 1.000 \\ 1.000 \end{bmatrix}.$$

If three-digit chopped arithmetic is used, the approximate solution vector is $\bar{x} = (0.991, 0.997, 1.000)^T$. Using double precision, we compute $A\bar{x}$ and the residual as

$$A\bar{x} = \begin{bmatrix} 5.24511 \\ 5.22246 \\ -2.59032 \end{bmatrix}, \qquad r = \begin{bmatrix} 0.0349 \\ -0.00246 \\ 0.0103 \end{bmatrix}.$$

We now solve $A\bar{e} = r$, again using three-digit precision, and get

$$\bar{e} = \begin{bmatrix} 0.00822 \\ 0.00300 \\ -0.00000757 \end{bmatrix}.$$

Finally, correcting $\bar{x}$ with $\bar{x} + \bar{e}$ gives almost exactly the correct solution:

$$\bar{x} + \bar{e} = \begin{bmatrix} 0.999 \\ 1.000 \\ 1.000 \end{bmatrix}.$$

In the general case, the iterations are repeated until the corrections are negligible. Since we want to make the solution of Eq. (2.39) as economical as possible, we should use an LU method to solve the original system and apply the LU to Eq. (2.39).

2.10 Iterative Methods

As opposed to the direct method of solving a set of linear equations by elimination, we now discuss iterative methods. In certain cases, these methods are preferred over the direct methods—when the coefficient matrix is sparse (has many zeros) they may be more rapid. They may be more economical in memory requirements of a computer. For hand computation they have the distinct advantage that they are self-correcting if an error is made; they may sometimes be used to reduce round-off error in the solutions computed by direct methods, as discussed earlier. They can also be applied to sets of nonlinear equations.

We illustrate one iterative method by a simple example:

$$
\begin{aligned}
6x_1 - 2x_2 + x_3 &= 11, \\
x_1 + 2x_2 - 5x_3 &= -1, \\
-2x_1 + 7x_2 + 2x_3 &= 5.
\end{aligned}
$$

(2.40)

The solution is $x_1 = 2, x_2 = 1, x_3 = 1$. We begin our iterative scheme by solving each equation for one of the variables, choosing, when possible, to solve for the variable with the largest coefficient:

$$
\begin{aligned}
x_1 &= 1.8333 + 0.3333x_2 - 0.1667x_3 && \text{(From line 1, Eq. 2.40)} \\
x_2 &= 0.7143 + 0.2857x_1 - 0.2857x_3 && \text{(From line 3, Eq. 2.40)} \\
x_3 &= 0.2000 + 0.2000x_1 + 0.4000x_2 && \text{(From line 2, Eq. 2.40)}
\end{aligned}
$$

We begin with some initial approximation to the value of the variables. (Each component might be taken equal to zero if no better initial estimates are at hand.) Substituting these approximations into the right-hand sides of the set of equations generates new approximations that, we hope, are closer to the true value. The new values are substituted in the right-hand sides to generate a second approximation, and the process is repeated until successive values of each of the variables are sufficiently alike. We indicate the iterative process on Eqs. (2.41), as follows, by putting superscripts on variables to indicate successive iterates. Thus our set of equations becomes

$$
\begin{aligned}
x_1^{(n+1)} &= 1.8333 + 0.3333x_2^{(n)} - 0.1667x_3^{(n)}, \\
x_2^{(n+1)} &= 0.7143 + 0.2857x_1^{(n)} - 0.2857x_3^{(n)}, \\
x_3^{(n+1)} &= 0.2000 + 0.2000x_1^{(n)} + 0.4000x_2^{(n)}.
\end{aligned}
$$

(2.41)

Starting with an initial vector of: $x^{(0)} = (0, 0, 0)$, we get:

Successive estimates of solution (Jacobi method)

	First	Second	Third	Fourth	Fifth	Sixth	$\cdots$	Ninth
x_1	0	1.833	2.038	2.085	2.004	1.994	$\cdots$	2.000
x_2	0	0.714	1.181	1.053	1.001	0.990	$\cdots$	1.000
x_3	0	0.200	0.852	1.080	1.038	1.001	$\cdots$	1.000

Note that this method is exactly the same as the method of fixed-point iteration for a single equation that was discussed in Chapter 1, but it is now applied to a set of equations; we see this if we write Eq. (2.41) in the form of

$$x^{(n+1)} = G(x^{(n)}) = b' - Bx^{(n)}, \tag{2.42}$$

which is identical to $x_{n+1} = g(x_n)$ as used in Chapter 1.

In the present context, of course, $x^{(n)}$ and $x^{(n+1)}$ refer to the nth and $(n+1)$st iterates of a vector rather than a simple variable, and g is a linear transformation rather than a nonlinear function. For the preceding example, we restate Eq. (2.40) in matrix form after interchanging equation lines 2 and 3:

$$Ax = b, \qquad \begin{bmatrix} 6 & -2 & 1 \\ -2 & 7 & 2 \\ 1 & 2 & -5 \end{bmatrix} \begin{bmatrix} x_1 \\ x_2 \\ x_3 \end{bmatrix} = \begin{bmatrix} 11 \\ 5 \\ -1 \end{bmatrix}. \tag{2.43}$$

Now, let $A = L + D + U$, where

$$L = \begin{bmatrix} 0 & 0 & 0 \\ -2 & 0 & 0 \\ 1 & 2 & 0 \end{bmatrix}, \qquad D = \begin{bmatrix} 6 & 0 & 0 \\ 0 & 7 & 0 \\ 0 & 0 & -5 \end{bmatrix}, \qquad U = \begin{bmatrix} 0 & -2 & 1 \\ 0 & 0 & 2 \\ 0 & 0 & 0 \end{bmatrix}.$$

Then Eq. (2.43) can be rewritten as

$$Ax = (L + D + U)x = b, \quad \text{or}$$
$$Dx = -(L + U)x + b, \quad \text{which gives}$$
$$x = -D^{-1}(L + U)x + D^{-1}b.$$

From this we have, identifying x on the left as the new iterate,

$$x^{(n+1)} = -D^{-1}(L + U)x^{(n)} + D^{-1}b. \tag{2.44}$$

In Eqs. (2.41), we see that

$$b' = D^{-1}b = \begin{bmatrix} 1.8333 \\ 0.7143 \\ 0.2000 \end{bmatrix},$$

$$B = D^{-1}(L + U) = \begin{bmatrix} 0 & -0.3333 & 0.1667 \\ -0.2857 & 0 & 0.2857 \\ -0.2000 & -0.4000 & 0 \end{bmatrix}.$$

The procedure we have just described is known as the *Jacobi method*, also called "the method of simultaneous displacements" because each of the equations is simultaneously changed by using the most recent set of x-values.

We can write the algorithm for the Jacobi iterative method as follows:

Algorithm for Jacobi Iteration

We assume that the system $Ax = b$ has been rearranged so that for each row of A the diagonal elements have magnitudes that are greater than the sum of the remaining elements in the corresponding row. That is,

$$|a_{i,i}| > \sum_{\substack{j=1 \\ j \neq i}}^{n} |a_{i,j}|, \qquad i = 1, 2, \ldots, n.$$

This is a sufficient condition for both this method and the one that follows to converge.

We begin with an initial approximation to the solution vector, which we store in the vector: old_x.

FOR $i = 1$ TO n
  $b[i] = b[i]/a[i, i]$
  $new\_x[i] = old\_x[i]$
  $a[i, j] = a[i, j]/a[i, i]; j = 1 \ldots n$ and $i <> j$ END FOR i;

REPEAT
  FOR $i = 1$ TO n
  $old\_x[i] = new\_x[i]$
  $new\_x[i] = b[i]$ END FOR i;

  FOR $i = 1$ TO n
    FOR $j = 1$ TO n
      IF $(j <> i)$ THEN
        $new\_x[i] = new\_x[i] - a[i, j] * old\_x[j]$

UNTIL new_x and old_x converge to each other;

Actually, the x-values of the next trial (new_x) are not used, even in part, until we have first found all its components. For instance, in the Jacobi method, even though we have computed the $new\_x[1]$, we still do not use this value in computing $new\_x[2]$, even though in nearly all cases the new values are better than the old, and should be used in preference to the poorer values. When this is done, the procedure is known as the *Gauss–Seidel method*. In this method our first step is to rearrange the set of equations by solving each equation for one of the variables in terms of the others, exactly as we have done in the Jacobi method. We then proceed to improve each x-value in turn, always using the most recent approximations to the values of the other variables. The rate of convergence is more rapid, as shown by reworking our earlier example (Eqs. 2.41, rearranged form of Eqs. 2.40):

Successive estimates of solution (Gauss–Seidel method)

	First	Second	Third	Fourth	Fifth	Sixth
x_1	0	1.833	2.069	1.998	1.999	2.000
x_2	0	1.238	1.002	0.995	1.000	1.000
x_3	0	1.062	1.015	0.998	1.000	1.000

These values were computed by using this iterative scheme:

$$x_1^{(n+1)} = 1.8333 + 0.3333x_2^{(n)} \quad - 0.1667x_3^{(n)},$$

$$x_2^{(n+1)} = 0.7143 + 0.2857x_1^{(n+1)} - 0.2857x_3^{(n)},$$

$$x_3^{(n+1)} = 0.2000 + 0.2000x_1^{(n+1)} - 0.4000x_2^{(n+1)},$$

beginning with $x^{(1)} = (0, 0, 0)^T$.

The algorithm for the Gauss–Seidel iteration is as follows:

Algorithm for Gauss–Seidel Iteration

We assume, as we did in the previous algorithm, that the system $Ax = b$ has been rearranged so that for each row of A the diagonal elements have magnitudes that are greater than the sum of the remaining elements in the corresponding row. That is,

$$|a_{i,i}| > \sum_{\substack{j=1 \\ j \neq i}}^{n} |a_{i,j}|, \qquad i = 1, 2, \ldots, n.$$

This is a sufficient condition for both this method and the previous one. As before we begin with an initial approximation to the solution vector, which we store in the vector, x.

```
FOR i = 1 TO n
   b[i] = b[i]/a[i, i];
   a[i, j] = a[i, j]/a[i, i]; j = 1 . . . n and i <> j   END FOR i;

WHILE Not (yet convergent)   DO

   FOR i = 1 TO n
      x[i] = b[i];
      FOR j = 1 TO n
         IF (j <> i) THEN
            x[i] = x[i] − a[i, j] * x[j] END; (* i, j *)
   END;   (* While *)
```

The Gauss–Seidel method will generally converge if the Jacobi method converges, and will do so more rapidly. However, we might still prefer the Jacobi method if we were running our program on parallel processors, since all n equations could be solved simultaneously at each iteration. In fact, other algorithms once thought "inefficient" may now become the algorithms of choice due to parallel processing.

The matrix formulation for the Gauss–Seidel method is almost like the one given in Eq. (2.44). Here $Ax = b$ can be rewritten as

$$(L + D)x = -Ux + b. \tag{2.45}$$

From this we have

$$x^{(n+1)} = -(L + D)^{-1}Ux^{(n)} + (L + D)^{-1}b. \tag{2.46}$$

The usefulness of the matrix notation will be more apparent in Chapter 6, when we study the eigenvalues of the matrices $D^{-1}(L + U)$ in Eq. (2.44) and $(L + D)^{-1}U$ in Eq. (2.46). The eigenvalues of these respective matrices affect the rates of convergence of both methods.

These iteration methods will not converge for all sets of equations, nor for all possible rearrangements of the equations. When the equations can be ordered so that each diagonal entry is larger in magnitude than the sum of the magnitudes of the other coefficients in that row (such a system is called *diagonally dominant*), the iteration will converge for any starting values. This is easy to visualize, because all the equations can be put in the form

$$x_i = \frac{b_i}{a_{ii}} - \frac{a_{i1}}{a_{ii}}x_1 - \frac{a_{i2}}{a_{ii}}x_2 - \cdots \tag{2.47}$$

The error in the next value of x_i will be the sum of the errors in all other x's multiplied by the coefficients of Eq. (2.47), and if the sum of the magnitudes of the coefficients is less than unity, the error will decrease as the iteration proceeds. The preceding convergence condition is a sufficient condition only; that is, if the condition holds, the system always converges, but sometimes the system converges even if the condition is violated.

The speed with which the iterations converge is obviously related to the degree of dominance of the diagonal terms, for the coefficients in Eq. (2.47) are then smaller and x_i is less affected by the errors in the other components. When the initial approximation is close to the solution vector, relatively few iterations are needed to get an acceptable solution.

2.11   The Relaxation Method

There is an iteration method that is more rapidly convergent than Gauss–Seidel and that can be used to advantage for hand calculations. Unfortunately, it is not well adapted to computer application. The method is due to a British engineer, Richard Southwell, and has been applied to a wide variety of problems. (Allen, 1954, is an excellent reference.) We discuss the method because of its historical importance and because it leads to an important acceleration technique called *overrelaxation*.

If we consider the Gauss–Seidel scheme, we realize that the order in which the equations are used is important. We should improve the x that is most in error, since, in the rearranged form, that variable does not appear on the right, and hence its own error will not affect the next iterate. By using that equation, then, we introduce lesser errors into the computation of the next iterate. The *method of relaxation* is a scheme that permits one to select the best equation to be used for maximum rate of convergence.

We illustrate the method by the following example. The original equations are

$$
\begin{aligned}
8x_1 + x_2 - x_3 &= 8, \\
2x_1 + x_2 + 9x_3 &= 12, \\
x_1 - 7x_2 + 2x_3 &= -4.
\end{aligned}
\tag{2.48}
$$

We again begin by a rearrangement of the equations, but different from that for the Gauss–Seidel or Jacobi methods. We transpose all the terms to one side, and then divide each equation by the negative of its largest coefficient. Equations (2.48) become

$$
\begin{aligned}
-x_1 - 0.125x_2 + 0.125x_3 + 1 &= 0, \\
-0.222x_1 - 0.111x_2 - x_3 + 1.333 &= 0, \\
0.143x_1 - x_2 + 0.286x_3 + 0.571 &= 0.
\end{aligned}
\tag{2.49}
$$

If we begin with some initial set of values and substitute in Eqs. (2.49), the equations will not be satisfied (unless, by chance, we have stumbled onto the solution); the left sides will not be zero, but some other value that we call the *residual* and denote by R_i. It is also convenient to reorder the equation so the -1 coefficients are on the diagonal. Equations (2.49) become, with these rearrangements,

$$
\begin{aligned}
-x_1 - 0.125x_2 + 0.125x_3 + 1 &= R_1, \\
0.143x_1 - x_2 + 0.286x_3 + 0.571 &= R_2, \\
-0.222x_1 - 0.111x_2 - x_3 + 1.333 &= R_3.
\end{aligned}
$$

For example, with $x_1 = 0, x_2 = 0, x_3 = 0$, we have

$$
R_1 = 1, \qquad R_2 = 0.571, \qquad R_3 = 1.333.
$$

The largest residual in magnitude, R_3, tells us that the third equation is most in error and should be improved first. The method gets its name "relaxation" from the fact that we make a change in x_3 to relax R_3 (the greatest residual) so as to make it zero. Observing the coefficients of the various equations, we see that increasing the value of x_3 by one, say, will decrease R_3 by one, will increase R_1 by 0.125, and increase R_2 by 0.286. To change R_3 from its initial value of 1.333 to zero, we should increase x_3 by that same amount.

We then select the new residual of greatest magnitude, and relax it to zero. We continue until all residuals are zero, and when this is true, the values of the x's will be at the exact solution. In implementing this method, there are some modifications that make the work easier. We illustrate in Fig. 2.2.

Eq. No.	x_1	x_2	x_3
1	-1	-0.125	0.125
2	0.143	-1	0.286
3	-0.222	-0.111	-1

x_1	R_1	x_2	R_2	x_3	R_3
0	~~1000~~	0	~~571~~	0	~~1333~~
	+167		+381		
	~~1167~~		~~952~~	+1333	~~0~~
					−259
+1167	~~0~~		~~1119~~	~~−259~~	
	−140		+167		
	~~−140~~	+1119	~~0~~		~~−383~~
					−124
	− 48		− 109		
	~~−189~~		~~−109~~	−383	~~0~~
−189			− 27		+ 42
	~~0~~		~~−136~~		~~42~~
	+ 17	−136			+ 15
	~~17~~		~~0~~		~~57~~
	+ 7		+ 16		
	~~24~~		~~16~~	+57	~~0~~
+24		+19	+ 3		− 5
	~~0~~		~~19~~		~~5~~
	− 2		~~0~~		− 2
	~~−2~~	+19	− 2		~~7~~
	− 1		~~−2~~	−7	~~0~~
	~~−3~~		0		+ 1
−3	~~0~~		~~−2~~		~~1~~
	0		~~0~~		0
	~~0~~	−2	~~0~~	+1	~~1~~
	0		0		0
999		1000		1001	

Check residuals: 1 -1 0

Figure 2.2 Solving a set of linear equations by relaxation

We make three double columns, one for each variable and for the residual of the equation in which that variable appears with -1 coefficient. The initial x values and the initial residuals are entered as the first row of the table. It is convenient to work entirely with integers by multiplying the initial x-values and residuals by 1000, and then to scale down the solution by dividing by 1000 at the end of the computations. We avoid fractions; if a fractional change in a variable is needed to relax to zero, we only relax to near zero.

In Fig. 2.2, we set down the increments to the x's but record the cumulative effect on the residuals. (The old values of the residuals are crossed out when replaced by a new value.) When the residuals are zero, we add the various increments to the initial value to get the final value. In this example, round-off errors cause an error of one in the third decimal.

It is important to make a final check by recomputing residuals at the end of the calculation to check for mistakes in arithmetic. The method is not usually programmed because searching on the computer for the largest residual is slow, adding enough execution time that the acceleration gives no net benefit. The search can be done rapidly by scanning the residuals in a hand calculation, however.

Southwell and his co-workers observed, for many situations, that relaxing the residuals to zero was less efficient than relaxing beyond zero (*overrelaxing*) or relaxing short of zero (*underrelaxing*). The reason this strategy is an improved one is that a zero residual doesn't stay zero; relaxing the residual of another equation affects the first residual, so it is appropriate to anticipate and allow for this by an appropriate under- or overrelaxation.

Table 2.1 shows that a significant improvement in the speed of convergence is obtained if R_1 is underrelaxed by 10% and R_3 is underrelaxed by 25%. Unfortunately, the optimum degree of under- or overrelaxation is not easily determined. In many problems, acceleration is obtained by overrelaxing rather than underrelaxing.

Table 2.1 Accelerated solution of linear equations by relaxation

Eq. No.	x_1	x_2	x_3
1	-1	-0.125	0.125
2	0.143	-1	0.286
3	-0.222	-0.111	-1

x_1	R_1	x_2	R_2	x_3	R_3
0	~~1000~~	0	~~571~~	0	~~1333~~
	+125		+286		
	~~1125~~		~~857~~	+1000	~~333~~
			+144		−225
+1013	~~112~~		~~1001~~		~~108~~
	−125				−111
	~~=13~~	+1001	~~0~~		~~=3~~
			−2		+3
−12	~~=1~~		~~=2~~		~~0~~
	+0				+0
	~~=1~~	−2	~~0~~		~~0~~
			+0		+0
−1	0		0		0
1000		999		1000	

Even though Southwell's relaxation method is not often used today, there is one aspect of it that has influence on the iterative solution of linear equations by computer. In using the Gauss–Seidel method, we can speed up the convergence by "overrelaxation," that is, by making the residuals go to the other side of zero instead of just relaxing to zero as in the first example. We can apply this technique to Gauss–Seidel iteration by modifying the algorithm.

The standard relationship for Gauss–Seidel iteration for the set of equations $Ax = b$, for variable x_i, can be written

$$x_i^{(k+1)} = \frac{1}{a_{ii}}\left(b_i - \sum_{j=1}^{i-1} a_{ij} x_j^{(k+1)} - \sum_{j=i+1}^{n} a_{ij} x_j^{(k)} \right), \qquad (2.50)$$

where the superscript $(k + 1)$ indicates that this is the $(k + 1)$st iterate. On the right side we use the most recent estimates of the x_j, which will be either $x_j^{(k)}$ or $x_j^{(k+1)}$.

An algebraically equivalent form for Eq. (2.50) is

$$x_i^{(k+1)} = x_i^{(k)} + \frac{1}{a_{ii}}\left(b_i - \sum_{j=1}^{i-1} a_{ij} x_j^{(k+1)} - \sum_{j=i}^{n} a_{ij} x_j^{(k)} \right),$$

because $x_i^{(k)}$ is both added to and subtracted from the right side. In this form, we see that Gauss–Seidel and Southwell's relaxation can have identical arithmetic: The term we add to $x_i^{(k)}$ to get $x_i^{(k+1)}$ is exactly the increment that relaxes the residual to zero. (Of course, we apply the relaxation to the x_i's in a different sequence in the two methods.) Overrelaxation can be applied to Gauss–Seidel if we will add to $x_i^{(k)}$ some multiple of the second term. It can be shown that this multiple should never be more than 2 in magnitude (to avoid divergence), and the optimum overrelaxation factor lies between 1.0 and 2.0. Our iteration equations take this form, where w is the *overrelaxation factor*:

$$x_i^{(k+1)} = x_i^{(k)} + \frac{w}{a_{ii}}\left(b_i - \sum_{j=1}^{i-1} a_{ij} x_j^{(k+1)} - \sum_{j=i}^{n} a_{ij} x_j^{(k)} \right). \qquad (2.51)$$

Table 2.2 shows how the convergence rate is influenced by the value of w for the system

$$\begin{bmatrix} -4 & 1 & 1 & 1 \\ 1 & -4 & 1 & 1 \\ 1 & 1 & -4 & 1 \\ 1 & 1 & 1 & -4 \end{bmatrix} x = \begin{bmatrix} 1 \\ 1 \\ 1 \\ 1 \end{bmatrix},$$

starting with an initial estimate of $x = 0$. The exact solution is

$$x_1 = -1, \qquad x_2 = -1, \qquad x_3 = -1, \qquad x_4 = -1.$$

Table 2.2   Acceleration of convergence of
Gauss–Seidel iteration

w, the overrelaxation factor	Number of iterations to reach error $< 1 \times 10^{-5}$
1.0	24
1.1	18
1.2	13
1.3	11 ←Minimum
1.4	14 of iterations
1.5	18
1.6	24
1.7	35
1.8	55
1.9	100+

We see that the optimum value for the overrelaxation factor is about $w = 1.3$ for this example. The optimum value will vary between 1.0 and 2.0, depending on the size of the coefficient matrix and the values of the coefficients. Overrelaxation is considered further in Chapter 7 in connection with methods to solve partial-differential equations. We also discuss there the question of finding w_{opt}.

2.12 Systems of Nonlinear Equations

As mentioned previously, the problem of finding the solution of a set of nonlinear equations is much more difficult than for linear equations. (In fact, some sets have no real solutions.) Consider the example of a pair of nonlinear equations:

$$x^2 + y^2 = 4,$$
$$e^x + y = 1. \tag{2.52}$$

Graphically, the solution to this system is represented by the intersections of the circle $x^2 + y^2 = 4$ with the curve $y = 1 - e^x$. Figure 2.3 shows that these are near $(-1.8, 0.8)$ and $(1, -1.7)$.

Newton's method can be applied to systems as well as to a single nonlinear equation. We begin with the forms

$$f(x, y) = 0,$$
$$g(x, y) = 0.$$

Let $x = r$, $y = s$ be a root, and expand both functions as a Taylor series about the point (x_i, y_i) in terms of $(r - x_i)$, $(s - y_i)$, where (x_i, y_i) is a point near the root:

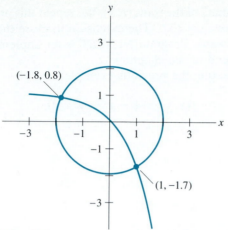

Figure 2.3

$$f(r,s) = 0 = f(x_i, y_i) + f_x(x_i, y_i)(r - x_i)$$
$$+ f_y(x_i, y_i)(s - y_i) + \cdots$$
$$g(r,s) = 0 = g(x_i, y_i) + g_x(x_i, y_i)(r - x_i)$$
$$+ g_y(x_i, y_i)(s - y_i) + \cdots$$

(2.53)

Truncating the series gives

$$\begin{bmatrix} 0 \\ 0 \end{bmatrix} = \begin{bmatrix} f(x_i, y_i) \\ g(x_i, y_i) \end{bmatrix} + \begin{bmatrix} f_x(x_i, y_i) & f_y(x_i, y_i) \\ g_x(x_i, y_i) & g_y(x_i, y_i) \end{bmatrix} \begin{bmatrix} r - x_i \\ s - y_i \end{bmatrix}.$$

(2.54)

We rewrite this to solve as the system of equations

$$\begin{bmatrix} f_x(x_i, y_i) & f_y(x_i, y_i) \\ g_x(x_i, y_i) & g_y(x_i, y_i) \end{bmatrix} \begin{bmatrix} \Delta x_i \\ \Delta y_i \end{bmatrix} = - \begin{bmatrix} f(x_i, y_i) \\ g(x_i, y_i) \end{bmatrix},$$

(2.55)

where $\Delta x_i = r - x_i$, and $\Delta y_i = s - y_i$.

We solve (2.55) by Gaussian elimination and then, if we set

$$\begin{bmatrix} x_{i+1} \\ y_{i+1} \end{bmatrix} = \begin{bmatrix} x_i \\ y_i \end{bmatrix} + \begin{bmatrix} \Delta x_i \\ \Delta y_i \end{bmatrix},$$

(2.56)

we get an improved estimate of the root, (r, s). We repeat this process with i replaced by $i + 1$ until f and g are close to 0. The extension to more than two simultaneous equations is straightforward. Program NLSYST is an implementation (see Computer Programs at the end of the chapter).

We illustrate by repeating the previous example:

$$f(x, y) = 4 - x^2 - y^2 = 0,$$
$$g(x, y) = 1 - e^x - y = 0.$$

The partials are

$$f_x = -2x, \qquad f_y = -2y,$$
$$g_x = -e^x, \qquad g_y = -1.$$

Beginning at $x_0 = 1, y_0 = -1.7$, we solve the equations

$$\begin{bmatrix} -2 & 3.4 \\ -2.7183 & -1.0 \end{bmatrix} \begin{bmatrix} \Delta x_0 \\ \Delta y_0 \end{bmatrix} = -\begin{bmatrix} 0.1100 \\ -0.0183 \end{bmatrix}, \qquad (2.57)$$

to get $\Delta x_0 = 0.0043, \Delta y_0 = -0.0298$. This then gives us $x_1 = 1.0043, y_1 = -1.7298$. The results already agree with the true value of the root within two in the fourth decimal place. Repeating the process once more produces $x_2 = 1.004169$, $y_2 = -1.729637$. The function values at (x_2, y_2) are approximately -0.0000001 and -0.00000001, respectively.

We can write Newton's method for a system of n equations by expanding Eq. (2.55). Thus we have

$$\begin{bmatrix} f_{1x} & f_{1y} & f_{1z} & \cdots \\ f_{2x} & f_{2y} & f_{2z} & \cdots \\ f_{3x} & f_{3y} & f_{3z} & \cdots \\ \vdots & \vdots & \vdots \\ f_{nx} & f_{ny} & f_{nz} & \cdots \end{bmatrix} \begin{bmatrix} \Delta x_i \\ \Delta y_i \\ \Delta z_i \\ \vdots \end{bmatrix} = -\begin{bmatrix} f_1 \\ f_2 \\ f_3 \\ \vdots \\ f_n \end{bmatrix} \qquad (2.58)$$

evaluated at $(x_i, y_i, z_i, \ldots)$. Solving this, we compute

$$x_{i+1} = x_i + \Delta x_i, \qquad y_{i+1} = y_i + \Delta y_i, \qquad z_{i+1} = z_i + \Delta z_i, \ldots \qquad (2.59)$$

In a computer program, it is awkward to introduce each of the partial-derivative functions (which often must be developed by hand, unless one has access to a program like *Mathematica* or DERIVE or MAPLE) to use in Eqs. (2.58). An alternative technique is to approximate these partials by recalculating the function with a small perturbation to each of the variables in turn:

$$(f_1)_x \doteq \frac{f_1(x + \delta, y, z, \ldots) - f_1(x, y, z, \ldots)}{\delta},$$

$$(f_1)_y \doteq \frac{f_1(x, y + \delta, z, \ldots) - f_1(x, y, z, \ldots)}{\delta},$$

$$\vdots$$

$$(f_i)_{x_j} \doteq \frac{f_i(x, y, z, \ldots, x_j + \delta, \ldots) - f_i(x, y, z, \ldots, x_j, \ldots)}{\delta}.$$

Similar relations are used for each variable in each function.* A computer program at the end of this chapter exploits this idea.

It is interesting to observe that Newton's method, as applied to a set of nonlinear equations, reduces the problem to solving a set of *linear equations* in order to determine the values that improve the accuracy of the estimates. This points out quite dramatically how linear and nonlinear problems vary in difficulty.

Newton's method has the advantage of converging quadratically, at least when we are near a root, but it is expensive in terms of function evaluations. For the preceding 2×2 system, there are six function evaluations at each step, while for a 3×3 system, there are twelve. For n simultaneous equations, the number of function evaluations is $n^2 + n$. We can see why this is rarely applied to large systems. It is good strategy, in all cases of simultaneous nonlinear equations, to reduce the number of equations as much as possible by solving for one variable in terms of the others and eliminating that one by substituting for it in the other equations. For example, we could attack the previous example as follows:

$$\begin{cases} 4 - x^2 - y^2 = 0 \\ 1 - e^x - y = 0 \end{cases} \quad \text{Solve for } y: y = 1 - e^x.$$

Substituting in the first equation, we get

$$4 - x^2 - (1 - e^x)^2 = 0,$$
$$3 - x^2 + 2e^x - e^{2x} = 0.$$

We then use the methods of Chapter 1.

When we must solve a larger system of nonlinear equations, a modification of Newton's method is often used. It converges less than quadratically but usually faster than linearly. Unfortunately, it may *diverge* unless we start fairly near to a root. In this method we do not recompute the matrix of partials at each step. Rather, we use the same matrix for several steps before recomputing it again. For a system of n equations we would recompute the matrix after every n steps. In this way we need

---

* Approximation of derivatives by such difference quotients is discussed in Chapter 4. If the limiting value (as $\delta \rightarrow 0$) of this ratio were used, we would have exactly the definition of a derivative. Since limits are not used, we have approximate values, as indicated by the dots over the equal signs.

only n function evaluations at each step, except when we occasionally have to update the matrix of partials, which then adds an additional n^2 function evaluations.

We first illustrate the method on our earlier 2×2 system. Reworking Eq. (2.57) produces

$$\begin{bmatrix} \Delta x_0 \\ \Delta y_0 \end{bmatrix} = -\begin{bmatrix} -2 & 3.4 \\ -2.7183 & -1.0 \end{bmatrix}^{-1} \begin{bmatrix} 0.110 \\ -0.0183 \end{bmatrix} = \begin{bmatrix} 0.0043 \\ -0.0298 \end{bmatrix};$$

$$\begin{bmatrix} x_1 \\ y_1 \end{bmatrix} = \begin{bmatrix} 1.0 \\ -1.7 \end{bmatrix} + \begin{bmatrix} 0.0043 \\ -0.0298 \end{bmatrix} = \begin{bmatrix} 1.0043 \\ -1.7298 \end{bmatrix};$$

$$\begin{bmatrix} \Delta x_1 \\ \Delta y_1 \end{bmatrix} = -\begin{bmatrix} -2 & 3.4 \\ -2.7183 & -1.0 \end{bmatrix}^{-1} \begin{bmatrix} 0.000827 \\ 0.000196 \end{bmatrix} = \begin{bmatrix} -0.000133 \\ 0.000165 \end{bmatrix};$$

$$x_2 = 1.004167, \qquad y_2 = -1.729635;$$

$$f(x_2, y_2) = -0.000011, \qquad g(x_2, y_2) = -0.000002.$$

In addition, we consider a different example. Here we have

$$e^x - y = 0,$$
$$xy - e^x = 0.$$

Eq. (2.55) for this system becomes

$$\begin{bmatrix} e^{x_i} & -1 \\ y_i - e^{x_i} & -x_i \end{bmatrix} \begin{bmatrix} \Delta x_i \\ \Delta y_i \end{bmatrix} = -\begin{bmatrix} f(x_i, y_i) \\ g(x_i, y_i) \end{bmatrix}.$$

Let $x_0 = 0.95, y_0 = 2.7$. Then solving the preceding with the inverse matrix

$$\begin{bmatrix} \Delta x_0 \\ \Delta y_0 \end{bmatrix} = -\begin{bmatrix} 2.5857 & -1 \\ 0.1143 & 0.95 \end{bmatrix}^{-1} \begin{bmatrix} -0.1143 \\ -0.0207 \end{bmatrix},$$

we get

$$x_1 = 1.00029, \qquad y_1 = 2.71575;$$

$$\begin{bmatrix} \Delta x_1 \\ \Delta y_1 \end{bmatrix} = -\begin{bmatrix} 2.5857 & -1 \\ 0.1143 & 0.95 \end{bmatrix}^{-1} \begin{bmatrix} 0.003325 \\ -0.002533 \end{bmatrix}, \qquad \text{we get}$$

$$x_2 = 1.000048, \qquad y_2 = 2.718445.$$

Since $n = 2$, we update our matrix of partials so that

$$\begin{bmatrix} \Delta x_2 \\ \Delta y_2 \end{bmatrix} = -\begin{bmatrix} 2.7184 & -1 \\ 0.00003 & 1.0000 \end{bmatrix}^{-1} \begin{bmatrix} -0.000033 \\ -0.000163 \end{bmatrix}, \qquad \text{we get}$$

$$x_3 = 1.000000, \qquad y_3 = 2.718282.$$

This agrees with the exact solution: $(1, e)$.

2.13 Theoretical Matters

In this section, we shall look a little deeper at the algorithm for Gaussian elimination of Section 2.4 for solving a system of n equations in n unknowns. We shall show that the algorithm is just a series of simple matrix multiplications that are important on their own. We will also show that there is a very big difference between a nearly singular matrix and an ill-conditioned matrix.

Interchanging Rows of a Matrix

Interchanging rows of a matrix can be done by multiplying by a *permutation matrix*. A permutation matrix, P, is an $n \times n$ matrix derived from the identity matrix, except that exactly two rows, i and j, are interchanged. If we want to indicate the particular rows, we might write $P(i \leftrightarrow j)$, but this task would be cumbersome. The idea becomes clear if we examine this concept for a 3×3 matrix. Define

$$P = \begin{bmatrix} 1 & 0 & 0 \\ 0 & 0 & 1 \\ 0 & 1 & 0 \end{bmatrix}, \qquad B = \begin{bmatrix} a & b & c \\ d & e & f \\ g & h & i \end{bmatrix}.$$

Observe that multiplying P times B interchanges rows 2 and 3 of B:

$$P * B = \begin{bmatrix} a & b & c \\ g & h & i \\ d & e & f \end{bmatrix}.$$

Conversely, if we change the order of multiplication, we get

$$B * P = \begin{bmatrix} a & c & b \\ d & f & e \\ g & i & h \end{bmatrix},$$

where now column 2 and column 3 of B have been interchanged. The identity matrix can be considered a permutation matrix that leaves all the rows and columns alone. Moreover, we see that for all permutation matrices,

$$P^{-1} = P.$$

Zeroing Elements Below the Diagonal

In Gaussian elimination, when we zeroed out the elements below the pivot element, we were actually doing the following:

$$H_1 B = \begin{bmatrix} 1 & 0 & 0 \\ -\frac{d}{a} & 1 & 0 \\ -\frac{g}{a} & 0 & 1 \end{bmatrix} * B = \begin{bmatrix} a & b & c \\ 0 & u & v \\ 0 & x & y \end{bmatrix}.$$

Similarly, if we set

$$H_2 = \begin{bmatrix} 1 & 0 & 0 \\ 0 & 1 & 0 \\ 0 & -\frac{x}{u} & 1 \end{bmatrix},$$

then it turns out that

$$H_2(H_1 B) = \begin{bmatrix} a & b & c \\ 0 & u & v \\ 0 & 0 & w \end{bmatrix}.$$

The matrices H_i, $i = 1, 2$, are referred to as simple *Householder transformations* on the ith column of B. The inverses to these transformations are also very simple. For instance, in our prior example,

$$H_1^{-1} = \begin{bmatrix} 1 & 0 & 0 \\ \frac{d}{a} & 1 & 0 \\ \frac{g}{a} & 0 & 1 \end{bmatrix}.$$

We shall now apply our discussion to the system of equations in Section 2.4, $Ax = b$ (Eq. 2.9), where

$$A = \begin{bmatrix} 0 & 2 & 0 & 1 \\ 2 & 2 & 3 & 2 \\ 4 & -3 & 0 & 1 \\ 6 & 1 & -6 & -5 \end{bmatrix}, \qquad b = \begin{bmatrix} 0 \\ -2 \\ -7 \\ 6 \end{bmatrix}.$$

Letting P_1 be the permutation that interchanges rows 1 and 4 and letting

$$H_1 = \begin{bmatrix} 1 & 0 & 0 & 0 \\ -0.3333 & 1 & 0 & 0 \\ -0.6667 & 0 & 1 & 0 \\ 0 & 0 & 0 & 1 \end{bmatrix},$$

we get

$$(H_1 P_1)A = \begin{bmatrix} 6 & 1 & -6 & -5 \\ 0 & 1.6667 & 5 & 3.6667 \\ 0 & -3.6667 & 4 & 4.3333 \\ 0 & 2 & 0 & 1 \end{bmatrix}, \tag{2.60}$$

which is found in Eq. (2.11a).

Now letting P_2 be the permutation that interchanges rows 2 and 3 of Eq. (2.60), and

$$H_2 = \begin{bmatrix} 1 & 0 & 0 & 0 \\ 0 & 1 & 0 & 0 \\ 0 & 0.4545 & 1 & 0 \\ 0 & 0.5455 & 0 & 1 \end{bmatrix},$$

then it is easy to see that

$$H_2 P_2(H_1 P_1 A) = \begin{bmatrix} 6 & 1 & -6 & -5 \\ 0 & -3.6667 & 4 & 4.3333 \\ 0 & 0 & 6.8182 & 5.6364 \\ 0 & 0 & 2.1818 & 3.3636 \end{bmatrix},$$

which is Eq. (2.13). Finally, if we let

$$H_3 = \begin{bmatrix} 1 & 0 & 0 & 0 \\ 0 & 1 & 0 & 0 \\ 0 & 0 & 1 & 0 \\ 0 & 0 & -0.3200 & 1 \end{bmatrix}$$

and P_3 be the identity matrix, since there are no row interchanges, we finally get

$$H_3 P_3(H_2 P_2 H_1 P_1 A) = \begin{bmatrix} 6 & 1 & -6 & -5 \\ 0 & -3.6667 & 4 & 4.3333 \\ 0 & 0 & 6.8182 & 5.6364 \\ 0 & 0 & 0 & 1.5600 \end{bmatrix}, \qquad \textbf{(2.61)}$$

which is the coefficient matrix part of Eq. (2.14).

To summarize: We wish to solve the system: $Ax = b$.
 Using a sequence of elementary matrices, we have

$$(H_3 P_3 H_2 P_2 H_1 P_1)A = (H_3 P_3 H_2 P_2 H_1 P_1)b.$$

It should be noted that the right-hand side of this equation corresponds to the last column in the augmented matrix of Eq. (2.14). It can be shown for this example—and it is true even in the general case—that we can rearrange the order of the simple transformations to get

$$H_3 P_3 H_2 P_2 H_1 P_1 A = H_3 H_2 H_1 P_3 P_2 P_1 A = U, \qquad \textbf{(2.62)}$$

where U is the upper-triangular matrix of Eq. (2.61). Then by multiplying both sides of the equation by the appropriate inverses, we get

$$A' = P_3 P_2 P_1 A = H_1^{-1} H_2^{-1} H_3^{-1} U = LU \qquad \textbf{(2.63)}$$

which for this 4×4 example is

$$\begin{bmatrix} 1 & 0 & 0 & 0 \\ 0.6667 & 1 & 0 & 0 \\ 0.3333 & -0.4545 & 1 & 0 \\ 0 & -0.5454 & 0.32 & 1 \end{bmatrix} * \begin{bmatrix} 6 & 1 & -6 & -5 \\ 0 & -3.6667 & 4 & 4.3333 \\ 0 & 0 & 6.8182 & 5.6364 \\ 0 & 0 & 0 & 1.5600 \end{bmatrix}.$$

In conclusion, the statements in Eqs. (2.62) and (2.63) are true for a general system of equations on which we use the Gaussian elimination of Section 2.4.

It also follows that

$$\det(A) = \det(P_n * \cdots * P_1) \det(L) \det(U)$$
$$= (-1)^k U(1,1) * \cdots * U(n,n),$$

where k = the number of row interchanges, the $U(i,i)$'s are the diagonal elements of U, and $\det(L) = 1$, since its diagonal elements are all 1.

Ill-Conditioned Systems

In Section 2.9, we encountered the notion of an ill-conditioned system, $Ax = b$. This notion is quite different from that of a singular or near singular system. For instance, if we had 10 equations in 10 unknowns in the form

$$0.1x_i = 1, \qquad \text{for } i = 1, \ldots, 10,$$

then the coefficient matrix, A, would just be a 10×10 matrix whose diagonal elements are all 0.1 for which $\det(A) = 10^{-10}$!. This would indicate a near singular matrix, but actually its condition number is 1 and the solution is obvious.

As indicated in Section 2.9, the condition number of a system is a magnification factor for errors in the input matrix A or in the right-hand side b. In Eq. (2.32) we saw how much the answers changed for just very small changes in the right-hand side b. In that example we were given

$$\begin{bmatrix} 1.01 & 0.99 \\ 0.99 & 1.01 \end{bmatrix} \begin{bmatrix} x \\ y \end{bmatrix} = \begin{bmatrix} 2.00 \\ 2.00 \end{bmatrix} \tag{2.64}$$

We thought of the system, $Ax = b$, as a linear system solver machine. In the example our inputs were just the right-hand sides, b. The outputs from the machine were the solutions to the system for the fixed coefficient matrix, A, and we drew a diagram:

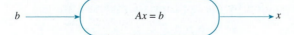

For three very "close" inputs—$b^{(1)} = (2, 2)^T, b^{(2)} = (2.02, 1.98)^T, b^{(3)} = (1.98, 2.02)^T$—we have in this very small example very "distant" outputs—namely, $x^{(1)} = (1, 1)^T, x^{(2)} = (2, 0)^T, x^{(3)} = (0, 2)^T$. We considered all this in Section 2.9. However, we did not complete our discussion of the condition number as a magnification factor. In this example, we see that

$$\|b^{(1)} - b^{(2)}\| = 0.02,$$

and $$\text{cond}(A) = \|A\|_\infty * \|A^{-1}\|_\infty = 100.$$

Observe that the norm of $(x^{(1)} - x^{(2)}) = \|x^{(1)} - x^{(2)}\|_\infty = 1$ is of the order of the product of the magnification factor (which is cond$(A) = 100$), and the norm of $(b_1^{(1)} - b_2^{(2)}) = 0.02$. A similar statement can be made for $b^{(1)}, b^{(3)}, x^{(1)}, x^{(3)}$.

Pivoting

Finally, a brief comment on pivoting. Gaussian elimination involves partial pivoting, *partial* meaning that at any time we are looking for the maximum magnitude in a particular column only. In contrast, full pivoting seeks the largest magnitude in the whole array at each stage and is more easily implemented in the Gauss–Jordan method. In any case, it should be pointed out that we would not need to use any pivoting if the computer could do arithmetic in infinite precision.

2.14   Using MAPLE

The MAPLE program is a very powerful mathematical package that allows the user to solve mathematical problems numerically, by symbolic methods, and graphics in two and three dimensions. MAPLE is very easy to use and allows the user to recall previous commands without having to rewrite them. The *Maple Technical Newsletter*, published by Springer-Verlag, is devoted to applications of MAPLE to solving real-world problems. In this section we look at the power of MAPLE in working with matrices and vectors and in solving systems of linear and nonlinear equations.

Defining Vectors and Matrices

Before applying MAPLE to linear algebra problems, we should first call the package: **with(linalg)**. Vectors can be defined in a number of ways. A vector with four components is declared as follows:

```
v2 := array([1, 3, -2, 5]);
```

Once the vectors have been defined, they may be added, subtracted, or multiplied. The dot-product and cross-product operations are done by the functions dotprod(v1, v2) and crossprod(v1, v2). Note that the semicolon is required at the end of the expression.

Matrices can be defined in a number of ways. Entering either

```
A := array([[1,2,3],[3,2,1],[1,0,3]]);
```

or

```
A := array(3,3,[1,2,3,3,2,1,1,0,3]);
```

will result in the following display:

$$\begin{bmatrix} 1 & 2 & 3 \\ 3 & 2 & 1 \\ 1 & 0 & 3 \end{bmatrix}.$$

Determinants and Matrix Operations

Once a square matrix has been defined, MAPLE will compute the determinant of the matrix M through the det(M) function. The result with the preceding matrix is -16, the correct value.

Two matrices can be added or subtracted and, if conformable, can be multiplied. The appropriate MAPLE notations are add(A, B) or evalm(A + B), multiply(A, B), or evalm(A &* B), where A, B are previously declared matrices. In the case of a vector, v, we write multiply(A, v) or evalm(A &* v).

Matrix M can be raised to the nth power by evalm(A^n) (when n is an integer). It is remarkable how fast the 50 power of a matrix can be computed, considering that it is done in exact arithmetic. The inverse is computed by writing inverse(A) or evalm(1/A).

Solving Linear Systems

MAPLE will find the solution of a system of linear equations. We first enter the elements of the augmented matrix (multiple right-hand sides are permitted). We then get the solution by using the reduced row echelon function rref(A) or gaussjord(A), where A is the augmented matrix.

Here is an example. If we have defined

```
A := array([[1,2,3,-2,4],[3,2,1,3,-1],[2,0,2,1,2]]);
```

$$A := \begin{bmatrix} 1 & 2 & 3 & -2 & 4 \\ 3 & 2 & 1 & 3 & -1 \\ 2 & 0 & 2 & 1 & 2 \end{bmatrix}$$

then

```
rref(A);
```

produces this result:

$$\begin{bmatrix} 1 & 0 & 0 & \frac{3}{2} & -\frac{3}{4} \\ 0 & 1 & 0 & -\frac{1}{4} & -\frac{1}{4} \\ 0 & 0 & 1 & -1 & \frac{7}{4} \end{bmatrix}.$$

Observe that the rref function also uses the Gauss–Jordan technique to reduce the coefficient matrix to the identity matrix and, at the same time, to reduce the right-hand-side columns so that they become the solution vector(s). MAPLE does have a linsolve(A, b) procedure.

There is another way to solve a system of linear equations that may be more convenient. For example,

```
solve({x + y + z = 2, y + 2*x + z = 3, x + y = 1}, {x, y, z});
```

gives the vector

$$\{x = 1, y = 0, z = 1\}$$

Nonlinear Systems

It is easy to solve a system of equations in MAPLE. For instance,

```
fsolve({x^2 − y = 1, x + y^2 = 2}, {x, y});
```

results in the following display of one of the answers:

$$\{x = -1.701644095, y = 1.926303220\}$$

One could also predefine the previous functions in fsolve. We first write

```
f := x^2 − y − 1 = 0;
```

```
g := x + y^2 − 2 = 0;
```

```
fsolve({f, g}, {x, y});
```

All this generates the same result. Now, if we rewrite the second equation,

```
h := y − sqrt(2 − x) = 0;
```

and then follow with

```
fsolve({f, h}, {x, y}, {x = 0 . . 2, y = 0 . . 2});
```

we get the other solution to the problem:

$$\{x = 1.345089393, y = .809265474\}.$$

The plot procedures allow us to see the graphs of the different functions for quick starting values. In the example just given, we need only to do the following to get the graph of the functions we want to solve:

```
f := x^2 - 1;

h := sqrt(2 - x);

with(plots);

plot({f, h}, x = -2..2);
```

Characteristic Polynomial

MAPLE has built-in functions to evaluate the characteristic polynomial of a matrix. Let's try the following:

```
A := matrix(4, 4, [1, 3, 4, 2, 3, 2, 4, 1, 4, 2, 1, 3, 2, 1, 4, 3]);
```

$$A := \begin{bmatrix} 1 & 3 & 4 & 2 \\ 3 & 2 & 4 & 1 \\ 4 & 2 & 1 & 3 \\ 2 & 1 & 4 & 3 \end{bmatrix}$$

```
charpoly(A, x);
```

produces

$$x^4 - 7x^3 - 33x^2 + 21x + 90$$

and

```
eigenvals(A);
```

has the output

$$10, -3, 3^{1/2}, -3^{1/2}$$

These four numbers are the eigenvalues of A.

In Chapter 6, we will devote more discussion to finding the eigenvalues and eigenvectors of a matrix.

2.15 Parallel Processing

We have mentioned that parallel processing can speed the operation of many numerical methods. In this section we describe how Gaussian elimination and Jacobi iteration can be efficiently performed in a parallel processing environment. We will show how much the performance can be improved in each case.

Gaussian Elimination

Recall how we achieve a solution to a system of linear equations through Gaussian elimination. We first perform a sequence of row reductions on the augmented matrix $(A : b)$ until the coefficient matrix A is in upper-triangular form. Then we employ back-substitution to find the solution.

To see how this can be done in a parallel processing environment, we must examine the row-reduction phase and the back-substitution phase in some detail.

We begin with the row-reduction phase: Consider the following example of the first stage of row reduction of a 4×4 system with one right-hand side:

$$
\begin{bmatrix}
1 & 2 & 1 & 3 & 4 \\
2 & 5 & 4 & 3 & 4 \\
1 & 4 & 2 & 3 & 3 \\
3 & 2 & 4 & 1 & 8
\end{bmatrix}
\begin{matrix}
\\
R_2 - (2/1) * R_1 \\
R_3 - (1/1) * R_1 \rightarrow \\
R_4 - (3/1) * R_1
\end{matrix}
\begin{bmatrix}
1 & 2 & 1 & 3 & 4 \\
0 & 1 & 2 & -3 & -4 \\
0 & 2 & 1 & 0 & -1 \\
0 & -4 & 1 & -8 & -4
\end{bmatrix}.
$$

Though each of these row reductions depends on the elements of row 1, they are completely independent of one another. For example, the elements of rows 2 and 4 play no part in the row operations performed on row 3. Thus the row reductions on rows 2, 3, and 4 can be computed simultaneously.

If we are computing in a parallel processing environment, we can take advantage of this independence by assigning each row-reduction task to a different processor:

$$
\begin{bmatrix}
1 & 2 & 1 & 3 & 4 \\
2 & 5 & 4 & 3 & 4 \\
1 & 4 & 2 & 3 & 3 \\
3 & 2 & 4 & 1 & 8
\end{bmatrix}
\begin{matrix}
\\
\rightarrow \text{Processor 1: } R_2 - (2/1) * R_1 \\
\rightarrow \text{Processor 2: } R_3 - (1/1) * R_1 \rightarrow \cdots \\
\rightarrow \text{Processor 3: } R_4 - (3/1) * R_1
\end{matrix}
$$

Suppose each row assignment statement requires 4 time units, one for each element in a row. Then the sequential algorithm performs this stage of the row reduction in 12 time units, whereas we need only 4 time units for the parallel algorithm. This example of parallel processing on the first stage of row reduction of a 4×4 system matrix generalizes to any row reduction in stage j of an $n \times n$ system matrix.

Recall that there are $n - 1$ row reduction stages in Gaussian elimination, one for each of the n columns of the coefficient matrix except for the last column. This suggests that we need $n - 1$ processors to do the reduction in parallel.* Also recall that each row-reduction stage j creates zeros in every cell below the diagonal in the jth column. The following two pseudocodes compare the use of a single processor with the use of n processors to perform the entire row-reduction phase of Gaussian elimination.

---

*Even so, we will need n processors in the final algorithm, as will be seen.

Algorithms for Row Reduction in Gaussian Elimination

Sequential processing (without pivoting):
FOR $j = 1$ TO $(n - 1)$
   FOR $i = (j + 1)$ TO n
     FOR $k = j$ TO $(n + 1)$
       $a[i, k] = a[i, k] - a[i, j] * a[j, k]$
     END (For k)
   END (FOR i)
END (FOR j)

Parallel processing:
FOR $i = 1$ TO $(n - 1)$ (Counts stages = columns)
   FOR $k = i$ TO $(n + 1)$ {ON PROCESSOR $j = i + 1$ TO n}
     $a[i, k] = a[i, k] - a[i, j] * a[j, k]$
   END (FOR k)
END (FOR i)

If we total the arithmetic operations required to carry out the reduction of an $n \times n$ coefficient matrix A to upper-triangular form, we find that the sequential algorithm requires $O(n^3)$ successive operations and that the parallel algorithm accomplishes the same task in $O(n^2)$ successive operations. (Operations done simultaneously are counted together.)

What happens if we have more than n processors? Can we speed up the reduction process even more? Suppose we increase the number of processors from n to, say, n^2. Can we effectively use this extra power? We examine two options: (1) We might hope to perform all n stages of the row reduction at once, creating all zeros below the diagonal simultaneously; or (2) we could perform row operations on all $(n + 1)$st elements of a row (in the augmented matrix) simultaneously, thereby reducing the time required for the reduction of each row from n time units to 1 time unit. Each of these options would decrease computing time by a factor of n, so we consider both as candidates for parallelization.

Of the two possibilities, the correct choice is clear. We cannot perform all n stages of the row reduction at once, since the computations at each stage depend on the preceding one. Thus we select option 2, and seek to improve on the time required to perform the reduction in a row. We examine the algorithm given for parallel processing but modified to agree with the algorithm of Section 2.4:

FOR $i = 1$ TO $(n - 1)$ (Counts stages = columns)
   FOR $k = i$ TO $(n + 1)$ {ON PROCESSOR $j = i + 1$ TO n}
     $a[i, k] = a[i, k] - a[i, j]/a[j, j] * a[j, k]$
   END (FOR k)
END (FOR i)

As in the case of row reductions, each iteration of the FOR k loop is independent and thus the operations can be done simultaneously. When we work on row m, we will do

ON PROCESSOR m:
  $a[i,m] = a[i,m] - a[i,j]/a[j,j] * a[j,m]$
  $\vdots$

ON PROCESSOR $n + 1$:
  $a[i, n + 1] = a[i, n + 1] - a[i,j]/a[j,j] * a[j, n + 1]$

where these processors are in addition to those we used before.

For reasons that will become clear later, we will assign processors to row 1 even though they are not used during the reduction phase. This additional parallelization reduces processing time in each row from $m + 1$ time units to 1 time unit. Notice, however, that this enhancement is useful to us only if we have $n + 1$ processors for each of the n rows, or $n^2 + n$ total processors.

The complete algorithm for the row-reduction phase of Gaussian elimination on $n^2 + n$ processors runs with only $O(n)$ successive steps. Imagine that each processor is labeled $\text{proc}(i,j), i = 1, \ldots, n, j = 1, \ldots, n + 1$. Then $\text{proc}(i,j)$ is responsible for all operations performed on element a_{ij} of the matrix $[A:b]$. We rewrite the algorithm for parallel processing to reflect this improvement:

```
FOR i = 2 TO n                         (Counts rows)
  {ON PROCESSOR(j, k)}
    a[j, k] = a[j, k] - a[j, i]/a[i, i] * a[i, k]
END (FOR i)
```

The row-reduction phase of Gaussian elimination leaves us with an upper-triangular coefficient matrix and an appropriately adjusted right-hand side. In the sequential algorithm we now find a solution using back-substitution. Before we consider parallelization of back-substitution, let us examine the activity of the processors during row reduction in greater detail.

As we have observed, the processors responsible for computations on the elements of row 1 sit idle during row reduction, since those elements of the matrix never change. In addition, after the first row-reduction stage in which zeros are placed in the first column, the processors for row 2 also sit idle. In fact, each stage of row reduction frees $n + 1$ processors.

It is natural to wonder if these idle processors could be employed in our algorithm. Indeed, they can. We use them to perform row reductions above the diagonal at the same time that corresponding row reductions occur below the diagonal. Thus at each stage j of the reduction, zeros appear in all but the diagonal element of the jth column.

This diagram illustrates our improved procedure, continuing the simple 4×4 example examined before and doing stages 3 and 4:

$$
\rightarrow \begin{bmatrix} 1 & 0 & -3 & 9 & 12 \\ 0 & 1 & 2 & -3 & -4 \\ 0 & 0 & -3 & 6 & 7 \\ 0 & 0 & 9 & -20 & -20 \end{bmatrix}
\begin{array}{l} R_1 - (3/3)*R_3 \\ R_2 - (-2/3)*R_3 \rightarrow \\ \\ R_4 - (9/-3)*R_3 \end{array}
\begin{bmatrix} 1 & 0 & 0 & 3 & 5 \\ 0 & 1 & 0 & 1 & \frac{2}{3} \\ 0 & 0 & -3 & 6 & 7 \\ 0 & 0 & 0 & -2 & 1 \end{bmatrix}
$$

$$
\begin{array}{l} R_1 - (3/-2)*R_4 \\ R_2 - (1/-2)*R_4 \\ R_3 - (6/-2)*R_4 \end{array}
\begin{bmatrix} 1 & 0 & 0 & 0 & \frac{13}{2} \\ 0 & 1 & 0 & 0 & \frac{7}{6} \\ 0 & 0 & -3 & 0 & 10 \\ 0 & 0 & 0 & -2 & 1 \end{bmatrix}
\rightarrow x = \begin{bmatrix} \frac{13}{2} \\ \frac{7}{6} \\ -\frac{10}{3} \\ -\frac{1}{2} \end{bmatrix}.
$$

The result of n such reductions—one for each column of the coefficient matrix A—is $[D : b']$, where D is a diagonal matrix and b' is an appropriately adjusted right-hand-side vector. Specifically, the solution x for $Dx = b'$ also satisfies $Ax = b$. This solution is the vector x whose elements are $x_i = b_i'/d_{ii}$ for $i = 1, \ldots, n$. We can use n processors to perform these n divisions simultaneously. Notice that the back-substitution phase of Gaussian elimination is no longer necessary! We find that the Gauss–Jordan procedure is preferred when doing parallel processing!

The parallel algorithm for n^2 processors required n time units for row reduction, and one additional time unit for division. Recall that the sequential algorithm required $O(n^3)$ time units. To understand the magnitude of the improvement in running time, consider that a solution achieved in 10 seconds via the parallel algorithm would require around 15 minutes via the sequential algorithm.*

Our final parallel algorithm for solving a system of linear equations more closely resembles the Gauss–Jacobi solution technique than it does Gaussian elimination. This is not surprising. It is not uncommon for good parallel algorithms to differ dramatically from their speediest sequential counterparts.

Problems in Using Parallel Processors

It is essential to mention some important concerns that have been neglected in the preceding discussion. When we actually implement this parallel algorithm, we must worry about four issues.

---

* This neglects the overhead of interprocessor communications.

1. The algorithm does not pivot. Thus our solution may not be as numerically stable as one obtained via a sequential algorithm with partial pivoting. In fact, if a zero appears on the diagonal at any stage of the reduction, we are in big trouble. Nonetheless, we elect to exclude pivoting because the searching it requires increases the running time considerably and partially nullifies our streamlined approach.

2. The coefficient matrix A is assumed to be nonsingular. It is easy to check for singularity at each stage of the row reduction, but such error-handling will more than double the running time of the algorithm.

3. We have ignored the communication and overhead time costs that are involved in parallelization. Because of these costs, it is probably more efficient to solve small systems of equations using a sequential algorithm.

4. Other, perhaps faster, parallel algorithms exist for solving systems of linear equations. One technique, which is easily derived from ours, involves computing A^{-1} via row operations and simply multiplying the right-hand side to get the solution $x = A^{-1}b$. Another technique requires computing the coefficients of the characteristic polynomial and then applying these coefficients in building A^{-1} from powers of A. This method finds a solution in only $[2 \log_2 n + O(\log n)]$ time units, but it requires $n^4/2$ processors to do so. In addition, it often leads to numeric instability.*

Despite these concerns, our algorithm is an effective approach to solving systems of linear equations in a parallel environment.

Iterative Solutions—The Jacobi Method

The method of simultaneous displacements (the Jacobi method) that was discussed in Section 2.10 is adapted very simply to a parallel environment. Recall that at each iteration of the algorithm a new solution vector $x^{(n+1)}$ is computed using only the elements of the solution vector from the previous iteration, $x^{(n)}$. In fact, the elements of the vector $x^{(n)}$ can be considered fixed with respect to the $(n + 1)$st iteration. Thus, though each element $x_i^{(n+1)}$ in the vector $x^{(n+1)}$ depends on the elements in $x^{(n)}$, these $x_i^{(n+1)}$ are independent of one another and can be computed simultaneously.

Suppose the solution vector x has m elements. Then each iteration of the Jacobi algorithm in a sequential environment requires m assignment statements. If we have m processors in parallel, these m assignment statements can be performed simultaneously, thereby reducing the running time of the algorithm by a factor of m.

Notice also that each assignment statement is a summation over approximately m terms. As demonstrated in Section 0.7, this summation can be performed in $\log_2 m$ time units with m parallel processors, compared to m time units for sequential

---

* JaJa (1992) describes these alternative algorithms in some detail.

addition. If m^2 processors are available, we can employ both of these parallelizations and reduce the time for each iteration of the Jacobi algorithm to $\log_2 m$ time units. This is a significant speedup over the sequential algorithm, which requires m^2 time units per iteration.

As was seen in Section 2.10, the actual running time of the algorithm (the number of iterations) depends on the degree of diagonal dominance of the coefficient matrix. Parallelization decreases only the time required for each iteration.

Because Gauss–Seidel iteration requires that the new iterates for each variable be used after they have been obtained, this method cannot be speeded up by parallel processing. Again, the preferred algorithm for sequential processing is not the best for parallel processing.

Chapter Summary

After working through this chapter, you should be able to

1. Handle the basic operations of matrices and vectors. You should know the definitions for triangular matrix, tridiagonal matrix, transpose, determinant, characteristic polynomial, and matrix inverse.

2. Use Gaussian elimination with partial pivoting and back-substitution to solve a system of equations, compute the determinant of the coefficient matrix, and find the LU decomposition. You should be able to compare this method with the Gauss–Jordan method.

3. Compute the number of additions, multiplications, and divisions required in the implementation of several of the algorithms in this chapter.

4. Be acquainted with another LU method for solving a system of equations. You should know how to use the LU decomposition to solve a system efficiently when there are multiple right-hand sides.

5. Understand the several ways for computing the determinant of a square matrix.

6. Know more than one way to find the inverse of a matrix and why using Gaussian elimination can be more efficient than the standard Gauss–Jordan technique.

7. Understand the concept of an ill-conditioned system and compare this to singular and near-singular systems.

8. Compute the various norms for vectors and matrices and evaluate the condition number of a matrix.

9. Solve certain specialized systems by iterative methods and explain why the use of parallel processors may make some methods preferred that would not be in sequential processing. You should know what is meant by *diagonal dominance*.

10. Use Newton's method to solve a system of nonlinear equations and tell how the number of function evaluations may be reduced.

11. Explain what is meant by a *parallel algorithm*.

12. Express the Gaussian elimination method as a product of simple matrices.

13. Use an example to show the importance of pivoting in getting a more accurate solution.

Computer Programs

In this section we include three computer programs: (1) a C program that solves a set of linear equations using Gaussian elimination, (2) a Pascal program that solves a diagonally dominant system using Gauss–Jacobi iteration, and (3) a FORTRAN program that gets the solution to a set of nonlinear equations with Newton's method.

Program 2.1 Gaussian Elimination

Figure 2.4(a) lists the driver program for Gaussian elimination (written in C). It is heavily commented to facilitate understanding; here is an outline of the logic:

1. After the usual include statements (observe that the subroutine file of Fig. 2.4(b) is one of these), some global variables are defined.
2. What follows is a procedure that prints out the augmented matrix of the system.
3. The main program comes next. It first fills the matrix of coefficients and the right-hand-side vector with zeros, then defines the coefficients and right-hand-side values, prints these, and finally calls the procedure that solves the system. Finally, the main program prints the *LU* decomposition of the system and the solution vector.

The output from the program is shown at the end of Fig. 2.4(a).

Figure 2.4(b) shows the two procedures that are called to solve a linear system. (These procedures are actually in an include file incorporated into the driver above.) The logic of these procedures should also be clear from the comments, but we summarize the steps.

First, procedure *elim* follows two steps to get the *LU* decomposition: (1) Some local variables are defined. (2) The *LU* decomposition of the coefficient matrix is computed, performing partial pivoting to improve accuracy. (Rows are actually interchanged during this step.) After rows are interchanged, the elements below the diagonal are reduced (storing the reduction factors to produce the *LU* matrix).

Procedure *solve* is also listed in Fig. 2.4(b). It uses the *LU* matrix constructed in *elim* to get the solution. Four steps are performed: (1) Local variables are defined. (2) The elements of the *b*-vector are rearranged to be in the same order as the rearranged rows of coefficients. (3) The *b*-vector is reduced by solving $Ly = b$. (4) Back-substitution is used to obtain x from $Ux = y$.

```
/*
                    *******************************
                    **                           **
                    **            PDEC.C          **
                    **                           **
                    *******************************
                              Chapter 2

           Gerald/Wheatley, APPLIED NUMERICAL ANALYSIS(fifth edition)
                            Addison-Wesley, 1994

       ----------------------------------------------------------------

           This program solves a system of equations using Gaussian
       elimination with partial pivoting and back-substitution according
       to the algorithm given in Section 2.4.
           The algorithm is implemented in two procedures: ELIM and
       SOLVE. The first procedure returns the LU decomposition of the
       coefficient matrix, and the second solves the system using both
       forward- and back-substitution.
       */

       #include <stdio.h>
       #include <math.h>
       #include "elim.c"

       float a[11][11],        /* the matrix of coefficients */
             b[11];            /* the right-hand side of the system */
       int   ipvt[11];         /* vector to keep track of row changes */
       int   n, i, j;

       printing(a, b)
         float a[][11], b[];
       {
         int i, j;

           printf("\n");
           for ( i = 1; i <= n; i++ )
             {
               for ( j = 1; j <= n; j++ )
                 printf("   %.4f  ", a[i][j]);
               printf("\t   %.4f  ", b[i]);
               printf("\n");
             }
       }

       main()        /* main program */
       {
       /*
       ----------------------------------------------------------------
           Fill matrix a and b with zeroes.
       */

           for ( i = 1; i <= 10; i++ )
             {
               b[i] = 0.0;
               for ( j = 1; j <= 10; j++ )
                 a[i][j] = 0.0;
             }

       /*
       ----------------------------------------------------------------
           Enter the number of equations, the matrix,a, and the
       right-hand side, b.
           This example is taken from Section 2.4.
```

Figure 2.4(a) Program 2.1a

Figure 2.4(a)   *(continued)*

```
*/
    n = 4;
    a[1][1] = 6; a[1][2] =  1; a[1][3] = -6; a[1][4] = -5;
    a[2][1] = 2; a[2][2] =  2; a[2][3] =  3; a[2][4] =  2;
    a[3][1] = 4; a[3][2] = -3;                a[3][4] =  1;
                 a[4][2] =  2;                a[4][4] =  1;
    b[1] = 6; b[2] = -2; b[3] = -7;

/*
---------------------------------------------------------------
    Write out the matrix of coefficients, a, and the right-hand
    side.
*/
    printf("\n\n\n");
    printf("\t\t\tTHE MATRIX A\t\t\tRIGHT HAND SIDE\n");
    printing(a,b);
    printf("\n\n");
/*
---------------------------------------------------------------
    We find the LU decomposition of the matrix A.
*/
    elim(a, n, ipvt);
/*
---------------------------------------------------------------
    Procedure solve now gets solution by forward/back-
    substitution.
*/
    solve(a, n, ipvt, b);
/*
---------------------------------------------------------------
    Write out the LU decomposition of A.
*/
    printf("\n\t\t THE LU DECOMPOSITION OF A\t\tSOLUTION
VECTOR\n");
    printing(a, b);
    printf("\n\n");
}
```

```
*******************************************************************

                        OUTPUT FOR PDEC.C


              THE MATRIX A                    RIGHT HAND SIDE

    6.0000      1.0000     -6.0000     -5.0000        6.0000
    2.0000      2.0000      3.0000      2.0000       -2.0000
    4.0000     -3.0000      0.0000      1.0000       -7.0000
    0.0000      2.0000      0.0000      1.0000        0.0000



           THE LU DECOMPOSITION OF A              SOLUTION VECTOR

    6.0000      1.0000     -6.0000     -5.0000       -0.5000
    0.6667     -3.6667      4.0000      4.3333        1.0000
    0.3333     -0.4545      6.8182      5.6364        0.3333
    0.0000     -0.5455      0.3200      1.5600       -2.0000
```

```
elim(a, n, ipvt)
  float a[][11];
  int   ipvt[], n;

/*
---------------------------------------------------------------

   ELIM: this procedure solves a set of linear equations and gives an
LU decomposition of the coefficient matrix. The Gaussian elimination
method is used, with partial pivoting and back-substitution.

      INPUT:   a  - the coefficient matrix
               n  - the number of equations
               ****************************
      OUTPUT:  a  - the LU decomposition of the matrix a
               ipvt - a vector containing the order of the rows
                      of the rearranged matrix due to pivoting.
*/


{
   float save, ratio, value, det;
   int   i, ipvtemp, nMinus1, iPlus1, j, l, kcol, jcol, jrow,
tempipvt;

/*
---------------------------------------------------------------
   Begin the LU decomposition
*/

   det = 1.0;
   nMinus1 = n-1;
   for ( i = 1; i <= n; i++ )
      ipvt[i] = i;
   for ( i = 1; i <= nMinus1; i++ )
   {
      iPlus1 = i+1; ipvtemp = i;
      for ( j = iPlus1; j <= n; j++ )            /* find the */
         if (fabs(a[ipvtemp][i]) < fabs(a[j][i]))   /* pivot */
            ipvtemp = j;                             /* row */
      if (ipvtemp != i)                           /* interchange */
      {
         tempipvt = ipvt[i];
         ipvt[i] = ipvt[ipvtemp];                 /* rows if  a[i,i] */
         for ( jcol = 1; jcol <= n; jcol++ )      /* is not the max */
         {
            save = a[i][jcol];                     /* in column. */
            a[i][jcol] = a[ipvtemp][jcol];
            a[ipvtemp][jcol] = save;
         }      /*  end of for jcol loop */
         ipvt[ipvtemp] = tempipvt;
         det = -det;
      }          /* end of if statement */

/*
---------------------------------------------------------------
      Now reduce all elements below the i'th row.
*/

      for ( jrow = iPlus1; jrow <= n; jrow++ )
         if (a[jrow][i] != 0.0)
         {
            a[jrow][i] = a[jrow][i]/a[i][i];
            for ( kcol = iPlus1; kcol <= n; kcol++ )
               a[jrow][kcol] = a[jrow][kcol] -
             a[jrow][i]*a[i][kcol];
```

Figure 2.4(b) Program 2.1b

Figure 2.4(b)   *(continued)*

```
      }      /* end of if statement */
    }        /* end of for i loop */
  }          /* end of elim procedure */

/*
----------------------------------------------------------------
                  PROCEDURE solve
*/

solve(a, n, ipvt, b)
  float a[][11], b[];
  int   n, ipvt[];

/*
----------------------------------------------------------------
   Making use of the LU decomposition of the matrix A, this
   procedure solves the system by forward- and back-substitution.

   INPUT   a - LU matrix from elim
           n - number of equations
           ipvt - a record of the rearrangement of the rows of a
                        from  elim
           b - right-hand side of the system of equations

   OUTPUT b - the solution vector
*/

{
  int    irow, jcol, i;
  float sum;
  float x[11];

/*
   Rearrange the elements of the b vector. Store them in the
   x vector.
*/

  for ( i = 1; i <= n; i++ )
    x[i] = b[ipvt[i]];

/*
----------------------------------------------------------------
   Solve using forward-substitution--Ly=b
*/
  for ( irow = 2; irow <= n; irow++ )
    {
      sum = x[irow];
      for ( jcol = 1; jcol <= (irow-1); jcol++ )
        sum = sum - a[irow][jcol]*x[jcol];
      x[irow] = sum;
    }                       /* end of for irow loop */
/*
----------------------------------------------------------------
   Solve by back-substution--Ux=y
*/

b[n] = x[n]/a[n][n];
for ( irow = (n-1); irow >= 1; irow-- )
  {
    sum = x[irow];
    for ( jcol = (irow+1); jcol <= n; jcol++ )
      sum = sum - a[irow][jcol]*b[jcol];
    b[irow] =  sum/a[irow][irow];
  }      /* end of irow loop */
}        /* end of solve procedure */
```

Program 2.2 Gauss–Jacobi Iteration

A driver program for doing Gauss–Jacobi iterations is shown in Fig. 2.5(a). The code for the procedure that does the work, shown in Fig. 2.5(b), is an include file within the driver program. These Pascal modules are fully commented. The program requires that the set of equations be in diagonally dominant form.

After declaring variables, the driver program calls a procedure to initialize the pertinent parameters, including the system matrix. It then calls procedures, listed in Fig. 2.5(b), that (1) display the original augmented matrix, (2) carry out the iterations until the system converges to meet a tolerance value or until the number of iterations reaches its limit, and (3) print the final answer or a message saying that the tolerance was not met.

Typical output is shown at the end of Fig. 2.5(a).

```
PROGRAM p_Gauss_Jacobi(INPUT,OUTPUT);
(*
                        Chapter 2

    Gerald/Wheatley, APPLIED NUMERICAL ANALYSIS(fifth edition)
                    Addison-Wesley, 1994
------------------------------------------------------------------

 This program calls the Procedure that uses the Gauss-Jacobi
 iterative method to compute the solution to the system of
 equations:  AX = b.
------------------------------------------------------------------
The global variables: a, b, x, n, niter, tol are defined in the
procedure, stored in the file:  GSITRN.PAS.

This program assumes the equations have been arranged so that
there is DIAGONAL DOMINANCE in the coefficient matrix.
*)

TYPE
    matrix = ARRAY[1..10, 1..10] OF REAL;
    vector = ARRAY[1..10] OF REAL;
VAR
    a      : matrix;
    b,x    : vector;

    n, nITER,
    iter        : INTEGER;
    tol         : REAL;

(*$I gjitrn.pas *)   (*  <===== INCLUDE FILE OPTION =====>  *)
(*
------------------------------------------------------------------

 INITIALIZE THE MATRIX A AND ITS RIGHT HAND SIDE B
*)
PROCEDURE initialize;
VAR
    i : INTEGER;
BEGIN
    n := 3;  nITER := 30;  tol := 0.00001;
```

Figure 2.5(a) Program 2.2a

Figure 2.5(a)   *(continued)*

```
    a[1,1] :=  6.0;   a[1,2] :=  -2.0;  a[1,3] :=   1.0;
    a[2,1] := -2.0;   a[2,2] :=   7.0;  a[2,3] :=   2.0;
    a[3,1] :=  1.0;   a[3,2] :=   2.0;  a[3,3] :=  -5.0;

    b[1] := 11.0;  b[2] :=  5.0;  b[3] := -1.0;
    FOR i := 1 TO n DO x[i] := 0.0
END;

(*
------------------------------------------------------------------
*)
BEGIN   (*  MAIN  *)

   WriteLn;
   WriteLn;
   WriteLn('*****************************************');
   WriteLn;
   WriteLn(' ':15, ' GAUSS-JACOBI ITERATIVE METHOD');
   WriteLn;
   WriteLn('*****************************************');
   WriteLn;
   initialize;
   write_out_matrix_and_right_hand_side;
   Gauss_Jacobi_iteration_on_a_and_b(iter);

   IF iter >= nITER THEN write_no_convergence
      ELSE write_final_answer;
   WriteLn
END.
```

```
*********************************************************

           GAUSS-JACOBI ITERATIVE METHOD

*********************************************************

  The given matrix a and its right-hand side b are:

     6.0 -2.0  1.0          11.00000
    -2.0  7.0  2.0           5.00000
     1.0  2.0 -5.0          -1.00000


    ******     0.3333   -0.1667        1.8333
    0.2857    ******    -0.2857        0.7143
    0.2000     0.4000    *****         0.2000


  0.00000  0.00000  0.00000   <====  FIRST ITERATION IN X,Y,Z
  1.83333  0.71429  0.20000
  2.03810  1.18095  0.85238
  2.08492  1.05306  1.08000
  2.00435  1.00141  1.03821
  1.99410  0.99033  1.00143
  1.99654  0.99790  0.99495
  2.00014  1.00045  0.99847
  2.00041  1.00048  1.00021
  2.00012  1.00006  1.00027
  1.99997  0.99996  1.00005
  1.99998  0.99998  0.99998
  2.00000  1.00000  0.99999
  2.00000  1.00000  1.00000   <====  LAST ITERATION IN X,Y,Z

  The solution to the equations is:

           2.00000
           1.00000
           1.00000
```

```
PROCEDURE Gauss_Jacobi_iteration_on_A_and_b(VAR iter : Integer);
(*
                    GJITRN.PAS
  ----------------------------------------------------------------
   This procedure obtains the solution to N linear equations by
   the GAUSS-JACOBI ITERATION METHOD. An initial approximation is
   sent to the procedure in the vector x. The solution, as
   approximated by the maximum change in any x component is less
   than TOL. If this cannot be accomplished in NITER iterations,
   a message is printed and the current approximation is
   returned.

   It is ASSUMED that equations are to be arranged so as to have
   the LARGEST VALUES ON THE DIAGONAL, i.e. diagonal dominance.
  ----------------------------------------------------------------
   GLOBAL VARIABLES

       a       - coefficient matrix with largest values on the diagonal.
       b       - right-hand-side vector.
       x       - on INPUT  the initial approximation to the solution,
               - on OUTPUT the solution.

       n       - the number of equations.
       nITER   - limit to the number of iterations.
       iter    - actual number of iterations to get final answer.

       tol     - test value to stop iterating on the solution.

  ----------------------------------------------------------------
*)
VAR
   old_x       : vector;
   save, xMAX  : Real;
   i, j        : Integer;
   continue    : BOOLEAN;

BEGIN
(*
  ----------------------------------------------------------------
  Make the diagonal elements equal to 1.0
*)
   WriteLn; WriteLn;
   FOR i := 1 TO n DO                                Begin
      save := a[i,i];
      b[i] := b[i]/save;
      FOR j := 1 TO n DO                             Begin
         a[i,j] := a[i,j]/save;
         If (j<>i) Then
                 Write( -a[i,j]:10:4)
             ELSE
                 Write( '     ******')     End;
      WriteLn( b[i]:15:4)                            End;
WriteLn; WriteLn;
(*
  ----------------------------------------------------------------
   Now we perform the iterations. Store max change
   on x values for testing against tol.
   WHILE tests the current number of iterations
 *)
   iter := 1;    continue := TRUE;

   WHILE (ITER < nITER) AND continue DO              Begin
      xMAX := 0.0;
      For i := 1 To n Do old_x[i] := x[i];
```

Figure 2.5(b) Program 2.2b

Figure 2.5(b)   *(continued)*

```
      FOR i := 1 TO n DO                      Begin
          save := x[i];
          x[i] := b[i];
          FOR j := 1 TO n DO
              IF j<>i
                  THEN x[i] := x[i] - a[i,j]*old_x[j];
          IF (ABS(x[i] - save) > xMAX ) THEN
              xMAX := ABS(x[i]-save)          End;

   iter := iter + 1;
      continue := xMAX > tol;
      For j := 1 TO n DO
              Write(outfile, old_x[j]:9:5);
      WriteLn(outfile);                              End;
END;
(*
------------------------------------------------------------
   PRINT OUT THE MATRIX AND THE RIGHT-HAND SIDE
*)

PROCEDURE write_out_matrix_and_right_hand_side;
VAR
   i,j  : Integer;
BEGIN
   WriteLn; WriteLn;
   WriteLn('The given matrix a and its right-hand side b are:');
   WriteLn;
   FOR i := 1 TO n DO                      BEGIN
      FOR j := 1 TO n DO
          WRITE( a[i,j]:5:1);
          WRITE(' ':10, b[i]:8:5);
          WRITELN                          END;

   WriteLn; WriteLn
END;
(*
------------------------------------------------------------
   PRINT OUT THE FINAL ANSWER
*)
PROCEDURE write_final_answer;
VAR
   i : Integer;
BEGIN
   WriteLn('  The solution to the equations is: ');
   WriteLn;
   FOR i := 1 TO n DO
      WriteLn(' ':10, x[i]:8:5);
   WriteLn
END;
(*
------------------------------------------------------------
   PRINT MESSAGE IF THERE IS NO CONVERGENCE
*)
PROCEDURE write_no_convergence;
BEGIN
   WriteLn;
   WriteLn(' THE TOLERANCE OF ', tol:10:7,
                      ' NOT MET IN ',
                   iter:3,' iterations.');
   WriteLn
END;
```

Program 2.3 Solving a Nonlinear System

A FORTRAN program that solves a nonlinear system is presented in Fig. 2.6. The driver program, written to solve a pair of equations, calls subroutine NLSYST, passing 10 parameters. These parameters include the name of the subroutine that computes function values, an array that initially sends starting values for the Newton iterations and then returns the final estimates, tolerance values to end iterating, and a parameter METHOD that tells whether to compute partial derivatives from their analytical expression or by finite differences.

This somewhat lengthy program lists the principal subroutine NLSYST, which does the work, and four subroutines that it uses: (1) FCN, which computes values for the equations being solved; (2) FCNJ, which computes values for the partial derivatives (either numerically or analytically), (3) LUD, which gets the *LU* decomposition of the linear equations used for improving the estimate of solution, and (4) SOLVE, which solves these linear equations.

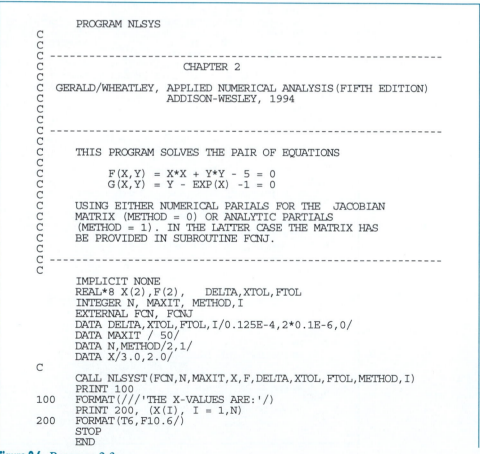

```
         PROGRAM NLSYS
C
C
C    ------------------------------------------------------------
C
C                         CHAPTER 2
C
C    GERALD/WHEATLEY, APPLIED NUMERICAL ANALYSIS(FIFTH EDITION)
C                     ADDISON-WESLEY, 1994
C
C
C    ------------------------------------------------------------
C
C       THIS PROGRAM SOLVES THE PAIR OF EQUATIONS
C
C           F(X,Y) = X*X + Y*Y - 5 = 0
C           G(X,Y) = Y - EXP(X) -1 = 0
C
C       USING EITHER NUMERICAL PARIALS FOR THE  JACOBIAN
C       MATRIX (METHOD = 0) OR ANALYTIC PARTIALS
C       (METHOD = 1). IN THE LATTER CASE THE MATRIX HAS
C       BE PROVIDED IN SUBROUTINE FCNJ.
C
C    ------------------------------------------------------------
C
         IMPLICIT NONE
         REAL*8 X(2),F(2),    DELTA,XTOL,FTOL
         INTEGER N, MAXIT, METHOD,I
         EXTERNAL FCN, FCNJ
         DATA DELTA,XTOL,FTOL,I/0.125E-4,2*0.1E-6,0/
         DATA MAXIT / 50/
         DATA N,METHOD/2,1/
         DATA X/3.0,2.0/
C
         CALL NLSYST(FCN,N,MAXIT,X,F,DELTA,XTOL,FTOL,METHOD,I)
         PRINT 100
100      FORMAT(///'THE X-VALUES ARE:'/)
         PRINT 200, (X(I), I = 1,N)
200      FORMAT(T6,F10.6/)
         STOP
         END
```

Figure 2.6 Program 2.3

Figure 2.6   *(continued)*

```
C
          SUBROUTINE NLSYST(FCN,N,MAXIT,X,F,DELTA,XTOL,FTOL,METHOD,I)
C
C      -------------------------------------------------------------
C
C      SUBROUTINE NLSYST :
C         THIS SUBROUTINE SOLVES A SYSTEM OF N NON-
C         LINEAR EQUATIONS BY NEWTON'S METHOD.
C         THE PARTIAL DERIVATIVES OF THE FUNCTIONS ARE
C         ESTIMATED BY DIFFERENCE QUOTIENTS WHEN A VARIABLE
C         IS PERTURBED BY AN AMOUNT EQUAL TO DELTA ( DELTA IS
C         ADDED ). THIS IS DONE FOR EACH VARIABL E IN EACH
C         FUNCTION. INCREMENTS TO IMPROVE THE ESTIMATES FOR
C         X-VALUES ARE COMPUTED FROM SOLVING
C         A SYSTEM OF EQUATIONS USING SUBROUTINES
C         SUBROUTINES LUD AND SOLVE
C
C      -------------------------------------------------------------
C
C      PARAMETERS ARE :
C
C      FCN     - SUBROUTINE THAT COMPUTES VALUES OF THE FUNCTIONS.
C                  IT MUST BE DECLARED EXTERNAL
C                  IN THE CALLING PROGRAM.
C      N       - THE NUMBER OF EQUATIONS
C      MAXIT   - LIMIT TO THE NUMBER OF ITERATIONS
C                  THAT WILL BE USED
C      X       - ARRAY TO HOLD THE X VALUES. INITIALLY
C                  THIS ARRAY HOLDS THE INITIAL GUESSES.
C                  IT RETURNS THE FINAL VALUES.
C      F       - AN ARRAY THAT HOLDS VALUES OF THE FUNCTIONS
C      DELTA   - A SMALL VALUE USED TO PERTURB  C
C                      THE X VALUES SO PARTIAL DERIVATIVES CAN
C                      BE COMPUTED BY DIFFERENCE QUOTIENT.
C      XTOL    - TOLERANCE VALUE FOR CHANGE IN X VALUES . WHEN THE
C                      LARGEST CHANGE IN ANY X MEETS XTOL, THE
C                      SUBROUTINE TERMINATES.
C      FTOL    - TOLERANCE VALUE ON F TO TERMINATE.
C      I       - RETURNS VALUES TO INDICATE HOW THE ROUTINE TERMINATED
C        I=1     XTOL WAS MET
C        I=2     FTOL WAS MET
C        I=-1    MAXIT EXCEEDED BUT TOLERANCES NOT MET
C        I=-2    VERY SMALL PIVOT ENCOUNTERED IN GAUSSIAN
C                      ELIMINATION
C                  STEP - NO RESULTS OBTAINED
C        I=-3    INCORRECT VALUE OF N WAS SUPPLIED - N MUST BE
C                  BETWEEN 2 AND 10
C      METHOD  - ALLOWS ONE TO COMPUTE THE JACOBIAN MATRIX IN
C                  SUBROUTINE FCNJ BY NUMERICAL OR
C                  ANALYTIC PARTIAL DERIVATIVES.
C        0       IMPLIES NUMERICAL PARTIALS ARE TO BE COMPUTED.
C        1       IMPLIES ANALYTIC PARTIALS. THIS REQUIRES THE USER
C                  TO INPUT THESE PARTIAL DERIVATIVES IN FCNJ.
C
C      -------------------------------------------------------------
C
          IMPLICIT NONE
          INTEGER N,  MAXIT, I
          REAL*8 X(N),F(N),DELTA,XTOL,FTOL
          REAL*8 A(10,10), XSAVE(10),FSAVE(10),B(10), DET
          INTEGER IPVT(10),IT,IVBL,ITEST,IFCN,IROW,JCOL,    METHOD
          COMMON A, XSAVE, FSAVE
          EXTERNAL FCN, FCNJ
C
```

Figure 2.6 *(continued)*

```
C     -----------------------------------------------------------
C
C     CHECK VALIDITY OF VALUE OF N
C
      IF ( N .LT. 2 .OR. N .GT. 10 ) THEN
        I = -3
        PRINT 1004, N
        RETURN
      END IF
C     -----------------------------------------------------------
C
C     BEGIN ITERATIONS - SAVE X VALUES, THEN GET F VALUES
C
      DO 100 IT = 1,MAXIT
        CALL FCN(X,F)
C     -----------------------------------------------------------
C
C     TEST F VALUES AND SAVE THEM
C
        ITEST = 0
        DO 20 IFCN = 1,N
          IF ( ABS(F(IFCN)) .GT. FTOL ) ITEST = ITEST + 1
20      CONTINUE
        IF ( I .EQ. 0 ) THEN
          PRINT 1000, IT,X
          PRINT 1001, F
        END IF
C     -----------------------------------------------------------
C
C     SEE IF FTOL IS MET. IF NOT, CONTINUE. IF SO, SET I = 2 AND
C       RETURN.
C
        IF ( ITEST .EQ. 0 ) THEN
          I = 2
          RETURN
        END IF
C     -----------------------------------------------------------
C
        CALL FCNJ(FCN,METHOD, B,N,X,F,DELTA)
C
        CALL LUD(A,N,IPVT,10, DET)
        CALL SOLVE(A,N,IPVT,B,10)
C
C     BE SURE THAT THE COEFFICIENT MATRIX IS NOT TOO ILL-CONDITIONED
C
        DO 70 IROW = 1,N
          IF ( ABS(A(IROW,IROW)) .LE. 1.0E-6 ) THEN
            I = -2
            PRINT 1003
            RETURN
          END IF
70      CONTINUE
C     -----------------------------------------------------------
C
C     APPLY THE CORRECTIONS TO THE X VALUES, SEE IF XTOL IS MET.
C
        ITEST = 0
        DO 80 IVBL = 1,N
          X(IVBL) = XSAVE(IVBL) + B(IVBL)
          IF ( ABS(B(IVBL)) .GT. XTOL ) ITEST = ITEST + 1
80      CONTINUE
```

Figure 2.6   *(continued)*

```
C
C     ------------------------------------------------------------
C
C   IF XTOL IS MET, PRINT LAST VALUES AND RETURN, ELSE DO ANOTHER
         ITERATION.
C
         IF ( ITEST .EQ. 0 ) THEN
             I = 1
             IF ( I .EQ. 0 ) PRINT 1002, IT,X
             RETURN
          END IF
100    CONTINUE
C
C     ------------------------------------------------------------
C
C   WHEN WE HAVE DONE MAXIT ITERATIONS, SET I = -1 AND RETURN
C
       I = -1
       RETURN
C
1000   FORMAT(/' AFTER ITERATION NUMBER',I3,' X AND F VALUES ARE'
      +         //10F13.5)
1001   FORMAT(/10F13.5)
1002   FORMAT(/' AFTER ITERATION NUMBER',I3,' X VALUES (MEETING',
      +        ' XTOL) ARE '//10F13.5)
1003   FORMAT(/' CANNOT SOLVE SYSTEM. MATRIX NEARLY SINGULAR.')
1004   FORMAT(/' NUMBER OF EQUATIONS PASSED TO NLSYST IS
      + INVALID.',
      +        ' MUST BE 1 < N < 11. VALUE WAS ',I3)
C
       END
C
C     ------------------------------------------------------------
C
C     SUBROUTINE FCN:
C         THE TWO NONLINEAR FUNCTION ARE DEFINED
C
C     ------------------------------------------------------------
       SUBROUTINE FCN(X,F)
       IMPLICIT NONE
       REAL*8 X(*),F(*)
       F(1) =  X(1)*X(1) + X(2)*X(2) - 5.0
       F(2) =  X(2) - EXP(X(1)) - 1.0
       RETURN
       END
C
C     ------------------------------------------------------------
C
C     SUBROUTINE FCNJ:
C         HERE THE MATRIX OF PARTIALS ARE COMPUTED:
C         METHOD:  0 - IMPLIES NUMERICAL PARTIALS
C                  1 - IMPLIES ANALYTIC PARTIALS
C                      WHICH  MUST BE SUPPLIED.
C
C     ------------------------------------------------------------
C
       SUBROUTINE FCNJ(FCN,METHOD,B, N,X,F,DELTA)
       INTEGER N, METHOD
       REAL*8   X(N), F(N), DELTA
       INTEGER IROW, JCOL, NP
       REAL*8 A(10,10), FSAVE(10), XSAVE(10), B(10)
       COMMON A,XSAVE,FSAVE
C
       NP = NP + 1
C
```

Figure 2.6   *(continued)*

```
            DO 10  IROW = 1,N
                FSAVE(IROW) = F(IROW)

                XSAVE(IROW) = X(IROW)
10          CONTINUE
C
            IF (METHOD .EQ. 0) THEN
C
C                       COMPUTE NUMERICAL PARTIALS
C
C  THIS DOUBLE LOOP COMPUTES THE PARTIAL DERIVATIVES OF EACH
C  FUNCTION FOR EACH VARIABLE AND STORES THEM
C  IN A COEFFICIENT ARRAY.
C
            DO 50 JCOL = 1,N
                X(JCOL) = XSAVE(JCOL) + DELTA
                CALL FCN(X,F)
                DO 40 IROW = 1,N
                  A(IROW,JCOL) = (F(IROW) - FSAVE(IROW)) / DELTA
40          CONTINUE
C
C  RESET X VALUES FOR NEXT COLUMN OF PARTIALS
C
                X(JCOL) = XSAVE(JCOL)
50          CONTINUE
            ELSE
C
C                       COMPUTE ANALYTIC PARTIALS
C
            A(1,1) = 2.0*X(1)
            A(2,1) = -EXP(X(1))
            A(1,2) = 2.0*X(2)
            A(2,2) = 1.0
C
            ENDIF
C
C  ------------------------------------------------------------
C
C  NOW WE PUT NEGATIVE OF F VALUES AS RIGHT-HAND SIDES
C
            DO 60 IROW = 1 , N
                B(IROW) = - FSAVE(IROW)
60          CONTINUE
            RETURN
            END
C
C               SUBROUTINE LUD
C
            SUBROUTINE LUD(A,N,IPVT,NDIM,DET)
C
C
C  ------------------------------------------------------------
C
C
C           SUBROUTINE LUD: RETURNS THE LU DECOMPOSITION OF THE
C                   COEFFICIENT MATRIX. THE METHOD IS BASED ON
C
C                   THE ALGORITHM PRESENTED IN SECTION 4.
C
C  INPUT:  A  -   THE COEFFICIENT MATRIX
C          N  -   THE NUMBER OF EQUATIONS
C          NDIM  -   THE MAXIMUM ROW DIMENSION OF A
C
C  OUTPUT: A  -   THE LU DECOMPOSITION OF THE MATRIX A
C                 THE ORIGINAL MATRIX A IS LOST
```

Figure 2.6   *(continued)*

```
C                 IPVT  -  A VECTOR CONTAINING THE ORDER OF THE ROWS
C                          OF THE REARRANGED MATRIX DUE TO PIVOTING
C          DET  -  THE DETERMINANT OF THE MATRIX. IT IS SET TO
C                    0 IF ANY PIVOT ELEMENT IS LESS THAN 0.00001.
C
C     ------------------------------------------------------------
C
      IMPLICIT NONE
      REAL*8 A(NDIM,N),  SAVE,RATIO,VALUE,DET
      INTEGER IPVT(N),N,NDIM,  I,IPVTMT,NLESS1,IPLUS1,J,L
      INTEGER KCOL,JCOL,JROW,TMPVT
      EXTERNAL FNCJ
C
      DET = 1.0
      NLESS1 = N - 1
      DO 10 I = 1,N
          IPVT(I) = I
10        CONTINUE
C
      DO 20 I = 1,NLESS1
          IPLUS1 = I+1
          IPVTMT = I
C
C         FIND PIVOT ROW
C
          DO 30 J = IPLUS1,N
              IF (ABS(A(IPVTMT,I)) .LT. ABS(A(J,I))) IPVTMT = J
30            CONTINUE
C
C     ------------------------------------------------------------
C
C         CHECK FOR SMALL PIVOT ELEMENT
C
C     ------------------------------------------------------------
C
          IF (ABS(A(IPVTMT,I)). LT. 1.0E-05) THEN
              DET = 0.0
              PRINT '(//)'
              PRINT *, '   MATRIX IS SINGULAR OR NEAR SINGULAR   '
              PRINT '(//)'
              RETURN
          ENDIF
C
C         INTERCHANGE ROWS IF NECESSARY
C
          IF (IPVTMT .NE. I) THEN
              TMPVT = IPVT(I)
              IPVT(I) = IPVT(IPVTMT)
              DO 40 JCOL = 1,N
                  SAVE = A(I,JCOL)
                  A(I,JCOL) = A(IPVTMT,JCOL)
                  A(IPVTMT,JCOL) = SAVE
40                CONTINUE
              IPVT(IPVTMT) = TMPVT
              DET = -DET
          END IF
C
C             REDUCE ALL ELEMENTS BELOW THE I'TH ROW
C
          DO 50 JROW = IPLUS1,N
              IF (A(JROW,I) .NE. 0.0) THEN
                  A(JROW,I) = A(JROW,I)/A(I,I)
```

Figure 2.6 *(continued)*

```
                         DO 60 KCOL = IPLUS1,N
                            A(JROW,KCOL) = A(JROW,KCOL) -
        +                               A(JROW,I)*A(I,KCOL)
60                             CONTINUE
                         END IF
50                    CONTINUE
C
20             CONTINUE
C
        IF (ABS(A(N,N)) .LT. 1.0E-5 ) THEN
             PRINT '(//)'
             PRINT *, '   MATRIX IS SINGULAR OR NEAR SINGULAR '
             PRINT '(//)'
             DET = 0.0


             RETURN
        END IF

C
C
C  ---------------------------------------------------------------
C
C    COMPUTE THE DETERMINANT OF THE MATRIX
C
C  ---------------------------------------------------------------
C
        DO 70 I = 1,N
             DET = DET * A(I,I)
70           CONTINUE
C
        RETURN
        END
C
C  -----------------------------------------------------------
C
C                  SUBROUTINE SOLVE
C
C  -----------------------------------------------------------
C
        SUBROUTINE SOLVE(A,N,IPVT,B,NDIM)
C
C  -----------------------------------------------------------
C
C    MAKING USE OF THE LU DECOMPOSITION OF THE MATRIX A, THIS
C    SUBROUTINE SOLVES THE SYSTEM BY FORWARD- AND BACK-SUBSTITUTION.
C
C    INPUT:    A - LU MATRIX FROM SUBROUTINE LUD
C              N - NUMBER OF EQUATIONS
C              IPVT - A RECORD OF THE REARRANGEMENT OF THE ROWS
C                     OR A FROM SUBROUTINE LUD
C              B - RIGHT-HAND-SIDE OF THE SYSTEM OF EQUATIONS
C
C    OUTPUT:  B - THE SOLUTION VECTOR
C
C  -----------------------------------------------------------
C
        IMPLICIT NONE
        REAL*8 A(NDIM,N), B(N), X(10), SUM
        INTEGER IPVT(N),N,NDIM,   IROW,JCOL,I
C
C  -----------------------------------------------------------
C
C    REARRANGE THE ELEMENTS OF THE B VECTOR. STORE THEM IN THE
C    X VECTOR.
C
```

Figure 2.6   *(continued)*

```
        DO 10 I = 1,N
           X(I)  = B(IPVT(I))
10         CONTINUE
C
C  ------------------------------------------------------------
C
C     SOLVE USING FORWARD SUBSTITUTION--LY = B
C
C  ------------------------------------------------------------
C
        DO 20 IROW = 2,N

           SUM = X(IROW)
           DO 30 JCOL = 1,(IROW-1)
               SUM = SUM - A(IROW,JCOL)*X(JCOL)
30             CONTINUE
           X(IROW) = SUM
20         CONTINUE
C
C  ------------------------------------------------------------
C
C     SOLVE BY BACK SUBSTITUTION--UX = Y
C
C  ------------------------------------------------------------
C
        B(N)  = X(N)/A(N,N)
         DO 40 IROW = (N-1),1,-1
           SUM = X(IROW)
           DO 50 JCOL = (IROW+1),N
               SUM = SUM - A(IROW,JCOL)*B(JCOL)
50             CONTINUE
           B(IROW) = SUM/A(IROW,IROW)
40         CONTINUE

        RETURN
        END

*******************************************************************
                    OUTPUT FOR NLSYST.F

*******************************************************************

 AFTER ITERATION NUMBER 1  X AND F VALUES ARE

        3.00000      2.00000

        8.00000     -19.08554

 AFTER ITERATION NUMBER 2  X AND F VALUES ARE

        2.02316      1.46525

        1.24016     -7.09696

 AFTER ITERATION NUMBER 3  X AND F VALUES ARE

        1.18226      2.20315

        1.25160     -2.05860
```

Figure 2.6 *(continued)*

```
AFTER ITERATION NUMBER 4  X AND F VALUES ARE

     0.56551        2.25006

     0.38258       -0.51029

AFTER ITERATION NUMBER 5  X AND F VALUES ARE

     0.26959        2.23942

     0.08768       -0.07001

AFTER ITERATION NUMBER 6  X AND F VALUES ARE

     0.20694        2.22739

     0.00407       -0.00252

AFTER ITERATION NUMBER 7  X AND F VALUES ARE

     0.20434        2.22671

     0.00000        0.00000

AFTER ITERATION NUMBER 8  X AND F VALUES ARE

     0.20434        2.22671

     0.00000        0.00000


THE X-VALUES ARE:

     0.204337

     2.226712
```

Exercises

Section 2.2

1. Given the matrices A, B, and the vectors x, y,

$$A = \begin{bmatrix} 2 & -1 & 3 & -4 \\ 0 & 3 & 4 & -1 \\ 2 & 5 & 5 & 4 \end{bmatrix}, \quad x = \begin{bmatrix} 1 \\ -3 \\ 8 \\ -2 \end{bmatrix},$$

$$B = \begin{bmatrix} 1 & -1 & 6 & -4 \\ 3 & 3 & 4 & 3 \\ 4 & 0 & 2 & -5 \end{bmatrix}, \quad y = \begin{bmatrix} -1 \\ 2 \\ -3 \\ 2 \end{bmatrix},$$

a. Find $3A, 2A + 4B, 2x - 3y$.
▶**b.** Find $A - B, Ax, By$.
▶**c.** Find $x^T y, xy^T$.
d. Find B^T.

2. Given the matrices

$$A = \begin{bmatrix} -3 & 1 & -2 \\ 2 & 3 & 0 \\ -1 & 2 & 3 \end{bmatrix}, \quad B = \begin{bmatrix} 3 & -2 & 5 \\ 2 & -4 & 1 \\ -4 & 1 & 6 \end{bmatrix},$$

▶**a.** Find BA, B^3, AA^T.
b. Find $\det(A), \det(B)$.

c. A square matrix can always be expressed as a sum of an upper-triangular matrix U and a lower-triangular matrix L. Find two different combinations of L's and U's such that $A = L + U$.

3. Given the matrices

$$A = \begin{bmatrix} 1 & -2 & 2 \\ 3 & 1 & 1 \\ 2 & 0 & 1 \end{bmatrix}, \quad B = \begin{bmatrix} -1 & -2 & 4 \\ 1 & 3 & -5 \\ 2 & 4 & -7 \end{bmatrix},$$

$$C = \begin{bmatrix} 2 & 1 & 2 \\ 1 & 0 & 4 \\ 1 & 1 & 3 \end{bmatrix},$$

a. Show that $AB = BA = I$, where I is the 3×3 identity matrix. We shall later identify B as the inverse of A.
b. Show that $AI = IA = A$.
c. Show that $AC \ne CA$ and also that $BC \ne CB$. In general, matrices do not commute under multiplication.
▶**d.** A square matrix can be expressed also as the sum of an upper-triangular matrix, a diagonal matrix, and a lower-triangular matrix. Express A as $L + D + U$.

4. Let

$$A = \begin{bmatrix} -2 & 8 \\ -1 & 7 \end{bmatrix}, \quad B = \begin{bmatrix} 8 & -1 & 1 \\ -2 & 2 & -1 \\ -2 & 4 & -3 \end{bmatrix}$$

a. Find the characteristic polynomials of both A and B.
▶**b.** Find the eigenvalues of both A and B.

5. Write as a set of equations:

$$\begin{bmatrix} 2 & 1 & 1 & -2 \\ 4 & 0 & 2 & 1 \\ 3 & 2 & 2 & 0 \\ 1 & 3 & 2 & 0 \end{bmatrix} \begin{bmatrix} x_1 \\ x_2 \\ x_3 \\ x_4 \end{bmatrix} = \begin{bmatrix} 0 \\ 8 \\ 7 \\ 3 \end{bmatrix}.$$

▶**6.** Write in matrix form:

$$3x - 6y + 2z = 15,$$
$$-4x + y - z = -2,$$
$$x - 3y + 7z = 22.$$

Section 2.3

7. a. Solve by back-substitution:

$$2x_1 - 3x_2 + x_3 = -11,$$
$$4x_2 - 3x_3 = -10,$$
$$2x_3 = 4.$$

▶**b.** Solve by forward-substitution:

$$5x_3 = 10,$$
$$3x_2 - 3x_3 = 3,$$
$$2x_1 - x_2 + 2x_3 = 7.$$

8. Solve the set of equations in Exercise 5.
9. Solve the set of equations in Exercise 6.
10. Solve the following (given as the augmented matrix):

$$\begin{bmatrix} 1 & 1 & -2 & | & 3 \\ 4 & -2 & 1 & | & 5 \\ 3 & -1 & 3 & | & 8 \end{bmatrix}.$$

▶**11.** Show that the following does not have a solution:

$$3x_1 + 2x_2 - x_3 - 4x_4 = 10,$$
$$x_1 - x_2 + 3x_3 - x_4 = -4,$$
$$2x_1 + x_2 - 3x_3 = 16,$$
$$-x_2 + 8x_3 - 5x_4 = 3.$$

12. If the right-hand side of Exercise 11 is $(2, 3, 1, 3)^T$, show that there are an infinite number of solutions.
▶**13.** Show that the set of left-hand sides of Exercise 11 are not independent vectors.

Section 2.4

14. Using Gaussian elimination with partial pivoting and back-substitution,
a. Solve the equations of Exercise 5.
b. Using part (a), find the determinant of the coefficient matrix.
c. What is the LU decomposition of the coefficient matrix? (Rows may have been interchanged.)
▶**15.** Using Gaussian elimination with partial pivoting and back-substitution,
a. Solve the equations of Exercise 10.
b. Using part (a), find the determinant of the coefficient matrix.
c. What is the LU decomposition of the coefficient matrix? (Rows may have been interchanged.)

▶**16. a.** Solve the system

$$2.51x_1 + 1.48x_2 + 4.53x_3 = 0.05,$$
$$1.48x_1 + 0.93x_2 - 1.30x_3 = 1.03,$$
$$2.68x_1 + 3.04x_2 - 1.48x_3 = -0.53,$$

by Gaussian elimination, carrying just three significant digits and chopping. Do not interchange rows. Observe that there is a small divisor in reducing the third equation.

b. Repeat part (a), but now use partial pivoting. Observe that there are no small divisors.

c. Substitute each set of answers into the original equations and observe that the left- and right-hand sides match much better with the answers to part (b). The solution, correct to six digits, is

$$x_1 = 1.45310, \quad x_2 = -1.58919, \quad x_3 = -0.27489.$$

17. Solve the systems of Exercises 5, 10, and 11 by the Gauss–Jordan method.

18. Augment the coefficient matrix with all three of the right-hand sides and get all three solutions simultaneously, given

$$A = \begin{bmatrix} 4 & 0 & -1 & 3 \\ 2 & 1 & -2 & 0 \\ 0 & 3 & 2 & -2 \\ 1 & 1 & 0 & 5 \end{bmatrix}, \quad b_1 = \begin{bmatrix} 0 \\ 1 \\ 4 \\ -2 \end{bmatrix},$$

$$b_2 = \begin{bmatrix} 0 \\ 0 \\ -2 \\ 4 \end{bmatrix}, \quad b_3 = \begin{bmatrix} 7 \\ -1 \\ 4 \\ -2 \end{bmatrix}.$$

19. a. For a general $n \times n$ matrix, show that steps 1–4 of Gaussian elimination take at most $n(n-1)(2n-1)/6 + n(n-1)$ multiplications/divisions. You will need to know that

$$1 + 2 + 3 + \cdots + n = \frac{n(n+1)}{2},$$

$$1^2 + 2^2 + 3^2 + \cdots + n^2 = \frac{n(2n+1)(n+1)}{6}.$$

b. Show also that the back-substitution part of Gaussian elimination takes $n(n+1)/2$ multiplications/divisions.

▶**c.** Verify that Gauss–Jordan takes about 50% more

operations than Gaussian elimination for the case of three equations. In this, add the number of adds, subtracts, multiplies, and divides.

▶**20.** Suppose we want to solve the system $Az = b$, where the a_{ij}, z_i, and b_i are complex numbers.

a. Show that this can be done using only real arithmetic. (*Hint: A can be written as $B + Ci$.*)

b. If one solves the system in a computer using a language that permits complex numbers, compare the amount of storage space needed compared to the amount if done as in part (a).

21. a. Show that the system

$$\begin{bmatrix} 2 + 2i & -1 + 2i \\ -3i & 3 - 2i \end{bmatrix} \begin{bmatrix} z_1 \\ z_2 \end{bmatrix} = \begin{bmatrix} 2 + 2i \\ 1 - 3i \end{bmatrix}$$

can be written as

$$\begin{bmatrix} 2 & -1 & -2 & -2 \\ 0 & 3 & 3 & 2 \\ 2 & 2 & 2 & -1 \\ -3 & -2 & 0 & 3 \end{bmatrix} \begin{bmatrix} x_1 \\ x_2 \\ y_1 \\ y_2 \end{bmatrix} = \begin{bmatrix} 2 \\ 1 \\ 2 \\ -3 \end{bmatrix}.$$

▶**b.** Solve the system of part (a), then find z_1 and z_2.

Section 2.5

▶**22.** Use Crout reduction to solve Exercise 10.

23. Use Crout reduction to solve Exercise 11.

24. Repeat Exercise 16, but this time use Crout reduction.

25. Suppose that we do not know all of the three right-hand sides of Exercise 18 in advance.

a. Solve $Ax = b_1$ by Gaussian elimination, getting the LU decomposition. Then use the LU to solve with the other two right-hand sides.

b. Repeat part (a), this time using Crout reduction.

26. Exercise 19 shows that the number of multiplications/divisions to solve a system of n equations is $O(n^3)$. If we already have the LU decomposition of the coefficient matrix, find how many multiplications/divisions are required to first get $Ly = b'$ and then solve from $Ux = b'$. Make a table that compares the total number of multiplications/divisions in the two cases, if $n = 5, 10, 20, 100$.

Section 2.6

27. Which of these matrices are singular?

a.
$$A = \begin{bmatrix} 3 & 2 & -1 \\ 0 & -1 & 4 \\ 6 & 3 & 2 \end{bmatrix}$$

b.
$$B = \begin{bmatrix} 1 & 0 & -2 & 3 \\ 3 & 1 & 1 & 4 \\ -1 & 0 & 2 & -1 \\ 4 & 2 & 6 & 0 \end{bmatrix}$$

c.
$$C = \begin{bmatrix} 1 & 0 & -2 & 3 \\ 3 & 1 & 1 & 4 \\ -1 & 0 & 2 & -1 \\ 4 & 3 & 6 & 0 \end{bmatrix}$$

28. a. Find values of x and y that make A singular:

$$A = \begin{bmatrix} 4 & 1 & 0 \\ 2 & -1 & 2 \\ x & y & -1 \end{bmatrix}.$$

b. Find values for x and y that make A nonsingular.

29. Matrix A in Exercise 27 is singular. Do its rows form linearly independent vectors? Find the values for the a_i in Eq. (2.28).
►**b.** Repeat part (a) for the elements of A considered as column vectors.

30. Do these sets of equations have a solution? Find a solution if it exists.

a.
$$\begin{cases} 3x - 2y + z = 2, \\ x - 3y + z = 5, \\ x + y - z = -5, \\ 3x \quad\quad + z = 0. \end{cases}$$

►**b.**
$$\begin{bmatrix} 1 & 1 & 0 \\ 0 & 1 & 1 \\ 1 & 0 & 1 \\ 1 & 1 & 1 \end{bmatrix} x = \begin{bmatrix} 1 \\ -2 \\ 0 \\ 4 \end{bmatrix}.$$

c.
$$\begin{bmatrix} 2 & 1 & 3 \\ -1 & 0 & 2 \\ 6 & 2 & 2 \end{bmatrix} x = \begin{bmatrix} 1 \\ 0 \\ 0 \end{bmatrix}.$$

d.
$$\begin{bmatrix} 2 & 1 & 3 \\ -1 & 0 & 2 \\ 6 & 2 & 2 \end{bmatrix} x = \begin{bmatrix} 1 \\ 0 \\ 2 \end{bmatrix}.$$

►**31.** The Hilbert matrix is a classic case of the pathological situation called "ill-conditioning." The 4×4 Hilbert matrix is

$$H = \begin{bmatrix} 1 & \frac{1}{2} & \frac{1}{3} & \frac{1}{4} \\ \frac{1}{2} & \frac{1}{3} & \frac{1}{4} & \frac{1}{5} \\ \frac{1}{3} & \frac{1}{4} & \frac{1}{5} & \frac{1}{6} \\ \frac{1}{4} & \frac{1}{5} & \frac{1}{6} & \frac{1}{7} \end{bmatrix}.$$

For the system $Hx = b$, with $b^T = [25/12, 77/60, 57/60, 319/420]$, the exact solution is $x^T = [1, 1, 1, 1]$.
a. Show that the matrix is ill-conditioned by showing that it is nearly singular.
b. Using only three significant digits (chopped) in your arithmetic, find the solution to $Hx = b$. Explain why the answers are so poor.
c. Using only three significant digits, but rounding, again find the solution and compare it to that obtained in part (b).
d. Using five significant digits in your arithmetic, again find the solution and compare it to those found in parts (b) and (c).

Section 2.7

32. Find the determinant of the matrix

$$\begin{bmatrix} 0 & 1 & -1 \\ 3 & 1 & -4 \\ 2 & 1 & 1 \end{bmatrix}$$

by row operations to make it (a) upper-triangular, (b) lower-triangular.

33. Find the determinant of the matrix

$$\begin{bmatrix} 1 & 4 & -2 & 1 \\ -1 & 2 & -1 & 1 \\ 3 & 3 & 0 & 4 \\ 4 & -4 & 2 & 3 \end{bmatrix}.$$

34. Invert the coefficient matrix in Exercise 5, and then use the inverse to generate the solution.

35. If the constant vector in Exercise 5 is changed to one with components $(1, 2, 4, 2)$, what is now the solution? Observe that the inverse obtained in Exercise 34 gives the answer readily.

36. Attempt to find the inverse of the coefficient matrix in Exercise 11. Note that a singular matrix has no inverse.

37. a. Find the determinant of the Hilbert matrix in Exercise 31. A small value of the determinant (when the matrix has elements of the order of unity) indicates ill-conditioning.
▶**b.** Find the inverse of the Hilbert matrix in Exercise 31. The inverse of an ill-conditioned matrix has some very large elements in comparison to the elements of the original matrix.

▶**38.** Both Gaussian elimination and the Gauss–Jordan method can be adapted to invert a matrix. In Section 2.7, we say "it is more efficient to use the Gaussian elimination algorithm." Verify this for the specific case of a 3×3 matrix by counting arithmetic operations for each method.

Section 2.8

39. For each of the following, evaluate the norms $\|*\|_p, p = 1, 2$, and ∞. (For a matrix, the 2-norm is the largest magnitude eigenvalue.)
  a. $x = [1.15, -2.3, 19.1, 2.0]$
  b. $y = [-2, 1, 0, 7, -11]$

$$\mathbf{c.}\ A = \begin{bmatrix} -2 & 8 & 0 \\ -1 & 7 & 0 \\ 0 & 3 & 2 \end{bmatrix}$$

$$\blacktriangleright \mathbf{d.}\ B = \begin{bmatrix} 8 & -2 & 1 \\ -2 & 2 & -1 \\ -2 & 4 & -3 \end{bmatrix}$$

  e. Find the norms of $B^2, A + B, AB$.
  f. Does the triangle inequality of Eq. (2.29) hold for $A + B$? for $x + y$?

▶**40.** Find the ∞-norm of the Hilbert matrix of Exercise 31.

41. Find the ∞-norm of the inverse of the Hilbert matrix of Exercise 31.

Section 2.9*

42. Consider the system $Ax = b$, where

$$A = \begin{bmatrix} 3.01 & 6.03 & 1.99 \\ 1.27 & 4.16 & -1.23 \\ 0.987 & -4.81 & 9.34 \end{bmatrix}, \quad b = \begin{bmatrix} 1 \\ 1 \\ 1 \end{bmatrix}.$$

▶**a.** Using double precision (or a calculator with 10 or more digits of accuracy), solve for x.
  b. Solve the system using three-digit (chopped) arithmetic for each arithmetic operation; call this solution $\bar{x}$.
  c. Compare x and $\bar{x}$, and compute $e = x - \bar{x}$. What is $\|e\|_2$?
  d. Is the system ill-conditioned? What evidence is there to support your conclusion?

43. Repeat Exercise 42, but change the element a_{33} to -9.34.

▶**44.** Suppose, in Exercise 42, that uncertainties of measurement give slight changes in some of the elements of A. Specifically, suppose a_{11} is 3.00 instead of 3.01 and a_{31} is 0.99 instead of 0.987. What change does this cause in the solution vector (using precise arithmetic)?

45. Compute the residuals for the imperfect solutions in 42(b) and 43(b). Use double precision in this computation.

46. What are the condition numbers of the coefficient matrices in Exercises 42 and 43? Use the 1-norms.

47. Verify Eq. (2.37), using the results in Exercises 42 and 43.

▶**48.** Verify Eq. (2.37), using the results in Exercises 43, 45, and 46.

49. Verify Eq. (2.38) for the results of Exercise 44.

50. Apply iterative improvement to the imperfect solution of Exercise 42.

▶**51.** Apply iterative improvement to the imperfect solution of Exercise 43.

Section 2.10

52. Solve Exercise 6 by the Jacobi method, beginning with the initial vector $(0, 0, 0)$. Compare the rate of convergence when Gauss–Seidel is used with the same starting vector.

53. Solve Exercise 6 by Gauss–Seidel iteration, beginning with approximate solution $(2, 2, -1)$.

▶**54.** The pair of equations

$$x_1 + 2x_2 = 3,$$
$$3x_1 + x_2 = 4,$$

---

* In certain exercises (42, 43, 44, 50, 51), imperfect solutions will result because low-precision arithmetic is used when the condition number is large. This exaggerates the condition number problem.

can be rearranged to give $x_1 = 3 - 2x_2, x_2 = 4 - 3x_1$. Apply the Jacobi method to this rearrangement, beginning with a vector very close to the solution $x^{(1)} = (1.01, 1.01)^T$, and observe divergence. Now apply Gauss–Seidel. Which method diverges more rapidly?

▶**55.** Solve the system

$$9x + 4y + z = -17,$$
$$x - 2y - 6z = 14,$$
$$x + 6y = 4,$$

a. Using the Gauss–Jacobi method.
b. Using the Gauss–Seidel method. How much faster is the convergence than in part (a)?

Section 2.11

▶**56.** Beginning with $(0, 0, 0)$, use relaxation to solve the system

$$6x_1 - 3x_2 + x_3 = 11,$$
$$2x_1 + x_2 - 8x_3 = -15,$$
$$x_1 - 7x_2 + x_3 = 10.$$

57. Solve the system in Exercise 55 by relaxation.

58. Relaxation is especially well adapted to problems like Exercise 18. Solve by the relaxation method, starting with the vector $[1, 1, 1, 1]$ which one obtains by inspection.

Section 2.12

▶**59.** Find the two intersections nearest the origin of the two curves $x^2 + x - y^2 = 1$ and $y - \sin x^2 = 0$.

60. Solve the system

$$x^2 + y^2 + z^2 = 9,$$
$$xyz = 1,$$
$$x + y - z^2 = 0,$$

by Newton's method to obtain the solution near $(2.5, 0.2, 1.6)$.

▶**61.** Solve by using Newton's method:

$$x^3 + 3y^2 = 21,$$
$$x^2 + 2y + 2 = 0.$$

Make sketches of the graphs to locate approximate values of the intersections.

62. Apply Eq. (2.58) to compute partials and solve this system by Newton's method:

$$xyz - x^2 + y^2 = 1.34,$$
$$xy - z^2 = 0.09,$$
$$e^x - e^y + z = 0.41.$$

There should be a solution near $(1, 1, 1)$.

63. At the end of Section 2.12, it is suggested that it would be more efficient to avoid recomputing the partials at each step of Newton's method for a non-linear system, doing it only after each nth step when there are n equations. Redo Exercises 59 and 62 using this modification. Compare the rate of convergence with that when the partials are recomputed at each step.

Section 2.13

64. Given matrix A, write the permutation matrix that does the following interchanges.

$$A = \begin{bmatrix} 4 & 7 & 3 & -2 \\ 0 & 0 & 2 & 1 \\ -6 & 1 & 1 & 0 \\ 1 & 0 & 1 & 8 \end{bmatrix}.$$

a. Row 3 with row 1
▶**b.** Row 1 with row 4
▶**c.** Column 2 with column 1
▶**d.** Row 2 with row 4 and column 4 with column 2 simultaneously

65. Confirm that $P^{-1} = P$ for each permutation matrix of Exercise 64.

66. Confirm Eq. (2.61) by first computing $H = H_3 P_3 H_2 P_2 H_1 P_1$ and then multiplying this times A.

67. Repeat Exercise 66, this time using

$$H = H_3 H_2 H_1 P_3 P_2 P_1.$$

Section 2.14

(Note that for Exercises 68–76, the matrix A can indicate either the coefficient matrix or the augmented matrix.)

68. Use MAPLE to solve Exercise 10.
a. Use linsolve.
b. Use gaussjord(A) and gausselim(A).

69. Use MAPLE to solve Exercise 2, parts (a) and (b).

70. Use MAPLE to solve Exercise 3.

71. Use MAPLE to solve the system of Exercise 10
  a. through reduced row echelon, rref(A).
  b. using the inverse matrix.
  c. with linsolve(A, b).

72. Use MAPLE to solve Exercise 4.

73. Use MAPLE to get the exact solution to the system given in Exercise 31.

74. Do Exercise 37 with MAPLE.

75. Plot the curves of Exercise 59 to get the approximations, then refine with fsolve.

76. Solve Exercise 61 using MAPLE.

Section 2.15

77. Show that solving an $n \times n$ system by Gaussian elimination requires these numbers of steps:
Making upper-triangular:

$(2n^3 + 3n^2 - 5n)/6$ multiplications/divisions

$(n^3 - n)/3$ additions/subtractions

Back-substitution:

$(n^2 + n)/2$ multiplications/divisions

$(n^2 - n)/2$ additions/subtractions

For a total of

$(n^3 + 3n^2 - n)/3$ multiplications/divisions

$(2n^3 + 3n^2 - 5n)/6$ additions/subtractions

78. Find the equivalent number of operations (as in Exercise 77) for the Gauss–Jordan method.

79. Develop an algorithm for inverting an $n \times n$ nonsingular matrix by parallel processing using approximately n^2 processors.

80. The final algorithm developed in Section 2.15 used $n^2 + n$ processors. Show how this can further be improved so that only $(n + 1)(n - 1) = n^2 - 1$ processors are needed.

81. Develop an algorithm for solving a system of n linear equations by Jacobi iteration using n^2 processors.

Applied Problems and Projects

82. In considering the movement of space vehicles, it is frequently necessary to transform coordinate systems. The standard inertial coordinate system has the N-axis pointed north, the E-axis pointed east, and the D-axis pointed toward the center of the earth. A second system is the vehicle's local coordinate system (with the i-axis straight ahead of the vehicle, the j-axis to the right, and the k-axis downward). We can transform the vector whose local coordinates are (i, j, k) to the inertial system by multiplying by transformation matrices:

$$\begin{bmatrix} n \\ e \\ d \end{bmatrix} = \begin{bmatrix} \cos a & -\sin a & 0 \\ \sin a & \cos a & 0 \\ 0 & 0 & 1 \end{bmatrix} \begin{bmatrix} \cos b & 0 & \sin b \\ 0 & 1 & 0 \\ -\sin b & 0 & \cos b \end{bmatrix} \begin{bmatrix} 1 & 0 & 0 \\ 0 & \cos c & -\sin c \\ 0 & \sin c & \cos c \end{bmatrix} \begin{bmatrix} i \\ j \\ k \end{bmatrix}.$$

Transform the vector $[2.06, -2.44, -0.47]^T$ to the inertial system if $a = 27°$, $b = 5°, c = 72°$.

83. a. Exercise 31 shows the pattern for Hilbert matrices. Find the condition number of the 9×9 Hilbert matrix.
  b. Suppose we have a system of nine linear equations whose coefficients are the 9×9 Hilbert matrix. Find the right-hand side (the b-vector) if the solution vector has ones for all components. Now increase the value of the first component of the b-vector by 1% and find the solution to the perturbed system. Which component of the solution vector is most changed?

84. Electrical engineers often must find the currents flowing and voltages existing in a complex resistor network. Here is a typical problem.

   Seven resistors are connected as shown, and voltage is applied to the circuit at points 1 and 6 (see Fig. 2.7). You may recognize the network as a variation on a Wheatstone bridge.

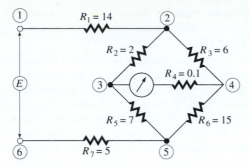

Figure 2.7

While we are especially interested in finding the current that flows through the ammeter, the computational method can give the voltages at each numbered point (these are called *nodes*) and the current through each of the branches of the circuit. Two laws are involved:

Kirchhoff's law: The sum of all currents flowing into a node is zero.

Ohm's law: The current through a resistor equals the voltage across it divided by its resistance.

We can set up eleven equations using these laws and from these solve for eleven unknown quantities (the four voltages and seven currents). If $V_1 = 5$ volts and $V_6 = 0$ volts, set up the eleven equations and solve to find the voltage at each other node and the currents flowing in each branch of the circuit.

85. Mass spectrometry analysis gives a series of peak height readings for various ion masses. For each peak, the height h_j is contributed to by the various constituents. These make different contributions c_{ij} per unit concentration p_i so that the relation

$$h_j = \sum_{i=1}^{n} c_{ij} p_i$$

holds, with n being the number of components present. Carnahan (1964) gives the values shown in Table 2.3 for c_{ij}:

Table 2.3

Peak number	Component				
	CH_4	C_2H_4	C_2H_6	C_3H_6	C_3H_8
1	0.165	0.202	0.317	0.234	0.182
2	27.7	0.862	0.062	0.073	0.131
3		22.35	13.05	4.420	6.001
4			11.28	0	1.110
5				9.850	1.684
6					15.94

If a sample had measured peak heights of $h_1 = 5.20$, $h_2 = 61.7$, $h_3 = 149.2$, $h_4 = 79.4$, $h_5 = 89.3$, and $h_6 = 69.3$, calculate the values of p_i for each component. The total of all the p_i values was 21.53.

86. The truss in Section 2.1 is called *statically determinate* because nine linearly independent equations can be established to relate the nine unknown values of the tensions in the members. If an additional cross brace is added, as sketched in Fig. 2.8, we have ten unknowns but still only nine equations can be written; we now have a statically *indeterminate* system. Consideration of the stretching or compression of the members permits a solution, however. We need to solve a set of equations that gives the displacements x of each pin, which is of the form $ASA^Tx = P$. We then get the tensions f by matrix multiplication: $SA^Tx = f$. The necessary matrices and vectors are

$$A = \begin{bmatrix} 0.7071 & 0 & 0 & -1 & -0.8660 & 0 & 0 & 0 & 0 & 0 \\ 0.7071 & 0 & 1 & 0 & 0.5 & 0 & 0 & 0 & 0 & 0 \\ 0 & 1 & 0 & 0 & 0 & -1 & 0 & 0 & 0 & -0.8660 \\ 0 & 0 & -1 & 0 & 0 & 0 & 0 & 0 & 0 & -0.5 \\ 0 & 0 & 0 & 0 & 0 & 0 & 1 & 0 & 0.7071 & 0.5 \\ 0 & 0 & 0 & 1 & 0 & 0 & 0 & 0 & -0.7071 & 0.8660 \\ 0 & 0 & 0 & 0 & 0.8660 & 1 & 0 & -1 & 0 & 0 \\ 0 & 0 & 0 & 0 & -0.5 & 0 & -1 & 0 & 0 & 0 \\ 0 & 0 & 0 & 0 & 0 & 0 & 0 & 1 & 0.7071 & 0 \end{bmatrix}.$$

S is a diagonal matrix with values (from upper left to lower right) of

$$\begin{array}{ccccc} 4255, & 6000, & 6000, & 3670, & 3000, \\ 3670, & 6000, & 6000, & 4255, & 3000. \end{array}$$

(These quantities are the values of aE/L, where a is the cross-sectional area of a member, E is the Young's modulus for the material, and L is the length.)

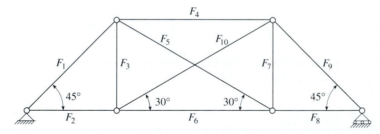

Figure 2.8

Solve the system of equations to determine the values of f for each of three loading vectors:

$$P_1 = [0, -1000, 0, 0, 500, 0, 0, -500, 0]^T,$$
$$P_2 = [1000, 0, 0, -500, 0, 1000, 0, -500, 0]^T,$$
$$P_3 = [0, 0, 0, -500, 0, 0, 0, -500, 0]^T.$$

87. For turbulent flow of fluids in an interconnected network (see Fig. 2.9), the flow rate V from one node to another is about proportional to the square root of the difference in pressures at the nodes. (Thus fluid flow differs from flow of electrical current in a network in that nonlinear equations result.) For the conduits in Fig. 2.9, find the pressure at each node. The values of b represent conductance factors in the relation $v_{ij} = b_{ij}(p_i - p_j)^{1/2}$.

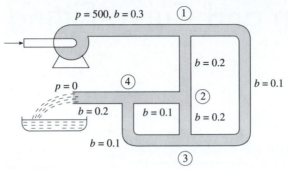

Figure 2.9

These equations can be set up for the pressures at each node:

$$\text{At node 1: } 0.3\sqrt{500 - p_1} = 0.2\sqrt{p_1 - p_2} + 0.1\sqrt{p_1 - p_3};$$
$$\text{node 2: } 0.2\sqrt{p_1 - p_2} = 0.1\sqrt{p_2 - p_4} + 0.2\sqrt{p_2 - p_3};$$
$$\text{node 3: } 0.1\sqrt{p_1 - p_3} + 0.2\sqrt{p_2 - p_3} = 0.1\sqrt{p_3 - p_4};$$
$$\text{node 4: } 0.1\sqrt{p_2 - p_4} + 0.1\sqrt{p_3 - p_4} = 0.2\sqrt{p_4 - 0}.$$

88. Translate each of the programs of this chapter to another computer language.

89. Implement the solutions to Exercise 55 with a spreadsheet.

3

Interpolation and Curve Fitting

Contents of This Chapter

Interpolation, the computing of values for a tabulated function at points not in the table, is historically a most important task. Many famous mathematicians have their names associated with procedures for interpolation: Gauss, Newton, Bessel, Stirling. The need to interpolate began with the early studies of astronomy when the motion of heavenly bodies was to be determined from periodic observations.

Today, students rarely have to interpolate for values of sines, logarithms, and other such nonalgebraic functions from tables. Their calculators and computers compute the values using techniques that are described in Chapter 10 of this book. Why then do we devote a lengthy chapter to a subject that seems almost obsolete? There are four reasons. First, interpolation methods are the basis for many other procedures that we will study: numerical differentiation and integration and solution methods for ordinary and partial-differential equations. Second, these methods demonstrate some important theory about polynomials and the accuracy of numerical methods. Third, the interpolating of polynomials serves as an excellent introduction to some techniques for drawing smooth curves. And, finally, history itself has a special fascination. This chapter compares several ways of doing interpolation and contrasts these procedures with several ways for fitting imprecise data and for drawing smooth curves.

3.1   An Interpolation Problem
Describes a typical situation when a professional numerical analyst would need to do interpolation.

3.2   Lagrangian Polynomials
Presents the Lagrangian polynomial—a straightforward, but computationally awkward, way to construct an interpolating polynomial. The cost

of determining the accuracy of the interpolated value is reduced by a variant, Neville's method.

3.3   Divided Differences

Describes these more efficient methods for constructing the interpolating polynomials and getting interpolated values. They readily allow changing the degree of the interpolating polynomial.

3.4   Evenly Spaced Data

Simplifies the problem of obtaining interpolating polynomials of varying degrees when the data are uniformly spaced, a common situation. Many of the applications of interpolation to other numerical procedures use the formulas of this section.

3.5   Interpolating with a Cubic Spline

Introduces spline curves, a newer and most important way to do interpolation. At the expense of added computations to find the polynomials, splines overcome some important problems that we find with ordinary interpolating polynomials.

3.6   Bezier Curves and B-Spline Curves

Describes modern techniques for constructing smooth curves that are widely used in computer graphics. They are not interpolating polynomials but are closely related to them.

3.7   Polynomial Approximation of Surfaces

Applies the techniques of the chapter to the more difficult problem of more than one independent variable. The section does not cover this topic in great detail.

3.8   Least-Squares Approximations

Considers the fitting of polynomials and other functions to data that are inexact. This classic problem must be faced when experimental data are interpreted.

3.9   Theoretical Matters

Presents the supporting theory behind many of the methods of the chapter.

3.10   Using *Mathematica*

Shows how this symbolic algebra program can help in obtaining interpolating and least-squares polynomials. *Mathematica* can also plot cubic spline curves and Bezier curves.

Chapter Summary

Is the usual review section to check your understanding.

Computer Programs

Gives examples of programs that implement some of the methods of the chapter.

3.1 An Interpolation Problem

Rita Laski and Ed Baker were at lunch in the cafeteria of Ruscon Engineering. Both had just started summer jobs and were excited at the prospect of finding out how their college training really applied in industry. Rita was explaining her first assignment.

"They have huge amounts of data on the performance of that new rocket. Telemetry signals are received every 10 sec, giving the position of the rocket as well as other information. My boss has asked me to look into how we can determine the position at intermediate times. In essence, it's a kind of interpolation problem."

"I see," said Ed, "something like what we did when we studied log tables in algebra. There we calculated the logs for intermediate values by assuming that a straight line went between the two values from the table."

"Exactly," Rita replied, "except that it isn't appropriate to assume that the positions are linear with time. Typically, the points look something like this when they are plotted." She drew on a paper napkin to illustrate her point. "As you can see, sometimes a signal is missed, like on the third point."

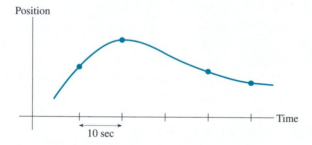

"You can see that a straight line is just impossible if we want to fit the data. And we can't just draw curves like I've done here because we want to get the intermediate values in the computer, and a computer can't look at a graph to find the values. What really bothers me is that Mr. Johnson, my boss, told me that we also must be able to get the intermediate values even when there is a maximum or minimum to the curve. Besides all this, we need a method of good efficiency because there are so many cases to handle."

In this chapter we will explore efficient techniques to interpolate, particularly for those situations where the data are far from linear. The principle that will be used is to fit a polynomial curve to the points. The problem is one of interest to many applications. Much of the development comes from work done by Newton and Kepler as they analyzed data on the positions of stars and planets.

In Chapters 1 and 2, we examined the question: "Given an explicit function of the independent variable x, what is the value of x corresponding to a certain value of the function?" In Chapter 1, x was a simple variable. In Chapter 2, x was a vector.

We now want to consider a question somewhat the reverse of this: "Given values of an unknown function corresponding to certain values of x, what is the behavior of the function?" We would like to answer the question "What is the function?" but this is always impossible to determine with a limited amount of data.

Our purpose in determining the behavior of a function, as evidenced by the sample of data pairs $[x, f(x)]$, is severalfold. We will wish to approximate other values of the function at values of x not tabulated (interpolation or extrapolation) and to estimate the integral of $f(x)$ and its derivative. The latter objectives will lead us into ways of solving ordinary- and partial-differential equations.

The strategy we will use in approximating unknown values of the function is straightforward. We will find a polynomial that fits a selected set of points $(x_i, f(x_i))$ and assume that the polynomial and the function behave nearly the same over the interval in question. Values of the polynomial then should be reasonable estimates of the values of the unknown function. When the polynomial is of the first degree, this leads to the familiar linear interpolation. We will be interested in polynomials of degree higher than the first, so we can approximate functions that are far from linear, or so we can get good values from a table with wider spacing.

However, there are problems with interpolating polynomials when the data are not "smooth," meaning that there are local irregularities. In such cases, a polynomial of high degree would be required to follow the irregularities, but we will find that such polynomials, while fitting to the irregularity, deviate widely at other regions where the function is smooth. One solution is to fit subregions of the data with different polynomials, but this method too is problematic in that the joins of the different polynomials are not continuous in their slope. To remedy this problem, special types of polynomials, called splines, are useful.

The study of splines leads to some other special forms of polynomials (Bezier curves and B-spline curves) that do not interpolate (they do not pass exactly through all of the points) but are very useful for sketching smooth curves.

We do not always want to find a polynomial that fits exactly to the data. Often the values we wish to fit are not exact, or they may come from a set of experimental measurements that are subject to error. Fitting the approximating polynomial in this instance fits to the errors in the data and we do not want to do that.

A technique called *least squares* is normally used in such cases. Based on statistical theory, this method finds a polynomial (or some other kind of approximating function) that is more likely to approximate the true values. In this chapter we will cover this method as well as methods for fitting exactly to the points.

3.2 Lagrangian Polynomials

In this section and the next three we assume that the given data are exact and represent values of some unknown function. If we desire to find a polynomial that passes through the same points as our unknown function, we could set up a system of equations involving the coefficients of the polynomial. For example, suppose we want to fit a cubic to these data:

x	$f(x)$
3.2	22.0
2.7	17.8
1.0	14.2
4.8	38.3
5.6	51.7

First, we need to select four points to determine our polynomial. (The maximum degree of the polynomial is always one less than the number of points.) Suppose we choose the first four points. If the cubic is $ax^3 + bx^2 + cx + d$, we can write four equations involving the unknown coefficients a, b, c, and d:

$$\text{when } x = 3.2: a(3.2)^3 + b(3.2)^2 + c(3.2) + d = 22.0,$$
$$\text{if } x = 2.7: a(2.7)^3 + b(2.7)^2 + c(2.7) + d = 17.8,$$
$$\text{if } x = 1.0: a(1.0)^3 + b(1.0)^2 + c(1.0) + d = 14.2,$$
$$\text{if } x = 4.8: a(4.8)^3 + b(4.8)^2 + c(4.8) + d = 38.3.$$

Solving these equations by the methods of the previous chapter gives us the polynomial. We can then estimate the values of the function at some value of x—say, $x = 3.0$—by substituting 3.0 for x in the polynomial.

For this example, the set of equations gives

$$a = -0.5275$$
$$b = 6.4952$$
$$c = -16.1177$$
$$d = 24.3499$$

and our polynomial is

$$-0.5275x^3 + 6.4952x^2 - 16.1177x + 24.3499.$$

At $x = 3.0$, the estimated value is 20.21.

We seek a better and simpler way of finding such interpolating polynomials. This procedure is awkward, especially if we want a new polynomial that is also made to fit at the point $(5.6, 51.7)$, or if we want to see what difference it would make to use a quadratic instead of a cubic. Furthermore, this technique leads to an ill-conditioned system of equations.*

---

* For this example, the system of four equations has a condition number of approximately 2700. Adding the last point to solve for the coefficients of a quartic polynomial causes the condition number to jump to near 100,000!

We will first look at one very straightforward approach—the Lagrangian polynomial. The Lagrangian polynomial is perhaps the simplest way to exhibit the existence of a polynomial for interpolation with unevenly spaced data. Data where the x-values are not equispaced often occur as the result of experimental observations or when historical data are examined.

Suppose we have a table of data with four pairs of x- and $f(x)$-values, with x_i indexed by variable i:

x	$f(x)$
x_0	f_0
x_1	f_1
x_2	f_2
x_3	f_3

Here we do not assume uniform spacing between the x-values, nor do we need the x-values arranged in a particular order. The x-values must all be distinct, however. Through these four data pairs we can pass a cubic. The Lagrangian form for this is

$$
P_3(x) = \frac{(x - x_1)(x - x_2)(x - x_3)}{(x_0 - x_1)(x_0 - x_2)(x_0 - x_3)} f_0 + \frac{(x - x_0)(x - x_2)(x - x_3)}{(x_1 - x_0)(x_1 - x_2)(x_1 - x_3)} f_1
$$
$$
+ \frac{(x - x_0)(x - x_1)(x - x_3)}{(x_2 - x_0)(x_2 - x_1)(x_2 - x_3)} f_2 + \frac{(x - x_0)(x - x_1)(x - x_2)}{(x_3 - x_0)(x_3 - x_1)(x_3 - x_2)} f_3 \quad \textbf{(3.1)}
$$

Note that this equation is made up of four terms, each of which is a cubic in x; hence the sum is a cubic. The pattern of each term is to form the numerator as a product of linear factors of the form $(x - x_i)$, omitting one x_i in each term, the omitted value being used to form the denominator by replacing x in each of the numerator factors. In each term, we multiply by the f_i corresponding to the x_i omitted in the numerator factors. The Lagrangian polynomial for other degrees of interpolating polynomials employs this same pattern of forming a sum of polynomials all of the desired degree; it will have $n + 1$ terms when the degree is n.

It is easy to see that the Lagrangian polynomial does in fact pass through each of the points used in its construction. For example, in the preceding equation for $P_3(x)$, let $x = x_2$. All terms but the third vanish because of a zero numerator, while the third term becomes just $(1) * f_2$. Hence $P_3(x_2) = f_2$. Similarly, $P_3(x_i) = f_i$ for $i = 0, 1, 3$.

EXAMPLE 3.1   Fit a cubic through the first four points of the preceding table and use it to find the interpolated value for $x = 3.0$.

$$P_3(3.0) = \frac{(3.0 - 2.7)(3.0 - 1.0)(3.0 - 4.8)}{(3.2 - 2.7)(3.2 - 1.0)(3.2 - 4.8)}(22.0)$$

$$+ \frac{(3.0 - 3.2)(3.0 - 1.0)(3.0 - 4.8)}{(2.7 - 3.2)(2.7 - 1.0)(2.7 - 4.8)}(17.8)$$

$$+ \frac{(3.0 - 3.2)(3.0 - 2.7)(3.0 - 4.8)}{(1.0 - 3.2)(1.0 - 2.7)(1.0 - 4.8)}(14.2)$$

$$+ \frac{(3.0 - 3.2)(3.0 - 2.7)(3.0 - 1.0)}{(4.8 - 3.2)(4.8 - 2.7)(4.8 - 1.0)}(38.3).$$

Carrying out the arithmetic, $P_3(3.0) = 20.21$.

Observe that we get the same result as before. The arithmetic in this method is tedious, although hand calculators are convenient for this type of computation. Writing a computer program that implements the method is not hard to do.

An interpolating polynomial, while passing through the points used in its construction, does not, in general, give exactly correct values when used for interpolation. The reason is that the underlying relationship is often not a polynomial of the same degree. We are therefore interested in the error of interpolation.

As shown in a later section of this chapter, the error term is given by the expression

$$E(x) = (x - x_0)(x - x_1) \cdots (x - x_n) \frac{f^{(n+1)}(\xi)}{(n + 1)!} \qquad \textbf{(3.2)}$$

with ξ in the smallest interval that contains $\{x, x_0, x_1, \ldots, x_n\}$.

The expression for error given in Eq. (3.2) is interesting but is not always extremely useful. This is because the actual function that generates the x_i, f_i values is often unknown; we obviously then do not know its $(n + 1)$st derivative. We can conclude, however, that if the function is "smooth," a low-degree polynomial should work satisfactorily. (A smooth function has small higher derivatives.) On the other hand, a "rough" function can be expected to have larger errors when interpolated. We can also conclude that extrapolation (applying the interpolating polynomial outside the range of x-values employed to construct it) will have larger errors than for interpolation. It also follows that the error is smaller if x is centered within the x_i, because this makes the product of the $(x - x_i)$ terms smaller.

Here is an algorithm for interpolation with a Lagrangian polynomial of degree N.

To interpolate for $f(x)$, given x and a set of $N + 1$ data pairs, $(x_i, f_i), i = 0, \ldots, N$:

Set SUM = 0.
DO FOR $I = 0$ to N:
    SET $P = 1$.

> DO FOR $J = 0$ to N:
>   IF $J \neq I$:
>   SET $P = P * (x - x(J))/(x(I) - x(J))$.
>   ENDDO (J).
>   SET SUM = SUM + $P * f_i$
> ENDDO (I).
> SUM is the interpolated value.

Neville's Method

The trouble with the standard Lagrangian polynomial technique is that we do not know which degree of polynomial to use. If the degree is too low, the interpolating polynomial does not give good estimates of $f(x)$. If the degree is too high, undesirable oscillations in polynomial values can occur. (More on this later, in the section on spline curves.)

Neville's method can overcome this difficulty. It essentially computes the interpolated value with polynomials of successively higher degree, stopping when the successive values are close together. The successive approximations are actually computed by linear interpolation from the intermediate values. The Lagrange formula for linear interpolation to get $f(x)$ from two data pairs, (x_1, f_1) and (x_2, f_2), is

$$f(x) = \frac{(x - x_2)}{(x_1 - x_2)} f_1 + \frac{(x - x_1)}{(x_2 - x_1)} f_2,$$

which can be written more compactly as

$$f(x) = \frac{(x - x_2) * f_1 + (x_1 - x) * f_2}{x_1 - x_2}. \tag{3.3}$$

We will use Eq. (3.3) in Neville's method.

If we examine Eq. (3.2) for the error term of Lagrangian interpolation, we see that the smallest error results when we use data pairs where the x_i's are closest to the x-value we are interpolating. Neville's method begins by arranging the given data pairs, (x_i, f_i), so the successive values are in order of the closeness of the x_i to x.

EXAMPLE 3.2 Suppose we are given these data:

x	$f(x)$
10.1	0.17537
22.2	0.37784
32.0	0.52992
41.6	0.66393
50.5	0.63608

and we want to interpolate for $x = 27.5$. We first rearrange the data pairs in order of closeness to $x = 27.5$:

i	$\|x - x_i\|$	x_i	$f_i = P_{i0}$
0	4.5	32.0	0.52992
1	5.3	22.2	0.37784
2	14.1	41.6	0.66393
3	17.4	10.1	0.17537
4	23.0	50.5	0.63608

Neville's method begins by renaming the f_i as P_{i0}. We build a table by first interpolating linearly between pairs of values for $i = 0, 1, i = 1, 2, i = 2, 3$, and so on. These values are written in a column to the right of the first P of each pair. The next column of the table is created by linearly interpolating from the previous column for $i = 0, 2, i = 1, 3, i = 2, 4$, and so on. The next column after this uses values for $i = 0, 3, i = 0, 4, \ldots$, and continues until we run out of data pairs.

Here is the Neville table for the preceding data:

i	x_i	P_{i0}	P_{i1}	P_{i2}	P_{i3}	P_{i4}
0	32.0	0.52992	0.46009	0.46200	0.46174	0.45754
1	22.2	0.37784	0.45600	0.46071	0.47901	
2	41.6	0.66393	0.44524	0.55843		
3	10.1	0.17537	0.37379			
4	50.5	0.63608				

The general formula for computing entries into the table is

$$P_{i,j} = \frac{(x - x_i) * P_{i+1,j-1} + (x_{i+j} - x) * P_{i,j-1}}{x_{i+j} - x_i}. \qquad (3.4)$$

Thus the values of P_{01} and P_{11} are computed by

$$P_{01} = \frac{(27.5 - 32.0) * 0.37784 + (22.2 - 27.5) * 0.52992}{22.2 - 32.0} = 0.46009,$$

$$P_{11} = \frac{(27.5 - 22.2) * 0.66393 + (41.6 - 27.5) * 0.37784}{41.6 - 22.2} = 0.45600.$$

Once we have the column of P_{i1}'s, we compute the next column. For example,

$$P_{22} = \frac{(27.5 - 41.6) * 0.37379 + (50.5 - 27.5) * 0.44524}{50.5 - 41.6} = 0.55843.$$

The remaining columns are computed similarly by using Eq. (3.4).

The top line of the table represents Lagrangian interpolates at $x = 27.5$ using polynomials of degree equal to the second subscript of the P's. Each of these polynomials uses the required number of data pairs, taking them as a set starting from the top of the table. (An exercise asks you to prove that the top line does represent Lagrangian interpolates with polynomials of increasing degree.)

The preceding data are for sines of angles in degrees and the correct value for $x = 27.5$ is 0.46175. Observe that the top line values get better and better until the last, when it diverges. This divergence becomes apparent when we notice that the successive values get closer to a constant value until the last one. (If the table is not arranged to center the x-value within the x_i, the convergence is not as quick.)

If we instead do this computation by hand, we can save computing time by computing, not the entire table, but only as much as required to get convergence to the desired number of decimal places. We therefore do only the computations needed to compute those top row values that are required. In a computer program it is hardly worth the added programming complications because the entire table is computed so quickly.

Parallel Processing

The several terms of a Lagrange polynomial, as shown in Eq. (3.1), can all be computed simultaneously with parallel processing. Each entry in the successive columns of the table for Neville's method can be computed simultaneously. An exercise asks you to determine the number of time steps that are saved.

3.3 Divided Differences

There are two disadvantages to using the Lagrangian polynomial method for interpolation. First, it involves more arithmetic operations than does the divided-difference method we now discuss. Second, and more importantly, if we desire to add or subtract a point from the set used to construct the polynomial, we essentially have to start over in the computations. Both the Lagrangian polynomials and Neville's method also must repeat all of the arithmetic if we must interpolate at a new x-value. The divided-difference method avoids all of this computation.

Actually, we will not get a polynomial different from that obtained by Lagrange's technique. As we will show in a later section, every nth-degree polynomial that passes through the same $n + 1$ points is identical. Only the way that the polynomial is expressed is different.

Our treatment of divided-difference tables assumes that function, $f(x)$, is known at several values for x:

$$
\begin{array}{cc}
x_0 & f_0 \\
x_1 & f_1 \\
x_2 & f_2 \\
x_3 & f_3
\end{array}
$$

We do not assume that the x's are evenly spaced or even that the values are arranged in any particular order (but some ordering may be advantageous).

Consider the nth-degree polynomial written in a special way:

$$
P_n(x) = a_0 + (x - x_0)a_1 + (x - x_0)(x - x_1)a_2 + \cdots
$$
$$
+ (x - x_0)(x - x_1)\cdots(x - x_{n-1})a_n. \tag{3.5}
$$

If we chose the a_i so that $P_n(x) = f(x)$ at the $n + 1$ known points, $(x_i, f_i), i = 0, \ldots, n$, then $P_n(x)$ is an interpolating polynomial. We will show that the a_i are readily determined by using what are called the *divided differences of the tabulated values*.

A special standard notation used for divided differences is

$$
f[x_0, x_1] = \frac{f_1 - f_0}{x_1 - x_0} = (f_0^{[1]}),
$$

called the *first divided difference between x_0 and x_1*. (We have added a nonstandard notation that is shorter.) The function

$$
f[x_1, x_2] = \frac{f_2 - f_1}{x_2 - x_1} = (f_1^{[1]})
$$

is the first divided difference between x_1 and x_2.

In general,

$$
f[x_s, x_t] = \frac{f_t - f_s}{x_t - x_s}
$$

is the first divided difference between x_s and x_t. (Observe that the order of the points is immaterial:

$$
f[x_s, x_t] = \frac{f_t - f_s}{x_t - x_s} = \frac{f_s - f_t}{x_s - x_t} = f[x_t, x_s].)
$$

Second- and higher-order differences are defined in terms of lower-order differences. For example,

$$
f[x_0, x_1, x_2] = \frac{f[x_1, x_2] - f[x_0, x_1]}{x_2 - x_0} = (f_0^{[2]});
$$

$$f[x_0, x_1, \ldots, x_n] = \frac{f[x_1, x_2, \ldots, x_n] - f[x_0, x_1, \ldots, x_{n-1}]}{x_n - x_0} = (f_0^{[n]}).$$

The concept is even extended to a zero-order difference:

$$f[x_s] = f_s = (f_s^{[0]}).$$

Using the standard notation, a divided-difference table is shown in symbolic form in Table 3.1. Table 3.2 shows specific numerical values. (These data are the same as in the table of Section 3.2.)

Table 3.1

x_i	f_i	$f[x_i, x_{i+1}]$	$f[x_i, x_{i+1}, x_{i+2}]$	$f[x_i, x_{i+1}, x_{i+2}, x_{i+3}]$
x_0	f_0			
x_1	f_1	$f[x_0, x_1]$	$f[x_0, x_1, x_2]$	$f[x_0, x_1, x_2, x_3]$
x_2	f_2	$f[x_1, x_2]$	$f[x_1, x_2, x_3]$	$f[x_0, x_1, x_2, x_4]$
x_3	f_3	$f[x_2, x_3]$	$f[x_2, x_3, x_4]$	
x_4	f_4	$f[x_3, x_4]$		

Table 3.2

x_i	f_i	$f[x_i, x_{i+1}]$	$f[x_i, \ldots, x_{i+2}]$	$f[x_i, \ldots, x_{i+3}]$	$f[x_i, \ldots, x_{i+4}]$
3.2	22.0				
2.7	17.8	8.400	2.856		
1.0	14.2	2.118	2.012	-0.528	0.256
4.8	38.3	6.342	2.263	0.0865	
5.6	51.7	16.750			

We are now ready to establish that the a_i of Eq. (3.5) are given by these divided differences. We write Eq. (3.5) with $x = x_0, x = x_1, \ldots, x = x_n$, giving

$$x = x_0: \quad P_n(x_0) = a_0,$$
$$x = x_1: \quad P_n(x_1) = a_0 + (x_1 - x_0)a_1,$$
$$x = x_2: \quad P_n(x_2) = a_0 + (x_2 - x_0)a_1 + (x_2 - x_0)(x_2 - x_1)a_2,$$
$$\vdots$$
$$x = x_n: \quad P_n(x_n) = a_0 + (x_n - x_0)a_1 + (x_n - x_0)(x_n - x_1)a_2 + \cdots$$
$$+ (x_n - x_0)\ldots(x_n - x_{n-1})a_n.$$

If $P_n(x)$ is to be an interpolating polynomial, it must match the table for all $n + 1$ entries:

$$P_n(x_i) = f_i \quad \text{for } i = 0, 1, 2, \ldots, n.$$

If the $P_n(x_i)$ in each equation is replaced by f_i, we get a triangular system, and each a_i can be computed in turn.

From the first equation,

$$a_0 = f_0 = f[x_0] \quad \text{makes} \quad P_n(x_0) = f_0.$$

If $a_1 = f[x_0, x_1]$, then

$$P_n(x_1) = f_0 + (x_1 - x_0)\frac{f_1 - f_0}{x_1 - x_0} = f_1.$$

If $a_2 = f[x_0, x_1, x_2]$, then

$$P_n(x_2) = f_0 + (x_2 - x_0)\frac{f_1 - f_0}{x_1 - x_0}$$

$$+ (x_2 - x_0)(x_2 - x_1)\frac{(f_2 - f_1)/(x_2 - x_1) - (f_1 - f_0)/(x_1 - x_0)}{x_2 - x_0}$$

$$= f_2.$$

One can show in similar fashion that each $P_n(x_i)$ will equal f_i if $a_i = f[x_0, x_1, \ldots, x_i]$. The interpolating polynomial that fits a divided difference table at $x = x_0$, $x_1, x_2, \ldots, x_n$ is then

$$P_n(x) = f_0^{[0]} + (x - x_0)f_0^{[1]} + (x - x_0)(x - x_1)f_0^{[2]}$$
$$+ (x - x_0)(x - x_1)(x - x_3)f_0^{[3]} + \cdots$$
$$+ (x - x_0)(x - x_1) \cdots (x - x_{n-1})f_0^{[n]}. \tag{3.6}$$

EXAMPLE 3.3   Write the interpolating polynomial of degree 3 that fits the data of Table 3.1 at all points from $x_0 = 3.2$ to $x_3 = 4.8$.

$$P_3(x) = 22.0 + 8.400(x - 3.2) + 2.856(x - 3.2)(x - 2.7)$$
$$- 0.528(x - 3.2)(x - 2.7)(x - 1.0).$$

What is the fourth-degree polynomial that fits at all five points? We only have to add one more term to $P_3(x)$:

$$P_4(x) = P_3(x) + 0.256(x - 3.2)(x - 2.7)(x - 1.0)(x - 4.8).$$

When this method is used for interpolation, we observe that nested multiplication can be used to cut down on the number of arithmetic operations, for example, for $x = 3$:

$$P_3(3) = \{[-0.528(3 - 1.0) + 2.586](3 - 2.7) + 8.400\}(3 - 3.2) + 22.0.$$

If we compute the interpolated value at $x = 3.0$ for each of the third-degree polynomials in Sections 3.2 and 3.3, we get the same result: $P_3(3.0) = 20.21$. This is not surprising, because all third-degree polynomials that pass through the same four points are identical. They may be of different appearance, but they can all be reduced to the same form.

An algorithm for constructing a divided difference table is

Given a set of $N + 1$ data pairs, $(x_i, f_i), i = 0, \ldots, n,$
    DO FOR $i = 1$ to n:
        DO FOR $j = 0$ to $n - i$:
            Compute $f_j^{[i]} = (f_{j+1}^{[i-1]} - f_j^{[i-1]})/(x_{j+i} - x_j)$
            and enter into column i of the table.
        ENDDO (j).
    ENDDO (i).

Observe that parallel processing can compute all entries in the successive columns simultaneously. If there are $N + 1$ data pairs and a full table is constructed, the number of time steps equals the number of new columns, N. Sequential processing would require $N(N + 1)/2$ steps.

Divided Difference for $f(x)$ a Polynomial

It is of interest to look at the divided differences for $f(x) = P_n(x)$. Suppose that $f(x)$ is the cubic

$$f(x) = 2x^3 - x^2 + x - 1.$$

Here is its divided-difference table:

x_i	$f(x_i)$	$f_i^{[1]}$	$f_i^{[2]}$	$f_i^{[3]}$	$f_i^{[4]}$	$f_i^{[5]}$
0.30	−0.7360	2.4800	3.0000	2.0000	0.0000	0.0000
1.00	1.0000	3.6800	3.6000	2.0000	0.0000	
0.70	−0.1040	2.2400	5.4000	2.0000		
0.60	−0.3280	8.7200	8.2000			
1.90	11.0080	21.0200				
2.10	15.2120					

Observe that the third divided differences are all the same. (It then follows that all higher divided differences will be zero.) We can take advantage of this fact by not using interpolating polynomials beyond the column where the values are essentially constant.

It is most important to also observe that the third derivative of a cubic polynomial is also a constant. (In this instance, $P^{(3)}(x) = 2*3! = 12$.) The relationship between divided differences and derivatives will be explored in detail in the next chapter. For now we just state that for an nth-degree polynomial, $P_n(x)$, whose highest power term has the coefficient a_n, the nth divided differences will always be equal to a_n. Since the nth derivative of this polynomial is equal to $a_n * n!$, the relationship between derivatives and divided differences seems to involve $n!$. We exploit this next.

Error of Interpolation

The error term for an interpolating polynomial derived from a divided-difference table is identical to that for the equivalent Lagrangian polynomial because, as we have already observed, all polynomials of degree n that match at $n + 1$ points are identical. That means that the error term associated with the nth-degree polynomial $P_n(x)$ of Eq. (3.5) is simply Eq. (3.2), which we repeat here:

$$E(x) = (x - x_0)(x - x_1) \cdots (x - x_n)\frac{f^{(n+1)}(\xi)}{(n+1)!}.$$

It is still not convenient to use this error expression, because the derivative of f that appears is unknown. However, if $f(x)$ is almost the same as some polynomial of degree n (and we will know that this is true because the nth divided differences will be almost constant), interpolating with an nth-degree polynomial should be nearly exact. The reason is that the $(n + 1)$st derivative of $f(x)$ will be nearly zero and the error of the nth-degree interpolating polynomial will be very small.

What if we use a lower-degree polynomial? The error should be larger. If $f(x)$ is a known function, we can use Eq. (3.2) to bound the error. Here is an example.

EXAMPLE 3.4   Here is a divided difference table for $f(x) = x^2 e^{-x/2}$:

x_i	$f(x_i)$	$f_i^{[1]}$	$f_i^{[2]}$	$f_i^{[3]}$	$f_i^{[4]}$
1.10	0.6981	0.8593	−0.1755	0.0032	0.0027
2.00	1.4715	0.4381	−0.1631	0.0191	
3.50	2.1287	−0.0511	−0.0657		
5.00	2.0521	−0.2877			
7.10	1.4480				

Find the error of the interpolates for $f(1.75)$ using polynomials of degrees one, two, and three.

The results are shown in Table 3.3, for which Eq. (3.6) was used to do the interpolations. (DERIVE helped in finding the derivatives and evaluating the maximum and minimum values within the intervals.) The error formula does bracket the actual errors, as expected. In this case, observe that the use of a cubic polynomial does not improve the accuracy. In part, this is because we do not have the x-value well centered within the tabulated values; also, the value of the derivative is not decreasing.

▲

Table 3.3 Errors of interpolation for $f(1.75)$

Degree	Interpolated value	Actual error	$f^{(t+1)}$ maximum	$f^{(t+1)}$ minimum	Upper bound	Lower bound
1	1.25668	0.01996	−0.3679	0.0594	0.0299	−0.00483
2	1.28520	−0.00856	−0.8661	0.1249	0.0059	−0.0408
3	1.28611	−0.00947	1.1398	−0.0359	0.0014	−0.0439

Error Estimation When $f(x)$ Is Unknown—The Next-Term Rule

Occasionally, almost always when dealing with experimental data, the function is unknown. Still, there is way to estimate the error of the interpolation. This is because the nth-order divided difference is itself an approximation for $f^{(n)}(x)/n!$, as will be demonstrated in the next chapter. What this means is that the error of the interpolation is given approximately by the value of the next term that would be added!

This most valuable rule for estimating the error of interpolation we call the *next-term rule*. It is easy to state and to use:

$$E_n(x) = \text{(approximately) the value of the next term that would be added to } P_n(x).$$

Here is how it works for the preceding example:

Degree	Exact error	Estimate from next-term rule
1	0.01996	0.02852
2	0.00856	0.00091
3	−0.00947	−0.00249

As you can see, the agreement is at least fair.

Interpolation Near the End of a Table

Thus far, we have assumed that the entries are indexed from the top to the bottom of the table. This would appear to indicate that our formulas do not work well for constructing polynomials from divided differences at the end of the table. Remember, however, that the ordering of the points is immaterial. We can just as well begin at the bottom and number the entries going upward, with no adjustment of Eq. (3.6) required. The table is really not changed at all, just the symbols that we use. We now use Eq. (3.6) with the newly indexed values.

Tables 3.4(a) and (b) compare the two different numbering schemes. The entries in the rows of Table 3.4(b) are exactly the same numbers as in the upward diagonals of Table 3.4(a).

Table 3.4(a)   Conventional divided-difference table

x_0	f_0	$f_0^{[1]}$	$f_0^{[2]}$	$f_0^{[3]}$	$f_0^{[4]}$
x_1	f_1	$f_1^{[1]}$	$f_1^{[2]}$	$f_1^{[3]}$	
x_2	f_2	$f_2^{[1]}$	$f_2^{[2]}$		
x_3	f_3	$f_3^{[1]}$			
x_4	f_4				

Table 3.4(b)   Divided-difference table indexed upwardly

x_4	f_4				
x_3	f_3	$f_3^{[1]}$			
x_2	f_2	$f_2^{[1]}$	$f_2^{[2]}$		
x_1	f_1	$f_1^{[1]}$	$f_1^{[2]}$	$f_1^{[3]}$	
x_0	f_0	$f_0^{[1]}$	$f_0^{[2]}$	$f_0^{[3]}$	$f_0^{[4]}$

3.4   Evenly Spaced Data

The problem of interpolation from tabulated data is considerably simplified if the values of the function are given at evenly spaced intervals of the independent variable. It is necessary here to arrange the data in a table with x-values in ascending order. In addition to columns for x and $f(x)$, we will tabulate differences of the functional values. Table 3.5 is a typical difference table. Each of the columns to the right of the $f(x)$ column is computed by calculating the difference between two values in the column to its left. (We do not divide by x-differences.)

Table 3.5   A difference table

x	$f(x)$	$\Delta f(x)$	$\Delta^2 f(x)$	$\Delta^3 f(x)$	$\Delta^4 f(x)$
0.0	0.000				
0.2	0.203	0.203			
0.4	0.423	0.220	0.017		
0.6	0.684	0.261	0.041	0.024	
0.8	1.030	0.346	0.085	0.044	0.020
1.0	1.557	0.527	0.181	0.096	0.052
1.2	2.572	1.015	0.488	0.307	0.211

Symbols that represent the entries in a difference table will be helpful in using the table of differences to determine coefficients in interpolating polynomials. It is conventional to let the letter h stand for the uniform difference in the x-values, $h = \Delta x$. Using subscripts to represent the order of the x and $f(x)$ values, we define the first differences of the function as

$$\Delta f_0 = f_1 - f_0, \quad \Delta f_1 = f_2 - f_1, \quad \Delta f_2 = f_3 - f_2, \quad \ldots, \quad \Delta f_i = f_{i+1} - f_i.$$

The second- and higher-order differences are similarly defined:*

$$\Delta^2 f_1 = \Delta(\Delta f_1) = \Delta(f_2 - f_1) = \Delta f_2 - \Delta f_1 = (f_3 - f_2) - (f_2 - f_1)$$
$$= f_3 - 2f_2 + f_1,$$
$$\Delta^2 f_i = f_{i+2} - 2f_{i+1} + f_i,$$
$$\Delta^3 f_1 = \Delta(\Delta^2 f_1) = f_4 - 3f_3 + 3f_2 - f_1, \tag{3.7}$$
$$\Delta^3 f_i = f_{i+3} - 3f_{i+2} + 3f_{i+1} - f_i,$$
$$\vdots \qquad\qquad \vdots$$
$$\Delta^n f_i = f_{i+n} - nf_{i+n-1} + \frac{n(n-1)}{2!}f_{i+n-2} - \frac{n(n-1)(n-2)}{3!}f_{i+n-3} + \cdots.$$

In Eqs. (3.7) and throughout this chapter, a subscript on f indicates the x-value at which it is evaluated: namely, $f_3 = f(x_3)$. The pattern of the coefficients in Eqs. (3.7) is the familiar array of coefficients in the binomial expansion. This fact we can prove most readily by symbolic methods, which we postpone until a later section. The second- and higher-order differences are generally obtained by differencing the previous differences, but Eqs. (3.7) show how any difference can be calculated directly from the functional values.

Table 3.6 illustrates the formation of a difference table using symbolic representation of the entries. While it might be natural always to call the first x in the table

Table 3.6 Difference table, using symbols

s	x	$f(x)$	Δf	$\Delta^2 f$	$\Delta^3 f$	$\Delta^4 f$
-2	x_{-2}	f_{-2}				
-1	x_{-1}	f_{-1}	Δf_{-2}			
0	x_0	f_0	Δf_{-1}	$\Delta^2 f_{-2}$		
1	x_1	f_1	Δf_0	$\Delta^2 f_{-1}$	$\Delta^3 f_{-2}$	$\Delta^4 f_{-2}$
2	x_2	f_2	Δf_1	$\Delta^2 f_0$	$\Delta^3 f_{-1}$	$\Delta^4 f_{-1}$
3	x_3	f_3	Δf_2	$\Delta^2 f_1$	$\Delta^3 f_0$	$\Delta^4 f_0$
4	x_4	f_4	Δf_3	$\Delta^2 f_2$	$\Delta^3 f_1$	

* The differences that we define here are called *forward differences*. Some texts define differences that are called *backward differences* (written as ∇f_i, defined as $\nabla f_i = f_i - f_{i-1}$) and *central differences* (written δf_i). Our treatment uses only the forward difference.

x_0, we will frequently wish to refer to only a portion of the data pairs of the table, so the "first" x loses its significance. We then arbitrarily choose the origin for the subscripted variable, because by the use of negative values we can refer to x-entries that precede x_0. We will set the variable s equal to the subscript on x; it then serves to index the x-values.

In hand computation of a difference table, great care should be exercised to avoid arithmetic errors in the subtractions—the fact that we subtract the upper entry from the lower adds a real source of confusion. One of the best ways to check for mistakes is to add the sum of the numbers in each column to the top entry in the column to its left. This sum should equal the bottom entry in the column to the left.

Since each entry in the difference table is the difference of a pair of numbers in the column to its left, we could recompute one of this pair if it should be erased. As a consequence, the entire table could be reproduced, given only one value in each column, if the table is extended to the highest possible order of differences.

When $f(x)$ behaves like a polynomial for the set of data given, the difference table has special properties. In Table 3.7, a function is tabulated over the domain $x = 1$ to $x = 6$, and $f(x)$ obviously behaves the same as x^3. (Note carefully that this does *not* imply that $f(x) \equiv x^3$; the value of $f(x)$ at $x = 7$ might well be 17 instead of $7^3 = 343$. We *only* know the values of $f(x)$ as given in the table.)

Table 3.7 Difference table for a function behaving like x^3

x	$f(x)$	Δf	$\Delta^2 f$	$\Delta^3 f$	$\Delta^4 f$
0	0				
1	1	1			
2	8	7	6	6	
3	27	19	12	6	0
4	64	37	18	6	0
5	125	61	24	6	0
6	216	91	30		

We observe that the third differences are constant. Consequently, the fourth and all higher differences will be zero. The fact that the nth-order differences of any nth-degree polynomial are constant is readily shown. To prove that these nth-order differences are constant, we first examine the differences of ax^n:

$$\Delta(ax^n) = a(x + h)^n - ax^n$$
$$= (ax^n + anx^{n-1}h + \cdots + ah^n) - ax^n$$
$$= (anh)x^{n-1} + \text{terms of lower degree in } x,$$
$$\Delta(anh\, x^{n-1}) = an(n - 1)h^2 x^{n-2} + \text{terms of lower degree.}$$

Noting that every time a difference is taken the leading term has a power of x one less than originally and that the difference of a constant term will be zero, we have for a polynomial

$$\Delta P_n(x) = \Delta(a_0 x^n + a_1 x^{n-1} + \cdots + a_{n-1} x + a_n)$$
$$= a_0 nh x^{n-1} + \text{terms of lower degree,}$$
$$\Delta^2 P_n(x) = a_0 n(n-1) h^2 x^{n-2} + \text{terms of lower degree,} \tag{3.8}$$
$$\vdots$$
$$\Delta^n P_n(x) = a_0 n(n-1)(n-2) \cdots (1) h^n x^{n-n} = a_0 n! h^n.$$

This shows not only that the nth difference is a constant, but that its value is $a_0 n! h^n$. For $P_3(x) = x^3$ (where $a_0 = 1$):

$$\Delta^3 P_3(x) = (1)(3!)(1)^3 = 6 \qquad \text{when } h = 1.$$

This is exactly what was found in Table 3.7. Note a similarity between the nth difference and the nth derivative of $P_n(x)$:

$$P_n^{(n)}(x) = a_0 n!,$$

where a_0 is the coefficient of x^n.

Differences Versus Divided Differences

Obviously, the table of function differences that we have been discussing is closely related to the table of divided differences of the previous section. Except for dividing function differences by a difference of x-values in the latter, these two tables are the same when the x-values are evenly spaced. To make this crystal clear, compare the tables for the simple case of $f(x) = 2x^3$ with $h = 0.5$, as shown in Tables 3.8(a) and (b).

Table 3.8(a) Table of function differences for $f(x) = 2x^3, h = 0.5$

x_i	f_i	Δf_i	$\Delta^2 f_i$	$\Delta^3 f_i$	$\Delta^4 f_i$	$\Delta^5 f_i$
0.00	0.00	0.25	1.50	1.50	0.00	0.00
0.50	0.25	1.75	3.00	1.50	0.00	0.00
1.00	2.00	4.75	4.50	1.50	0.00	
1.50	6.75	9.25	6.00	1.50		
2.00	16.00	15.25	7.50			
2.50	31.25	22.75				
3.00	54.00					

Table 3.8(b) Table of divided differences for $f(x) = 2x^3, h = 0.5$

x_i	f_i	$f_i^{[1]}$	$f_i^{[2]}$	$f_i^{[3]}$	$f_i^{[4]}$	$f_i^{[5]}$
0.00	0.00	0.50	3.00	2.00	0.00	0.00
0.50	0.25	3.50	6.00	2.00	0.00	0.00
1.00	2.00	9.50	9.00	2.00	0.00	
1.50	6.75	18.50	12.00	2.00		
2.00	16.00	30.50	15.00			
2.50	31.25	45.50				
3.00	54.00					

As expected, the columns of third differences are constant in both tables. For divided differences, this constant is equal to just 2, the coefficient of x^3. For the difference table, it is equal to that coefficient times $(3!)(h^3)$, or $2*6*0.5^3 = 1.5$.

For first differences, the divided differences are equal to the function differences divided by h (0.5 here). Second divided differences are equal to second function differences divided by $(h)(2h)$ (0.5 in this example). Third divided differences are equal to third function differences divided by $(h)(2h)(3h)$ (0.75 in this instance). The pattern should now be clear:

$$f_i^{[n]} = \frac{\Delta^n f_i}{n! h^n}.$$

If the values are not evenly spaced, a comparison is impossible because the table of function differences is not defined.

This difference between the two kinds of tables has a great effect on the relation between differences and derivatives, a topic that we explore in the next chapter.

Interpolating Polynomials Based on Difference Tables

When the function that is tabulated behaves like a polynomial (and this we can tell by observing that its nth-order differences are constant or nearly so), we can approximate it by the polynomial it resembles. Difference tables are another way to write the nth-degree polynomial that passes through $n + 1$ pairs of points, $(x_i, f_i), i = 0, 1, \ldots, n$. Note that such a polynomial is unique—there is only one polynomial of degree n passing through $n + 1$ points, as we have previously observed.

One of the easiest ways to write a polynomial that passes through a group of equispaced points is the Newton–Gregory forward polynomial; we write it in terms of the index s:

$$P_n(x_s) = f_0 + s\Delta f_0 + \frac{s(s-1)}{2!}\Delta^2 f_0 + \frac{s(s-1)(s-2)}{3!}\Delta^3 f_0 + \cdots$$

$$= f_0 + \binom{s}{1}\Delta f_0 + \binom{s}{2}\Delta^2 f_0 + \binom{s}{3}\Delta^3 f_0 + \binom{s}{4}\Delta^4 f_0 + \cdots \qquad \textbf{(3.9)}$$

In this equation we have used the notation $\binom{s}{n}$, the number of combinations of s things taken n at a time, which is the same as the factorial ratios also shown. Referring to Table 3.6, we now observe that $P_n(x)$ does match the table at all the data pairs $(x_i, f_i), i = 0, 1, 2, \ldots, n$. When $s = 0$, $P_n(x_0) = f_0$. If $s = 1$,

$$P_n(x_1) = f_0 + \Delta f_0 = f_0 + f_1 - f_0 = f_1.$$

If $s = 2$,

$$P_n(x_2) = f_0 + 2\Delta f_0 + \Delta^2 f_0 = f_2.$$

Similarly we can demonstrate that $P_n(x)$ formed according to Eq. (3.9) matches at all $n + 1$ points.*

In Eq. (3.9), note that differences all fall on a descending diagonal line beginning at f_0, in Table 3.6.

We have previously observed that if, over the domain from x_0 to x_n, $P_n(x)$ and $f(x)$ have the same values at tabulated values of x, it is reasonable to assume that they will be nearly the same at intermediate points. This assumption is the basis for the use of $P_n(x)$ as an interpolating polynomial. We again emphasize that $f(x)$ and $P_n(x)$ will, in general, not be the same function. Hence there is some error to be expected in the estimate from such interpolation. We use the polynomial in Eq. (3.9) as an interpolating polynomial by letting s take on nonintegral values. This extends the definition of s so that, for any value of x,

$$s = \frac{x - x_0}{h}.$$

EXAMPLE 3.5   Write a Newton–Gregory forward polynomial of degree 3 that fits Table 3.5 for the four points at $x = 0.4$ to $x = 1.0$. Use it to interpolate for $f(0.73)$.

To make the polynomial fit as specified, we must index the x's so that $x_0 = 0.4$. It follows then that $f_0 = 0.423$, $\Delta f_0 = 0.261$, $\Delta^2 f_0 = 0.085$, and $\Delta^3 f_0 = 0.096$. We compute s:

$$s = \frac{x - x_0}{h} = \frac{0.73 - 0.4}{0.2} = 1.65.$$

---

*This demonstration is not a proof, of course. We give a proof in another section.

Applying these to Eq. (3.9) with terms through $\Delta^3 f_0$ to give a cubic, we have

$$f(0.73) = 0.423 + (1.65)(0.261) + \frac{(1.65)(0.65)}{2}(0.085)$$

$$+ \frac{(1.65)(0.65)(-0.35)}{6}(0.096)$$

$$= 0.423 + 0.4306 + 0.0456 - 0.0060$$

$$= 0.893. \tag{3.10}$$

The function tabulated in Table 3.5 is $\tan x$, for which the true value is 0.895 at $x = 0.73$. We hence see that there is an error in the third decimal place. We should anticipate some error because, since the third differences are far from constant, our cubic polynomial is not a perfect representation of the function. Even so, our interpolating polynomial gave a fair estimate, certainly better than linear interpolation, which gives 0.911.

Even though the fourth differences are also not constant, we would hope for some improvement if we approximated $f(x)$ by a fourth-degree polynomial. We can just add one more term to Eq. (3.10) to do this:

$$\binom{s}{4}\Delta^4 f_0 = \frac{(1.65)(0.65)(-0.35)(-1.35)}{4!}(0.211) = 0.0045,$$

$$f(0.73) = 0.893 + 0.0045 = 0.898.$$

Normally the higher-degree polynomial is better, but in this instance, adding another term does not improve our estimate; in fact it worsens it slightly. This is due to round-off errors in the original values in this instance.

The domain over which an interpolating polynomial agrees with the function is most readily found by working backward from the last difference that is included in Eq. (3.9), drawing imaginary diagonals to the left between the entries. The x-values included between this fan of diagonals is the domain of the interpolating polynomial. It will always be found to include one more x-entry than the degree of the polynomial.

Again we remark that the error term is the same as that derived for the Lagrangian polynomials. We also observe that the values used should be such that the x-value is centered as well as possible among those used to construct the interpolating polynomial. (There are other ways to construct the polynomial that ease the problem of keeping the x-value centered—see the next subsection.)

As previously stated, one way to estimate the error of the interpolated value is the *next-term rule*: The next term that we would add to the interpolating polynomial approximates the error of the estimate. We can test this rule by looking at Eq. (3.10). The last term added is -0.006; the actual error of the quadratic to which this is added is -0.004. In the section on theory, we will show why this rule applies and discuss the error of interpolation in further detail.

An algorithm for computing a difference table is

Given $N + 1$ data pairs, (x_i, f_i), $i = 0 .. N$,
   Relabel the f_i as $\Delta^0 f_i$.
   DO FOR $i = 1$ to N:
      DO FOR $j = 0$ to $N - i$:
         Compute $\Delta^i f_j = \Delta^{i-1} f_{j+1} - \Delta^{i-1} f_j$
         and enter into column i of the table.
      ENDDO (j).
   ENDDO (i).

We again observe that parallel processing can compute each column of a difference table in a single time step.

Other Interpolating Polynomials

Several other forms of interpolating polynomials can be generated from a difference table. The Newton–Gregory backward polynomial uses differences from a diagonal row that rises upward and to the right (in contrast to the Newton–Gregory forward polynomial that uses a diagonal row that *descends* to the right). Other interpolating polynomials use different paths through the difference table. The Gauss forward polynomial takes a zigzag path horizontally through the table, the first step an upward one. The Gauss backward polynomial also proceeds in a zigzag fashion but takes its first step downward. Stirling's and Bessel's formulas use a horizontal path, employing averages of differences that lie above and below the path when a column is traversed between the differences. Each of these polynomials has a different way of computing the coefficients.

There are special advantages to polynomials that use horizontal paths. By starting the path at an entry point nearest to x, the value for which we are interpolating, the value of x remains centered among the tabular values that are used. As previously observed, this decreases the magnitude of the error term.

There is less interest today in such interpolating polynomials because computers are able to compute values of functions rather than to interpolate from tabulated values. For more information on these, see previous editions of this text.

3.5   Interpolating with a Cubic Spline

There is another way to fit a polynomial to a set of data. A cubic spline fits a "smooth curve" to the points, borrowing from the idea of a draftman's spline. This device is a flexible rod, bent to conform to the points (and usually held in place by weights).

This device is better than a French curve, for how the French curve is placed to draw the curve is very subjective.

The spline curve can be of varying degrees. Suppose we have the following set of $n + 1$ data points (which need not be evenly spaced):

$$(x_i, y_i), \qquad i = 0, 1, 2, \ldots, n.$$

In general, we fit a set of nth-degree polynomials between each pair of adjacent points, $g_i(x)$, from x_i to x_{i+1}. If the degree of the spline is one (just straight lines between the points), the "curve" would appear as shown in the accompanying figure. The problem with this *linear spline* is that the slope is discontinuous at the points. Splines of degree greater than one do not have this problem. An interpolating polynomial that passes through all the points also is continuous in its slope. But higher-degree interpolating polynomials have another problem. Here is an example of how bad the situation can be.

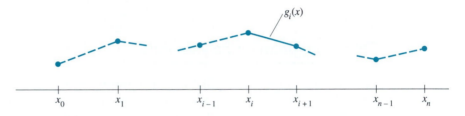

Let $f(x) = \cos^{10}(x)$ over the interval $[-2, 2]$. Fit polynomials of degrees 2, 4, 6, and 8 that match $f(x)$ at equispaced points on the interval. Figure 3.1 shows that none of the polynomials fit well to $f(x)$. The problem is that $f(x)$ is nearly flat except for the "bump" between about $x = -1$ and $x = 1$. The flat portions require $P_n(x)$ to have zeros outside of $[-1, 1]$, creating the oscillations. One remedy to the problem is to fit different polynomials to subregions of $f(x)$. Fitting first-degree polynomials is equivalent to a linear spline.

Figure 3.2 shows a much better fit using a quadratic between $x = -.65$ and $x = .65$ with $P(x) = 0$ outside this region. Although the fit is better (and we could improve it further), there are discontinuities in the slopes at the joins of the polynomials and $f(x)$ has no such breaks in its slope. Spline curves of degree greater than one have a continuous slope.

The drafting spline bends according to the laws of beam flexure so that both the slope and curvature are continuous. Our mathematical spline curve must use polynomials of degree three (or more) to match this behavior. While splines can be of any degree, *cubic splines* are by far the most popular; only these will be described.

We will fit a succession of cubic splines to successive intervals. These polynomials will have the same slope and curvature at the points where they join. We do not require that the intervals be of the same width.

At the endpoints of the data set on which we fit $f(x)$ with spline curves, there is no "joining" polynomial. This means that the slope and curvature there is not constrained. This point will be covered later in the development.

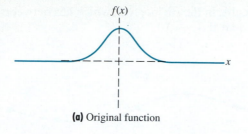

(a) Original function

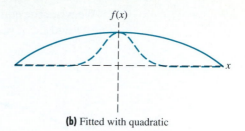

(b) Fitted with quadratic

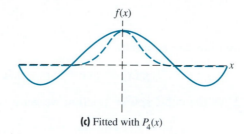

(c) Fitted with $P_4(x)$

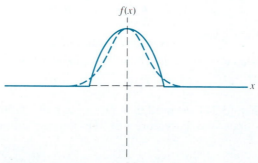

(d) Fitted with $P_6(x)$

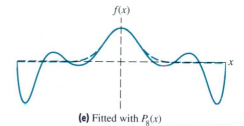

(e) Fitted with $P_8(x)$

Figure 3.1

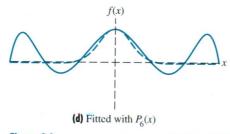

Figure 3.2

We write the equation for a cubic in the ith interval, which lies between the points (x_i, y_i) and (x_{i+1}, y_{i+1}). It looks like this:

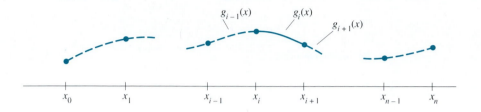

and has the equation

$$g_i(x) = a_i(x - x_i)^3 + b_i(x - x_i)^2 + c_i(x - x_i) + d_i. \tag{3.11}$$

Thus the cubic spline function we want is of the form

$$g(x) = g_i(x) \text{ on the interval } [x_i, x_{i+1}], \quad \text{for } i = 0, 1, \dots, n - 1$$

and meets these conditions:

$$g_i(x_i) = y_i, \quad i = 0, 1, \dots, n - 1 \quad \text{and} \quad g_{n-1}(x_n) = y_n; \tag{3.12a}$$

$$g_i(x_{i+1}) = g_{i+1}(x_{i+1}), \quad i = 0, 1, \dots, n - 2; \tag{3.12b}$$

$$g_i'(x_{i+1}) = g_{i+1}'(x_{i+1}), \quad i = 0, 1, \dots, n - 2; \tag{3.12c}$$

$$g_i''(x_{i+1}) = g_{i+1}''(x_{i+1}), \quad i = 0, 1, \dots, n - 2. \tag{3.12d}$$

[Equations (3.12) say that the cubic spline fits to each of the points (3.12a), is continuous (3.12b), and is continuous in slope and curvature, (3.12c) and (3.12d), throughout the region spanned by the points.]

If there are $n + 1$ points, the number of intervals and the number of $g_i(x)$'s are n. There are thus four times n unknowns, which are the $\{a_i, b_i, c_i, d_i\}$ for $i = 0, 1, \dots, n - 1$. Equation (3.12a) immediately gives

$$d_i = y_i, \quad i = 0, 1, \dots, n - 1. \tag{3.13}$$

Equation (3.12b) then gives

$$y_{i+1} = a_i(x - x_i)^3 + b_i(x - x_i)^2 + c_i(x - x_i) + y_i$$
$$= a_i h_i^3 + b_i h_i^2 + c_i h_i + y_i, \quad i = 0, 1, \dots, n - 1. \tag{3.14}$$

[In the last part of Eq. (3.14), we used $h_i = (x_{i+1} - x_i)$, the width of the ith interval.]

To relate the slopes and curvatures of the joining splines, we differentiate Eq. (3.11):

$$g_i'(x) = 3a_i h_i^2 + 2b_i h_i + c_i, \tag{3.15}$$

$$g_i''(x) = 6a_i h_i + 2b_i, \quad \text{for } i = 0, 1, \dots, n - 1. \tag{3.16}$$

Development is simplified if we write the equations in terms of the second derivatives—that is, if we let $S_i = g_i''(x_i)$ for $i = 0, 1, \ldots, n - 1$ and $S_n = g_{n-1}''(x_n)$.

From Eq. (3.16), we have

$$S_i = 6a_i(x_i - x_i) + 2b_i$$
$$= 2b_i;$$
$$S_{i+1} = 6a_i(x_{i+1} - x_i) + 2b_i$$
$$= 6a_i h_i + 2b_i.$$

Hence we can write

$$b_i = \frac{S_i}{2}, \tag{3.17}$$

$$a_i = \frac{S_{i+1} - S_i}{6h_i}. \tag{3.18}$$

We substitute the relations for a_i, b_i, d_i given by Eqs. (3.13), (3.17), and (3.18) into Eq. (3.11) and then solve for c_i:

$$y_{i+1} = \left(\frac{S_{i+1} - S_i}{6h_i}\right)h_i^3 + \frac{S_i}{2}h_i^2 + c_i h_i + y_i;$$

$$c_i = \frac{y_{i+1} - y_i}{h_i} - \frac{2h_i S_i + h_i S_{i+1}}{6}.$$

We now invoke the condition that the slopes of the two cubics that join at (x_i, y_i) are the same. For the equation in the ith interval, Eq. (3.12c) becomes, with $x = x_i$,

$$y_i' = 3a_i(x_i - x_i)^2 + 2b_i(x_i - x_i) + c_i = c_i.$$

In the previous interval, from x_{i-1} to x_i, the slope at its right end will be

$$y_i' = 3a_{i-1}(x_i - x_{i-1})^2 + 2b_{i-1}(x_i - x_{i-1}) + c_{i-1}$$
$$= 3a_{i-1} h_{i-1}^2 + 2b_{i-1} h_{i-1} + c_{i-1}.$$

Equating these, and substituting for a, b, c, d their relationships in terms of S and y, we get

$$y_i' = \frac{y_{i+1} - y_i}{h_i} - \frac{2h_i S_i + h_i S_{i+1}}{6}$$

$$= 3\left(\frac{S_i - S_{i-1}}{6h_{i-1}}\right)h_{i-1}^2 + 2\left(\frac{S_{i-1}}{2}\right)h_{i-1} + \frac{y_i - y_{i-1}}{h_{i-1}} - \frac{2h_{i-1} S_{i-1} + h_{i-1} S_i}{6}.$$

Simplifying this equation, we get

$$h_{i-1} S_{i-1} + (2h_{i-1} + 2h_i)S_i + h_i S_{i+1} = 6\left(\frac{y_{i+1} - y_i}{h_i} - \frac{y_i - y_{i-1}}{h_{i-1}}\right)$$

$$= 6(f[x_i, x_{i+1}] - f[x_{i-1}, x_i]). \tag{3.19}$$

Equation (3.19) applies at each internal point, from $i = 1$ to $i = n - 1$, there being $n + 1$ total points. This gives $n - 1$ equations relating the $n + 1$ values of S_i. We get two additional equations involving S_0 and S_n when we specify conditions pertaining to the end intervals of the whole curve. To some extent, these end conditions are arbitrary. Four* alternative choices are often used:

1. Take $S_0 = 0$ and $S_n = 0$. This makes the end cubics approach linearity at their extremities. This condition, called a *natural spline*, matches precisely to the drafting device. This technique is used very frequently.
2. Another often used condition is to force the slopes at each end to assume specified values. When that information is not known, the slope might be estimated from the points. If $f'(x_0) = A$ and $f'(x_n) = B$, we use these relations (note that divided differences are employed):

$$\text{At left end:} \quad 2h_0 S_0 + h_1 S_1 = 6(f[x_0, x_1] - A).$$

$$\text{At right end:} \quad h_{n-1} S_{n-1} + 2h_n S_n = 6(B - f[x_{n-1}, x_n]).$$

3. Take $S_0 = S_1, S_{n-1} = S_{n-2}$. This is equivalent to assuming that the end cubics approach parabolas at their extremities.
4. Take S_0 as a linear extrapolation from S_1 and S_2, and S_n as a linear extrapolation from S_{n-1} and S_{n-2}. Only this condition gives cubic spline curves that match exactly to $f(x)$ when $f(x)$ is itself a cubic. For condition 4, we use these relations:

$$\text{At left end:} \quad \frac{S_1 - S_0}{h_0} = \frac{S_2 - S_1}{h_1}, \qquad S_0 = \frac{(h_0 + h_1)S_1 - h_0 S_2}{h_1}.$$

$$\text{At right end:} \quad \frac{S_n - S_{n-1}}{h_{n-1}} = \frac{S_{n-1} - S_{n-2}}{h_{n-2}}, \tag{3.20}$$

$$S_n = \frac{(h_{n-2} + h_{n-1})S_{n-1} - h_{n-1}S_{n-2}}{h_{n-2}}.$$

Relation 1, where $S_0 = 0$ and $S_n = 0$, is called a *natural spline*. It is often felt that this flattens the curve too much at the ends; in spite of this, it is frequently used. Relation 4 frequently suffers from the other extreme, giving too much curvature in the end intervals. Probably the best end condition to use is condition 2, provided reasonable estimates of the derivative are available.

If we write the equation of $S_1, S_2, \ldots, S_{n-1}$ (Eq. (3.19)) in matrix form, we get

---

* A fifth condition is sometimes encountered—a function is periodic and the data cover a full period. In this case, $S_0 = S_n$ and the slopes are also the same at the first and last points.

$$
\begin{bmatrix}
h_0 & 2(h_0 + h_1) & h_1 & & & & \\
 & h_1 & 2(h_1 + h_2) & h_2 & & & \\
 & & h_2 & 2(h_2 + h_3) & h_3 & & \\
 & & & & \ddots & & \\
 & & & & h_{n-2} & 2(h_{n-2} + h_{n-1}) & h_{n-1}
\end{bmatrix}
\begin{bmatrix}
S_0 \\ S_1 \\ S_2 \\ S_3 \\ \vdots \\ S_{n-1} \\ S_n
\end{bmatrix}
$$

$$
= 6
\begin{bmatrix}
f[x_1, x_2] & - & f[x_0, x_1] \\
f[x_2, x_3] & - & f[x_1, x_2] \\
f[x_3, x_4] & - & f[x_2, x_3] \\
\vdots & & \\
f[x_{n-1}, x_n] & - & f[x_{n-2}, x_{n-1}]
\end{bmatrix}.
$$

In this matrix array there are only $n - 1$ equations, but $n + 1$ unknowns. We can eliminate two unknowns (S_0 and S_n) using the relations that correspond to the end-condition assumptions. In the first three cases, this reduces the S vector to $n - 1$ elements, and the coefficient matrix becomes square, of size ($n - 1 \times n - 1$). Furthermore, the matrix is always tridiagonal (even in case 4), and hence is solved speedily and can be stored economically. A program is given at the end of this chapter that creates this tridiagonal matrix and augments it with the proper right-hand-side vector, then solves for the values of S in each interval. You will remember that S is the second derivative of the cubic in each interval, so there is more work to do to get the actual cubic that makes the interpolating curve.

For each end condition, the coefficient matrices become

Condition 1 $S_0 = 0, S_n = 0$:

$$
\begin{bmatrix}
2(h_0 + h_1) & h_1 & & & \\
h_1 & 2(h_1 + h_2) & h_2 & & \\
 & h_2 & 2(h_2 + h_3) & h_3 & \\
 & & & \ddots & \\
 & & & h_{n-2} & 2(h_{n-2} + h_{n-1})
\end{bmatrix}
$$

Condition 2 $f'(x_0) = A$ and $f'(x_n) = B$:

$$
\begin{bmatrix}
2h_0 & h_1 & & & \\
h_0 & 2(h_0 + h_1) & h_1 & & \\
 & h_1 & 2(h_1 + h_2) & h_2 & \\
 & & & \ddots & \\
 & & & h_{n-2} & 2h_{n-1}
\end{bmatrix}
$$

Condition 3 $S_0 = S_1, S_n = S_{n-1}$:

$$
\begin{bmatrix}
(3h_0 + 2h_1) & h_1 & & & \\
h_1 & 2(h_1 + h_2) & h_2 & & \\
& h_2 & 2(h_2 + h_3) & h_3 & \\
& & & \ddots & \\
& & & h_{n-2} & (2h_{n-2} + 3h_{n-1})
\end{bmatrix}
$$

Condition 4 S_0 and S_n are linear extrapolations:

$$
\begin{bmatrix}
\dfrac{(h_0 + h_1)(h_0 + 2h_1)}{h_1} & \dfrac{h_1^2 - h_0^2}{h_1} & & & \\
h_1 & 2(h_1 + h_2) & h_2 & & \\
& h_2 & 2(h_2 + h_3) & h_3 & \\
& & & \ddots & \\
& & \dfrac{h_{n-2}^2 - h_{n-1}^2}{h_{n-2}} & \dfrac{(h_{n-1} + h_{n-2})(h_{n-1} + 2h_{n-2})}{h_{n-2}}
\end{bmatrix}
$$

With condition 4, after solving the set of equations, we must compute S_0 and S_n using Eq. (3.20). For conditions 1, 2, and 3, no computations are needed. For each of the first three cases, the right-hand-side vector is the same; it is given in Eq. (3.19). If the data are evenly spaced, the matrices reduce to a simple form.

After the S_i values are obtained, we get the coefficients a_i, b_i, c_i, and d_i for the cubics in each interval. From these we can compute points on the interpolating curve.

$$a_i = \frac{S_{i+1} - S_i}{6h_i} \, ;$$

$$b_i = \frac{S_i}{2} \, ;$$

$$c_i = \frac{y_{i+1} - y_i}{h_i} - \frac{2h_i S_i + h_i S_{i+1}}{6} \, ;$$

$$d_i = y_i.$$

EXAMPLE 3.6   Fit the data of Table 3.9 with a natural cubic spline curve, and evaluate the spline values $g(0.66)$ and $g(1.75)$. [The true relation is $f(x) = 2e^x - x^2$.] We see that $h_0 = 1.0$, $h_1 = 0.5$, and $h_2 = 0.75$. The divided differences that we can use to get the right-hand sides of our equations are $f[0, 1] = 2.4366$, $f[1, 1.5] = 4.5536$, and $f[1.5, 2.25] = 9.5995$.

Table 3.9

x	$f(x)$
0.0	2.0000
1.0	4.4366
1.5	6.7134
2.25	13.9130

For a natural cubic spline, we use end condition 1 and solve

$$\begin{bmatrix} 3.0 & 0.5 \\ 0.5 & 2.5 \end{bmatrix} \begin{bmatrix} S_1 \\ S_2 \end{bmatrix} = \begin{bmatrix} 12.7020 \\ 30.2754 \end{bmatrix},$$

giving $S_1 = 2.2920$ and $S_2 = 11.6518$. ($S_0 = S_3 = 0$, of course.) Using these S's, we compute the coefficients of the individual cubic splines to arrive at

i	Interval	$g_i(x)$
0	$[0.0, 1.0]$	$0.3820(x - 0)^3 + 0(x - 0)^2 + 2.0546(x - 0) + 2.0000$
1	$[1.0, 1.5]$	$3.1199(x - 1)^3 + 1.146(x - 1)^2 + 3.2005(x - 1) + 4.4366$
2	$[1.5, 2.25]$	$-2.5893(x - 1.5)^3 + 5.8259(x - 1.5)^2 + 6.6866(x - 1.5) + 6.7134$

Figure 3.3 shows the cubic spline curve. (You should verify that these equations satisfy all the conditions that were given for cubic spline curves.)

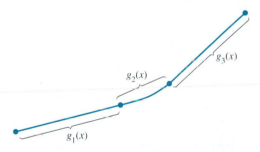

$g_1(x) = 0.3820(x - 0)^3 + 0(x - 0)^2 + 2.0546(x - 0) + 2.0000$
$g_2(x) = 3.1199(x - 1)^3 + 1.146(x - 1)^2 + 3.2005(x - 1) + 4.4366$
$g_3(x) = -2.5893(x - 1.5)^3 + 5.8259(x - 1.5)^2 + 6.6866(x - 1.5) + 6.7134$

Figure 3.3

We use g_0 to find $g(0.66)$: It is 3.4659. (True = 3.4340)

We use g_2 to find $g(1.75)$: It is 8.7087. (True = 8.4467)

Some observations on this example: (a) We were given four points that define three intervals, (b) on each of the three intervals a $g(x)$ is defined, and (c) since each g has four coefficients, we must evaluate 12 unknown coefficients. But by introducing the S's, we only had to solve two equations!

EXAMPLE 3.7   The data in the following table are from astronomical observations of a type of variable star called a *Cepheid variable* and represent variations in its apparent magnitude with time:

Time	0.0	0.2	0.3	0.4	0.5	0.6	0.7	0.8	1.0
Apparent magnitude	0.302	0.185	0.106	0.093	0.240	0.579	0.561	0.468	0.302

Use each of the four end conditions to compute cubic splines, and compare the values interpolated from each spline function at intervals of time of 0.05.

The augmented matrices whose solutions give values for $S_1, S_2, \ldots, S_7$ are shown in Table 3.10. A computer program was used to obtain the results shown in Table 3.11.

Table 3.10

Condition 1

Matrix coefficients are

—	0.60	0.10	−1.23
0.10	0.40	0.10	3.96
0.10	0.40	0.10	9.60
0.10	0.40	0.10	11.52
0.10	0.40	0.10	−21.42
0.10	0.40	0.10	−4.50
0.10	0.60	—	0.60

Condition 2

Matrix coefficients are

—	0.40	0.10	0.00
0.20	0.60	0.10	−1.23
0.10	0.40	0.10	3.96
0.10	0.40	0.10	9.60
0.10	0.40	0.10	11.52
0.10	0.40	0.10	−21.42
0.10	0.40	0.10	−4.50
0.10	0.60	0.20	0.60
0.10	0.40	—	0.00

Table 3.10 (*Cont.*)

	Condition 3		
Matrix coefficients are			
—	0.80	0.10	−1.23
0.10	0.40	0.10	3.96
0.10	0.40	0.10	9.60
0.10	0.40	0.10	11.52
0.10	0.40	0.10	−21.42
0.10	0.40	0.10	−4.50
0.10	0.80	—	0.60

	Condition 4		
Matrix coefficients are			
—	1.20	−0.30	−1.23
0.10	0.40	0.10	3.96
0.10	0.40	0.10	9.60
0.10	0.40	0.10	11.52
0.10	0.40	0.10	−21.42
0.10	0.40	0.10	−4.50
−0.30	1.20	—	0.60

Table 3.11

t	Values, condition 1	Values, condition 2*	Values, condition 3	Values, condition 4
0.00	0.302	0.302	0.302	0.302
0.05	0.278	0.276	0.282	0.297
0.10	0.252	0.250	0.256	0.271
0.15	0.222	0.221	0.224	0.231
0.20	0.185	0.185	0.185	0.185
0.25	0.143	0.143	0.142	0.141
0.30	0.106	0.106	0.106	0.106
0.35	0.087	0.087	0.088	0.088
0.40	0.093	0.093	0.093	0.093
0.45	0.133	0.133	0.133	0.133
0.50	0.240	0.240	0.240	0.240
0.55	0.424	0.424	0.424	0.424
0.60	0.579	0.579	0.579	0.579
0.65	0.608	0.608	0.608	0.608
0.70	0.561	0.561	0.561	0.561
0.75	0.511	0.511	0.511	0.511
0.80	0.468	0.468	0.468	0.468
0.85	0.426	0.426	0.426	0.430
0.90	0.385	0.385	0.384	0.392
0.95	0.343	0.343	0.343	0.350
1.00	0.302	0.302	0.302	0.302

*Note that in the values for condition 2 we used forward and backward differences to approximate the slope at either end of the curve; that is, $V'(0.0) = -0.585$ and $V'(1.0) = -0.830$.

A graph of the four solutions is shown in Fig. 3.4. The points are so close to one another that we must magnify the portions near the ends to see the differences. In the central part of the curve, between $t = 0.2$ and 0.8, none differ by more than 0.001.

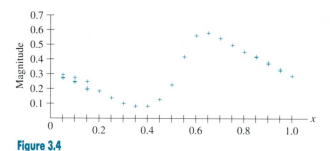

Figure 3.4

EXAMPLE 3.8   Cubic splines that fit the function $f(x) = \cos^{10}(x)$ at $x = -2, -1, -0.5, 0, 0.5, 1$, and 2 were constructed. (This is the same function that showed how poorly ordinary interpolating polynomials worked.) Figure 3.5 compares the cubic spline interpolates with the true function. The two plots are barely distinguishable.

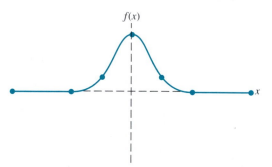

Figure 3.5

3.6 Bezier Curves and B-Spline Curves

In addition to the splines we have studied in the previous section, there are others that are important. In particular, Bezier curves and B-splines are widely used in computer graphics and computer-aided design. These two types of curves are not really interpolating splines, since the curves do not normally pass through all of the points. In this respect they show some similarity to least-squares curves that are discussed in a later section. However, both Bezier curves and B-splines have the important property of staying within the polygon determined by the given points.

We will be more explicit about this property later. In addition, these two new spline curves have a nice geometric property in that in changing one of the points we change only one portion of the curve, a "local" effect. For the cubic spline curve of the previous section, changing just one point has a "global" effect, in that the curve from the first to the last point is affected. Finally, for the cubic splines just studied the points were given data points. For the two curves we study in this section the points in question are more likely "control" points we select to determine the shape of the curve we are working on.

For simplicity, we consider mainly the cubic version of these two curves. In what follows, we will express $y = f(x)$ in parametric form. The parametric form represents a relation between x and y by two other equations, $x = F_1(u), y = F_2(u)$. The independent variable u is called the *parameter*. For example, the equation for a circle can be written, with θ as the parameter, as

$$x = r \cos(\theta),$$
$$y = r \sin(\theta).$$

When y and x are expressed in terms of a parameter u, $(x(u), y(u)), 0 \le u \le 1$, defines a set of points (x, y), associated with the values of u. We discuss Bezier curves before B-splines. Bezier curves are named after the French engineer, P. Bezier of the Renault Automobile Company. He developed them in the early 1960s to fill a need for curves whose shape can be readily controlled by changing a few parameters. Bezier's application was to construct pleasing surfaces for car bodies.

Suppose we are given a set of control points, $p_i = (x_i, y_i), i = 0, 1, \ldots, n$. (These points are also referred to as *Bezier points*.) Figure 3.6 is an example.

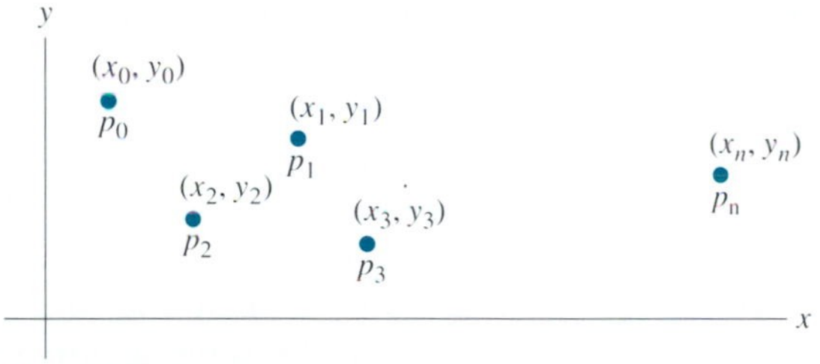

Figure 3.6

These points could be chosen on a computer screen, using either a mouse or a track ball. The points do not necessarily progress from left to right. We treat the coordinates of each point as a two-component vector,

$$p_i = \begin{pmatrix} x_i \\ y_i \end{pmatrix}.$$

The set of points, in parametric form, is

$$P(u) = \begin{pmatrix} x(u) \\ y(u) \end{pmatrix}, \qquad 0 \le u \le 1.$$

The nth-degree Bezier polynomial determined by $n + 1$ points is given by

$$P(u) = \sum_{i=0}^{n} \binom{n}{i}(1 - u)^{n-i}u^i p_i, \quad \text{where}$$

$$\binom{n}{i} = \frac{n!}{i!(n - i)!}.$$

(The preceding formula really represents two other scalar equations, one for x_i and the other for y_i.) For $n = 2$, this would give the quadratic equation defined by three points, p_0, p_1, p_2:

$$P(u) = (1)(1 - u)^2 p_0 + 2(1 - u)(u)p_1 + (1)u^2 p_2,$$

since, for $n = 2$ and $i = 0, 1, 2$, we have $\binom{2}{0} = 1, \binom{2}{1} = 2, \binom{2}{2} = 1$. The preceding equation represents the pair of equations

$$x(u) = (1 - u)^2 x_0 + 2(1 - u)(u)x_1 + u^2 x_2,$$
$$y(u) = (1 - u)^2 y_0 + 2(1 - u)(u)y_1 + u^2 y_2.$$

Observe that, if $u = 0$, $x(0)$ is identical to x_0 and similarly for $y(0)$. If $u = 1$, the point referred to is (x_2, y_2). As u takes on values between 0 and 1, a curve is traced that goes from the first point to the third of the set. Ordinarily the curve will not pass through the central point of the three. (If they are collinear, the curve is the straight line through them all.) In effect, the points of the second-degree Bezier curve have coordinates that are weighted sums of the coordinates of the three points that are used to define it. From another point of view, one can think of the Bezier equations as weighted sums of three polynomials in u, where the weighting factors are the coordinates of the three points. Applying the general defining equation for $n = 3$, we get the cubic Bezier polynomial that we shall consider in some detail. The properties of other Bezier polynomials are the same as for the cubic. Here is the Bezier cubic:

$$x(u) = (1 - u)^3 x_0 + 3(1 - u)^2 ux_1 + 3(1 - u)u^2 x_2 + u^3 x_3,$$
$$y(u) = (1 - u)^3 y_0 + 3(1 - u)^2 uy_1 + 3(1 - u)u^2 y_2 + u^3 y_3.$$

Observe again that $(x(0), y(0)) = p_0$ and $(x(1), y(1)) = p_3$, and that the curve will not ordinarily go through the intermediate points. As illustrated in the example curves in Fig. 3.7, changing the intermediate "control" points changes the shape of the curve. The examples are in Fig. 3.7(a) through 3.7(e). The first three of these show Bezier curves defined by one group of four points.

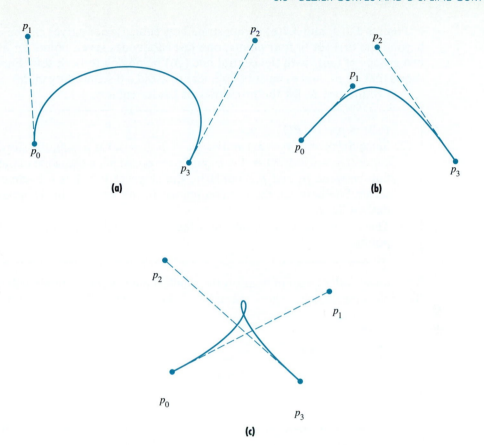

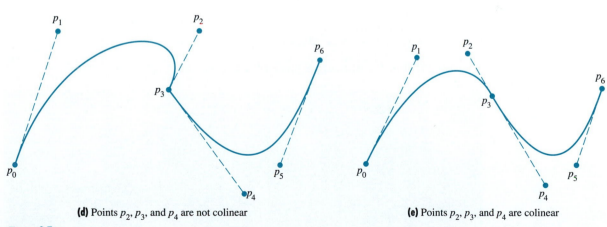

(d) Points p_2, p_3, and p_4 are not colinear

(e) Points p_2, p_3, and p_4 are colinear

Figure 3.7

Figures 3.7(d) and 3.7(e) demonstrate how cubic Bezier curves can be continued beyond the first set of four points; one just subdivides seven points (p_0 to p_6) into two groups of four, with the central one (p_3) belonging to both sets. Figure 3.7(e) shows that p_2, p_3, and p_4 must be collinear to avoid a discontinuity in the slope at p_3.

It is of interest to list the properties of Bezier cubics:

1. $P(0) = p_0$, $P(1) = p_3$.
2. Since $dx/du = 3(x_1 - x_0)$ and $dy/du = 3(y_1 - y_0)$ at $u = 0$, the slope of the curve at $u = 0$ is $dy/dx = (y_1 - y_0)/(x_1 - x_0)$, which is the slope of the secant line between p_0 and p_1. Similarly, the slope at $u = 1$ is the same as the secant line between the last two points. In the figures, this is indicated by dashed lines.
3. The Bezier curve is contained in the convex hull determined by the four points.

The *convex hull* of a set of points is the smallest convex set containing all the points. The following sketches show examples of the convex hull of four points.

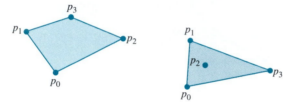

It is often convenient to represent the Bezier curve in matrix form. For Bezier cubics, this is

$$P(u) = [u^3, u^2, u, 1] \begin{bmatrix} -1 & 3 & -3 & 1 \\ 3 & -6 & 3 & 0 \\ -3 & 3 & 0 & 0 \\ 1 & 0 & 0 & 0 \end{bmatrix} \begin{bmatrix} p_0 \\ p_1 \\ p_2 \\ p_3 \end{bmatrix}$$

$$= u^T M_2 p.$$

An algorithm for drawing a piece of a Bezier curve is

Given four points, $(x_i, y_i), i = 0, \ldots, 3$:

  DO FOR $u = 0$ TO 1 STEP 0.01:
    Compute
      $x = (1 - u)^3 x_0 + 3(1 - u)^2 u x_1 + 3(1 - u)u^2 x_2 + u^3 x_3,$
      $y = (1 - u)^3 y_0 + 3(1 - u)^2 u y_1 + 3(1 - u)u^2 y_2 + u^3 y_3.$
    Plot (x, y).
  ENDDO.

To continue the curve, repeat this process for the next set of four points, beginning with point p_3.

B-Spline Curves

We now discuss B-splines. These curves are like Bezier curves in that they do not ordinarily pass through the given data points. They can be of any degree, but we will concentrate on the cubic form. Cubic B-splines resemble the ordinary cubic splines of the previous section, in that a separate cubic is derived for each pair of points in the set. However, the B-spline need not pass through any of the set that are used in its definition.

We begin the description by stating the formula for a cubic B-spline in terms of parametric equations whose parameter is u.

Given the points $p_i = (x_i, y_i), i = 0, 1, \ldots, n$, the cubic B-spline for the interval $(p_i, p_{i+1}), i = 1, 2, \ldots, n - 1$, is

$$B_i(u) = \sum_{k=-1}^{2} b_k p_{i+k}, \quad \text{where}$$

$$b_{-1} = \frac{(1 - u)^3}{6},$$

$$b_0 = \frac{u^3}{2} - u^2 + \frac{2}{3}, \tag{3.21}$$

$$b_1 = -\frac{u^3}{2} + \frac{u^2}{2} + \frac{u}{2} + \frac{1}{6},$$

$$b_2 = \frac{u^3}{6}, \quad 0 \le u \le 1.$$

As before, p_i refers to the point (x_i, y_i); it is a two-component vector. The coefficients, the b_k's, serve as a basis and do not change as we move from one set of points to the next. Observe that they can be considered weighting factors applied to the coordinates of a set of four points. The weighted sum, as u varies from 0 to 1, generates the B-spline curve.

If we write out the equations for x and y from Eq. (3.21), we get

$$x_i(u) = \frac{1}{6}(1 - u)^3 x_{i-1} + \frac{1}{6}(3u^3 - 6u^2 + 4)x_i$$

$$+ \frac{1}{6}(-3u^3 + 3u^2 + 3u + 1)x_{i+1} + \frac{1}{6}u^3 x_{i+2};$$

$$y_i(u) = \frac{1}{6}(1 - u)^3 y_{i-1} + \frac{1}{6}(3u^3 - 6u^2 + 4)y_i$$

$$+ \frac{1}{6}(-3u^3 + 3u^2 + 3u + 1)y_{i+1} + \frac{1}{6}u^3 y_{i+2}.$$

Note the notation here: $x_i(u)$ is a function (of u) and x_i, y_i are the components of the point p_i. (The end portions are a special situation that we discuss later.)

As we have said, the u-cubics act as weighting factors on the coordinates of the four successive points to generate the curve. For example, at $u = 0$, the weights applied are 1/6, 2/3, 1/6, and 0. At $u = 1$, they are 0, 1/6, 2/3, and 1/6. These values vary throughout the interval from $u = 0$ to $u = 1$. As an exercise, you are asked to graph these factors. This will give you a visual impression of how the weights change with u.

Let us now examine two B-splines determined from a set of exactly four points. Figures 3.8(a) and 3.8(b) show the effect of varying just one of the points. As you would expect, when p_2 is moved upward and to the left, the curve tends to follow; in fact, it is pulled to the opposite side of p_1. You may be surprised to see that the curve is never very close to the two intermediate points, though it begins and ends at positions somewhat adjacent. It will be helpful to think of the curve generated from the defining equation for B_1 as associated with a curve that goes from near p_1 to p_2. It is also helpful to remember that points p_0, p_1, p_2, and p_3 are used to get B_1.

Figure 3.8

Since a set of four points is required to generate only a portion of the B-spline, that associated with the two inner points, we must consider how to get the B-spline for more than four points as well as how to extend the curve into the region outside of the middle pair. We use a method analogous to the cubic splines of Section 3.5 marching along one point at a time, forming new sets of four. We abandon the first of the old set when we add the new one.

The conditions that we want to impose on the B-spline are exactly the same as for ordinary splines—continuity of the curve and its first and second derivatives. It turns out that the equations for the weighting factors (the u-polynomials, the b_k) are such that these requirements are met. Figure 3.9 shows how three successive parts of a B-spline might look.

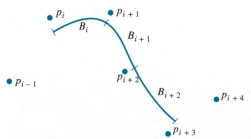

Figure 3.9 Successive B-splines joined together

We can summarize the properties of B-splines as

1. Like the cubic splines of Section 3.5, B-splines are pieced together so they agree at their joints in three ways:

   a. $B_i(1) = B_{i+1}(0) = \dfrac{p_i + 4p_{i+1} + p_{i+2}}{6}$,

   b. $B_i'(1) = B_{i+1}'(0) = \dfrac{-p_i + p_{i+2}}{2}$,

   c. $B_i''(1) = B_{i+1}''(0) = p_i - 2p_{i+1} + p_{i+2}$.

   The subscripts here refer to the portions of the curve and the points in Fig. 3.9.

2. The portion of the curve determined by each group of four points is within the convex hull of these points.

Now we consider how to generate the ends of the joined B-spline. If we have points from p_0 to p_n, we already can construct B-splines B_1 through B_{n-2}. We need B_0 and B_{n-1}. Our problem is that, using the procedure already defined, we would need additional points outside the domain of the given points. We probably also want to tie down the curve in some way—having it start and end at the extreme points of the given set seems like a good idea. How can we do this?

First, we can add more points without creating artificiality by making the added points coincide with the given extreme points. If we add not just a single fictitious point at each end of the set, but two at each end, we will find that the new curves not only join properly with the portions already made, but start and end at the extreme points as we wanted. (It looks like we have added two extra portions, but reflection shows these are degenerate, giving only a single point.)

In summary—we add fictitious points p_{-2}, p_{-1}, p_{n+1}, and p_{n+2}, with the first two identical with p_0 and the last two identical with p_n. (There are other methods to handle the starting and ending segments of B-splines that we do not cover.)

The matrix formulation for cubic B-splines is helpful. Here it is:

$$
B_i(u) = \frac{1}{6}[u^3, u^2, u, 1]\begin{bmatrix} -1 & 3 & -3 & 1 \\ 3 & -6 & 3 & 0 \\ -3 & 0 & 3 & 0 \\ 1 & 4 & 1 & 0 \end{bmatrix}\begin{bmatrix} p_{i-1} \\ p_i \\ p_{i+1} \\ p_{i+2} \end{bmatrix}
$$

$$
= \frac{u^T M_b p}{6} . \tag{3.22}
$$

This applies on the interval $[0, 1]$ and for the points (p_i, p_{i+1}).

We conclude this section by looking at several examples of B-splines. The five parts of Fig. 3.10 show B-splines that are defined by the same set of points as the Bezier curves in Fig. 3.7. There are significant differences.

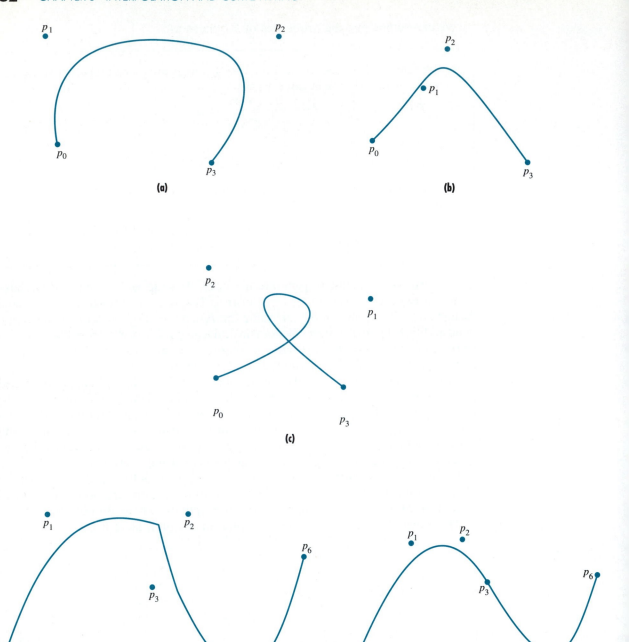

Figure 3.10

B-splines differ from Bezier curves in three ways:

1. For a B-spline, the curve does not begin and end at the extreme points.
2. The slopes of the B-splines do not have any simple relationship to lines drawn between the points.
3. The endpoints of the B-splines are in the vicinity of the two intermediate given points, but neither the x- nor the y-coordinates of these endpoints normally equal the coordinates of the intermediate points.

An algorithm for drawing a B-spline curve is

Given $n + 1$ points, $p_i = (x_i, y_i), i = 0, \ldots, n$:
    Set $p_{-1} = p_{-2} = p_0$.
    Set $p_{n+1} = p_{n+2} = p_n$.
DO FOR $i = 0$ to $n - 1$:
    DO FOR $u = 0$ TO 1 STEP 0.01:
    Compute
$$x = (1 - u)^3/6x_{i-1} + (3u^3 - 6u^2 + 4)/6x_i$$
$$+ (-3u^3 + 3u^2 + 3u + 1)/6x_{i+1} + u^3/6x_{i+2},$$
$$y = (1 - u)^3/6y_{i-1} + (3u^3 - 6u^2 + 4)/6y_i$$
$$+ (-3u^3 + 3u^2 + 3u + 1)/6y_{i+1} + u^3/6y_{i+2}.$$
    Plot (x, y).
    ENDDO (u).
ENDDO (i).

3.7 Polynomial Approximation of Surfaces

When a function z is a polynomial function* of two variables x and y—say, of degree 3 in x and of degree 2 in y—we would have

$$z = f(x, y) = a_0 + a_1 x + a_2 y + a_3 x^2 + a_4 xy + a_5 y^2 + a_6 x^3$$
$$+ a_7 x^2 y + a_8 xy^2 + a_9 x^3 y + a_{10} x^2 y^2 + a_{11} x^3 y^2. \tag{3.23}$$

---

* We approximate a nonpolynomial function by a polynomial that agrees with the function, just as we have done with a function of one variable.

Such a function describes a surface; (x, y, z) is a point on it. The functional relation is seen to involve many terms. If we are concerned with four independent variables (three space dimensions plus time, say), even low-degree polynomials would be quite intractable. Except for special purposes, such as when we need an explicit representation, perhaps to permit ready differentiation at an arbitrary point, we can avoid such complications by handling each variable separately. We will treat only this case.

Note the immediate simplification of Eq. (3.23) if we let y take on a constant value, say, $y = c$. Combining the y factors with the coefficients, we get

$$z\big|_{y=c} = b_0 + b_1 x + b_2 x^2 + b_3 x^3.$$

This will be our attack in interpolating at the point (a, b) in a table of two variables—hold one variable constant—say, $y = y_1$—and the table becomes a single-variable problem. The preceding methods then apply to give $f(a, y_1)$. If we repeat this at various values of y, $y = y_2, y_3, \ldots, y_n$, we will get a table with x constant at the value $x = a$ and with y varying. We then interpolate at $y = b$.

EXAMPLE 3.9   Estimate $f(1.6, 0.33)$ from the values in Table 3.12. Use quadratic interpolation in the x-direction and cubic interpolation for y. We select one of the variables to hold constant, say, x. (This choice is arbitrary since we would get the same result, except for differences due to round-off, if we had chosen to hold y constant.) We decide to interpolate for y within the three rows of the table at $x = 1.0, 1.5,$ and 2.0, since the desired value at $x = 1.6$ is most nearly centered within this set. We choose y-values of $0.2, 0.3, 0.4,$ and 0.5 so that $y = 0.33$ is centralized. The shading in Table 3.12 shows the region of fit for our polynomials.

Table 3.12   Tabulation of a function of two variables $z = f(x, y)$

x \ y	0.1	0.2	0.3	0.4	0.5	0.6
0.5	0.165	0.428	0.687	0.942	1.190	1.431
1.0	0.271	0.640	1.003	1.359	1.703	2.035
1.5	0.447	0.990	1.524	2.045	2.549	3.031
2.0	0.738	1.568	2.384	3.177	3.943	4.672
2.5	1.216	2.520	3.800	5.044	6.241	7.379
3.0	2.005	4.090	6.136	8.122	10.030	11.841
3.5	3.306	6.679	9.986	13.196	16.277	19.198

We may either use divided differences or derive the interpolated values using difference tables. Let us use the latter method, since the data are evenly spaced. This gives Table 3.13.

Table 3.13

	y	z	Δz	$\Delta^2 z$	$\Delta^3 z$
	0.2	0.640			
			0.363		
	0.3	1.003		−0.007	
$x = 1.0$			0.356		−0.005
	0.4	1.359		−0.012	
			0.344		
	0.5	1.703			
	0.2	0.990			
			0.534		
	0.3	1.524		−0.013	
$x = 1.5$			0.521		−0.004
	0.4	2.045		−0.017	
			0.504		
	0.5	2.549			
	0.2	1.568			
			0.816		
	0.3	2.384		−0.023	
$x = 2.0$			0.793		−0.004
	0.4	3.177		−0.027	
			0.766		
	0.5	3.943			

We need the subtables from $y = 0.2$ to $y = 0.5$, since, for a cubic interpolation, four points are required. Using any convenient formula (remember that all cubics that agree at four points are identical), we get Table 3.14. In the last tabulation we carry one extra decimal to guard against round-off errors. Interpolating again, we get $z = 1.8406$, which we report as $z = 1.841$.

Table 3.14

	x	z	Δz	$\Delta^2 z$
	1.0	1.1108		
			0.5710	
$y = 0.33$	1.5	1.6818		0.3717
			0.9427	
	2.0	2.6245		

The function tabulated in Table 3.12 is $f(x, y) = e^x \sin y + y - 0.1$, so the true value is $f(1.6, 0.33) = 1.8350$. Our error of -0.006 occurs because quadratic interpolation for x is inadequate in view of the large second difference. In retrospect, it would have been better to use quadratic interpolation for y, since the third differences of the y-subtables are small, and let x take on a third-degree relationship.

It is instructive to observe which of the values in Table 3.12 entered into our computation. The shaded rectangle covers these values. This is the "region of fit" for the interpolating polynomial that we have used. The principle of choosing values so that the point at which the interpolating polynomial is used is centered in the region of fit obviously applies here in exact analogy to the one-way table situation. It also applies to tables of three and four variables in the same way. Of course, the labor of interpolating in such multidimensional cases soon becomes burdensome.

A rectangular region of fit is not the only possibility. We may change the degree of interpolation as we subtabulate the different rows or columns. Intuitively, it would seem best to use higher-degree polynomials for the rows near the interpolating point, decreasing the degree as we get farther away. The coefficient of the error term, when this is done, will be found to be minimized thereby, though for multidimensional interpolating polynomials the error term is quite complex. The region of fit will be diamond-shaped when such tapered degree functions are used.

We may adapt the Lagrangian form of interpolating polynomial to the multidimensional case also. It is perhaps easiest to employ a process similar to the preceding example. Holding one variable constant, we write a series of Lagrangian polynomials for interpolation at the given value of the other variable, and then combine these values in a final Lagrange form. The net result is a Lagrangian polynomial in which the function factors are replaced by Lagrangian polynomials. The resulting expression for the above example would be

$$
\frac{(y - 0.3)(y - 0.4)(y - 0.5)}{(0.2 - 0.3)(0.2 - 0.4)(0.2 - 0.5)}
$$
$$
\times \left[\frac{(x - 1.5)(x - 2.0)}{(1.0 - 1.5)(1.0 - 2.0)} (0.640) + \frac{(x - 1.0)(x - 2.0)}{(1.5 - 1.0)(1.5 - 2.0)} (0.990) + \frac{(x - 1.0)(x - 1.5)}{(2.0 - 1.0)(2.0 - 1.5)} (1.568) \right]
$$

$$
+ \frac{(y - 0.2)(y - 0.4)(y - 0.5)}{(0.3 - 0.2)(0.3 - 0.4)(0.3 - 0.5)}
$$
$$
\times \left[\frac{(x - 1.5)(x - 2.0)}{(1.0 - 1.5)(1.0 - 2.0)} (1.003) + \frac{(x - 1.0)(x - 2.0)}{(1.5 - 1.0)(1.5 - 2.0)} (1.534) + \frac{(x - 1.0)(x - 1.5)}{(2.0 - 1.0)(2.0 - 1.5)} (2.384) \right]
$$

$$
+ \frac{(y - 0.2)(y - 0.3)(y - 0.5)}{(0.4 - 0.2)(0.4 - 0.3)(0.4 - 0.5)} \tag{3.24}
$$
$$
\times \left[\frac{(x - 1.5)(x - 2.0)}{(1.0 - 1.5)(1.0 - 2.0)} (1.359) + \frac{(x - 1.0)(x - 2.0)}{(1.5 - 1.0)(1.5 - 2.0)} (2.045) + \frac{(x - 1.0)(x - 1.5)}{(2.0 - 1.0)(2.0 - 1.5)} (3.177) \right]
$$

$$
+ \frac{(y - 0.2)(y - 0.3)(y - 0.4)}{(0.5 - 0.2)(0.5 - 0.3)(0.5 - 0.4)}
$$
$$
\times \left[\frac{(x - 1.5)(x - 2.0)}{(1.0 - 1.5)(1.0 - 2.0)} (1.703) + \frac{(x - 1.0)(x - 2.0)}{(1.5 - 1.0)(1.5 - 2.0)} (2.549) + \frac{(x - 1.0)(x - 1.5)}{(2.0 - 1.0)(2.0 - 1.5)} (3.943) \right].
$$

The equation is easy to write, but its evaluation by hand is laborious. If one is writing a computer program for interpolation in such multivariate situations, the Lagrangian form is recommended. There is a special advantage in that equal spacing in the table is not required. The Lagrangian form is also perhaps the most straightforward way to write out the polynomial as an explicit function.

When the given points are not evenly spaced, Lagrangian polynomials or the method of divided differences would be used for interpolation. With the latter, exactly the same principle is involved—hold one variable constant while subtables of divided differences are constructed, then combine the interpolated values from these subtables into a new table.

Parallel processing can save many time steps in the preceding computations. Each value in the column of differences of Tables 3.13 and 3.14 can be computed at the same time. (We must wait for the interpolations from Table 3.13 to be completed before we do Table 3.14, of course.) Every factor of Eq. (3.23) can be evaluated in parallel.

Using Cubic Splines, Bezier Surfaces, and B-Spline Surfaces

Another alternative is to use cubic splines for interpolation in multivariate cases. Here again it is perhaps best to hold one variable constant while constructing one-way splines, then combine the results from these in the second phase. The computational effort would be significant, however.

Interpolating for values of functions of two independent variables can also be thought of as constructing a surface that is defined by the given points. Rather than finding values on a surface that contains the given points, we can construct surfaces that are analogous to Bezier curves and B-spline curves where the surface does not normally contain the given points.

So far we have been able to interpolate on simple surfaces where we are given z as a function of x and y. Suppose now we are given a set of points, $p_i = \{(x_i, y_i, z_i),$ $i = 0, \ldots, n\}$, and we wish to fit a surface to those points. This would be the case if we were trying to draw a mountain, an airplane, or a teapot. But first we consider the representation of more general surfaces. Let $p = (x, y, z)$ be any point on the surface. Then the coordinates of each point are represented as the equations

$$x = x(u, v),$$
$$y = y(u, v),$$
$$z = z(u, v),$$

where u, v are the independent variables that range over a given set of values and x, y, z are the dependent variables. This is a slight change of notation from the first part of this section.

An example of this would be the equations of a sphere of radius r about the origin: $(0, 0, 0)$. Here any point on the surface of the sphere is given by

$$x = r \cos(u) \sin(v),$$
$$y = r \sin(u) \sin(v),$$
$$z = r \cos(v),$$

where u ranges in value from 0 to 2π and v ranges from 0 to π. Figure 3.11 illustrates this.

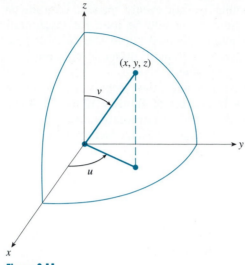

Figure 3.11

We will only describe constructing a B-spline surface. (A most interesting and informative description of Bezier surfaces can be found in Crow, 1987. See also Pokorny and Gerald, 1989.)

From the previous section, we know that a cubic B-spline curve segment starting near the point p_i to near the point p_{i+1} is determined by the four points

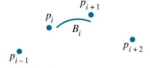

where $p_i(u) = (x_i(u), y_i(u))$ in two dimensions, or $p_i(u) = (x_i(u), y_i(u), z_i(u))$ if we had been working in three dimensions. The segment was then extended by introducing p_{i+3}, deleting p_{i-1}, and generating the curve for $0 \leq u \leq 1$.

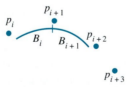

The process is continued until we have B_{n-2}. Finally, the first and last segments are generated by starting with p_0, p_0, p_0, p_1 and ending with p_{n-1}, p_n, p_n, p_n.

In an analogous manner the interpolating B-spline surface patch depends on 16 points, as Fig. 3.12 shows. Here $p_{i,j} = (x_{i,j}, y_{i,j}, z_{i,j})$, a point in E^3. This patch is generated by computing the points $p_{i,j}(u, v)$, for $0 \leq u \leq 1$ and $0 \leq v \leq 1$. Here we have changed the subscripts on the points $p_{i,j}$ so as to fit into matrix notation.

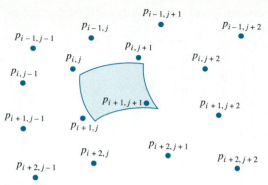

Figure 3.12

For simplicity, we will consider only the x-coordinate in detail. Comparable formulations hold for the y- and z-coordinates. The simplest formulation for $x_{ij}(u, v)$ is based on the matrix formulation of Eq. (3.22) and is given by

$$x_{ij}(u, v) = \frac{1}{36} [u^3, u^2, u, 1] M_b X_{i,j} M_b^T \begin{bmatrix} v^3 \\ v^2 \\ v \\ 1 \end{bmatrix}, \tag{3.25}$$

where $X_{i,j}$ is the 4×4 matrix

$$\begin{bmatrix} x_{i-1,j-1} & x_{i-1,j} & x_{i-1,j+1} & x_{i-1,j+2} \\ x_{i,j-1} & x_{i,j} & x_{i,j+1} & x_{i,j+2} \\ x_{i+1,j-1} & x_{i+1,j} & x_{i+1,j+1} & x_{i+1,j+2} \\ x_{i+2,j-1} & x_{i+2,j} & x_{i+2,j+1} & x_{i+2,j+2} \end{bmatrix},$$

which are just the x-coordinates of the 16 points of Fig. 3.12. The matrix M_b is the matrix we saw before in Eq. (3.22):

$$M_b = \begin{bmatrix} -1 & 3 & -3 & 1 \\ 3 & -6 & 3 & 0 \\ -3 & 0 & 3 & 0 \\ 1 & 4 & 1 & 0 \end{bmatrix}.$$

The y and z equations are then obtained merely by substituting the corresponding matrices $Y_{i,j}$ and $Z_{i,j}$, which are formed from the y and z components of the 16 points. Since each of these equations is cubic in u and v, they are referred to as *bicubic equations*. The coordinates of the points on a patch are given by

$$x(u, v) = \frac{1}{36} [u^3, u^2, u, 1] M_b X_{i,j} M_b^T [v^3, v^2, v, 1]^T,$$

$$y(u, v) = \frac{1}{36} [u^3, u^2, u, 1] M_b Y_{i,j} M_b^T [v^3, v^2, v, 1]^T,$$

$$z(u, v) = \frac{1}{36} [u^3, u^2, u, 1] M_b Z_{i,j} M_b^T [v^3, v^2, v, 1]^T,$$

as u and v range between 0 and 1. It is easily verified that the weights applied to each of the 16 points are

$$
\begin{bmatrix}
1 & 4 & 1 & 0 \\
4 & 16 & 4 & 0 \\
1 & 4 & 1 & 0 \\
0 & 0 & 0 & 0
\end{bmatrix}
\quad \text{At } p_{i,j}(u, v) \text{ (for } u = 0, v = 0\text{), and}
$$

$$
\begin{bmatrix}
0 & 0 & 0 & 0 \\
0 & 1 & 4 & 1 \\
0 & 4 & 16 & 4 \\
0 & 1 & 4 & 1
\end{bmatrix}
\quad \text{At } p_{i,j}(u, v) \text{ (for } u = 1, v = 1\text{)}
$$

where each (i, j)th element is the coefficient for the corresponding point in Fig. 3.12. In effect, these matrices are templates that overlay the points shown in Fig. 3.12.

The surface patch is extended by adding another row or column of points and deleting a corresponding row or column of points. One should verify that the current and previous patches are connected smoothly along the edge where they join. An initial or final patch can be obtained by repeating a corner, as was suggested for the B-spline curve. This will ensure that the patch actually starts or ends at a point. For the surface we would repeat a point nine times, instead of three times as was done for the curve.

For a more detailed and informative discussion of interpolating curves and surfaces, the reader should consult Pokorny and Gerald (1989).

3.8 Least-Squares Approximations

Suppose we wish to fit a curve to an approximate set of data, such as from the determination of the effects of temperature on a resistance by students in their physics laboratory. They have recorded the temperature and resistance measurements as shown in Fig. 3.13, where the graph suggests a linear relationship. We want to suitably determine the constants a and b in the equation relating resistance R and temperature T,

$$
R = aT + b, \tag{3.26}
$$

so that in subsequent use the resistance can be predicted at any temperatures. The line as sketched by eye represents the data fairly well, but if we replotted the data and asked someone else to draw a line, rarely would exactly the same line be obtained. One of our requirements for fitting a curve to data is that the process be *unambiguous*. We would also like, in some sense, to minimize the deviations of the points from the line. The deviations are measured by the distances from the points to the line—how these distances are measured depends on whether both variables are subject to error. We will assume that the error of reading the temperatures in Fig. 3.13 is negligible, so that all the errors are in the resistance measurements, and we will use vertical distances. (If both were subject to error, we might use perpendicular distances and would modify the following. In this way the problem also becomes considerably more complicated. We will treat only the simpler case.)

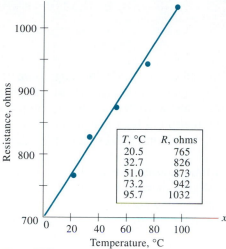

T, °C	R, ohms
20.5	765
32.7	826
51.0	873
73.2	942
95.7	1032

Figure 3.13

We might first suppose we could minimize the deviations by making their sum a minimum, but this is not an adequate criterion. Consider the case of only two points (Fig. 3.14). Obviously, the best line passes through each, but any line that passes through the midpoint of the segment connecting them has a sum of errors equal to zero.

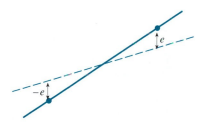

Figure 3.14

Then what about making the sum of the magnitudes of the errors a minimum? This also is inadequate, as the case of three points shows (Fig. 3.15). Assume that two of the points are at the same x-value (which is not an abnormal situation since frequently experiments are duplicated). The best line will obviously pass through the average of the duplicated tests. However, any line that falls between the dotted lines shown will have the same sum of the magnitudes of the vertical distances. Since we wish an unambiguous result, we cannot use this as a basis for our work.

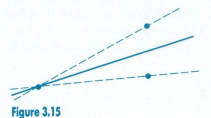

Figure 3.15

We might accept the criterion that we make the magnitude of the maximum error a minimum (the so-called *minimax* criterion), but for the problem at hand this is rarely done.* This criterion is awkward because the absolute-value function has no derivative at the origin, and it also is felt to give undue importance to a single large error. The usual criterion is to minimize the sum of the *squares* of the errors, the "least-squares" principle.†

In addition to giving a unique result for a given set of data, the least-squares method is also in accord with the *maximum-likelihood* principle of statistics. If the measurement errors have a so-called normal distribution and if the standard deviation is constant for all the data, the line determined by minimizing the sum of squares can be shown to have values of slope and intercept that have maximum likelihood of occurrence.

Let Y_i represent an experimental value, and let y_i be a value from the equation

$$y_i = ax_i + b,$$

where x_i is a particular value of the variable assumed to be free of error. We wish to determine the best values for a and b so that the y's predict the function values that correspond to x-values. Let $e_i = Y_i - y_i$. The least-squares criterion requires that

$$
\begin{aligned}
S &= e_1^2 + e_2^2 + \cdots + e_N^2 \\
&= \sum_{i=1}^{N} e_i^2 \\
&= \sum_{i=1}^{N} (Y_i - ax_i - b)^2
\end{aligned}
$$

be a minimum. N is the number of (x, Y)-pairs. We reach the minimum by proper choice of the parameters a and b, so they are the "variables" of the problem. At a minimum for S, the two partial derivatives $\partial S/\partial a$ and $\partial S/\partial b$ will both be zero.

---

* We will use this criterion in a later section of this book.
† The various criteria for a "best fit" can be described by minimizing a norm of the error vector. Relate each criterion to its corresponding vector norm to review the definition of such norms.

Hence, remembering that the x_i and Y_i are data points unaffected by our choice of values for a and b, we have

$$\frac{\partial S}{\partial a} = 0 = \sum_{i=1}^{N} 2(Y_i - ax_i - b)(-x_i),$$

$$\frac{\partial S}{\partial b} = 0 = \sum_{i=1}^{N} 2(Y_i - ax_i - b)(-1).$$

Dividing each of these equations by -2 and expanding the summation, we get the so-called *normal equations*

$$a \sum x_i^2 + b \sum x_i = \sum x_i Y_i,$$

$$a \sum x_i + bN = \sum Y_i. \tag{3.27}$$

All the summations in Eq. (3.27) are from $i = 1$ to $i = N$. Solving these equations simultaneously gives the values for slope and intercept a and b.

For the data in Fig. 3.13 we find that

$$N = 5, \quad \sum T_i = 273.1, \quad \sum T_i^2 = 18{,}607.27, \quad \sum R_i = 4438,$$

$$\sum T_i R_i = 254{,}932.5.$$

Our normal equations are then

$$18{,}607.27a + 273.1b = 254{,}932.5,$$

$$273.1a + 5b = 4438.$$

From these we find $a = 3.395$, $b = 702.2$, and hence write Eq. (3.26) as

$$R = 702 + 3.39T.$$

Nonlinear Data

In many cases, of course, data from experimental tests are not linear, so we need to fit to them some function other than a first-degree polynomial. Popular forms that are tried are the exponential forms

$$y = ax^b$$

or

$$y = ae^{bx}.$$

We can develop normal equations for these analogously to the preceding development for a least-squares line by setting the partial derivatives equal to zero. Such nonlinear simultaneous equations are much more difficult to solve* than linear equations. Thus the exponential forms are usually linearized by taking logarithms before determining the parameters:

$$\ln y = \ln a + b \ln x$$

or

$$\ln y = \ln a + bx.$$

We now fit the new variable $z = \ln y$ as a linear function of $\ln x$ or x as described earlier. Here we do not minimize the sum of squares of the deviations of Y from the curve, but rather the deviations of $\ln Y$. In effect, this amounts to minimizing the squares of the percentage errors, which itself may be a desirable feature. An added advantage of the linearized forms is that plots of the data on either log-log or semilog graph paper show at a glance whether these forms are suitable by whether a straight line represents the data when so plotted.

In cases when such linearization of the function is not desirable, or when no method of linearization can be discovered, graphical methods are frequently used; one merely plots the experimental values and sketches in a curve that seems to fit well. Special forms of graph paper, in addition to log-log and semilog, may be useful (probability, log-probability, and so on). Transformation of the variables to give near linearity, such as by plotting against $1/x, 1/(ax + b), 1/x^2$, and other polynomial forms of the argument may give curves with gentle enough changes in slope to allow a smooth curve to be drawn. S-shaped curves are not easy to linearize; the Gompertz relation

$$y = ab^{cx}$$

is sometimes employed. The constants a, b, and c are determined by special procedures. Another relation that fits data to an S-shaped curve is

$$\frac{1}{y} = a + be^{-x}.$$

In awkward cases, subdividing the region of interest into subregions with a piecewise fit in the subregions can be used.

The objection to the graphical technique is its *lack of uniqueness*. Two individuals will usually not draw the same curve through the points. One's judgment is frequently distorted by one or two points that deviate widely from the remaining data. Often one tends to pay too much attention to the extremities in comparison to the points in the central parts of the region of interest.

Further problems are caused if we wish to integrate or differentiate the function. Our discussion of least-squares polynomials is one solution to these difficulties.

---

* They were treated in Chapter 2.

Because polynomials can be readily manipulated, fitting such functions to data that do not plot linearly is common. We now consider this case. It will turn out that the normal equations are linear for this situation, which is an added advantage. In the development, we use n as the degree of the polynomial and N as the number of data pairs. Obviously, if $N = n + 1$, the polynomial passes exactly through each point and the methods discussed earlier in this chapter apply, so we will always have $N > n + 1$ in the following.

We assume the functional relationship

$$y = a_0 + a_1 x + a_2 x^2 + \cdots + a_n x^n, \tag{3.28}$$

with errors defined by

$$e_i = Y_i - y_i = Y_i - a_0 - a_1 x_i - a_2 x_i^2 - \cdots - a_n x_i^n.$$

We again use Y_i to represent the observed or experimental value corresponding to x_i, with x_i free of error. We minimize the sum of squares,

$$S = \sum_{i=1}^{N} e_i^2 = \sum_{i=1}^{N} (y_i - a_0 - a_1 x_1 - a_2 x_i^2 - \cdots - a_n x_i^n)^2.$$

At the minimum, all the partial derivatives $\partial S/\partial a_0, \partial S/\partial a_1, \ldots, \partial S/\partial a_n$ vanish. Writing the equations for these gives $n + 1$ equations:

$$\frac{\partial S}{\partial a_0} = 0 = \sum_{i=1}^{N} 2(Y_i - a_0 - a_1 x_i - \cdots - a_i x_i^n)(-1),$$

$$\frac{\partial S}{\partial a_1} = 0 = \sum_{i=1}^{N} 2(Y_i - a_0 - a_1 x_i - \cdots - a_i x_i^n)(-x_i),$$

$$\vdots \qquad \vdots$$

$$\frac{\partial S}{\partial a_n} = 0 = \sum_{i=1}^{N} 2(Y_i - a_0 - a_1 x_i - \cdots - a_n x_i^n)(-x_i^n).$$

Dividing each by -2 and rearranging gives the $n + 1$ normal equations to be solved simultaneously:

$$a_0 N + a_1 \sum x_i + a_2 \sum x_i^2 + \cdots + a_n \sum x_i^n = \sum Y_i,$$

$$a_0 \sum x_i + a_1 \sum x_i^2 + a_2 \sum x_i^3 + \cdots + a_n \sum x_i^{n+1} = \sum x_i Y_i,$$

$$a_0 \sum x_i^2 + a_1 \sum x_i^3 + a_2 \sum x_i^4 + \cdots + a_n \sum x_i^{n+2} = \sum x_i^2 Y_i, \tag{3.29}$$

$$\vdots \qquad\qquad\qquad\qquad\qquad\qquad \vdots$$

$$a_0 \sum x_i^n + a_1 \sum x_i^{n+1} + a_2 \sum x_i^{n+2} + \cdots + a_n \sum x_i^{2n} = \sum x_i^n Y_i.$$

Putting these equations in matrix form shows an interesting pattern in the coefficient matrix.

$$
\begin{bmatrix}
N & \sum x_i & \sum x_i^2 & \sum x_i^3 & \ldots \sum x_i^n \\
\sum x_i & \sum x_i^2 & \sum x_i^3 & \sum x_i^4 & \ldots \sum x_i^{n+1} \\
\sum x_i^2 & \sum x_i^3 & \sum x_i^4 & \sum x_i^5 & \ldots \sum x_i^{n+2} \\
\vdots & & & & \\
\sum x_i^n & \sum x_i^{n+1} & \sum x_i^{n+2} & \sum x_i^{n+3} & \ldots \sum x_i^{2n}
\end{bmatrix}
a =
\begin{bmatrix}
\sum Y_i \\
\sum x_i Y_i \\
\sum x_i^2 Y_i \\
\vdots \\
\sum x_i^n Y_i
\end{bmatrix}.
\tag{3.30}
$$

All the summations in Eqs. (3.29) and (3.30) run from 1 to N.

Solving large sets of linear equations is not a simple task. Methods for this are the subject of Chapter 2. These particular equations have an added difficulty in that they have the undesirable property known as *ill-conditioning*. Its result is that round-off errors in solving them cause unusually large errors in the solutions, which of course are the desired values of the coefficients a_i in Eq. (3.28). Up to $n = 4$ or 5, the problem is not too great (that is, double-precision arithmetic in computer solutions is only desirable, not essential), but beyond this point special methods are needed. Such special methods use orthogonal polynomials in an equivalent form of Eq. (3.28). We will not pursue this matter further,* although we will treat one form of orthogonal polynomials in a later chapter in connection with representation of functions. From the point of view of the experimentalist, functions more complex than fourth-degree polynomials are rarely needed, and when they are, the problem can often be handled by fitting a series of polynomials to subsets of the data.

The matrix of Eq. (3.30) is called the *normal matrix* for the least-squares problem. There is another matrix that corresponds to this, called the *design matrix*. It is of the form

$$
A =
\begin{bmatrix}
1 & 1 & 1 & \ldots & 1 \\
x_1 & x_2 & x_3 & \ldots & x_N \\
x_1^2 & x_2^2 & x_3^2 & \ldots & x_N^2 \\
\vdots & & & & \vdots \\
x_1^n & x_2^n & x_3^n & \ldots & x_N^n
\end{bmatrix}.
$$

It is easy to show that AA^T is just the coefficient matrix of Eq. (3.30). It is also easy to see that Ay, where y is the column vector of Y-values, gives the right-hand side of Eq. (3.30). (You ought to try this for, say, a 3×3 case to reassure yourself.) This means that we can rewrite Eq. (3.30) in matrix form, as

$$
AA^T a = Ba = Ay.
$$

---

* Ralston (1965) is a good source of further information. The ill-conditioning problem, though very real, is often academic, since seldom is a degree above 4 or 5 needed to give a curve that fits the data with adequate precision.

Usually we would use Gaussian elimination to solve the system, but because B has special properties, we can use other methods that avoid the problem of ill-conditioning that already has been pointed out.

1. The matrix $B = AA^T$ is symmetric and positive semidefinite. An $n \times n$ matrix is said to be positive semidefinite if, for every n-component vector, $x^T M x \geq 0$. If we add the condition that $x^T M x = 0$ only if x is the zero vector, M is said to be positive definite. (You should show that B is positive semidefinite and symmetric.)

2. In linear algebra, it is shown that B can be diagonalized by an orthogonal matrix P:

$$PBP^T = PAA^TP^T = D,$$

   where the diagonal elements of D are the eigenvalues of B. Note that orthogonality implies that $PP^T = I$, the identity matrix.

3. Since B is positive semidefinite, all of its eigenvalues are nonnegative. This means that we can define a matrix S as

$$S = \sqrt{D}, \quad \text{or} \quad S^2 = D.$$

   The diagonal elements of S are called the singular values of A.

4. We can rewrite Eq. (3.30) and its solution as follows:

$$AA^Ta = P^TDPa = PS(PS)^Ta = Ay,$$
$$a = PD^{-1}P^T Ay.$$

This last eliminates having to multiply out AA^T and, by extending this approach, leads to an important method for solving Eq. (3.30) called *singular-value decomposition*. (See Press, *Numerical Recipes*, 1986, on this topic.)

We illustrate the use of Eqs. (3.29) to fit a quadratic to the data of Table 3.15. Figure 3.16 shows a plot of the data. (The data are actually a perturbation of the relation $y = 1 - x + 0.2x^2$. It will be of interest to see how well we approximate this function.) To set up the normal equations, we need the sums tabulated in Table 3.15.

Table 3.15   Data to illustrate curve-fitting

x_i	0.05	0.11	0.15	0.31	0.46	0.52	0.70	0.74	0.82	0.98	1.17
Y_i	0.956	0.890	0.832	0.717	0.571	0.539	0.378	0.370	0.306	0.242	0.104

$$\sum x_i = 6.01 \qquad\qquad N = 11$$
$$\sum x_i^2 = 4.6545 \qquad\qquad \sum Y_i = 5.905$$
$$\sum x_i^3 = 4.1150 \qquad\qquad \sum x_i Y_i = 2.1839$$
$$\sum x_i^4 = 3.9161 \qquad\qquad \sum x_i^2 Y_i = 1.3357$$

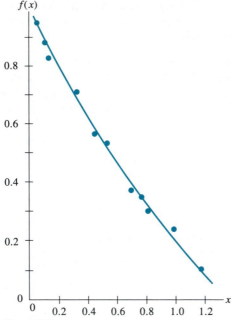

Figure 3.16

A calculator can give these as accumulated totals directly. We need to solve the set of equations

$$11a_0 + 6.01a_1 + 4.6545a_2 = 5.905,$$
$$6.01a_0 + 4.6545a_1 + 4.1150a_2 = 2.1839,$$
$$4.6545a_0 + 4.1150a_1 + 3.9161a_2 = 1.3357.$$

The result is $a_0 = 0.998$, $a_1 = -1.018$, $a_2 = 0.225$, so the least-squares method gives

$$y = 0.998 - 1.018x + 0.225x^2.$$

Compare this to $y = 1 - x + 0.2x^2$. We do not expect to reproduce the coefficients exactly because of the errors in the data.

What Degree Polynomial Should be Used?

In the general case, we may wonder what degree of polynomial should be used. As we use higher-degree polynomials, we of course will reduce the deviations of the points from the curve until, when the degree of the polynomial, n, equals $N - 1$, there is an exact match (assuming no duplicate data at the same x-value) and we have an interpolating polynomial. The answer to this problem is found in statistics. One increases the degree of approximating polynomial as long as there is a statistically significant decrease in the variance, σ^2, which is computed by

$$\sigma^2 = \frac{\sum e_i^2}{N - n - 1}.$$ (3.31)

For the preceding example, when the degree of the polynomial made to fit the points is varied from 1 to 7, we obtain the results shown in Table 3.16.

The criterion of Eq. (3.31) chooses the optimum degree as 2. This is no surprise, in view of how the data were constructed. It is important to realize that the numerator of Eq. (3.31), the *sum of the deviations squared* of the points from the curve, should continually decrease as the degree of the polynomial is raised. It is the denominator of Eq. (3.31) that makes σ^2 increase as we go above the optimum degree. In this example, this behavior is observed for $n = 3$. Above $n = 3$, a second effect sets in. Due to ill-conditioning, the coefficients of the least-squares polynomials are determined with poor precision. This modifies the expected increases of the values of σ^2.

Table 3.16

Degree	Equation	σ^2 (Eq. 3.30)	$\sum e^2$
1	$y = 0.952 - 0.760x$	0.0010	0.0092
2	$y = 0.998 - 1.018x + 0.225x^2$	0.0002	0.0018
3	$y = 1.004 - 1.079x + 0.351x^2 - 0.069x^3$	0.0003	0.0018
4	$y = 0.998 - 0.838x - 0.522x^2 + 1.040x^3$ $- 0.454x^4$	0.0003	0.0016
5	$y = 1.031 - 1.704x + 4.278x^2 - 9.477x^3$ $+ 9.394x^4 - 3.290x^5$	0.0001	0.0007
6	$y = 1.038 - 1.910x + 5.952x^2 - 15.078x^3$ $+ 18.277x^4 - 9.835x^5 + 1.836x^6$	0.0002	0.0007
7	$y = 1.032 - 1.742x + 4.694x^2 - 11.898x^3$ $+ 16.645x^4 - 14.346x^5 + 8.141x^6 - 2.293x^7$	0.0002	0.0007

Before leaving this section, we illustrate how to apply these methods to more complicated functions.

EXAMPLE 3.10 The results of a wind tunnel experiment on the flow of air on the wing tip of an airplane provide the following data:

R/C: 0.73, 0.78, 0.81, 0.86, 0.875, 0.89, 0.95, 1.02, 1.03, 1.055, 1.135, 1.14, 1.245, 1.32, 1.385, 1.43, 1.445, 1.535, 1.57, 1.63, 1.755;

V_θ/V_∞: 0.0788, 0.0788, 0.064, 0.0788, 0.0681, 0.0703, 0.0703, 0.0681, 0.0681, 0.079, 0.0575, 0.0681, 0.0575, 0.0511, 0.0575, 0.049, 0.0532, 0.0511, 0.049, 0.0532, 0.0426;

where R is the distance from the vortex core, C is the aircraft wing chord, V_θ is the vortex tangential velocity, and V_∞ is the aircraft free-stream velocity. Let $x = R/C$ and $y = V_\theta/V_\infty$. We would like our curve to be of the form

$$g(x) = \frac{A}{x}(1 - e^{-\lambda x^2}),$$

and our least-squares equations become

$$S = \sum_{i=1}^{21} (Y_i - g(x_i))^2$$

$$= \sum_{i=1}^{21} \left(Y_i - \frac{A}{x_i}(1 - e^{-\lambda x_i^2}) \right)^2.$$

Setting $S_A = S_\lambda = 0$ gives the following equations:

$$\sum_{i=1}^{21} \left(\frac{1}{x_i} \right)(1 - e^{-\lambda x_i^2})\left(Y_i - \frac{A}{x_i}(1 - e^{-\lambda x_i^2}) \right) = 0,$$

$$\sum_{i=1}^{21} x_i(e^{-\lambda x_i^2})\left(Y_i - \frac{A}{x_i}(1 - e^{-\lambda x_i^2}) \right) = 0.$$

When this system of nonlinear equations is solved, we get

$$g(x) = \frac{0.07618}{x}(1 - e^{-2.30574x^2}).$$

For these values of A and λ, $S = 0.000016$. The graph of this function is presented in Fig. 3.17.

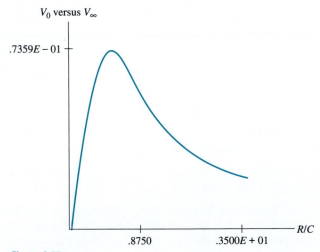

Figure 3.17

Here is an algorithm for obtaining a least-squares polynomial:

Given N data pairs $(x_i, Y_i), i = 1, \ldots, N$:
Form the coefficient matrix, M, with $N + 1$ rows (r) and $N + 1$ columns (c), by:

$$\text{Set } M_{rc} = \sum_{i=1}^{N} x_i^{r+c-2}.$$

Form the right-hand side vector b, with $N + 1$ rows (r), by:

$$\text{Set } b_r = \sum_{i=1}^{N} x_i^{r-1} Y_i.$$

Solve the linear system $Ma = b$ to get the coefficients in

$$y = a_0 + a_1 x + a_2 x^2 + \cdots + a_n x^n,$$

which is the desired polynomial that fits the data.

3.9 Theoretical Matters

In this section we assemble several items of theory whose conclusions were referred to earlier in this chapter.

Why Polynomials?

Throughout this chapter the emphasis has been on fitting polynomials to functions. Why choose polynomials over other functions? The answer lies in the Weierstrass approximation theorem. This states:

If $f(x)$ is continuous on a finite interval $[a, b]$, there exists a polynomial $P_n(x)$ of degree n such that

$$|f(x) - P_n(x)| < \text{ERROR},$$

throughout the interval $[a, b]$, for any given ERROR > 0. (The degree required of $P_n(x)$ is a function of ERROR.)

What this means is that we can achieve what is called *uniform approximation*—the maximum error in $[a, b]$ is bounded rather than some average of the errors.

A proof of this theorem can be found in Ralston (1965).

Identical Polynomials

We have written interpolating polynomials in several forms: with Lagrangian polynomials, using divided differences, using terms from a difference table (if the data are equispaced), or in the standard form:

$$P_n(x) = a_0 + a_1 x + a_2 x^2 + \cdots + a_n x^n.$$

We now show that every polynomial of degree n that has the same value at $n + 1$ distinct points is identical.

First, the conclusion seems intuitively true because the $n + 1$ data pairs are exactly enough to determine the $n + 1$ coefficients of the polynomial, and, since every expression of the polynomial can be reduced to the standard form, they must be identical.

A more formal and compelling proof is by contradiction:

Suppose there are two different polynomials of degree n that agree at $n + 1$ distinct points. Call these $P_n(x)$ and $Q_n(x)$, and write their difference:

$$D(x) = P_n(x) - Q_n(x),$$

where $D(x)$ is a polynomial of at most degree n. But since P and Q match at the $n + 1$ points, their difference $D(x)$ is equal to zero at all $n + 1$ of these x-values; that is, $D(x)$ is a polynomial of degree n at most but has $n + 1$ distinct zeros. But this is impossible unless $D(x)$ is identically zero. Hence $P_n(x)$ and $Q_n(x)$ are not different—they must be the same polynomial.

A most important consequence of this uniqueness property of interpolating polynomials is that their error terms are also identical (though we may want to express the error term in different forms). We only have to derive the error term for one form of interpolating polynomial to have the error term for all forms of interpolating polynomials.

The Error Term for Interpolation

We begin the development of an expression for the error of $P_n(x)$, an nth-degree interpolating polynomial, by writing the error function in a form that has the known property that it is zero at the $n + 1$ points, from x_0 through x_n, where $P_n(x)$ and $f(x)$ are the same. We call this function $E(x)$:

$$E(x) = f(x) - P_n(x) = (x - x_0)(x - x_1) \cdots (x - x_n)g(x).$$

The $n + 1$ linear factors give $E(x)$ the zeros we know it must have, and $g(x)$ accounts for its behavior at values other than at $x_0, x_1, \ldots, x_n$. Obviously, $f(x) - P_n(x) - E(x) = 0$, so

$$f(x) - P_n(x) - (x - x_0)(x - x_1) \cdots (x - x_n)g(x) = 0. \tag{3.32}$$

To determine $g(x)$, we now use the interesting mathematical device of constructing an auxiliary function (the reason for its special form becomes apparent as the development proceeds). We call this auxiliary function $W(t)$, and define it as

$$W(t) = f(t) - P_n(t) - (t - x_0)(t - x_1) \cdots (t - x_n)g(x).$$

Note in particular that x has *not* been replaced by t in the $g(x)$ portion. (W is really a function of both t and x, but we are only interested in variations of t.) We now examine the zeros of $W(t)$.

Certainly at $t = x_0, x_1, \ldots, x_n$, the W function is zero ($n + 1$ times), but it is also zero if $t = x$ by virtue of Eq. (3.32). There are then a total of $n + 2$ values of t that make $W(t) = 0$. We now impose the necessary requirements on $W(t)$ for the *law of mean value* to hold. $W(t)$ must be continuous and differentiable. If this is so, there is a zero to its derivative $W'(t)$ between each of the $n + 2$ zeros of $W(t)$, a total of $n + 1$ zeros. If $W''(t)$ exists, and we suppose it does, there will be n zeros of $W''(t)$, and likewise $n - 1$ zeros of $W'''(t)$, and so on, until we reach $W^{(n+1)}(t)$, which must have at least one zero in the interval that has x_0, x_n, or x as endpoints. Call this value of $t = \xi$. We then have

$$W^{(n+1)}(\xi) = 0 = \frac{d^{n+1}}{dt^{n+1}}[f(t) - P_n(t) - (t - x_0) \cdots (t - x_n)g(x)]_{t=\xi}$$

$$= f^{(n+1)}(\xi) - 0 - (n + 1)!g(x). \tag{3.33}$$

The right-hand side of Eq. (3.33) occurs because of the following arguments. The $(n + 1)$st derivative of $f(t)$, evaluated at $t = \xi$, is obvious. The $(n + 1)$st derivative of $P_n(t)$ is zero because every time any polynomial is differentiated its degree is reduced by one, so that the nth derivative is of degree zero (a constant) and its $(n + 1)$st derivative is zero. We apply the same argument to the $(n + 1)$st-degree polynomial in t that occurs in the last term—its $(n + 1)$st derivative is a constant that results from the t^{n+1} term and is $(n + 1)!$. Of course $g(x)$ is independent of t and goes through the differentiations unchanged. The form of $g(x)$ is now apparent:

$$g(x) = \frac{f^{(n+1)}(\xi)}{(n + 1)!}, \quad \xi \text{ between } (x_0, x_n, x)$$

The conditions on $W(t)$ that are required for this development (continuous and differentiable $n + 1$ times) will be met if $f(x)$ has these same properties, because $P_n(x)$ is continuous and differentiable. We now have our error term:

$$E(x) = (x - x_0)(x - x_1) \cdots (x - x_n)\frac{f^{(n+1)}(\xi)}{(n + 1)!}, \tag{3.34}$$

with ξ on the smallest interval that contains $\{x, x_0, x_1, \ldots, x_n\}$.

If we modify this, expressing it in terms of $s = (x - x_0)/h$, it becomes more compatible with the Newton–Gregory forward polynomials. Remembering that

$$x_1 = x_0 + h, \qquad x_2 = x_0 + 2h, \qquad \ldots,$$

so that

$$(x - x_0) = sh, \qquad (x - x_1) = sh - h = (s - 1)h,$$
$$(x - x_2) = sh - 2h = (s - 2)h, \qquad \ldots,$$

we find that Eq. (3.34) becomes

$$E(x_s) = \frac{(s)(s - 1)(s - 2)\cdots(s - n)}{(n + 1)!} h^{n+1} f^{(n+1)}(\xi)$$

$$= \binom{s}{n + 1} h^{n+1} f^{(n+1)}(\xi), \quad \xi \text{ between } (x_0, x_n, x_s) \qquad (3.35)$$

Referring again to Eq. (3.9), we observe that the next term after the last one included in an nth-degree Newton–Gregory forward interpolating polynomial is $\binom{s}{n+1}\Delta^{n+1}f_0$. We can get the error term of Eq. (3.35) by substituting $h^{n+1}f^{(n+1)}(\xi)$ for the $(n + 1)$st difference. This is true for all interpolating polynomials, not just the Newton–Gregory forward one. It is true for polynomials based on divided differences because of the relationship between divided differences and function differences that was pointed out in Section 3.4.

This is the justification for our *next-term rule*.

Centering the x-Value

We have remarked several times that centering the x-value within the range of the x_i used to construct the interpolating polynomial will give better results. This is easy to see. Consider a set of data pairs with these x-values:

i	x_i
0	0.2
1	0.4
2	0.5
3	0.7
4	0.8
5	1.0

Suppose that $x = 0.63$ and that we want to develop a second-degree polynomial (that will utilize three data pairs). It is obvious that the $(x - x_i)$ factors in the error term will have the smallest product (in magnitude) if we choose x_i-values of 0.5, 0.7, and 0.8, since these three x_i's are nearest to $x = 0.63$. (There may be unusual situations when the derivative portion of the error term contradicts this rule, but such an event is so rare that we usually ignore it, particularly when we do not know the size of the derivative term.)

Recognizing which data pairs are nearest to x is easier when the data are equispaced. It is particularly hard when the data pairs are not arranged in order of the x_i.

We can further demonstrate this rule by observing the errors for second-degree interpolating polynomials based on the data for $f(x) = \sin(x)$ when we vary the range of x_i's:

i-values	x_i-values	Estimate of $f(0.8)$	Actual error
0, 1, 2	0.1, 0.5, 0.9	0.71445	2.904E-3
1, 2, 3	0.5, 0.9, 1.3	0.71895	−1.593E-3
2, 3, 4	0.9, 1.3, 1.7	0.71450	2.854E-3
3, 4, 5	1.3, 1.7, 2.1	0.75083	−3.347E-2

The results agree with our rule: $x = 0.8$ is best centered when we take values at $i = 1, 2, 3$, and the error is smaller when that is done.

In Summary

To summarize, then, the error of polynomial interpolation is reduced by making its range as symmetrical as possible about the point of interpolation and by choosing a higher degree of polynomial, up to the point where round-off or the effect of local irregularities causes offsetting errors. But another factor has the greatest importance of all—the step size h. With a given set of tabulated data, we may not be able to do much about the step size. However, if we are designing a new set of tables, the step size may be open to our selection; it is then advantageous to make it small. (This makes our tables more bulky and adds proportionately to the computational effort, of course.)

This effect of h is so important that a special notation is often used to focus attention on it. We write

$$\text{Error} = O(h^n)$$

if there is a constant K such that if h is "small enough," and

$$|\text{Error}| \le Kh^n$$

where $h > 0$ and where K is some constant not equal to zero. The expression $O(h^n)$ is read "order of h to the nth power." For example, the error of a quadratic interpolating polynomial whose range is (x_0, x_2) would be of order h *cubed* because

$$\text{Error} = \frac{(s)(s - 1)(s - 2)}{6} h^3 f'''(\xi), \quad \text{or} \quad \text{Error} = O(h^3).$$

As h gets small, $f'''(\xi) \rightarrow f'''(x_0)$ because ξ is squeezed between x_0 and x_2, and hence approaches a constant value.

A Contradiction?

As we have pointed out, one way to improve the accuracy of a polynomial that approximates a known function is to use closely spaced points of agreement. But that is not always true! We saw in Section 3.5 that some functions, those that are smooth everywhere except for a local "bump," are difficult to approximate with ordinary interpolating polynomials.

Here is another example:

$$\text{Let} \quad f(x) = \frac{1}{1 + 25x^2} \quad \text{on} \quad [-1, 1].$$

Now, suppose we find interpolating polynomials that fit to $f(x)$ at uniformly spaced points on $[-1, 1]$. It turns out that the maximum error does not decrease as we increase the degree of the interpolating polynomials! The following table gives, for polynomials of odd degree, the approximate value of the maximum error.

Degree	Maximum error	Degree	Maximum error
3	0.7070	13	1.0700
5	0.4327	15	2.1041
7	0.2572	17	4.2190
9	0.3001	19	8.8526
11	0.5567		

From the data, it is clear that the maximum errors are increasing after degree $= 7$. (This phenomenon occurs after degree $= 2$ for polynomials of even degree.)

Why such behavior, particularly in view of the Weierstrass theorem? Because the theorem does not apply if we constrain the matching with $f(x)$ to occur at evenly spaced points! In Chapter 10, we will find that fitting the function at special points does cause the maximum error to decrease with the degree.

Coefficients for $\Delta^n f_i$—Symbolic Operators

In Section 3.4, Eq. (3.7) exhibited $\Delta^n f_i$ as an alternating sum of function values with coefficients like those for a binomial expansion. To establish the truth of these relations, we can use a convenient method, called the *symbolic operator* method.

Recall that we used Δ as a *forward-differencing operator*. We supplement this with a backward-differencing operator ∇ and a *stepping* operator E. These are defined by their actions on a function:

$$\Delta f(x_0) = f(x_0 + h) - f(x_0),$$
$$\Delta^2 f(x_0) = \Delta[\Delta f(x_0)] = \Delta f(x_0 + h) - \Delta f(x_0),$$
$$\nabla f(x_0) = f(x_0) - f(x_0 - h),$$
$$\nabla^2 f(x_0) = \nabla[\nabla f(x_0)] = \nabla f(x_0) - \nabla f(x_0 - h),$$

$$Ef(x_0) = f(x_0 + h),$$
$$E^2f(x_0) = E[Ef(x_0)] = Ef(x_0 + h) = f(x_0 + 2h),$$
$$E^nf(x_0) = f(x_0 + nh).$$

These obvious relationships exist:

$$\Delta f(x_0) = Ef(x_0) - f(x_0) = (E - 1)f(x_0),$$
$$\nabla f(x_0) = f(x_0) - E^{-1}f(x_0) = (1 - E^{-1})f(x_0).$$

(3.36)

We abstract from Eq. (3.36) the symbolic operator relations:

$$\Delta = E - 1,$$
$$\nabla = 1 - E^{-1}.$$

(3.37)

Equations (3.37) are really meaningless, for neither side is defined as it stands. What they signify is that the effect of Δ when operating on a function is the same as the effect of operating with $(E - 1)$, and that ∇ and $(1 - E^{-1})$ have the same effect on a function, though all these quantities are without significance standing alone. We must apply them to a function to interpret them. This is no different, really, from other operational symbols that we use, such as $\sqrt{}$ for square root and d/dx for differentiation.

Since all the operators represented by Δ, ∇, and E are linear operators (the effect on a linear combination of functions is the same as the linear combination of the operator acting on the functions), the laws of algebra are obeyed in relationships between them. This means we can manipulate the relations of Eq. (3.37) by algebraic transformations, and then interpret the results by letting them operate on a function. For example, by raising $\Delta = E - 1$ to the nth power, we have

$$\Delta^n = (E - 1)^n = E^n - nE^{n-1} + \binom{n}{2}E^{n-2} - \binom{n}{3}E^{n-3} + \cdots,$$

so

$$\Delta^nf(x_0) = \left[E^n - nE^{n-1} + \binom{n}{2}E^{n-2} - \cdots \right]f(x_0)$$
$$= f(x_0 + nh) - nf[x_0 + (n - 1)h] + \binom{n}{2}f[x_0 + (n - 2)h] - \cdots,$$

or

$$\Delta^nf_0 = f_n - nf_{n-1} + \binom{n}{2}f_{n-2} - \cdots.$$

(3.38)

Equation (3.38) is a proof of the alternating-sign, binomial-coefficient formula given in Eq. (3.7).

We can develop interesting relations between Δ and ∇, such as

$$E\nabla = E(1 - E^{-1}) = E - 1 = \Delta,$$
$$E^n\nabla^n = \nabla^nE^n = \Delta^n,$$
$$\Delta^nf_0 = \nabla^nE^nf_0 = \nabla^nf_n.$$

This illustrates the fact that a given difference entry in a table can be interpreted as either a forward or backward difference of the appropriate f-values. (This was already obvious from inspection of a difference table.)

Here is another example of the method: We can very easily derive the Newton–Gregory forward formula:

$$E = 1 + \Delta, \qquad E^s = (1 + \Delta)^s,$$

$$f_s = E^s f_0 = (1 + \Delta)^s f_0 = \left[1 + s\Delta + \binom{s}{2}\Delta^2 + \binom{s}{3}\Delta^3 + \cdots \right] f_0$$

$$= f_0 + s\Delta f_0 + \binom{s}{2}\Delta^2 f_0 + \binom{s}{3}\Delta^3 f_0 + \cdots .$$

We will use this symbolic operator method to derive other formulas in later chapters.

3.10 Using *Mathematica*

The *Mathematica* program is perhaps the most extensive of all current computer algebra systems. It is a very large program (maybe the correct word is "huge") with 843 built-in functions, called "objects" in the *Mathematica* documentation. In addition to the built-in objects, *Mathematica* comes with almost 150 other "standard packages"—functions that are created using the powerful programming capabilities that are incorporated in the system.

Mathematica is available for several computer platforms—Macintosh, PCs (DOS and Windows versions), Unix-based systems, among others. The computational parts of these different versions are identical, but there are different "front ends" (user interfaces) for the various versions; these interfaces take advantage of the special capabilities of the particular computer system.

Because of its size, *Mathematica* requires memory and hard disk capacities that minimum computer systems do not provide. It also takes several seconds to load, even on a fast personal computer. Because of its size and complexity, *Mathematica* is not an easy program to learn, although doing the most common and simple things is not hard.

Mathematica can carry out many of the numerical procedures of this chapter. Functions that create interpolating polynomials, plot spline and Bezier curves, and obtain least-squares fits to data points are available. We describe how each of these procedures is done in this system through examples. These examples show that *Mathematica* displays the results of computations in a dialog: $In[n]$ identifies the user input and $Out[n]$ labels the response, with the value of n increasing for successive requests.

Most of these illustrative examples can also be done in MAPLE as well as DERIVE.

Difference Tables

Mathematica has no built-in procedure for computing differences of a set of function values (this function is not necessary, because the interpolating polynomial can be obtained without them). Still, it is easy to get the differences and instructive to do so. We start by defining a list of function values, in [1], and then we use this list in a **Table** operation to construct the differences in [2]. (**Table** creates a new list that, in this case, contains the differences of the original list of function values.) Request [3] asks for the display to be a column of values:

```
In[1]:=
  f = {.203,  .423,  .684,  1.030,  1.557,  2.572}
Out[1]=
  {0.203,  0.423,  0.684,  1.03,  1.557,  2.572}
In[2]:=
  dif = Table [ %[[i+i]] − %[[i]], {i, Length[f] − 1}  ]
Out[2]=
  {0.22,  0.261,  0.346,  0.527,  1.015}
In[3]:=
  ColumnForm [ dif ]
Out[3]=
  0.22
  0.261
  0.346
  0.527
  1.015
```

Observe that *Mathematica* uses curly brackets to enclose a list of values, which are then treated as a single entity. We name the list "f" and then obtain the differences of "f" through the **Table** operation. Its parameters need some explanation.

The % symbol refers to the previous result. (%% would refer to the result before the previous one, %%% to the one before that, and %n refers to the result of request [n].) We obtain the ith element of the list through %[[i]] and the (i + 1)st element through %[[i + 1]]. The last parameter, { i, Length[f] − 1 }, causes i to run from 1 to the length of list f, minus 1.

Interpolation

There are at least two ways to develop an interpolating polynomial that passes through each of a set of points. The most obvious is the function **InterpolatingPolynomial**. (*Mathematica* almost always uses very long names for its operations to avoid ambiguity.) We begin with a set of data points and then ask for the polynomial:

```
In[4]:=
  data = { {1,1},{2,4},{4,3},{5,4} }
```

```
Out[4]=
  {{1, 1}, {2, 4}, {4, 3}, {5, 4}}
In[5]:=
  poly = InterpolatingPolynomial [ data, x ]
Out[5]=
```

$$1 + (3 + (-(\frac{7}{6}) + \frac{5\ (-4 + x)}{12})\ (-2 + x))\ (-1 + x)$$

The result from [5] can be put into more familiar form with a request to expand (to multiply out the terms):

```
In[6]:=
  Expand [ % ]
Out[6]=
```

$$-(\frac{23}{3}) + \frac{37\ x}{3} - \frac{49\ x^2}{12} + \frac{5\ x^3}{12}$$

We can evaluate the polynomial as follows:

```
In[7]:=
  poly /. x→3.5
Out[7]=
  3.34375
```

In the last operation, the symbol /. means to evaluate "poly," and we specify that this evaluation is for $x = 3.5$. The result, 3.34375, is the interpolated value (from a cubic interpolating polynomial, since there are four data points) at $x = 3.5$.

A second way to get an interpolating polynomial is through **Interpolation**. This develops a so-called **InterpolatingFunction**:

```
In[8]:=
  interp = Interpolation [ data ]
Out[8]=
  InterpolatingFunction[{1, 5}, <>]
```

Observe that the interpolating function is not displayed explicitly. We do learn that its range is from $x = 1$ to $x = 5$. We also evaluate it in a different way, as shown in request [9] below. The result is identical to the previous interpolating polynomial because Interpolation creates cubic interpolating polynomials by default. We can change this default (see [10]):

```
In[9]:=
  interp [ 3.5 ]
Out[9]=
  3.34375
In[10]:=
  interp2 = Interpolation [ data, InterpolationOrder→2 ]
Out[10]=
  InterpolatingFunction[{1, 5}, <>]
```

```
In[11]:=
  interp2 [ 3.5 ]
Out[11]=
  4.125
In[12]:=
  interp2 [ 4.5 ]
Out[12]=
  3.375
```

This last interpolating function created quadratics through successive triples of the data points. You should verify the truth of this statement.

Interpolating on a Surface

If we give **Interpolation** a set of values defined on a grid of (x, y) points, we can interpolate on the surface defined by the points (which are in three-dimensional space). Here is an example where the points are actually values of $f(x, y) = 3x - y^2$:

```
In[13]:=
  tabl5 =
  {{2, 1, 5}, {2, 1.5, 3.75}, {2, 2., 2.}, {2, 2.5, −0.25},
   {3, 1, 8}, {3, 1.5, 6.75}, {3, 2., 5.}, {3, 2.5, 2.75},
   {4, 1, 11}, {4, 1.5, 9.75}, {4, 2., 8.}, {4, 2.5, 5.75},
   {5, 1, 14}, {5, 1.5, 12.75}, {5, 2., 11.}, {5, 2.5, 8.75}};
In[14]:=
  interp2 = Interpolation [ tabl5 ]
Out[14]=
  InterpolatingFunction[{{2, 5}, {1, 2.5}}, <>]
In[15]:=
  interp2 [3,2]
Out[15]=
  5.
In[16]:=
  interp2 [ 3.2,1.7 ]
Out[16]=
  6.71
In[17]:=
  3*3.2 − 1.7^2
Out[17]=
  6.71
```

Some explanations are in order. When we terminated the input in [13] with a semicolon, the output was suppressed. In [15] we evaluated the interpolating function at $x = 3, y = 2$, getting the point-value at $(3, 2)$, where $f(x, y) = 5$. In [16], we asked for $f(3.2, 1.7)$ and got 6.71. This was verified in [17].

Least-Squares Fitting of Data

Mathematica has a built-in function to do least-squares fitting of data. The desired relation does not have to be linear, and more than one independent variable is permitted. We work with a list of data points and supply a pattern for the equation to be fitted. How this is done can be understood from some examples.

EXAMPLE 3.11 Fit a least-squares line to $(20.5, 765), (32.7, 826), (51.0, 873), (73.2, 942), (95.7, 1032)$. (These are the data for the first example in Section 3.8 of this chapter.) Here is how we do this:

```
In[18]:=
   d1 = {{20.5,765},{32.7,826},{51.0,873},{73.2,942},{95.7,1032}
Out[18]=
   {{20.5, 765}, {32.7, 826}, {51., 873}, {73.2, 942},
    {95.7, 1032}}
In[19]:=
   Fit [ d1, {1,t}, t]
Out[19]=
   702.172 + 3.39487 t
```

The result in [19] agrees with the result in Section 3.8.

EXAMPLE 3.12 We reproduce the second example in the Section 3.8 with

```
In[20]:=
   d2 = {{.05,.956},{.11,.890},{.15,.832},{.31,.717},
   {.46,.571},{.52,.539},{.70,.378},{.74,.370},{.82,.306},
   {.98,.242},{1.17,.104}}
Out[20]=
   {{0.05, 0.956}, {0.11, 0.89}, {0.15, 0.832}, {0.31, 0.717},
    {0.46, 0.571}, {0.52, 0.539}, {0.7, 0.378}, {0.74, 0.37},
    {0.82, 0.306}, {0.98, 0.242}, {1.17, 0.104}}
In[21]:=
   Fit [ d2, (1,x,x^2}, x ]
Out[21]=
   0.997968 - 1.01804 + 0.224682 x²
```

EXAMPLE 3.13 In this example we fit to the values of a known function: $f(x, y) = 2x^2 + 3y - 1$. We fit a function of two variables in this fashion:

```
In[22]:=
   d3 = {{2,3,16},{3,2,23},{4,1,34}}
Out[22]=
   {{2, 3, 16}, {3, 2, 23}, {4, 1, 34}}
In[23]:=
   Fit [ d3, {1,x^2,y}, {x,y} ]
Out[23]=
   -1. + 2. x² + 3.y
```

In this example we have exactly enough data to compute the coefficients so the result, $2x^2 + 3y - 1$, is the exact answer. (We then see that this is still another way to get an interpolating polynomial—supply exactly enough data to compute the coefficients through a least-squares fit.) If we supply more data points than parameters in the function being fit, a least-squares evaluation of the coefficients results.

▲

Plotting

Mathematica is notable for its plotting capabilities. Both two- and three-dimensional plots are easy to create and can be displayed with or without grid lines and with or without labels on the axes and for the plot as a whole. (There are still more options.)

We might want to plot an interpolating polynomial to see how well the polynomial matches the function (that is, if we know the function!). This next compares an interpolating polynomial to the function itself. We take $f(x) = xe^{-x/2}$, and begin with a table of data pairs:

x	$f(x)$
−0.5	−0.6420
2.3	0.7283
5.7	0.3297
7.2	0.1967

We will fit a cubic to these four points and draw its graph, then draw the graph for $f(x)$, and, finally, superimpose the two graphs. Here is how we do this in *Mathematica*:

```
In[24]:=
   f[x_] :=x*E^(-x/2)
```

```
In[25]:=
   x_f = {{-.5, f[-.5]},(2.3, f[2.3]},{5.7, f[5.7]},{7.2, f[7.2]}}
Out[25]=
   {{-0.5, -0.642013}, {2.3, 0.728265}, {5.7, 0.329713},
    {7.2, 0.196731}}
In[26]:=
   InterpolatingPolynomial [ %,x ]
Out[26]=
   -0.642013 + (0.489385 + (-0.0978397 + 0.0134636(-5.7 + x))
      (-2.3 + x)) (0.5 + x)
In[27]:=
   pltpoly = Plot [ %, {x,-1,8} ]
Out[27]=
   -Graphics-
```

```
In[28]:=
   pltfx = Plot [ x*E^(-x/2), {x,-1,8} ]
Out[28]=
   -Graphics-
```

```
In[29]:=
   Show [ pltpoly, pltfx ]
Out[29]=
   -Graphics-
```

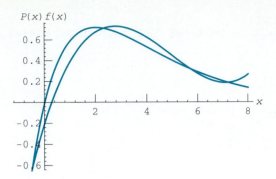

Spline Curves

We can plot cubic spline curves and Bezier curves by using one of *Mathematica*'s standard packages. Before we use the package, we must load it; we do this in [31]. We create the cubic spline in [32] and plot it in [33]. The straight lines connect the data points.

This same package also allows us to create Bezier curves. In [34], we combine the creation and plotting of the Bezier curve defined by our four points:

```
In[30]:=
    spdata = {{1,1},{2,4},{3,3},{4,4}}
Out[30]=
    {{1, 1}, {2, 4}, {3, 3}, {4, 4}}
In[31]:=
    <<Graphics'Spline'
In[32]:=
    splin = Spline [spdata, Cubic]
Out[32]=
    Spline[{{1, 1}, {2, 4}, {3, 3}, {4, 4}}, Cubic, <>]
In[33]:=
    Show[Graphics[{Line[spdata], splin}]]
Out[33]=
    -Graphics-
```

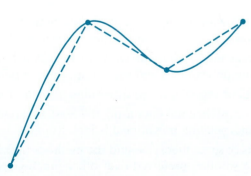

```
In[34]:=
  Show[Graphics[{Line[spdata], Spline[spdata, Bezier]}]]
Out[34]=
  -Graphics-
```

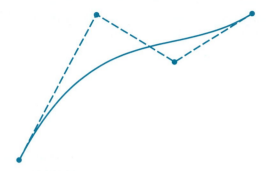

Chapter Summary

If you fully understand this chapter, you are now able to

1. Construct interpolating polynomials for any set of data by using the Lagrangian formulation or by using divided differences. You can explain why the divided-difference technique is preferred, and you can use either method to interpolate from a given set of data pairs.

2. Understand Neville's method for interpolation, and explain how it only needs to do a succession of linear interpolations to accomplish the task.

3. Write and use an expression for the error of the interpolate. You know the difficulty and limitations of such error estimates. You can explain and use the next-term rule.

4. Build a difference table from a set of evenly spaced data, and construct an interpolating polynomial from the table. You are aware of several variants on the Newton–Gregory forward polynomial and why these variants may be preferred.

5. Show a situation where the error of interpolation increases with the degree of polynomial, and tell why this occurs. You know about methods to overcome the problem.

6. Set up the equations for a cubic spline and explain how the end portions can be established. You know why a cubic spline is an improved method for interpolation.

7. Construct Bezier curves and B-spline curves. You can explain how these curves differ from the interpolating polynomials and some important applications for them.

8. Apply the techniques of this chapter to three-dimensional problems.

9. Fit a line or curve to experimental data using the least-squares method. You know why the use of an interpolating polynomial is inappropriate.

10. Explain the elements of some theory behind the methods of this chapter, including why polynomials are so often preferred over other functions, that the nth-degree polynomial through $n + 1$ points is unique, how the error term for interpolation can

be derived, why the data pairs should be chosen to make the x-value centered within them, why the next-term rule can estimate the error, and what the "$O(\ )$" expression means with respect to errors.

11. Use *Mathematica* to get an interpolating polynomial and to fit a set of experimental data with a least-squares curve.

12. Write and use computer programs to carry out the procedures of the chapter.

Computer Programs

This section includes three programs. The first, written in C, uses divided differences to formulate a polynomial that passes through a set of points and then evaluates the polynomial to approximate the function at a given point. The second, a Pascal program, draws a B-spline curve that is defined from a set of (x, y)-values. The last is a FORTRAN program that finds a sequence of polynomials that are fitted to a set of points by the least-squares procedure.

Program 3.1 Divided-Difference Program

Figure 3.18 is a listing of the C program. The coordinates of the points that define the interpolating polynomial are set up in the main procedure (which comes at the end of the figure). The coefficients of the polynomial are then determined by a call to procedure *ddcoef* that appears in the listing just after the global declarations. These coefficients are just the divided differences that are computed in the procedure.

After printing the coefficients, seven evenly spaced interpolated values are printed through calls to procedure *ddvalue*. The program's output is shown at the end of Fig. 3.18.

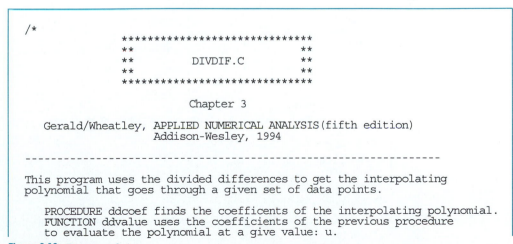

```
/*
                    ******************************
          **                              **
          **              DIVDIF.C         **
          **                              **
          ******************************

                          Chapter 3

          Gerald/Wheatley, APPLIED NUMERICAL ANALYSIS(fifth edition)
                        Addison-Wesley, 1994

    ----------------------------------------------------------------

    This program uses the divided differences to get the interpolating
    polynomial that goes through a given set of data points.

       PROCEDURE ddcoef finds the coefficents of the interpolating polynomial.
       FUNCTION ddvalue uses the coefficients of the previous procedure
       to evaluate the polynomial at a give value: u.
```

Figure 3.18 Program 3.1

Figure 3.18 *(continued)*

```
*/

#include <stdio.h>
#include <math.h>
    float x[10], y[10],    /* the given data points */
            dd[10];           /* the vector of coefficients for p(x) */
    float u;
    int   i, j, n;
/*
        *** ddcoef ***
*/

ddcoef(x, y, dd, n)
  float x[], y[], dd[];
  int    n;

/*
-----------------------------------------------------------------
        INPUT: x,y -  the given data points.
                 n   -  the number of data points
        OUTPUT: dd -  the coefficients of p(x), i.e.f[,,,]
*/

{
    int    i, j, k;
    float  temp1, temp2;


    for ( i = 1; i <= n; ++i )
        dd[i] = y[i];
    for ( j = 2; j <= n; ++j )
    {

        temp1 = dd[j-1];
        printf("\n");

        for ( k = j; k <= n; ++k )
        {
            temp2 = dd[k];
            dd[k] = (dd[k] - temp1)/(x[k] - x[k-j+1]);
            temp1 = temp2;
        }           /* end of for k loop */
    }               /* end of for j loop */
}                   /* end of ddcoef */

/*
        *** ddvalue ***
*/

float ddvalue(u)
  float u;
/*
-----------------------------------------------------------------
        INPUT:  u - the x-value
        OUTPUT: ddvalue - the corresponding y-value, i.e. p(u)
*/

{
    float sum;
    int    i;


    sum = 0.0;

    /* Compute value by nested multiplication from highest term */
```

Figure 3.18   *(continued)*

```
    for ( i = n; i >= 2; --i )

        sum = (sum + dd[i]) * (u - x[i-1]);

    sum = sum + dd[1];
    return(sum);
}

        /* end of function ddvalue */

main()          /* main program */
{
/*
-----------------------------------------------------------------
   set up five data points
*/

    n = 5;
    x[1] = 3.2; x[2] = 2.7; x[3] = 1.0;
    x[4] = 4.8; x[5] = 5.6;
    y[1] = 22.0; y[2] = 17.8;  y[3] = 14.2;
    y[4] = 38.3; y[5] = 51.7;

/*
-----------------------------------------------------------------
     GENERATE THE COEFFICIENTS FOR THE POLYNOMIAL
*/

    ddcoef(x, y, &dd, n);       /* compute coefficients */
    printf("\n");
    printf("The coefficients for the polynomial are:\n\n");
    for ( i = 1; i <= n; ++i )
       printf(" %.5f ", dd[i]);
    printf("\n\n");
    printf("\t\t\t*********\n\n");
    printf("\t\t  U\t\t   P(U)\t\t\n\n");

/*
-----------------------------------------------------------------
     SET UP TABLE OF VALUES FROM 1 TO 7
*/

    u = 1.0;
    do
      {
        printf(" \t\t%.3f\t\t %.5f\n", u, ddvalue(u));
        u = u + 0.2;
      }
    while ( u < 5.601 );
    printf("\n\n");
    printf("\t\t\t*********\n\n");
}


***********************************************************

                OUTPUT FOR DIVDIF.C

The coefficients for the polynomial are:

 22.00000   8.40000   2.85562   -0.52748   0.25584

                   *********
```

Figure 3.18   *(continued)*

U	P(U)
1.000	14.20000
1.200	12.89775
1.400	12.25424
1.600	12.16433
1.800	12.53272
2.000	13.27390
2.200	14.31221
2.400	15.58181
2.600	17.02667
2.800	18.60060
3.000	20.26723
3.200	22.00000
3.400	23.78221
3.600	25.60695
3.800	27.47714
4.000	29.40553
4.200	31.41471
4.400	33.53706
4.600	35.81480
4.800	38.30000
5.000	41.05450
5.200	44.15002
5.400	47.66806
5.600	51.69998

****

Program 3.2 The B-Spline Program

This Pascal program contains commands for drawing a curve (as a sequence of short straight lines); these commands may have to be modified, depending on the graphics package incorporated in the Pascal compiler.

The program is listed in Fig. 3.19, beginning with the usual declarations of global variables. One variable is f, the name of a file that contains the data points. The program includes five major procedures that carry out the operation:

1. Procedure initialize reads in the control points that are used to create the B-spline curve. It also displays these points.
2. Procedure set_blending_functions() evaluates the parametric equations for a B-spline at the endpoints of the short straight lines that will create the curve. These parametric equations are equivalent to those in Eq. (3.21) but in a different form.
3. Procedure start_b_spline() starts the curve. The first point is used three times to fasten the curve to the first point.
4. Procedure curve_abs_2() extends the curve by adding the next point to the group of four that defines a curve segment, draws the segment, and resets the control points for the next segment. These operations are done by calls to three secondary procedures: put_in_sm(), make_curve(), and next_section().
5. Procedure end_B_spline completes the last segment, duplicating the last point to tie the curve to it.

The main procedure, seen at the end of the figure, first makes calls to activate the graphics commands and then makes a call to the initializing procedure. Following this a box is drawn in which the curve will be displayed.

The actual curve is drawn by calls to procedures (2), (3), (4), and (5). (Procedure (3), curve_abs_2, is repeated in a loop.)

```
PROGRAM cubic_B_spline(Input,Output,f);
(*
        ****************************************
        **                                    **
        **              BSPLINE.PAS           **
        **                                    **
        ****************************************
                        Chapter 3

Gerald/Wheatley, APPLIED NUMERICAL ANALYSIS (fifth edition)
                    Addison-Wesley, 1994
------------------------------------------------------------------------

This program implements the algorithm of S. Harrington, but in a more
simplified form. It is assumed that one has the graphics commands,
such as:

        Procedure draw_line(x1,y1,  x2,y2 : Integer);
        Procedure plotPoint(x,y  : Integer);
        gotoXY(x,y : Integer);
        GraphON;
        clearGRAPHICS;
*)

Const
        max_number_of_lines = 20;   (* Each curve segment consists
                                          of 20 straight lines *)
        max_number_of_points = 20;
Type
        matrix = Array[1..4,1..max_number_of_lines] Of Real;
        vector = Array[1..max_number_of_points] Of Real;
        vector4 = Array[1..4] Of Real;
Var
        blend           : matrix;

        xsm,ysm                     (* set of four points to create current
                                        segment of B-spline curve           *)
                : vector4;
        ax,ay           (* data points as stored in the array  *)
                                : vector;
        x0,y0           : Real;

        lines_per_section, i,
        number_of_points                : Integer;

        f  :  TEXT; (* data file that stores the number of points and
                        the data points (x,y) in  screen coordinates. *)
(*
        INITIALIZE THE BLENDING FUNCTIONS
------------------------------------------------------------
This procedure evaluates the B-spline basis functions at the
number_of_points of u on the interval [0,1].
*)
```

Figure 3.19 Program 3.2

Figure 3.19 *(continued)*

```
Procedure set_blending_functions(number_of_lines : Integer);
Var
        i,j                     : Integer;
        u, u_cube, u_square,
        u_minus1_cube           : Real;
Begin

        For i := 1 To number_of_lines  Do                      Begin
          u := i/number_of_lines;
          u_square := u*u;
          u_cube    := u*u_square;
          u_minus1_cube :=  (u - 1.0)*Sqr(u - 1.0);

          blend[1,i] := u_minus1_cube/6.0;
          blend[4,i] := u_cube/6.0;
          blend[3,i] := (-u_cube/2.0 + u_square/2.0 + u/2.0 + 1.0/6.0);
          blend[2,i] := (u_cube/2.0 - u_square + 2.0/3.0);
                                                                       End;
End;  (* Initialize Blending Functions  *)


(*
        Put_in_Sm
------------------------------------------------------------
This procedure places a new sample point into the -sm arrays.
*)

Procedure put_in_sm(x, y : Real);
Begin
        xsm[4] := x;  ysm[4] := y;
End;


(*
        Make_Curve
------------------------------------------------------------
This procedure fills in a section of the curve.
*)
Procedure make_curve(Var  b : matrix);
Var
        i,j     : Integer;
        x,y     : Real;
Begin
        For j := 1 To lines_per_section      Do                      Begin
          x := 0.0;  y := 0.0;
          For i := 1 To 4 Do          Begin
              x := x + xsm[i]*b[i,j];
              y := y + ysm[i]*b[i,j]          End;
          draw_line(Round(x0), Round(y0), Round(x), Round(y) );
          x0 := x;  y0 := y                                          End;
End;    (* make_curve *)


(*
        NEXT_SECTION
------------------------------------------------------------
After a section of the curve has been drawn, we shift the sample points
so that our blending functions can be applied to the next section.
*)
Procedure next_section;
Var
        i : Integer;
Begin
        For i := 1 To 3 Do             Begin
            xsm[i] := xsm[i+1];
            ysm[i] := ysm[i+1]  End
End;
(*
```

Figure 3.19   *(continued)*

```
          CURVE_ABS_2
----------------------------------------------------------------
This procedure extends the curve by taking a new sample point as its
argument and storing it into arrays.
*)
Procedure curve_abs_2(x,y  : Real);
Begin
        put_in_sm(x,y);
        make_curve(blend);
        next_section
End;                                    (* curve_abs_2 *)
(*
          START_B_SPLINE
----------------------------------------------------------------
We require the first four sample points to start the curve. s

Procedure start_B_spline(ax,ay  : vector); (* first 4 points *)
Var
        i  :  Integer;
Begin
        For i := 1 To 3 Do               Begin
           xsm[i]  := ax[1];
           ysm[i]  := ay[1];             End;

        xsm[4]  := ax[2]; ysm[4]  := ay[2];
        make_curve(blend);
        next_section
End;                                    (* start_B_spline *)


(*
          END_B_SPLINE
----------------------------------------------------------------
This procedure terminates the B-spline curve
*)
Procedure end_B_spline;
Begin
        put_in_sm(ax[number_of_points], ay[number_of_points]);
        make_curve(blend);
        next_section;
        put_In_sm(ax[number_of_points], ay[number_of_points]);
        make_curve(blend)
End;                                 (* end_B_spline *)

(*
          INITIALIZE
----------------------------------------------------------------
This procedure reads in the control points from a data file.
*)

Procedure initialize;
Var
        i : Integer;
Begin
        WriteLn; WriteLn;
        WriteLn(' ':20, ' Control Points ');
        WriteLn;
        WriteLn(' X':24, ' Y':5);
        WriteLn;
        lines_per_section := 10;

        ReadLn(f, number_of_points);
        For i := 1 To number_of_points Do              Begin
           ReadLn(f, ax[i], ay[i]);
           WriteLn(ax[i]:25:0, ay[i]:7:0)              End
        x0 := ax[1]; y0 := ay[i];
```

Figure 3.19 *(continued)*

```
            For i := 1 To number_of_points  Do
                               plotPoint(Trunc(ax[i]), Trunc(ay[i]))
    End;

    Begin           (* MAIN  *)
            gotoXY(1,1);
            ClrScr;
            GraphON;  clearGRAPHICS;

            Assign(f, 'b:bspline2.dat');
            Reset(f);
            While Not Eof(f) Do                        Begin
                initialize;
    (*
            Create a box about the display
    *)
            draw_line(0,0,0,389);
            draw_line(0,389,511,389);
            draw_line(511,389,511,0);
            draw_line(511,0,0,0);

            set_blending_functions(lines_per_section);

            start_b_spline(ax,ay);

            For i := 3 To number_of_points Do
                curve_abs_2(ax[i], ay[i]);

            end_B_spline;                          End; (* While *)
            Close(f)
    End.
```

Program 3.3 The Least-Squares Program

Our third program, listed in Fig. 3.20, fits least-squares polynomials to a set of data pairs. This FORTRAN program is designed to either read in the data pairs from the console or get them from DATA statements. The version here does the latter by commenting out the READ statement. (To have the program read the data pairs, we would change the comment to the DATA statements.)

After obtaining the data pairs and the range of degrees of polynomials to compute (the maximum degree cannot exceed one less than the number of data pairs, of course, and this is checked), the program proceeds to set up the matrix of normal equations. These are written out. After this, the *LU* decomposition of the coefficient matrix is computed through a call to a subroutine, LUDCMP.

For each degree requested, beginning with the lowest, a right-hand-side vector is set up for the next degree of polynomial and the system of equations is solved through a call to a second subroutine, SOLNQ. After each of these calls, the coefficients of the polynomials are printed as well as the value of BETA (the variance, Eq. (3.31)). From this last value, the user can determine the best degree of least-squares polynomial to employ. Typical output is shown at the end of Fig. 3.20.

```
      PROGRAM LEASQR
C
C
C -------------------------------------------------------------------
C                      CHAPTER 3
C
C GERALD/WHEATLEY, APPLIED NUMERICAL ANALYSIS(FIFTH EDITION)
C                  ADDISON-WESLEY, 1994
C
C -------------------------------------------------------------------
C
C THIS PROGRAM IS USED IN FITTING A POLYNOMIAL TO A SET OF DATA.
C THE PROGRAM READS IN N PAIRS OF X AND Y VALUES AND COMPUTES
C THE COEFFICIENTS OF THE NORMAL EQUATIONS FOR THE
C LEAST-SQUARES METHOD.
C
C -------------------------------------------------------------------
C
C     PARAMETERS ARE :
C
C     X, Y  - ARRAY OF X AND Y VALUES
C     N     - NUMBER OF DATA PAIRS
C     MS,MF - THE RANGE OF DEGREE OF POLYNOMIALS TO BE COMPUTED
C             THE MAXIMUM DEGREE IS 9.
C     A     - AUGMENTED ARRAY OF THE COEFFICIENTS OF THE NORMAL EQUATIONS.
C     C     - ARRAY OF COEFFICIENTS OF THE LEAST-SQUARES POLYNOMIALS.
C
C -------------------------------------------------------------------
      REAL*8 X(100),Y(100),C(100),A(10,11),XN(100),SUM,BETA
      INTEGER N,MS,MF,MFP1,MFP2,I,J,IM1,IPT,ICOEF,JCOEF
C -------------------------------------------------------------------
C
C READ IN N, THEN THE X AND Y VALUES.
C
C     READ *, N,( X(I),Y(I), I = 1,N )
C
      DATA N/11/
      DATA X/0.05,0.11,0.15,0.31,0.46,0.52,0.7,0.74,0.82,
     *    0.98,1.17,89*0.0/
      DATA Y/0.956, 0.89, 0.832, 0.717, 0.571, 0.539, 0.378,
     *    0.37, 0.306, 0.242, 0.104, 89*0.0/
C
C READ IN MS,MF. THE PROGRAM WILL FIND COEFFICIENTS FOR EACH
C DEGREE OF POLYNOMIAL FROM DEGREE MS TO DEGREE MF.
C
C     READ *, MS,MF
C
      DATA MS,MF/1,7/
C -------------------------------------------------------------------
C
C COMPUTE MATRIX OF COEFFICIENTS AND R.H.S. FOR MF DEGREE.
C HOWEVER, FIRST CHECK TO SEE IF MAX DEGREE REQUESTED IS TOO
C LARGE. IT CANNOT EXCEED N-1. IF IT DOES, REDUCE TO EQUAL N-1
C AND PRINT MESSAGE.
C
      IF ( MF .GT. (N-1) ) THEN
        MF = N - 1
        PRINT 200, MF
      END IF
  5   MFP1 = MF + 1
      MFP2 = MF + 2
C
C -------------------------------------------------------------------
C
```

Figure 3.20 Program 3.3

Figure 3.20 *(continued)*

```
C    PUT ONES INTO A NEW ARRAY. THIS WILL HOLD THE POWERS OF
C    THE X VALUES AS WE PROCEED.
C
        DO 10 I = 1,N
          XN(I) = 1.0
10      CONTINUE
C
C    -------------------------------------------------------------
C
C    COMPUTE FIRST COLUMN AND N+1 ST COLUMN OF A.
C        I MOVES DOWN THE ROWS,
C        J SUMS OVER THE N VALUES.
C
        DO 30 I = 1,MFP1
          A(I,1) = 0.0
          A(I,MFP2) = 0.0
          DO 20 J = 1,N
            A(I,1) = A(I,1) + XN(J)
            A(I,MFP2) = A(I,MFP2) + Y(J)*XN(J)
            XN(J) = XN(J) * X(J)
20        CONTINUE
30      CONTINUE
C
C    -------------------------------------------------------------
C
C    COMPUTE THE LAST ROW OF A.
C        I MOVES ACROSS THE COLUMNS,
C        J SUMS OVER THE N VALUES.
C
        DO 50 I = 2,MFP1
          A(MFP1,I) = 0.0
          DO 40 J = 1,N
            A(MFP1,I) = A(MFP1,I) + XN(J)
            XN(J) = XN(J) * X(J)
40        CONTINUE
50      CONTINUE
C    -------------------------------------------------------------
C
C    NOW FILL IN THE REST OF THE A MATRIX.
C        I MOVES DOWN THE ROWS,
C        J MOVES ACROSS THE COLUMNS.
C
        DO 70 J = 2,MFP1
          DO 60 I = 1,MF
            A(I,J) = A(I+1,J-1)
60        CONTINUE
70      CONTINUE
C
C    -------------------------------------------------------------
C
C    WRITE OUT THE MATRIX OF NORMAL EQUATIONS.
C
        PRINT '(///)'
        PRINT *, '          THE NORMAL MATRIX IS:  '
        PRINT '(/)'
        PRINT 201, ((A(I,J), J=1,MFP2), I=1,MFP1)
        PRINT '(//)'
C
C    NOW CALL A SUBROUTINE TO SOLVE THE SYSTEM. DO THIS FOR EACH
C    DEGREE FROM MS TO MF. GET THE LU DECOMPOSITION OF A.
C
        CALL LUDCMQ(A,MFP1,10)
C
C    -------------------------------------------------------------
C
```

Figure 3.20   *(continued)*

```
C  RESET THE R.H.S. INTO C. WE NEED TO DO THIS FOR EACH DEGREE.
C
      MSP1 = MS + 1
      DO 95 I = MSP1,MFP1
        DO 90 J = 1,I
          C(J) = A(J,MFP2)
90      CONTINUE
        CALL SOLNQ(A,C,I,10)
        IM1 = I - 1
C
C  -----------------------------------------------------------------
C
C  NOW WRITE OUT THE COEFFICIENTS OF THE LEAST-SQUARES POLYNOMIAL.
C
      PRINT 202, IM1, ( C(J), J=1,I)
C
C  -----------------------------------------------------------------
C
C  COMPUTE AND PRINT THE VALUE OF BETA = SUM OF DEV SQUARED
C    DIVIDED BY ( N - M - 1 ).
      BETA = 0.0
      DO 94 IPT = 1,N
        SUM = 0.0
        DO 93 ICOEF = 2,I
          JCOEF = I - ICOEF + 2
          SUM = ( SUM + C(JCOEF) ) * X(IPT)
93      CONTINUE

        SUM = SUM + C(1)
        BETA = BETA + ( Y(IPT) - SUM )**2
94      CONTINUE
      BETA = BETA / (N - I)
      PRINT 203, BETA
95    CONTINUE
C
C  -----------------------------------------------------------------
C
200   FORMAT(//' DEGREE OF POLYNOMIAL CANNOT EXCEED N - 1.',/
     +       '   REQUESTED MAXIMUM DEGREE TOO LARGE - ',
     +       'REDUCED TO ',I3)
201   FORMAT(1X,9F8.2)
202   FORMAT(/' FOR DEGREE OF ',I2,' COEFFICIENTS ARE'//
     +       ' ',5X,11F9.3)
203   FORMAT(9X,' BETA IS ',F10.5//)
      STOP
      END
C
      SUBROUTINE LUDCMQ(A,N,NDIM)
C
C  -----------------------------------------------------------------
C
C  SUBROUTINE LUDCMQ :
C      THIS SUBROUTINE FORMS THE LU EQUIVALENT OF THE SQUARE
C      COEFFICIENT MATRIX A. THE LU IS RETURNED IN THE A
C      MATRIX SPACE. THE UPPER TRIANGULAR MATRIX U HAS ONES
C      ON ITS DIAGONAL - THESE VALUES ARE NOT INCLUDED IN
C      THE RESULT.
C  -----------------------------------------------------------------
C
      REAL*8 A(NDIM,NDIM),SUM
      INTEGER N,NDIM,I,J,JM1,IM1,K
C
C  -----------------------------------------------------------------
C
      DO 30 I = 1,N
      DO 30 J = 2,N
```

Figure 3.20 *(continued)*

```
              SUM = 0.0
              IF ( J .LE. I ) THEN
                 JM1 = J - 1
                 DO 10 K = 1,JM1
                    SUM = SUM + A(I,K)*A(K,J)
10               CONTINUE
                 A(I,J) = A(I,J) - SUM
              ELSE
                 IM1 = I - 1
                 IF ( IM1 .NE. 0 ) THEN
                    DO 20 K = 1,IM1
                       SUM = SUM  + A(I,K)*A(K,J)
20                  CONTINUE
                 END IF
C
C   ------------------------------------------------------------
C
C  TEST FOR SMALL VALUE ON THE DIAGONAL
C
25                IF ( ABS(A(I,I)) .LT. 1.0E-10 ) THEN
                     PRINT 100, I
                     RETURN
                  ELSE
                     A(I,J) = ( A(I,J) - SUM ) / A(I,I)
                  END IF
              END IF
30         CONTINUE
           RETURN
C
100    FORMAT(' REDUCTION NOT COMPLETED BECAUSE SMALL VALUE',
      +         ' FOUND FOR DIVISOR IN ROW ',I3)
       END


       SUBROUTINE SOLNQ(A,B,N,NDIM)
C
C   ------------------------------------------------------------
C
C      SUBROUTINE SOLNQ :
C          THIS SUBROUTINE FINDS THE SOLUTION TO A SET OF N LINEAR
C          EQUATIONS THAT CORRESPONDS TO THE RIGHT-HAND SIDE
C          VECTOR B. THE A MATRIX IS THE LU DECOMPOSITION
C          EQUIVALENT TO THE COEFFICIENT MATRIX OF THE
C          ORIGINAL EQUATIONS, AS PRODUCED BY LUDCMQ. THE
C          SOLUTION VECTOR IS RETURNED IN THE B VECTOR.
C
C   ------------------------------------------------------------
C
       REAL*8 A(NDIM,NDIM),B(NDIM),SUM
       INTEGER N,NDIM,I,IM1,K,J,NMJP1,NMJP2
C   ------------------------------------------------------------
C
C  DO THE REDUCTION STEP
C
       B(1) = B(1) / A(1,1)
       DO 20 I = 2,N
          IM1 = I - 1
          SUM = 0.0
          DO 10 K = 1,IM1
             SUM = SUM + A(I,K)*B(K)
10        CONTINUE
          B(I) = ( B(I) - SUM ) / A(I,I)
20     CONTINUE
C
C   ------------------------------------------------------------
C
```

Figure 3.20 *(continued)*

```
C  NOW WE ARE READY FOR BACK SUBSTITUTION. REMEMBER THAT THE
C  ELEMENTS OF U ON THE DIAGONAL ARE ALL ONES.
C
       DO 40 J = 2,N
         NMJP2 = N - J + 2
         NMJP1 = N - J + 1
         SUM = 0.0
         DO 30 K = NMJP2,N
           SUM = SUM + A(NMJP1,K)*B(K)
30       CONTINUE
         B(NMJP1) = B(NMJP1) - SUM
40     CONTINUE
       RETURN
       END
```

```
********************************************************************
                   OUTPUT FOR LEASQR.F

THE NORMAL MATRIX IS:

11.00    6.01    4.65    4.11    3.92    3.92    4.07    4.34    5.90
 6.01    4.65    4.11    3.92    3.92    4.07    4.34    4.72    2.18
 4.65    4.11    3.92    3.92    4.07    4.34    4.72    5.22    1.34
 4.11    3.92    3.92    4.07    4.34    4.72    5.22    5.84    1.00
 3.92    3.92    4.07    4.34    4.72    5.22    5.84    6.59     .83
 3.92    4.07    4.34    4.72    5.22    5.84    6.59    7.50     .74
 4.07    4.34    4.72    5.22    5.84    6.59    7.50    8.57     .70
 4.34    4.72    5.22    5.84    6.59    7.50    8.57    9.84     .68

                   ************

FOR DEGREE OF  1 COEFFICIENTS ARE
        .952    -.760
      BETA IS     .00102


FOR DEGREE OF  2 COEFFICIENTS ARE
        .998   -1.018     .225
      BETA IS     .00023


FOR DEGREE OF  3 COEFFICIENTS ARE
       1.004   -1.079     .351    -.069
      BETA IS     .00026


FOR DEGREE OF  4 COEFFICIENTS ARE
        .988    -.837    -.527   1.046    -.456
      BETA IS     .00027


FOR DEGREE OF  5 COEFFICIENTS ARE
       1.037   -1.824    4.895  -10.753   10.537   -3.659
      BETA IS     .00013


FOR DEGREE OF  6 COEFFICIENTS ARE
       1.041   -1.946    5.886  -14.081   15.818   -7.599    1.112
      BETA IS     .00017


FOR DEGREE OF  7 COEFFICIENTS ARE
       1.060   -2.562   12.441  -44.819   88.979  -99.704   59.448  -14.607
      BETA IS     .00021
```

Exercises

Section 3.2

1. Write the Lagrangian interpolating polynomial that passes through each point:

x	-1.1	2.1	3.5
y	1.3	2.5	-1.6

Plot the points, and sketch the parabola that passes through them.

▶**2.** Given the four points $(1, 0)$, $(-2, 16)$, $(3, -2)$, $(-1, 1)$, write the cubic in Lagrangian form that passes through them. Multiply out each term to express in standard form, $ax^3 + bx^2 + cx + d$.

3. Given that $\ln(1) = 0$, $\ln(2) = 0.69315$, and $\ln(5) = 1.60944$, interpolate with a Lagrangian polynomial for the natural logarithms of each integer from 1 to 10. Tabulate these together with the error at each point.

4. If $e^{0.2}$ is approximated by Lagrangian interpolation among the values of $e^0 = 1$, $e^{0.1} = 1.1052$, and $e^{0.3} = 1.3499$, find the maximum and minimum estimates of the error. Compare to the actual error.

5. Repeat Exercise 4, but this time extrapolate to get $e^{0.4}$.

▶**6.** Given the following data, compute the Neville table. From this get interpolates for $f(3)$, using polynomials of degrees 2, 3, and 4.

x	$f(x)$
2	3.4899
5	21.7889
6	31.3585
-1	0.8726
-2	3.4899

7. Construct the Neville table for Exercise 4 to approximate $e^{0.2}$. Does the estimate from a polynomial of degree 2 agree with that in Exercise 4? What if we were to use linear interpolation? Would this be the same?

8. Repeat Exercise 5, this time using Neville's method.

9. Show that the entries in the top line of the Neville table of Exercise 7 do in fact represent the results of interpolating for $e^{0.2}$ with polynomials of increasing degrees.

10. Suppose a Neville table for the following n data points is computed with parallel processors. How many fewer time steps are required compared to doing it with a single CPU?
  a. For $n = 8$
  b. For $n = 16$
  c. For $n = 11$

Section 3.3

11. Construct a divided-difference table from these data:

x	$f(x)$
0.5	1.0025
-0.2	1.3940
0.7	1.0084
0.1	1.1221
0.0	1.1884

12. Repeat Exercise 2, except this time use divided differences. Compare the polynomial in standard form with that obtained in Exercise 2.

13. Repeat Exercise 4, but now use divided differences.

14. Use the divided difference table of Exercise 11 to estimate $f(0.15)$, using
  a. a polynomial of degree 2 through the first three points.
  b. a polynomial of degree 2 through the last three points.
  c. a polynomial of degree 3 through the first four points.
  d. a polynomial of degree 3 through the last four points.
  e. a polynomial of degree 4.
  f. Why are the results different?

▶**15.** In Exercise 14, which three points are best to use for constructing the quadratic if we want
  a. $f(0.15)$?
  b. $f(-0.1)$?
  c. $f(1.2)$?

16. Repeat Exercise 5, this time using divided differences.

17. The function in Exercise 11 is unknown, but that does not hinder our use of the table for interpolation. Interpolate with a cubic polynomial that passes through the first four points to get $f(0.2)$. Estimate the error from the next-term rule.

▶18. In Exercise 13, you estimated a value for $e^{0.2}$. How does the actual error compare to the error bounds from Eq. (3.2)?

Section 3.4

19. Complete the difference table for the following data:

x	1.20	1.25	1.30	1.35	1.40	1.45	1.50
$f(x)$	0.1823	0.2231	0.2624	0.3001	0.3365	0.3716	0.4055

20. In Exercise 19, what degree of polynomial is required to exactly fit to all seven data pairs? What lesser-degree polynomial will nearly fit the data? Justify your answer.

21. Form a difference table for $f(x) = 2x^3 - 4x^2 - 2x + 3$ for the interval $[-1, 1]$ with a spacing of 0.2. (If you do this by hand, remember that nested multiplication is more efficient.) Are the third differences a constant as expected? Is the value of this constant equal to $a_0 n! h^n$?

▶22. Using the data in Exercise 19, compute the value of $\Delta^3 f_0$, if x_0 is the second entry in the table ($x_0 = 1.25$), directly from the f-values, not from the table.

23. Without computing the divided-difference table for the data in Exercise 19, what is $f[x_0, x_1, x_2, x_3]$ if x_0 is 1.25? Compute this in two ways, first from the table values, then from the answer to Exercise 22.

24. Use a Newton–Gregory interpolating polynomial of degree 3 to estimate the value of $f(1.37)$ from the data of Exercise 19. Select the best point to call x_0. Estimate the error by the next-term rule.

25. Repeat Exercise 24, but now get $f(0.77)$. Is the estimated error larger than in Exercise 24? If so, explain.

▶26. The following table is already computed. Use a Newton–Gregory interpolating polynomial of degree 2 to estimate $f(0.203)$, taking $x_0 = 0.125$. Then add one term to get $f(0.203)$ from a third-degree polynomial. Estimate the errors of each from the next-term rule.

x	$f(x)$	Δf	$\Delta^2 f$	$\Delta^3 f$	$\Delta^4 f$
0.125	0.79168				
		−0.01834			
0.250	0.77334		−0.01129		
		−0.02963		0.00134	
0.375	0.74371		−0.00995		0.00038
		−0.03958		0.00172	
0.500	0.70413		−0.00823		0.00028
		−0.04781		0.00200	
0.625	0.65632		−0.00623		
		−0.05404			
0.750	0.60228				

27. Repeat Exercise 26, but now get $f(0.612)$, taking $x_0 = 0.375$.

28. What would be the answers to Exercise 27 if we took $x_0 = 0.125$?

29. Use the following data to find a value for $y(0.54)$ using a cubic that fits at $x = 0.3, 0.5, 0.7$, and 0.9.

x	y	Δy	$\Delta^2 y$	$\Delta^3 y$
0.1	0.003			
		0.064		
0.3	0.067		0.017	
		0.081		0.002
0.5	0.148		0.019	
		0.100		0.003
0.7	0.248		0.022	
		0.122		0.004
0.9	0.370		0.026	
		0.148		0.005
1.1	0.518		0.031	
		0.179		
1.3	0.697			

▶30. What is the minimum degree of polynomial that will exactly fit all seven pairs of data in Exercise 29? (Answer is *not* sixth-degree.)

31. Construct a divided-difference table for the data in Exercise 29. How do the values compare to those in the given table?

32. The precision of $f(x)$ data has a considerable effect on a table of differences. Demonstrate this fact by recomputing the table of Exercise 26 after rounding to three decimal places. Repeat this calculation, but chop after three places.

Section 3.5

33. For $f(x)$ as defined here, find polynomials of degrees 2, 3, 4, and 5 that fit $f(x)$ at equally spaced points in

[−1,1]. Plot these values and observe that the fit is poor.

$$f(x) = \begin{cases} 0, & -1 \ < x < -0.25 \\ 1 - |4x|, & -0.25 < x < \ 0.25 \\ 0, & \ 0.25 < x < \ 1.0 \end{cases}$$

▶**34.** Find the coefficient matrix and the right-hand side for fitting a cubic spline curve to the following data. Use linearity end conditions (condition 1).

x	f(x)
0.15	0.1680
0.27	0.2974
0.76	0.7175
0.89	0.7918
1.07	0.8698
2.11	0.9972

35. Solve the set of equations in Exercise 34, and then determine the coefficients of the various cubics. Plot the cubic spline curve. Compare the interpolates at $x = 0.33$, $x = 0.92$, and $x = 2.05$ with the tabulated values for $ERF(x)$ (the so-called error function).

36. Repeat Exercises 34 and 35, this time for each of the other end conditions. How different are the interpolates, and which end condition gives the least average error?

37. Fit a natural cubic spline to the function in Exercise 33, matching to the function at five equally spaced points between −1 and 1. Plot the cubic spline curve and compare it to the plots of the polynomials of Exercise 33, particularly with the fourth-degree polynomial.

38. Repeat Exercise 37, but use end conditions 3 and 4. Plot these points and compare to the plot from end condition 1.

▶**39.** Repeat Exercise 38, but now force the end slopes to be zero.

40. If the data given are periodic and cover one period, the first and last points will have identical f-values and the beginning and ending slopes will be the same. Develop the relations that give a cubic spline curve for such periodic data.

41. The data in Example 3.7 are from the kind of periodic data referred to in Exercise 40. Use the relation developed in Exercise 40 to get the periodic cubic

spline curve. Which of the results of Example 3.7 are closest to this spline?

Section 3.6

42. Show that the matrix forms of the equation for Bezier and B-spline curves are equivalent to the algebraic equations given in Section 3.6.

▶**43.** Write the matrix form of the equations for Bezier and B-spline curves of order 4.

44. Prove that the convex hull does enclose all the points for both Bezier and B-spline curves. (*Hint:* Use the fact that any point p in the convex hull for the set of points $\{p_0, p_1, \ldots, p_n\}$ can be written as $p = \sum a_i p_i$ where all the a's are nonnegative and their sum is one.)

▶**45.** The slopes at the ends of the cubic B-spline curve seem to be the same as the slopes between adjacent points. Is this true?

46. Suppose that we have constructed a connected B-spline curve and then one of the points is changed. What part(s) of the connected curve is (are) affected? Are Bezier curves similar in this respect? What about a cubic spline? Do the terms *local control* and *global control* apply to the phenomena you observe?

47. A higher-degree B-spline curve is a natural extension of our cubic B-splines. What about reducing the degree to give a quadratic B-spline curve? What assumptions are reasonable for such quadratics?

48. Compute and then graph the cubic Bezier curve defined by this set of points.

Point	x	f(x)
1	100	100
2	50	150
3	200	150
4	50	50
5	100	200
6	50	100
7	200	100
8	50	50
9	100	200
10	100	100

49. By letting u vary from 0 to 1, plot the weighting factors that produce a Bezier curve.

50. Repeat Exercise 49, this time for a B-spline curve.

Section 3.7

51. In Section 3.7, it is asserted that the order in which the interpolation is done does not matter. Verify that this is true by interpolating within the data of Table 3.12 to find values at $y = 0.33$ (for x constant at 1.0, 1.5, and 2.0), using cubic interpolation with y-values of 0.2, 0.3, 0.4, and 0.5. Then interpolate from these to get $f(1.6, 0.33)$, and compare to the value 1.841 obtained in the text.

▶**52.** In Example 3.9, after the computations were completed, it was observed that a cubic in x and a quadratic in y would be preferred. Do this to obtain an estimate of $f(1.6, 0.33)$ and compare it to the true value, 1.8350. Use the best "region of fit."

53. Example 3.9 used a rectangular region of fit when a more nearly circular region should be advantageous. Interpolate from the data of Table 3.12 to evaluate $f(1.62, 0.31)$ by a set of polynomials that fit at $x = 1.5$ and 1.6 when y is 0.2 or 0.4, and fits at $x = 0.5$ to 2.5 when y is 0.3. Do this by forming a set of difference tables. This is awkward to do if we begin with x held constant, but there is no problem if we begin with y held constant.

▶**54.** Interpolate for $f(3.32, 0.71)$ from the following data. Use cubics in each direction and with the best region of fit. Since the x- and y-values are not evenly spaced, you will need to use Lagrangian polynomials or divided differences.

x \ y	0.1	0.4	0.6	0.9	1.2
1.1	1.100	0.864	0.756	0.637	0.550
3.0	8.182	6.429	5.625	4.737	4.091
3.7	12.445	9.779	8.556	7.205	6.223
5.2	24.582	19.314	16.900	14.232	12.291
6.5	38.409	30.179	26.406	22.237	19.205

55. Find a value at $(3.7, 0.6)$ on the B-spline surface constructed from the 16 points in the upper-left corner of the data of Exercise 54.

56. A Bezier surface can be constructed in a manner similar to that for a B-spline surface. Repeat Exercise 55, but for a Bezier surface.

Section 3.8

57. The "least-squares" line for the data in Fig. 3.13 has the equation

$$R = 702.2 + 3.395T.$$

A line drawn by eye was $R = 700 + 3.5T$. Compute the deviations of the actual data from each of these lines. Then compare the sum of the squares of these deviations. Observe that, even though the maximum deviation is about the same for each, the sum of squares differs significantly. How do the average errors compare?

58. Show that the point (X, Y), where X is the mean of all x-values and Y is the mean of all y-values, falls on the least-squares line. Often a change of variable is made to put the point at the origin, thus reducing the magnitude of the numbers worked with, which is an advantage if the least-squares line is computed by hand.

▶**59.** Find the least-squares line that fits the following data, assuming that the x-values are free from error. (The data are actually $y = 3x + 2$ plus a random variation.)

x	y
1	5.04
2	8.12
3	10.64
4	13.18
5	16.20
6	20.04

60. Repeat Exercise 59, but now consider all the errors to be in the x-values with y free of error. Modify the normal equations to get the least-squares line $x = ay + b$. Observe that this is not the same line obtained in Exercise 59.

▶**61.** Multivariate analysis finds a function of more than one independent variable. Suppose that z is a function of both x and y. Find the normal equations to fit

$$z = ax + by + c$$

and then use the following data to fit the least-squares plane to them.

x	0	1.2	2.1	3.4	4.0	4.2	5.6	5.8	6.9
y	0	0.5	6.0	0.5	5.1	3.2	1.3	7.4	10.2
z	1.2	3.4	−4.6	9.9	2.4	7.2	14.3	3.5	1.3

62. Draw the straight line between $(2, 3)$ and $(6, 5)$. Write its equation in the form $y = ax + b$. How much does this line shift to give the least-squares line that fits to these two points and a third point, if the third point is

a. $(4, 5)$?
b. $(4, 2)$?
c. $(5, 5)$?
d. $(3, 2)$?

63. In Section 3.8, the statement is made that $A * A^T$, where A is the design matrix given in Section 3.8, is equal to the coefficient matrix of Eq. (3.30). Show that this is true, and show that $A * y$ where y is the column vector of y-values is the same as the right-hand side of Eq. (3.30).

64. Show that $A * A^T$, where A is the design matrix, is symmetric and positive semidefinite.

65. Observe that the following data seem to be fit by a curve $y = ae^{bx}$ by plotting on semilog paper and noting that the points appear to fall on a straight line. (The data are the solubilities of n-butane in anhydrous hydrofluoric acid at high pressures and were used in the design of petroleum refineries.) Find values of a and b from the plot.

Temperature, °F	Solubility, wt. %
77	2.4
100	3.4
185	7.0
239	11.1
285	19.6

66. Plot the data of Exercise 65 on ordinary graph paper, and observe that the data are nonlinear.

▶67. Find the least-square values for the parameters of $y = ae^{bx}$ by fitting the data of Exercise 65 to the relation $\ln(y) = \ln(a) + bx$.

68. It is suspected (from theoretical considerations) that the rate of flow from a fire hose is proportional to some power of the pressure at the nozzle. Do the following data confirm the speculation? What is the least-squares value of the exponent? (Assume the pressure data are more accurate.)

Flow, gallons per minute	94	118	147	180	230
Pressure, psi	10	16	25	40	60

69. Plot the data of Exercise 68 on log-log paper, and observe that they nearly fall on a line of slope 2. That means that a quadratic would be a good function for

fitting them. Find the least-squares values for the constants in

$$\text{Flow} = aP^2 + bP + c, \quad \text{where } P = \text{pressure}.$$

70. The data in Exercise 59, while actually perturbations from a straight line, plot better along a curve because of the accidental occurrence of three negative deviations in succession. Fit a quadratic to the data. How do the sums of squares of the deviations compare between the line and the curve?

71. If the degree of the least-squares polynomial is exactly one less than the number of points, the polynomial passes through all of the points; that is, it is an interpolating polynomial. Fit a fifth-degree polynomial to the six points of Exercise 59. Now plot this polynomial and compare it to plots of the least-squares line of Exercise 59 and the least-squares quadratic of Exercise 70. Compare the computed y-values at $x = 1.5, 2.5, 3.5, 4.5,$ and 5.5 from the three relations.

72. How do the maximum and minimum values of the slopes of the three curves of Exercise 71 compare to the true slope of 3? What does this mean with respect to getting the slope of experimental data?

▶73. The following data seem to fit to a cubic equation but determine by least squares the optimum degree.

x	$f(x)$	x	$f(x)$
0.1	1.9	9.4	−3.1
1.1	7.9	11.1	−13.0
1.6	24.9	11.4	−28.7
2.4	24.9	12.2	−39.5
2.5	34.9	13.2	−48.6
4.1	42.7	14.1	−40.2
5.2	29.7	15.6	−51.6
6.1	49.8	16.1	−30.5
6.6	36.1	17.6	−34.6
7.1	23.7	17.9	−16.4
8.2	13.0	19.1	−13.4
9.1	20.5	20.0	−1.1

74. Repeat Exercise 73, but this time use every other point $(x = 0.1, 1.6, 2.5, \dots)$. Do you get the same results? Repeat again but with the other half of the points.

75. The data of Exercise 73 suggest a function of the form $y = A + B * \sin(Cx)$. How can least squares

be used to determine the coefficients? What difficulties will there be in solving the normal equations? Suppose that it were known that $C = \pi/10$. Would this make it easier to get the values of A and B?

Section 3.9

76. Write $x^3 - 2x^2 + x - 7$ in at least two different ways that do not look identical.

77. Find bounds to the errors of each of the results of Exercise 3, and compare them to the actual errors.

▶**78.** Find bounds to the errors when each of the polynomials of Exercise 33 is used to estimate $f(0.1)$. Compare to the actual errors.

79. Make a table for $f(x) = e^{-x}(x^2)$ for $x = 0.2, 0.3, 0.6$, 0.9, and 1.0. Construct quadratic interpolating polynomials using three successive points beginning at $x = 0.2$, $x = 0.3$, and $x = 0.6$. What are the errors if we use each of these to estimate $f(0.5)$? Compare to the bounds for the errors.

80. Fit $f(x) = 2x - 1 + \exp(-10x^2)$ between $x = -1$ and $x = 1$ with polynomials of degrees 2, 3, 4, 5, and 6 that match to $f(x)$ at equally spaced points. Compare the graphs of these to the plot of $f(x)$.

▶**81.** Use the symbolic operator method to develop the Newton–Gregory backward-interpolating polynomial. (This is similar to Eq. (3.9), but it uses function values that precede f_0.)

Section 3.10

Use *Mathematica* or another symbolic algebra program to solve Exercises 82–89.

82. Construct the divided-difference table for the data of Exercise 11.

83. This is more challenging: Construct the Neville table for the data of Exercise 6.

84. Obtain the cubic interpolating polynomial that fits to the data of Exercise 2.

▶**85.** Solve Exercise 51 with a computer algebra program.

86. Solve Exercise 61 with a computer algebra program.

▶**87.** Solve Exercise 67 with a computer algebra program.

88. Plot the cubic spline curve that passes through the points of Exercise 34.

89. Plot the Bezier curve that is defined by the data of Exercise 48.

Applied Problems and Projects

90. S.H.P. Chen and S.C. Saxena report experimental data for the emittance of tungsten as a function of temperature [*Ind. Eng. Chem. Fund.* 12, 220 (1973)]. Their data follow. They found that the equation

$$e(T) = 0.02424\left(\frac{T}{303.16}\right)^{1.27591}$$

correlated the data for all temperatures accurately to three digits. What degree of interpolating polynomial is required to match to their correlation at points midway between the tabulated temperatures? Discuss the pros and cons of polynomial interpolation in comparison to using their correlation.

$T, °K$	300	400	500	600	700	800	900	1000	1100
e	0.024	0.035	0.046	0.058	0.067	0.083	0.097	0.111	0.125

$T, °K$	1200	1300	1400	1500	1600	1700	1800	1900	2000
e	0.140	0.155	0.170	0.186	0.202	0.219	0.235	0.252	0.269

91. In studies of radiation-induced polymerization, a source of gamma rays was employed to give measured doses of radiation. However, the dosage varied with position in the apparatus, with these figures being recorded:

Position, in. from base point	0	0.5	1.0	1.5	2.0	3.0	3.5	4.0
Dosage, 10^5 rads/hr	1.90	2.39	2.71	2.98	3.20	3.20	2.98	2.74

For some reason, the reading at 2.5 in. was not reported, but the value of radiation there is needed. Fit interpolating polynomials of various degrees to the data to supply the missing information. What do you think is the best estimate for the dosage level at 2.5 in.?

92. Studies of the kinetics of elution of copper compounds from ion-exchange resins gave the following data. The normality of the leaching liquid was the most important factor in determining the diffusivity. The data were obtained at convenient values of normality; we desire a table of D for integer values of normality ($N = 0.0, 1.0, 2.0, 3.0, 4.0, 5.0$). Use the data to construct such a table.

N	$D \times 10^6$, cm²/sec	N	$D \times 10^6$, cm²/sectr
0.0521	1.65	0.9863	3.12
0.1028	2.10	1.9739	3.06
0.2036	2.27	2.443	2.92
0.4946	2.76	5.06	2.07

93. Experiment with the placing of intermediate points in Exercise 37 to see whether you can reduce the average error of the cubic spline curve.

94. When the steady-state heat-flow equation is solved numerically, temperatures $u(x,y)$ are calculated at the nodes of a gridwork constructed in the domain of interest. (This is the content of Chapter 7.) When a certain problem was solved, the values given in the following table were obtained. This procedure does not give the temperatures at points other than the nodes of the grid; if they are desired, one can interpolate to find them. Use the data to estimate the values of the temperature at the points $(0.7, 1.2)$, $(1.6, 2.4)$, and $(0.65, 0.82)$.

x \ y	0.0	0.5	1.0	1.5	2.0	2.5
0.0	0.0	5.00	10.00	15.00	20.00	25.00
0.5	5.00	7.51	10.05	12.70	15.67	20.00
1.0	10.00	10.00	10.00	10.00	10.00	10.00
1.5	15.00	12.51	9.95	7.32	4.33	0.0
2.0	20.00	15.00	10.00	5.00	0.00	−5.00

95. Star S in the Big Dipper (Ursa Major) has a regular variation in its apparent magnitude. Leon Campbell and Laizi Jacchia give data for the mean light curve of this star in their book *The Story of Variable Stars* (Blakeston, 1941). A portion of these data is given here.

Phase	−110	−80	−40	−10	30	80	110
Magnitude	7.98	8.95	10.71	11.70	10.01	8.23	7.86

The data are periodic in that the magnitude for phase $= -120$ is the same as for phase $= +120$. The spline functions discussed in Section 3.5 do not allow for periodic behavior. For a periodic function, the slope and second derivatives are the same at the two endpoints. Taking this into account, develop a spline that interpolates the preceding data.

Other data given by Campbell and Jacchia for the same star are

Phase	−100	−60	−20	20	60	100
Magnitude	8.37	9.40	11.39	10.84	8.53	7.89

How well do interpolants based on your spline function agree with this second set of observations?

96. Develop the matrices to make Eq. (3.25) generate points on a Bezier surface. Show that this passes through the 12 points on the borders of the group of 16 in Fig. 3.12. How could a Bezier surface be created that passes through the innermost set of four points?

97. A fictitious chemical experiment produces seven data points:

t	−1	−0.96	−0.86	−0.79	0.22	0.5	0.930
y	−1	−0.151	0.894	0.986	0.895	0.5	−0.306

a. Plot the points and interpolate a smooth curve by intuition.
b. Plot the unique sixth-degree polynomial that interpolates these points.
c. Use a spline program to evaluate enough points to plot this curve.
d. Compare your results with the graph in Fig. 3.21.

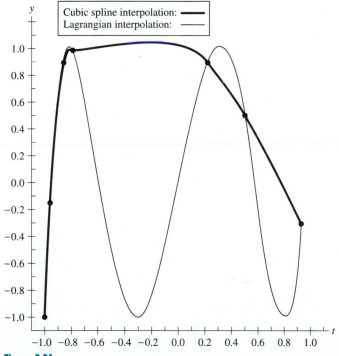

Figure 3.21

4

Numerical Differentiation and Numerical Integration

Contents of This Chapter

Chapter 4 tells you how to estimate the derivative or get the integral of a function by numerical methods. Numerical procedures differ from the analytical methods of calculus in that they can be applied to functions known only as a table of values as well as to a function stated explicitly. Although computers can be programmed to do the symbolic manipulations of formal integration techniques (DERIVE, MAPLE, and *Mathematica* are such programs), computers more often use the numerical methods to perform differentiation and integration. The material of this chapter has obvious applications to the solving of differential equations numerically, the subject of Chapters 5 and 6. This chapter also explains the theory behind the "next-term rule."

4.4   Extrapolation Techniques

Allow you to get an improved estimate of the value of a derivative that is equivalent to using a formula based on a higher-degree polynomial. These techniques are often used in computer programs. The Richardson extrapolation procedure has wide application beyond its use for derivatives.

4.5   Newton–Cotes Integration Formulas

Are derived by integrating an interpolation formula analogous to the development of differentiation formulas. The three most important of these formulas for integration are based on polynomials of degrees 1, 2, and 3, with the integration interval the same as the region of fit for the polynomial.

4.6   The Trapezoidal Rule—A Composite Formula

Is another name for the first of the Newton–Cotes formulas when applied to a succession of evenly spaced data points. It is very widely used in computer programs. When the trapezoidal rule is used for a known function, its results may be extrapolated, a technique known as Romberg integration.

4.7   Simpson's Rules

Are the composite rules obtained when the Newton–Cotes formulas based on polynomials of degrees 2 and 3 are applied to a succession of evenly spaced intervals. These rules are more accurate than the trapezoidal rule, particularly so for Simpson's $\frac{1}{3}$ rule, based on a quadratic interpolating polynomial. Extrapolation may be used here also.

4.8   Other Ways to Derive Integration Formulas

Explains the application of symbolic methods to obtain integration formulas and a second method, the method of undetermined coefficients. This last technique has many applications.

4.9   Gaussian Quadrature

Is a procedure for getting the integral of a known function that uses fewer function evaluations for a required accuracy than do the previous methods. Because the cost of computer time is usually related directly to the number of function evaluations, this procedure is widely used in modern computer programs. You are also introduced to Legendre polynomials, a representative of the very important *orthogonal polynomials*.

4.10   Adaptive Integration

Can reduce the number of function evaluations when Simpson's $\frac{1}{3}$ rule is used, by properly selecting the size of intervals within subregions of the integration region. A kind of binary search is used. An interesting bookkeeping problem is involved.

4.11 Multiple Integrals

Can also be evaluated numerically by extending the methods for a single integral.

4.12 Multiple Integration with Variable Limits

Shows how to apply the methods to this more complicated type of problem.

4.13 Applications of Cubic Splines

Cubic splines often can give more accurate values for the derivatives and the integral but cannot be used for derivatives higher than the second.

4.14 Theoretical Matters

Discusses several important theoretical items that were used without proof in earlier sections or that merit expansion on the previous treatment.

4.15 Using *Mathematica*

Illustrates the ability of this symbolic algebra program to perform differentiation and integration. The program is able to obtain both derivatives and integrals analytically as well as with numerical methods.

4.16 Parallel Processing

Contains remarks on how the use of processors in parallel can speed the operations of the chapter.

Chapter Summary

Lets you evaluate your understanding of the material.

Computer Programs

Gives examples of how some of the methods of this chapter can be implemented on the computer.

4.1 Getting Derivatives and Integrals Numerically

Rita and Ed were again at lunch in the cafeteria of Ruscon Engineering.

"How did you make out on the interpolation problem?" Ed asked.

"It went all right," Rita replied. "In fact, we have computer programs now to interpolate either with polynomials or with cubic splines. My boss now wants me to work on some extensions."

"What do you mean, extensions?" Ed asked.

"Well, we need the time derivative of the position. That's the velocity, of course. And some calculations that lead to the fuel consumption require us to integrate a function when we know the values only at discrete times."

"I should think that would follow directly from what you did before," Ed said. "If you have a polynomial that gives you the position as a function of time, can't you just differentiate or integrate that polynomial?"

"Very true," Rita replied. "The catch is that we never really develop the polynomial; we just work with the position values and their differences. My boss warns me also that estimating the errors of derivatives and integrals determined from function values that are known only at discrete points in time is pretty tricky. I suspect that there are ways to tackle these problems that are similar to the methods for interpolation."

Rita is right; there are methods for getting derivatives and integrals from a table of function values, and these methods resemble those for interpolation. In this chapter we will explore these methods. These same methods are also useful to estimate the value of the derivative or integral when $f(x)$ is known.

4.2 Derivatives from Difference Tables

We can use a divided-difference table to estimate values for derivatives. Recall that the interpolating polynomial of degree n that fits at points $p_0, p_1, p_3, \ldots, p_n$ is, in terms of divided differences,

$$
\begin{aligned}
f(x) = P_n(x) + \text{error} \\
= f[x_0] + f[x_0, x_1](x - x_0) \\
+ f[x_0, x_1, x_2](x - x_0)(x - x_1) \\
+ \cdots + f[x_0, x_1, \ldots, x_n] \prod_{i=0}^{n-1} (x - x_i) \\
+ \text{error}
\end{aligned}
\tag{4.1}
$$

(We may sometimes use an alternate notation for the divided difference: $f_0^{[n]}$.)

If $P_n(x)$ matches well to $f(x)$, we should get a polynomial that approximates the derivative, $f'(x)$, by differentiating it. Recall that the derivative of a product of n terms is a sum of n of these product terms with one member of each term in the sum replaced by its derivative. For example,

$$
\frac{d}{dx} (u * v * w) = u' * v * w + u * v' * w + u * v * w'
$$

or, for the products in $P_n(x)$ of Eq. (4.1),

$$
\begin{aligned}
\frac{d}{dx} \prod_{i=0}^{n-1} (x - x_i) = \sum_{i=0}^{n-1} \frac{(x - x_0)(x - x_1) \cdots (x - x_{n-1})}{(x - x_i)} \\
= \sum_{i=0}^{n-1} \prod_{\substack{j=0 \\ j \neq i}}^{n-1} (x - x_j).
\end{aligned}
$$

Carrying this out, we get this approximation for $f'(x)$:

$$
\begin{aligned}
P_n'(x) = f[x_0, x_1] + f[x_0, x_1, x_2][(x - x_1) + (x - x_0)] \\
+ \cdots + f[x_0, x_1, \ldots, x_n] \sum_{i=0}^{n-1} \frac{(x - x_0)(x - x_1) \cdots (x - x_{n-1})}{(x - x_i)}.
\end{aligned}
\tag{4.2}
$$

To get the error term for Eq. (4.2), we have to differentiate the error term for $P_n(x)$:

$$\text{Error} = (x - x_0)(x - x_1) \cdots (x - x_n) \frac{f^{(n+1)}(\xi)}{(n + 1)!}. \qquad (4.3)$$

When this error term is differentiated, we will find a sum that has in one of its terms

$$\frac{d}{dx} [f^{(n+1)}(\xi)],$$

which is impossible to evaluate because ξ depends on x in an unknown way. However, if we take $x = x_i$ (where x_i is one of the tabulated points), the difficult term drops out and we get this expression for the error:

Error of the approximation to $f'(x)$, when $x = x_i$, is

$$\text{Error} = \left[\prod_{\substack{j=0 \\ j \neq i}}^{n} (x_i - x_j) \right] \frac{f^{(n+1)}(\xi)}{(n + 1)!}, \qquad \xi \text{ in } [x, x_0, x_n]. \qquad (4.4)$$

Observe that the error is not zero even when x is a tabulated value, although the interpolating polynomial agrees with $f(x)$ at this point. In fact, the error of the derivative is less at some x-values between the points.

It is not surprising that the next-term rule applies here as it did for interpolating polynomials. We use a known polynomial function in the following example to show that the procedure works correctly.

EXAMPLE 4.1   Let $f(x) = x^2 - x + 1$, and tabulate for $x = 0, 2, 3, 5, 6$ (five points). Here is the divided difference table:

i	x_i	f_i	$f_i^{[1]}$	$f_i^{[2]}$	$f_i^{[3]}$	$f_i^{[4]}$
0	0	1	1	1	0	0
1	2	3	4	1	0	
2	3	7	7	1		
3	5	21	10			
4	6	31				

(In this table, we have used our nonstandard notation for the divided differences, thus simplifying the writing of equations that follow.) Observe that the second differences are constant and the third are zero, as we expect for a quadratic polynomial.

What is the estimated value for the derivative if x is 4.1 and we use a cubic interpolating polynomial that starts at $i = 1$? Here is the arithmetic:

$$f'(x) = f_1^{[1]} + f_1^{[2]}[(x - x_2) + (x - x_1)]$$
$$+ f_1^{[3]}[(x - x_2)(x - x_3) + (x - x_1)(x - x_3) + (x - x_1)(x - x_2)]. \qquad (4.5)$$

Substituting values gives

$$f'(4.1) = 4 + 1*[(4.1 - 3) + (4.1 - 2)]$$
$$+ 0*[(4.1 - 3)(4.1 - 5) + (4.1 - 2)(4.1 - 5) + (4.1 - 2)(4.1 - 3)]$$
$$= 4 + 1*(1.1 + 2.1)$$
$$+ 0*[(1.1)(-0.9) + (2.1)(-0.9) + (2.1)(1.1)]$$
$$= 7.2,$$

which is the exact answer, as we expect since $f(x)$ is a quadratic and the cubic between $x = 2$ and $x = 6$ is actually a quadratic. In fact, if we expand Eq. (4.5), we get $f'(x) = 2x - 1$, the analytical expression for the derivative. (If the tabulated function is not a polynomial, we would not get exact answers, of course.)

If we were to use an interpolating polynomial of degree 1 (linear interpolation) to get the derivative polynomial, starting at point $x = x_i$, we would have just

$$f_1'(x) = f[x_i, x_{i+1}] + \text{error} = f_i^{[1]} + \text{error}.$$

For $x = 4.1$, if we take $i = 2$ (the best value, why?), we estimate $f'(4.1)$ as 7. The true value is 7.2, so the error is 0.2. Let us use the next-term rule to estimate the error. The next term is

$$f_2^{[2]}[(x - x_3) + (x - x_2)] = 1*[(4.1 - 5) + (4.1 - 3)] = 0.2,$$

which is precisely the actual error!

As we saw in the last chapter, if we want to estimate the derivative for an x-value near the end of the table, we appear to be severely limited in the degree of interpolating polynomial. We can overcome this limitation by reordering i-values, putting them in reverse order. Our formulas still work correctly, but we must remember to go diagonally upward to get the values for a given value of i.

Since an interpolating polynomial fits better to the function if the x-values used in its construction are such that the x-value for getting the derivative is centered within them, we should choose the starting point (the i-value) to make this true. (If the x-values are in order, our task is easier.)

An Algorithm to Obtain an Estimate of the Derivative from a Divided-Difference Table

Given $n + 1$ data pairs, $(x_i, f_i), i = 0, \ldots, n$:

(Do the table)
DO FOR $i = 0$ TO n STEP 1:
  SET $f(i, 0) = f(i)$,
ENDDO (FOR i).
DO FOR $j = 1$ TO n STEP 1:
  DO FOR $i = 0$ TO $n - j$ STEP 1:
    $f(i, j) = [f(i + 1, j - 1) - f(i, j - 1)]/[x(i + j) - x(i)]$,
  ENDDO (FOR i).

```
ENDDO (FOR j).
(Get user inputs)
INPUT:
   x = x-value,
   i = i-value to start,
   ND = degree of polynomial.
(Compute derivative)
SET SUM = 0.
DO FOR j = ND TO 2 STEP −1:
   SET SUMP = 0.
   DO FOR k = 0 TO j − 1 STEP 1:
      SET p = 1.
      DO FOR l = 0 TO j − 1 STEP 1:
         IF l = k THEN:
            p = p * [x − x(i + 1)]
         ENDIF.
      ENDDO (FOR l).
      SET SUMP = SUMP + p.
   ENDDO (FOR k).
   SET SUM = SUM + f(i, 1).
ENDDO (FOR j).
(Display result)
DISPLAY SUM as derivative value at x.
```

(The computations of the derivative value can be repeated with new user inputs.)

Evenly Spaced Data

When the data are evenly spaced, we can use a table of function differences to construct the interpolating polynomial. We write this in terms of $s = (x - x_i)/h$:

$$P_n(s) = f_i + s\Delta f_i + \frac{s(s-1)}{2!}\Delta^2 f_i + \frac{s(s-1)(s-2)}{3!}\Delta^3 f_i$$

$$+ \cdots + \prod_{j=0}^{n-1}(s-j)\frac{\Delta^n f_i}{n!} + \text{error};$$

$$\text{Error} = \left[\prod_{j=0}^{n}(s-j)\right]\frac{f^{(n+1)}(\xi)}{(n+1)!}, \qquad \xi \text{ in } [x, x_1, \ldots, x_n].$$

(In this formula, i is the index value where we enter the difference table.)

The derivative of $P_n(s)$ should approximate $f'(x)$. We do exactly the same as we did for the polynomial constructed from a divided-difference table, getting

$$\frac{d}{dx} P_n(s) = \frac{d}{ds} P_n(s) \frac{ds}{dx}$$

$$= \frac{1}{h}\left[\Delta f_i + \sum_{j=2}^{n}\left\{ \sum_{k=0}^{j-1} \prod_{\substack{\ell=0 \\ \ell \neq k}}^{j-1} (s - \ell) \right\} \frac{\Delta^j f_i}{j!} \right]. \tag{4.6}$$

(The $1/h$ factor comes from $ds/dx = d/dx\,(x - x_i)/h = 1/h$.)

Again, the error term involves an unknown quantity unless x is one of the tabulated values. When $x = x_i$, $s = 0$. In this case, we get this analog of Eq. (4.4) when an interpolating polynomial of degree n is used:

$$\begin{array}{c} \text{Error} \\ (\text{when } x = x_i) \end{array} = \frac{(-1)^n h^n}{n + 1} f^{(n+1)}(\xi), \qquad \xi \text{ in } [x_1, \ldots, x_n]. \tag{4.7}$$

Equation (4.6) is a formula for estimating derivatives from a table of differences that we enter at index value i. Here is an example (that again uses a known function, $f(x) = e^x$).

EXAMPLE 4.2   Estimate the value of $f'(3.3)$ with a cubic polynomial that is created if we enter the table at $i = 2$, given this difference table:

i	x_i	f_i	Δf_i	$\Delta^2 f_i$	$\Delta^3 f_i$	$\Delta^4 f_i$	$\Delta^5 f_i$
0	1.30	3.669	3.017	2.479	2.041	1.672	1.386
1	1.90	6.686	5.496	4.520	3.713	3.058	2.504
2	2.50	12.182	10.016	8.233	6.771	5.562	
3	3.10	22.198	18.249	15.004	12.333		
4	3.70	40.447	33.253	27.337			
5	4.30	73.700	60.590				
6	4.90	134.290					

For our example, $h = 0.6$ and we take $x_i = 2.5$ (the best choice). With $x = 3.3$, we have $s = (3.3 - 2.5)/0.6 = 4/3$. Here is the polynomial (it will be of degree 2, of course):

$$P_2(x) = \frac{1}{h}\left\{ \Delta f_2 + \frac{1}{2!}\left[\sum_{k=0}^{1} \prod_{\substack{\ell=0 \\ \ell \neq k}}^{1} (s - \ell) \right]\Delta^2 f_2 + \frac{1}{3!}\left[\sum_{k=0}^{2} \prod_{\substack{\ell=0 \\ \ell \neq k}}^{2} (s - \ell) \right]\Delta^3 f_2 \right\}$$

$$= \frac{1}{0.6}\left\{ 10.016 + \frac{1}{2}\left[\left(\frac{4}{3} - 1\right) + \left(\frac{4}{3} - 0\right)\right] * 8.233 \right.$$

$$\left. + \frac{1}{6}\left[\left(\frac{4}{3} - 1\right)\left(\frac{4}{3} - 2\right) + \left(\frac{4}{3} - 0\right)\left(\frac{4}{3} - 2\right) + \left(\frac{4}{3} - 0\right)\left(\frac{4}{3} - 1\right)\right] * 6.771 \right\}$$

$$= 27.875 \qquad (\text{versus } 27.113, \text{ the exact value of } f'(3.3)).$$

If we use the next-term rule to estimate the error of the derivative based on a cubic interpolating polynomial, we get a value of 0.315 compared to the actual error of 0.238. This is not an unreasonable estimate.

▲

Figure 4.1 plots the polynomial that estimates the derivative (the curve that is nearly straight), the true derivative curve, and the points where the interpolating polynomial fits the function. (In this special case, these points fall on the derivative curve because, for $f(x) = e^x$, the function and its derivative are the same.) Observe that the line for the derivative polynomial does not agree with the exact derivative curve at the points where the interpolating polynomial agrees with the function but that there are two places where the derivative polynomial does agree with the exact value for $f'(x)$.

A higher-degree polynomial would estimate the values for the derivative more exactly.

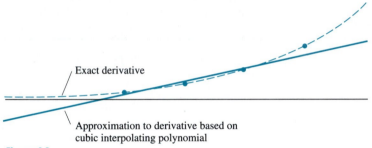

Exact derivative

Approximation to derivative based on cubic interpolating polynomial

Figure 4.1

Simpler Formulas

Equation (4.6) is awkward to use when we do hand computations (but a computer program, though using several nested loops, has no problem). If we stipulate that the x-value must be in the difference table, the computation is simplified considerably. Thus, for an estimate of $f'(x_i)$, we get

$$f'(x) = \frac{1}{h}\left[\Delta f_i - \frac{1}{2}\Delta^2 f_i + \frac{1}{3}\Delta^3 f_i - \cdots \pm \frac{1}{n}\Delta^n f_i \right]_{x=x_i} \qquad (4.8)$$

because, at $x = x_i$, $s = 0$ and

$$\frac{1}{n!}\prod_{j=0}^{n-1}(s - j) = \frac{(-1)^{n-1}}{n}.$$

Equation (4.8) is easy to use for hand computations because the multipliers of the differences are so simple. There are other consequences as well. Suppose we write the error term as an "order of" expression. If we use just one term of Eq. (4.8) (this means that we are linearly interpolating, using a polynomial of degree 1), we have

$$f'(x_i) = \frac{1}{h}[\Delta f_i] - \frac{1}{2}hf''(\xi), \qquad \text{error is } O(h).$$

With two terms,

$$f'(x_i) = \frac{1}{h}\left[\Delta f_i - \frac{1}{2}\Delta^2 f_i\right] + \frac{1}{3}h^2 f^{(3)}(\xi), \qquad \text{error is } O(h^2).$$

In general, with n terms, the error is $O(h^n)$. As already pointed out, we also can estimate the size of the error from the next-term rule.

The derivative formula of Eq. (4.8) is called a *forward-difference approximation* because the differences all involve f-values that lie forward in the table from f_i. Suppose we use a second-degree polynomial that matches the table at x_i, x_{i+1}, and x_{i+2} but evaluate it for $f'(x_{i+1})$, using $s = 1$. Eq. (4.6) then becomes

$$f'(x_{i+1}) = \frac{1}{h}\left[\Delta f_i + \frac{1}{2}\Delta^2 f\right] + O(h^2).$$

Rewriting in terms of the f-values, we have

$$f'(x_{i+1}) = \frac{1}{h}\left[(f_{i+1} - f_i) + \frac{1}{2}(f_{i+2} - 2f_{i+1} + f_i)\right] + \text{error}$$

$$= \frac{1}{h}\frac{f_{i+2} - f_i}{2} + \text{error}, \qquad \text{(4.9)}$$

$$\text{error} = -\frac{1}{6}h^2 f^{(3)}(\xi) = O(h^2).$$

Equation (4.9) is called a *central-difference formula* because the x-value is centered within the range of x-values used in its construction. It is most important to note that we get an improved order of error even though only two function values are involved. The coefficient of the error term is also smaller, an added benefit. This formula is widely used to estimate the derivative both from tabulated values and even when the function is known.

EXAMPLE 4.3   For $f(x) = xe^{-x/2}$, estimate the value of $f'(0.3)$ using $h = 0.1$, 0.05, and 0.025. Compare the results from Eqs. (4.8) and (4.9). In each case, compute the actual error.

We use these two formulas to get the estimates:

From Eq. (4.8) $f'(0.3) = \dfrac{1}{h}[f(0.3 + h) - f(0.3)]$.

From Eq. (4.9) $f'(0.3) = \dfrac{1}{2h}[f(0.3 + h) - f(0.3 - h)]$.

Table 4.1 shows the results.

Table 4.1

h	Estimate of $f'(0.3)$	Error	
0.1	0.69280	−0.0388	
0.05	0.71195	−0.0196	From Eq. (4.8)
0.025	0.72171	−0.0099	
0.1	0.73262	0.00102	
0.05	0.73186	0.00026	From Eq. (4.9)
0.025	0.73167	0.00006	

The results from the forward-difference formula have errors much greater than those from central differences. The former are about halved as h is halved (successive errors are decreased by a factor of 1.98), while the errors from the preferred formula decrease in proportion to h^2 (successive errors are decreased by a factor of 3.98). These errors are also much smaller.

▲

4.3 Higher-Order Derivatives

We can differentiate the formulas of Section 4.2 to obtain formulas for higher-order derivatives, but we prefer to use a simpler method. The *symbolic operator method* was described in Section 3.9, where it was used to derive formulas including the Newton–Gregory forward-interpolating polynomial. We will use it to develop formulas for derivatives of higher order based on evenly spaced data.

Recall that there we used several important operators that were interrelated:

$$\text{Difference operator:} \quad \Delta f(x_0) = f(x_0 + h) - f(x_0),$$
$$\text{Or:} \quad \Delta f(x_i) = \Delta f_i = f_{i+1} - f_i.$$
$$\text{Stepping operator:} \quad E f(x_0) = f(x_0 + h),$$
$$\text{Or:} \quad E f_i = f_{i+1}.$$
$$E^n f_i = f_{i+n}$$
$$\text{Relation between } E \text{ and } \Delta: \quad E = 1 + \Delta.$$

We will also use the symbol D to represent the differentiation operator: $D(f) = df/dx$, $D^n(f) = d^n/dx^n(f)$.

Let us begin with

$$f_{i+s} = E^s f_i,$$

where $s = (x - x_i)/h$. If we take the derivative as if this were an algebraic expression, we get

$$Df_{i+s} = \frac{d}{dx} f(x_{i+s}) = \frac{d}{dx} (E^s f_i)$$

$$= \frac{1}{h} \frac{d}{ds} (E^s f_i) = \frac{1}{h} (\ln E) E^s f_i.$$

If $s = 0$, we get

$$Df_i = f'_i = \frac{1}{h} (\ln E) f_i = \frac{1}{h} \ln(1 + \Delta) f_i, \qquad \textbf{(4.10)}$$

which means that D and Δ are related by

$$D = \frac{1}{h} \ln(1 + \Delta).$$

Now we will use a Maclaurin expansion for $\ln(1 + \Delta)$ to transform Eq. (4.10) to

$$f'_i = \frac{1}{h}\left(\Delta f_i - \frac{1}{2}\Delta^2 f_i + \frac{1}{3}\Delta^3 f_i - \frac{1}{4}\Delta^4 f_i + \cdots \right), \qquad \textbf{(4.11)}$$

which is identical to Eq. (4.8). We have derived the formula for the derivative of $f(x)$ at $x = x_i$ through symbol manipulation. Now the second derivative of $f(x)$ is $f''(x) = D^2 f$, for which we can get the operator by multiplying the operator of Eq. (4.11) by itself. Further, we can get a formula for the nth derivative by multiplying n times!

Here are the results for $D^2 f_i$ if we perform the multiplication:

$$f''_i = \frac{1}{h^2}\left(\Delta^2 f_i - \Delta^3 f_i + \frac{11}{12}\Delta^4 f_i - \frac{5}{6}\Delta^5 f_i + \cdots \right). \qquad \textbf{(4.12)}$$

Formulas for higher derivatives we leave as an exercise. A most important consequence of such expansions is that the first term is

$$D^n f_i = \frac{1}{h^n} \Delta^n f_i, \qquad \textbf{(4.13)}$$

which we have used previously in the next-term rule. An even better approximation to $f^{(n)}(x_i)$ when n is even is $\Delta^n f_{i-n/2}$, which centers x_i within the range of data points.

Divided Differences

If the data are evenly spaced, a divided-difference table is really the same as a table of function differences except that the h-value is included in the entries. This relation holds

$$\frac{\Delta^n f_i}{(h^n)(n!)} = f_i^{[n]} = f[x_i, x_{i+1}, \ldots, x_{i+n}],$$

which means that the nth divided difference and the nth derivative are related by another expression we have used:

$$f_i^{[n]} = f[x_i, x_{i+1}, \ldots, x_{i+n}] = \frac{D^n f(x_i)}{n!} \approx \frac{f^{(n)}(x_i)}{n!}. \tag{4.14}$$

EXAMPLE 4.4   Given Table 4.2, estimate the second derivative at $x = 1.3$ using one, two, and three terms of Eq. (4.12). Also estimate the errors of each.

Table 4.2

i	x_i	f_i	Δf_i	$\Delta^2 f_i$	$\Delta^3 f_i$	$\Delta^4 f_i$	$\Delta^5 f_i$
0	1.10	3.24075	−0.27596	−0.02885	0.00700	0.00039	−0.00020
1	1.30	2.96479	−0.30481	−0.02185	0.00739	0.00019	−0.00021
2	1.50	2.65999	−0.32666	−0.01446	0.00758	−0.00002	
3	1.70	2.33333	−0.34112	−0.00688	0.00756		
4	1.90	1.99221	−0.34800	0.00067			
5	2.10	1.64421	−0.34733				
6	2.30	1.29688					

Using one term,

$$f''(1.3) = \frac{1}{h^2}(\Delta^2 f_1) = \frac{1}{0.04} * (-0.02185) = -0.5463.$$

Using two terms,

$$f''(1.3) = \frac{1}{h^2}(\Delta^2 f_1 - \Delta^3 f_1)$$

$$= \frac{1}{0.04} * (-0.02185 - 0.00739) = -0.7310.$$

Using three terms,

$$f''(1.3) = \frac{1}{h^2}\left(\Delta^2 f_1 - \Delta^3 f_1 + \frac{11}{12}\Delta^4 f_1\right)$$

$$= \frac{1}{0.04} * \left[-0.02185 - 0.00739 + \frac{11}{12}(0.00019)\right]$$

$$= -0.7267.$$

We can estimate the errors in each case from the next-term rule. Doing so gives −0.1848, 0.0044, and 0.0044 for the successive results. (We would expect the errors to decrease in the last two, but round-off errors are beginning to affect the computations.)

The function tabulated in Table 4.2 is $4 * \sin(x)/x$ for which the true value of $f''(1.3)$ is −0.72243. (A symbolic algebra program was of great help in getting this result!) We then compute the actual errors as -0.1761, 0.0086, and 0.0043. The estimates are certainly reasonable, though not exact.

Central-Difference Formulas

In the previous section, a central-difference formula for the derivative was shown to be more accurate than a forward-difference formula. Let us use the symbolic method to derive the central-difference formula.

It is convenient to define a *backward-difference operator*:

$$\nabla f_0 = f_0 - f_{-1}, \qquad \nabla f_i = f_i - f_{i-1},$$

which is related to the stepping operator, E, by

$$E^{-1} = 1 - \nabla.$$

If we use this relation in Eq. (4.10), we get

$$\frac{d}{dx} f(x)_{x=x_i} = f_i' = \frac{1}{h}(\ln E)f_i = \frac{1}{h}\left[\ln\left(\frac{1}{1-\nabla}\right)\right]f_i, \qquad \textbf{(4.15)}$$

which shows that another symbolic relation for the derivative operator, D, is

$$D = \frac{1}{h}\ln\left(\frac{1}{1-\nabla}\right).$$

Expanding the logarithm term gives a formula for the derivative in terms of backward differences:

$$f_i' = \frac{1}{h}\left(\nabla + \frac{1}{2}\nabla^2 + \frac{1}{3}\nabla^3 + \frac{1}{4}\nabla^4 + \cdots\right)f_i. \qquad \textbf{(4.16)}$$

Suppose we take the average of the derivative values given by Eqs. (4.11) and (4.16) by adding the two and dividing by 2:

$$f_i' = \frac{1}{2h}\left[(\Delta + \nabla) + \frac{1}{3}(\Delta^3 + \nabla^3) + \frac{1}{5}(\Delta^5 + \nabla^5) + \cdots\right]f_i. \qquad \textbf{(4.17)}$$

The first term of this formula, rewritten in terms of function values, is

$$f_i' = \frac{(f_{i+1} - f_i) + (f_i - f_{i-1})}{2h} = \frac{1}{h}\frac{f_{i+1} - f_{i-1}}{2},$$

which is our central-difference formula. We can get central-difference formulas for the derivative based on higher-degree interpolating polynomials by using more terms of Eq. (4.17).

As before, we can get the central-difference operators for higher derivatives by multiplying the operator of Eq. (4.17) by itself. It turns out that this technique gives very complicated expressions. Since only the first term of the expression is commonly used, we can take a simpler approach.

We observed that a quadratic polynomial that fits from points $i - 1$ to $i + 1$ has its central point at i [which we can call point (i)]. We used this fact to get a central-difference formula for the first derivative (Eq. 4.9) from the forward-difference formula of Eq. (4.8).

We can take a similar approach for the second derivative. We write Eq. (4.13) for point $(i - 1)$ but interpret it as the second derivative at point (i):

$$f''(x_i) = \frac{1}{h^2} (\Delta^2 f_{i-1}) = \frac{f_{i+1} - 2f_i + f_{i-1}}{h^2}. \qquad (4.18)$$

Equation (4.18) gives an approximation of error $O(h^2)$.

EXAMPLE 4.5   Apply Eq. (4.18) to estimate $f''(1.3)$ (i.e., at $i = 1$) from Table 4.2.
We have

$$f''(1.3) = \frac{1}{h^2} \Delta^2 f_0 = \frac{-0.02885}{0.2^2} = -0.7213,$$

which is much closer to the true value of -0.72243 than even with three terms of the forward-difference formula used in Example 4.4.

▲

There are other ways to obtain formulas for the various derivatives from a difference table. Appendix B explains the method of undetermined coefficients. Earlier editions of this book described the use of a "lozenge diagram," a device to determine the coefficients in formulas such as Eq. (4.17) but with more terms. These diagrams can be built up from one known formula such as Eq. (4.8) or (4.12) to give a device that works for any desired formula, including some that we have not discussed here.

An Algorithm to Compute First and Second Derivatives from Central-Difference Formulas

Given a function $f(x)$:

(Get user inputs)
INPUT:
   x = x-value,
   h = delta-x value.
(Compute derivatives)
SET $d_1 = [f(x + h) - f(x - h)]/(2h)$.
SET $d_2 = [f(x + h) - 2f(x) + f(x - h)]/h^2$.
(Display results)
DISPLAY d_1 and d_2 as estimates of $f'(x), f''(x)$.

(The computations of the derivative values can be repeated with new user inputs.)

4.4 Extrapolation Techniques

There is another way to improve the accuracy of the estimates of derivatives from a table of evenly spaced values. This technique is equivalent to using formulas based on higher-degree polynomials without explicitly finding the formula. This method is best explained through an example.

EXAMPLE 4.6 Estimate the value of $f'(2.4)$ and $f''(2.4)$ from Table 4.3, for which the x's are evenly spaced.

Table 4.3

i	x_i	f_i
0	2.0	0.123060
1	2.1	0.105706
2	2.2	0.089584
3	2.3	0.074764
4	2.4	0.061277
5	2.5	0.049126
6	2.6	0.038288
7	2.7	0.028722
8	2.8	0.020371
9	2.9	0.013164
10	3.0	0.007026

We will use central-difference formulas, Eqs. (4.9) and (4.18), both of which have an error of $O(h^2)$.

We begin with computations for $f'(2.4)$. Eq. (4.9) gives

$$f'(2.4) = \frac{f(2.5) - f(2.3)}{2(0.1)} = \frac{0.0491256 - 0.0747636}{0.2},$$ (4.19)

$$f'(2.4)[\text{exact}] = -0.12819 + C_1(0.1)^2,$$

where we show the exact value as the estimate plus the error, a quantity that is proportional to h^2. (C_1 is the proportionality constant.)

Suppose we repeat the calculation, this time using $h = 0.2$. (We can't make h smaller using only a table of values. It may seem that we are losing ground, but be patient.) This computation gives

$$f'(2.4) = \frac{f(2.6) - f(2.2)}{2(0.2)} = \frac{0.038288 - 0.089584}{0.4},$$ (4.20)

$$f'(2.4)[\text{exact}] = -0.12824 + C_2(0.2)^2.$$

The values of C_1 and C_2 in Eqs. (4.19) and (4.20) are not usually identical, but we will assume they are the same. (Each of the C's actually involves the value of $f^{(3)}(\xi)$, where the ξ's are not identical but should not differ by much. It can be shown that we make an error of $O(h^4)$ in taking the two values as equal.) Based on this assumption that the C's are equal, we can solve the two equations to eliminate the C's:

$$f'(2.5)[\text{exact}] = -0.12819 + \frac{1}{(0.2/0.1)^2} - 1\,[-0.12819 - (-0.12824)]$$

$$= -0.12817 + O(h^4).$$

The computation is an instance of a general rule: Given two estimates of a value that have errors of $O(h^n)$, where the h's are in the ratio of 2 to 1, we can extrapolate to a better estimate of the exact value as follows:

Better estimate = more accurate

$$+ \frac{1}{2^n - 1} (\text{more accurate} - \text{less accurate}). \quad \textbf{(4.21)}$$

(The more accurate value is the one computed with the smaller value for h.)

We can repeat the scheme using h-values of 0.2 and 0.4. Table 4.4 shows the result as "first-order extrapolations." The first-order extrapolations have errors of $O(h^4)$ so the second-order extrapolation comes from

$$f'(2.4)[\text{exact}] = -0.12817 + \frac{1}{2^4 - 1}[-0.12817 - (-0.12820)]$$

$$= -0.12817 + O(h^6).$$

Table 4.4

h	Initial estimate	First-order extrapolation	Second-order extrapolation
0.1	−0.12819		
		−0.12817	
0.2	−0.12824		−0.12817
		−0.12820	
0.4	−0.12836		

Since there was no change in the fifth decimal place from the first extrapolation, we are tempted to say that the improved estimate is good to that many places. This conclusion is correct, except that round-off in the original data can affect the results and invalidate our conclusion. [The data were constructed from $f(x) = e^{-x}\sin(x)$, and $f'(2.4)$ is −0.128171.)

Second-Derivative Computations

We proceed in exactly the same manner to get an improved estimate of $f''(2.4)$. The initial estimates are as follows:

For $h = 0.1$,

$$f''(2.4) = \frac{0.049126 - 2(0.061277) + 0.074764}{(0.1)^2} = 0.13360.$$

For $h = 0.2$,

$$f''(2.4) = \frac{0.038288 - 2(0.061277) + 0.089584}{(0.2)^2} = 0.13295.$$

For $h = 0.4$,

$$f''(2.4) = \frac{0.020371 - 2(0.061277) + 0.123060}{(0.4)^2} = 0.13048.$$

Table 4.5 shows these values in the initial estimates column. The first- and second-order improvements are obtained from these initial estimates in exactly the same way as for the first derivatives in Table 4.4.

Table 4.5

h	Initial estimate	First-order extrapolation	Second-order extrapolation
0.1	0.13360		
		0.13382	
0.2	0.13295		0.13382
		0.13377	
0.4	0.13048		

Again we are tempted to say that $f''(2.4)$ equals 0.13382 to five decimal places, but in this instance round-off errors have made the conclusion not true. [The exact value of $f''(2.4)$ is 0.13379.]

Additional extrapolations could be made, using the fact that the error of the second-order extrapolation is $O(h^6)$, but it is not recommended to go that far because of round-off effects.

Richardson Extrapolations

We can apply this same technique when we want to differentiate a known function numerically. In this application, known as Richardson's extrapolation method, we can make the h-values smaller rather than use larger values as is required when the function is known only as a table.

We begin at some arbitrarily selected value of h and compute $f'(x)$ from

$$f'(x) = \frac{f(x + h) - f(x - h)}{2h}.$$

We then compute a second value for $f'(x)$ with h half as large. From these two computations, we extrapolate using Eq. (4.21). This improved value has an error of $O(h^4)$. Normally, one builds a table by continuing with higher-order extrapolations with the h-value halved at each stage.

EXAMPLE 4.7   Build a Richardson table for $f(x) = x^2 \cos(x)$ to evaluate $f'(1)$. Start with $h = 0.1$. Repeat for $f'(2)$.

Tables 4.6 and 4.7 show the results. The exact answers are $f'(1) = 0.2391336$ and $f'(2) = -5.3017770$. The Richardson technique indicates convergence when two successive values on the same line are the same. Because single precision was used in computing these tables, convergence is to values slightly different from the exact values, to $f'(1) = 0.239132$ and $f'(2) = -5.301808$.

Table 4.6   Richardson table starting with $h = 0.1$ for derivative at $x = 1$

0.226736			
0.236031	0.239129		
0.238358	0.239133	0.239134	
0.238938	0.239132	0.239132	0.239132

Table 4.7   Richardson table starting with $h = 0.1$ for derivative at $x = 2$

-5.296478			
-5.300449	-5.301773		
-5.301454	-5.301789	-5.301790	
-5.301719	-5.301807	-5.301808	-5.301808

An Algorithm to Compute a Table of Richardson Extrapolations for Computing the Derivative

Given a function, $f(x)$:

(Get user inputs)
INPUT
    x = x-value,
    h = h-value to start,
    MAXST = maximum number of stages,

TOL = tolerance value for termination.
$d(0, 1) = 0$
(Compute lines of the table)
DO FOR ST = 0 TO MAXST STEP 1:
  SET $d(ST, 0) = [f(x + h) - f(x - h)]/(2h)$.
  DO FOR j = 1 TO ST STEP 1:
    SET $d(ST, j) = d(ST, j - 1) + [d(ST, j - 1)$
      $- d(ST - 1, j - 1)]/(2^{2j} - 1)$,
  ENDDO (FOR j).
  IF $|d(ST, ST) - d(ST, ST - 1)|$ < TOL THEN:
    EXIT
  ENDIF.
  SET $h = h/2$.
ENDDO (FOR ST).

On termination, the last computed value is the extrapolated estimate of the derivative.

For convenience, we collect formulas for computing derivatives here.

Formulas for Computing Derivatives

Formulas for the first derivative:

$$f'(x_0) = \frac{f_1 - f_0}{h} + O(h)$$

$$f'(x_0) = \frac{f_1 - f_{-1}}{2h} + O(h^2) \qquad \text{Central difference}$$

$$f'(x_0) = \frac{-f_2 + 4f_1 - 3f_0}{2h} + O(h^2)$$

$$f'(x_0) = \frac{-f_2 + 8f_1 - 8f_{-1} + f_{-2}}{12h} + O(h^4) \quad \text{Central difference}$$

Formulas for the second derivative:

$$f''(x_0) = \frac{f_2 - 2f_1 + f_0}{h^2} + O(h)$$

$$f''(x_0) = \frac{f_1 - 2f_0 + f_{-1}}{h^2} + O(h^2) \qquad \text{Central difference}$$

$$f''(x_0) = \frac{-f_3 + 4f_2 - 5f_1 + 2f_0}{h^2} + O(h^2)$$

$$f''(x_0) = \frac{-f_2 + 16f_1 - 30f_0 + 16f_{-1} - f_{-2}}{12h^2} + O(h^4) \quad \text{Central difference}$$

Formulas for the third derivative:

$$f'''(x_0) = \frac{f_3 - 3f_2 + 3f_1 - f_0}{h^3} + O(h)$$

$$f'''(x_0) = \frac{f_2 - 2f_1 + 2f_{-1} - f_{-2}}{2h^3} + O(h^2) \quad \text{Averaged difference}$$

Formulas for the fourth derivative:

$$f^{iv}(x_0) = \frac{f_4 - 4f_3 + 6f_2 - 4f_1 + f_0}{h^4} + O(h)$$

$$f^{iv}(x_0) = \frac{f_2 - 4f_1 + 6f_0 - 4f_{-1} + f_{-2}}{h^4} + O(h^2) \quad \text{Central difference}$$

4.5   Newton–Cotes Integration Formulas

The usual strategy in developing formulas for numerical integration is similar to that for numerical differentiation. We pass a polynomial through points defined by the function, and then integrate this polynomial approximation to the function. This permits us to integrate a function known only as a table of values. When the values are equispaced, our familiar Newton–Gregory forward polynomial is a convenient starting point, so

$$\int_a^b f(x)\, dx \doteq \int_a^b P_n(x_s)\, dx. \tag{4.22}$$

The formula we get from Eq. (4.22) will not be exact because the polynomial is not identical with $f(x)$. We get an expression for the error by integrating the error term of $P_n(x_s)$:

$$\text{Error} = \int_a^b \binom{s}{n+1} h^{n+1} f^{(n+1)}(\xi)\, dx.$$

There are various ways that we can employ Eq. (4.22). The interval of integration (a, b) can match the range of fit of the polynomial, (x_0, x_n). In this case, we get the Newton–Cotes formulas; these are a set of integration rules corresponding to the varying degrees of the interpolating polynomial. The first three, with the degree of the polynomial 1, 2, or 3, are particularly important, and we discuss them at length in the following sections.

If the degree of the polynomial is too high, errors due to round-off and local irregularities can cause a problem. This explains why it is only the lower-degree Newton–Cotes formulas that are often used.

The range of the polynomial and the interval of integration do not have to be the same. If the interval of integration extends outside the range of fit, however, we are extrapolating, and this incurs larger errors. If we only desire to get the integral of a known function, we will normally avoid extrapolation. As we will see in the next chapter, however, integrating the polynomial outside its range of fit leads to some important methods for solving differential equations. Using an interval of integration that is a subset of the points at which the polynomial agrees with the function also has special application in solving differential equations numerically.

The utility of numerical integration extends beyond the need to integrate a function known only as a table of values. Most computer programs for integration of functions whose form is known use these numerical techniques rather than the analytical methods of the calculus. Although programs like DERIVE and *Mathematica* do use analytical procedures, in a large number of cases no closed form for the integral exists. Numerical integration applies regardless of the complexity of the integrand or the existence of a closed form for the integral.

Let us now develop our three important Newton–Cotes formulas. During the integration, we will need to change the variable of integration from x to s, since our polynomials are expressed in terms of s. Observe that $dx = h\,ds$.

For $n = 1$,

$$\int_{x_0}^{x_1} f(x)\,dx \doteq \int_{x_0}^{x_1} (f_0 + s\Delta f_0)\,dx = h\int_{s=0}^{s=1} (f_0 + s\Delta f_0)\,ds$$

$$= hf_0 s\Big]_0^1 + h\Delta f_0 \frac{s^2}{2}\Big]_0^1 = h\left(f_0 + \frac{1}{2}\Delta f_0\right)$$

$$= \frac{h}{2}[(2f_0 + (f_1 - f_0)] = \frac{h}{2}(f_0 + f_1). \tag{4.23}$$

$$\text{Error} = \int_{x_0}^{x_1} \frac{s(s-1)}{2} h^2 f''(\xi)\,dx = h^3 f''(\xi_1)\int_0^1 \frac{s^2 - s}{2}\,ds$$

$$= h^3 f''(\xi_1)\left(\frac{s^3}{6} - \frac{s^2}{4}\right)\Big]_0^1 = -\frac{1}{12} h^3 f''(\xi_1), \qquad x_0 < \xi_1 < x_1. \tag{4.24}$$

The details of getting the error term here and below are spelled out in a later section.

For $n = 2$,

$$\int_{x_0}^{x_2} f(x)\,dx \doteq \int_{x_0}^{x_2}\left(f_0 + s\Delta f_0 + \frac{s(s-1)}{2}\Delta^2 f_0\right)dx$$

$$= h\int_0^2 \left(f_0 + s\Delta f_0 + \frac{s(s-1)}{2}\Delta^2 f_0\right)ds$$

$$= hf_0 s\Big]_0^2 + h\Delta f_0 \frac{s^2}{2}\Big]_0^2 + h\Delta^2 f_0\left(\frac{s^3}{6} - \frac{s^2}{4}\right)\Big]_0^2$$

$$= h\left(2f_0 + 2\Delta f_0 + \frac{1}{3}\Delta^2 f_0\right) = \frac{h}{3}(f_0 + 4f_1 + f_2), \tag{4.25}$$

with an error term of

$$\text{Error} = -\frac{1}{90} h^5 f^{iv}(\xi), \qquad x_0 < \xi < x_2. \tag{4.26}$$

Similarly, for $n = 3$, we find

$$\int_{x_0}^{x_3} f(x)\,dx = \int_{x_0}^{x_3} P_3(x_s)\,dx = \frac{3h}{8}(f_0 + 3f_1 + 3f_2 + f_3). \tag{4.27}$$

$$\text{Error} = -\frac{3}{80} h^5 f^{iv}(\xi_1), \qquad x_0 < \xi_1 < x_3. \tag{4.28}$$

In summary, the basic Newton–Cotes formulas are

$$\int_{x_0}^{x_1} f(x)\,dx = \frac{h}{2}(f_0 + f_1) - \frac{1}{12} h^3 f''(\xi),$$

$$\int_{x_0}^{x_2} f(x)\,dx = \frac{h}{3}(f_0 + 4f_1 + f_2) - \frac{1}{90} h^5 f^{iv}(\xi),$$

$$\int_{x_0}^{x_3} f(x)\,dx = \frac{3h}{8}(f_0 + 3f_1 + 3f_2 + f_3) - \frac{3}{80} h^5 f^{iv}(\xi).$$

An important item to observe is that the error terms for both $n = 2$ and $n = 3$ are $O(h^5)$. This means that the error of integration using a quadratic is similar to the integral using a cubic; it is a consequence of finding the integral of the next term equal to zero in deriving the error term of Eq. (4.26). Note also that the coefficient in Eq. (4.26)—$(-\frac{1}{90})$—is smaller than that in Eq. (4.28)—$(-\frac{3}{80})$. The formula based on a quadratic is unexpectedly more accurate.

This phenomenon is true of all the even-order Newton–Cotes formulas; each has an order of h in its error term the same as for the formula of next higher order. This fact suggests that the even-order rules are especially useful.

4.6 The Trapezoidal Rule—A Composite Formula

The first of the Newton–Cotes formulas, based on approximating $f(x)$ on (x_0, x_1) by a straight line, is also called the *trapezoidal rule*. We have derived it by integrating $P_1(x_s)$, but the familiar and simple trapezoidal rule can also be considered to be an adaptation of the definition of the definite integral as a sum. To evaluate $\int_a^b f(x)\,dx$, we subdivide the interval from a to b into n subintervals, as in Fig. 4.2. The area under the curve in each subinterval is approximated by the trapezoid formed by replacing the curve by its secant line drawn between the endpoints of the curve. The integral is then approximated by the sum of all the trapezoidal areas. (If we found the limiting value of this sum as the widths of the intervals approach zero, we would

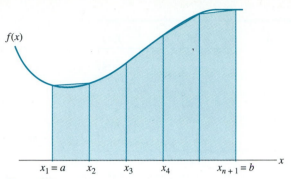

Figure 4.2

have the exact value of the integral, but in numerical integration the widths are finite.) There is no necessity to make the subintervals equal in width, but our formula is simpler if this is done. Let h be the constant Δx. Since the area of a trapezoid is its average height times the base, for each subinterval,

$$\int_{x_i}^{x_{i+1}} f(x)\,dx \doteq \frac{f(x_i) + f(x_{i+1})}{2}(\Delta x) = \frac{h}{2}(f_i + f_{i+1}), \tag{4.29}$$

and for $[a, b]$ subdivided into n subintervals of size h,

$$\int_a^b f(x)\,dx \doteq \sum_{i=1}^{n} \frac{h}{2}(f_i + f_{i+1}) = \frac{h}{2}(f_1 + f_2 + f_2 + f_3 + \cdots + f_n + f_{n+1});$$

$$\int_a^b f(x)\,dx = \frac{h}{2}(f_1 + 2f_2 + 2f_3 + \cdots + 2f_n + f_{n+1}). \tag{4.30}$$

Equation (4.29) is identical to Eq. (4.23). Equation (4.30) is called the *composite trapezoidal rule:* It lets us apply the formula over an extended region where $f(x)$ is far from linear by applying the procedure to subintervals in which it can be approximated by linear segments. The formula is beautifully simple, and its applicability to unequally spaced values is useful in finding the integral of an experimentally determined function. It is obvious from Fig. 4.2 that the method is subject to large errors unless the subintervals are small, for replacing a curve by a straight line is hardly accurate.

EXAMPLE 4.8   Suppose we wished to integrate the function tabulated in Table 4.8 over the interval from $x = 1.8$ to $x = 3.4$. The composite trapezoidal rule gives

$$\int_{1.8}^{3.4} f(x)\,dx = \frac{0.2}{2}[6.050 + 2(7.389) + 2(9.025) + 2(11.023) + 2(13.464)$$

$$+ 2(16.445) + 2(20.086) + 2(24.533) + 29.964] = 23.9944.$$

The data in Table 4.8 are for $f(x) = e^x$, so the true value of the integral is $e^{3.4} - e^{1.8} = 23.9144$. We are off in the second decimal.

Table 4.8

x	$f(x)$	x	$f(x)$
1.6	4.953	2.8	16.445
1.8	6.050	3.0	20.086
2.0	7.389	3.2	24.533
2.2	9.025	3.4	29.964
2.4	11.023	3.6	36.598
2.6	13.464	3.8	44.701

We have previously shown the error of the trapezoidal rule as Eq. (4.24). We repeat it here:

$$\text{Local error of trapezoidal rule} = -\frac{1}{12}h^3 f''(\xi_1), \qquad x_0 < \xi_1 < x_1.$$

This error, it should be emphasized, is the error of only a single step, and is hence called the *local error*. We normally apply the trapezoidal formula to a series of subintervals to get the integral over a large interval from $x = a$ to $x = b$. We are interested in the total error, which is called the *global error*.

To develop the formula for global error of the trapezoidal rule, we note that it is the sum of the local errors:

$$\text{Global error} = -\frac{1}{12}h^3[f''(\xi_1) + f''(\xi_2) + \cdots + f''(\xi_n)]. \qquad \textbf{(4.31)}$$

In Eq. (4.31) each of the values of ξ_i is found in the n successive subintervals. If we assume that $f''(x)$ is continuous on (a, b), there is some value of x in (a, b)—say, $x = \xi$—at which the value of the sum in Eq. (4.31) is equal to $n \cdot f''(\xi)$. Since $nh = b - a$, the global error becomes

$$\text{Global error of trapezoidal rule} = -\frac{1}{12}h^3 n f''(\xi) = \frac{-(b-a)}{12}h^2 f''(\xi) = O(h^2). \qquad \textbf{(4.32)}$$

The fact that the global error is $O(h^2)$ while the local error is $O(h^3)$ is reasonable, since, for example, if h is halved the number of subintervals is doubled, so we add together twice as many errors.

When the function $f(x)$ is known, Eq. (4.32) permits us to estimate the error of numerical integration by the trapezoidal rule. In applying this equation we bracket the error by calculating with the maximum and the minimum values of $f''(x)$ on the interval $[a, b]$.

For this example, our error expression gives these estimates:

$$\text{Error} = -\frac{1}{12} h^3 n f''(\xi), \qquad 1.8 \le \xi \le 3.4,$$

$$= -\frac{1}{12}(0.2)^3(8) \begin{Bmatrix} e^{1.8} & (\min) \\ e^{3.4} & (\max) \end{Bmatrix} = \begin{Bmatrix} -0.0323 & (\min) \\ -0.1598 & (\max) \end{Bmatrix}.$$

Alternatively,

$$\text{Error} = -\frac{1}{12}(0.2)^2(3.4 - 1.8) \begin{Bmatrix} e^{1.8} & (\min) \\ e^{3.4} & (\max) \end{Bmatrix} = \begin{Bmatrix} -0.0323 \\ -0.1598 \end{Bmatrix}.$$

The actual error was -0.080.

If we had not known the form of the function for which we have tabulated values, we would have estimated $h^2 f''(\xi)$ from the second differences.

An Algorithm for Composite Trapezoidal Rule Integration

Given a function $f(x)$:

(Get user inputs)
INPUT:
   a, b = endpoints of interval,
   n = number of intervals.
(Do the integration)
SET $h = (b - a)/n$.
SET SUM = 0.
DO FOR $i = 1$ TO $n - 1$ STEP 1:
   SET $x = a + h * i$,
   SET SUM = SUM + $2 * f(x)$
ENDDO (FOR i).
SET SUM = SUM + $f(a) + f(b)$.
SET ANS = SUM $* h/2$.

The value of the integral is given by ANS.

Romberg Integration

We can improve the accuracy of the trapezoidal rule integral by a technique that is similar to Richardson extrapolation. This technique is known as *Romberg integration*.

Since the integral determined with the trapezoidal method has an error of $O(h^2)$, we can combine two estimates of the integral that have h-values in a $2:1$ ratio by Eq. (4.21), which we repeat here:

$$\text{Better estimate} = \text{more accurate} + \frac{1}{2^n - 1}(\text{more accurate} - \text{less accurate}). \quad (4.33)$$

When we apply this equation to get the integral of a known function, we begin with an arbitrary value for h in Eq. (4.30). A second estimate is then made with the value of h halved. From these two estimates we extrapolate to get an improved estimate using Eq. (4.33). This has an error of $O(h^4)$.

Obviously, this can be extended to produce a table of successively better estimates. When we find that the values converge, we have the best estimate that we can make in the light of round-off error. As shown in Section 4.14, each new extrapolation has error orders that increase: $O(h^4), O(h^6), O(h^8), \ldots$.

We can reduce the number of computations because, when h is halved, all of the old points at which the function was evaluated to get Eq. (4.30) appear in the new computation and we thus can avoid repeating the evaluations. Figure 4.3 illustrates this point.

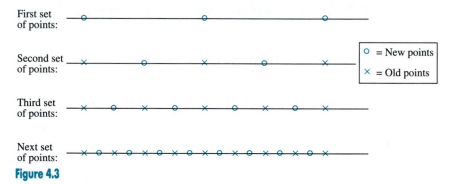

Figure 4.3

This next example shows how the *Romberg table* appears for the function $f(x) = e^{-x^2}$ integrated between the limits of 0.2 and 1.5. This integral has no closed form solution. It is closely related to the *error function*, a quantity that is so important in statistics and other branches of applied mathematics that values have been tabulated.

EXAMPLE 4.9   Use Romberg integration to find the integral of e^{-x^2} between the limits of $a = 0.2$ and $b = 1.5$. Take the initial subinterval size as $h = (b - a)/2 = 0.65$.

Our first estimate is

$$\text{Integral} = \frac{h}{2}[f(a) + 2f(a + h) + f(b)]$$

$$= \frac{0.65}{2}[e^{-0.2^2} + 2e^{-0.85^2} + e^{-1.5^2}]$$

$$= 0.66211.$$

The next estimate uses $h = 0.65/2 = 0.325$:

$$\text{Integral} = \frac{h}{2}[f(a) + 2f(a + h) + 2f(a + 2h) + 2f(a + 3h) + f(b)]$$

$$= \frac{0.325}{2}[e^{-0.2^2} + 2e^{-0.525^2} + 2e^{-0.85^2} + 2e^{-1.175^2} + e^{-1.5^2}]$$

$$= 0.65947.$$

Observe that only two new function evaluations appear in the second estimate. We now extrapolate:

$$\text{Improved} = 0.65947 + \frac{1}{3}[0.65947 - 0.66211]$$

$$= 0.65859.$$

Table 4.9 exhibits the calculations when we repeat the estimations, halving the h-value each time.

Table 4.9   Romberg table of integrals over interval
from 0.2 to 1.5 with an initial $h(O)$ of 0.65

0.66211			
0.65947	0.65859		
0.65898	0.65881	0.65882	
0.65886	0.65882	0.65882	0.65882

The Romberg Method for a Tabulated Function

We can apply the Romberg method to integrate a function known only as a table of evenly spaced function values, but now we cannot make h smaller. Instead, we use estimates of the integral with h doubled each time, just as we did to improve the estimates of derivatives in Section 4.4. Here is an example.

EXAMPLE 4.10   Use the data of Table 4.8 to get the integral between the limits of $x = 1.8$ and $x = 3.4$. Begin with $h = 0.4$.

Our first estimate is

$$\text{Integral} = \frac{0.4}{2}[6.050 + 2(9.025) + 2(13.464) + 2(20.086) + 29.964]$$

$$= 24.2328.$$

We now use $h = 0.8$ to get the next estimate:

$$\text{Integral} = \frac{0.8}{2}[6.050 + 2(13.464) + 29.964]$$

$$= 25.1768.$$

We then extrapolate with Eq. (4.32):

$$\text{Improved} = 24.2328 + \frac{1}{3}[24.2328 - 25.1768]$$
$$= 23.9181.$$

If we had started with $h = 0.2$, we would have had the results shown in Table 4.10, where an additional extrapolation is made to give the value 23.9147. (The exact value of the integral is 23.9144. Round-off errors make it impossible to reach the exact value from these tabulated function values.)

Table 4.10 Romberg table of integrals over interval from 1.8 to 3.4 from tabulated values

$h = 0.2$	23.9944	23.9149	23.9147
$h = 0.4$	24.2328	23.9181	
$h = 0.8$	25.1768		

The Romberg method is applicable to a wide class of functions—to any Riemann–integrable function, in fact. Smoothness and continuity are not required.

An Algorithm for Romberg Integration

Given a function, $f(x)$:

(Get user inputs)
INPUT:
  a, b = endpoints of interval,
  NST = number of stages.
(Do the integration)
SET $h = (b - a)/2$.
(Do the lines of the table)
SET SUM $= f(a) + 2*f(a + h) + f(b)$.
SET NTGRL$(0, 0) = $ SUM$*h/2$. (First value)
SET $d = 2*h$. (d is distance between added points)
DO FOR ST $= 1$ TO NST STEP 1:
  SET $h = h/2$.
  SET $d = d/2$.
  DO FOR $i = 1$ TO 2ST STEP 1:
    SET $x = a - h + d*i$.
    SET SUM $= $ SUM $+ 2*f(x)$.
  ENDDO (FOR i).
  SET NTGRL(ST, 0) $= $ SUM$*h/2$.

(Now extrapolate)
    DO FOR j = 1 TO ST STEP 1:
        SET NTGRL(ST, j) = NTGRL(ST, j − 1) +
            [NTGRL(ST, j − 1) − NTGRL(ST − 1, j − 1)]/(2^j − 1).
    ENDDO (FOR j).
ENDDO (FOR ST).

The last computed value is the estimate of the integral.

An alternative stopping criterion is when two successive computations in a line differ by less than some tolerance value.

4.7 Simpson's Rules

The composite Newton–Cotes formulas based on quadratic and cubic interpolating polynomials are known as *Simpson's rules*. The first, based on a quadratic, is known as *Simpson's $\frac{1}{3}$ rule* and that based on a cubic is known as *Simpson's $\frac{3}{8}$ rule*. These names come from the coefficients of the formulas.

Simpson's $\frac{1}{3}$ Rule

The second-degree Newton–Cotes formula integrates a quadratic over two intervals of equal width. (We shall call these intervals *panels*.) We build the composite rule from Eq. (4.25), which we repeat here:

$$f(x)\,dx = \frac{h}{3}[f_0 + 4f_1 + f_2]. \tag{4.34}$$

This very popular formula has a local error of $O(h^5)$:

$$\text{Error} = -\frac{h^5}{90}f^{(4)}(\xi).$$

We build upon this formula to get a composite rule that is applied to a subdivision of the interval of integration into n panels (n must be even):

$$f(x)\,dx = \frac{h}{3}[f(a) + 4f_1 + 2f_2 + 4f_3 + 2f_4 + \cdots + 4f_{n-1} + f(b)] \tag{4.35}$$

with an error term of

$$\text{Error} = -\frac{(b-a)}{180}h^4f^{(4)}(\xi).$$

The order of the global error changes to $O(h^4)$. The denominator in the error term changes to 180 because we integrate over pairs of panels (meaning that the local rule is applied $n/2$ times). The fact that the error is $O(h^4)$ is of special importance.

EXAMPLE 4.11 Use Simpson's $\frac{1}{3}$ rule to evaluate the integral of e^{-x^2} over the interval 0.2 to 1.5, using $2, 4, 6, \ldots$ subdivisions until the values converge to five decimal places.

Table 4.11 shows the results. The exact answer to five decimal places is 0.65882.

Table 4.11

n	Estimate of integral
2	0.65181
4	0.65860
6	0.65878
8	0.65881
10	0.65882

If we compare these values to those in the Romberg table in Table 4.9, we will see that the result with four panels agrees with the first Romberg extrapolation and that the result with eight panels agrees with the second Romberg extrapolation. We then see that doing a Romberg extrapolation of results from the trapezoidal rule gives an identical result to the Simpson $\frac{1}{3}$ rule integral with the same number of panels. For comparison, the trapezoidal rule without extrapolation requires about 50 subintervals to get five-decimal-place accuracy.

We can do a Romberg-type extrapolation to the Simpson's rule values of Table 4.11. Since the errors are of $O(h^4)$, we compute from the results with $n = 4$ and 8:

$$\text{Improved} = \text{more accurate} + \frac{1}{15}[\text{more} - \text{less}]$$

$$= 0.65881 + \frac{1}{15}[0.65881 - 0.65860] = 0.65882,$$

which is the correct value to five decimal places. This extrapolated result has an error of $O(h^6)$, so the next extrapolation would have a multiplier in the correction term of 1/63.

Algorithm for Simpson's $\frac{1}{3}$ Rule Integration

Given a function, $f(x)$:

(Get user inputs)
INPUT:
    a, b = endpoints of interval.
    n = number of intervals (n must be even).

(Do the integration)
SET $h = (b - a)/n$.
SET SUM = 0.
DO FOR $i = 1$ TO $n/2$ STEP 1:
    SET $x = a - h + 2*h*i$,
    SET SUM = SUM + $4*f(x)$,
    IF $i \neq n/2$ THEN:
        SET SUM = SUM + $2*f(x + h)$
    ENDIF.
ENDDO (FOR i).
SET SUM = SUM + $f(a) + f(b)$.
SET ANS = SUM $* h/3$.

The value of the integral is given by ANS.

Using Tabulated Values

We can apply Simpson's $\frac{1}{3}$ rule to a table of evenly spaced function values in an obvious way. If the number of intervals from the table is not even, we can do a subinterval at one end or the other with the trapezoidal rule and the rest with Simpson's rule. (We should select the end subinterval for applying the trapezoidal rule where the function is more nearly linear.)

EXAMPLE 4.12   Apply Simpson's $\frac{1}{3}$ rule to the data of Table 4.12. What difference results if we apply the trapezoidal rule at the left end rather than the right end?

Table 4.12

x	$f(x)$
0.7	0.64835
0.9	0.91360
1.1	1.16092
1.3	1.36178
1.5	1.49500
1.7	1.55007
1.9	1.52882
2.1	1.44513

If we use the values from $x = 0.7$ to $x = 1.9$ (six panels) in Simpson's $\frac{1}{3}$ rule, we get 1.51938 for this integral. We add to this the trapezoidal integral from $x = 1.9$ to $x = 2.1$, which is 0.29740, to get 1.81678 for the integral from $x = 0.7$ to $x = 2.1$.

Conversely, if we select the first interval to integrate by the trapezoidal rule and the remainder by Simpson's $\frac{1}{3}$ rule, we add 0.15620 to 1.66142 to get 1.81762 for the integral over the entire range.

The exact value for the integral over the entire range is 1.81759; the second choice is the better one. Our criterion seems to work for choosing the end for applying the trapezoidal rule. (Looking at the first differences shows that the function is more nearly linear at the start.)

Simpson's $\frac{3}{8}$ Rule

The composite rule based on fitting four points with a cubic leads to Simpson's $\frac{3}{8}$ rule. We start with Eq. (4.27), which integrates over three panels:

$$
\begin{aligned}
f(x)\,dx &= \frac{3h}{8}\,[f_0 + 3f_1 + 3f_2 + f_3] \\
&\quad - \frac{3}{80}\,h^5 f^{(4)}(\xi).
\end{aligned}
\tag{4.36}
$$

(We remark again that the local order of error, $O(h^5)$, is the same as for Simpson's $\frac{1}{3}$ rule and that the coefficient is larger.)

We apply this to sets of triple panels to get the composite rule:

$$
\begin{aligned}
f(x)\,dx &= \frac{3h}{8}\,[f(a) + 3f_1 + 3f_2 + 2f_3 + 3f_4 + 3f_5 + 2f_6 \\
&\quad + \cdots + 2f_{n-3} + 3f_{n-2} + 3f_{n-1} + f(b)]
\end{aligned}
\tag{4.37}
$$

with an error term of

$$
\text{Error} = -\frac{(b-a)}{80}\,h^4 f^{(4)}(\xi).
$$

We apply this just as we did with the $\frac{1}{3}$ rule. If the function is known, we divide the interval of integration into n panels where n is divisible by 3.

EXAMPLE 4.13   Apply Simpson's $\frac{3}{8}$ rule to the same function as in Example 4.11 $[f(x) = e^{-x^2}]$, integrating from $x = 0.2$ to $x = 1.5$. Compare the results with $3, 6, 9, \ldots$ panels.

Table 4.13 shows the results. Comparing these results with those in Table 4.11, we see that the correct result was not achieved until $n = 12$, whereas with Simpson's $\frac{1}{3}$ rule, it took only 8 subdivisions.

Table 4.13

n	Estimate of integral
3	0.65593
6	0.65872
9	0.65881
12	0.65882

One might wonder why Simpson's $\frac{3}{8}$ rule is ever used if it is less efficient than the $\frac{1}{3}$ rule. One useful application is the calculation of a tabulated function with an odd number of panels by doing the first (or the last) three with the $\frac{3}{8}$ rule and the rest with the $\frac{1}{3}$ rule. As an example, we apply this to the values of Table 4.12.

If we use the $\frac{3}{8}$ rule over points 0 through 3 and then the $\frac{1}{3}$ rule, the values are $0.61753 + 1.20015 = 1.81768$. With the other choice, we add 0.91326 to 0.90445 to get 1.81771. There is little difference, although the former is slightly better.

We summarize the rules for integration:

Formulas for Integration (Uniform spacing, $\Delta x = h$)

Trapezoidal rule:

$$\int_a^b f(x)\, dx = \frac{h}{2}(f_1 + 2f_2 + 2f_3 + \cdots + 2f_n + f_{n+1})$$

$$- \frac{(b-a)}{12} h^2 f''(\xi), \qquad a \le \xi \le b$$

Simpson's $\frac{1}{3}$ rule:

$$\int_a^b f(x)\, dx = \frac{h}{3}(f_1 + 4f_2 + 2f_3 + 4f_4 + 2f_5 + \cdots + 4f_n + f_{n+1})$$

$$- \frac{(b-a)}{180} h^4 f^{iv}(\xi), \qquad a \le \xi \le b$$

(requires an even number of panels)

Simpson's $\frac{3}{8}$ rule:

$$\int_a^b f(x)\, dx = \frac{3h}{8}(f_1 + 3f_2 + 3f_3 + 2f_4 + 3f_5 + 3f_6 + \cdots + 3f_n + f_{n+1})$$

$$- \frac{(b-a)}{80} h^4 f^{iv}(\xi), \qquad a \le \xi \le b$$

(requires a number of panels divisible by 3)

4.8 Other Ways to Derive Integration Formulas

It is interesting to examine other ways of deriving integration formulas than by integrating the interpolating polynomial. One way is to use symbolic methods, similar to those of Section 4.3 and Section 3.9.

In terms of the stepping operator E,

$$f(x_s) = f_s = E^s f_0.$$

Multiplying by $dx = h\, ds$ and integrating from x_0 to x_1 ($s = 0$ to $s = 1$), we obtain

$$\int_{x_0}^{x_1} f(x)\, dx = h \int_0^1 E^s f_0\, ds = \left[\frac{hE^s}{\ln E} f_0 \right]_0^1 = \frac{h(E-1)}{\ln E} f_0.$$

Let $E = 1 + \Delta$ and expand $\ln(1 + \Delta)$ as a power series:

$$\ln(1 + \Delta) = \Delta - \frac{1}{2}\Delta^2 + \frac{1}{3}\Delta^3 - \frac{1}{4}\Delta^4 + \cdots .$$

On dividing Δ by this series, we get

$$\int_{x_0}^{x_1} f(x)\, dx = \frac{h\Delta}{\Delta - \dfrac{1}{2}\Delta^2 + \dfrac{1}{3}\Delta^3 - \dfrac{1}{4}\Delta^4 + \cdots} f_0$$

$$= h\left(f_0 + \frac{1}{2}\Delta f_0 - \frac{1}{12}\Delta^2 f_0 + \frac{1}{24}\Delta^3 f_0 - \cdots \right). \tag{4.38}$$

The coefficients are considerably easier to get by this technique than by the term-by-term integration of Section 4.5.

Equation (4.38) is not a Newton–Cotes formula unless we use only the first two terms, making the interval of integration and the range of fit of the polynomial agree. When n terms are used, the formula represents a polynomial of degree n, fitting from x_0 to x_n, but integrated only from x_0 to x_1. This is especially useful in connection with differential-equation methods.

We can develop the formula for an nth-degree interpolating polynomial integrated over two panels, from x_0 to x_2, in a similar fashion:

$$\int_{x_0}^{x_2} f(x)\, dx = \left[\frac{hE^s}{\ln E} f_0 \right]_0^2 = \frac{h(E^2 - 1)}{\ln E} f_0 = \frac{h(E + 1)(E - 1)}{\ln E} f_0.$$

Again letting $E = 1 + \Delta$ so $E - 1 = \Delta$ and $E + 1 = 2 + \Delta$, and dividing Δ by the series for $\ln(1 + \Delta)$, we get

$$\int_{x_0}^{x_2} f(x)\, dx = h(2 + \Delta)\left(f_0 + \frac{1}{2}\Delta f_0 - \frac{1}{12}\Delta^2 f_0 + \frac{1}{24}\Delta^3 f_0 - \cdots \right)$$

$$= h\left(2f_0 + \Delta f_0 - \frac{1}{6}\Delta^2 f_0 + \frac{1}{12}\Delta^3 f_0 - \cdots \right.$$

$$\left. + \Delta f_0 + \frac{1}{2}\Delta^2 f_0 - \frac{1}{12}\Delta^3 f_0 + \cdots \right)$$

$$= h\left(2f_0 + 2\Delta f_0 + \frac{1}{3}\Delta^2 f_0 + 0 - \cdots \right). \tag{4.39}$$

Obviously the method can be extended. One reason that we might desire formulas such as Eqs. (4.38) and (4.39) is for constructing formulas that permit integration over m panels based on polynomials of degree n.

We now present still another interesting method of deriving formulas that can be applied to a variety of situations including the development of integration formulas. It may be called the method of *undetermined coefficients*.* We express the formula as a sum of $n + 1$ terms with unknown coefficients, and then evaluate the coefficients

---

* Appendix B describes this method in more detail, illustrating it with several examples.

by requiring that the formula be exact for all polynomials of degree n or less. We illustrate it here by finding Simpson's $\frac{1}{3}$ rule by this other technique. Express the integral as a weighted sum of three equispaced function values:

$$\int_{-1}^{1} f(x)\, dx = af(-1) + bf(0) + cf(+1). \tag{4.40}$$

The symmetrical interval of integration simplifies the arithmetic. We stipulate that the function is to be evaluated at three equally spaced intervals, the two end values and the midpoint. Since the formula contains three terms, we can require it to be correct for all polynomials of degree 2 or less. If that is true, it certainly must be true for the three special cases of $f(x) = x^2$, $f(x) = x$, and $f(x) = 1$. We rewrite Eq. (4.40) three times, applying each definition of $f(x)$ in turn:

$$f(x) = 1: \qquad \int_{-1}^{1} dx = 2 = a(1) + b(1) + c(1) = a + b + c;$$

$$f(x) = x: \qquad \int_{-1}^{1} x\, dx = 0 = a(-1) + b(0) + c(1) = -a + c;$$

$$f(x) = x^2: \qquad \int_{-1}^{1} x^2\, dx = \frac{2}{3} = a(1) + b(0) + c(1) = a + c.$$

Solving the three equations simultaneously gives $a = \frac{1}{3}, b = \frac{4}{3}, c = \frac{1}{3}$. Here the spacing between points was unity; obviously the integral is proportional to $\Delta x = h$. We then get Simpson's $\frac{1}{3}$ rule:

$$\int_{-h}^{h} f(x)\, dx = h \left[\frac{1}{3} f(-h) + \frac{4}{3} f(0) + \frac{1}{3} f(h) \right].$$

4.9 Gaussian Quadrature

Our previous formulas for numerical integration were all predicated on evenly spaced x-values; this means the x-values were predetermined. With a formula of three terms, then, there were three parameters, the coefficients (weighting factors) applied to each of the functional values. A formula with three parameters corresponds to a polynomial of the second degree, one less than the number of parameters. Gauss observed that if we remove the requirement that the function be evaluated at predetermined x-values, a three-term formula will contain six parameters (the three x-values are now unknowns, plus the three weights) and should correspond to an interpolating polynomial of degree 5. Formulas based on this principle are called *Gaussian quadrature formulas*. They can be applied only when $f(x)$ is known explicitly, so that it can be evaluated at any desired value of x.

We will determine the parameters in the simple case of a two-term formula containing four unknown parameters:

$$\int_{-1}^{1} f(t) \doteq af(t_1) + bf(t_2).$$

The method is the same as that illustrated in the previous section, by determining unknown parameters. We use a symmetrical interval of integration to simplify the arithmetic, and call our variable t. (This notation agrees with that of most authors. Since the variable of integration is only a dummy variable, its name is unimportant.) Our formula is to be valid for any polynomial of degree 3; hence it will hold if $f(t) = t^3$, $f(t) = t^2$, $f(t) = t$, and $f(t) = 1$:

$$f(t) = t^3: \qquad \int_{-1}^{1} t^3\, dt = 0 = at_1^3 + bt_2^3;$$

$$f(t) = t^2: \qquad \int_{-1}^{1} t^2\, dt = \frac{2}{3} = at_1^2 + bt_2^2;$$

$$f(t) = t: \qquad \int_{-1}^{1} t\, dt = 0 = at_1 + bt_2;$$

$$f(t) = 1: \qquad \int_{-1}^{1} dt = 2 = a + b.$$

(4.41)

Multiplying the third equation by t_1^2, and subtracting from the first, we have

$$0 = 0 + b[t_2^3 - t_2 t_1^2] = b(t_2)(t_2 - t_1)(t_2 + t_1).$$ (4.42)

We can satisfy Eq. (4.42) by either $b = 0$, $t_2 = 0$, $t_1 = t_2$ or $t_1 = -t_2$. Only the last of these possibilities is satisfactory, the others being invalid, or else reduce our formula to only a single term, so we choose $t_1 = -t_2$. We then find that

$$a = b = 1,$$

$$t_2 = -t_1 = \sqrt{\frac{1}{3}} = 0.5773,$$

$$\int_{-1}^{1} f(t)\, dt \doteq f(-0.5773) + f(0.5773).$$

It is remarkable that adding these two values of the function gives the exact value for the integral of any cubic polynomial over the interval from -1 to 1.

Suppose our limits of integration are from a to b, and not -1 to 1 for which we derived this formula. To use the tabulated Gaussian quadrature parameters, we must change the interval of integration to $(-1, 1)$ by a change of variable. We replace the given variable by another to which it is linearly related according to the following scheme:

If we let

$$x = \frac{(b - a)t + b + a}{2} \qquad \text{so that } dx = \left(\frac{b - a}{2}\right) dt,$$

then

$$\int_{a}^{b} f(x)\, dx = \frac{b - a}{2} \int_{-1}^{1} f\left(\frac{(b - a)t + b + a}{2}\right) dt.$$

EXAMPLE 4.14 Evaluate $I = \int_0^{\pi/2} \sin x \, dx$. (Obviously, $I = 1.0$, so we can readily see the error of our estimate.)

To use the two-term Gaussian formula, we must change the variable of integration to make the limits of integration from -1 to 1.

Let

$$x = \frac{(\pi/2)t + \pi/2}{2}, \qquad \text{so } dx = \frac{\pi}{4}\,dt.$$

Observe that when $t = -1$, $x = 0$; when $t = 1$, $x = \pi/2$. Then

$$I = \frac{\pi}{4}\int_{-1}^{1} \sin\left(\frac{\pi t + \pi}{4}\right) dt.$$

The Gaussian formula calculates the value of the new integral as a weighted sum of two values of the integrand, at $t = -0.5773$ and at $t = 0.5773$. Hence,

$$I = \frac{\pi}{4}[(1.0)(\sin(0.10566\pi) + (1.0)(\sin(0.39434\pi))]$$

$$= 0.99847.$$

The error is 1.53×10^{-3}.

▲

The power of the Gaussian method derives from the fact that we need only two functional evaluations. If we had used the trapezoidal rule, which also requires only two evaluations, our estimate would have been $(\pi/4)(0.0 + 1.0) = 0.7854$, an answer quite far from the mark. Simpson's $\frac{1}{3}$ rule requires three functional evaluations and gives $I = 1.0023$, with an error of -2.3×10^{-3}, somewhat greater than for Gaussian quadrature.

Gaussian quadrature can be extended beyond two terms. The formula is then given by

$$\int_{-1}^{1} f(t) \, dt \doteq \sum_{i=1}^{n} w_i f(t_i), \qquad \text{for } n \text{ points.} \tag{4.43}$$

This formula is *exact* for functions $f(t)$ that are polynomials of degree $2n - 1$ or less! Moreover, by extending the method we used previously for the 2-point formula, for each n we obtain a system of $2n$ equations:

$$w_1 t_1^k + \cdots + w_n t_n^k = \begin{cases} 0, & \text{for } k = 1, 3, 5, \ldots, 2n - 1; \\ \dfrac{2}{k + 1}, & \text{for } k = 0, 2, 4, \ldots, 2n - 2. \end{cases}$$

This approach is obvious. However, this set of equations, obtained by writing $f(t)$

as a succession of polynomials, is not easily solved. We wish to indicate an approach that is easier than the methods for a nonlinear system that we used in Chapter 2.

It turns out that the t_i's for a given n are the roots of the nth-degree Legendre polynomial. The Legendre polynomials are defined by recursion:

$$(n + 1)L_{n+1}(x) - (2n + 1)xL_n(x) + nL_{n-1}(x) = 0,$$
$$\text{with } L_0(x) = 1, \quad L_1(x) = x.$$

Then $L_2(x)$ is

$$L_2(x) = \frac{3xL_1(x) - (1)L_0(x)}{2} = \frac{3}{2}x^2 - \frac{1}{2};$$

here roots are $\pm\sqrt{\frac{1}{3}} = \pm 0.5773$, precisely the t-values for the two-term formula.

By using the recursion relation, we find

$$L_3(x) = \frac{5x^3 - 3x}{2},$$

$$L_4(x) = \frac{35x^4 - 30x^2 + 3}{8}, \qquad \text{and so on.}$$

The methods of Chapter 1 allow us to find the roots of these polynomials. After they have been determined, the set of equations analogous to Eqs. (4.41) can easily be solved for the weighting factors because the equations are linear with respect to these unknowns.

Table 4.14 lists the zeros of Legendre polynomials up to degree 5, giving values that we need for Gaussian quadrature where the basic polynomial is up to degree 9. For example, $L_3(x)$ has zeros at $x = 0$, $+0.77459667$, and -0.77459667.

Before continuing with an example of the use of Gaussian quadrature, it is of interest to summarize the properties of Legendre polynomials.

1. The Legendre polynomials are *orthogonal* over the interval $[-1, 1]$. That is,

$$\int_{-1}^{1} L_n(x) L_m(x) \, dx \begin{cases} = 0 \text{ if } n \neq m; \\ > 0 \text{ if } n = m. \end{cases}$$

   This is a property of several other important functions, such as $\{\cos(nx), n = 0, 1, \ldots\}$. Here we have

$$\int_{0}^{2\pi} \cos(mx) \cos(nx) \, dx \begin{cases} = 0 \text{ if } n \neq m; \\ > 0 \text{ if } n = m. \end{cases}$$

   In this case we say that these functions are orthogonal over the interval $[0, 2\pi]$.

2. Any polynomial of degree n can be written as a sum of the Legendre polynomials:

$$P_n(x) = \sum_{i=0}^{n} c_i L_i(x).$$

Table 4.14 Values for Gaussian quadrature

Number of terms	Values of t	Weighting factor	Valid up to degree
2	−0.57735027	1.0	3
	0.57735027	1.0	
3	−0.77459667	0.55555555	5
	0.0	0.88888889	
	−0.77459667	0.55555555	
4	−0.86113631	0.34785485	7
	−0.33998104	0.65214515	
	0.33998104	0.65214515	
	0.86113631	0.34785485	
5	−0.90617975	0.23692689	9
	−0.53846931	0.47862867	
	0.0	0.56888889	
	0.53846931	0.47862867	
	0.90617975	0.23692689	

3. The n roots of $L_n(x) = 0$ lie in the interval $[-1, 1]$.

Using these properties, we are able to show that Eq. (4.43) is exact for polynomials of degree $2n - 1$ or less.

The weighting factors and t-values for Gaussian quadrature have been tabulated. (Love, 1966, gives values for up to 200-term formulas.) We are content to give a few of the values in Table 4.14.

We illustrate the three-term formula with an example.

EXAMPLE 4.15 Evaluate $I = \int_{0.2}^{1.5} e^{-x^2} \, dx$ using the three-term Gaussian formula,

$$x = \frac{(1.5 - 0.2)t + 1.5 + 0.2}{2} = 0.65t + 0.85.$$

Then

$$I = \frac{1.5 - 0.2}{2} \int_{-1}^{1} e^{-(0.65t+0.85)^2} \, dt$$

$$= 0.65[0.555 \ldots e^{-[0.65(-0.774\ldots)+0.85]^2} + 0.888 \ldots e^{-[0.65(0.0)+0.85]^2}$$

$$+ 0.555 \ldots e^{-[0.65(0.774\ldots)+0.85]^2}]$$

$$= 0.65860 \quad \text{(compare to exact value 0.65882)}$$

A four-term Gaussian quadrature gives 0.65883, off only in the last digit from the correct value. Compare this result, which required only four function evaluations, to the result of Example 4.11, where Simpson's $\frac{1}{3}$ rule required eight evaluations to achieve the same accuracy.

4.10 Adaptive Integration

The trapezoidal rule and Simpson's $\frac{1}{3}$ rule are often used to find the integral of $f(x)$ over a fixed interval $[a, b]$ using a uniform value for Δx. When $f(x)$ is a known function, we can choose the value for $\Delta x = h$ arbitrarily. The problem is that we do not know a priori what value to choose for h to attain a desired accuracy. Romberg-type integration is a way to find the necessary h. We start with two panels, $h = h_1 = (b - a)/2$, and apply one of the formulas. Then we let $h_2 = h_1/2$ and apply the formula again, now with four panels, and compare the results. If the new value is sufficiently close, we terminate and use a Richardson extrapolation to further reduce the error. If the second result is not close enough to the first, we again halve h and repeat the procedure. We continue in this way until the last result is close enough to its predecessor.

We illustrate this obvious procedure with an example.

EXAMPLE 4.16   Integrate $f(x) = 1/x^2$ over the interval $[0.2, 1]$ using Simpson's $\frac{1}{3}$ rule. Use a tolerance value of 0.02 to terminate the halving of $h = \Delta x$. From calculus, we know that the exact answer is 4.0.

We introduce a special notation that will be used throughout this section.

$S_n[a, b]$ = the computed value using Simpson's $\frac{1}{3}$ rule with $\Delta x = h_n$ over $[a, b]$.

If we use this notation, the composite Simpson rule becomes

$$I(f) = S_n[a, b] - \frac{(b - a)}{180} h_n^4 f^{iv}(\xi), \qquad a < \xi < b.$$

Using this with $h_1 = (1.0 - 0.2)/2 = 0.4$, we compute $S_1[0.2, 1.0]$. We continue halving h, $h_{n+1} = h_n/2$, computing its corresponding $S_{n+1}[a, b]$ until $|S_{n+1} - S_n| < 0.02$, the tolerance value. The following table shows the results.

n	h_n	S_n	$|S_{n+1} - S_n|$
1	0.4	4.948148	
			0.761111
2	0.2	4.187037	
			0.162756
3	0.1	4.024281	
			0.022117
4	0.05	4.002164	
			0.002010
5	0.025	4.000154	

From the table we see that, at $n = 5$, we have met the tolerance criterion, since $|S_5 - S_4| < 0.02$. A Romberg extrapolation gives

$$RS[a,b] = S_5 + \frac{S_5 - S_4}{15} = 4.00002.$$

(We use $RS[a,b]$ to represent the Romberg extrapolation from Simpson's rule.)

The Adaptive Scheme

The disadvantage of this technique is that the value of h is the same over the entire interval of integration, while the behavior of $f(x)$ may not require such uniformity. Consider Fig. 4.4. It is obvious that, in the subinterval $[c,b]$, h can be much larger than in subinterval $[a,c]$, where the curve is much less smooth. We could subdivide the entire interval $[a,b]$ nonuniformly by personal intervention after examining the graph of $f(x)$. We prefer to avoid such intervention.

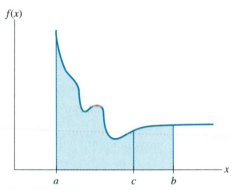

Figure 4.4

Adaptive integration automatically allows for different h's on different subintervals of $[a,b]$, choosing values adequate for a specified accuracy. We do not specify where the size change for h occurs; this can occur anywhere within it. We use something like a binary search to locate the point where we should change the size of h. Actually, the total interval $[a,b]$ may be broken into several subintervals, with different values for h within each of them. This depends on the tolerance value, TOL, and the nature of $f(x)$.

To describe this strategy, we repeat the preceding example to find the integral of $f(x) = 1/x^2$ between $x = 0.2$ and $x = 1$. We choose a value for TOL of 0.02, and do the computations in double precision to minimize the effects of round-off.

We begin as before by specifying just two subintervals in $[a,b]$. The first computation is a Simpson integration over $[0.2, 1]$ with $h_1 = 0.4$. The result, which we call

$S_1[0.2, 1]$, is 4.94814815. The next step is to integrate over each half of $[0.2, 1]$ but with h half as large, $h_2 = 0.2$. We get

$$S_2[0.2, 0.6] = 3.51851852 \quad \text{and} \quad S_2[0.6, 1] = 0.66851852.$$

We now test the accuracy of our initial computations by seeing whether the difference between $S_1[0.2, 1]$ and the sum of $S_2[0.2, 0.6]$ and $S_2[0.6, 1]$ is greater than TOL. (Actually we compare the magnitude of this difference.)

$$S_1[0.2, 1] - (S_2[0.2, 0.6] + S_2[0.6, 1]) = 0.7611111$$

Since this result is greater than TOL $= 0.02$, we must use a smaller value for h.

We continue by applying the strategy to one-half of the original interval. We arbitrarily choose the right half and compute $S_2[0.6, 1]$ with $h = h_2 = (1 - 0.6)/2 = 0.2$, comparing it to $S_3[0.6, 0.8] + S_3[0.8, 1]$ (both of these use $h_3 = h_2/2 = 0.1$). We also halve the value for TOL, getting

$$S_2[0.6, 1] - (S_3[0.6, 0.8] + S_3[0.8, 1]) = 0.66851852 - (0.41678477 + 0.25002572)$$
$$= 0.66851852 - 0.66681049$$
$$= 0.001708 \quad \text{versus TOL} = 0.01.$$

Since this passes the test, we take advantage of the results that we have available and do a Richardson extrapolation to get

$$RS[0.6, 1] = 0.66681049$$
$$+ \frac{1}{15}(0.66681049 - 0.66851852)$$
$$= 0.66669662.$$

We now move to the next adjacent subinterval, $[0.2, 0.6]$, and repeat the procedure. We compute

$$S_2[0.2, 0.6] = 3.51851852, \quad \text{with } h_2 = 0.2;$$
$$S_3[0.2, 0.4] = 2.52314815; \quad S_3[0.4, 0.6] = 0.83425926;$$
$$S_2[0.2, 0.6] - (S_3[0.2, 0.4] + S_3[0.4, 0.6]) = 0.161111 \quad \text{versus TOL} = 0.01,$$

which fails, so we proceed to another level with the right half:

$$S_3[0.4, 0.6] = 0.83425926, \quad \text{with } h_3 = 0.1;$$
$$S_4[0.4, 0.5] = 0.50005144; \quad S_4[0.5, 0.6] = 0.33334864;$$
$$S_3[0.4, 0.6] - (S_4[0.4, 0.5] + S_4[0.5, 0.6]) = 0.000859 \quad \text{versus TOL} = 0.005,$$

which passes. We extrapolate:

$$RS[0.4, 0.6] = 0.8333428.$$

The next adjacent interval is $[0.2, 0.4]$. For this we use TOL $= 0.005$. We find that this does not meet the criterion, so we next do $[0.3, 0.4]$. We do meet the TOL level of 0.0025:

$$S_4[0.3, 0.4] = 0.83356954, \quad \text{with } h_4 = 0.05;$$

$$S_5[0.3, 0.35] = 0.47620166; \quad S_5[0.35, 0.4] = 0.35714758;$$

$$S_4[0.3, 0.4] - S_5[0.3, 0.35] + S_5[0.35, 0.4]) = 0.000220 \quad \text{versus TOL} = 0.0025,$$

which passes, so

$$RS[0.3, 0.4] = 0.83333492.$$

Our last subinterval is $[0.2, 0.3]$. We find that we again meet the test. We give only the extrapolated result

$$RS[0.2, 0.3] = 1.666686.$$

Adding all of the RS-values gives the final answer:

$$\text{Integral over } [0.2, 1] = 4.00005957.$$

By employing adaptive integration, we reduced the number of function evaluations from 33 to 17.

Bookkeeping and the Avoidance of Repeating Function Evaluations

It should be obvious that we were recomputing many of the values of $f(x)$ in the previous integration. We can avoid these recalculations if we store these computations in such a way as to retrieve them appropriately. We also need to keep track of the current subinterval, the previous subintervals that we return to, and the appropriate value for h and TOL for each subinterval. The mechanism for storing these quantities is a *stack*, a data structure that is a last-in first-out device that resembles a stack of dishes in a restaurant. Actually, we use just a two-dimensional array of seven columns and as many rows as levels that we wish to accommodate. (Often a large number of levels is provided—say, 200—even though we hardly ever need so many.)

After an initial calculation to get $h_1 = (b - a)/2, c = a + h_1, f(a), f(c), f(b)$, and $S_1[a, b]$, we store a set of seven values: $a, f(a), f(c), f(b), h, \text{TOL}, S[a, b]$. We retrieve these values into variables that represent these quantities and continue with the first stage of the computations.

Whenever the test fails after computing for the current subinterval, we store two sets of values in two rows of the seven columns:

$$\text{First row:} \quad a, f(a), f(d), f(c), h_n, \text{TOL}, S[a, c],$$

$$\text{Next row:} \quad c, f(c), f(e), f(b), h_n, \text{TOL}, S[c, b] \leftarrow \text{TOP},$$

where the letters a, d, c, e, b refer to points in the last subinterval that are evenly spaced from left to right in that order. We also use a pointer to the last row stored. It is named TOP to indicate it is the "top" of the stack (even though it points to the last row stored as we normally view an array). Whenever we store a set of values, we add one to TOP; whenever we retrieve a set of values, we subtract one so that TOP always points to the row that is next available for retrieval.

We begin each iteration by retrieving the row of quantities pointed to by TOP (the one above labeled "Next row"). In this way, we can reuse the previously computed function values to get values for computing the rightmost remaining subinterval. (Observe that the next subinterval begins at the c-value for the last subinterval.)

The following algorithm implements the adaptive integration scheme that we have described.

Algorithm for Computing $I(f) = \int_a^b f(x)\,dx$

Set: VALUE = 0.0,
Evaluate: $h_1 = (b - a)/2, c = a + h_1, Fa = f(a)$,
$\quad\quad Fc = f(c), Fb = f(b), Sab = S_1(a, b)$.
STORE$(a, Fa, Fc, Fb, h_1, \text{TOL}, Sab)$.
SET: TOP = 1.
REPEAT
RETRIEVE$(a, Fa, Fc, Fb, h_1, \text{TOL}, Sab)$.
SET: TOP = TOP − 1.
Evaluate: $h_2 = h_1/2, d = a + h_2, e = a + 3h_2, Fd = f(d)$,
$\quad\quad Fe = f(e)$,
$\quad\quad Sac = S_2(a, c), Scb = S_2(c, b), S_2(a, b) = Sac + Scb$.
IF $|S_2(a, b) − S_1(a, b)| < \text{TOL}$ THEN
$\quad$ Compute $RS(a, b)$,
$\quad$ VALUE = VALUE + $RS(a, b)$,
ELSE
$\quad h_1 = h_2, \text{TOL} = \text{TOL}/2$,
$\quad$ SET: TOP = TOP + 1,
$\quad$ STORE$(a, Fa, Fd, Fc, h_1, \text{TOL}, Sac)$,
$\quad$ SET: TOP = TOP + 1,
$\quad$ STORE$(c, Fc, Fe, Fb, h_1, \text{TOL}, Scb)$
UNTIL TOP = 0.
$I(f) = $ VALUE.

4.11 Multiple Integrals

We consider first the case when the limits of integration are constants. In the calculus we learned that a double integral may be evaluated as an iterated integral; in other words, we may write

$$\iint\limits_A f(x, y)\,dA = \int_a^b \left(\int_c^d f(x, y)\,dy \right) dx = \int_c^d \left(\int_a^b f(x, y)\,dx \right) dy. \tag{4.44}$$

In Eq. (4.44) the rectangular region A is bounded by the lines

$$x = a, \qquad x = b, \qquad y = c, \qquad y = d.$$

In computing the iterated integrals, we hold x constant while integrating with respect to y (vice versa in the second case).

Adapting the numerical integration formulas developed earlier in this chapter for integration with respect to just one independent variable is quite straightforward. Recall that any one of the integration formulas is just a linear combination of values of the function, evaluated at varying values of the independent variable. In other words, a quadrature formula is just a weighted sum of certain functional values. The inner integral is written then as a weighted sum of function values with one variable held constant. We then add together a weighted sum of these sums. If the function is known only at the nodes of a rectangular grid through the region, we are constrained to use these values. The Newton–Cotes formulas are a convenient set to employ. There is no reason why the same formula must be used in each direction, although it is often particularly convenient to do so.

EXAMPLE 4.17   We illustrate this technique by evaluating the integral of the function of Table 4.15 over the rectangular region bounded by

$$x = 1.5, \qquad x = 3.0, \qquad y = 0.2, \qquad y = 0.6.$$

Let us use the trapezoidal rule in the x-direction and Simpson's $\frac{1}{3}$ rule in the y-direction. (Since the number of panels in the x-direction is not even, Simpson's $\frac{1}{3}$ rule does not apply readily.) It is immaterial which integral we evaluate first. Suppose we start with y constant:

$$y = 0.2: \qquad \int_{1.5}^{3.0} f(x,y)\, dx = \int_{1.5}^{3.0} f(x, 0.2)\, dx = \frac{h}{2}(f_1 + 2f_2 + 2f_3 + f_4)$$

$$= \frac{0.5}{2}[0.990 + 2(1.568) + 2(2.520) + 4.090]$$

$$= 3.3140;$$

$$y = 0.3: \qquad \int_{1.5}^{3.0} f(x, 0.3)\, dx = \frac{0.5}{2}[1.524 + 2(2.384) + 2(3.800) + 6.136]$$

$$= 5.0070.$$

Similarly, at

$$y = 0.4, \qquad I = 6.6522;$$
$$y = 0.5, \qquad I = 8.2368;$$
$$y = 0.6, \qquad I = 9.7435.$$

We now sum these in the y-direction according to Simpson's rule:

$$f(x,y)\, dx = \frac{0.1}{3}[3.3140 + 4(5.0070) + 2(6.6522) + 4(8.2368) + 9.7435]$$

$$= 2.6446$$

Table 4.15 Tabulation of a function of two variables, $u = f(x, y)$

x \ y	0.1	0.2	0.3	0.4	0.5	0.6
0.5	0.165	0.428	0.687	0.942	1.190	1.431
1.0	0.271	0.640	1.003	1.359	1.703	2.035
1.5	0.447	0.990	1.524	2.045	2.549	3.031
2.0	0.738	1.568	2.384	3.177	3.943	4.672
2.5	1.216	2.520	3.800	5.044	6.241	7.379
3.0	2.005	4.090	6.136	8.122	10.030	11.841
3.5	3.306	6.679	9.986	13.196	16.277	19.198

(In this example our answer does not check well with the analytical value of 2.5944 because the x-intervals are large. We could improve our estimate somewhat by fitting a higher-degree polynomial than the first to provide the integration formula. We can even use values outside the range of integration for this, using the techniques of Section 4.8 to derive the formulas.)

The previous example shows that double integration by numerical means reduces to a double summation of weighted function values. The calculations we have just made could be written in the form

$$\int f(x, y)\, dx\, dy = \sum_{j=1}^{m} v_j \sum_{i=1}^{n} w_i f_{ij}$$

$$= \frac{\Delta y}{3} \frac{\Delta x}{2} [(f_{1,1} + 2f_{2,1} + 2f_{3,1} + f_{4,1})$$

$$+ 4(f_{1,2} + 2f_{2,2} + 2f_{3,2} + f_{4,2})$$

$$+ \cdots + (f_{1,5} + 2f_{2,5} + 2f_{3,5} + f_{4,5})].$$

It is convenient to write this in pictorial operator form, in which the weighting factors are displayed in an array that is a map to the location of the functional values to which they are applied.

$$\int f(x, y)\, dx\, dy = \frac{\Delta y}{3} \frac{\Delta x}{2} \begin{bmatrix} 1 & 4 & 2 & 4 & 1 \\ 2 & 8 & 4 & 8 & 2 \\ 2 & 8 & 4 & 8 & 2 \\ 1 & 4 & 2 & 4 & 1 \end{bmatrix} f_{i,j}. \tag{4.45}$$

We interpret the numbers in the array of Eq. (4.45) in this manner: We use the values 1, 4, 2, 4, and 1 as weighting factors for functional values in the top row of the portion of Table 4.15 that we integrate over (values where $x = 1.5$ and y varies from 0.2 to 0.6). Similarly, the second column of the array in Eq. (4.45) represents weighting factors that are applied to a column of function values where $y = 0.4$ and x varies from 1.5 to 3.0. Observe that the values in the pictorial operator of Eq. (4.45) follow immediately from the Newton–Cotes coefficients for single-variable integration.

Other combinations of Newton–Cotes formulas give similar results. It is probably easiest for hand calculation to use these pictorial integration operators. Pictorial integration is readily adapted to any desired combination of integration formulas. Except for the difficulty of representation beyond two dimensions, this operator technique also applies to triple and quadruple integrals.

There is an alternative representation to such pictorial operators that is easier to translate into a computer program. We also derive it somewhat differently. Consider the numerical integration formula for one variable

$$\int_{-1}^{1} f(x) \, dx \doteq \sum_{i=1}^{n} a_i f(x_i). \tag{4.46}$$

We have seen in Section 4.9 that such formulas can be made exact if $f(x)$ is any polynomial of a certain degree. Assume that Eq. (4.46) holds for polynomials up to degree s.*

We now consider the multiple integral formula

$$\int_{-1}^{1} \int_{-1}^{1} \int_{-1}^{1} f(x, y, z) \, dx \, dy \, dz \stackrel{?}{=} \sum_{i=1}^{n} \sum_{j=1}^{n} \sum_{k=1}^{n} a_i a_j a_k f(x_i, y_j, z_k). \tag{4.47}$$

We wish to show that Eq. (4.47) is exact for all polynomials in x, y, and z up to degree s. Such a polynomial is a linear combination of terms of the form $x^\alpha y^\beta z^\gamma$, where α, β, and γ are nonnegative integers whose sum is equal to s or less. If we can prove that Eq. (4.47) holds for the general term of this form, it will then hold for the polynomial.

To do this we assume that

$$f(x, y, z) = x^\alpha y^\beta z^\gamma.$$

Then, since the limits are constants and the integrand is factorable,

$$I = \int_{-1}^{1} \int_{-1}^{1} \int_{-1}^{1} x^\alpha y^\beta z^\gamma \, dx \, dy \, dz$$
$$= \left(\int_{-1}^{1} x^\alpha \, dx \right) \left(\int_{-1}^{1} y^\beta \, dy \right) \left(\int_{-1}^{1} z^\gamma \, dz \right).$$

Replacing each term according to Eq. (4.46), we get,

$$I = \left(\sum_{i=1}^{n} a_i x_i^\alpha \right) \left(\sum_{j=1}^{n} a_j y_j^\beta \right) \left(\sum_{k=1}^{n} a_k z_k^\gamma \right) = \sum_{i=1}^{n} a_i x_i^\alpha \sum_{j=1}^{n} a_j y_j^\beta \sum_{k=1}^{n} a_k z_k^\gamma. \tag{4.48}$$

We need now an elementary rule about the product of summations. We illustrate it for a simple case. We assert that

$$\left(\sum_{i=1}^{3} u_i \right) \left(\sum_{j=1}^{2} v_j \right) = \sum_{i=1}^{3} \left(\sum_{j=1}^{2} u_i v_j \right)$$
$$= \sum_{i=1}^{3} \sum_{j=1}^{2} u_i v_j.$$

---

* For Newton–Cotes formulas, $s = n - 1$ for n even and $s = n$ for n odd. For Gaussian quadrature formulas, $s = 2n - 1$, and the x_i will be unevenly spaced.

The last equality is purely notational. We prove the first by expanding both sides:

$$\left(\sum_{i=1}^{3} u_i\right)\left(\sum_{j=1}^{2} v_j\right) = \sum_{i=1}^{3} u_i \sum_{j=1}^{2} v_j$$

$$= (u_1 + u_2 + u_3)(v_1 + v_2)$$

$$= u_1 v_1 + u_1 v_2 + u_2 v_1 + u_2 v_2 + u_3 v_1 + u_3 v_2;$$

$$\sum_{i=1}^{3}\sum_{j=1}^{2} u_i v_j = (u_1 v_1 + u_1 v_2) + (u_2 v_1 + u_2 v_2) + (u_3 v_1 + u_3 v_2).$$

On removing parentheses, we see the two sides are the same. Using this principle, we can write Eq. (4.48) in the form

$$I = \sum_{i=1}^{n}\sum_{j=1}^{n}\sum_{k=1}^{n} a_i a_j a_k x_i^{\alpha} y_j^{\beta} z_k^{\gamma}, \tag{4.49}$$

which shows that the questioned equality of Eq. (4.47) is valid, and we can write a program for a triple integral by three nested DO loops. The coefficients a_i are chosen from any numerical integration formula. If the three one-variable formulas corresponding to Eq. (4.47) are not identical, an obvious modification of Eq. (4.49) applies. In some cases a change of variable is needed to correspond to Eq. (4.46).

If we are evaluating a multiple integral numerically where the integrand is a known function, our choice of the form of Eq. (4.46) is wider. Of higher efficiency than the Newton–Cotes formulas is Gaussian quadrature. Since it also fits the pattern of Eq. (4.46), the formula of Eq. (4.49) applies. We illustrate this with a simple example.

EXAMPLE 4.18   Evaluate

$$I = \int_{0}^{1}\int_{-1}^{0}\int_{-1}^{1} yze^x \, dx \, dy \, dz$$

by Gaussian quadrature using a three-term formula for x and two-term formulas for y and z. We first make the changes of variables to adjust the limits for y and z to $(-1, 1)$:

$$y = \frac{1}{2}(u - 1), \qquad dy = \frac{1}{2} du;$$

$$z = \frac{1}{2}(v + 1), \qquad dz = \frac{1}{2} dv.$$

Our integral becomes

$$I = \frac{1}{16}\int_{-1}^{1}\int_{-1}^{1}\int_{-1}^{1} (u - 1)(v + 1)e^x \, dx \, du \, dv.$$

The two- and three-point Gaussian formulas are, from Section 4.9:

$$\int_{-1}^{1} f(x)\, dx = (1)f(-0.5774) + (1)f(0.5774),$$

$$\int_{-1}^{1} f(x)\, dx = \left(\frac{5}{9}\right)f(-0.7746) + \left(\frac{8}{9}\right)f(0) + \left(\frac{5}{9}\right)f(0.7746).$$

The integral is then

$$I = \frac{1}{16}\sum_{i=1}^{2}\sum_{j=1}^{2}\sum_{k=1}^{3} a_i\, a_j\, b_k (u_i + 1)(v_j - 1)e^{x_k},$$

$$a_1 = 1, \qquad a_2 = 1,$$

$$b_1 = \frac{5}{9}, \qquad b_2 = \frac{8}{9}, \qquad b_3 = \frac{5}{9},$$

and values of u, v, and x as above.

A few representative terms of the sum are

$$
\begin{aligned}
I = \frac{1}{16}\Bigg[&(1)(1)\left(\frac{5}{9}\right)(-0.5774 + 1)(-0.5774 - 1)e^{-0.7446} \\
+ &(1)(1)\left(\frac{8}{9}\right)(-0.5774 + 1)(-0.5774 - 1)e^{0} \\
+ &(1)(1)\left(\frac{5}{9}\right)(-0.5774 + 1)(-0.5774 - 1)e^{0.7746} \\
+ &(1)(1)\left(\frac{5}{9}\right)(0.5774 + 1)(-0.5774 - 1)e^{-0.7746} \\
+ &\cdots \Bigg].
\end{aligned}
$$

On evaluating, we get $I = -0.58758$. The analytical value is

$$-\frac{1}{4}(e - e^{-1}) = -0.58760.$$

4.12 Multiple Integration with Variable Limits

If the limits of integration are not constant, so that the region in the x, y-plane upon which the integrand $f(x, y)$ is to be summed is not rectangular, we must modify the procedure of Section 4.11. We consider a simple example to illustrate. Evaluate

$$\iint f(x, y)\, dy\, dx$$

over the region bounded by the lines $x = 0, x = 1, y = 0$, and the curve $y = x^2 + 1$.

The region is sketched in Fig. 4.5. If we draw vertical lines spaced at $\Delta x = 0.2$ apart, shown as dashed lines in Fig. 4.5, it is obvious that we can approximate the inner integral at constant x-values along any one of the vertical lines (including $x = 0$ and $x = 1$). If we use the trapezoidal rule with five panels for each of these, we get the series of sums

$$S_1 = \frac{h_1}{2}(f_a + 2f_b + 2f_c + 2f_d + 2f_e + f_f),$$

$$S_2 = \frac{h_2}{2}(f_g + 2f_h + 2f_i + 2f_j + 2f_k + f_l),$$

$$S_3 = \frac{h_3}{2}(f_m + 2f_n + \cdots),$$

$$\vdots$$

$$S_6 = \frac{h_6}{2}(f_u + 2f_v + 2f_w + 2f_x + 2f_y + f_z).$$

The subscripts here indicate the values of the function at the points so labeled in Fig. 4.5. The values of the h_i are not the same in the preceding equations, but in each they are the vertical distances divided by 5. The combination of these sums to give an estimate of the double integral will then be

$$\text{Integral} = \frac{0.2}{2}(S_1 + 2S_2 + 2S_3 + 2S_4 + 2S_5 + S_6).$$

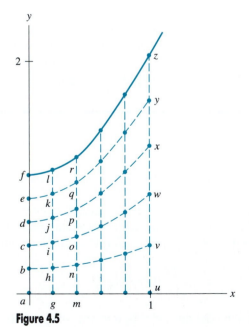

Figure 4.5

To be even more specific, suppose that $f(x,y) = xy$. Then

$$S_1 = \frac{1.0/5}{2}(0 + 0 + 0 + 0 + 0 + 0) = 0,$$

$$S_2 = \frac{1.04/5}{2}(0 + 0.0832 + 0.1664 + 0.2496 + 0.3328 + 0.208) = 0.1082,$$

$$S_3 = \frac{1.16/5}{2}(0 + 0.1856 + 0.3712 + 0.5568 + 0.7424 + 0.464) = 0.2691,$$

$$S_4 = \frac{1.36/5}{2}(0 + 0.3264 + 0.6528 + 0.9792 + 1.3056 + 0.816) = 0.5549,$$

$$S_5 = \frac{1.64/5}{2}(0 + 0.5248 + 1.0496 + 1.5744 + 2.0992 + 1.312) = 1.0758,$$

$$S_6 = \frac{2.0/5}{2}(0 + 0.8 + 1.6 + 2.4 + 3.2 + 2.0) = 2.0;$$

$$\text{Integral} = \frac{0.2}{2}(0 + 0.2164 + 0.5382 + 1.1098 + 2.1516 + 2.0)$$

$$= 0.6016 \qquad \text{versus analytical value of } 0.583333.$$

The extension of this to more complicated regions and the adaptation to the use of Simpson's rule should be obvious. If the functions that define the region are not single-valued, we must divide the region into subregions to avoid the problem, but we must also do this when we integrate analytically.

The previous calculations were not very accurate because the trapezoidal rule has relatively large errors. Gaussian quadrature should be an improvement, even using fewer points within the region. Let us use 3-point quadrature in the x-direction and 4-point quadrature in the y-direction. As in Section 4.9, we must change the limits of integration:

$$\int_0^1 \int_0^{x^2+1} x\,y\,dy\,dx$$

to

$$\frac{1}{4}\int_{-1}^1 \int_{-1}^1 \frac{s+1}{2}\left[\frac{(x^2(s)+1)t + (x^2(s)+1)}{2}\right]dt\,ds$$

in which we make the following substitutions:

$$x = \frac{s+1}{2} \qquad y = \frac{(x^2(s)+1)t + (x^2(s)+1)}{2}.$$

The integral is approximated by the sum

$$\sum_{i=1}^3 \sum_{j=1}^4 w_i\,W_j\,f(s_i, t_j),$$

where the w_i's, W_j's, s_i's, and t_j's are the values taken from Table 4.14. Using that table, we set $w_1 = 0.55555555$, $w_3 = w_1$, and $w_2 = 0.88888889$; we set $s_1 = -0.77459667$, $s_3 = -s_1$, and $s_2 = 0.0$. The values for the W_j's and t_j's are obtained in the same way. For each fixed i, $i = 1, 2, 3$, let S_i be the corresponding value obtained using Gaussian quadrature for a fixed s_i, where $S_i = \sum_{j=1}^{4} W_j f(s_i, t_j)$.

The following intermediate values are easily verified:

$$S_1 = (0.00279158 + 0.02487506 + 0.05050174 + 0.03741447) = 0.11558285,$$

$$S_2 = (0.01886891 + 0.16813600 + 0.34135240 + 0.25289269) = 0.78125000,$$

$$S_3 = (0.06845742 + 0.61000649 + 1.23844492 + 0.91750833) = 2.83441716.$$

We sum these values as follows:

$$\frac{w_1 S_1 + w_2 S_2 + w_3 S_3}{4} = 0.58333334,$$

which agrees with the exact answer to seven places. In this case we used only 12 evaluations of the function (exceptionally simple to do here, but usually more costly) compared to the 36 used with the trapezoidal rule.

To keep track of the intermediate computations, it is convenient to use a template such as

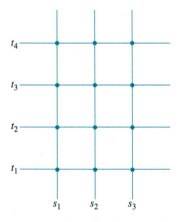

and to compute the S_i's along the verticals. The points (s_i, t_i) within the region are often called *Gauss points*.

4.13 Applications of Cubic Splines

In addition to their obvious use for interpolation, splines (Chapter 3) can be used for finding derivatives and integrals of functions, even when the function is known only as a table of values. The smoothness of splines, because of the requirement that each portion have the same first and second derivatives as its neighbor where they join, can give improved accuracy in some cases.

For the cubic spline that approximates $f(x)$, we can write, for the interval $x_i \leq x \leq x_{i+1}$,

$$f(x) = a_i(x - x_i)^3 + b_i(x - x_i)^2 + c_i(x - x_i) + d_i,$$

where the coefficients are determined as in Section 3.5. The method outlined in that section computes S_i and S_{i+1}, the values of the second derivative at each end of the subinterval. From these S-values and the values of $f(x)$, we compute the coefficients of the cubic:

$$a_i = \frac{S_{i+1} - S_i}{6(x_{i+1} - x_i)},$$

$$b_i = \frac{S_i}{2},$$

$$c_i = \frac{f(x_{i+1}) - f(x_i)}{x_{i+1} - x_i} - \frac{2(x_{i+1} - x_i)S_i + (x_{i+1} - x_i)S_{i+1}}{6},$$

$$d_i = f(x_i).$$

Approximating the first and second derivatives is straightforward; we estimate these as the values of the derivatives of the cubic:

$$f'(x) = 3a_i(x - x_i)^2 + 2b_i(x - x_i) + c_i, \tag{4.50}$$

$$f''(x) = 6a_i(x - x_i) + 2b_i. \tag{4.51}$$

At the $n + 1$ points x_i where the function is known and the spline matches $f(x)$, these formulas are particularly simple:

$$f'(x_i) \doteq c_i,$$

$$f''(x_i) \doteq 2b_i.$$

(We note that a cubic spline is not useful for approximating derivatives of order higher than the second. A higher degree of spline function would be required for these values.)

Approximating the integral of $f(x)$ over the n intervals where $f(x)$ is approximated by the spline is similarly straightforward:

$$\int_{x_1}^{x_{n+1}} f(x)\, dx = \int_{x_1}^{x_{n+1}} P_3(x)\, dx$$

$$= \sum_{i=1}^{n} \left[\frac{a_i}{4}(x - x_i)^4 + \frac{b_i}{3}(x - x_i)^3 + \frac{c_i}{2}(x - x_i)^2 + d_i(x - x_i) \right]_{x_i}^{x_{i+1}}$$

$$= \sum_{i=1}^{n} \left[\frac{a_i}{4}(x_{i+1} - x_i)^4 + \frac{b_i}{3}(x_{i+1} - x_i)^3 + \frac{c_i}{2}(x_{i+1} - x_i)^2 + d_i(x_{i+1} - x_i) \right].$$

If the intervals are all the same size, $(h = x_{i+1} - x_i)$, this equation becomes

$$\int_{x_1}^{x_{n+1}} f(x)\,dx = \frac{h^4}{4} \sum_{i=1}^{n} a_i + \frac{h^3}{3} \sum_{i=1}^{n} b_i + \frac{h^2}{2} \sum_{i=1}^{n} c_i + h \sum_{i=1}^{n} d_i.$$

We illustrate the use of splines to compute derivatives and integrals by a simple example.

EXAMPLE 4.19 Compute the integral and derivatives of $f(x) = \sin \pi x$ over the interval $0 \le x \le 1$ from the spline that fits at $x = 0, 0.25, 0.5, 0.75$, and 1.0. (See Table 4.16.) We use end condition 1: $S_1 = 0, S_5 = 0$. Solving for the coefficients of the cubic spline, we get the results shown in Table 4.17.

Table 4.16

i, point number	x	$f(x)$
1	0	0
2	0.25	0.7071
3	0.50	1.0000
4	0.75	0.7071
5	1.0	0

Table 4.17

i	x	S_i	a_i	b_i	c_i	d_i
1	0	0	−4.8960	0	3.1340	0
2	0.25	−7.3440	−2.0288	−3.6720	2.2164	0.7071
3	0.50	−10.3872	2.0288	−5.1936	0	1.000
4	0.75	−7.3440	4.8960	−3.6720	−2.2164	0.7071

The estimated values for $f'(x)$ and $f''(x)$ computed with Eqs. (4.50) and (4.51) are shown in Table 4.18. The errors of these estimates from the exact values $(f'(x) = \pi \cos(\pi x)$ and $f''(x) = -\pi^2 \sin(\pi x))$ are shown in the last two columns.

In general, the cubic spline gives good estimates of the derivatives, the maximum error being 2.5% for the first derivative and 5.0% for the second.

It is of interest to compare these values with estimates of the derivatives from a fourth-degree interpolating polynomial that fits $f(x)$ at the same five points. Table 4.19 exhibits these estimates. For the first derivative, the spline curve gives better results near the ends of the range for $f(x)$; the polynomial gives better results near the midpoint. Both are very good in this example.

Comparison of estimates for the second derivative shows a similar relationship, except for the fourth-degree polynomial, which is very bad at the endpoints.

Table 4.18 Estimates of $f'(x)$ and $f''(x)$ from a cubic spline

x	$f'(x)$	$f''(x)$	Error in $f'(x)$	Error in $f''(x)$
0.00	3.1344	0.0000	0.007146	0.000000
0.05	3.0977	−1.4689	0.005191	−0.075053
0.10	2.9876	−2.9378	0.000275	−0.112090
0.15	2.8039	−4.4067	−0.004766	−0.074028
0.20	2.5469	−5.8756	−0.005287	0.074363
0.25	2.2164	−7.3445	0.005053	0.365600
0.30	1.8340	−7.9529	0.012627	−0.031778
0.35	1.4211	−8.5613	0.005155	−0.232546
0.40	0.9778	−9.1698	−0.007015	−0.216781
0.45	0.5041	−9.7782	−0.012668	0.030114
0.50	−0.0000	−10.3866	0.000000	0.517038
0.55	−0.5041	−9.7782	0.012668	0.030113
0.60	−0.9778	−9.1698	0.007015	−0.216781
0.65	−1.4211	−8.5613	−0.005155	−0.232547
0.70	−1.8340	−7.9529	−0.012628	−0.031778
0.75	−2.2164	−7.3445	−0.005053	0.365598
0.80	−2.5469	−5.8756	0.005287	0.074362
0.85	−2.8039	−4.4067	0.004766	−0.074028
0.90	−2.9876	−2.9378	−0.000275	−0.112088
0.95	−3.0977	−1.4689	−0.005190	−0.075051
1.00	−3.1344	0.0000	−0.007146	0.000003

Table 4.19 Estimates of $f'(x)$ and $f''(x)$ from $P_4(x)$

x	$f'(x)$	$f''(x)$	Error in $f'(x)$	Error in $f''(x)$
0.00	3.0849	1.1505	0.056643	−1.150496
0.05	3.0894	−0.9358	0.013513	−0.608130
0.10	2.9950	−2.8025	−0.007196	−0.247359
0.15	2.8128	−4.4496	−0.013630	−0.031101
0.20	2.5537	−5.8771	−0.012126	0.075874
0.25	2.2288	−7.0849	−0.007321	0.106083
0.30	1.8489	−8.0732	−0.002311	0.088523
0.35	1.4251	−8.8418	0.001151	0.047960
0.40	0.9684	−9.3909	0.002436	0.004319
0.45	0.4897	−9.7203	0.001778	−0.027803
0.50	−0.0000	−9.8301	−0.000000	−0.039509
0.55	−0.4897	−9.7203	−0.001779	−0.027803
0.60	−0.9684	−9.3909	−0.002437	0.004321
0.65	−1.4251	−8.8418	−0.001152	0.047961
0.70	−1.8489	−8.0732	0.002311	0.088524
0.75	−2.2288	−7.0849	0.007320	0.106084
0.80	−2.5537	−5.8771	0.012125	0.075876
0.85	−2.8128	−4.4496	0.013630	−0.031100
0.90	−2.9950	−2.8025	0.007196	−0.247358
0.95	−3.0894	−0.9358	−0.013513	−0.608130
1.00	−3.0849	1.1505	−0.056643	−1.150496

We readily compute the integral from the cubic spline:

$$\int_0^1 f(x)\,dx = \frac{(0.25)^4}{4}(0) + \frac{(0.25)^3}{3}(-12.5376) + \frac{(0.25)^2}{2}(3.1340)$$

$$+ 0.25(2.4142)$$

$$= 0.6362 \quad (\text{exact} = 0.6366;\ \text{error} = +0.0004).$$

For this example, the accuracy is significantly better than that obtained with Simpson's $\frac{1}{3}$ rule, which gives 0.6381, error $= -0.0015$.

This example illustrates the fact that getting derivatives numerically is more difficult than getting integrals by numerical methods.

▲

4.14 Theoretical Matters

This section assembles several theoretical topics that are important for numerical differentiation and integration.

Error Terms for Derivatives

In Section 4.2, the error terms for derivative formulas were computed by differentiating the error term for the corresponding interpolation polynomial. It is of interest to get expressions for the errors in a different way.

Consider this Taylor series:

$$f(x_i + h) = f(x_i) + f'(x_i)(h) + f''(\xi)\frac{h^2}{2!}, \qquad \xi \text{ in } [x_i, x_i + h],$$

where the last term is the error of the series. Solve for $f'(x_i)$:

$$f'(x_i) = \frac{f(x_i + h) - f(x_i)}{h} + \text{error},$$

$$\text{error} = -f''(\xi)\frac{h}{2} = O(h), \qquad \xi \text{ in } [x_i, x_i + h],$$

which is the forward-difference formula for the first derivative, which we have written with an alternative notation:

$$f_i' = \frac{f_{i+1} - f_i}{h} - f''(\xi)\frac{h}{2}.$$

Similarly, from the two Taylor series for $f(x_i + h)$ and $f(x_i - h)$, we can solve for the central-difference formula:

$$f(x_i + h) = f(x_i) + f'(x_i)(h) + f''(x_i)\frac{h}{2} + f'''(\xi_1)\frac{h^3}{3!},$$

$$f(x_i - h) = f(x_i) - f'(x_i)(h) + f''(x_i)\frac{h}{2} - f'''(\xi_2)\frac{h^3}{3!},$$

$$f_i' = \frac{f_{i+1} - f_{i-1}}{2h} + \text{error},$$

(4.52)

$$\text{error} = -f'''(\xi)\frac{h^2}{6} = O(h^2), \qquad \xi \text{ in } [x_{i-1}, x_{i+1}].$$

In this we have used the intermediate value theorem to rewrite the error in terms of ξ. This formula is sometimes called a *three-point formula* because three points, at x_i, x_{i+1}, and x_{i-1}, are used (even though the central one does not appear).

Proceeding in a similar fashion, we can obtain additional formulas for derivatives from

$$f(x_i \pm nh) = f_i \pm f_i'(nh) + f_i''\frac{(nh)^2}{2!} \pm f_i'''\frac{(nh)^3}{3!} + \cdots, \qquad n = 1, 2, 3, \ldots.$$

Three-point forward-difference formula for f':

$$f_i' = \frac{-f_{i+2} + 4f_{i+1} - 3f_i}{2h} + \text{error},$$

$$\text{error} = f'''(\xi)\frac{h^2}{3} = O(h^2).$$

Five-point forward-difference formula for f':

$$f_i' = \frac{-3f_{i+4} + 16f_{i+3} - 36f_{i+2} + 48f_{i+1} - 25f_i}{12h} + \text{error},$$

$$\text{error} = f^{(5)}(\xi)\frac{h^4}{5} = O(h^4).$$

Five-point central difference formula for f':

$$f_i' = \frac{-f_{i+2} + 8f_{i+1} - 8f_{i-1} + f_{i-2}}{12h} + \text{error},$$

$$\text{error} = f^{(5)}(\xi)\frac{h^4}{30} = O(h^4).$$

Three-point forward-difference formula for f'':

$$f_i'' = \frac{f_{i+2} - 2f_{i+1} + f_i}{h^2} + \text{error},$$

$$\text{error} = -f'''(\xi)(h) = O(h).$$

Three-point central-difference formula for f'':

$$f_i'' = \frac{f_{i+1} - 2f_i + f_{i-1}}{h^2} + \text{error},$$

$$\text{error} = -f^{(4)}(\xi)\frac{h^2}{12} = O(h^2).$$

Five-point forward-difference formula for f'':

$$f_i'' = \frac{11f_{i+4} - 56f_{i+3} + 114f_{i+2} - 104f_{i+1} + 35f_i}{12h^2} + \text{error},$$

$$\text{error} = -f^{(5)}(\xi)\frac{5h^3}{6} = O(h^3).$$

Five-point central-difference formula for f'':

$$f_i' = \frac{-f_{i+2} + 16f_{i+1} - 30f_i + 16f_{i-1} - f_{i-2}}{12h^2} + \text{error},$$

$$\text{error} = f^{(6)}(\xi)\frac{h^4}{90} = O(h^4).$$

Error Terms in Richardson Extrapolations

In Section 4.4, improved estimates of $f'(x)$ were obtained from central-difference estimates, with the value of h differing by a factor of two. The rationale was based on the assumption that the errors were proportional to h^2 (when the error was just $O(h^2)$). We can establish the validity of the extrapolation by beginning with more terms of the Taylor series that gave Eq. (4.52):

$$f_i' = \frac{f_{i+1} - f_{i-1}}{2h} + a_1 h^2 + a_2 h^4 + a_3 h^6 + \cdots, \qquad (4.53)$$

where the added terms represent the error. If we let $F(h)$ be the first term on the right and rearrange, we have

$$F(h) = f_i' - a_1 h^2 - a_2 h^4 - a_3 h^6 - \cdots. \qquad (4.54)$$

Now, if h is halved, we have

$$F\left(\frac{h}{2}\right) = f_i' - a_1\left(\frac{h}{2}\right)^2 - a_2\left(\frac{h}{2}\right)^4 - a_3\left(\frac{h}{2}\right)^6 - \cdots. \qquad (4.55)$$

The Richardson extrapolation adds $\frac{1}{3}$ of the difference between Eq. (4.55) and Eq. (4.54) to Eq. (4.55) to give

$$G(h) = f_i' + \frac{1}{4}a_2 h^4 + \frac{5}{16}a_3 h^6 + \cdots,$$

and we have improved our estimate to one with an error of $O(h^4)$.

Obviously, we can repeat this step, but now we need a multiplier of 1/15 to eliminate the first added term. Then we can extrapolate again, using multipliers of $1/(2^n - 1)$, where n is the exponent on h in the term being eliminated.

This scheme is commonly used for any computation with an error of the form of Eq. (4.53). In particular, it is valid for trapezoidal rule integrals—Romberg integration.

Error Terms for Newton–Cotes Formulas

The Newton–Cotes integration formulas can be obtained by integrating a Newton–Gregory forward-interpolating polynomial over one, two, or three equal subdivisions. The corresponding error term comes from integrating the error term of the polynomial. In Section 4.5 the error term for the trapezoidal rule was shown to be

$$\text{Error of trapezoidal rule} = \frac{-h^3 f''(\xi)}{12}.$$

It is instructive to carry more terms in the error expression. We begin with $P(x)$:

$$P(x) = f_i + s\Delta f_i + s(s - 1)\frac{\Delta^2 f_i}{2!} + s(s - 1)(s - 2)\frac{\Delta^3 f_i}{3!} + \cdots, \quad \textbf{(4.56)}$$

where $s = (x - x_i)/h$.

The integral of the first two terms from x_i to $x_i + h$ gives the trapezoidal rule, as we have seen. An expression for the error is obtained by integrating the remaining terms:

$$\text{Error} = -h^3 \frac{\Delta^2 f_i}{12} + h^4 \frac{\Delta^3 f_i}{24} - 19h^5 \frac{\Delta^4 f_i}{720} + 3h^6 \frac{\Delta^5 f_i}{160} - \cdots.$$

(A symbolic algebra program was very helpful in evaluating these integrals.) We observe that using only the first term gives the normal expression for the error:

$$\text{Error} = -\frac{h^3 f''(\xi)}{12} = O(h^3).$$

Let us do the same for Simpson's $\frac{1}{3}$ rule. The rule is given by the integral of the first three terms of Eq. (4.56) between the limits of x_i to $x_i + 2h$ and the error expression by integrating the rest of the polynomial between the same limits. Carrying out the integrations gives

$$\text{Error} = 0 - h^5 \frac{\Delta^4 f_i}{90} + h^6 \frac{\Delta^5 f_i}{90} - 37h^7 \frac{\Delta^6 f_i}{3780} + \cdots.$$

Because the first term of this error expression is zero, Simpson's $\frac{1}{3}$ rule has an unexpectedly small error. If we replace $\Delta^4 f_i$ with its equivalent in terms of derivatives, we get the normal expression for the error:

$$\text{Error} = -\frac{h^5 f^{(4)}(\xi)}{90} = O(h^5).$$

Simpson's $\frac{3}{8}$ rule does not have a zero for the first term in the error expression, so its error term is the same, $O(h^5)$.

Error Terms for the Composite Rules

When we apply the Newton–Cotes formula to a succession of intervals, the error is the sum of the errors of the individual error terms. We refer to the error of the composite rule as a *global error*, and to the error of the single application as the *local error*. For example, suppose we use the trapezoidal rule to integrate from $x = a$ to $x = b$, subdividing the interval of integration into n equal subintervals of size h. We then have

$$\text{Global error} = \frac{-h^3[f''(\xi_1) + f''(\xi_2) + \cdots + f''(\xi_n)]}{12}.$$

If $f''(x)$ is continuous on $[a, b]$, there is a value, ξ, for which the sum of the n terms can be written as

$$\text{Global error} = \frac{-nh^3 f''(\xi)}{12} = \frac{-(b - a)h^2 f''(\xi)}{12} = O(h^2)$$

because $(b - a) = nh$.

The reduction in order of error occurs for any of the composite Newton–Cotes rules. Both of Simpson's rules then have global errors of $O(h^4)$.

The Next-Term Rule

We have made much of the fact that the error term for differentiation, integration, and even interpolation can be easily written using the *next-term rule*. By this we mean that the term after the last one used in the formula can be converted into the error term by either of these replacements:

$$\Delta^n f_i = h^n f^{(n)}(x_i) \qquad \text{or} \qquad f_i^{[n]} = \frac{f^{(n)}(x_i)}{n!}.$$

The basis for this is the development of Section 4.3.

Round-Off and Accuracy of Derivatives

It is important to realize the effect of round-off error on the accuracy of derivatives computed by finite-difference formulas. As we have seen, decreasing the value of h or increasing the degree of the interpolating polynomial increases the accuracy of our derivative formulas. (These procedures reduce the truncation error.)

As h is reduced, however, we are required to subtract function values that are almost equal, and this, we have seen, incurs a large error due to round-off. The increase in round-off error as h gets smaller causes the best accuracy to occur at some intermediate point. The data in Table 4.20 demonstrate this phenomenon. The test was made with both single-precision arithmetic (giving 6–7 significant figure accuracy) and double-precision (14–15 figures). The function was $f(x) = e^x$, at $x = 0$. The true value of both $f'(x)$ and $f''(x)$ is 1.0, of course. Central-difference approximations were employed.

In the single-precision test, the optimum value of h was only 0.01 for the first derivative, and the best accuracy of the second derivative was at $h = 0.1$, a surprisingly large value. When the round-off errors were reduced by using double-precision, essentially exact results were obtained for the first derivative throughout the range of h values. For the second derivative, $h = 0.1 \times 10^{-2}$ or $h = 0.1 \times 10^{-3}$ gives the best accuracy. Note how large the error grows; it is especially marked for second derivatives when h becomes much smaller than the optimum. Higher derivatives than the second will be expected to show an even more exaggerated influence of round-off.

Table 4.20 Derivatives computed by central-difference formulas $f(x) = e^x$, at $x = 0$

h	$f'(x)$	$f''(x)$
Results with single precision:		
0.1E 00	0.1001663E 01	0.1000761E 01
0.1E–01	0.1000001E 01	0.9959934E 00
0.1E–02	0.9999569E 00	0.8940693E 00
0.1E–03	0.9959930E 00	−0.8344640E 02
0.1E–04	0.9775158E 00	−0.4768363E 04
0.1E–05	0.9834763E 00	−0.5960462E 05
0.1E–06	0.5960463E 00	−0.1192093E 08
0.1E–07	0.2980230E 01	−0.5960458E 09
0.1E–08	0.2980229E 02	−0.5960459E 11
0.1E–09	0.2980229E 03	−0.5960456E 13
Results with double precision:		
0.1E 00	0.1001667E 01	0.1000834E 01
0.1E–01	0.1000016E 01	0.1000008E 01
0.1E–02	0.1000000E 01	0.1000000E 01
0.1E–03	0.1000000E 01	0.9999999E 00
0.1E–04	0.1000000E 01	0.9999989E 00
0.1E–05	0.1000000E 01	0.1000031E 01
0.1E–06	0.9999999E 01	0.9894834E 00
0.1E–07	0.9999999E 00	−0.4163322E 00
0.1E–08	0.9999999E 00	−0.4163319E 02
0.1E–09	0.1000000E 01	0.2775545E 04

Computer models vary widely in the number of digits of accuracy they provide, due to the different number of bits provided for the storage of floating-point values and also, to a lesser degree, due to the format of their floating-point representations. Another factor that affects the accuracy of functional values is the algorithm used in their evaluation. Chapter 10 gives some insight into this. Different software packages that are used with computer operating systems employ different algorithms, which are of varying accuracy. Further, as Chapter 10 will show, the error of approximating the function differs with the value of the argument.

No analytical way of predicting the effects of round-off error seems to exist. The best we can do is to warn budding numerical analysts so they will be properly wary when computing derivatives numerically. In the words of many authorities, *the process of differentiating is basically an unstable one*—meaning that small errors made during the process cause greatly magnified errors in the final result.

Differentiation of "noisy" data encounters a similar problem. If the data being differentiated are from experimental tests, or are observations subject to errors of measurement, the errors so influence the derivative values calculated by numerical procedures that they may be meaningless. The usual recommendation is to smooth the data first, using methods that are discussed in Chapter 3. Passing a cubic spline through the points and then getting the derivative of this approximation to the data has become quite popular. A least-squares curve may also be used. The strategy involved is straightforward—we don't try to represent the function by one that fits exactly to the data points, since this fits to the errors as well as to the trend of the information. Rather, we approximate with a smoother curve that we hope is closer to the truth than the data themselves. The problem, of course, is how much smoothing should be done. One can go too far and "smooth" beyond the point where only errors are eliminated.

A final situation should be mentioned. Some functions, or data from a series of tests, are inherently "rough." By this we mean that the function values change rapidly; a graph would show sharp local variations. When the derivative values of the function incur rapid changes, a sampling of the information may not reflect them. In this instance, the data indicate a smoother function than actually exists. Unless enough data are at hand to show the local variations, valid values of the derivatives just cannot be obtained. The only solution is more data, especially near the "rough" spots. And then we are beset by problems of accuracy of the data!

Fortunately this problem does not occur with numerical integration. As you have seen, all the integration formulas add function values together. Since the errors can be positive or negative and the probability for each is the same, errors tend to cancel out. That means that integration is a smoothing process. We assess integration as *inherently stable*.

Errors in Multiple Integration and Extrapolations

The error term of a one-variable quadrature formula is an additive one just like the other terms in the linear combination (although of special form). It would seem reasonable that it would go through the multiple summations in a similar fashion, so we should expect error terms for multiple integration that are analogous to the one-dimensional case. We illustrate that this is true for double integration using the trapezoidal rule in both directions, with uniform spacings, choosing n intervals in the x-direction and m in the y-direction.

From Section 4.6 we have

$$\text{Error of} \int_a^b f(x)\,dx = -\frac{b-a}{12}h^2 f''(\xi) = O(h^2),$$

$$h = \Delta x = \frac{b-a}{n}.$$

In developing Romberg integration, we found that the error term could be written as

$$\text{Error} = O(h^2) = Ah^2 + O(h^4) \doteq Ah^2 + Bh^4,$$

where A is a constant and the value of B depends on a fourth derivative of the function. Appending this error term to the trapezoidal rule, we get (equivalent to Eq. 4.54):

$$\int_a^b f(x,y)\,dx \bigg|_{y=y_j} = \frac{h}{2}(f_{0,j} + 2f_{1,j} + 2f_{2,j} + \cdots + f_{n,j}] + A_j h^2 + B_j h^4.$$

Summing these in the y-direction and retaining only the error terms, we have

$$\int_c^d \int_a^b f(x,y)\,dx\,dy = \frac{k}{2}\frac{h}{2}\sum_{i=0}^{n}\sum_{j=0}^{m} a_i a_j f_{i,j} + \frac{k}{2}(A_0 + 2A_1 + 2A_2 + \cdots + A_m)h^2$$

$$+ \frac{k}{2}(B_0 + 2B_1 + 2B_2 + \cdots + B_m)h^4 + \bar{A}k^2 + \bar{B}k^4, \qquad \textbf{(4.57)}$$

$$k = \Delta y = \frac{d-c}{m}.$$

In Eq. (4.57), $\bar{A}$ and $\bar{B}$ are the coefficients of the error term for y. The coefficients A and B for the error terms in the x-direction may be different for each of the $(m+1)$ y-values, but each of the sums in parentheses in Eq. (4.57) is $2n$ times some average value of A or B, so the error terms become

$$\text{Error} = \frac{k}{2}(nA_{av})h^2 + \frac{k}{2}(nB_{av})h^4 + \bar{A}k^2 + \bar{B}k^4. \qquad \textbf{(4.58)}$$

Since both Δx and Δy are constant, we may take $\Delta y = k = \alpha \Delta x = \alpha h$, where $\alpha = \Delta y/\Delta x$, and Eq. (4.58) can be written, with $nh = (b-a)$,

$$\text{Error} = \left(\frac{b-a}{2}A_{av}\alpha\right)h^2 + \left(\frac{b-a}{2}B_{av}\alpha\right)h^4 + A\alpha^2 h^2 + B\alpha^4 h^4$$

$$= K_1 h^2 + K_2 h^4.$$

Here K_2 will depend on fourth-order partial derivatives. This confirms our expectation that the error term of double integration by numerical means is of the same form as for single integration.

Since this is true, a Romberg integration may be applied to multiple integration, whereby we extrapolate to an $O(h^4)$ estimate from two trapezoidal computations at a 2-to-1 interval ratio. From two such $O(h^4)$ computations we may extrapolate to one of $O(h^6)$ error.

4.15 Using *Mathematica*

Getting Derivatives

Mathematica has no built-in capabilities for doing numerical differentiation, but it is able to get derivatives analytically, even for very complicated expressions. These can be of any order, and differentiation can apply to only one of several variables, giving partial derivatives.

The command format to obtain the derivative of $f(x)$ is

$$D [f(x), x];$$

and to get, say, the eighth derivative:

$$D [f(x), \{x, 8\}].$$

Observe that the command to differentiate employs only the single letter D, rather than spelling out the word, in contrast to *Mathematica*'s usual style.

Here are two examples involving the exponential function, represented in *Mathematica* by E (it must be uppercase):

```
In[1]:=
   D[E^x/x, x]
Out[1]=
```
$$-(\frac{E^x}{x^2}) + \frac{E^x}{x}$$
```
In[2]:=
   D[E^x/x, {x,8}]
Out[2]=
```
$$\frac{40320\ E^x}{x^9} - \frac{40320\ E^x}{x^8} + \frac{20160\ E^x}{x^7} - \frac{6720\ E^x}{x^6} + \frac{1680\ E^x}{x^5} -$$
$$\frac{336\ E^x}{x^4} + \frac{56\ E^x}{x^3} - \frac{8\ E^x}{x^2} + \frac{E^x}{x}$$

In [2], we asked for the eighth derivative; we got the answer in a fraction of a second!

If we really want to do a numerical integration—say, for data known only through a table—we could write a simple program to obtain the difference table (as in Section 3.10) and then use these values in one of the formulas of this chapter. We prefer to take a simpler approach: Fit an interpolating polynomial and then differentiate that.

EXAMPLE 4.20   In Section 4.13, we wanted to compare values of the first and second derivatives of $f(x) = \sin(\pi x)$ from a cubic spline and from the fourth-degree interpolating polynomial that agrees with $f(x)$ at x-values of 0, 0.25, 0.5, 0.75, and 1. Use *Mathematica* to get the interpolating polynomial, and then the expressions for the derivatives. Evaluate these at $x = 0.63$.

We first define a vector of the data pairs (in [3]), then ask for the interpolating polynomial (in [4]), and finally, in [5], differentiate. This gives us a third-degree polynomial:

```
In[3]:=
   data = {{0, 0},{.25, .707107}, {.5, 1}, {.75, .707107}, {1,0}}
Out[3]=
   {{0,  0},  {0.25,  0.707107},  {0.5,  1},  {0.75,  0.707107},
    {1,  0}}
In[4]:=
   poly = InterpolatingPolynomial[data, x]
Out[4]=
   (2.82843 + (-3.31371 + (-1.8301 + 3.6602 (-0.75 + x))
     (-0.5 + x)) (-0.25 + x)) x
In[5]:=
   D[poly, x]
Out[5]=
   2.82843 + (-3.31371 + (-1.8301 + 3.6602 (-0.75 + x))
     (-0.5 + x)) (-0.25 + x) +
   (-3.31371 + (-1.8301 + 3.6602 (-0.75 + x)) (-0.5 + x) +
     (-1.8301 + 3.6602 (-0.75 + x) + 3.6602 (-0.5 + x))
       (-0.25 + x)) x
```

In [6], we evaluate the derivative for $x = 0.63$, getting -1.24575 as the result. (Compare to the exact answer, -1.247677.)

```
In[6]:=
   % /. x → .63
Out[6]=
   -1.24575
```

In [7], we request the second derivative and, in [8], evaluate this second-degree polynomial for $x = 0.63$. The result, -9.08781, is not as close to the exact value, -9.05787, as was the first derivative value:

```
In[7]:=
   D[poly, {x,2}]
Out[7]=
   -6.62742 + 2 (-1.8301 + 3.6602 (-0.75 + x)) (-0.5 + x) +
     2 (-1.8301 + 3.6602 (-0.75 + x) + 3.6602 (-0.5 + x))
       (-0.25 + x) + (-3.6602 + 7.32041 (-0.75 + x) +
         7.32041 (-0.5 + x) + 7.32041 (-0.25 + x) x
In[8]:=
   % /. x → .63
Out[8]=
   -9.08781
```

If we prefer the polynomials that approximate the derivates to be expressed in more customary form, we can expand them as in [9] and [10]:

In[9]:=
  Expand[%5]
Out[9]=
  $3.08495 + 1.15051\ x - 21.9612\ x^2 + 14.6408\ x^3$
In[10]:=
  Expand[%7]
Out[10]=
  $1.15051 - 43.9224\ x + 43.9224\ x^2$

Doing Integrations

Mathematica uses an extensive set of rules to compute analytical integrations. The format of the command is

$$\texttt{Integrate [f(x), x]}$$

to obtain the indefinite integral of $f(x)$. (This command can also do multiple integrations.) If we want the definite integral between limits of $x = a$ and $x = b$, the format changes to

$$\texttt{Integrate [f(x), \{x, a, b\}].}$$

Here are two examples.

EXAMPLE 4.21   Integrate $f(x) = e^{-x}(x^2 + 1)$. First get the indefinite integral, then the definite integral between $x = 0.2$ and $x = 1.5$.
Entries [11] and [12] carry out the computations:

In[11]:=
  Integrate[E^$-$x*(x^2 + 1), x]
Out[11]=
  $$\frac{-3 - 2\ x - x^2}{E^x}$$
In[12]:=
  Integrate[E^$-$x*(x^2+1), (x,$-$.3,2.7})
Out[12]=
  2.30669

EXAMPLE 4.22   Some integrals cannot be represented in terms of ordinary functions, but *Mathematica* has some special capabilities that permit the integration of functions normally thought to be impossible to carry out. If we raise e to the power $-x^2$, a function important in statistical computations, most people would say there is no closed-form solution. But *Mathematica* can do this:

```
In[13]:=
   Integrate[E^-x^2, x]
Out[13]=
   Sqrt[Pi] Erf[x]
   ───────────────
         2
```

We see in [13] that the result is expressed in terms of Erf(x), the so-called error function. If we request the definite integral between $x = 0.2$ and $x = 1.5$, *Mathematica* gives the result in terms of Pi (*Mathematica*'s term for pi):

```
In[14]:=
   Integrate[E^-x^2, {x, .2,1.5}]
Out[14]=
   0.371701 Sqrt[Pi]
```

If we prefer this in a numerical form, the *N* function will evaluate it ([15]):

```
In[15]:=
   N[%]
Out[15]=
   0.658823
```

▲

Some functions are beyond *Mathematica*'s ability to obtain analytically. When this is the case ([16]), we see that the response is just an echo of our input (though in different form):

```
In[16]:=
   Integrate[E^-x^2/Sin[x^2], x]
Out[16]=
                 Csc[x²]
   Integrate[─────────, x]
                  E^x²
```

Mathematica also computes numerical integrations. We use this command for such numerical integrations:

$$\text{NIntegrate [f(x), \{x, a, b\}];}$$

Here is an example of a really hard problem:

```
In[17]:=
   NIntegrate[E^-x^2/Sin[x^2], {x,0,1}]
   NIntegrate::slwcon:
      Numerical integration converging too slowly; suspect one of the
         following: singularity, oscillatory integrand, or insufficient
         WorkingPrecision.
```

```
NIntegrate::ncvb:
   NIntegrate failed to converge to prescribed accuracy after
      7 recursive bisections in x near x = 4.36999 10⁻⁵⁷.
Out[17]=
   1.2 10³⁴⁹⁸
```

The difficulty in the previous example is that the integrand has a singularity at $x = 0$. *Mathematica* does the best it can but warns that it has trouble carrying out the request. We can make the problem much easier to handle by avoiding the value of $x = 0$. But observe that the answer in [18] is very different from that in [17].

```
In[18]:=
   NIntegrate[E^−x^2/Sin[x^2], {x, .0001,1}]
Out[18]=
   9998.17
```

4.16 Parallel Processing

Our discussion in this section will concern only those cases when parallel processing can speed the computations of numerical differentiation and integration. As pointed out in Chapter 3, parallel processing can obtain the columns of difference tables individually and speed up the creation of the table thereby.

In formulas like Eq. (4.9), the product terms and their sum can be speeded up by parallel processing, but since we normally use only a few factors and terms, the overhead makes parallel processing a minor advantage. This is also true for Eqs. (4.12) and (4.18). Each column of a Richardson table (Section 4.4) can be computed in parallel, but usually we do not do them in this way. Rather we alternate between column and row computations to determine when to stop the extrapolation.

The Newton–Cotes formulas of Section 4.5 also have too few terms to be good candidates for parallel processing.

When we integrate using the composite trapezoidal and Simpson's rules, parallel processing can be advantageous. Each of the function evaluations can be done at the same time (and there may be parallelization possibilities within these), and the summations can be speeded up as well. This opportunity applies to Romberg integration.

Gaussian quadrature is similar but, since we generally use only a few terms, the speedup is marginal.

Since adaptive integration employs Simpson's rule, the preceding remarks apply to this method. The same is true for multiple integration because this technique is also just successive applications of a trapezoidal or Simpson rule.

Getting integrals and derivatives from cubic splines cannot profit much from parallel processing because there are not enough terms to justify its use. Of course, parallel processors can be effective in computing the elements of the system matrix and in solving the system (as explained in Chapter 2) to give the coefficients of the individual cubics.

In summary, except for doing integrations with many subintervals, the advantages of parallel processing are not significant for the operations of Chapter 4.

Chapter Summary

If you are not able to do the following, we advise you to restudy portions of this chapter. You should now be able to

1. Develop formulas for differentiation and integration from an interpolating polynomial.

2. Get an expression for error of the formula you have developed.

3. Derive formulas by the symbolic method and from a Taylor series.

4. Apply these formulas to obtain derivatives and integrals and estimate the error in the result.

5. Use extrapolation techniques to get improved estimates of derivatives and integrals. You should be able to explain why this extrapolation works.

6. Evaluate an integral using Gaussian quadrature. You know why this method obtains the integral with fewer function evaluations than with the other integration techniques. You can develop a Legendre polynomial from the recursion formula and tell what is meant by an orthogonal polynomial.

7. Explain why adaptive integration is a more efficient way to obtain the integrals. You can use this technique.

8. Apply the various integration methods to evaluate improper and indefinite integrals and multiple integrals.

9. Use a cubic spline to obtain derivatives and integrals. You know the limitations of the cubic spline for getting higher derivatives.

10. Discuss the principles behind the error terms for the various formulas of the chapter.

11. Explain why numerical differentiation is called an unstable process and why numerical integration is not an unstable process.

12. Use *Mathematica* to do integration and differentiation.

13. Employ computer programs (perhaps even write your own) to carry out differentiation and integration.

Computer Programs

This section presents three programs that exemplify how some of the algorithms of this chapter can be implemented. Three different languages are used: (1) BASIC (a version called GWBASIC, very similar to BASICA), to get derivatives from a table; (2) Ada, which does Gaussian quadrature, and (3) FORTRAN, which carries out adaptive integration.

Program 4.1 Derivatives from a Table

Figure 4.6 lists a BASIC program that uses central-difference formulas for the first and second derivatives. The program does not represent good programming practice; it is similar to what an inexperienced programmer might do. Although the program performs the job satisfactorily, its logic is hard to follow. This is in part due to the lack of good control structures in this obsolete version of BASIC and its inability to have blank lines.

Observe also that not having to declare variables before they are used avoids what may be considered to be a tedious chore but does not help the reader to understand what is going on. It would be preferred to explain in REMark statements the names and purpose of the variables. (We deliberately did not do this to emphasize the point. You should rewrite this program to eliminate its deficiencies.)

Here is how the program gets derivatives:

1. A difference table is constructed from the x-values in line 60 and the $y = f(x)$-values in line 70. The computations are done in lines 110–160, and the table is printed in lines 170–230.
2. A value for x where $f'(x)$ and $f''(x)$ are to be computed is obtained from the user in lines 240–250.
3. The program uses the central-difference approximations of Eqs. (4.9) and (4.18) to compute derivative values. Since these formulas work only when the x-value is one of the tabulated values, a test (lines 260–280) is made to assure this. It is also required that there are function values spaced one Δx before and after the x-value, so the user is told if this is not true in lines 290 and 300. If this requirement is not met, the user is asked for another x-value.
4. If there are enough function values available, approximations based on a fourth-degree polynomial are also computed. Lines 380–410 perform this task.
5. Lines 340–350 and 380–410 display the result. The program loops to get the derivatives at a new x-value, so the user must terminate the program with the ubiquitous "CONTROL-C."

Figure 4.6 also shows some typical output when the program was run.

```
5     REM                        CHAPTER 4
1     REM           GERALD/WHEATLEY, APPLIED NUMERICAL ANALYSIS
15    REM                     (FIFTH EDITION)
20    REM
25    REM  THIS PROGRAM GETS DERIVATIVES FROM A EQUISPACED TABLE
30    REM          USING CENTRAL DIFFERENCES
35    REM    N = NO OF PTS, PUT X-VALUES IN 60, Y'S IN 70
40    N = 7
50    DIM X(N), Y(N), DIF(N,5)
60    DATA 1.3, 1.5, 1.7, 1.9, 2.1, 2.3, 2.5
70    DATA 3.669, 4.482, 5.474, 6.686, 8.166, 9.974, 12.182
80    REM SET UP THE ARRAYS
90    FOR I = 1 TO N: READ X(I): NEXT I
100   FOR I = 1 TO N: READ Y(I): NEXT I
110   REM CREATE THE TABLE OF DIFFERENCES
```

Figure 4.6 Program 4.1

Figure 4.6 *(continued)*

```
120    O = 5: IF N<5 THEN O = N-1
130    FOR I = 1 TO N: DIF(I,1) = Y(I): NEXT I
140    FOR I = 2 TO O: FOR J = 1 TO N-I+1
150    DIF(J,I) = (DIF(J+1,I-1)-DIF(J,I-1))
160    NEXT J: NEXT I
170    REM PRINT THE TABLE
180    FOR I = 1 TO N
190    PRINT USING " ##.### ";X(I);
200    FOR J = 1 TO N-I+1
210    IF J>5 THEN GOTO 230
220    PRINT USING " ##.### ";DIF(I,J);
230    NEXT J: PRINT: NEXT I: PRINT
240    REM GET X-VALUE FOR DERIV AND DEGREE OF POLYN
250    INPUT "WHAT X-VALUE FOR DERIV ";XX
260    REM BE SURE XX IS IN TABLE
270    SS = 0: FOR I = 1 TO N: IF XX = X(I) THEN SS = I
275    NEXT I
280    IF SS = 0 THEN PRINT "YOUR VALUE NOT IN TABLE": GOTO 250
290    IF SS < 2 THEN PRINT "X-VALUE TOO SMALL": GOTO 250
300    IF SS = N THEN PRINT "X-VALUE TOO LARGE": GOTO 250
310    H = X(2)-X(1)
320    D1 = (DIF(SS,2)+DIF(SS-1,2))/2/H
330    D2 = DIF(SS-1,3)/H/H
340    PRINT "USING POLY OF DEGREE 2, DERIV IS ";D1
350    PRINT "     THE SECOND DERIVATIVE IS ";D2
360    IF SS < 3 THEN GOTO 250
370    IF SS > N-2 THEN GOTO 250
380    PRINT "USING POLY OF DEGREE 4, DERIV IS ";
390    PRINT D1 - (DIF(SS-2,4)+DIF(SS-1,4))/12/H
400    PRINT "     THE SECOND DERIVATIVE IS ";
410    PRINT D2 - DIF(SS-2,5)/12/H/H
420    GOTO 250


****************************************************************
                   OUTPUT FOR DER.BAS

     1.300    3.669   0.813   0.179   0.041   0.007
     1.500    4.482   0.992   0.220   0.048   0.012
     1.700    5.474   1.212   0.268   0.060   0.012
     1.900    6.686   1.480   0.328   0.072
     2.100    8.166   1.808   0.400
     2.300    9.974   2.208
     2.500   12.182


WHAT X-VALUE FOR DERIV ? 1.6
YOUR VALUE NOT IN TABLE
WHAT X-VALUE FOR DERIV ? 1.7
USING POLY OF DEGREE 2, DERIV IS  5.509999
        THE SECOND DERIVATIVE IS  5.499992
USING POLY OF DEGREE 4, DERIV IS  5.472915
        THE SECOND DERIVATIVE IS  5.485406
WHAT X-VALUE FOR DERIV ? 2.5
X-VALUE TOO LARGE
WHAT X-VALUE FOR DERIV ? 2.3
USING POLY OF DEGREE 2, DERIV IS  10.04
        THE SECOND DERIVATIVE IS  10.00001
WHAT X-VALUE FOR DERIV ? 1.9
USING POLY OF DEGREE 2, DERIV IS  6.73
        THE SECOND DERIVATIVE IS  6.700012
USING POLY OF DEGREE 4, DERIV IS  6.685
        THE SECOND DERIVATIVE IS  6.675016
WHAT X-VALUE FOR DERIV ?
Break in 250
Ok
```

Program 4.2 Gaussian Quadrature

Figure 4.7 lists a program written in Ada that performs Gaussian integration with formulas of two, three, or four terms. (The Ada language is important because it is the standard for software used by the Department of Defense.) The program begins by declaring variables that define the limits for the integration and the number of terms to be used. Functions are specified that define the integrand and that carry out the computations. The computation is simple: A weighted sum is computed at the proper "Gauss points." (A CASE statement is employed to use the specified number of terms.) The output from the program is given at the end of Fig. 4.7.

```
with TEXT_IO, FLOAT_MATH_FUNCTIONS;
use TEXT_IO, FLOAT_MATH_FUNCTIONS;
--------------------------------------------------------------------
--                        Chapter 4
--
-- Gerald/Wheatley. Applied Numerical Analysis, (fifth edition)
--                        Addison-Wesley, 1994
--
-- This program implements the Gaussian quadrature method
-- presented in Chapter 4. However, it can handle only 2,3,4
-- terms depending on the parameter, nterms.
--------------------------------------------------------------------
procedure MAIN is
    package I_IO is new INTEGER_IO (INTEGER);
    use I_IO;
    package F_IO is new FLOAT_IO (FLOAT);
    use F_IO;

    A      : FLOAT   := 0.2; -- lower bound
    B      : FLOAT   := 1.5; -- upper bound
    NTERMS : INTEGER := 4;   -- number of terms to be used
    --------------------------------------------------------------
    function F (X : FLOAT) return FLOAT is

    begin
        return (1.0 / EXP (X * X));
    end F;
    --------------------------------------------------------------
    -- Input:
    --      a, b    : lower and upper bounds of definite interval.
    --      nterms : number of terms to be used.
    -- Output: the value of the integral.
    --------------------------------------------------------------
    function GAUSSQ (A, B   : FLOAT;
                     NTERMS : INTEGER) return FLOAT is
        type TEMP_ARRAY is array (1 .. 4) of FLOAT;
        T,             -- vector of T-values in [-1,1]
        W : TEMP_ARRAY; -- vector of weights
        BPLUSA,        -- (b + a) / 2
        BLESSA,        -- (b - a) / 2
        SUM : FLOAT;
    begin -- gaussq
        case NTERMS is
            when 2 =>
                T (1) := - 0.5773502691;
                T (2) := - T (1);
                W (1) := 1.0;
                W (2) := 1.0;
```

Figure 4.7 Program 4.2

Figure 4.7 *(continued)*

```
         when 3 =>
            T (1) := - 0.774596692;
            T (2) := 0.0;
            T (3) := - T (1);
            W (1) := 0.55555555556;
            W (2) := 0.88888888889;
            W (3) := W (1);
         when 4 =>
            T (1) := - 0.8611363116;
            T (2) := - 0.3399810436;
            T (3) := - T (2);
            T (4) := - T (1);
            W (1) := 0.3478548451;
            W (2) := 0.6521451549;
            W (3) := W (2);
            W (4) := W (1);
         when others =>
            null;
      end case;
      -- Compute integral from above values
      BLESSA := (B - A) / 2.0;
      BPLUSA := (B + A) / 2.0;
      SUM    := 0.0;
      for I in 1 .. NTERMS loop
         SUM := SUM + W (I) * F (BLESSA * T (I) + BPLUSA);
      end loop;
      return BLESSA * SUM;
   end GAUSSQ;
begin -- main
   PUT ("The value of the integral is: ");
   PUT (GAUSSQ (A, B, NTERMS), 4, 7, 0);
   NEW_LINE;
end MAIN;
===============================================================
                 Output for GAUSSQDT.ADA
         The value of the integral is: 0.6588290
===============================================================
```

Program 4.3 Adaptive Integration

The third program of this chapter uses FORTRAN to carry out adaptive integration. It follows the algorithm of Section 4.10 quite closely but also counts the number of times that the function is evaluated. A DATA statement in the driver program sets values for the limits of integration and tolerance. A call to subroutine ADAPT is made to do the integration. This uses two other subroutines to store and retrieve values (PUSH and POP), and a function subprogram to compute function values. A listing of the program appears in Fig. 4.8. Output from the program is shown at the end of the figure.

```
      PROGRAM PADAPT
C
C -----------------------------------------------------------
C                         CHAPTER 4
C
C  GERALD/WHEATLEY, APPLIED NUMERICAL ANALYSIS (FIFTH EDITION)
C                   ADDISON-WESLEY, 1994
C
C -----------------------------------------------------------
```

Figure 4.8 Program 4.3

Figure 4.8 *(continued)*

```
C
C               THIS PROGRAM IMPLEMENTS THE ADAPTIVE SIMPSON
C               INTEGRATION PRESENTED IN THIS CHAPTER
C
C      -------------------------------------------------------------
C
       REAL F, A,B,TOL, ANSWER
       INTEGER  MAXCNT, FCNCNT
       EXTERNAL F
       COMMON MAXCNT, FCNCNT
C
       DATA A,B,TOL/0.2, 1.0, 0.050/
C
C
C      -------------------------------------------------------------
C
C
C          MAXCNT: MAXIMUN NUMBER OF FUNCTION CALLS
C          FCNCNT: ACTUAL NUMBER OF FUNCTION CALLS
C
C      -------------------------------------------------------------
C
       MAXCNT = 500
       FCNCNT = 0
C
C
       CALL ADAPT(F, A,B,TOL,  ANSWER)
C
C          PRINT ANSWER = INTEGRAL(F) FROM A TO B
C
       PRINT 100, ANSWER
        PRINT 200, FCNCNT
100    FORMAT(///,10X,'THE INTEGRAL VALUE IS ', F10.6/)
200    FORMAT(10X,'THE NUMBER OF FUNCTION CALLS = ', I4///)
       STOP
       END
C
C          SUBROUTINE ADAPT
C
       SUBROUTINE ADAPT(F,   A,B,TOL,   ANSWER)
C
C
C      -------------------------------------------------------------
C
C
       THIS SUBROUTINE USES THE ADAPTIVE SIMPSON
C      METHOD AS PRESENTED IN THIS CHAPTER.
C
C
C      -------------------------------------------------------------
C
       REAL A,B,ANSWER,TOL,
     +     H1,H2,C,D,E,FA,FB,FC,FD,FE,
     +     S1AB,S2AB,S2AC,S2CB,  STORE
       REAL STACK(7,100)
       INTEGER COUNT, TOP , MAXCNT,FCNCNT
C
       COMMON MAXCNT, FCNCNT
       COMMON /SUB/STACK, TOP
C
       STORE = 0.0
       COUNT = 0
       H1 = (B-A)/2.0
       C = A + H1
       FA = F(A)
       FB = F(B)
       FC = F(C)
       S1AB = H1*(FA + 4.0*FC + FB)/3.0
       TOP = 0
```

Figure 4.8 *(continued)*

```
          CALL  PUSH(A,FA,FC,FB,H1,TOL,S1AB)
C
10        CONTINUE
              COUNT = COUNT + 1
              CALL POP(A,FA,FC,FB,H1,TOL,S1AB)
              H2 = H1/2.0
              D = A + H2
              E = A + 3.0*H2
              C = A + H1
              B = A + 2.0*H1
              FD = F(D)
              FE = F(E)
              S2AC = H2*(FA + 4.0*FD + FC)/3.0
              S2CB = H2*(FC + 4.0*FE + FB)/3.0
C
              S2AB = S2AC + S2CB
C
              IF (ABS(S2AB - S1AB) .LT. TOL )  THEN
                  STORE = STORE + S2AB + (S2AB - S1AB)/15.0
              ELSE
                  H1 = H2
                  TOL = TOL/2
                  CALL PUSH(A,FA,FD,FC,H1,TOL,S2AC)
                  CALL PUSH(C,FC,FE,FB,H1,TOL,S2CB)
              END IF
C
              IF (TOP .GT. 0  .AND. COUNT .LT. MAXCNT) GO TO 10
C
C             RETURN ANSWER
C
          ANSWER = STORE
          RETURN
          END
C
C --------------------------------------------------------------
C
C
C         SUBROUTINE PUSH WHICH STORES THE VALUES IN A STACK
C
C --------------------------------------------------------------
C
          SUBROUTINE PUSH(LPT,LVAL,MIDVAL,RTVAL,STPSZE, TOL,INTGRL)
C
          REAL LPT,LVAL,MIDVAL,RTVAL,STPSZE,TOL,INTGRL
          REAL STACK(7,100)
          INTEGER TOP
          COMMON /SUB/ STACK,TOP
C
              TOP = TOP + 1
              STACK(1,TOP) = LPT
              STACK(2,TOP) = LVAL
              STACK(3,TOP) = MIDVAL
              STACK(4,TOP) = RTVAL
              STACK(5,TOP) = STPSZE
              STACK(6,TOP) = TOL
              STACK(7,TOP) = INTGRL
          RETURN
          END
C
C --------------------------------------------------------------
C
C         SUBROUTINE PUSH RETURNS THE STORED VALUES FROM THE STACK
C
C
C --------------------------------------------------------------
C
          SUBROUTINE POP(LPT,LVAL,MIDVAL,RTVAL,STPSZE, TOL,INTGRL)
C
```

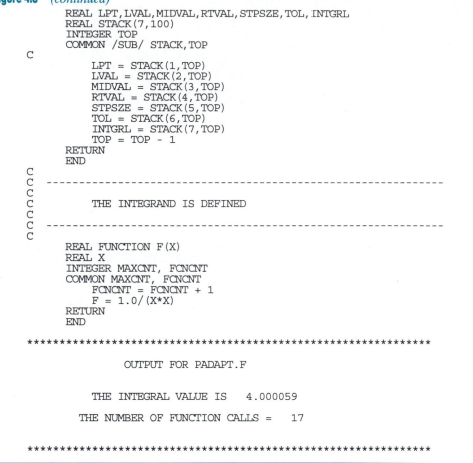

Figure 4.8 *(continued)*

```
          REAL LPT,LVAL,MIDVAL,RTVAL,STPSZE,TOL,INTGRL
          REAL STACK(7,100)
          INTEGER TOP
          COMMON /SUB/ STACK,TOP
C
          LPT = STACK(1,TOP)
          LVAL = STACK(2,TOP)
          MIDVAL = STACK(3,TOP)
          RTVAL = STACK(4,TOP)
          STPSZE = STACK(5,TOP)
          TOL = STACK(6,TOP)
          INTGRL = STACK(7,TOP)
          TOP = TOP - 1
       RETURN
       END
C
C     -----------------------------------------------------------
C
C         THE INTEGRAND IS DEFINED
C
C     -----------------------------------------------------------
C
       REAL FUNCTION F(X)
       REAL X
       INTEGER MAXCNT, FCNCNT
       COMMON MAXCNT, FCNCNT
          FCNCNT = FCNCNT + 1
          F = 1.0/(X*X)
       RETURN
       END

  ****************************************************************

                  OUTPUT FOR PADAPT.F

          THE INTEGRAL VALUE IS    4.000059

          THE NUMBER OF FUNCTION CALLS =    17


  ****************************************************************
```

Exercises

Section 4.2

1. For the following divided-difference table, compute a value for $f'(0.242)$ from a quadratic polynomial that fits the table at $i = 1, 2,$ and 3 (the best choice?).

i	x_i	f_i	$f_i^{[1]}$	$f_i^{[2]}$	$f_i^{[3]}$
0	0.15	0.1761	2.4355	−5.7505	15.3476
1	0.21	0.3222	1.9754	−3.9088	8.7492
2	0.23	0.3617	1.7409	−2.9464	5.9642
3	0.27	0.4314	1.4757	−2.2307	
4	0.32	0.5051	1.2973		
5	0.35	0.5441			

2. Estimate the error in the answer to Exercise 1 from the next-term rule. The function actually is $1 + \log_{10}(x)$. Compare the estimate to the actual error.

▶**3.** Repeat Exercise 1 but this time for $f'(x)$ at $x = 0.21, 0.22, 0.23, 0.24, 0.25, 0.26,$ and 0.27. Plot the estimates and compare to a graph of the true values. Make another plot of the errors versus x. At what point is the error smallest?

4. Use Eq. (4.4) to find bounds for the errors at $x = 0.21, 0.23,$ and 0.27 in Exercise 3. Do these bounds bracket the errors found in Exercise 3?

5. Use the next-term rule to estimate the error in

Exercise 3. Compare these errors with the actual errors. Are the estimates always larger?

6. Using the table of Exercise 1, get $f'(0.242)$ from quadratics that begin with i-values of
   a. $i = 0$.
   b. $i = 2$.
   c. $i = 3$.
   Do these results confirm that using $i = 1$ is the best starting value?

7. What degree of polynomial gives the most accurate value of $f'(0.242)$ from the table of Exercise 1?

▶**8.** The following difference table is for $f(x) = x + \sin(x)/3$. Use it to find
   a. $f'(0.72)$ from a cubic polynomial.
   b. $f'(1.33)$ from a quadratic.
   c. $f'(0.50)$ from a fourth-degree polynomial.
   In each part, choose the best starting i-value.

i	x_i	f_i	Δf_i	$\Delta^2 f_i$	$\Delta^3 f_i$	$\Delta^4 f_i$
0	0.30	0.3985	0.2613	−0.0064	−0.0022	0.0003
1	0.50	0.6598	0.2549	−0.0086	−0.0018	0.0004
2	0.70	0.9147	0.2464	−0.0104	−0.0014	0.0005
3	0.90	1.1611	0.2360	−0.0118	−0.0010	
4	1.10	1.3971	0.2241	−0.0128		
5	1.30	1.6212	0.2113			
6	1.50	1.8325				

9. What are the errors in Exercise 8? For part (c), use Eq. (4.7) to find bounds on the error. Do these bounds bracket the actual error?

10. Use the next-term rule to estimate the errors for each part of Exercise 8. Compare these errors with the actual errors. Is the estimate always larger?

▶**11.** Use the central-difference formula, Eq. (4.9), to compute $f'(0.50)$ from the data of Exercise 8. Find the maximum and minimum estimates of the error, and compare them to the actual error.

12. Repeat Exercise 11, this time using a forward-difference formula obtained by using only the first term of Eq. (4.8). Also compute the error by the next-term rule.

13. Repeat Exercise 12, but now use two terms of Eq. (4.8).

14. Use a central-difference formula to estimate $f'(0.66)$ if $f(x) = e^x/(x - 2)$. Also compute with one term of Eq. (4.8), the forward-difference formula. Compare the errors of each to the actual error for
   a. $h = 0.1$.

b. $h = 0.01$.
c. $h = 0.001$.

15. Repeat Exercise 14, but now round all calculated values to three decimal places to see the effect of round-off.

Section 4.3

16. Multiply the operator of Eq. (4.11) by itself to get formulas for the third and fourth derivatives of $f(x_i)$. (A symbolic algebra program can help in doing the multiplications!) Confirm that the first term of these formulas matches Eq. (4.13).

17. Use the formulas of Exercise 16 to compute the third and fourth derivatives of $f(x) = e^{-x/2}$ at $x = 0.3$ using $h = 0.05$.

▶**18.** Use one, two, and three terms of Eq. (4.12) to estimate the second derivative of $f(x) = e^x/(x - 2)$ at $x = 0.5$ with $h = 0.05$. Estimate the errors by the next-term rule, and compare to the actual errors.

19. Derive a central-difference approximation to $f^{iv}(x)$ in a way similar to that of Eq. (4.17). What order of error will it have? Can you do the same for the third derivative?

20. Use the formula of Exercise 19 to estimate the fourth derivative of $e^x/(x - 2)$ at $x = 0.5$ with $h = 0.05$. Repeat with $h = 0.025$. Do the errors of these two estimates agree with the order of the error?

21. Another way to derive formulas for derivatives is from a Taylor-series expansion of $f(x)$ about the point x_0. For example, if we now subtract the series for $x = x_0 - h$ from that where $x = x_0 + h$ and solve for $f'(x_0)$, we get the central-difference formula as well as its error term and order of error. Carry out this operation to confirm this.

▶**22.** Repeat Exercise 21 to get the forward-difference formula for $f'(x_0)$. Repeat again to get a backward-difference formula in terms of x_0 and $x_0 - h$.

23. Use the Taylor-series technique of Exercise 21 to derive both central- and forward-difference formulas for the second derivative together with their error terms.

Section 4.4

24. Use the extrapolation technique of Section 4.4 to get $f'(0.90)$ from the table of Exercise 8 to an $O(h^6)$ accuracy.

25. Show that the first extrapolation for $f'(x_0)$ with h-values differing by 2 to 1 is the same as the formula

$$f_0' = \frac{1}{H}\left(\frac{\Delta f_{-1} + \Delta f_0}{2} - \frac{1}{6}\frac{\Delta^3 f_{-2} + \Delta^3 f_{-1}}{2}\right),$$

where H is the smaller of the h's.

26. Consider whether extrapolations similar to that of Section 4.4 can be used for unevenly spaced data and a divided-difference table. (Taylor-series expansions as described in Exercise 21 may be helpful.) If you succeed in getting a formula for $f'(x)$, use it to estimate $f'(0.27)$ for the data in Exercise 1. What order of error results?

▶27. Apply Richardson extrapolation to get $f'(0.32)$ accurate to five significant figures for $f(x) = \sin^2(x/2)$, starting with $h = 0.1$ and using central differences. When the extrapolations agree to five significant figures, are they that accurate?

28. Repeat Exercise 27, but now for $f''(0.32)$.

▶29. Can Richardson extrapolation be applied to forward-difference approximations? If you think they can, repeat Exercise 27 but this time with forward differences.

30. Write a computer program to implement the algorithm of Section 4.4 to create a Richardson table.

Section 4.5

31. Use the Newton–Cotes formula for $n = 1$ applied twice to estimate the integral of $f(x) = e^{-x} * \sin(x)$ over $[1, 3]$ with each application having $h = 1$. Compare the actual error with the error bounds computed from Eq. (4.24) applied to each subinterval.

32. Repeat Exercise 31, this time using the Newton–Cotes formula for $n = 2$ applied once to the whole interval ($h = 2$). Use Eq. (4.26) to get the error bounds.

33. Repeat Exercise 31, but now use the Newton–Cotes formula for $n = 3$ applied once to the whole interval ($h = 2/3$). Use Eq. (4.28) to get the error bounds.

▶34. Derive the Newton–Cotes formulas for $n = 4$ and $n = 5$.

35. Beginning with the divided-difference form of the interpolating polynomial, rederive the New-

ton–Cotes formulas for $n = 1, 2$, and 3. (Assume, of course, that the points are evenly spaced.)

Section 4.6

36. The following table has values for $f(x)$. Integrate between $x = 1.0$ and $x = 1.8$ using the trapezoidal rule (Eq. 4.30) with
   a. $h = 0.1$.
   b. $h = 0.2$.
   c. $h = 0.4$.

x	$f(x)$
1.0	1.543
1.1	1.669
1.2	1.811
1.3	1.971
1.4	2.151
1.5	2.352
1.6	2.577
1.7	2.828
1.8	3.107

▶37. The function tabulated in Exercise 36 is $\cosh(x)$. What are the errors in parts (a), (b), and (c)? How closely are these proportional to h^2? What other errors are present besides the truncation error?

38. In Example 4.8, the integral of e^x between $x = 1.8$ and $x = 3.4$ was computed with the trapezoidal rule using $h = 0.2$ and has an error of -0.08. It is shown that this falls within the bounds given by Eq. (4.32). But if we don't know what $f(x)$ is, we cannot use that equation. Estimate the error using the second differences of the tabulated data.

39. If we want the integral of Example 4.8 correct to five decimal places (error < 0.000005), how small must h be?

40. Find the integral of $f(x)$ between $x = 0$ and $x = 2$ for

x	$f(x)$
0.00	1.0000
0.12	0.8869
0.53	0.5886
0.87	0.4190
1.08	0.3396
1.43	0.2393
2.00	0.1353

▶**41.** Use Romberg integration to evaluate the integral of $f(x) = 1/x$ between $x = 1$ and $x = 2$. Carry six decimals, and continue until there is no change in the fifth place. Compare to the analytical value, $\ln(2)$.

42. Apply Romberg extrapolation to the results of Exercise 36.

43. Extrapolate to the limit to get the integral of the following data between 0 and 1.00:

x	$f(x)$
0.00	0.3989
0.25	0.3867
0.50	0.3521
0.75	0.3011
1.00	0.2420

Section 4.7

44. Repeat Exercise 36, but use Simpson's $\frac{1}{3}$ rule.

45. Use the error expression for Eq. (4.35) to find bounds on the error of Exercise 44. Does the actual error fall within these bounds? [The function tabulated in Exercise 36 is $\cosh(x)$].

▶**46.** Evaluate the integral of e^x between $x = 0$ and $x = 1$ with a value of h small enough to guarantee five-decimal-place accuracy. What is the maximum size for h?

47. Show that one extrapolation of the trapezoidal rule is identical to Simpson's $\frac{1}{3}$ rule with a comparable value for h.

48. Compute the integral of $f(x) = \sin(x)/x$ between $x = 0$ and $x = 1$ using Simpson's $\frac{1}{3}$ rule with $h = 0.5$ and then with $h = 0.25$. (Remember that the limit of $\sin(x)/x$ at $x = 0$ is 1.) From these two results, extrapolate to get a better result. What is the order of the error after the extrapolation? Compare your answer with the true answer.

49. Repeat Exercise 46 but use Simpson's $\frac{3}{8}$ rule.

50. The function tabulated here is given at points that divide seven panels. That means that Simpson's $\frac{1}{3}$ rule does not fit, but we can use a combination of the $\frac{1}{3}$ and $\frac{3}{8}$ rules. Compare the results when the $\frac{3}{8}$ rule is applied at different subintervals within [3.0, 6.5]. Which of these gives the smallest error? [The function of the table is $f(x) = 3 + \cos(x)/(x - 1.5)$.]

x	$f(x)$
3.0	2.34000
3.5	2.53177
4.0	2.73854
4.5	2.92973
5.0	3.08105
5.5	3.17717
6.0	3.21337
6.5	3.19532

▶**51.** Simpson's $\frac{1}{3}$ rule, although based on passing a quadratic through three evenly spaced points, actually gives the exact answer if $f(x)$ is a cubic. The implication is that the area under any cubic between $x = a$ and $x = b$ is identical to the area of a parabola that matches the cubic at $x = a$, $x = b$, and $x = (a + b)/2$. Prove this.

Section 4.8

52. Use the symbolic method to derive a formula for three-panel integration, using a polynomial of degree m.

53. Integration from x_0 to x_1, as in Eq. (4.38), is an arbitrary choice taken for convenience only. If the limits were taken from x_{-1} to x_0, a similar formula for one-panel integration would result, but different coefficients would be obtained. Carry out the integrations.

54. Perform computations similar to those in Exercise 53, but get a two-panel integration formula from x_{-1} to x_1.

▶**55.** Use the method of undetermined coefficients to derive the trapezoidal rule.

56. Repeat Exercise 55 but for Simpson's $\frac{3}{8}$ rule.

57. Use the method of undetermined coefficients to derive the central-difference formulas, Eqs. (4.9) and (4.17).

Section 4.9

▶**58.** Evaluate the integral of e^x between $x = 0$ and $x = 1$ using three-term Gaussian quadrature. Compare the accuracy of your answer with that of Exercise 46. If you get five-decimal-place accuracy, how many fewer function evaluations are required with Gaussian quadrature?

59. Evaluate the integral of $\sin(x)/x$ between $x = 0$ and $x = 1$ with a four-term Gaussian formula. What is the error? How small must h be with Simpson's $\frac{1}{3}$ rule to match this accuracy?

▶60. By using Gaussian formulas of increasing complexity, determine how many terms are needed to evaluate the integral of $x^3 * \sin(x^2)e^{x-3}$ over the interval $[-1.5, 2.7]$ to get accuracy to six significant figures.

61. An n-term Gaussian formula assumes that a polynomial of degree $2n - 1$ is used to fit the function between $x = a$ and $x = b$. Does this mean that the error is the same as for a Newton–Cotes integration formula based on a polynomial of degree $2n - 1$?

62. The t-values given in Table 4.14 are the zeros of Legendre polynomials that can be constructed as shown in Section 4.9. Using any method from Chapter 1, confirm the values for $n = 3$, 4, and 5.

Section 4.10

63. Use a calculator to verify the computations in Example 4.16.

64. Redo Example 4.16 using an adaptive trapezoidal rule method and obtaining the same accuracy. How many more function evaluations would be required if the adaptive scheme were not used?

▶65. Evaluate the integral of $x * \exp(-2x^2)$ between $x = 0$ and $x = 2$ using a tolerance value, TOL, sufficiently small as to get an answer within 0.1% of the true answer, 0.249916.

66. Most programs for adaptive integration will compute the appropriate step size if they use the procedure of Section 4.10. However, in some cases, this leads to significant errors. For instance, the integral of $\sin^2(16x)$ between $x = 0$ and $x = \pi/2$ is $\pi/4$, but it is easy to see that the values of $S_1[0, \pi/2]$ and $S_2[0, \pi/2]$ both equal zero, where $h_1 = \pi/4$ and $h_2 = \pi/8$.

How can we solve this problem correctly with the adaptive method of Section 4.10? (It is interesting to know that the HP-15C calculator avoids this error.)

67. Implement the algorithm for adaptive integration in a computer program.

Section 4.11

68. In Example 4.17 the statement is made, "It is immaterial which integral we evaluate first." Confirm this statement by repeating Example 4.17, but this time integrating with respect to y first and holding x constant.

69. Write pictorial operators similar to that of Eq. (4.45) for
   ▶a. Simpson's $\frac{1}{3}$ rule in the x-direction and the trapezoidal rule in the y-direction.
   b. Simpson's $\frac{1}{3}$ rule in both directions.
   c. Simpson's $\frac{3}{8}$ rule in both directions.
   ▶d. What conditions are placed on the number of panels in both directions by parts (a), (b), and (c)?

70. Since Simpson's $\frac{1}{3}$ rule is exact when $f(x)$ is a cubic, evaluation of the following triple integral should be exact. Confirm by evaluating both numerically and analytically. Use Eq. (4.49) adapted for this integral.

$$\int_0^1 \int_0^2 \int_{-1}^0 x^3 y z^2 \, dx \, dy \, dz$$

71. Draw a pictorial operator that represents the formula used in Exercise 70. You may want to do this with three widely separated planes such as:

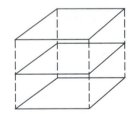

72. Evaluate the following integral, and compare your answers to the analytical solution. Use $h = 0.1$ in both directions in parts (a) and (b),
   a. using the trapezoidal rule in both directions.
   b. using Simpson's $\frac{1}{3}$ rule in both directions.
   c. using Gaussian quadrature, three-term formulas in both directions.

$$\int_{-0.2}^{1.4} \int_{0.4}^{2.6} e^y \cos(x) \, dy \, dx$$

73. Solve Exercise 72 by performing the trapezoidal rule integrations first with $h = 0.2$ (in both directions), then with $h = 0.1$, and extrapolate. The answer should match parts (b) of the exercise. Does it?

▶74. Integrate with varying values of Δx and Δy using the trapezoidal rule in both directions, and show that the error decreases about in proportion to h^2:

$$\int_0^1 \int_0^1 (x^2 + y^2)\, dx\, dy.$$

Section 4.12

75. Integrate $\int\int \cos(x) \sin(2y)\, dx\, dy$ over the region defined by the portion of a unit circle that lies in the first quadrant. Integrate first with respect to x holding y constant, using $h = 0.25$. Subdivide the vertical lines into four panels.
   a. Use the trapezoidal rule.
   b. Use Simpson's $\frac{1}{3}$ rule.

76. The order of integration in multiple integrations can usually be changed. Repeat Exercise 75, but now integrate first with respect to y. Compare the answer with that of Exercise 75.

77. Integrate $\exp(-x^2 y^2)$ over the region bounded by the two parabolas $y = x^2$ and $y = 2x^2 - 1$. Observe that, since the integrand is an even function and the region is symmetrical about the y-axis, the integration may be over half the area and then doubled. Choose reasonable values for the number of panels in each direction and use any formula you prefer.

78. Use Gaussian quadrature to solve Exercise 75, with
   ▶**a.** three-term formulas.
   b. four-term formulas.

79. In performing finite element analysis (described in Chapters 7 and 8), integrations over a rectangular region with sides parallel to the axes are required. Gaussian quadrature is a preferred method. If two-term formulas are employed, we need to evaluate the integrand at only four points within the rectangle. These points are called "Gauss points." Where are the Gauss points for the rectangle with opposite corners at $(2.1, 3.7)$ and $(5.2, 7.0)$?

Section 4.13

80. The following table is for $f(x) = 1/(x + 2)$. Find values for $f'(x)$ and $f''(x)$ at $x = 1.5$, 2.0, and 2.5 from cubic spline functions that approximate $f(x)$. Compare to the true values to determine the errors. Also compare to derivative values computed from central-difference formulas.

a. Use end condition 1.
b. Use end condition 3.
c. Use end condition 4.

x	1.0	1.5	2.0	2.5	3.0
$f(x)$	0.333	0.286	0.250	0.222	0.200

81. Plot the values of $f'(x)$ and $f''(x)$ from the cubic splines of Exercise 80 on $[1.0, 3.0]$, and compare to plots of the true values.

82. The comparisons in Exercise 80 may favor the cubic spline because they are based on cubic polynomials, whereas the central-difference formulas are based on quadratics. Repeat Exercise 80, but now use interpolating polynomials of degrees 3 and 4.

▶83. Repeat Exercise 80, but this time use cubic splines that have the correct slopes at the ends, condition 2.

▶84. Integrate sech(x) over $[0, 2]$ by integrating the natural cubic spline curve (end condition 1) that fits at five evenly spaced points on $[0, 2]$. Compare the result to the analytical value. Also compare to the integral from Simpson's $\frac{1}{3}$ rule.

85. Repeat Exercise 83 using end conditions 2, 3, and 4. For condition 2, use the analytical values for $f'(x)$.

86. Repeat Exercise 84, but now force the values of $f''(x)$ at the ends to the analytical values of the second derivative of sech(x).

Section 4.14

87. Use the Taylor series method to derive expressions for $f'(x)$ and $f''(x)$ and their error terms using f-values that precede f_0. (These are called backward-difference formulas.)

88. Repeat Exercise 87, but now use the symbolic operator method.

89. Repeat Exercise 87, but this time use the method of undetermined coefficients. (See Appendix B.)

90. Develop the expanded form for the error terms of the Newton–Cotes integration formulas for $n = 4$ and $n = 5$.

91. Prove that *all* Newton–Cotes integration formulas of even order have error terms of order 1 greater than expected.

▶92. Demonstrate by varying the size of h that the local error for Simpson's $\frac{1}{3}$ rule is $O(h^5)$ and that the global error is $O(h^4)$. Repeat using several choices for $f(x)$.

93. Demonstrate with several different polynomials that the two-term Gaussian quadrature formula is exact for all polynomials of degree 3 or less.

94. Assume that $f(x) = \sin(x^2/2)$. Make a table similar to Table 4.20 to show that the error of $f'(x)$ and $f''(x)$ does not continually decrease as h is made smaller.

95. Repeat Exercise 94 but for forward-difference formulas. Are the best values for h the same?

96. Repeat Exercise 94, but now use the trapezoidal rule to integrate between $x = 0$ and $x = 0.4$. How does the optimum h-value compare with that for differentiation?

97. Repeat Exercise 96, but now use Simpson's $\frac{1}{3}$ rule.

Section 4.15

98. Use *Mathematica* or some other symbolic algebra program to get the analytical values for the derivatives in

a. Exercise 2.
b. Exercise 9.
c. Exercise 14.
d. Exercise 27.
e. Exercise 45.
f. Exercise 83.
g. Exercise 86.

99. Use *Mathematica* or some other symbolic algebra program to get the analytical values for the integrals in

a. Exercise 31.
b. Exercise 37.
c. Exercise 46.
d. Exercise 48.
e. Exercise 50.
f. Exercise 59.
g. Exercise 60.
h. Exercise 70.
i. Exercise 72.

▶100. Use *Mathematica* or some other symbolic algebra program to do the multiplications in Exercise 16.

101. Use *Mathematica* or some other symbolic algebra program to get the zeros in Exercise 62.

Applied Problems and Projects

102. Differential thermal analysis is a specialized technique that can be used to determine transition temperatures and the thermodynamics of chemical reactions. It has special application in the study of minerals and clays. Vold [*Anal. Chem.* 21, 683 (1949)] describes the technique. In this method, the temperature of a sample of the material being studied is compared to the temperature of an inert reference material when both are heated simultaneously under identical conditions. The furnace housing the two materials is normally heated so that its temperature T_f increases (approximately) linearly with time (t), and the difference in temperatures (ΔT) between the sample and the reference is recorded. Some typical data are

t, min	0	1	2	3	4	5	6	7
ΔT, °F	0.00	0.34	1.86	4.32	8.07	13.12	16.80	18.95
T_f, °F	86.2	87.8	89.4	91.0	92.7	94.3	95.9	97.5

t	8	9	10	11	12	13	14	15	16
ΔT	18.07	16.69	15.26	13.86	12.58	11.40	10.33	8.95	6.46
T_f	99.2	100.8	102.3	103.9	105.5	107.1	108.6	110.2	111.8

t	17	18	19	20	21	22	23	24	25
ΔT	4.65	3.37	2.40	1.76	1.26	0.88	0.63	0.42	0.30
T_f	113.5	115.1	116.8	118.4	120.0	121.6	123.2	124.9	126.5

The ΔT values increase to a maximum, then decrease, due to the heat evolved in an exothermic reaction. One item of interest is the time (and furnace temperature) when the reaction is complete. Vold shows that the logarithm of ΔT should decrease linearly after the reaction is over; while the chemical reaction is occurring, the data depart from this linear relation. She used a graphical method to find this point. Perform numerical computations to find, from the preceding data, the time and the furnace temperature when the reaction terminates. Compare the merits of doing it graphically or numerically.

103. The temperature difference data in Problem 102 can be used to compute the heat of reaction. To do this, the integral of the values of ΔT is required, from the point where the reaction begins (which is at the point where ΔT becomes nonzero) to the time when the reaction ceases, as found in Exercise 102. Determine the value of the required integral. Which of the methods of this chapter should give the best value for the integral?

104. *Fugacity* is a term used by engineers to describe the available work from an isothermal process. For an ideal gas, the fugacity f is equal to its pressure P, but for real gases,

$$\ln \frac{f}{P} = \int_0^P \frac{C-1}{P}\, dp,$$

where C is the experimentally determined *compressibility factor*. For methane, values of C are

P (atm)	C	P (atm)	C
1	0.9940	80	0.3429
10	0.9370	120	0.4259
20	0.8683	160	0.5252
40	0.7043	250	0.7468
60	0.4515	400	1.0980

Write a program that reads in the P and C values and uses them to compute and print f corresponding to each pressure given in the table. Assume that the value of C varies linearly between the tabulated values (a more precise assumption would fit a polynomial to the tabulated C values). The value of C approaches 1.0 as P approaches 0.

105. The stress developed in a rectangular bar when it is twisted can be computed if one knows the values of a torsion function U that satisfies a certain partial-differential equation. Chapter 7 describes a numerical method that can determine values of U. To compute the stress, it is necessary to integrate $\iint U\, dx\, dy$ over the rectangular region for which the data given here apply. Determine the stress. (You may be able to simplify the integration because of the symmetry in the data.)

x \ y	0.0	0.2	0.4	0.6	0.8	1.0	1.2
0.0	0	0	0	0	0	0	0
0.2	0	2.043	3.048	3.354	3.048	2.043	0
0.4	0	3.123	4.794	5.319	4.794	3.123	0
0.6	0	3.657	5.686	6.335	5.686	3.657	0
0.8	0	3.818	5.960	6.647	5.960	3.818	0
1.0	0	3.657	5.686	6.336	5.686	3.657	0
1.2	0	3.123	4.794	5.319	4.794	3.123	0
1.4	0	2.043	3.048	3.354	3.048	2.043	0
1.6	0	0	0	0	0	0	0

106. Make a critical comparison of the accuracy of Newton–Cotes integration formulas compared to Gaussian quadrature. Test the formulas for a variety of functions for which you can calculate the integrals analytically. Select some functions that are smooth, some that have sharp changes in value, and some with periodic behavior.

107. Write a general-purpose subroutine that performs Gaussian quadrature. You should have the subroutine change the limits of the integration appropriately and call a function subprogram to compute function values, with the name of the function subprogram passed as an argument. It should also receive as an argument the degree of the formula to be employed.

108. The data in Exercise 80 exhibit round-off in three of the entries. By recalculating the spline function, with more and less accurate values of these entries, determine how the values of S are affected by the precision of the data. Also calculate how the precision affects the estimates of $f'(x)$ and $f''(x)$. Compare the effects of the precision of the original data, using a cubic spline, with the effects of precision on the values of $f'(x)$ and $f''(x)$ when central-difference formulas are used.

5

Numerical Solution of Ordinary Differential Equations

Contents of This Chapter

The subject of ordinary differential equations not only is one of the most beautiful parts of mathematics, but it is also an essential tool for modeling many physical situations: spring–mass systems, resistor–capacitor–inductance circuits, bending of beams, chemical reactions, pendulums, the motion of a rotating mass around another body, and so forth. These equations have also demonstrated their usefulness in ecology and economics. The predator–prey problem has become a classic example of differential equations.

The prominence of ordinary differential equations in applied mathematics is due to the fact that most scientific laws are more readily expressed in terms of rates of change. For example,

$$\frac{du}{dt} = -0.27(u - 60)^{5/4}$$

is an equation describing (approximately) the rate of change of temperature u of a body losing heat by natural convection with constant-temperature surroundings. This is termed a first-order differential equation because the highest-order derivative is the first.

If the equation contains derivatives of nth order, it is said to be an nth-order differential equation. For example, a second-order equation describing the oscillation of a weight acted upon by a spring, with resistance to motion proportional to the square of the velocity, might be

$$\frac{d^2x}{dt^2} + 4\left(\frac{dx}{dt}\right)^2 + 0.6x = 0,$$

where x is the displacement and t is time.

The solution to a differential equation is the function that satisfies the differential equation and that also satisfies certain initial conditions on the function. In solving a differential equation analytically, we usually find a general solution containing arbitrary constants and then evaluate the arbitrary constants so that the expression agrees with the initial conditions. For an nth-order equation, n independent initial conditions must usually be known.* The analytical methods are limited to certain special forms of equations; elementary courses normally treat only linear equations with constant coefficients when the degree of the equation is higher than the first. Neither of these examples is linear.

Numerical methods have no such limitations to only standard forms. We obtain the solution as a tabulation of the values of the function at various values of the independent variable, however, and not as a functional relationship. We must also pay a price for our ability to solve practically any equation, in that we must recompute the entire table if the initial conditions are changed.

Our procedure will be to explore several methods of solving first-order equations, and then to show how these same methods can be applied to systems of simultaneous first-order equations and to higher-order differential equations. We will use for our typical first-order equation the form

$$\frac{dy}{dx} = f(x, y),$$

$$y(x_0) = y_0.$$

5.1 The Spring–Mass Problem—A Variation

Outlines the solution of an example of the spring–mass problem that differs from the usual case in that the spring is not in line with the motion of the weight. This leads to a nonlinear second-order differential equation.

5.2 Taylor-Series Method

Is a straightforward adaptation of classic calculus to develop the solution as an infinite series. The catch is that a computer usually cannot be programmed to construct the terms and one doesn't know how many terms should be used.

5.3 Euler and Modified Euler Methods

Are simple to use but subject to error unless the step size Δx is made very small.

5.4 The Runge–Kutta Methods

Are very popular because of their good efficiency; they are used in most computer programs for differential equations. They are single-step methods, as are the Euler methods. In this section we compare the methods presented so far.

---

* In later chapters we will study differential equations that are subject to boundary conditions as well as initial conditions.

Chapter Summary

Lists the important topics of this chapter.

Computer Programs

Gives the listing and description of two implementations of the algorithms of this chapter.

Parallel Processing

As you will see when you study the algorithms of this chapter, numerical methods for solving ordinary differential equations (posed as an initial value problem) do not lend themselves to parallel processing. We quote from a recent conference summary (*SIAM News*, November 1991):

> Attempts to apply parallelism to ordinary differential equations are met with an inherent difficulty—the initial value problem propagates information forward in time and therefore has an essentially serial character. If a large system is to be integrated, it is possible to divide the equations among processors and use parallelism across space, but issues of synchronization and communication can become very important. Because these issues are to some extent problem-dependent, little progress has been made in developing a general theory. . . .
>
> The progress that has been made is in low-degree parallelism, in which a few processors are applied to different parts of the same step using Runge–Kutta-like methods.

The report pointed out that research in this area is continuing and that there are several ways in which parallelism can be profitable.

5.1   The Spring–Mass Problem—A Variation

Every student of differential equations has studied the spring–mass problem, in which the motion of a weight that is suspended from a spring is determined. We will look at a variation of this problem. Instead of hanging from a spring, the mass moves horizontally along a frictionless bar and is connected by a spring to a support located centrally below the bar. Figure 5.1 shows this system.

Suppose that the spring has an unstretched length of $\sqrt{10} \doteq 3.162$ meters, the spring constant, k, is 100 N/m (newtons per meter), and the mass, m, is 3 kg. It is easy to see that the equilibrium position of the weight is at $x = 1.0$ m (or -1.0 m). Let x be the distance to the mass from the center of the bar and y be the length of the spring when the mass is at point x. Let y_0 be the unstretched length of the spring ($y_0 = \sqrt{10}$ meters). Then the differential equation that describes the motion of the mass is this second-order differential equation:

$$\frac{d^2x}{dt^2} = -\frac{k}{m}x\left(1 - \frac{y_0}{\sqrt{x^2 + 9}}\right).$$

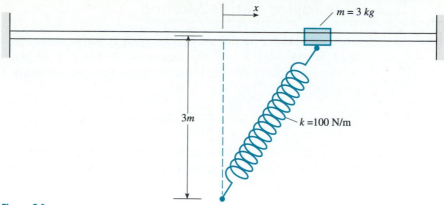

Figure 5.1

The solution depends on how far we place the weight to the right or left initially, x_0, and on the initial velocity of the mass, x_0'. We could solve this problem analytically, but our emphasis will be on doing it numerically.

While we postpone the description of methods to solve such problems numerically until Section 5.11, we will show a plot of the motion. Figure 5.2 shows two different solutions, part (a) with $x_0 = 1.4$ and part (b) with $x_0 = 2.5$. (The initial velocity is zero for both.) Observe in part (a) that the spring is stretched only slightly, and that the mass does not cross the center point of the bar.

These plots were generated by an important tool for engineers: a program called ACSL.

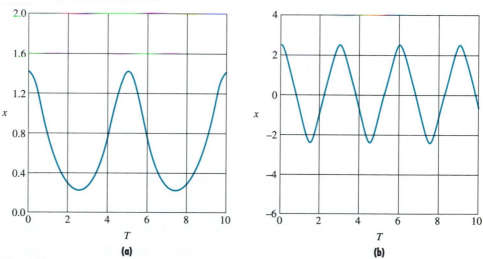

Figure 5.2

5.2 The Taylor-Series Method

The first method we discuss is not strictly a numerical method, but it is sometimes used in conjunction with the numerical schemes, is of general applicability, and serves as an introduction to the other techniques we will study. Consider the example problem

$$\frac{dy}{dx} = -2x - y, \qquad y(0) = -1. \tag{5.1}$$

(This particularly simple example is chosen to illustrate the method so that you can readily check the computational work. The analytical solution,

$$y(x) = -3e^{-x} - 2x + 2$$

is obtained immediately by application of standard methods and will be compared with our numerical results to show the error at any step.)

We develop the relation between y and x by finding the coefficients of the Taylor series in which we expand y about the point $x = x_0$:

$$y(x) = y(x_0) + y'(x_0)(x - x_0) + \frac{y''(x_0)}{2!}(x - x_0)^2 + \frac{y'''(x_0)}{3!}(x - x_0)^3 + \cdots.$$

If we let $x - x_0 = h$, we can write the series as

$$y(x) = y(x_0) + y'(x_0)h + \frac{y''(x_0)}{2}h^2 + \frac{y'''(x_0)}{6}h^3 + \cdots. \tag{5.2}$$

Since $y(x_0)$ is our *initial condition*, the first term is known from the initial condition $y(0) = -1$. (Since the expansion is about the point $x = 0$, our Taylor series is actually the Maclaurin series in this example.)

We get the coefficient of the second term by substituting $x = 0$, $y = -1$ in the equation for the first derivative, Eq. (5.1):

$$y'(x_0) = y'(0) = -2(0) - (-1) = 1.$$

We get the second- and higher-order derivatives by successively differentiating the equation for the first derivative. Each of these derivatives is evaluated corresponding to $x = 0$ to get the various coefficients:

$$y''(x) = -2 - y', \qquad y''(0) = -2 - 1 = -3,$$
$$y'''(x) = -y'', \qquad y'''(0) = 3,$$
$$y^{iv}(x) = -y''', \qquad y^{iv}(0) = -3.$$

(Getting the derivatives is deceptively easy in this example. You need only compare this to another function—say, $f(x, y) = x/(y - x^2)$, with $y(0) = 1$—to see how quickly the differentiation can become messy.) We then write our series solution for y, letting $x = h$ be the value at which we wish to determine y:

$$y(h) = -1 + 1.0h - 1.5h^2 + 0.5h^3 - 0.125h^4 + \text{error}.$$

The solution to the differential equation, $dy/dx = -2x - y$, where $y(0) = -1$, is then given by Table 5.1. As we computed the last two entries, we were doubtful about their accuracy without having used more terms of the Taylor series, because the successive terms were decreasing less and less rapidly. Thus we need more terms than we have calculated to get four-decimal-place accuracy.

Table 5.1

x	y	y, analytical
0.0	-1.00000	-1.00000
0.1	-0.91451	-0.91451
0.2	-0.85620	-0.85619
0.3	-0.82251	-0.82245
0.4	-0.81120	-0.81096
0.5	-0.82031	-0.81959

When a convergent Taylor series is truncated, the error is simple to express. We merely take the next term and evaluate the derivative at the point $x = \xi, 0 < \xi < h$, instead of at the point $x = x_0$. This is exactly what we did to write error terms in the previous chapters. The error term of the Taylor series after the h^4 term is

$$\text{Error} = \frac{y^{(v)}(\xi)}{5!} h^5, \qquad 0 < \xi < h.$$

However, this cannot be computed, because evaluating the derivative at $x = \xi$ is impossible with ξ unknown, and even bounding it in the interval $[0, h]$ is impossible because the derivatives are known only at $x = 0$ and not at $x = h$.

Numerical analysis is sometimes termed an art instead of a science because, in situations like this, the number of Taylor-series terms to be included is a matter of judgment and experience. We normally truncate the Taylor series when the contribution of the last term is negligible to the number of decimal places to which we are working. However, this is correct only when the succeeding terms become small rapidly enough—in some cases the sum of the many neglected small terms is significant.

The Taylor series is easily applied to a higher-order equation. For example, if we are given

$$y'' = 3 + x - y^2, \qquad y(0) = 1, \qquad y'(0) = -2,$$

we can find the derivative terms in the Taylor series as follows:

$y(0)$, and $y'(0)$ are given by the initial conditions.

$y''(0)$ comes from substitution into the differential equation from $y(0)$ and $y'(0)$.

$y'''(0)$ and higher derivatives are found by differentiating the equation for the previous order of derivative and substituting previously computed values.

5.3 Euler and Modified Euler Methods

As we have seen in the previous section, the Taylor-series method may be awkward to apply if the derivatives become complicated and in this case the error is difficult to determine. Of course, many computer packages, including MAPLE and DERIVE, can use symbolic methods to differentiate a function. In fact, even a calculator like the HP-48S can differentiate a function, but the analytic derivatives often grow very complicated.

We know that the error in a Taylor series will be small if the step size h is small. In fact, if we make h small enough, we may only need a few terms of the Taylor-series expansion for good accuracy. The Euler method follows this idea to the extreme for first-order differential equations—it uses only the first two terms of the Taylor series! Suppose that we have chosen h small enough that we may truncate after the first-derivative term. Then

$$y(x_0 + h) = y(x_0) + y'(x_0) + \frac{y''(\xi)h^2}{2}, \qquad x_0 < \xi < x_0 + h,$$

where we have written the usual form of the error term for the truncated Taylor series.

In using this equation, the value of $y(x_0)$ is given by the initial condition and $y'(x_0)$ is evaluated from $f(x_0, y_0)$ which is given by the differential equation, $dy/dx = f(x, y)$. It is necessary to use this method iteratively, advancing the solution to $x = x_0 + 2h$ after $y(x_0 + h)$ has been computed, and then on to $x = x_0 + 3h$, and so on. Adopting a subscript notation for the successive y-values and representing the error by the order relation, we may write the algorithm for the Euler method in the form[*]

$$y_{n+1} = y_n + hy'_n + O(h^2). \tag{5.3}$$

As an example, we apply this to the previous equation

$$\frac{dy}{dx} = -2x - y, \qquad y(0) = -1,$$

where the computation can be done rather simply. It is convenient to arrange the work as in Table 5.2. Here, we take $h = 0.1$.

Table 5.2

x_n	y_n	y'_n	hy'_n
0.0	−1.00000	1.00000	0.10000
0.1	−0.90000	0.70000	0.07000
0.2	−0.83000	0.43000	0.04300
0.3	−0.78700	0.18700	0.01870
0.4	−0.76830	−0.03170	

(Analytical answer is −0.81096, error is −0.04266.)

[*] This error is just the local error. Over many steps, the global error becomes $O(h)$.

Each of the y_n values is computed using Eq. (5.3), adding hy_n' and y_n of the previous line. Comparing the last result to the analytical answer $y(0.40) = -0.81096$, we see that there is only one-decimal-place accuracy, even though we have advanced the solution only four steps! To gain four-decimal-place accuracy, we must reduce the error by more than 400-fold. Since the global error is about proportional to h, we will need to reduce the step size about 426-fold, to <0.00024.

The trouble with this most simple method is its lack of accuracy, requiring an extremely small step size. Figure 5.3 suggests how we might improve this method with just a little additional effort.

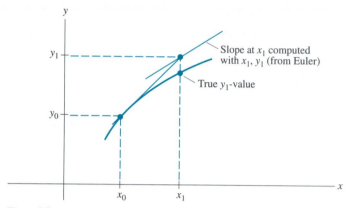

Slope at x_1 computed with x_1, y_1 (from Euler)

True y_1-value

Figure 5.3

In the simple Euler method, we use the slope at the beginning of the interval, y_n', to determine the increment to the function. This technique would be correct only if the function were linear. What we need instead is the correct average slope within the interval. This can be approximated by the mean of the slopes at both ends of the interval.

Suppose we use the arithmetic average of the slopes at the beginning and end of the interval to compute y_{n+1}:

$$y_{n+1} = y_n + h\frac{y_n' + y_{n+1}'}{2}. \tag{5.4}$$

This should give us an improved estimate for y at x_{n+1}. However, we are unable to employ Eq. (5.4) directly, since the derivative is a function of both x and y and we cannot evaluate y_{n+1}' with the true value of y_{n+1} unknown. The modified Euler method works around this problem by estimating or "predicting" a value of y_{n+1} by the simple Euler relation, Eq. (5.3). It then uses this value to compute y_{n+1}', giving an improved estimate (a "corrected" value) for y_{n+1}. Since the value of y_{n+1}' was computed using the predicted value, of less than perfect accuracy, we might want to recorrect the y_{n+1} value as many times as will make a significant difference. (However, if more than one or two recorrections are required, it is more efficient to reduce the step size or use a different method.)

We will illustrate the modified Euler method, which we also call the Euler predictor–corrector method, on the problem we previously treated. Table 5.3 shows

Table 5.3

x_n	y_n	hy_n'	$y_{n+1,p}$	$hy_{n+1,p}'$	hy_{av}'	$y_{n+1,c}$
0.0	−1.0000	0.1000	−0.9000	0.0700	0.0850	−0.9150
0.1	−0.9150	0.0715	−0.8435	0.0444	0.0579	−0.8571
0.2	−0.8571	0.0457	−0.8114	0.0211	0.0334	−0.8237
0.3	−0.8237	0.0224	−0.8013	0.0001	0.0112	−0.8124
0.4	−0.8124	0.0012	−0.8112	−0.0189	−0.0088	−0.8212
0.5	−0.8212					

$$[y(0.5) = -0.8196, \text{ the analytical value}]$$

all the steps in going from $x = 0.0$ to $x = 0.5$. To show the method as simply as possible, we did not recorrect at any step in the table.

However, in computing from $x = 0.0$ to $x = 0.2$, we will get improved results if we recorrect. The next display shows the results of recorrecting. (Lines 2 and 4 are where we recorrect.)

x_n	y_n	hy_n'	$y_{n+1,p}$	$hy_{n+1,p}'$	hy_{av}'	$y_{n+1,c}$
0.0	−1.0000	0.1000	−0.9000	0.0700	0.0850	−0.9150
			−0.9150	0.0715	0.0858	−0.9142
0.1	−0.9142	0.0714	−0.8428	0.0443	0.0578	−0.8564
			−0.8564	0.0456	0.0585	−0.8557
0.2	−0.8557					

We work as before, but recorrect when there is a significant difference between $y_{n+1,p}$ and $y_{n+1,c}$, that is, the predicted and corrected values for the next y_{n+1}. It is obvious that this recorrected value, −0.9142, is closer to the exact value, $y(0.1) = -0.9145$. The corrected value at $x = 0.2$, −0.8557, is also closer to the exact value. However, we would be far better off to use a more efficient method such as the Runge–Kutta methods of Section 5.4 rather than using many recorrections in the modified Euler method. It is true, however, that this method is somewhat more efficient than the Euler method of Eq. (5.3).

We can find the error of the modified Euler method by comparing it with the Taylor series:

$$y_{n+1} = y_n + y_n'h + \frac{1}{2}y_n''h^2 + \frac{y'''(\xi)}{6}h^3, \qquad x_n < \xi < x_n + h.$$

Replace the second derivative by the forward-difference approximation for y'', $(y_{n+1}' - y_n')/h$, which has error of $O(h)$, and write the error term as $O(h^3)$:

$$y_{n+1} = y_n + h\left(y_n' + \frac{1}{2}\left[\frac{y_{n+1}' - y_n'}{h} + O(h)\right]h\right) + O(h^3),$$

$$y_{n+1} = y_n + h\left(y_n' + \frac{1}{2}y_{n+1}' - \frac{1}{2}y_n'\right) + O(h^3),$$

$$y_{n+1} = y_n + h\left(\frac{y_n' + y_{n+1}'}{2}\right) + O(h^3).$$

This shows that the error of one step of the modified Euler method is $O(h^3)$. This is the local error. There is an accumulation of errors from step to step, so that the error over the whole range of application, the so-called global error, is $O(h^2)$. This seems intuitively reasonable, since the number of steps into which the interval is subdivided is proportional to $1/h$; hence the order of error is reduced to $O(h^2)$ on continuing the technique. We shall treat the accumulation of errors for the simple Euler method more fully in Section 5.9.

(We take note of the fact that the MAPLE package can solve our example problem analytically as well as numerically.)

5.4 Runge–Kutta Methods

The two numerical methods of the last section, though not very impressive, serve as a good introduction to our next procedures. While we can improve the accuracy of these two methods by taking smaller step sizes, much greater accuracy can be obtained more efficiently by a group of methods named after two German mathematicians, Runge and Kutta. They developed algorithms that solve a differential equation efficiently and yet are the equivalent of approximating the exact solution by matching the first n terms of the Taylor-series expansion. We will consider only the fourth- and fifth-order Runge–Kutta methods, even though there are higher-order methods. Actually, the modified Euler method of the last section is a second-order Runge–Kutta method.

To impart some idea of how the Runge–Kutta methods are developed, we will show the derivation of a simple second-order method. Here, the increment to the y is a weighted average of two estimates of the increment which we call k_1 and k_2. Thus for the equation $dy/dx = f(x, y)$,

$$
\begin{aligned}
y_{n+1} &= y_n + ak_1 + bk_2, \\
k_1 &= hf(x_n, y_n), \\
k_2 &= hf(x_n + \alpha h, y_n + \beta k_1).
\end{aligned}
\tag{5.5}
$$

We can think of the values k_1 and k_2 as estimates of the change in y when x advances by h, because they are the product of the change in x and a value for the slope of the curve, dy/dx. The Runge–Kutta methods always use the simple Euler estimate as the first estimate of Δy; the other estimate is taken with x and y stepped up by the fractions α and β of h and of the earlier estimate of $\Delta y, k_1$. Our problem is to devise a scheme of choosing the four parameters, a, b, α, β. We do so by making Eq. (5.5) agree as well as possible with the Taylor-series expansion, in which the y-derivatives are written in terms of f, from $dy/dx = f(x, y)$,

$$
y_{n+1} = y_n + hf(x_n, y_n) + \frac{h^2}{2} f'(x_n, y_n) + \cdots .
$$

An equivalent form, since $df/dx = f_x + f_y\, dy/dx = f_x + f_y f$, is

$$y_{n+1} = y_n + h f_n + h^2\left(\frac{1}{2} f_x + \frac{1}{2} f_y f\right)_n. \tag{5.6}$$

[All the derivatives in Eq. (5.6) are evaluated at the point (x_n, y_n).] We now rewrite Eq. (5.6) by substituting the definitions of k_1 and k_2:

$$y_{n+1} = y_n + ahf(x_n, y_n) + bhf[x_n + \alpha h, y_n + \beta h f(x_n, y_n)]. \tag{5.7}$$

To make the last term of Eq. (5.7) comparable to Eq. (5.6), we expand $f(x, y)$ in a Taylor series in terms of x_n, y_n, remembering that f is a function of two variables,* retaining only first derivative terms:

$$f[x_n + \alpha h, y_n + \beta h f(x_n, y_n)] \doteq (f + f_x \alpha h + f_y \beta h f)_n. \tag{5.8}$$

On the right side of both Eqs. (5.6) and (5.8) f and its partial derivatives are all to be evaluated at (x_n, y_n).

Substituting from Eq. (5.8) into Eq. (5.7), we have

$$y_{n+1} = y_n + ahf_n + bh(f + f_x \alpha h + f_y \beta h f)_n,$$

or, rearranging,

$$y_{n+1} = y_n + (a + b)hf_n + h^2(\alpha b f_x + \beta b f_y f)_n. \tag{5.9}$$

Equation (5.9) will be identical to Eq. (5.6) if

$$a + b = 1, \qquad \alpha b = \frac{1}{2}, \qquad \beta b = \frac{1}{2}.$$

Note that only three equations need to be satisfied by the four unknowns. We can choose one value arbitrarily (with minor restrictions); hence we have a set of second-order methods. For example, taking $a = \frac{2}{3}$, we have $b = \frac{1}{3}, \alpha = \frac{3}{2}, \beta = \frac{3}{2}$. Other choices give other sets of parameters that agree with the Taylor-series expansion. If we take $a = \frac{1}{2}$, the other variables are $b = \frac{1}{2}, \alpha = 1, \beta = 1$. This last set of parameters gives the modified Euler algorithm that we have previously discussed; the modified Euler method is a special case of a second-order Runge–Kutta method.

Fourth-order Runge–Kutta methods are most widely used and are derived in similar fashion. Greater complexity results from having to compare terms through h^4, and gives a set of 11 equations in 13 unknowns. The set of 11 equations can be solved with 2 unknowns being chosen arbitrarily. The most commonly used set of values leads to the algorithm

---

* Appendix A will remind readers of this expansion.

$$y_{n+1} = y_n + \frac{1}{6}(k_1 + 2k_2 + 2k_3 + k_4),$$

$$k_1 = hf(x_n, y_n),$$

$$k_2 = hf\left(x_n + \frac{1}{2}h, y_n + \frac{1}{2}k_1\right),$$

$$k_3 = hf\left(x_n + \frac{1}{2}h, y_n + \frac{1}{2}k_2\right),$$

$$k_4 = hf(x_n + h, y_n + k_3).$$

(5.10)

Using Eqs. (5.10) to apply the Runge–Kutta fourth-order to our problem, $dy/dx = -2x - y, y(0) = -1$ with $h = 0.1$, we obtain the results shown in Table 5.4. The results here are very impressive compared to those given in Table 5.1, where we computed the values using the terms of the Taylor series up to the h^4 term. Table 5.4 agrees to five decimals with the analytical result—illustrating a further gain in accuracy with less effort than with the Taylor series method of Section 5.2—and certainly is better than the Euler or modified Euler methods.

Table 5.4

x_n	y_n	k_1	k_2	k_3	k_4	Average
0.0	−1.0000	0.1000	0.0850	0.0858	0.0714	0.0855
0.1	−0.9145	0.0715	0.0579	0.0586	0.0456	0.0583
0.2	−0.8562	0.0456	0.0333	0.0340	0.0222	0.0337
0.3	−0.8225	0.0222	0.0111	0.0117	0.0011	0.0115
0.4	−0.8110	0.0011	−0.0090	−0.0085	−0.0181	−0.0086
0.5	−0.81959					

[$y(0.5) = -0.81959$, the analytical value]

Figure 5.4 illustrates the four slope values that are combined in the four k's of the Runge-Kutta method.

The local error term for the fourth-order Runge–Kutta method is $O(h^5)$; the global error would be $O(h^4)$. It is computationally more efficient than the modified Euler method because, while four evaluations of the function are required per step rather than two, the steps can be manyfold larger for the same accuracy. The Runge–Kutta techniques have been very popular, especially the fourth-order method just presented. Since going from second to fourth order was so beneficial, we may wonder whether we should use a still higher order of formula. Higher-order (fifth, sixth, and so on) Runge–Kutta formulas have been developed and can be used to advantage in determining a suitable size for h, as we will see.

One way to determine whether the Runge–Kutta values are sufficiently accurate

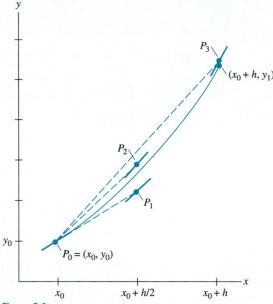

Figure 5.4

is to recompute the value at the end of each interval with the step size halved. If only a slight change in the value of y_{n+1} occurs, the results are accepted; if not, the step must be halved again until the results are satisfactory. This procedure is very expensive, however. For instance, to implement Eq. (5.10) this way, we would need an additional seven function evaluations to determine the accuracy of our y_{n+1}. The best case then would require $4 + 7 = 11$ function evaluations to go from (x_n, y_n) to (x_{n+1}, y_{n+1}).

A different approach uses two Runge–Kutta methods of different orders. For instance, we could use one fourth-order and one fifth-order method to move from (x_n, y_n) to (x_{n+1}, y_{n+1}). We would then compare our results at y_{n+1}. The Runge–Kutta–Fehlberg method, now one of the most popular of these methods, does just this. Only six functional evaluations (versus eleven) are required, and we also have an estimate of the error (the difference of the two y's at $x = x_{n+1}$):

An Algorithm for the Runge–Kutta–Fehlberg Method

$$k_1 = h \cdot f(x_n, y_n),$$

$$k_2 = h \cdot f\left(x_n + \frac{h}{4}, y_n + \frac{k_1}{4}\right),$$

$$k_3 = h \cdot f\left(x_n + \frac{3h}{8}, y_n + \frac{3k_1}{32} + \frac{9k_2}{32}\right),$$

$$k_4 = h \cdot f\left(x_n + \frac{12h}{13}, y_n + \frac{1932k_1}{2197} - \frac{7200k_2}{2197} + \frac{7296k_3}{2197} \right),$$

$$k_5 = h \cdot f\left(x_n + h, y_n + \frac{439k_1}{216} - 8k_2 + \frac{3680k_3}{513} - \frac{845k_4}{4104} \right),$$

$$k_6 = h \cdot f\left(x_n + \frac{h}{2}, y_n - \frac{8k_1}{27} + 2k_2 - \frac{3544k_3}{2565} + \frac{1859k_4}{4104} - \frac{11k_5}{40} \right);$$

$$\hat{y}_{n+1} = y_n + \left(\frac{25k_1}{216} + \frac{1408k_3}{2565} + \frac{2197k_4}{4104} - \frac{k_5}{5} \right), \text{ with global error } O(h^4),$$

$$y_{n+1} = y_n + \left(\frac{16k_1}{135} + \frac{6656k_3}{12825} + \frac{28561k_4}{56430} - \frac{9k_5}{50} + \frac{2k_6}{55} \right),$$

with global error $O(h^5)$;

$$\text{Error, } E \doteq \frac{k_1}{360} - \frac{128k_3}{4275} - \frac{2197k_4}{75240} + \frac{k_5}{50} + \frac{2k_6}{55}. \tag{5.11}$$

The basis for the Runge–Kutta–Fehlberg scheme is to compute two Runge–Kutta estimates for the new value of $\hat{y}_{n+1}$ but of different orders of errors. Thus, instead of comparing estimates of y_{n+1} for h and $h/2$, we compare the estimates $\hat{y}_{n+1}$ and y_{n+1} using fourth- and fifth-order (global) Runge–Kutta formulas. Moreover, both equations make use of the same k's, so only six function evaluations are needed versus the previous eleven. In addition, one can increase or decrease h depending on the value of the estimated error. As our estimate for the new y_{n+1}, we use the fifth-order (global) estimate.

As an example, we once more solve $dy/dx = -2x - y, y(0) = -1$ with $h = 0.1$, using the Runge–Kutta–Fehlberg method:

$$k_1 = 0.1,$$
$$k_2 = 0.0925000,$$
$$k_3 = 0.0889609,$$
$$k_4 = 0.0735157,$$
$$k_5 = 0.0713736,$$
$$k_6 = 0.0853872,$$

$$\hat{y}_1 = -0.914512212, \qquad y_1 = -0.914512251, \qquad \text{Error, } E = -0.000000040.$$

The exact value is $y(0.1) = -0.914512254$. Thus on the first step, y_1 agrees with the exact answer to 8 decimal places with only two additional function evaluations. Moreover, we have the value E to adjust our step size for the next iteration. Of course, we would use the more accurate y_{n+1} for the next step. This algorithm is well documented and implemented in the FORTRAN program, RKF45 (Forsythe, Malcolm, and Moler, 1977).

A summary and comparison of the four numerical methods we have studied for solving $y' = f(x, y)$ is presented in Table 5.5.

Table 5.5

Method	Estimate of slope over x-interval	Global error	Local error	Evaluations of $f(x,y)$ per step
Euler	Initial value	$O(h)$	$O(h^2)$	1
Modified Euler	Arithmetic average of initial and final predicted slope	$O(h^2)$	$O(h^3)$	2
Runge–Kutta (fourth-order)	Weighted average of four values	$O(h^4)$	$O(h^5)$	4
Runge–Kutta–Fehlberg	Weighted average of six values	$O(h^5)$	$O(h^6)$	6

Table 5.6 shows that the global errors in column 3 of Table 5.5 hold empirically. Once again, we will consider the example, $dy/dx = -2x - y, y(0) = -1$, where we solve this problem from $x = 0$ to $x = 0.4$ in a variety of step sizes and their corresponding errors. We will then look at the ratios of the consecutive step sizes.

Table 5.6

		Error in value computed at $x = 0.4$			Ratios of successive errors	
h	Euler	Modified Euler	Runge–Kutta 4th	Euler	Modified Euler	Runge–Kutta 4th
0.4000	2.11E-01	2.90E-02	2.40E-04			
0.2000	9.10E-02	6.42E-03	1.27E-05	3.3	4.5	18.9
0.1000	4.27E-02	1.44E-03	7.29E-07	2.1	4.5	17.4
0.0500	2.07E-02	3.48E-04	4.37E-08	2.1	4.1	16.7
0.0250	1.02E-02	8.54E-05	2.76E-09	2.0	4.1	15.8
0.0125	5.06E-03	2.11E-05	1.65E-10	2.0	4.0	16.7

In Table 5.6, we obtain the second row in this way: For a step size of $h = 0.2$, we compute the errors in the values for y at $x = 0.4$ using the three methods indicated at the top of columns two through four. We write down the values of the differences between the computed value and the exact value. The last three columns represent the ratio between the previous error (larger step size h) and the current. For instance, the 3.3 in the second row is the ratio of 2.11E-01/9.10E-2 for the errors from Euler's method for $h = 0.4$ and $h = 0.2$. We do the same for the modified Euler method and the Runge–Kutta fourth-order method in columns six and seven. We see that as h gets smaller, the last three columns approach the ratios of 2.0, 4.0, and 16.0. This is what we expect, since these three methods are, respectively, $O(h)$, $O(h^2)$, and $O(h^4)$ and since at each stage the step size is halved.

We end this section by showing the Runge–Kutta–Merson method, another fourth-order method even though five different k's must be computed. It can be seen from the formula that the order is given, not by the number of k's, but by the global error.

$$k_1 = h \cdot f(x_n, y_n),$$

$$k_2 = h \cdot f\left(x_n + \frac{h}{3}, y_n + \frac{k_1}{3}\right),$$

$$k_3 = h \cdot f\left(x_n + \frac{h}{3}, y_n + \frac{k_1}{6} + \frac{k_2}{6}\right),$$

$$k_4 = h \cdot f\left(x_n + \frac{h}{2}, y_n + \frac{k_1}{8} + \frac{3k_3}{8}\right),$$ (5.12)

$$k_5 = h \cdot f\left(x_n + h, y_n + \frac{k_1}{2} - \frac{3k_3}{2} + 2k_4\right);$$

$$y_{n+1} = y_n + \frac{(k_1 + 4k_4 + k_5)}{6} + O(h^5);$$

$$\text{Error, } E \doteq \frac{1}{30}(2k_1 - 9k_3 + 8k_4 - k_5).$$

As we have already indicated, there are methods that use Runge–Kutta formulas of orders 5, 6, and higher. In fact, the IMSL routine DVERK uses formulas or orders 5 and 6 that were developed by J. H. Verner. In this case, the method uses eight function evaluations.

Although the Runge–Kutta method has been very popular in the past, it has its limitations in solving certain types of differential equations. However, for a large class of problems the methods presented in this chapter produce some very stunning results. Also the technique introduced by Fehlberg in comparing two different orders rather than halving step sizes increases the efficiency of the Runge–Kutta methods.

The methods so far discussed are called single-step methods. They use only the information at (x_n, y_n) to get to (x_{n+1}, y_{n+1}). In the next sections we examine methods that utilize past information from previous points to get (x_{n+1}, y_{n+1}).

5.5 Multistep Methods

Runge–Kutta-type methods (which include Euler and modified Euler as special cases) are called single-step methods because they use only the information from the last step computed. In this they have the ability to perform the next step with a different step size and are ideal for beginning the solution where only the initial conditions are available. After the solution has begun, however, there is additional information available about the function (and its derivative) if we are wise enough to retain it in the memory of the computer. A multistep method is one that takes advantage of this fact.

The principle behind a multistep method is to utilize the past values of y and/or y' to construct a polynomial that approximates the derivative function, and extrapolate this into the next interval. Most methods use equispaced past values to make the construction of the polynomial easy. The Adams method is typical. The number of past points that are used sets the degree of the polynomial and is therefore responsible for the truncation error. The order of the method is equal to the power of h in the global error term of the formula, which is also equal to one more than the degree of the polynomial.

To derive the relations for the Adams method, we write the differential equation $dy/dx = f(x, y)$ in the form

$$dy = f(x, y)\, dx,$$

and we integrate between x_n and x_{n+1}:

$$\int_{x_n}^{x_{n+1}} dy = y_{n+1} - y_n = \int_{x_n}^{x_{n+1}} f(x, y)\, dx.$$

To integrate the term on the right, we approximate $f(x, y)$ as a polynomial in x, deriving this by making it fit at several past points. If we use three past points, the approximating polynomial will be a quadratic. If we use four points, it will be a cubic. The more points we use, the better accuracy (until round-off interferes, of course).

Suppose that we fit a second-degree polynomial through the last three points: $(x_n, f_n), (x_{n-1}, f_{n-1}), (x_{n-2}, f_{n-2})$. We could use an interpolating polynomial formula, but, for variety, we will use *Mathematica* and the InterpolatingPolynomial function (see Section 3.10). When we perform the operation, we get a quadratic approximation to the derivative function (but displayed differently):

$$f(x, y) = \frac{1}{2} h^2 (f_n - 2f_{n-1} + f_{n-2}) x^2$$

$$+ \frac{1}{2} h (3f_n - 4f_{n-1} + f_{n-2}) x + f_n.$$

Now we again use *Mathematica* to integrate between the limits of $x = x_n$ and $x = x_{n+1}$. The result is a formula for the increment in y (again displayed differently):

$$y_{n+1} - y_n = \frac{h}{12} (23f_n - 16f_{n-1} + 5f_{n-2}),$$

and we have the formula to advance y:

$$y_{n+1} = y_n + \frac{h}{12} [23f_n - 16f_{n-1} + 5f_{n-2}] + O(h^4). \tag{5.13}$$

Observe that Eq. (5.13) resembles the single-step formulas of the previous sections in that the increment to y is a weighted sum of the derivatives times the step size, but differs in that past values are used rather than estimates in the forward direction.

EXAMPLE 5.1 We illustrate the use of Eq. (5.13) to calculate $y(0.6)$ for $dy/dx = -2x - y, y(0) = -1$. We compute good values for $y(0.2)$ and $y(0.4)$ using a single-step method. In this case we obtain these values using the Runge–Kutta–Fehlberg method with $h = 0.2$. These values are given in Table 5.7.

Table 5.7

x	y	y, exact	$f(x,y)$
0.0	−1.0000000	−1.0000000	1.0000000
0.2	−0.8561921	−0.8561923	0.4561921
0.4	−0.8109599	−0.8109601	0.0109599

Then, from Eq. (5.13), we have

$$y(0.6) = -0.81096 + \frac{0.2}{12}[23(0.01096) - 16(0.45619) + 5(1.0)]$$

$$= -0.84508.$$

Comparing our result with the exact solution (-0.84643), we find that the computed value has an error of 0.00135. We can reduce the size of the error by doing the calculations with a smaller step size of 0.1. We use the fifth-order values of the Runge–Kutta–Fehlberg method once again to obtain the values in Table 5.8.

Table 5.8

x	y	y, exact	$f(x,y)$
0.0	−1.00000	−1.00000	1.00000
0.1	−0.91451	−0.91451	0.71451
0.2	−0.85619	−0.85619	0.45619
0.3	−0.82245	−0.82245	0.22245
0.4	−0.81096	−0.81096	0.01096
0.5	−0.81959	−0.81959	−0.18041

Using Eq. (5.13) again with the values for $f(x,y)$ at $x = 0.3, x = 0.4, x = 0.5$, from Table 5.8, we recompute $y(0.6)$:

$$y(0.6) = -0.81959 + \frac{0.1}{12}[23(-0.18041) - 16(0.01096) + 5(0.22245)]$$

$$= -0.84636,$$

which has an error of 0.00007.

In Section 5.7 we will see that we can reduce the error by using more past points for fitting a polynomial. In fact, when the derivation is done for four points to get a cubic approximation to $f(x, y)$, the following is obtained:

$$y_{n+1} = y_n + \frac{h}{24}[55f_n - 59f_{n-1} + 37f_{n-2} - 9f_{n-3}] + O(h^5). \qquad \textbf{(5.14)}$$

We repeat Example 5.1 with $h = 0.1$ to compute $y(0.6)$, now using values from Table 5.8 for $x = 0.2$, $x = 0.3$, $x = 0.4$, and $x = 0.5$, and computing

$$y(0.6) = -0.81959 + \frac{0.1}{24}[55(-0.18041) - 59(0.01096)$$

$$+ 37(0.22245) - 9(0.45619)]$$

$$= -0.84644.$$

The error of this computation has been reduced to 0.00001. We summarize the results of these two formulas in Table 5.9.

Table 5.9

Number of points used	Estimate of $y(0.6)$	Error $(h = 0.1)$
3	−0.8463626	0.000072
4	−0.8464420	0.000007

5.6 Milne's Method

The method of Milne is a multistep approach that first predicts a value for y_{n+1} by extrapolating values for the derivative. It differs from the Adams method in that it integrates over more than one interval. The past values that we require may have been computed by the Runge–Kutta method, or possibly by the Taylor-series method. In the Milne method, we suppose that four equispaced starting values of y are known, at the points x_n, x_{n-1}, x_{n-2}, and x_{n-3}. We can employ quadrature formulas to integrate as follows:

$$\frac{dy}{dx} = f(x, y),$$

$$\int_{x_{n-3}}^{x_{n+1}} \left(\frac{dy}{dx} \right) dx = \int_{x_{n-3}}^{x_{n+1}} f(x, y)\, dx = \int_{x_{n-3}}^{x_{n+1}} P_2(x)\, dx; \qquad \textbf{(5.15)}$$

therefore

$$y_{n+1} - y_{n-3} = \frac{4h}{3}(2f_n - f_{n-1} + 2f_{n-2}) + \frac{28}{90}h^5 y^v(\xi_1), \qquad x_{n-3} < \xi_1 < x_{n+1}.$$

We integrate the function $f(x, y)$ by replacing it with a quadratic interpolating polynomial that fits at the three points, where $x = x_n, x_{n-1}$, and x_{n-2}, and integrating according to the methods of Chapter 4.* Note that we extrapolate in the integration by one panel both to the left and to the right of the region of fit. Hence the error is larger, because of the extrapolation, than it would be with only interpolation.

With the value of y_{n+1} we can calculate f_{n+1} reasonably accurately. In Milne's method, we use Eq. (5.15) as a predictor formula and then correct with

$$\int_{x_{n-1}}^{x_{n+1}} \left(\frac{dy}{dx}\right) dx = \int_{x_{n-1}}^{x_{n+1}} f(x, y)\, dx = \int_{x_{n-1}}^{x_{n+1}} P_2(x)\, dx,$$

$$y_{n+1,c} - y_{n-1} = \frac{h}{3}(f_{n+1} + 4f_n + f_{n-1}) - \frac{h^5}{90}y^v(\xi_2), \qquad x_{n-1} < \xi_2 < x_{n+1}.$$

(5.16)

In Eq. (5.16), the polynomial P_2 is not identical to that in Eq. (5.15) because they do not fit the function at the same three points. In Eq. (5.16) the polynomial fits at x_{n+1}, x_n, and x_{n-1}. Note the changed range of integration and the smaller coefficient in the error term of Eq. (5.16) because the polynomial is not extrapolated; f_{n+1} is calculated using y_{n+1} from the predictor formula. The integration formula is the familiar Simpson's $\frac{1}{3}$ rule, since we integrate a quadratic over two panels within the region of fit. In this method we do not try to recorrect y_{n+1}.

We illustrate once again with our familiar simple problem, $dy/dx = -2x - y$, $y(0) = -1$, where the first four values are calculated using the Runge–Kutta fifth-order formula. See Table 5.10.

Table 5.10

x	y	$f(x, y)$	
0.0	-1.0000000	1.0000000	
0.1	-0.9145122	0.7145123	
0.2	-0.8561923	0.4561923	
0.3	-0.8224547	0.2224547	
0.4	(-0.8109678) predicted		
	(-0.8109596) corrected	$(-0.8109601$ exact$)$	
0.5	(-0.8195969) predicted		
	(-0.8195906) corrected	$(-0.8195920$ exact$)$	

---

* The formula can also be derived by the method of undetermined coefficients, as shown in Appendix B.

Normally, the values of y_{n+1} from the predictor and the corrector do not agree. Consideration of the error terms of Eqs. (5.15) and (5.16) suggests that the true value should usually lie between the two values and closer to the corrector value. While ξ_1 and ξ_2 are not necessarily the same value, they lie in similar intervals. If we assume that the values of $y^v(\xi_1)$ and $y^v(\xi_2)$ are equal, the error in the corrector formula is 1/28 times the error in the predictor formula. Hence the difference between the predictor and corrector formula is about 29 times the error in the corrected value. This is frequently used as a criterion of accuracy for Milne's method.* The ease with which we can monitor the error is a particular advantage of predictor–corrector methods. We are able to know immediately whether the step size is too big to give the desired degree of accuracy. This is in strong contrast to the Runge–Kutta method of the previous section (but not of Runge–Kutta–Fehlberg).

Milne's method is simple and has a good error term, $O(h^5)$, for local error. It is subject to an instability problem in certain cases, however, in which the errors do not tend to zero as h is made smaller. This unexpected phenomenon is discussed next. Because of possible instability, another method, a modification of that of Adams, is more widely used than Milne's method.

It is sufficient to show the instability of Milne's method for one simple case. We will be able to draw the necessary conclusions from this. Consider the differential equation $dy/dx = Ay$, where A is a constant. The general solution is $y = ce^{Ax}$. If $y(x_0) = y_0$ is the initial condition that the solution must satisfy, $c = y_0 e^{-Ax_0}$. Hence, letting y_n be the value of the function when $x = x_n$, the analytical solution is

$$y_n = y_0 e^{A(x_n - x_0)}. \tag{5.17}$$

If we solve the differential equation by the method of Milne, we have, from the corrector formula,

$$y_{n+1} = y_{n-1} + \frac{h}{3}(y'_{n+1} + 4y'_n + y'_{n-1}).$$

Letting $y'_n = Ay_n$, from the original differential equation, and rearranging, we get

$$y_{n+1} = y_{n-1} + \frac{h}{3}(Ay_{n+1} + 4Ay_n + Ay_{n-1}),$$

$$\left(1 - \frac{hA}{3}\right)y_{n+1} - \frac{4hA}{3}y_n - \left(1 + \frac{hA}{3}\right)y_{n-1} = 0. \tag{5.18}$$

We would like to solve this equation for y_n in terms of y_0 to compare to Eq. (5.17). Equation (5.18) is a second-order linear difference equation that can be solved in a manner analogous to that for differential equations. The solution is

$$y_n = C_1 Z_1^n + C_2 Z_2^n, \tag{5.19}$$

---

* A further criterion as to whether the single correction normally used in Milne's method is adequate is discussed in Section 5.9.

where Z_1, Z_2 are the roots of the quadratic

$$\left(1 - \frac{hA}{3}\right)Z^2 - \frac{4hA}{3}Z - \left(1 + \frac{hA}{3}\right) = 0. \tag{5.20}$$

[The reader should check that Eq. (5.19) is a solution of Eq. (5.18) by direct substitution.] For simplification, let $hA/3 = r$; the roots of Eq. (5.20) are then

$$Z_1 = \frac{2r + \sqrt{3r^2 + 1}}{1 - r},$$

$$Z_2 = \frac{2r - \sqrt{3r^2 + 1}}{1 - r}. \tag{5.21}$$

We are interested in comparing the behavior of Eqs. (5.19) and (5.17) as the step size h becomes small. As $h \to 0$, $r \to 0$, and $r^2 \to 0$ even faster. Neglecting the $3r^2$ terms in comparison to the constant 1 under the radical in Eq. (5.21) gives

$$Z_1 \doteq \frac{2r + 1}{1 - r} = 1 + 3r + O(r^2) = 1 + Ah + O(h^2),$$

$$Z_2 \doteq \frac{2r - 1}{1 - r} = -1 + r + O(r^2) = -\left(1 - \frac{Ah}{3}\right) + O(h^2). \tag{5.22}$$

The last results are obtained by dividing the fractions. We now compare Eq. (5.22) with the Maclaurin series of the exponential function,

$$e^{hA} = 1 + hA + O(h^2),$$

$$e^{-hA/3} = 1 - \frac{hA}{3} + O(h^2).$$

We see that, for $h \to 0$,

$$Z_1 \simeq e^{-hA}, \qquad Z_2 \simeq e^{-hA/3}.$$

Hence the Milne solution is represented by

$$y_n = C_1(e^{hA})^n + C_2(e^{-hA/3})^n = C_1 e^{A(x_n - x_0)} + C_2 e^{-A(x_n - x_0)/3}. \tag{5.23}$$

In Eq. (5.23), we have used $x_n - x_0 = nh$. The solution consists of two parts. The first term obviously agrees with the analytical solution, Eq. (5.17). The second term, called a *parasitic term*, will die out as x_n increases if A is a positive constant, but if A is negative, it will grow exponentially with x_n. Note that we get this peculiar behavior independent of h; smaller step size is of no benefit in eliminating the error.

A numerical example that demonstrates the instability of Milne's method is given in the next section. Such a demonstration by a numerical example is less conclusive than the analytical approach just given, but is much easier to grasp.

5.7 The Adams–Moulton Method

A method that does not have the same instability problem as the Milne method, but is about as efficient, is the Adams–Moulton method. It also assumes a set of starting values already calculated by some other technique. Here we take a cubic through four points, from x_{n-3} to x_n, and integrate over one step, from x_n to x_{n+1}. This is the same as the method described in Section 5.5. We assume that $dy/dx = f(x,y)$.

Thus, getting our formula for the cubic interpolating polynomial by the methods explained in Chapter 3 (or through a computer algebra system) and then integrating from x_n to x_{n+1}, we get the Adams–Moulton *predictor formula*:

$$\text{Predictor:}$$
$$y_{n+1} = y_n + \frac{h}{24}(55f_n - 59f_{n-1} + 37f_{n-2} - 9f_{n-3}) + \frac{251}{720}h^5 y^v(\xi_1). \quad \textbf{(5.24)}$$

(The last part of Eq. (5.24) is the error.) After we have computed a tentative value for y at x_{n+1}, we use it together with the y-values at x_n, x_{n-1}, and x_{n-2} to construct another cubic polynomial that we integrate from x_n to x_{n+1}. This gives the Adams–Moulton *corrector formula* (again with the last term being the error):

$$\text{Corrector:}$$
$$y_{n+1} = y_n + \frac{h}{24}(9f_{n+1} + 19f_n - 5f_{n-1} + f_{n-2}) - \frac{19}{720}h^5 y^v(\xi_2). \quad \textbf{(5.25)}$$

(These formulas can also be developed with the method of undetermined coefficients, as shown in Appendix B.)

We illustrate the Adams–Moulton method using our earlier example, $dy/dx = -2x - y, y(0) = -1$. Using Eqs. (5.24) and (5.25), we construct Table 5.11.

Table 5.11

x	y	$f(x,y)$	
0.0	-1.0000000	1.0000000	
0.1	-0.9145122	0.7145123	
0.2	-0.8561923	0.4561923	
0.3	-0.8224547	0.2224547	
0.4	(-0.8109687) predicted		
	(-0.8109652) corrected		(-0.8109601) exact
0.5	(-0.8195978) predicted		
	(-0.8195905) corrected		(-0.8195920) exact

Here is how the entries in the table were obtained. By the predictor formula of (5.24), we get

$$y(0.4) = -0.8224547 + \frac{0.1}{24}[55(0.2224547) - 59(0.4561923)$$
$$+ 37(0.7145123) - 9(1.0)]$$
$$= -0.8109687.$$

Then $f(0.4, -0.8109687)$ is computed, getting -0.0109688, and we use the corrector formula of Eq. (5.25) to get

$$y(0.4) = -0.8224547 + \frac{0.1}{24}[9(-0.0109688) + 19(0.2224547)$$
$$- 5(0.4561923) + 0.7145123]$$
$$= -0.8109652.$$

The computations are continued in the same manner to get $y(0.5)$. The corrected value almost agrees to five decimals with the predicted value. Comparing error terms of Eqs. (5.24) and (5.25) and assuming that the two fifth-derivative values are equal, we see that the true value should lie between the predicted and corrected values, with the error in the corrected value being about

$$\frac{19}{251 + 19} \quad \text{or} \quad \frac{1}{14.2}$$

times the difference between the predicted and corrected values. A frequently used criterion for accuracy of the Adams–Moulton method with four starting values is that the corrected value is not in error by more than 1 in the last place if the difference between predicted and corrected values is less than 14 in the last decimal place.[*] If this degree of accuracy is inadequate, we know that h is too large.

Changing the Step Size

When the predicted and corrected values agree to as many decimals as the desired accuracy, we can save computational effort by increasing the step size. We can conveniently double the step size, after we have seven equispaced values, by omitting every second one. When the difference between predicted and corrected values reaches or exceeds the accuracy criterion, we should decrease step size. If we interpolate two additional y-values with a fourth-degree polynomial, where the error will be $O(h^5)$, consistent with the rest of our work, we can readily halve the step size.[†]

---

[*] The convergence criterion of Section 5.9 should also be met.

[†] An alternative but computationally more expensive way to get intermediate points would be to compute them with Runge–Kutta formulas.

Convenient formulas for this are

$$y_{n-1/2} = \frac{1}{128}[35y_n + 140y_{n-1} - 70y_{n-2} + 28y_{n-3} - 5y_{n-4}],$$

$$y_{n-3/2} = \frac{1}{64}[-y_n + 24y_{n-1} + 54y_{n-2} - 16y_{n-3} + 3y_{n-4}]. \tag{5.26}$$

Use of these values with y_n, y_{n-1} gives four values of the function at intervals of $\Delta x = h/2$.

The efficiency of both the Adams–Moulton and Milne methods is about twice that of the Runge–Kutta–Fehlberg and Runge–Kutta methods. Only two function evaluations are needed per step for the former methods, whereas six or four are required with the single-step alternatives. All have similar error terms. Change of step size with the multistep methods is considerably more awkward, however.

Stability Considerations

As we have stressed, the advantage of the Adams–Moulton method over that of Milne is that it is a stable method rather than an unstable one. An analytical proof of the stability of Adams–Moulton for the equation $y' = Ay$ is similar to the analysis of Section 5.6—an equation results with parasitic terms that do die out as h gets small regardless of the sign of A. Such a treatment is not entirely satisfying because we can't prove stability by examining only one case (proving that an assertion is *not* true is always easier because we need to find only one counterexample).

It is much clearer to compare the stability of Adams–Moulton with Milne through a numerical example. The table in Fig. 5.5 presents results from a computer program that solved $y' = -10y, y(0) = 1$ (for which the analytical solution is $y = e^{-10x}$) over the interval $x = 0$ to $x = 2$. The first part of the table is computed with $h = 0.04$. In the second part, for which only partial output is given, $h = 0.004$.

Observe that the results with the Milne method grow to have very large relative errors. In fact, near $x = 2.0$, the relative error is practically 100%. The oscillatory behavior of the solution is also characteristic of instability. Even with the smaller h, the Milne method blows up. The errors are even greater near $x = 2.0$ in spite of the smaller step size. (When x is small the Milne solution is considerably more accurate when $h = 0.004$. Results are not shown for this.)

The results by the Adams–Moulton method do not show this anomalous behavior. The relative error, while growing to some degree, still stays manageable (8% at $x = 2.0$ with $h = 0.04$, $<0.1\%$ at $x = 2.0$ with $h = 0.004$). The expected decrease of error with decrease in step size is realized. The oscillation of values we notice in the results by Milne is absent. In sum, we conclude that the Adams–Moulton method gives good results, particularly at the smaller value of h, while the results of the Milne method are hopeless.

It is worth remarking that we usually don't have analytical answers to compare with our numerical results. Observing oscillatory behavior in itself does not mean instability, because the correct solution may *be* oscillatory. For this reason, the usual practice in numerical analysis is to entirely avoid methods that are sometimes unstable, even though they might be more accurate in some instances.

Figure 5.5   Comparison of relative error in results obtained by the Milne and Adams–Moulton methods

Values with $h = 0.04$	Solution using Milne method		Solution using Adams–Moulton method	
x	y	Error	y	Error
0.000000E 00	0.100000E 01	0.000000E 00	0.100000E 01	0.000000E 00
0.400000E–01	0.670320E 00	0.000000E 00	0.670320E 00	0.000000E 00
0.800000E–01	0.449329E 00	0.000000E 00	0.449329E 00	0.000000E 00
0.120000E 00	0.301195E 00	0.000000E 00	0.301195E 00	0.000000E 00
0.160000E 00	0.201667E 00	0.229836E–03	0.201552E 00	0.344634E–03
0.200000E 00	0.135271E 00	0.640750E–04	0.134873E 00	0.462234E–03
0.240000E 00	0.904572E–01	0.260949E–03	0.902753E–01	0.442863E–03
0.280000E 00	0.607594E–01	0.507981E–04	0.604129E–01	0.397373E–03
0.320000E 00	0.405497E–01	0.212613E–03	0.404273E–01	0.335068E–03
0.360000E 00	0.273069E–01	0.169165E–04	0.270547E–01	0.269119E–03
0.400000E 00	0.181599E–01	0.155795E–03	0.181052E–01	0.210475E–03
0.440000E 00	0.122874E–01	−0.100434E–04	0.121160E–01	0.161357E–03
0.480000E 00	0.811849E–02	0.111297E–03	0.810813E–02	0.121657E–03
0.520000E 00	0.554193E–02	−0.253431E–04	0.542601E–02	0.905804E–04
0.560000E 00	0.361742E–02	0.804588E–04	0.363111E–02	0.667726E–04
0.599999E 00	0.251041E–02	−0.316396E–04	0.242996E–02	0.488090E–04
0.639999E 00	0.160174E–02	0.598307E–04	0.162614E–02	0.354277E–04
0.679999E 00	0.114626E–02	−0.324817E–04	0.108822E–02	0.255615E–04
0.719999E 00	0.700667E–03	0.459237E–04	0.728243E–03	0.183482E–04
0.759999E 00	0.530960E–03	−0.305050E–04	0.487344E–03	0.131114E–04
0.799999E 00	0.299215E–03	0.362506E–04	0.326133E–03	0.933232E–05
0.839999E 00	0.252197E–03	−0.273281E–04	0.218250E–03	0.661935E–05
0.879999E 00	0.121497E–03	0.292376E–04	0.146054E–03	0.468052E–05
0.919999E 00	0.124879E–03	−0.238383E–04	0.977399E–04	0.330036E–05
0.959999E 00	0.437888E–04	0.239405E–04	0.654081E–04	0.232129E–05
0.999999E 00	0.658748E–04	−0.204745E–04	0.437714E–04	0.162897E–05
0.104000E 01	0.106330E–04	0.197998E–04	0.292920E–04	0.114074E–05
0.108000E 01	0.378257E–04	−0.174260E–04	0.196024E–04	0.797314E–06
0.112000E 01	−0.280455E–05	0.164789E–04	0.131180E–04	0.556327E–06
0.116000E 01	0.239183E–04	−0.147521E–04	0.877864E–05	0.387549E–06
0.120000E 01	−0.762395E–05	0.137682E–04	0.587471E–05	0.269573E–06
0.124000E 01	0.165678E–04	−0.124492E–04	0.393138E–05	0.187252E–06
0.128000E 01	−0.876950E–05	0.115303E–04	0.263090E–05	0.129902E–06
0.132000E 01	0.123371E–04	−0.104865E–04	0.176061E–05	0.900100E–07
0.136000E 01	−0.842895E–05	0.966946E–05	0.117821E–05	0.622986E–07
0.140000E 01	0.965523E–05	−0.882369E–05	0.788466E–06	0.430742E–07
0.144000E 01	−0.755810E–05	0.811550E–05	0.527645E–06	0.297529E–07

Figure 5.5 (*continued*)

	Solution using Milne method		Solution using Adams–Moulton method	
Values with $h = 0.04$ x	y	Error	y	Error
0.148000E 01	0.779361E–05	−0.741997E–05	0.353103E–06	0.205325E–07
0.152000E 01	−0.656401E–05	0.681446E–05	0.236298E–06	0.141571E–07
0.156000E 01	0.640522E–05	−0.623733E–05	0.158132E–06	0.975331E–08
0.160000E 01	−0.561102E–05	0.572355E–05	0.105823E–06	0.671417E–08
0.164000E 01	0.531755E–05	−0.524211E–05	0.708171E–07	0.461910E–08
0.168000E 01	−0.475746E–05	0.480803E–05	0.473911E–07	0.317556E–08
0.172000E 01	0.443906E–05	−0.440517E–05	0.317144E–07	0.218119E–08
0.176000E 01	−0.401659E–05	0.403931E–05	0.212234E–07	0.149759E–08
0.180000E 01	0.371683E–05	−0.370160E–05	0.142028E–07	0.102763E–08
0.184000E 01	−0.338345E–05	0.339366E–05	0.950460E–08	0.704578E–09
0.188000E 01	0.311713E–05	−0.311028E–05	0.636053E–08	0.482931E–09
0.192000E 01	−0.284671E–05	0.285129E–05	0.425650E–08	0.330768E–09
0.196000E 01	0.261645E–05	−0.261337E–05	0.284847E–08	0.226487E–09
0.200000E 01	−0.239358E–05	0.239564E–05	0.190621E–08	0.155008E–09

Values near $x = 2.0$ with $h = 0.004$

0.191192E 01	−0.734619E–05	0.735117E–05	0.496895E–08	0.425615E–11
0.191592E 01	0.744385E–05	−0.743908E–05	0.477411E–08	0.405009E–11
0.191992E 01	−0.752346E–05	0.752805E–05	0.458691E–08	0.392220E–11
0.192392E 01	0.762249E–05	−0.761808E–05	0.440706E–08	0.379785E–11
0.192792E 01	−0.770495E–05	0.770919E–05	0.423425E–08	0.368061E–11
0.193192E 01	0.780545E–05	−0.780138E–05	0.406822E–08	0.350298E–11
0.193592E 01	−0.789078E–05	0.789469E–05	0.390870E–08	0.339284E–11
0.193992E 01	0.799286E–05	−0.798910E–05	0.375544E–08	0.328626E–11
0.194392E 01	−0.808104E–05	0.808465E–05	0.360819E–08	0.318456E–11
0.194792E 01	0.818480E–05	−0.818133E–05	0.346671E–08	0.303002E–11
0.195192E 01	−0.827585E–05	0.827918E–05	0.333078E–08	0.293365E–11
0.195592E 01	0.838139E–05	−0.837819E–05	0.320017E–08	0.284017E–11
0.195992E 01	−0.847532E–05	0.847839E–05	0.307469E–08	0.274958E–11
0.196392E 01	0.858274E–05	−0.857979E–05	0.295413E–08	0.261657E–11
0.196792E 01	−0.867956E–05	0.868240E–05	0.283830E–08	0.253331E–11
0.197192E 01	0.878896E–05	−0.878623E–05	0.272701E–08	0.245226E–11
0.197592E 01	−0.888869E–05	0.889132E–05	0.262008E–08	0.237388E–11
0.197992E 01	0.900017E–05	−0.899765E–05	0.251735E–08	0.225930E–11
0.198392E 01	−0.910284E–05	0.910526E–05	0.241864E–08	0.218714E–11
0.198792E 01	0.921648E–05	−0.921415E–05	0.232380E–08	0.211720E–11
0.199192E 01	−0.932212E–05	0.932435E–05	0.223268E–08	0.204925E–11
0.199592E 01	0.943801E–05	−0.943586E–05	0.214514E–08	0.195066E–11
0.199992E 01	−0.954665E–05	0.954871E–05	0.206103E–08	0.188805E–11

5.8 Multivalued Methods

Although the multistep methods of Sections 5.6 and 5.7 have the advantage of advancing the solution of a differential equation with fewer function evaluations than Runge–Kutta methods for the same accuracy, they have two serious disadvantages. First, they require the computation of several previous steps by some single-step method. Second, they also require that these steps be at evenly spaced x-values; then the methods always advance the solution by these same equal step sizes.

The combination of these two problems with multistep methods may impose more closely spaced computations on the preliminary single-step computations than would otherwise be required to attain the required accuracy. The second problem means that changing the step size to maintain the required accuracy is awkward and computationally expensive.

One way to avoid both of these disadvantages uses a different multistep approach called a *multivalued method*. Such methods are normally posed in predictor–corrector form and are an interesting blend of analytical and numerical procedures. We present a typical multivalued method as an example of modern developments in numerical analysis. Our development is in stages.

We will solve the standard form of a first-order ordinary differential equation:

$$\frac{dy}{dx} = f(x, y), \qquad y(x_0) = y_0.$$

We begin by computing a predicted value for y at $x = x_0 + h$, where h is the step size. The prediction is through a Taylor-series expansion for $y(x)$. If we write our formula with terms through h^3, we get for $y_1 = y(x_1) = y(x_0 + h)$:

$$y_{1,p} = y_0 + hy_0' + \frac{h^2}{2}y_0'' + \frac{h^3}{6}y_0''' + \text{error}, \tag{5.27}$$

where the error term involves h^4 and a fourth-derivative value. We have added p to the subscript of y_1 to emphasize that this is a predicted quantity. Since we know y_0 and $y' = f(x, y)$ from the statement of the problem, we can develop the multipliers of the h's in Eq. (5.27) and hence compute $y_{1,p}$.

As will be seen later, we will also need predicted values for the derivatives in Eq. (5.27), which we get from Taylor-series expansions:

$$y_{1,p}' = y_0' + hy_0'' + \frac{h^2}{2}y_0''',$$

$$y_{1,p}'' = y_0'' + h y_0''', \tag{5.28}$$

$$y_{1,p}''' = y_0'''.$$

(Observe that we do not use terms beyond those in y'''.)

Actually, we will need the values of Eq. (5.28) multiplied by a power of h equal to the order of the derivative, so we write

$$h * y'_{1,p} = h * y'_0 + h^2 * y''_0 + \frac{h^3}{2} * y'''_0,$$

$$h^2 * y''_{1,p} = h^2 * y''_0 + h^3 * y'''_0, \tag{5.29}$$

$$h^3 * y'''_{1,p} = h^3 * y'''_0.$$

An example will clarify the prediction step.

EXAMPLE 5.2   Compute the predicted value for $y(x_0 + 0.1)$ for

$$\frac{dy}{dx} = f(x,y) = -2x - y, \qquad y(0) = -1.$$

The derivative values that we need in Eq. (5.27) are easy to get (we already computed these in Section 5.2):

$$y' = f(x,y) = -2x - y, \qquad y'_0 = 2(0) - (-1) = 1,$$
$$y'' = f'(x,y) = -2 - y', \qquad y''_0 = -2 - (1) = -3,$$
$$y''' = [f'(x,y)]' = -y'', \qquad y'''_0 = -(-3) = 3.$$

Substituting into Eq. (5.27) with $h = 0.1$ gives

$$y_{1,p} = -1 + (0.1) * (1) + \frac{0.01}{2} * (-3) + \frac{0.001}{6} * (3)$$

$$= -0.9145.$$

▲

Correcting the Predicted y-value

In the multivalued method, we correct the predicted value by adding a correction term. While there are many candidates for this correction, we choose one based on the difference in two estimates of the increment to y_0 to get $y(x)$ at $x = x_1 = x_0 + h$. These estimates of the increment are h times a value for the slope between x_0 and x_1. The two slope-values that we use in these estimates are at $x = x_1$, the end of the interval.

We have one approximation to the slope at $x = x_1$: it is $y'_{1,p}$, from Eq. (5.28). We get the other from the analytical expression for the slope, $f(x,y) = f(x_1,y_1)$. Even though we do not know y_1 (the true value of y at $x = x_1$), we will proceed as if it were known. So we write for the corrected y_1:

$$y_1 = y_{1,p} + r_1[h * f_1 - h * y'_{1,p}], \tag{5.30}$$

where r_1 is the fraction of the difference that we add. (We can change the value of r_1 to get variations of the method.) Although not readily apparent, y_1 appears on both sides of Eq. (5.30); it is contained in f_1. In a specific problem, we can solve for y_1 from the equation, although this may not always be easy to do. An example will show this.

EXAMPLE 5.3 Correct the predicted value for $y(0.1)$ of Example 5.2 using $r_1 = \frac{3}{8}$.
We need $y'_{1,p}$ and f_1. The first comes from Eq. (5.28):

$$y'_{1,p} = y'_0 + h * y''_0 + \frac{h^2}{2} * y'''_0$$

$$= 1 + (0.1)(-3) + \frac{0.01}{2}(3)$$

$$= 0.7150.$$

Our second expression for y'_1 is $f(x_1, y_1) = -2(x_0 + h) - y_1$. We substitute these into Eq. (5.30), getting

$$y_1 = y_{1,p} + r_1[h * f_1 - h * y'_{1,p}]$$
$$= y_{1,p} + r_1[h * (-2x_1 - y_1) - h * y'_{1,p}]$$
$$= -0.9145 + \frac{3}{8}[(0.1)(-2)(0.1) - (0.1) * (+y_1) - (0.1) * 0.7150] \quad \text{(5.31)}$$

We now solve Eq. (5.31) for y_1, an easy task in this example. (We might have to solve by some numerical procedure in other cases.) This gives for the corrected value for y_1:

$$y_{1,c} = \frac{y_{1,p} - r_1 h[y'_{1,p} - (-2x_1)]}{1 + r_1 h}$$

$$= \frac{-0.9145 - (3/8)(0.1)[0.7150 + (2)(0.1)]}{1 + (3/8)(0.1)}$$

$$= -0.914518.$$

The change on correction is small in this example, suggesting that we could have used a larger value for h.

Advancing the Solution

We have seen how to compute one new value for y by the multivalue method. We could repeat everything from a new starting value for x and y: x_1 and $y_{1,c}$. However, we can avoid evaluating the functions again (which can be expensive) if we apply corrections to the derivatives in Eq. (5.27) rather than recompute them. To do this, we write a column vector, Y_n, with these components:

$$Y_n = \begin{bmatrix} y_n \\ h * y_n' \\ (h^2/2) * y_n'' \\ (h^3/6) * y_n''' \end{bmatrix}.$$

We define a 4×4 matrix, B, as

$$B = \begin{bmatrix} 1 & 1 & 1 & 1 \\ 0 & 1 & 2 & 3 \\ 0 & 0 & 1 & 3 \\ 0 & 0 & 0 & 1 \end{bmatrix}. \tag{5.32}$$

Now we multiply B times Y_n, getting Y_{n+1}. Carrying out this step, we find that the elements of Y_{n+1} are precisely the Taylor-series expansions for the derivatives of y at $x = x_{n+1}$ multiplied by powers of h, as seen in Eq. (5.29).

We use the $Y_{n+1} = BY_n$ as a set of predicted values that we will correct. The procedure for correcting the first element (y_{n+1}) was shown in Eq. (5.31) for $x = x_0 + h$. Writing this for the step beyond x_n (but for our particular example), we have

$$y_{n+1,c} = \frac{y_{n+1,p} + r_1[h(-2x_{n+1}) - hy_{n+1,p}']}{1 + r_1 h}.$$

We observe that $y_{n+1,p}$ is just the first element of BY and that $hy_{n+1,p}'$ is the second element, so we modify the correcting equation to

$$y_{n+1,c} = \frac{Y_{n+1,p}[1] + r_1[h(-2x_{n+1}) - Y_{n+1,p}[2]]}{1 + r_1 h},$$

where the bracketed numbers after the Y's indicate the element position.

We need to correct the other elements of the Y-vector. We apply the same correction as for y_{n+1} but with different weighting factors. We will use these values in the R-vector:

$$R = \left[\frac{3}{8}, 1, \frac{3}{4}, \frac{1}{6} \right]^T.$$

Our equations for correcting the other elements of Y_{n+1} are then

$$\begin{aligned} Y_{n+1,c}[4] &= Y_{n+1,p}[4] + r_4 * (h * f(x_{n+1}, y_{n+1,c}) - Y_{n+1,p}[2]), \\ Y_{n+1,c}[3] &= Y_{n+1,p}[3] + r_3 * (h * f(x_{n+1}, y_{n+1,c}) - Y_{n+1,p}[2]), \\ Y_{n+1,c}[2] &= Y_{n+1,p}[2] + r_2 * (h * f(x_{n+1}, y_{n+1,c}) - Y_{n+1,p}[2]). \end{aligned} \tag{5.33}$$

(We computed these in reverse order so that the value of $Y_{n+1,p}[2]$ would not be changed too soon.)

Another example, using the same equation as before, will show how we advance the solution.

EXAMPLE 5.4   Use the multivalue method to find $y(0.3)$, starting at $x = 0$ with $h = 0.1$, given that

$$\frac{dy}{dx} = -2x - y, \qquad y(0) = -1.$$

$$Y_0 = \begin{bmatrix} y_0 \\ h*y_0' \\ (h^2/2)*y_0'' \\ (h^3/6)*y_0''' \end{bmatrix} = \begin{bmatrix} -1 \\ (0.1)(1) \\ (0.01/2)(-3) \\ (0.001/6)(3) \end{bmatrix} = \begin{bmatrix} -1 \\ 0.1 \\ -0.015 \\ 0.0005 \end{bmatrix}.$$

Using matrix B (Eq. (5.32)), we compute BY_0 to get $Y_{1,p}$, the predicted values. We get

$$Y_{1,p} = \begin{bmatrix} -0.9145 \\ 0.0715 \\ -0.0135 \\ 0.0005 \end{bmatrix}.$$

Now we find the corrected value for y_1, using Eq. (5.31). We did this before, in Example 5.3, so we know the result is -0.914481. We proceed to correct the other elements of $Y_{1,p}$, using Eqs. (5.33). We show the computation for $Y_{n+1}[4]$; the others are similar (but they use the appropriate element from the r-vector):

$$Y_{1,c}[4] = Y_{1,p}[4] + r_4 * [h*f(x_1, y_{1,c}) - Y_{1,p}[2]]$$

$$= 0.0005 + \frac{1}{6}[0.1(0.71448) - 0.0715]$$

$$= 0.000492.$$

The corrected y at $x = 0.1$ is then

$$Y_{1,c} = \begin{bmatrix} -0.914518 \\ 0.071452 \\ -0.013536 \\ 0.000492 \end{bmatrix}.$$

To get y-values at $x = 0.2$ and $x = 0.3$, we continue the same scheme. Table 5.12 summarizes the computations.

Table 5.12

| Y_1 | $x = 0.2$ | | $x = 0.3$ | |
	$Y_{2,p}$	$Y_{2,c}$	$Y_{3,p}$	$Y_{3,c}$
-0.914518	-0.856110	-0.856199	-0.822363	-0.822460
0.071452	0.045855	0.045620	0.022504	0.022246
-0.013536	-0.012060	-0.012237	-0.010879	-0.011072
0.000492	0.000492	0.000453	0.000453	0.000410

We have stated that different values can be used in the r-vector. It is important to know that proper choices can give predictor–corrector algorithms that are the same as the Adams–Bashforth and other well-known algorithms.

It should be relatively obvious that we can change the step size at any time. We only need to adjust the Y-vector to reflect the new value for h before we make the next step. We leave it as an exercise to find the diagonal matrix that performs this adjustment.

5.9 Convergence Criteria

In Section 5.3, we recorrected in the modified Euler method until no further change in y_{n+1} resulted. Usually this requires one more calculation than would otherwise be needed if we could predict whether the recorrection would make a significant change. In the methods of Milne and Adams–Moulton, we usually do not recorrect, but use a value of h small enough that this is unnecessary. We now look for a criterion to show how small h should be in the Adams–Moulton method, for $dy/dx = f(x,y)$, so that recorrections are not necessary. Let

$$y_p = \text{value of } y_{n+1} \text{ from predictor formula,}$$

$$y_c = \text{value of } y_{n+1} \text{ from corrector formula,}$$

$$y_{cc}, y_{ccc}, \text{ etc.} = \text{values of } y_{n+1} \text{ if successive recorrections are made,}$$

$$y_\infty = \text{value to which successive recorrections converge,}$$

$$D = y_c - y_p.$$

The change of y_c by recorrecting would be

$$
y_{cc} - y_c = \left(y_n + \frac{h}{24}(9y_c' + 19y_n' - 5y_{n-1}' + y_{n-2}') \right)
$$

$$
- \left(y_n + \frac{h}{24}(9y_p' + 19y_n' - 5y_{n-1}' + y_{n-2}') \right)
$$

$$
= \frac{9h}{24}(y_c' - y_p'). \tag{5.34}
$$

In Eq. (5.34) we have used the subscript p or c to denote which y-value is used in evaluating the derivative at $x = x_{n+1}$. We now manipulate the difference $(y_c' - y_p')$:

$$
y_c' - y_p' = f(x_{n+1}, y_c) - f(x_{n+1}, y_p) = \frac{f(x_{n+1}, y_c) - f(x_{n+1}, y_p)}{(y_c - y_p)}(y_c - y_p)
$$

$$
= f_y(\xi_1)D, \qquad \xi_1 \text{ between } y_c \text{ and } y_p, \text{ with } D = y_c - y_p.
$$

Hence,

$$y_{cc} - y_c = \frac{9hD}{24} f_y(\xi_1)$$

is the difference on recorrecting. If recorrected again, the result is

$$y_{ccc} - y_{cc} = \frac{9h}{24}(y'_{cc} - y'_c)$$

$$= \frac{9h}{24} f_y(\xi_2) \cdot (y_{cc} - y_c)$$

$$= \frac{9h}{24} f_y(\xi_2) \left[\frac{9hD}{24} f_y(\xi_1) \right]$$

$$= \left(\frac{9h}{24} \right)^2 [f_y(\xi)]^2 D, \qquad \xi \text{ between } y_c \text{ and } y_{cc}. \tag{5.35}$$

In Eq. (5.35), we need to impose the restrictions that f be continuous and $f_y(\xi_2)$ have the same sign; ξ lies between the extremes of y_p, y_c, and y_{cc}.

On further recorrections we will have a similar relation. We get y_∞ by adding all the corrections of y_p together:

$$y_\infty = y_p + (y_c - y_p) + (y_{cc} - y_c) + (y_{ccc} - y_{cc}) + \cdots$$

$$= y_p + D + \frac{9hf_y(\xi)}{24} D + \left(\frac{9hf_y(\xi)}{24} \right)^2 D + \left(\frac{9hf_y(\xi)}{24} \right)^3 D + \cdots.$$

The increment to y_p is a geometric series; so, if the ratio is less than unity, which is necessary if a geometric series is to have a sum,

$$y_\infty = y_p + \frac{D}{1 - r}, \qquad r = \frac{9hf_y(\xi)}{24}, \qquad \xi \text{ between } y_p \text{ and } y_\infty.$$

Hence, unless

$$|r| = \frac{h|f_y(\xi)|}{24/9} \doteq \frac{h|f_y(x_n, y_n)|}{24/9} < 1,$$

the successive recorrections diverge. Our first convergence criterion is

$$h < \frac{24/9}{|f_y(x_n, y_n)|}. \tag{5.36}$$

If we wish to have y_c and y_∞ the same to within one in the Nth decimal place, then

$$y_\infty - y_c = \left(y_p + \frac{D}{1 - r} \right) - (y_p + D) = \frac{rD}{1 - r} < 10^{-N}.$$

If $r \ll 1$, the fraction

$$\frac{r}{1 - r} \doteq r;$$

and a second convergence criterion, which ensures that the first corrected value is adequate (that is, it will not be changed in the Nth decimal place by further corrections), is

$$D \cdot 10^N < \left| \frac{1}{r} \right| \doteq \frac{24/9}{h \, |f_y(x_n, y_n)|}. \tag{5.37}$$

For the Adams–Moulton method we have the following three criteria. If all are met, the corrected value should be good to N decimals.

$$\text{Convergence criteria:} \begin{cases} h < \dfrac{24/9}{|f_y|}, \\[2mm] D \cdot 10^N < \dfrac{24/9}{h|f_n|}; \end{cases}$$

$$\text{Accuracy criterion: } D \cdot 10^N < 14.2.$$

Similar criteria for the Milne method are derived in the same way. They are

$$\text{Convergence criteria:} \begin{cases} h < \dfrac{3}{|f_y|}, \\[2mm] D \cdot 10^N < \dfrac{3}{h|f_y|}; \end{cases} \tag{5.38}$$

$$\text{Accuracy criterion: } D \cdot 10^N < 29.$$

These criteria are for a single first-order equation only. A similar analysis for a system is much more complicated.

We illustrate the use of these criteria with an example. Given the equation $dy/dx = \sin x - 3y, y(0) = 1$. In the neighborhood of the point $(1, 0.3)$, what maximum value of h is permitted if we wish to compute by (1) Milne's method, (2) the Adams–Moulton method, and get accuracy to five decimals? How close must the predictor and corrector values be?

1. For Milne:

$$f_y = -3, \qquad \text{so} \qquad h < \frac{3}{|-3|} = 1$$

to ensure convergence if we were to recorrect y_c in the Milne method. This requirement is not severe—we certainly would choose Δx smaller than this to give information throughout the range of x-values from 0 to 1. Suppose h were taken as 0.2. Then

$$D < \frac{3}{h|f_y|10^N} = \frac{3}{0.2|-3|10^5} = 5.0 \times 10^{-5}.$$

The difference between y_p and y_c cannot exceed this value; if it does, recorrections will be needed. This is more severe than $D < 29 \times 10^{-5}$ required for accuracy to one in the fifth decimal for y_c. We should monitor a Milne program to be sure that the difference between y_p and y_c does not exceed 5×10^{-5}. If it should exceed this value, we would need to reduce h so that this criterion could be met.

2. For Adams–Moulton, we calculate

$$h < \frac{24/9}{|-3|} = 0.89.$$

If $h = 0.2$,

$$D < \frac{24/9}{0.2|-3|10^5} = 4.4 \times 10^{-5}.$$

Again, this difference in y_p and y_c is more severe than $D < 14.2 \times 10^{-5}$ and should be used to control the value of h.

5.10 Errors and Error Propagation

Our previous error analyses have examined the error of a single step only, the so-called *local truncation error* of the methods. Since all practical applications of numerical methods to differential equations involve many steps, the accumulation of these errors, termed the *global truncation error*, is important. We remember that there are several sources of error in a numerical calculation in addition to the truncation error.

Original Data Errors

If the initial conditions are not known exactly (or must be expressed inexactly as a terminated decimal number), the solution will be affected to a greater or lesser degree, depending on the sensitivity of the equation. Highly sensitive equations are said to be subject to *inherent instability*.

Round-Off Errors

Since we can carry only a finite number of decimal places, our computations are subject to inaccuracy from this source, no matter whether we round or whether we chop off. Carrying more decimal places in the intermediate calculations than we require in the final answer is the normal practice to minimize this, but in lengthy calculations this is a source of error that is extremely difficult to analyze and control. Furthermore, in a computer program, if we use double precision, we require a longer execution time and also more storage to hold the more precise values. If these values are for a large array, the memory space needed may exceed that available to the program. This type of error is especially acute when two nearly equal quantities are subtracted.

Truncation Errors of the Method

These are the types of error we have been discussing, because we use truncated series for approximation in our work, when an infinite series is needed for exactness. The choice of method is our best control here, with suitable selection of h.

In addition to these three types of error, when we solve differential equations numerically we must be concerned about the propagation of previous errors through the subsequent steps. Since we use the end values at each step as the starting values for the next one, it is as if incorrect original data were distorting the later values. (Round-off would almost always produce error even if our method were exact.) This effect we now examine, but only for the very simple case of the Euler method. This will show how such error studies are made, as well as suggest how difficult the analysis of more practical methods is.

We consider the first-order equation $dy/dx = f(x, y), y(x_0) = y_0$. Let

$$Y_n = \text{calculated value at } x_n,$$

$$y_n = \text{true value at } x_n,$$

$$e_n = y_n - Y_n = \text{error in } Y_n; \ y_n = Y_n + e_n.$$

By the Euler algorithm,

$$Y_{n+1} = Y_n + hf(x_n, Y_n).$$

By Taylor series,

$$y_{n+1} = y_n + hf(x_n, y_n) + \frac{h^2}{2} y''(\xi_n), \qquad x_n < \xi_n < x_n + h,$$

$$e_{n+1} = y_{n+1} - Y_{n+1} = y_n - Y_n + h[f(x_n, y_n) - f(x_n, Y_n)] + \frac{h^2}{2} y''(\xi_n)$$

$$= e_n + h \frac{f(x_n, y_n) - f(x_n, Y_n)}{y_n - Y_n} (y_n - Y_n) + \frac{h^2}{2} y''(\xi_n) \tag{5.39}$$

$$= e_n + hf_y(x_n, \eta_n)e_n + \frac{h^2}{2} y''(\xi_n), \qquad \eta_n \text{ between } y_n, Y_n.$$

In Eq. (5.39), we have used the mean-value theorem, imposing continuity and existence conditions on $f(x, y)$ and f_y. We suppose, in addition, that the magnitude of f_y is bounded by the positive constant K in the region of x, y-space in which we are interested.* Hence,

$$e_{n+1} \leq (1 + hK)e_n + \frac{1}{2} h^2 y''(\xi_n). \tag{5.40}$$

---

* This is essentially the same as the Lipschitz condition, which will guarantee existence and uniqueness of a solution. See Section 5.14.

Here $y(x_0) = y_0$ is our initial condition, which we assume free of error. Since $Y_0 = y_0$, $e_0 = 0$:

$$e_1 \leq (1 + hK)e_0 + \frac{1}{2}h^2 y''(\xi_0) = \frac{1}{2}h^2 y''(\xi_0),$$

$$e_2 \leq (1 + hK)\left[\frac{1}{2}h^2 y''(\xi_0)\right] + \frac{1}{2}h^2 y''(\xi_1) = \frac{1}{2}h^2[(1 + hK)y''(\xi_0) + y''(\xi_1)].$$

Similarly,

$$e_3 \leq \frac{1}{2}h^2[(1 + hK)^2 y''(\xi_0) + (1 + hK)y''(\xi_1) + y''(\xi_2)],$$

$$e_n \leq \frac{1}{2}h^2[(1 + hK)^{n-1} y''(\xi_0) + (1 + hK)^{n-2} y''(\xi_1) + \cdots + y''(\xi_{n-1})].$$

If $f_y \leq K$ is positive, the truncation error at every step is propagated to every later step after being amplified by the factor $(1 + hf_y)$ each time. Note that as $h \to 0$, the error at any point is just the sum of all the previous errors. If the f_y are negative and of magnitude such that $|hf_y| < 2$, the errors are propagated with diminishing effect.

We now show that the accumulated error after n steps is $O(h)$; that is, the global error of the simple Euler method is $O(h)$. We assume, in addition, that y'' is bounded, $|y''(x)| < M, M > 0$. Equation (5.40) becomes, after taking absolute values,

$$|e_{n+1}| \leq (1 + hK)|e_n| + \frac{1}{2}h^2 M.$$

Now we compare to the second-order difference equation:

$$Z_{n+1} = (1 + hK)Z_n + \frac{1}{2}h^2 M,$$

$$Z_0 = 0.$$

(5.41)

Obviously the values of Z_n are at least equal to the magnitudes of $|e_n|$. The solution to Eq. (5.41) is (check by direct substitution)

$$Z_n = \frac{hM}{2K}(1 + hK)^n - \frac{hM}{2K}.$$

The Maclaurin expansion of e^{hK} is

$$e^{hK} = 1 + hK + \frac{(hK)^2}{2} + \frac{(hK)^3}{6} + \cdots,$$

so that

$$1 + hK < e^{hK} \qquad (K > 0),$$

$$Z_n < \frac{hM}{2k}(e^{hK})^n - \frac{hM}{2K} = \frac{hM}{2K}(e^{nhK} - 1)$$

$$= \frac{hM}{2K}(e^{(x_n - x_0)K} - 1) = O(h). \qquad \textbf{(5.42)}$$

It follows that the global error e_n is $O(h)$. (This result can be derived without difference equations.)

5.11 Systems of Equations and Higher-Order Equations

We have so far treated only the case of a first-order differential equation. Most differential equations that are the mathematical model for a physical problem are of higher order, or even a *set* of simultaneous higher-order differential equations. For example,

$$\frac{w}{g}\frac{d^2x}{dt^2} + b\frac{dx}{dt} + kx = f(x, t)$$

represents a vibrating system in which a linear spring with spring constant k restores a displaced mass of weight w against a resisting force whose resistance is b times the velocity. The function $f(x, t)$ is an external forcing function acting on the mass.

An analogous second-order equation describes the flow of electricity in a circuit containing inductance, capacitance, and resistance. The external forcing function in this case represents the applied electromotive force. Compound spring–mass systems and electrical networks can be simulated by a system of such second-order equations.

We first show how a higher-order differential equation can be reduced to a system of simultaneous first-order equations. We then show that these can be solved by an application of the methods previously studied. We treat here initial-value problems only, for which n values of the functions or derivatives (with n equal to the order of the system) are all specified at the same (initial) value of the independent variable. When some of the conditions are specified at one value of the independent variable and others at a second value, we call it a *boundary-value problem*. Methods of solving these are discussed in the next chapter.

By solving for the second derivative, we can normally express a second-order equation as

$$\frac{d^2x}{dt^2} = f\left(t, x, \frac{dx}{dt}\right), \qquad x(t_0) = x_0, \qquad x'(t_0) = x_0'. \qquad \textbf{(5.43)}$$

The initial value of the function x and its derivative are generally specified. We convert this to a pair of first-order equations by the simple expedient of defining the derivative as a second function. Then, since $d^2x/dt^2 = (d/dt)(dx/dt)$,

$$\frac{dx}{dt} = y, \qquad x(t_0) = x_0,$$

$$\frac{dy}{dt} = f(t, x, y), \qquad y(t_0) = x_0'.$$

This pair of first-order equations is equivalent to the original Eq. (5.43). For even higher orders, each of the lower derivatives is defined as a new function, giving a set of n first-order equations that correspond to an nth-order differential equation. For a system of higher-order equations, each is similarly converted, so that a larger set of first-order equations results. Thus the nth-order differential equation

$$y^{(n)} = f(x, y, y', \ldots, y^{(n-1)}),$$
$$y(x_0) = A_1, \qquad y'(x_0) = A_2, \qquad \ldots, \qquad y^{(n-1)}(x_0) = A_n$$

is converted into a system of n first-order differential equations by letting $y_1 = y$ and

$$y_1' = y_2,$$
$$y_2' = y_3,$$
$$\vdots$$
$$y_{n-1}' = y_n,$$
$$y_n' = f(x, y_1, y_2, \ldots, y_n);$$

with initial conditions

$$y_1(x_0) = A_1, \qquad y_2(x_0) = A_2, \qquad \ldots, \qquad y_n(x_0) = A_n.$$

We now illustrate the application of the various methods to the pair of first-order equations:

$$\frac{dx}{dt} = xy + t, \qquad x(0) = 1,$$

$$\frac{dy}{dt} = ty + x, \qquad y(0) = -1.$$

$$(5.44)$$

Taylor-Series Method

We need the various derivatives $x', x'', x''', \ldots, y', y'', y''', \ldots$, all evaluated at $t = 0$:

$x' = xy + t,$	$x'(0) = (1)(-1) + 0 = -1$
$y' = ty + x,$	$y'(0) = (0)(-1) + 1 = 1,$
$x'' = xy' + x'y + 1,$	$x''(0) = (1)(1) + (-1)(-1) + 1 = 3,$
$y'' = y + ty' + x',$	$y''(0) = -1 + (0)(1) - 1 = -2,$
$x''' = x'y' + xy'' + x''y + x'y',$	$x'''(0) = -7,$
$y''' = y' + y' + ty'' + x'',$	$y'''(0) = 5,$

and so on;   and so on;

$$x(t) = 1 - t + \frac{3}{2}t^2 - \frac{7}{6}t^3 + \frac{27}{24}t^4 - \frac{124}{120}t^5 + \cdots,$$

$$y(t) = -1 + t - t^2 + \frac{5}{6}t^3 - \frac{13}{24}t^4 + \frac{47}{120}t^5 + \cdots.$$

$$(5.45)$$

At $t = 0.1$, $x = 0.9139$ and $y = -0.9092$.

Equations (5.45) are the solution to the set (5.44). Note that we need to alternate between the functions in getting the derivatives; for example, we cannot get $x''(0)$ until $y'(0)$ is known; we cannot get $y'''(0)$ until $x''(0)$ is known. After we have obtained the coefficients of the Taylor-series expansions in Eq. (5.45), we can evaluate x and y at any value of t, but the error will depend on how many terms we employ.

Euler Predictor-Corrector

We apply the predictor to each equation; then the corrector can be used. Again note that we work alternately with the two functions.

Take $h = 0.1$. Let p and c subscripts indicate predicted and corrected values, respectively:

$$x_p(0.1) = 1 + 0.1[(1)(-1) + 0] = 0.9,$$

$$y_p(0.1) = -1 + 0.1[(0)(-1) + 1] = -0.9,$$

$$x_c(0.1) = 1 + 0.1\left(\frac{-1 + [(0.9)(-0.9) + 0.1]}{2}\right) = 0.9145,$$

$$y_c(0.1) = -1 + 0.1\left(\frac{1 + [(0.1)(-0.9) + 0.9145]}{2}\right) = -0.9088.$$

In computing $x_c(0.1)$, we used the x_p and y_p. In computing $y_c(0.1)$ after $x_c(0.1)$ is known, we have a choice between x_p and x_c. There is an intuitive feel that one should use x_c, with the idea that one should always use the best available values. This does not always expedite convergence, probably due to compensating errors. Here we have used the best values to date. Recorrecting in the obvious manner gives

$$x(0.1) = 0.9135,$$

$$y(0.1) = -0.9089.$$

We can now advance the solution another step if desired, by using the computed values at $t = 0.1$ as the starting values. From this point we can advance one more step, and so on for any value of t. The errors will be the combination of local truncation error at each step plus the propagated error resulting from the use of inexact starting values.

Runge–Kutta–Fehlberg

Again there is an alternation between the x and y calculations. In applying this method, one always uses the previous k-value in incrementing the function values and the value of h to increment the independent variable. As in the previous calculations, we oscillate between computations for x and for y; for example, we do $k_{1,x}$, then $k_{1,y}$, before doing $k_{2,x}$, and so on.

Keeping in mind that the equations are

$$\frac{dx}{dt} = f(t,x,y) = xy + t, \qquad x(0) = 1,$$

$$\frac{dy}{dt} = g(t,x,y) = ty + x, \qquad y(0) = -1,$$

the k-values for x and y are

for x:

$$\begin{aligned}
k_{1,x} &= hf(0,1,-1) \\
&= 0.1[(1)(-1) + 0] \\
&= -0.1;
\end{aligned}$$

$$\begin{aligned}
k_{2,x} &= hf(0.025, 0.975, -0.975) \\
&= 0.1[(0.975)(-0.975) + 0.025] \\
&= -0.092562;
\end{aligned}$$

$$\begin{aligned}
k_{3,x} &= hf(0.038, 0.965, -0.964) \\
&= 0.1[(0.965)(-0.964) + 0.038] \\
&= -0.089226;
\end{aligned}$$

$$\begin{aligned}
k_{4,x} &= hf(0.092, 0.919, -0.915) \\
&= 0.1[(0.919)(-0.915) + 0.092] \\
&= -0.074892;
\end{aligned}$$

$$\begin{aligned}
k_{5,x} &= hf(0.1, 0.913, -0.908) \\
&= 0.1[(0.913)(-0.908) + 0.1] \\
&= -0.072904;
\end{aligned}$$

$$\begin{aligned}
k_{6,x} &= hf(0.05, 0.954, -0.953) \\
&= 0.1[(0.954)(-0.953) + 0.05] \\
&= -0.085868.
\end{aligned}$$

for y:

$$\begin{aligned}
k_{1,y} &= hg(0,1,-1) \\
&= 0.1[(0)(-1) + 1] \\
&= 0.1;
\end{aligned}$$

$$\begin{aligned}
k_{2,y} &= hg(0.025, 0.975, -0.975) \\
&= 0.1[(0.025)(-0.975) + 0.975] \\
&= 0.095062;
\end{aligned}$$

$$\begin{aligned}
k_{3,y} &= hg(0.038, 0.965, -0.964) \\
&= 0.1[(0.038)(-0.964) + 0.095] \\
&= 0.092845;
\end{aligned}$$

$$\begin{aligned}
k_{4,y} &= hg(0.092, 0.919, -0.915) \\
&= 0.1[(0.092)(-0.915) + 0.919] \\
&= 0.083461;
\end{aligned}$$

$$\begin{aligned}
k_{5,y} &= hg(0.1, 0.913, -0.908) \\
&= 0.1[(0.1)(-0.908) + 0.913] \\
&= 0.082178;
\end{aligned}$$

$$\begin{aligned}
k_{6,y} &= hg(0.05, 0.954, -0.953) \\
&= 0.1[(0.05)(-0.953) + 0.954] \\
&= 0.090628.
\end{aligned}$$

Then using the fifth-order formula, we get

$$\begin{aligned}
x(0.1) &= 1 + (-0.01185 - 0.046307 - 0.037905 + 0.013123 - 0.003122) \\
&= 0.913936; \\
y(0.1) &= -1 + (0.01185 + 0.048185 + 0.042242 - 0.014792 + 0.003296) \\
&= -0.909217.
\end{aligned}$$

Extending the Taylor-series solution even further shows that the Runge–

Kutta–Fehlberg values are correct to more than five decimals, while the modified Euler values are correct to only three, so $h = 0.1$ may be too large for that method.

Advancing the solution by the Runge–Kutta–Fehlberg method will again involve using the computed values of x and y as the initial values for another step. The errors here will be much less than those for the Euler predictor–corrector method.

Adams–Moulton

After getting four starting values, we proceed with the algorithm of Eqs. (5.24) and (5.25), again alternately computing x and then y (see Table 5.13).

Table 5.13

	t	x	x'	t	y	y'
Starting values	0.000	1.0	−1.0	0.00	−1.0	1.0
	0.025	0.9759	−0.9271	0.025	−0.9756	0.9515
	0.050	0.9536	−0.8582	0.050	−0.9524	0.9060
	0.075	0.9330	−0.7929	0.075	−0.9303	0.8632
Predicted	0.10	(0.9139)	(−0.7310)	0.10	(−0.9092)	(0.8230)
Corrected		0.9139			−0.9092	

In the computations we first get predicted values of x and y:

$$x(0.1) = 0.9330 + \frac{0.025}{24}[55(-0.7929) - 59(-0.8582) + 37(-0.9271) - 9(-1.0)]$$

$$= 0.913937;$$

$$y(0.1) = -0.9303 + \frac{0.025}{24}[55(0.8632) - 59(0.9060) + 37(0.9515) - 9(1.0)]$$

$$= -0.909217.$$

After getting x' and y' at $t = 0.1$, using $x(0.1)$ and $y(0.1)$, we then correct:

$$x(0.1) = 0.9330 + \frac{0.025}{24}[9(-0.7310) + 19(-0.7929) - 5(-0.8582) + (-0.9271)]$$

$$= 0.913936;$$

$$y(0.1) = -0.9303 + \frac{0.025}{24}[9(0.8230) + 19(0.8632) - 5(0.9060) + (0.9515)]$$

$$= -0.909217.$$

The close agreement of predicted and corrected values indicates six-decimal-place accuracy.

In this method, as we advance the solution to larger values of t, the comparison between predictor and corrector values tells us whether the step size needs to be changed.

Milne

This method can be applied to a system of first-order equations exactly analogously to the application of Adams–Moulton. We do not illustrate it because it is subject to instability.

5.12 Comparison of Methods/Stiff Equations

Comparison of Methods

It is appropriate that we summarize the various methods that have been discussed in this chapter and compare them. Table 5.14 compares the accuracy, effort required, stability, and other features of the methods.

Table 5.14 Comparison of methods for differential equations

Method	Type	Local error	Global error	Function evaluations/step	Stability	Ease of changing step size	Recommended?
Modified Euler	Single-step	$O(h^3)$	$O(h^2)$	2	Good	Good	No
Fourth-order Runge–Kutta	Single-step	$O(h^5)$	$O(h^4)$	4	Good	Good	Yes
Runge–Kutta–Fehlberg	Single-step	$O(h^6)$	$O(h^5)$	6	Good	Good	Yes
Milne	Multistep	$O(h^5)$	$O(h^4)$	2	Poor	Poor	No
Adams–Moulton	Multistep	$O(h^5)$	$O(h^4)$	2	Good	Poor	Yes

The data in Table 5.14 lead us to draw the usual conclusion about the best scheme for solving a differential equation of higher order, or a system of N first-order equations: We begin with a fourth-order Runge–Kutta to get a total of four values for each of the functions (this also allows us to compute four values for each of the derivatives), and then advance the solution with Adams–Moulton. At each step after employing Adams–Moulton, we check* the accuracy and adjust the step size when appropriate.

There is still a problem during the starting phase when Runge–Kutta is being used. For instance, when a Runge–Kutta–Fehlberg method (Section 5.4) is used to start the solution for a multistep method that needs equispaced function values, an additional restriction is imposed; the step size must be uniform, which may mean that closer spaced values may need to be computed than are required to meet the accuracy criterion alone.

A method due to Hamming has been widely accepted. It is available through subroutines in some FORTRAN libraries. It begins the solution with a fourth-order Runge–Kutta, and then continues with a predictor–corrector. The equations employed are

$$y_{i+1,p} = y_{i-3} + \frac{4h}{3}(2f_i - f_{i-1} + 2f_{i-2}),$$

$$y_{i+1,m} = y_{i+1,p} - \frac{112}{121}(y_{i,p} - y_{i,c}),$$

$$y_{i+1,c} = \frac{1}{8}[9y_i - y_{i-2} + 3h(f_{i+1,m} + 2f_i - f_{i-1})],$$

$$y_{i+1} = y_{i+1,c} + \frac{9}{121}(y_{i+1,p} - y_{i+1,c}).$$

(5.46)

The predictor equation is the Milne predictor. Before correcting, the estimate of y_{i+1} is modified using the difference between the predicted and corrected values in the *previous* interval (this is omitted in the first interval, since these are not available). A corrector formula is used that depends on two previous y-values, though heavily weighted in favor of the last one. Finally, an adjustment is made based on the error estimate computed from the difference between predicted and corrected values. Note that, while two additional equations are employed in each step compared to the predictor–corrector methods previously described, only two evaluations of the derivative function are needed, the same as before. For many applications, it is the evaluation of the derivative function that is costly in computer time—the two extra algebraic steps don't count for much.

The special merits of Hamming's method are stability combined with good accuracy. Like Milne's and Adams', the method has a local error of $O(h^5)$ and a global error of $O(h^4)$.

Gear (1967) has proposed a predictor–corrector method that has an $O(h^6)$ local error but uses only three previous steps rather than the four previous steps employed by Adams–Moulton and Milne. It obtains its high order of error by using recorrected values of the function and derivative values. The formulas are

---

* Ordinarily a weighted average of the N errors is monitored. Alternatively, the maximum error is controlled.

$$y_{n+1,p} = -18y_n + 9y_{n-1,c_1} + 10y_{n-2,c_2} + 9hy'_n + 18hy'_{n-1} + 3hy'_{n-2},$$

$$hy'_{n+1,p} = -57y_n + 24y_{n-1,c_1} + 33y_{n-2,c_2} + 24hy'_n + 57hy'_{n-1} + 10hy'_{n-2},$$

$$F = hy'_{n+1,p} - hf(x_{n+1}, y_{n+1,p}),$$

$$y_{n+1} = y_{n+1,p} - \frac{95}{288}F,$$

$$y_{n,c_1} = y_n + \frac{3}{160}F,$$

$$y_{n-1,c_2} = y_{n-1,c_1} - \frac{11}{1440}F,$$

$$hy'_{n+1} = hy'_{n+1,p} - F.$$

This method is stable and is applicable to systems of first-order differential equations. In addition, Gear (1971) has a listing and a complete description of a subroutine called DIFSUB, which includes both the Adams predictor–corrector method and Gear's stiff methods. This subroutine is also the basis of an IMSL subroutine DGEAR, which contains Adams methods up to order 12 and methods for stiff differential equations.

Stiff Equations

A stiff equation results from phenomena with widely differing time scales. For example, the general solution of a differential equation may involve sums or differences of terms of the form ae^{ct}, be^{dt}, where both c and d are negative but c is much smaller than d. In such cases, using a small value for the step size can introduce enough round-off errors to cause instability.

An example is the following:

$$x' = 1195x - 1995y, \qquad x(0) = 2,$$
$$y' = 1197x - 1997y, \qquad y(0) = -2. \tag{5.47}$$

The analytical solution of Eq. (5.47) is

$$x(t) = 10e^{-2t} - 8e^{-800t}, \qquad y(t) = 6e^{-2t} - 8e^{-800t}.$$

Observe that the exponents are all negative and of very different magnitude, qualifying this as a stiff equation. Suppose we solve Eq. (5.47) by the simple Euler method with $h = 0.1$, applying just one step. The iterations are

$$x_{i+1} = x_i + hf(x_i, y_i) = x_i + 0.1(1195x_i - 1995y_i),$$
$$y_{i+1} = y_i + hg(x_i, y_i) = y_i + 0.1(1197x_i - 1997y_i).$$

This gives $x(0.1) = 640, y(0.1) = 636$, while the exact values are $x(0.1) = 8.187$ and $y(0.1) = 4.912$. Such a result is typical (though here exaggerated) for stiff equations.

One solution to this problem is to use an implicit method rather than an explicit one. All the methods so far discussed have been explicit, meaning that the new values, x_{i+1} and y_{i+1}, are computed in terms of the previous ones, x_i and y_i. The implicit form of Euler's method is

$$x_{i+1} = x_i + hf(x_{i+1}, y_{i+1}),$$
$$y_{i+1} = y_i + h(x_{i+1}, y_{i+1}). \tag{5.48}$$

If the derivative functions $f(x, y)$ and $g(x, y)$ are nonlinear, this is difficult to solve. However, in Eq. (5.47) they are linear. Solving Eq. (5.47) by use of Eq. (5.48) we have

$$x_{i+1} = x_i + 0.1(1195x_{i+1} - 1995y_{i+1}),$$
$$y_{i+1} = y_i + 0.1(1197x_{i+1} - 1997y_{i+1}).$$

Since the system is linear, we can write

$$\begin{bmatrix} x_{i+1} \\ y_{i+1} \end{bmatrix} = \begin{bmatrix} (1 - 1195(0.1)) & 1995(0.1) \\ -1197 & (1 + 1997(0.1)) \end{bmatrix}^{-1} \begin{bmatrix} x_i \\ y_i \end{bmatrix}$$

which has the solution $x(0.1) = 8.23, y(0.1) = 4.90$, reasonably close to the analytical values.

In summary, our results for the solution of Eq. (5.47) are

	$x(0.1)$	$y(0.1)$
Exact	8.19	4.91
Euler		
Explicit	640	636
Implicit	8.23	4.90

If the step size is very small, we can get good results from the simpler Euler after the first step. With $h = 0.0001$, the table of results becomes

	$x(0.0001)$	$y(0.0001)$
Exact	2.61	-1.39
Euler		
Explicit	2.64	-1.36
Implicit	2.60	-1.41

but this would require 1000 steps to reach $t = 0.1$, and round-off errors would be large.

If we anticipate some material from the next chapter, we can give a better description of stiffness as well as indicate the derivation of the general solution to Eq. (5.47). We rewrite Eq. (5.47) in matrix form:

$$\begin{bmatrix} x \\ y \end{bmatrix}' = A \begin{bmatrix} x \\ y \end{bmatrix}, \qquad \text{where } A = \begin{bmatrix} 1195 & -1995 \\ 1197 & -1997 \end{bmatrix}.$$

The general solution, in matrix form, is

$$\begin{bmatrix} x \\ y \end{bmatrix} = ae^{-2t}v_1 + ce^{-800t}v_2,$$

where

$$v_1 = \begin{bmatrix} 5 \\ 3 \end{bmatrix} \qquad \text{and} \qquad v_2 = \begin{bmatrix} 1 \\ 1 \end{bmatrix}.$$

You can easily verify that $Av_1 = -2v_1$ and $Av_2 = -800v_2$. In Chapter 6 we will see that this means that v_1 is an eigenvector of A and that -2 is the corresponding eigenvalue. Similarly, v_2 is an eigenvector of A with the corresponding eigenvalue of -800. (In Chapter 6 you will learn additional methods to find the eigenvectors and eigenvalues of a matrix.)

A stiff equation can be defined in terms of the eigenvalues of the matrix A that represents the right-hand sides of the system of differential equations. When the eigenvalues of A have real parts that are negative and differ widely in magnitude as in this example, the system is stiff. In the case of a nonlinear system

$$\begin{bmatrix} x_1 \\ x_2 \\ \vdots \\ x_n \end{bmatrix}' = \begin{bmatrix} f_1(x_1, x_2, \ldots, x_n) \\ f_2(x_1, x_2, \ldots, x_n) \\ \vdots \\ f_n(x_1, x_2, \ldots, x_n) \end{bmatrix},$$

one must consider the Jacobian matrix whose terms are $\partial f_i / \partial x_j$. See Gear (1971) for more information.

5.13 Using MAPLE

MAPLE has a built-in function, called dsolve, with several options that make it very flexible and powerful for solving a single equation or a system of equations, both numerically and analytically.

In several sections of Chapter 5, we solved this first-order differential equation numerically,

$$\frac{dy}{dx} = -2x - y, \qquad y(0) = -1,$$

and used the analytical solution to compare to the numerical results.

MAPLE can generate the analytical solution for this example problem by symbolic methods. To solve the preceding example, we first define a variable that will store the equation:

```
deq := diff(y(x),x) + y(x) + 2*x = 0;
dsolve( {deq, y(0) = -1 }, y(x));
```

The program comes back with the answer:

$$y(x) = -2x + 2 - 3 \exp(-x)$$

Since MAPLE has excellent plot routines for two and three dimensions, we illustrate them with this first example. The plot programs are called by

```
with(plots);
G := dsolve( {deq, y(0) = -1 }, y(x));
plot(rhs(G), x = 0..0.4);
```

The last command calls for a plot of the right-hand side of

$$G := y(x) = -2x + 2 - 3 \exp(-x)$$

from $x = 0$ to $x = 0.4$.

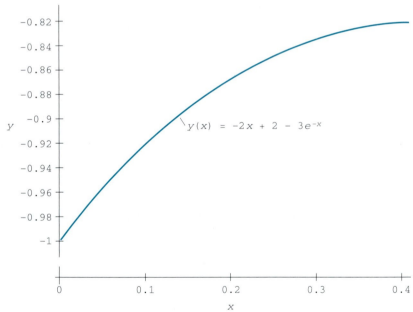

Solution to $y' = -2x - y$

The Taylor-Series Method

A Taylor-series expansion for a first-order equation is readily obtained with the function dsolve. We only need to add the series option. Let's apply this to our example

$$\frac{dy}{dx} = -2x - y, \qquad y(0) = -1.$$

Since we have already defined this in deq, we call dsolve:

```
dsolve( {deq, y(0) = -1}, y(x), series);
```

The resulting fifth-degree polynomial is obtained:

$$y(x) = -1 + x - 3/2\ x^2 + 1/2\ x^3 - 1/8\ x^4 + 1/40\ x^5$$

which is precisely the result in Section 5.2.

This same operator, dsolve, can be used to solve a system of equations both by series, as before, and numerically. (Remember that we solve higher-order equations by reducing them to a system of first-order equations.) The format is the same as before. Let us apply this to the example in Section 5.11, which is

$$\frac{dx}{dt} = xy + t, \qquad x(0) = 1,$$

$$\frac{dy}{dt} = ty + x, \qquad y(0) = -1.$$

First we define two variables to hold both the system of equations and the dependent variables:

```
sys := diff(x(t),t) = x(t)*y(t) + t,diff(y(t),t) = t*y(t) + x(t);
fcns := {x(t), y(t) };
```

The call to dsolve again includes the series option:

```
dsolve({sys,x(0) = 1,y(0) = -1}, fcns, series);
```

The series solution is then produced:

$$\{1 - t + 3/2t^2 - 7/6t^3 + 9/8t^4 - 31/30t^5 + 0(t^6),$$
$$-1 + t - t^2 + 5/6t^3 - 13/24t^4 + 47/120t^5 + 0(t^6)\}$$

To get a plot of these two solutions, we now need only to define

```
u := 1 - t + 3/2*t^2 - 7/6*t^3 + 9/8*t^4 - 31/30*t^5;
v := -1 + t - t^2 + 5/6*t^3 - 13/24*t^4 + 47/120*t^5;
```

Then,

```
plot({u,v}, t = 0..0.4);
```

produces the plot for the two equations on the interval from $x = 0$ to $x = 0.5$, which is identical to the result in Section 5.11.

The Runge–Kutta–Fehlberg Method

The dsolve program also contains an option that allows us to solve our problem numerically by calling on the Runge–Kutta–Fehlberg 4–5 order method. It is designed to handle a system of first-order differential equations as well as a single equation.

Once more we illustrate the use of the MAPLE procedure dsolve in solving the first-order equations in Section 5.10. However, this time we change our option from series to numeric and must indicate at which values of t we wish to evaluate $(x(t), y(t))$. To do so, MAPLE allows us to write a short Pascal-like "for loop."

We first define the two equations and the initial values:

```
d1 := {D(x)(t) = x(t)*y(t) + t, D(y)(t) = t*y(t) + x(t)};
init1 := {x(0) = 0, y(0) = -1};
```

Next we define a vector function, $F(t) = (x(t), y(t))$, which is to evaluate and store the values, $(x(t), y(t))$, for each t:

```
F := dsolve (d1 union init1, {x(t),y(t)}, numeric);
```

Then we define the values of t from the initial t_0 to final t_f and the step size, h:

```
for t from 0 by 0.1 to 0.3 do F(t) od;
```

The following results are produced:

```
                 0, 1., -1.
      .1000000000, .9139363455, -.9092171620
      .2000000000, .8521861364, -.8340893659
      .3000000000, .8106335047, -.7710932710
```

The system of first-order equations in Section 5.10 is solved through the Runge–Kutta–Fehlberg procedure by dsolve.

MAPLE does echo our input in a more mathematical way so that we can verify what we are typing in. However, for brevity, we omitted this feature of MAPLE from our example.

5.14 Theoretical Matters

We have included much of the theoretical material in this chapter in previous sections, but we will mention one or two items here. We begin with a discussion of the Lipschitz condition and then discuss more on stability.

The Lipschitz Condition

In the current chapter we have studied a number of numerical techniques to solve the differential equation

$$\frac{dy}{dx} = f(x, y), \qquad \text{with } y(x_0) = y_0. \tag{5.49}$$

We solved Eq. (5.49) by generating values for $y(x)$ at discrete points along the x-axis (or the t-axis if t is the independent variable). In a mathematics course we study analytical methods of solving the problem. Unfortunately, it often happens that the analytical solution is so complicated that it is more efficient to solve it numerically.

In any case, we should examine the condition under which Eq. (5.49) really has a solution and whether the solution is unique. The *Lipschitz condition* is involved in the answer to these questions. A definition of this condition is:

Let $f(x,y)$ be defined and continuous on a region R that contains the point (x_0, Y_0). We assume that the region is a closed and bounded rectangle. Then $f(x,y)$ is said to satisfy the Lipschitz condition if:

There is an $L > 0$ so that for all x, y_1, y_2 in R, we have

$$|f(x,y_1) - f(x,y_2)| < L|y_1 - y_2|.$$

If $f(x,y)$ satisfies this condition, it can be shown that there is a solution to Eq. (5.49) and that this solution is unique.

For most of the problems encountered in practice, $f(x,y)$ is continuous over region R; when this is true, we are guaranteed the existence of a solution. We can write this solution in the form

$$y(x) = Y_0 + \int_{x_0}^{x} f(t, y(t))\, dt.$$

If, in addition, $f(x,y)$ satisfies the Lipschitz condition, we then are guaranteed that the solution is unique. In the actual examples that we have considered, $f(x,y)$ is not only continuous but $f_y(x,y)$ is bounded and continuous in the region of interest.

The example that we have used many times,

$$y' = -2x - y,$$

satisfies the Lipschitz condition, since

$$|f(x,y) - f(x,z)| = L|y - z|, \qquad L = 1 > 0.$$

From this we can infer that the solution we found was the unique one.

Stability Reconsidered

In Section 5.6, we considered the stability of a particular method, Milne's method. The case of stability of methods for differential equations is similar to the case for systems of equations. Just as before, we can consider our system as a machine that gives different final values from different initial conditions. A picture of that machine is

Once again, we find that the idea of an unstable equation is one that gives widely separated outputs from inputs that are quite close together.

For instance, if the solution contains the expression e^{ax}, with $a > 0$, we expect the solutions to separate as a increases. In the case of a system of equations, we are interested in the Jacobian matrix of the system. In matrix terms, this Jacobian is

$$(\partial f_i/\partial y_j)_{i,j}, \qquad j = 1, \ldots, n,$$
$$y_i' = f_i(x, y_1, y_2, \ldots, y_n) \qquad \text{for } i = 1, 2, \ldots, n.$$

The eigenvalues of the matrix are involved. If at least one of the eigenvalues has a large positive real part, the system is unstable. (A good reference on this is Kahaner, Moler, and Nash, 1989).

We might contrast this to our earlier discussion of a stiff system. In the case of stiffness, the real parts of all the eigenvalues of the matrix are negative and at least one is much smaller than the rest. In general, stiff equations are stable.

Chapter Summary

If you understand the material in this chapter, you should be able to

1. Use the Taylor-series procedure to solve first- or second-order differential equations. You can explain why this method is not generally used.

2. Solve first-order equations and systems of first-order equations by any of these methods:
   a. Simple Euler and modified Euler
   b. Fourth-order Runge–Kutta and Runge–Kutta–Fehlberg
   c. Multistep methods, such as Adams–Moulton
   d. Multivalued methods

3. Explain the difference between a single-step method and a multistep method.

4. Rewrite a higher-order differential equation or system of equations in terms of an equivalent system of first-order equations.

5. Tell why the global error is different from the local error.

6. Explain the terms efficiency, accuracy, stability, and convergence as applied to the procedures of this chapter. In particular, you know why Milne's method is rejected.

7. Understand what is meant by the term *stiff differential equation* and give examples. You can solve this type of equation by an implicit method.

8. State the condition a differential equation must meet to have a unique solution.

Computer Programs

This section presents two programs that solve ordinary differential equations: (1) a BASIC program that advances the solution from three or four starting values by the Adams method, and (2) a program written in C that solves a system of first-order equations using the Runge–Kutta–Fehlberg method.

Program 5.1 The Adams Method

Figure 5.6 lists the BASIC program. It uses QuickBASIC, version 4.5. The logic is quite simple:

1. A data file is read to obtain the number of past points (3 or 4) and the $x, f(x)$ values. (These are evenly spaced.)
2. The number of past points and the step size for x is determined from the data.
3. Depending on the number of past points, one of two subroutines is called to compute and display five new values using the Adams method. A function subprogram is employed to compute values for the derivative function.
4. The user is asked whether to compute more points. If the response is positive, these points are computed and displayed and the request is repeated. If the user's response is negative, the program terminates.

```
' *****************************************************

'               PROGRAM ADAMS.BAS

' *****************************************************

'                   Chapter 5

'  Gerald/Wheatley, APPLIED NUMERICAL ANALYSIS (fifth edition)
'              Addison-Wesley, 1994


'  This program gets the solution to a first-order
'    differential equation by the Adams method.

'  It can use either three or four past points to compute
'    the next point. Starting values (that have been generated
'    by some single-step method) are read from a file named
'    ADAMS.DTA. The first item in that file is the number of
'    past points that are to be used. The items that follow
'    are the starting data pairs, x, y(x). Each item is
'    delimited with a comma (or end of line).

'  This solves dy/dx = -2x - y,  y(0) = -1

'  Variables used are:
'        x()       array of x-values
'        y()       array of y-values
'        f()       array of f(x)-values, f(x) = dy/dx
'        i,j       loop control indices
'         N        number of past point to be used
'    response$     user request to repeat computations
'        delx      the spacing between x-values

' Subprogram declarations
DECLARE SUB sub3 (x!(), y!(), f!(), delx)   'Computes from 3 points
DECLARE SUB sub4 (x!(), y!(), f!(), delx)   'Computes from 4 points
DECLARE FUNCTION fcn! (x!, y!)              'Computes f(x)

' Open file and read in the data
OPEN "ADAMS.DTA" FOR INPUT AS #1
INPUT #1, N
```

Figure 5.6   Program 5.1

Figure 5.6 *(continued)*

```
' See if N = 3 or 4
IF N < 3 OR N > 4 THEN
  PRINT "Invalid value for N in file"
  STOP
END IF

' Get the past x- and y-values
FOR i = 1 TO N
  INPUT #1, x(i), y(i)
NEXT i

' Find delx
delx = x(N) - x(N - 1)

' Compute dy/dx = f(x)-values for past data
FOR i = 1 TO N
  f(i) = fcn(x(i), y(i))
NEXT i

' Print a heading and the initial data

CLS
PRINT "   x    "; "       y     "; "      dy/dx"
PRINT
FOR i = 1 TO N
  PRINT USING "##.### "; x(i);
  PRINT USING "###.######  "; y(i); f(i)
NEXT i
PRINT

' While "response$ = Y" do a set of five iterations

response$ = "Y"

DO WHILE response$ = "Y"

  FOR i = 1 TO 5
    IF N = 3 THEN
      CALL sub3(x(), y(), f(), delx)
    ELSE
      CALL sub4(x(), y(), f(), delx)
    END IF

' Compute f-value at new point
    f(N + 1) = fcn(x(N + 1), y(N + 1))

' Print the new values
      PRINT USING "##.### "; x(N + 1);
      PRINT USING "###.######  "; y(i); f(i)

' Reset the arrays for next computation
      FOR j = 1 TO N
        x(j) = x(j + 1)
        y(j) = y(j + 1)
        f(i) = f(i + 1)
      NEXT j
  NEXT i

' See if user wants more computations
    INPUT "Enter Y if you want another five computations ", response$
    IF response$ = "y" THEN response$ = "Y"

LOOP  ' End of WHILE loop
```

Figure 5.6   *(continued)*

```
FUNCTION fcn (x, y)
' This computes a new value for dy/dx

   fcn = -2 * x - y
END FUNCTION

SUB sub3 (x(), y(), f(), delx)
'   This computes a new value for y from 3 past values
   y(4) = y(3) + delx / 12 * (23 * f(3) - 16 * f(2) + 5 * f(1))
   x(4) = x(3) + delx
END SUB


SUB sub4 (x(), y(), f(), delx)
' This computes a new y-value from 4 past values
   y(5) = y(4) + delx / 24 * (55 * f(4) - 59 * f(3) + 37 * f(2) - 9 * f(1))
   x(5) = x(4) + delx
END SUB


****************************************************************

        OUTPUT FOR ADAMS.BAS

     x          y         dy/dx

  0.000   -1.000000    1.000000
  0.200   -0.856192    0.456192
  0.400   -0.810960    0.010960
  0.600   -1.000000    1.000000
  0.800   -0.810960    0.456192
  1.000   -0.924510    0.010960
  1.200   -1.277456   -1.122544
  1.400    0.000000    0.000000
Enter Y if you want another five computations y
  1.600   -0.885216    0.456192
  1.800   -1.669697    0.010960
  2.000   -2.491281   -1.114784
  2.200   -2.851074   -1.548927
  2.400    0.000000    0.000000
Enter Y if you want another five computations
```

Program 5.2 The Runge–Kutta–Fehlberg Method

The second program of this chapter, a C program that uses the Runge–Kutta–Fehlberg method on a system of N first-order equations, is listed in Fig. 5.7. The system has t for its independent variable, and the dependent variables are an x-array.

After the definitions for global variables, a number of procedures are listed in Fig. 5.7:

1. Procedure *initialize*, which sets up the initial values
2. Procedure *getDERIVS*, which evaluates the derivative functions for a particular time-value
3. Procedure *TheLargerOf*, which returns the larger of two values
4. Procedure *computeERROR*, which estimates the error from the last set of k's

5. Procedure *ComputeNewX*, which computes the next set of *x*-values
6. Procedure *rkf*, which actually solves the problem by calls to the other procedures. At each new time step, the estimated error is examined. If it exceeds a tolerance value, the step size is cut in half; if less than one-tenth of the tolerance value, the step size is doubled. Whenever the tolerance value is exceeded, a message is displayed.

At the end of Fig. 5.7 is the listing of the main procedure that computes and displays successive values for the solution from *t* = tINIT to tFINAL. In this implementation, the analytical values are also displayed.

```
/*
                    *******************************
                    **                           **
                    **           RKFSYST.C        **
                    **                           **
                    *******************************
                              Chapter 5

        Gerald/Wheatley, APPLIED NUMERICAL ANALYSIS(Fifth edition)
                       Addison-Wesley, 1994

    ----------------------------------------------------------------

        This program solves a system of N first order differential
        equations by the RUNGE-KUTTA-FEHLBERG method. The equations
        are of the form:
            DX1/DT = F1(T,X)
            DX2/DT = F2(T,X), ETC.

        The particular problem solved here is the second order
        differential equation:

            Y'' = 2 - 4Y*Y/SQR(SIN(x)) ,    1 <= x <= 2.
            where Y(1) = sin(1)*sin(1)
                  Y'(1) = 2*sin(1)*cos(1)

        The Analytical solution to this equation is:
            Y(x) = sin(x)*sin(x).

        The computed and the analytical solution is printed out at
        each time step.

*/
#include <stdio.h>
#include <math.h>

#define   maxN       11    /* max number of first order equations */
#define   hSTART     0.1
#define   tol        0.00005
#define   MaxCount   10
#define   tINIT      1.0        /* initial time */
#define   tFINAL     2.0        /* final time at last interval */

float x0[maxN],      /* state vector at beginning of interval */
      xend[maxN],    /* state vector at end of interval */
      f[maxN],           /* array that holds the values of the derivatives   */
      h,          /* adjustable step size */
      t0,      /* initial time value at beginning of interval */
      finalTime;     /* value equal to: tFINAL - hSTART/2.0 */
```

Figure 5.7 Program 5.2

Figure 5.7   *(continued)*

```
float workarray[7][maxN],    /* a two-dimensional array to store the k values */
      tEND, error;           /* tEND is t[i+1] = t[i] + hSTART */

int   i_rkf, row, col, count_rkf;
int   numberOFequations,     /* actual number of differential equations */
      i, count;

/*
--------------------------------------------------------------
   This Procedure sets up all the initial values.
*/

initialize()
  {
    numberOFequations = 2;

    x0[1] = sin(1.0)*sin(1.0);
    x0[2] = 2.0*sin(1.0)*cos(1.0);

    t0 = tINIT;
    h = hSTART;
  }             /* end of initialize */

getDERIVS(x0, f, t0)
  float x0[], f[], t0;

/*
--------------------------------------------------------
This procedure evaluates the first-order equations for
the given time and state values.

   HERE:  dx1/dt = x2;   dx2/dt := 2 - 4*x1*x1/(sin(t))**2
*/
  {
    f[1] = x0[2];
    f[2] = 2.0 - 4.0*(x0[1]*x0[1])/(sin(t0)*sin(t0));
  }             /* end of getDERIVS */

/*
--------------------------------------------------------------
   This function returns the larger of two real numbers.
*/

float TheLargerOf(a, b)
float a, b;
  {
    if (a > b)
      return(a);
    else
      return(b);
  }                 /* end of TheLargerOf */

/*
--------------------------------------------------------------
   This procedure computes the estimated error from the new
   ki's (workarray) in getting the next state vector.
*/

computeERROR()
  {
    float temp;

    error = 0.0;
    for ( i = 1; i <= numberOFequations; ++i )
      {
```

Figure 5.7 *(continued)*

```
      temp = fabs(workarray[1][i]/360.0
                    - 128.0*workarray[3][i]/4275.0
                    - 2197.0*workarray[4][i]/75240.0
                    + workarray[5][i]/50.0 +
                    + 2.0*workarray[6][i]/55.0);
      error = TheLargerOf(error,temp);
    }                           /* end of the for i loop */
  }                                 /* end of computeERROR */



/*
----------------------------------------------------------------
This procedure updates the state vector from time[i] to time[i+1].
*/

ComputeNewX()
  {
   int i;

    for ( i = 1; i <= numberOFequations; ++i )
      xend[i] = x0[i] + 25.0*workarray[1][i]/216.0
                +1408.0*workarray[3][i]/2565.0
                +2197.0*workarray[4][i]/4104.0
                -workarray[5][i]/5.0;
  }                             /* end of ComputeNewX */

/*
----------------------------------------------------------------
    This Procedure solves a system of differential equations using
    the 4th/5th order method of Runge-Kutta-Fehlberg from t[i] to
    t[i+1]. Use is made of the error estimate to adjust the stepsize,
    h-value, for the next subinterval.
*/
        rkf(t0, h, x0, xend)
        float t0, h, x0[], xend[];
  {
   tEND = t0 + hSTART;
   count_rkf = 0;
   while ((t0 < tEND) && (count_rkf < MaxCount))
     {
        count_rkf = count_rkf + 1;
/*
----------------------------------------------------------------
   Get first estimate of the delta x's
*/
   getDERIVS(x0,f,t0);
   for ( i_rkf = 1; i_rkf <= numberOFequations; ++i_rkf )
     {
       workarray[1][i_rkf] = h*f[i_rkf];
       xend[i_rkf] = x0[i_rkf] + workarray[1][i_rkf]/4.0;
     }
/*
----------------------------------------------------------------
   Get second estimate. The xend vector holds the x values.
*/
   getDERIVS(xend,f,t0+h/4.0);
   for ( i_rkf = 1; i_rkf <= numberOFequations; ++i_rkf )
     {
       workarray[2][i_rkf] = h*f[i_rkf];
       xend[i_rkf] = x0[i_rkf] + (workarray[1][i_rkf]*3.0
                        + workarray[2][i_rkf]*9.0)/32.0;
     }
/*
----------------------------------------------------------------
   Repeat for third estimate
```

Figure 5.7 *(continued)*

```
*/
    getDERIVS(xend,f, t0+3.0*h/8.0);
    for ( i_rkf = 1; i_rkf <= numberOFequations; ++i_rkf )
       {
         workarray[3][i_rkf] = h*f[i_rkf];
         xend[i_rkf] = x0[i_rkf] + (workarray[1][i_rkf]*1932.0
                        - workarray[2][i_rkf]*7200
                        + workarray[3][i_rkf]*7296)/2197.0;
       }
/*
  -------------------------------------------------------------------
    Fourth estimates, i.e. k4's
*/
    getDERIVS(xend,f, t0+12.0*h/13.0);
    for ( i_rkf = 1; i_rkf <= numberOFequations; ++i_rkf )
       {
         workarray[4][i_rkf] = h*f[i_rkf];
         xend[i_rkf] = x0[i_rkf] + 439.0*workarray[1][i_rkf]/216.0
                        - 8.0*workarray[2][i_rkf]
                        + 3680.0*workarray[3][i_rkf]/513.0
                        -845.0*workarray[4][i_rkf]/4104.0;
       }
/*
  -------------------------------------------------------------------
    Fifth estimates
*/
    getDERIVS(xend,f,t0+h);
    for ( i_rkf = 1; i_rkf <= numberOFequations; ++i_rkf )
       {
         workarray[5][i_rkf] = h*f[i_rkf];
         xend[i_rkf] = x0[i_rkf]   - 8.0*workarray[1][i_rkf]/27.0
                        + 2.0*workarray[2][i_rkf]
                        -3544.0*workarray[3][i_rkf]/2565.0
                        +1859.0*workarray[4][i_rkf]/4104.0
                        -11.0*workarray[5][i_rkf]/40.0;
       }
/*
  -------------------------------------------------------------------
    Sixth estimates
*/
    getDERIVS(xend,f,t0+h/2.0);
    for ( i_rkf = 1; i_rkf <= numberOFequations; ++i_rkf )
      workarray[6][i_rkf] = h*f[i_rkf];
/*
  -------------------------------------------------------------------
    Compute error estimate
*/
    error = 0.0;
    computeERROR();
    if (error < tol)
       {
         t0 = t0 + h;
         ComputeNewX();
         for ( i_rkf = 1; i_rkf <= numberOFequations; ++i_rkf )
            x0[i_rkf] = xend[i_rkf];
       }
    if (error > tol)
       {
         printf("\n\n");
         printf("\t\t FOR A STEPSIZE OF%.4f\t",h);
            printf(" THE ERROR ESTIMATE IS  %.6f", h,error);
         printf("\n");
         h = h/2.0;
       }
    if (error < h*tol/10.0)
       h = 2.0*h;
```

Figure 5.7 *(continued)*

```
        if (t0 + h > tEND)
            h = tEND - t0;
        }       /* end of while loop */
    }           /* end of rkf procedure */

    main()              /* begin of main */
    {
        initialize();
        count = 0;
        printf("\n\t AT TIME T =  %.2f\n", t0);
        for ( i = 1; i <= numberOFequations; ++i )
            printf("\t\t   X[ %d ] = %.6f\n", i, x0[i]);
        printf("\n");

        finalTime = tFINAL - hSTART/2.0;
        while ((t0 < finalTime) && (count < MaxCount))
        {
            rkf(t0,h, x0,xend);
            t0 = t0 + hSTART;
            printf("\n\t AT TIME T =  %.2f\n ", t0);
            for ( i = 1; i <= numberOFequations; ++i )
                printf("\t\t   X[ %d ] = %.6f\n", i, xend[i]);
            printf("\t\t\t   EXACT IS : %.6f   FOR X[1]\n\n ",
    (sin(t0)*sin(t0)));
        }               /* end of while loop */
    }                       /* end of the rkfsyst.c program */

    ******************************************************************

                    OUTPUT FOR RKFSYST.C

            AT TIME T =  1.00

                    X[ 1 ] =  0.708073

                    X[ 2 ] =  0.909297

            AT TIME T =  1.10

                    X[ 1 ] =  0.794251

                    X[ 2 ] =  0.808496

                            EXACT IS :  0.794251   FOR X[1]

            AT TIME T =  1.20

                    X[ 1 ] =  0.868698

                    X[ 2 ] =  0.675462

                            EXACT IS :  0.868697   FOR X[1]

            AT TIME T =  1.30

                    X[ 1 ] =  0.928445

                    X[ 2 ] =  0.515498

                            EXACT IS :  0.928444   FOR X[1]

            AT TIME T =  1.40

                    X[ 1 ] =  0.971112

                    X[ 2 ] =  0.334983

                            EXACT IS :  0.971111   FOR X[1]
```

Figure 5.7 *(continued)*

```
AT TIME T =  1.50
                    X[ 1 ] =  0.994997
                    X[ 2 ] =  0.141113
                            EXACT IS :  0.994996    FOR X[1]
AT TIME T =  1.60
                    X[ 1 ] =  0.999148
                    X[ 2 ] =  -0.058383
                            EXACT IS :  0.999147    FOR X[1]
AT TIME T =  1.70
                    X[ 1 ] =  0.983399
                    X[ 2 ] =  -0.255551
                            EXACT IS :  0.983399    FOR X[1]
AT TIME T =  1.80
                    X[ 1 ] =  0.948378
                    X[ 2 ] =  -0.442532
                            EXACT IS :  0.948379    FOR X[1]
AT TIME T =  1.90
                    X[ 1 ] =  0.895481
                    X[ 2 ] =  -0.611869
                            EXACT IS :  0.895484    FOR X[1]
AT TIME T =  2.00
                    X[ 1 ] =  0.826818
                    X[ 2 ] =  -0.756813
                            EXACT IS :  0.826822    FOR X[1]
```

Exercises

Section 5.2

1. ▸a. Solve the differential equation

$$\frac{dy}{dx} = x + y + xy, \qquad y(0) = 1$$

by Taylor-series expansion to get the value of y at $x = 0.1$ and at $x = 0.5$. Use terms through x^5.

b. Do the same for

$$\frac{dy}{dx} = x + y, \qquad y(0) = 1.$$

(Analytical solution is $y(x) = 2e^x - x - 1$.)

c. Do the same for

$$\frac{dy}{dx} = \frac{2x}{y} - xy, \qquad y(0) = 1.$$

(Analytical solution is $y(x) = \sqrt{2 - e^{-x^2}}$.)

2. The general solution to a differential equation normally defines a family of curves. For the differential equation

$$\frac{dy}{dx} = x^2 y^2,$$

find, using the Taylor-series method, the particular curve that passes through $(1, 0)$. Also find the curve through $(0, 1)$. Compare to the analytical solutions.

▶**3.** Use the Taylor-series method to get y at $x = 0.2(0.2)0.6$,* given that

$$y'' = xy, \qquad y(0) = 1, \qquad y'(0) = 1.$$

4. A spring system has resistance to motion proportional to the square of the velocity, and its motion is described by

$$\frac{d^2x}{dt^2} + 0.1\left(\frac{dx}{dt}\right)^2 + 0.6x = 0.$$

If the spring is released from a point that is a unit distance above its equilibrium point, $x(0) = 1$, $x'(0) = 0$, use the Taylor-series method to write a series expression for the displacement as a function of time, including terms up to t^6.

Section 5.3

▶**5.** Use the simple Euler method to solve for $y(0.1)$ from

$$\frac{dy}{dx} = x + y + xy, \qquad y(0) = 1,$$

with $h = 0.01$. Comparing your result to the value determined by Taylor series in Exercise 1, estimate how small h would need to be to obtain four-decimal accuracy.

6. Solve the differential equation

$$\frac{dy}{dx} = \frac{x}{y}, \qquad y(0) = 1,$$

by the simple Euler method with $h = 0.1$, to get $y(1)$. Then repeat with $h = 0.2$ to get another estimate of $y(1)$. Extrapolate these results, assuming errors are proportional to step size, and then compare them to the analytical result. (Analytical result is $y^2 = 1 + x^2$.)

▶**7.** Repeat Exercise 5 but with the modified Euler method with $h = 0.025$, so that the solution is obtained after four steps. Comparing with the result of Exercise 5, about how much less effort is it to solve this problem to four decimals with the modified Euler method in comparison to the simple Euler method?

8. Find the solution to

$$\frac{dy}{dt} = y^2 + t^2, \qquad y(1) = 0, \qquad \text{at } t = 2,$$

by the modified Euler method, using $h = 0.1$. Repeat with $h = 0.05$. From the two results, estimate the accuracy of the second computation.

▶**9.** Solve $y' = \sin x + y$, $y(0) = 2$, by the modified Euler method to get y at $x = 0.1(0.1)0.5$.

10. A sky diver jumps from a plane, and during the time before the parachute opens, the air resistance is proportional to the $\frac{3}{2}$ power of the diver's velocity. If it is known that the maximum rate of fall under these conditions is 80 mph, determine the diver's velocity during the first 2 sec of fall using the modified Euler method with $\Delta t = 0.2$. Neglect horizontal drift and assume an initial velocity of zero.

Section 5.4

11. Solve Exercise 5 by the Runge–Kutta method but with $h = 0.1$ so that the solution is obtained in only one step. Carry five decimals, and compare the accuracy and amount of work required with this method against the simple and modified Euler techniques in Exercises 5 and 7.

12. Solve Exercise 8 by the Runge–Kutta method, using $h = 0.2, 0.1$, and 0.05.

▶**13.** Determine y at $x = 0.2(0.2)0.6$ by the Runge–Kutta technique, given that

$$\frac{dy}{dx} = \frac{1}{x + y}, \qquad y(0) = 2.$$

14. Using the conditions of Exercise 10, determine how long it takes for the jumper to reach 90% of his or her maximum velocity, by integrating the equation using the Runge–Kutta technique with $\Delta t = 0.5$ until the velocity exceeds this value, and then interpolating. Then use numerical integration on the velocity values to determine the distance the diver falls in attaining $0.9v_{max}$.

15. It is not easy to know the accuracy with which the function has been determined by either the Euler methods or the Runge–Kutta method. A possible way to measure accuracy is to repeat the problem with a smaller step size, and compare results. If the

---

* This notation means for $x = 0.2$ through $x = 0.6$ with increments of 0.2.

two computations agree to n decimal places, one then assumes the values are correct to that many places. Repeat Exercise 14 with $\Delta t = 0.3$, which should give a global error about one-eighth as large, and by comparing results, determine the accuracy in Exercise 14. (Why do we expect to reduce the error eightfold by this change in Δt?)

16. Repeat Exercises 11, 12, and 13 using the fifth-order equation of (5.11) of the Runge–Kutta–Fehlberg method.

17. Solve Exercises 1, 6, and 9 using Runge–Kutta–Fehlberg.

▶**18.** Solve $y' = 2x^2 - y, y(0) = -1$ by the Runge–Kutta–Fehlberg method to $x = 2.0$.

19. Solve the equation in Exercise 10 by Runge–Kutta–Fehlberg.

Section 5.5

▶**20.** The equation

$$\frac{dy}{dx} = f(x, y) = 2x(y - 1), \qquad y(0) = 0,$$

has initial values as follows:

x	y	f
0	0	0
0.1	−0.01005	−0.20201
0.2	−0.04081	−0.41632
0.3	−0.09417	−0.65650
0.4	−0.17351	−0.93881

a. Using the Adams procedure described in Section 5.5, compute $y(0.5)$ by fitting a quadratic through the last three values of $f(x, y)$. Compare to the exact answer: −0.28403.
b. Repeat, using the last four points to fit a cubic.
c. Repeat again, but use five points to fit a quartic.

21. For the differential equation

$$\frac{dy}{dt} = y - t^2, \qquad y(0) = 1,$$

starting values are known:

$$y(0.2) = 1.2186, \qquad y(0.4) = 1.4682,$$
$$y(0.6) = 1.7379.$$

Use the Adams method, fitting cubics with the last four (y, t) values and advance the solution to $t = 1.2$. Compare to the analytical solution.

22. For the equation

$$\frac{dy}{dt} = t^2 - t, \qquad y(1) = 0,$$

the analytical solution is easy to find:

$$y = \frac{t^3}{3} - \frac{t^2}{2} + \frac{1}{6}.$$

If we use three points in the Adams method, what error would we expect in the numerical solution? Confirm your expectation by performing the computations.

Section 5.6

23. For the differential equation

$$\frac{dy}{dx} = y - x^2, \qquad y(0) = 1,$$

starting values are known:

$$y(0.2) = 1.2186, \qquad y(0.4) = 1.4682,$$
$$y(0.6) = 1.7379.$$

Use the Milne method to advance the solution to $x = 1.2$. Carry four decimals and compare to the analytical solution.

▶**24.** For the differential equation $dy/dx = x/y$, the following values are given:

x	y
0	$\sqrt{1}$
1	$\sqrt{2}$
2	$\sqrt{5}$
3	$\sqrt{10}$

To how many decimal places will Milne's method give the value at $x = 4$? How many decimal places must be carried in the starting values of y to ensure this accuracy?

25. For the equation $y' = y \sin \pi x, y(0) = 1$, get starting values by the Runge–Kutta–Fehlberg method for $x = 0.2(0.2)0.6$, and advance the solution to $x = 1.0$ by Milne's method.

▶**26.** Continue the results of Exercise 13 to $x = 2.0$ by the method of Milne. If you find that the corrector formula reproduces the predictor values, double the value of h after sufficient values are available.

Section 5.7

27. Solve Exercise 21 using the Adams–Moulton method.

28. Repeat Exercise 25 using the Adams–Moulton method.

▶**29.** For the equation

$$\frac{dy}{dx} = x^3 + y^2, \qquad y(0) = 0,$$

using $h = 0.2$, compute three new values by the Runge–Kutta method (four decimals). Then advance to $x = 1.4$ using the Adams–Moulton method. If you find the accuracy criterion is not met, use Eqs. (5.26) to interpolate additional values so that four-place accuracy is maintained.

30. Derive the interpolation formulas given in Eqs. (5.26).

Section 5.8

31. Use the multivalued method to find $y(x_0 + h)$ for each of the following equations. First get the predicted value, then correct using the method of solving Eq. (5.30).
   a. $dy/dx = xy - 2y$, $y(1) = 2$, $h = 0.1$
   b. $dy/dx = -2x - y$, $y(0) = -1$, $h = 0.1$
   c. $dy/dx = -10xy + 10y$, $y(0) = 1$, $h = 0.05$
   d. $dy/dx = 2x^2 - y$, $y(0) = -1$, $h = 0.1$

32. Confirm that $Y_{n+1,p} = BY_n$ gives the Taylor-series expansion for y, hy', $(h^2/2)y''$, and $(h^3/6)y'''$. Matrix B is given in Eq. (5.32).

33. Confirm the values in Table 5.12.

▶**34.** For each equation in Exercise 31, advance the solution to $y(x_0 + 3h)$.

35. Use the multivalued method to compute the starting values for Exercise 21. Do these values agree with those that are given?

36. Solve Exercise 1, parts (a) and (b), and Exercise 21 using the multivalued method of Section 5.8.

▶**37.** The multivalued method can use other vectors R. Solve Exercise 34 with this method for
   a. $R = [0, 1, 3/4, 1/6]$.
   b. $R = [6/11, 1, 6/11, 1/11]$.

38. Repeat Exercise 35 but with each of the R's of Exercise 37.

Section 5.9

▶**39.** Given the linear differential equation $dy/dx = y \sin x$.
   a. What is the maximum value of h that ensures convergence of the Adams–Moulton method when continuing applications of the corrector formula are made?
   b. If an h one-tenth of this maximum value is used, how close must the predictor and corrector values be so that recorrections are not required?
   c. In terms of the maximum h in part (a), what size of h is implied in the accuracy criterion, $D \cdot 10^n < 14.2$?

40. Repeat Exercise 39 for the differential equation

$$\frac{dy}{dx} = x^3 + y^2, \qquad y(0) = 0,$$

in the neighborhood of the point $(1.0, 0.15)$.

41. Repeat Exercise 39, using the Milne method. For part (c), the accuracy criterion is $D \cdot 10^n < 29$.

42. Derive Eq. (5.38).

43. Derive convergence criteria similar to Eqs. (5.36) and (5.37) for the Euler predictor–corrector method. Why can one not derive an accuracy criterion similar to those for the methods of Milne and Adams–Moulton?

Section 5.10

44. Estimate the propagated error at each step when the equation

$$\frac{dy}{dx} = x + y, \qquad y(0) = 1,$$

is solved by the simple Euler method with $h = 0.02$, for $x = 0(0.02)0.1$. Compare to the actual errors.

▶**45.** Follow the propagated error between $x = 1$ and $x = 1.6$ when the simple Euler method is used to solve

$$\frac{dy}{dx} = xy^2, \qquad y(1) = 1.$$

Take $h = 0.1$. Compare to the actual errors at each step. The analytical solution is $y = 2/(3 - x^2)$.

46. We can derive the global error (Eq. (5.42)) for Euler's method without making use of the second-order difference equation. With the same assumptions about M and K and using the fact that the series

$$1 + s + s^2 + \cdots + s^n = \frac{s^{n+1} - 1}{s - 1},$$

show that

$$e_n \leq \frac{hM}{2K} (e^{(x_n - x_0)K} - 1).$$

(*Hint:* Let $s = 1 + hK$.)

Section 5.11

47. The mathematical model of an electrical circuit is given by the equation

$$0.5 \frac{d^2Q}{dt^2} + 6 \frac{dQ}{dt} + 50Q = 24 \sin 10t,$$

with $Q = 0$ and $I = dQ/dt = 0$ at $t = 0$. Express as a pair of first-order equations.

▶**48.** In the theory of beams it is shown that the radius of curvature at any point is proportional to the bending moment:

$$EI \frac{y''}{[1 + (y')^2]^{3/2}} = M(x),$$

where y is the deflection of the neutral axis. In the usual approach, $(y')^2$ is neglected in comparison to unity, but if the beam has appreciable curvature, this is invalid. For the cantilever beam for which $y(0) = y'(0) = 0$, express the equation as a pair of simultaneous first-order equations.

49. The motion of the compound spring system as sketched in Fig. 5.8 is given by the solution of the pair of simultaneous equations

$$m_1 \frac{d^2 y_1}{dt^2} = -k_1 y_1 - k_2(y_1 - y_2),$$

$$m_2 \frac{d^2 y_2}{dt^2} = k_2(y_1 - y_2),$$

where y_1 and y_2 are the displacements of the two masses from their equilibrium positions. The initial conditions are

$$y_1(0) = A, \quad y_1'(0) = B, \quad y_2(0) = C, \quad y_2'(0) = D.$$

Express as a set of first-order equations.

▶**50.** Solve the pair of simultaneous equations

$$\frac{dx}{dt} = xy + t, \quad x(0) = 0, \quad \frac{dy}{dt} = x - t, \quad y(0) = 1,$$

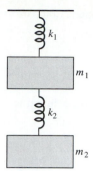

Figure 5.8

by the modified Euler method for $t = 0.2(0.2)0.6$. (Carry three decimals rounded.) Recorrect until reproduced to three decimals.

51. Advance the solution of Exercise 50 to $x = 1.0$ ($h = 0.2$) by the Adams–Moulton method.

52. Repeat Exercise 51 but use Milne's method.

▶**53.** Find y at $x = 0.6$, given that

$$y'' = yy', \quad y(0) = 1, \quad y'(0) = -1.$$

Begin the solution by the Taylor-series method, getting

$$y(0.1), \quad y(0.2), \quad y(0.3).$$

Then advance to $x = 0.6$ employing the Adams–Moulton technique with $h = 0.1$ on the equivalent set of first-order equations.

54. Express the third-order equation

$$y''' + ty'' - ty' - 2y = t, \quad y(0) = y''(0) = 0, \quad y'(0) = 1,$$

as a set of first-order equations and solve at $t = 0.2, 0.4, 0.6$ by the Runge–Kutta method ($h = 0.2$).

55. Using the Adams–Moulton method with $h = 0.2$, advance the solution of Exercise 54 to $t = 1.0$. Estimate the accuracy of the value of y at $t = 1.0$.

Section 5.12

▶**56.** For a resonant spring system with a periodic forcing function, the differential equation is

$$\frac{d^2x}{dt^2} + 64x = 16 \cos 8t, \quad x(0) = x'(0) = 0.$$

Determine the displacement at $t = 0.1(0.1)0.8$ by the method of Eq. (5.46), getting the starting values by any other method of this chapter. Compare to the analytical solution $t \sin 8t$.

57. For the first-order equation

$$\frac{dy}{dt} - 12y = \frac{3}{2}t^2 - 12t + 6t^3 - 1, \qquad y(0) = a,$$

it is not difficult to verify that the solution is

$$y = ae^{-12t} + \frac{t^3}{2} - t.$$

If $a = 0$ (so $y(0) = 0$), the solution reduces to a simple cubic in t. Still, accumulated round-off errors act as though the initial value of y is not exactly zero, causing the exponential term to appear when it should not. This makes the problem a stiff equation. Assume that $y(0) = 0$ and demonstrate that you do not get the analytical answer when you integrate from $t = 0$ to $t = 2$ using the Runge–Kutta fourth-order method with a step size of 0.005.

Section 5.13

58. Use MAPLE to solve these differential equations. Get the analytical solutions with dsolve.
a. $y' = x^2 + xy, \quad y(1) = 2$
b. $x' = \sin(t), \quad x(0) = 1$
c. $y' + x + y = 2, \quad y(x_0) = y_0$

59. Use the plot program in MAPLE to graph the solutions to the equations in Exercise 58, a and b.

60. $y' = xy^2 + xy - y^2 - y$ is what is called a "separable equation." Use MAPLE to get the general solution.

▶61. Make MAPLE get the Taylor-series solution to the equations in Exercise 58 and evaluate at $x_0 + 2$.

62. Repeat Exercise 61, but use Runge–Kutta–Fehlberg

with $h = 0.2$. Compare the solutions to those in Exercise 61 and to the analytical answers.

63. Use MAPLE to get the fifth-order Taylor-series solution, and evaluate at $x = 2$ for $y'' + 2y' = x^2 + y, y(0) = 1, y'(0) = -1$.

64. Repeat Exercise 63, but use the Runge–Kutta–Fehlberg method with $h = 0.05$.

▶65. Use the plot program in MAPLE to graph the solution to Exercise 63 from $x = 0$ to $x = 2$.

Section 5.14

66. Which of these satisfy the Lipschitz conditions?
▶a. $y' = x^2 - y^2$ on the unit square
▶b. $y' = x^2/y$ on the rectangle with corners at $(-2, 4)$, $(4, -1)$
c. $x' = tx$ on the rectangle with corners at $(1, 5)$, $(5, 1)$

67. Given $y = 1/(x - 1)$ so that $y' = -1/(x - 1)^2$. Over what x-intervals do the Lipschitz conditions hold?

▶68. The Lipschitz conditions are weaker than the requirement that $\partial f / \partial y$ is continuous and bounded. Find an example where the Lipschitz conditions hold but the stronger condition does not.

▶69. By making a slight change in the initial condition, determine the "amplification factor," the ratio of the change in $y(2)$ to the change of $y(0)$ for these equations. Which of them would you call unstable?
a. $y' = 1 - x + 4y, \quad y(0) = 5$
b. $y' = 1 - x, \quad y(0) = 5$
c. $y' = -4y, \quad y(0) = 5$
d. $y' = 4y, \quad y(0) = 5$

70. For the equation of Exercise 56, compute the eigenvalues for the Jacobian (see Section 2.4). Is the system unstable?

Applied Problems and Projects

71. For the mechanical system of Fig. 5.9, where the wheel has a radius of r_2 and a mass of m_2,
a. show that the differential equations for motion are

$$(m_1 + 0.5m_2)\frac{d^2x_1}{dt^2} - 0.5m_2\frac{d^2x_2}{dt^2} + k_1x_1 = 0,$$

$$-0.5m_2\frac{d^2x_1}{dt^2} + 1.5m_2\frac{d^2x_2}{dt^2} + k_2x_2 = 0;$$

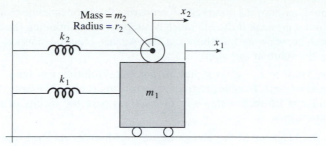

Figure 5.9

b. for $m_1 = m_2 = 5$, $k_1 = k_2 = 15$, $x_1(0) = 1.5$, and with $x_2 = dx_1/dt = dx_2/dt$ all equal to zero at $t = 0$, integrate from $t = 0$ to $t = 10$. Verify several points on the graphs of Fig. 5.10.

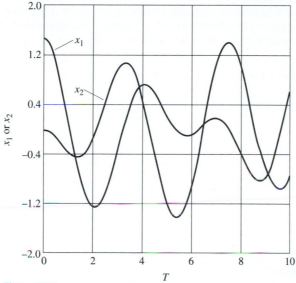

Figure 5.10

72. In finding the deflection of a beam, the y' term is often neglected (see Exercise 48). (The analytical method is easy if y' is neglected in the equation.) What is the difference in the calculated values of the maximum deflection when the nonlinear relationship (Exercise 48) is used in comparison to the simpler linear equation? For light loads, the error is negligible, of course; only for heavier ones is there a need to use the nonlinear equation. At what value of the load is the error equal to 1% of the true value?

73. Write a program that solves a first-order initial-value problem by the Adams–Moulton method, calling the RKF subroutine of the text to obtain starting values. As each new point

is computed, monitor the accuracy by comparing the error estimate to a tolerance parameter, printing out a message if the tolerance is not met, but continuing the solution anyway. This might best be done by incorporating the Adams–Moulton method in a subroutine that advances the solution one step.

74. The equation $y' = 1 + y^2, y(0)$, has the analytical solution $y = \tan x$. The tangent function is infinite at $x = \pi/2$. Use the program you wrote in Problem 73 to solve this equation between $x = 0$ and $x = 1.6$ with a step size of 0.1. Compare the results of the program with the analytical solution.

75. What is the behavior of the Runge–Kutta fourth-order method when used over $x = 0$ to $x = 1.6$ on the equation of Problem 74? How does it compare with the multistep methods? How does its behavior change with step size?

76. Enlarge on Problem 73 by modifying your program so that the step size is halved if the error estimate exceeds the tolerance value. Use Eq. (5.26) to interpolate for the new values needed when this is done. If the error estimate is very small, say 1/20 times the tolerance, your program should double the step size. (Some programs that provide for doubling the step size keep track of how many uniformly spaced values are available and defer the doubling until there are seven. You may wish to do this, but it is a tricky bit of programming. Alternatively, one could always compute two additional values after discovering that the step size can be increased; this guarantees that there are at least seven equispaced values. Another method would be to extrapolate to obtain a value at t_{n-5} when we find, at t_{n+1}, that the error is 1/20 of the tolerance.)

77. In an electrical circuit (Fig. 5.11) containing resistance, inductance, and capacitance (and every circuit does), the voltage drop across the resistance is iR (i is current in amperes, R is resistance in ohms), across the inductance it is $L(di/dt)$ (L is inductance in henries), and across the capacitance it is q/C (q is charge in the capacitor in coulombs, C is capacitance in farads). We then can write, for the voltage difference between points A and B,

$$V_{AB} = L\frac{di}{dt} + Ri + \frac{q}{C}.$$

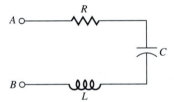

Figure 5.11

Differentiating with respect to t and remembering that $dq/dt = i$, we have a second-order differential equation:

$$L\frac{d^2i}{dt^2} + R\frac{di}{dt} + \frac{1}{C}i = \frac{dV}{dt}.$$

If the voltage V_{AB} (which has previously been 0 V) is suddenly brought to 15 V (let us say, by connecting a battery across the terminals) and maintained steadily at 15 V (so $dV/dt = 0$), current will flow through the circuit. Use an appropriate numerical method to determine how the current varies with time between 0 and 0.1 sec if $C = 1000$ μf, $L = 50$ mH, and $R = 4.7$

ohms; use Δt of 0.002 sec. Also determine how the voltage builds up across the capacitor during this time. You may want to compare the computations with the analytical solution.

78. Repeat Problem 77 but let the voltage source be a 60-Hz sinusoidal input:

$$V_{AB} = 15 \sin(120\pi t).$$

How closely does the voltage across the capacitor resemble a sine wave during the last full cycle of voltage variation?

79. After the voltages have stabilized in Problem 77 (15 V across the capacitor), the battery is shorted so that the capacitor discharges through the resistance and inductor. Follow the current and the capacitor voltages for 0.1 sec, again with $\Delta t = 0.002$ sec. The oscillations of decreasing amplitude are called *damped oscillations*. If the calculations are repeated but with the resistance value increased, the oscillations will be damped out more quickly; at $R = 14.14$ ohms the oscillations should disappear; this is called *critical damping*. Perform numerical computations with values of R increasing from 4.7 to 22 ohms to confirm that critical damping occurs at 14.14 ohms.

80. In Chapter 2, Problem 84, electrical networks were discussed, but these were simple ones with resistance only. A more realistic situation is to have RLC circuits combined into a network. Using either the current or voltage laws of Kirchhoff, we can set up a set of simultaneous equations, but now these are simultaneous differential equations. For example, for the circuit in Fig. 5.12, which has two voltage sources, we have

$$(L_1 + L_3)\frac{d^2i_1}{dt^2} + (R_1 + R_3)\frac{di_1}{dt} + \left(\frac{1}{C_1} + \frac{1}{C_3}\right)i_1 - L_3\frac{d^2i_2}{dt^2} - R_3\frac{di_2}{dt} - \frac{1}{C_3}i_2 = e_1'(t),$$

$$(L_2 + L_3)\frac{d^2i_2}{dt^2} + (R_2 + R_3)\frac{di_2}{dt} + \left(\frac{1}{C_2} + \frac{1}{C_3}\right)i_2 - L_3\frac{d^2i_1}{dt^2} - R_3\frac{di_1}{dt} - \frac{1}{C_3}i_1 = e_2'(t).$$

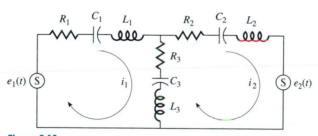

Figure 5.12

In the equations, i_1 and i_2 represent the currents in each of the loops. Solve the equations for i_1 and i_2 between $t = 0$ and $t = 0.2$ sec, if

$$e_1(t) = 100 \sin(120\pi t), \qquad e_2(t) = 0,$$

given that

$$R_1 = 22 \text{ ohms}, \qquad C_1 = C_2 = C_3 = 10 \text{ } \mu\text{f},$$
$$R_2 = 4.7 \text{ ohms}, \qquad L_1 = 2.5 \text{ mH},$$
$$R_3 = 47 \text{ ohms}, \qquad L_2 = L_3 = 0.5 \text{ mH}.$$

81. A Foucault pendulum is one free to swing in both the x- and y-directions. It is frequently displayed in science museums to exhibit the rotation of the earth, which causes the pendulum to swing in directions that continuously vary. The equations of motion are

$$\ddot{x} - 2\omega \sin\psi \dot{y} + k^2 x = 0,$$
$$\ddot{y} + 2\omega \sin\psi \dot{x} + k^2 y = 0,$$

when damping is absent (or compensated for). In these equations the dots over the variable represent differentiation with respect to time. Here ω is the angular velocity of the earth's rotation (7.29×10^{-5} sec^{-1}), ψ is the latitude, $k^2 = g/\ell$ where ℓ is the length of the pendulum. How long will it take a 10-m-long pendulum to rotate its plane of swing by 45° at the latitude where you live? how long if located in Quebec, Canada?

82. Condon and Odishaw (1967) discuss Duffing's equation for the flux ϕ in a transformer. This nonlinear differential equation is

$$\ddot{\phi} + \omega_0^2 \phi + b\phi^3 = \frac{\omega}{N} E \cos\omega t.$$

In this equation, $E \sin\omega t$ is the sinusoidal source voltage and N is the number of turns in the primary winding, while ω_0 and b are parameters of the transformer design. Make a plot of ϕ versus t (and compare to the source voltage) if $E = 165$, $\omega = 120\pi$, $N = 600$, $\omega_0^2 = 83$, and $b = 0.14$. For approximate calculations, the nonlinear term $b\phi^3$ is sometimes neglected. Evaluate your results to determine whether this makes a significant error in the results.

83. Ethylene oxide is an important raw material for the manufacture of organic chemicals. It is produced by reacting ethylene and oxygen together over a silver catalyst. Laboratory studies gave the equation shown.

   It is planned to use this process commercially by passing the gaseous mixture through tubes filled with catalyst. The reaction rate varies with pressure, temperature, and concentrations of ethylene and oxygen, according to this equation:

$$r = 1.7 \times 10^6 e^{-9716/T}\left(\frac{P}{14.7}\right) C_E^{0.328} C_O^{0.672},$$

where

r = reaction rate (units of ethylene oxide formed per lb of catalyst per hr),

T = temperature, °K (°C + 273),

P = absolute pressure (lb/in.^{2}),

C_E = concentration of ethylene,

C_O = concentration of oxygen.

Under the planned conditions, the reaction will occur, as the gas flows through the tube, according to the equation

$$\frac{dx}{dL} = 6.42r,$$

where

x = fraction of ethylene converted to ethylene oxide,

L = length of reactor tube (ft).

The reaction is strongly exothermic, so that it is necessary to cool the tubular reactor to prevent overheating. (Excessively high temperatures produce undesirable side reactions.) The reactor will be cooled by surrounding the catalyst tubes with boiling coolant under pressure so that the tube walls are kept at 225°C. This will remove heat proportional to the temperature difference between the gas and the boiling water. Of course, heat is generated by the reaction. The net effect can be expressed by this equation for the temperature change per foot of tube, where B is a design parameter:

$$\frac{dT}{dL} = 24{,}302r - B(T - 225).$$

For preliminary computations, it has been agreed that we can neglect the change in pressure as the gases flow through the tubes; we will use the average pressure of $P = 22$ lb/in.^{2} absolute. We will also neglect the difference between the catalyst temperature (which should be used to find the reaction rate) and the gas temperature. You are to compute the length of tubes required for 65% conversion of ethylene if the inlet temperature is 250°C. Oxygen is consumed in proportion to the ethylene converted; material balances show that the concentrations of ethylene and oxygen vary with x, the fraction of ethylene converted, as follows:

$$C_E = \frac{1 - x}{4 - 0.375x},$$

$$C_O = \frac{1 - 1.125x}{4 - 0.375x}.$$

The design parameter B will be determined by the diameter of tubes that contain the catalyst. (The number of tubes in parallel will be chosen to accommodate the quantities of materials flowing through the reactor.) The tube size will be chosen to control the maximum temperature of the reaction, as set by the minimum allowable value of B. If the tubes are too large in diameter (for which the value of B is small), the temperatures will run wild. If the tubes are too *small* (giving a large value to B), so much heat is lost that the reaction tends to be quenched. In your studies, vary B to find the least value that will keep the maximum temperature below 300°C. Permissible values for the parameter B are from 1.0 to 10.0.

   In addition to finding how long the tubes must be, we need to know how the temperature varies with x and with the distance along the tubes. To have some indication of the controllability of the process, you are also asked to determine how much the outlet temperature will change for a 1°C change in the inlet temperature, using the value of B already determined.

84. An ecologist has been studying the effects of the environment on the population of field mice. Her research shows that the number of mice born each month is proportional to the number of females in the group and that the fraction of females is normally constant in any group. This implies that the number of births per month is proportional to the total population.

   She has located a test plot for further research, which is a restricted area of semiarid land. She has constructed barriers around the plot so mice cannot enter or leave. Under the conditions of the experiment, the food supply is limited, and it is found that the death rate is affected as a result, with mice dying of starvation at a rate proportional to some power of the population. (She also hypothesizes that when the mother is undernourished, the babies have less chance for survival and that starving males tend to attack one another, but these factors are only speculation.)

   The net result of this scientific analysis is the following equation, with N being the number of mice at time t (with t expressed in months.) She has come to you for help in solving the equation; her calculus doesn't seem to apply.

$$\frac{dN}{dt} = aN - BN^{1.7}, \qquad \text{with } B \text{ given by Table 5.15}$$

Table 5.15

t	B	t	B
0	0.0070	5	0.0013
1	0.0036	6	0.0028
2	0.0011	7	0.0043
3	0.0001	8	0.0056
4	0.0004		

As the season progresses, the amount of vegetation varies. She accounts for this change in the food supply by using a "constant" B that varies with the season.

If 100 mice were initially released into the test plot and if $a = 0.9$, estimate the number of mice as a function of t, for $t = 0$ to $t = 8$.

6

Boundary-Value Problems and Characteristic-Value Problems

Contents of This Chapter

In all differential equations of order greater than 1, two or more values must be known to evaluate the constants in the solution function that satisfies the differential equation. In the problems discussed in Chapter 5, these several values were all specified at the same value of the independent variable, generally at the start, or the *initial value*. For that reason, such problems are termed *initial-value problems*.

For an important class of problems, the several known values of the function or its derivatives are not all known at the same point, but rather at two different values of the independent variable. Because these values of the independent variable are usually at the endpoints (or *boundaries*) of some domain of interest, problems with this type of condition are classed as *boundary-value problems*. Determining the deflection of a simply supported beam is a typical example, where the conditions specified are the deflections and second derivatives of the elastic curve at the supports. Heat-flow problems fall in this class when temperatures or temperature gradients are given at two points. A special case of the boundary-value problem occurs in vibration problems.

We will study three different methods for solving boundary-value problems. In addition, material on a special kind of boundary-value problem—the characteristic-value problem—is presented.

6.1 The "Shooting Method"

Applies the techniques of Chapter 5 to boundary-value problems by assuming values needed to make them initial-value problems. This method involves a degree of trial and error, but for a linear second-order problem, no more than two trials are required.

6.2 Solution Through a Set of Equations

Uses finite-difference approximations for the derivatives, allowing one to write a set of equations whose unknowns are values for the dependent variable(s) at several points within the domain. The solution to this system gives a solution to the original boundary-value problem.

6.3 Derivative Boundary Conditions

Require a modification of the method in Section 6.2; this modification is straightforward, but it does require an artificial extension of the domain.

6.4 Rayleigh–Ritz, Collocation, and Galerkin Methods

Use a technique founded on the calculus of variations. A function that approximates the true solution is found by optimizing a *functional*, an entity related to the original differential equation.

6.5 The Finite-Element Method

Is a modern method of solving boundary-value problems that applies the methods of Section 6.4 to subintervals of the domain, called elements. By working with small elements, the approximating function can be a low-degree polynomial. The development is involved but straightforward.

6.6 Characteristic-Value Problems

Describes this important special class of boundary-value problems, which have a solution only for certain characteristic values of a parameter, called the *eigenvalues*. In this section, you learn more about both eigenvalues and eigenvectors, essential matrix-related quantities that have application in many fields.

6.7 Eigenvalues by Iteration—The Power Method

Describes a method for getting eigenvalues and eigenvectors that is well adapted to computers.

6.8 Eigenvalues by the QR Method

Tells how to get all the eigenvalues of a matrix all at once. It carries out what are called *similarity transformations*, which triangularize the matrix to put the eigenvalues on the diagonal while preserving their values.

6.9 Applications of Eigenvalues

Shows that the eigenvalues of a matrix tell whether an iterative method for solving sets of equations will converge. Eigenvalues and eigenvectors also give the solution to simultaneous linear-differential equations.

6.10 Theoretical Matters

Presents several items of theoretical interest when we solve boundary-value problems and characteristic-value problems. Some of the theory behind the power method is given.

6.11 Using DERIVE

Describes the use of this "mathematical assistant" to aid in solving boundary-value problems and getting eigenvalues and eigenvectors.

Chapter Summary

Is provided to test your knowledge of the chapter's topics.

Computer Programs

Gives a few programming examples that implement the methods of this chapter.

Parallel Processing

Parallel processing is possible with several methods of this chapter, but nothing new is introduced. In Section 6.1, we solve an initial-value problem with two sets of parameters for which the computations can be done simultaneously. But this process involves a very low degree of parallelism, as was pointed out in Chapter 5.

In Sections 6.2, 6.3, 6.5, 6.6, and 6.8, sets of simultaneous linear equations are solved for which parallel processing can be effective, as you learned in Chapter 2. The power method of Section 6.7 is inherently serial because each step depends on the previous one. However, within each step a matrix multiplication is done, and these computations can be speeded up with parallel processors.

6.1 The "Shooting Method"

Suppose we wish to solve the second-order boundary-value problem

$$\frac{d^2x}{dt^2} - \left(1 - \frac{t}{5}\right)x = t, \qquad x(1) = 2, \qquad x(3) = -1. \tag{6.1}$$

Note that if $x'(1)$ were given in addition to $x(1)$, this would be an initial-value problem. There is a way to adapt our previous methods to this problem, as illustrated in Fig. 6.1. We know the value of x at $t = 1$, and at $t = 3$, as given by the dots. The curve that represents x between these two points is desired. We anticipate that some such curve, such as the dotted line, exists;* its slope and curvature are interrelated to x and t by differential Eq. (6.1).

If we assume the slope of the curve at $t = 1$—say, $x'(1) = -1.5$—we could solve the equation as an initial-value problem using this assumed value. The test of our assumption is whether we calculate x at $t = 3$ to match the known value, $x(3) = -1$. In Table 6.1 we show the results of this computation employing a computer program

---

*The existence of a solution to a boundary-value problem cannot be taken for granted, however. For example, the problem $y'' + y = 0, y(0) = 1, y(\pi) = 0$ has no solution.

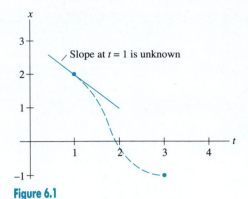

Figure 6.1

that implements the modified Euler method with $\Delta t = 0.2$. Since the result of this gives $x(3) = 4.811$, and not the desired $x(3) = -1$, we assume another value for $x'(1)$ and repeat. Since this first calculated value of $x(3)$ is too high, we assume a smaller value for the slope, say, $x'(1) = -3.0$. The second computation gives $x(3) = 0.453$, which is better but still too high. After these two trials, we linearly interpolate (here extrapolate) for a third trial. At $x'(1) = -3.500$, we get the correct value of $x(3)$. In some problems more attempts are needed to get the correct solution.

The method we have illustrated is called the *shooting method* because it resembles an artillery problem. The artillery officer sets the elevation of the gun and fires a preliminary round at the target. After successive shots have straddled the target, the officer zeroes in on it by using intermediate values of the gun's elevation. This method corresponds to our using assumed values of the initial slope and interpolating based on how close we come to $x(3)$.

Table 6.1 Solving a second-order equation by the shooting method:

$$x'' = t + \left(1 - \frac{t}{5}\right)x, \quad x(1) = 2, \quad x(3) = -1$$

Time	Distance	Velocity	Distance	Velocity	Distance	Velocity
$\Delta t = 0.2$:	Assume $x'(1) = -1.5$		Assume $x'(1) = -3.0$		Assume $x'(1) = -3.500$	
1.00	2.000	−1.500	2.000	−3.000	2.000	−3.500
1.20	1.751	−0.987	1.499	−2.510	1.348	−3.018
1.40	1.605	−0.478	0.991	−2.068	0.787	−2.599
1.60	1.561	0.043	0.619	−1.655	0.305	−2.221
1.80	1.625	0.594	0.328	−1.252	−0.104	−1.867
2.00	1.803	1.186	0.118	−0.844	−0.443	−1.521
2.20	2.105	1.832	−0.007	−0.417	−0.712	−1.167
2.40	2.542	2.542	−0.045	0.040	−0.908	−0.794
2.60	3.128	3.324	0.013	0.539	−1.026	−0.391
2.80	3.880	4.185	0.175	1.087	−1.060	0.054
3.00	4.811	5.128	0.453	1.693	−1.000	0.547

You may wonder whether it was only a lucky accident that gave correct results on our third trial, using the initial slope $x'(1) = -3.500$, extrapolated from our earlier guesses. This desirable result will always be true when the boundary-value problem is *linear*, as in this case. (A differential equation is linear when the coefficients of each derivative term and the function are not functions of x. They may be functions of t, as in this example, however.) The reason is that, if there are two different functions of x that satisfy the differential equation (but corresponding to different boundary conditions, of course), then a linear combination of them is also a solution. A proof of this statement will be found in Section 6.10.

With two functions x_1 and x_2 both having $x(1) = 2.0$ at the left boundary and differing at $t = 3.0$, we can take a linear combination, so the correct value of $x(3.0)$ results.

Let G1 = first guess at initial slope;

Let G2 = second guess at initial slope;

Let R1 = first result at endpoint (using G1);

Let R2 = second result at endpoint (using G2).

With D = the desired value at the endpoint, we can write

$$\text{Extrapolated estimate for initial slope} = G1 + \frac{G2 - G1}{R2 - R1}(D - R1).$$

In the preceding example, we get

$$-1.5 + \frac{(-3.0) - (-1.5)}{0.453 - 4.811}(-1.0 - 4.811) = -3.500.$$

Using a linear combination of two assumed initial slopes to get the correct slope goes further than just allowing us to interpolate from the two initial guesses of $x'(1)$ to predict exactly the correct value of $x'(1)$ that is needed to give agreement with the right-hand boundary condition. This relationship is true *throughout* the interval as well. This means that each value of x could be calculated by taking the proper combination of the values obtained from the earlier calculations. This proper combination is the same as that which gives the correct value of x at $t = 3$. (The true curve for x versus t will be intermediate between two curves that bracket the final condition.)

In our example,

$$c_1 x_1 + c_2 x_2 = \text{true values.}$$

At left end,

$$t = 1: \qquad c_1(2.0) + c_2(2.0) = 2.0.$$

At right end,

$$t = 3: \qquad c_1(4.811) + c_2(0.453) = -1.0.$$

From the first equation, $c_2 = 1 - c_1$, and from the second, $c_1 = -0.3334$, giving $c_2 = 1.3334$. We can calculate the correct values of x using these values. For example,

$$\text{at } t = 2.0: \qquad -0.3334(1.803) + 1.3334(0.118) = -0.443.$$

We therefore conclude that a second-order linear boundary-value problem can always be solved by doing only two computations using two different guesses for the initial slope.

With higher-order linear boundary-value problems, the situation is somewhat more difficult, although with a third-order equation we can choose the starting point at that boundary where two of the three conditions are specified. (If this means working backward from the right-hand end, Δt will be negative.) For a fourth-order problem, with two conditions specified at each boundary, we seem to have a double trial-and-error situation. However, a combination of solutions also is a solution in this case, permitting us to get the correct solution by combining four trial computations. In these higher-order problems, some of the conditions we must match are values of the derivatives, requiring us to estimate the derivatives of our computed functions using approximation methods of Chapter 4—and these methods are more subject to errors.

When the differential equation is nonlinear (the coefficients F, G, and/or H in $x'' + Fx' + Gx = H$ are functions of x or x'), the interpolation between two assumed values for the initial slope does not usually give the correct value. Consider

$$x'' - \left(1 - \frac{t}{5}\right)(x')(x) = t, \qquad x(1) = 2, \qquad x(3) = -1.$$

The presence of the $(x')(x)$-term makes it nonlinear. In this case, the values in Table 6.2 result, using the modified Euler method with $\Delta t = 0.2$.

Only when the assumed values are very near to the correct initial slope do we get the correct value by interpolating linearly. (If the computer program that solves the differential equation is available on a personal computer, the preceding trial-and-error search can be conducted quickly and conveniently. This was the case with the data in Table 6.2.)

Note that we are faced with a root-finding problem, in that the value of $x(3)$ is a function of $x'(1)$. The methods of Chapter 1 then apply, although Newton's method requires the *slope* of this function, which is not readily available. Muller's method, applied after three estimates $(x'(1) = -1.5, -3.0, -2.213)$ immediately gave the correct value for the initial slope.

After we know the correct value for $x'(1)$, we solve the problem as an initial-value problem, getting the results shown in Table 6.3.

Table 6.2 Values for a nonlinear boundary-value problem by the shooting method:

$$x'' = t + \left(1 - \frac{t}{5}\right)xx', \quad x(1) = 2, \quad x(3) = -1$$

Assumed value for $x'(1)$	Calculated value for $x(3)$
−1.5	−0.016
−3.0	−2.085
−2.213*	−1.271
−2.0	−0.972
−1.8	−0.642
−2.017*	−0.998
−2.01	−0.987
−2.02	−1.002
−2.018*	−1.000

* Interpolated linearly from two previous values.

Table 6.3

t	x	x'
1.000	2.000	−2.018
1.200	1.557	−2.407
1.400	1.053	−2.632
1.600	0.527	−2.625
1.800	0.026	−2.383
2.000	−0.408	−1.959
2.200	−0.747	−1.431
2.400	−0.977	−0.867
2.600	−1.094	−0.307
2.800	−1.101	0.237
3.000	−1.000	0.775

A Summary of the Shooting Method (We Give Only an Outline.)

To solve a boundary-value problem, create an initial-value problem by assuming a sufficient number of initial values. (Choose the right-hand end of the interval as the "initial point," and use a negative value of h if this requires fewer assumed conditions.) Solve this initial-value problem and compare the computed values with the given conditions at the other boundary.

Repeat the solution with varying values of the assumed conditions until agreement is attained at the other boundary. Methods for solving nonlinear systems can be employed to estimate the new assumed values for the successive iterations in an efficient manner.

The shooting method is often quite laborious. Especially with problems of fourth and higher order, the necessity to assume two or more conditions at the starting point (and match with the same number of conditions at the end) is slow and tedious.

6.2 Solution Through a Set of Equations

We have seen in Chapter 4 how the derivatives of a function can be approximated by finite-difference quotients. If we replace the derivatives in a differential equation by such expressions, we convert it to a difference equation whose solution is an approximation to the solution of the differential equation. This method is sometimes preferred over the shooting method discussed earlier. Consider the same linear example as before:

$$\frac{d^2x}{dt^2} - \left(1 - \frac{t}{5}\right)x = t, \qquad x(1) = 2, \qquad x(3) = -1. \tag{6.2}$$

Central-difference approximations to derivatives are more accurate than forward or backward approximations [$O(h^2)$ versus $O(h)$], so we replace the derivatives with

$$\left.\frac{dx}{dt}\right|_{t=t_i} = \frac{x_{i+1} - x_{i-1}}{2h} + O(h^2),$$

$$\left.\frac{d^2x}{dt^2}\right|_{t=t_i} = \frac{x_{i+1} - 2x_i + x_{i-1}}{h^2} + O(h^2).$$

The quantity h is the constant difference in t-values. Substituting these equivalences into Eq. (6.2) and rearranging, we get

$$\frac{x_{i+1} - 2x_i + x_{i-1}}{h^2} - \left(1 - \frac{t_i}{5}\right)x_i = t_i,$$

$$x_{i-1} - \left[2 + h^2\left(1 - \frac{t_i}{5}\right)\right]x_i + x_{i+1} = h^2 t_i. \tag{6.3}$$

In Eqs. (6.3) we have replaced x with x_i and t with t_i, since these values correspond to the point at which the difference quotients represent the derivatives.* Our problem now reduces to solving Eqs. (6.3) at points in the interval from $t = 1$ to $t = 3$. Let us subdivide the interval into a number of equal subintervals. For example, if $h = \Delta t = \frac{1}{2}$, the points

$$t_1 = 1, \qquad t_2 = 1.5, \qquad t_3 = 2, \qquad t_4 = 2.5, \qquad \text{and} \qquad t_5 = 3$$

subdivide the interval into four subintervals.

We write the difference equation in (6.3) for each of the three intermediate points where x is unknown, giving

at $t = t_2 = 1.5$: $\quad x_1 - \left[2 + \left(\frac{1}{2}\right)\left(\frac{1}{2}\right)\left(1 - \frac{1.5}{5}\right)\right]x_2 + x_3 = \left(\frac{1}{2}\right)\left(\frac{1}{2}\right)(1.5);$

at $t = t_3 = 2.0$: $\quad x_2 - \left[2 + \left(\frac{1}{2}\right)\left(\frac{1}{2}\right)\left(1 - \frac{2.0}{5}\right)\right]x_3 + x_4 = \left(\frac{1}{2}\right)\left(\frac{1}{2}\right)(2.0);$

at $t = t_4 = 2.5$: $\quad x_3 - \left[2 + \left(\frac{1}{2}\right)\left(\frac{1}{2}\right)\left(1 - \frac{2.5}{5}\right)\right]x_4 + x_5 = \left(\frac{1}{2}\right)\left(\frac{1}{2}\right)(2.5).$

---

* Our procedure is to subdivide the interval from the initial to the final value of t into equal subintervals, replacing the differential equation with a difference equation at each of the discrete points where the function is unknown.

We know $x_1 = 2$ and $x_5 = -1$; hence the difference equation is not written corresponding to $t = 1$ or $t = 3$. Substituting the values $x_1 = 2$ and $x_5 = -1$ and simplifying, we have, in matrix form,

$$\begin{bmatrix} -2.175 & 1 & 0 \\ 1 & -2.150 & 1 \\ 0 & 1 & -2.125 \end{bmatrix} \begin{bmatrix} x_2 \\ x_3 \\ x_4 \end{bmatrix} = \begin{bmatrix} -1.625 \\ 0.5 \\ 1.625 \end{bmatrix}.$$

Solving, we get

$$x_2 = 0.552, \qquad x_3 = -0.424, \qquad \text{and} \qquad x_4 = -0.964.$$

These are reasonably close to the values obtained by the shooting method. We would expect some significant error because our step size h is large and the finite differences will be poor approximations to the derivatives. If we take $h = 0.2$ (10 subdivisions) and write the approximating equations, we get

$$\begin{bmatrix} -2.0304 & 1 & 0 & 0 & 0 & 0 & 0 & 0 & 0 \\ 1 & -2.0288 & 1 & 0 & 0 & 0 & 0 & 0 & 0 \\ 0 & 1 & -2.0272 & 1 & 0 & 0 & 0 & 0 & 0 \\ 0 & 0 & 1 & -2.0256 & 1 & 0 & 0 & 0 & 0 \\ 0 & 0 & 0 & 1 & -2.0240 & 1 & 0 & 0 & 0 \\ 0 & 0 & 0 & 0 & 1 & -2.0224 & 1 & 0 & 0 \\ 0 & 0 & 0 & 0 & 0 & 1 & -2.0208 & 1 & 0 \\ 0 & 0 & 0 & 0 & 0 & 0 & 1 & -2.0192 & 1 \\ 0 & 0 & 0 & 0 & 0 & 0 & 0 & 1 & -2.0176 \end{bmatrix} x = \begin{bmatrix} -1.952 \\ 0.056 \\ 0.064 \\ 0.072 \\ 0.080 \\ 0.088 \\ 0.096 \\ 0.104 \\ 1.112 \end{bmatrix}. \quad \textbf{(6.4)}$$

We observe that the system is tridiagonal and therefore speedy to solve and also economical of memory space to store the coefficients. This will be true even if there are many equations, because we only use x_{i-1}, x_i, x_{i+1} in any equation to replace x' or x''. This is one reason why the finite-difference method is widely used to solve second-order linear boundary-value problems.

When a computer program designed for tridiagonal systems was used, we got the values shown in Table 6.4, where we also display the results by the shooting method from Section 6.1 for comparison.

The values obtained with $\Delta t = 0.2$ are more accurate. We might anticipate that the error would decrease proportionately to h^2 since the approximations to the derivatives are of $O(h^2)$ error. If we use the values from the shooting method as the exact values for comparison, the errors are reduced only about threefold (comparing the values at $t = 2$) instead of sixfold. However, the shooting method calculations are themselves imperfect, adding an element of doubt to our observation. To observe how the errors decrease when Δt decreases, we look at another example, one where the exact solution is known.

Table 6.4 Results for the linear boundary-value problem:

$$x'' = t + \left(1 - \frac{t}{5}\right)x, \qquad x(1) = 2, \quad x(3) = -1$$

t	Values from finite-difference method	Values from shooting method
1.0	2.000	2.000
1.2	1.351	1.348
1.4	0.792	0.787
1.6	0.311	0.305
1.8	−0.097	−0.104
2.0	−0.436	−0.443
2.2	−0.705	−0.712
2.4	−0.903	−0.908
2.6	−1.022	−1.026
2.8	−1.058	−1.060
3.0	−1.000	−1.000

EXAMPLE 6.1   Solve

$$\frac{d^2y}{dx^2} = y, \qquad y(1) = 1.1752, \qquad y(3) = 10.0179.$$

(The analytical solution is $y = \sinh x$, to which we can compare our estimates of the function.)

Although it is not strictly necessary, it is common to normalize the function to the interval $(0, 1)$.* This we can do by the change of variable

$$x = (b - a)t + a,$$

where (a, b) is the original interval to be normalized. Letting $x = 2t + 1$, we write

$$\frac{dy}{dx} = \frac{dy}{dt}\frac{dt}{dx} = \frac{1}{2}\frac{dy}{dt},$$

$$\frac{d^2y}{dx^2} = \frac{d}{dt}\left(\frac{1}{2}\frac{dy}{dt}\right)\frac{dt}{dx} = \frac{1}{2^2}\frac{d^2y}{dt^2},$$

and the problem becomes

$$\frac{d^2y}{dt^2} = 4y, \qquad y(0) = 1.1752, \qquad y(1) = 10.0179.$$

---

* Perhaps the most important reason is to make a computer program more general. Normalizing is really a scaling of the interval so we have an immediate understanding of whether a given value of h is "small" or "large." Also, when we compute the LU equivalent of the coefficient matrix to solve the system, it is then of more general applicability.

Replacing the second derivative by the central-difference approximation, and thus converting the differential equation to a difference equation, we have

$$\frac{y_{i+1} - 2y_i + y_{i-1}}{h^2} = 4y_i, \qquad i = 1, 2, 3, \dots, n.$$

Subdividing the interval $(0, 1)$ into four parts, $h = 0.25$, and writing the difference equation at the three internal points where y is unknown, we have the set of equations

$$\begin{bmatrix} -2.25 & 1 & 0 \\ 1 & -2.25 & 1 \\ 0 & 1 & -2.25 \end{bmatrix} \begin{bmatrix} y_2 \\ y_3 \\ y_4 \end{bmatrix} = \begin{bmatrix} -1.1752 \\ 0 \\ -10.0179 \end{bmatrix}. \tag{6.5}$$

In Eqs. (6.5) we have used $y_1 = 1.1752$, $y_5 = 10.0179$. Again, a tridiagonal matrix of coefficients occurs. The solution to this set of equations is given in Table 6.5.

Table 6.5

t	x	Calculated y	Exact value, $\sinh x$	Error
0	1.0	(1.1752)	1.1752	—
0.25	1.5	2.1467	2.1293	−0.0174
0.50	2.0	3.6549	3.6269	−0.0280
0.75	2.5	6.0768	6.0502	−0.0266
1.0	3.0	(10.0179)	10.0179	—

If we recalculate this problem using varying values of h, we get the results shown in Table 6.6 for y at $x = 2.0$ (equivalent to $t = 0.5$).

Table 6.6

Number of equations	$h = \Delta t$ for normalized interval	Result	Error
1	0.50	3.7310	−0.1041
3	0.25	3.6549	−0.0280
7	0.125	3.6340	−0.0071
∞	0	3.6269	(exact)

We observe an approximate fourfold decrease in the error when the step size is halved, meaning an $O(h^2)$ error does exist. When we know how errors vary with step size, we can perform a Richardson extrapolation, using results for which the step size is halved:

$$\begin{matrix} \text{Improved} \\ \text{value} \end{matrix} = \begin{matrix} \text{More} \\ \text{accurate} \end{matrix} + \frac{1}{3}\left[\left(\begin{matrix} \text{More} \\ \text{accurate} \end{matrix}\right) - \left(\begin{matrix} \text{Less} \\ \text{accurate} \end{matrix}\right)\right].$$

In this instance we get $3.6340 + \frac{1}{3}(3.6340 - 3.6549) = 3.6270$, when results with $\Delta t = \frac{1}{4}$ and $\Delta t = \frac{1}{8}$ are combined, off by only 1 in the fourth decimal place.

▲

When the differential equation underlying a boundary-value problem is nonlinear, this method of finite differences runs into a problem in that the resulting system of equations is nonlinear. Consider the nonlinear problem solved by the shooting method in Section 6.1:

$$\frac{d^2x}{dt^2} - \left(1 - \frac{t}{5}\right)\left(\frac{dx}{dt}\right)(x) = t, \qquad x(1) = 2, \qquad x(3) = -1.$$

When finite differences are substituted for the derivatives, we obtain an equivalent set of difference equations as follows:

$$x_{i-1} - \left[2 + \frac{h}{2}\left(1 - \frac{t_i}{5}\right)(x_{i+1} - x_{i-1})\right]x_i + x_{i+1} = h^2 t_i, \qquad i = 1, 2, 3, \ldots, n, \quad \textbf{(6.6)}$$

which is to be written at each point t_i at which x is unknown. In the middle term, products of the x's occur, exhibiting nonlinearity in the difference equations as a consequence of the nonlinearity in the differential equation. The standard elimination method fails for these equations.

In such cases, iteration techniques can be employed. Suppose we obtain some estimates of the x-vector, the solution to the equations. We could use these values as a means of approximating the coefficient of x_i in Eq. (6.6). (Since the values are multiplied by $h/2$, we might hope that errors of the estimate would be diluted in their effect, especially if h is fairly small, in comparison to the number 2, which should dominate in the coefficient of x_i.)

When this was done, using $h = 0.2$ and the initial x_i values estimated from $x = 3.5 - 1.5t$ [a linear relation between $x(1) = 2$ and $x(3) = -1$], the values in Table 6.7 resulted.

Further iterations after the eighth did not change the values. In Table 6.7 we have listed the results by the shooting method for comparison. In both techniques, an iterative process was used; in the finite-difference method we iterate on the approximations to x, whereas in the shooting method we use successive estimates to the initial slope. For this example, about the same number of iterations were required and roughly the same accuracy was achieved. (The correct solution is intermediate between the results for the two methods shown in Table 6.7.)

The choice between the finite-difference method and the shooting method for nonlinear boundary-value problems is not clearcut. The choice depends on whether a reasonably good initial estimate for the x-vector is available and how strongly nonlinear the equation is. When the nonlinearity in the set of equations is not diluted

Table 6.7 Successive approximations to finite-difference equations for

$$x'' - \left(1 - \frac{t}{5}\right)(x')(x) = t, \quad x(1) = 2, \quad x(3) = -1$$

			Approximations to x-vector			
t	Initial estimate	After 1 iteration	After 2 iterations	After 4 iterations	After 8 iterations	Results from shooting method
1.0	2.000	(2.000)	(2.000)	(2.000)	(2.000)	(2.000)
1.2	1.700	1.431	1.601	1.558	1.552	1.557
1.4	1.400	0.845	1.110	1.048	1.039	1.053
1.6	1.100	0.279	0.583	0.513	0.503	0.527
1.8	0.800	−0.235	0.077	0.006	−0.005	0.026
2.0	0.500	−0.668	−0.362	−0.430	−0.440	−0.408
2.2	0.200	−0.997	−0.704	−0.766	−0.775	−0.747
2.4	−0.100	−0.204	−0.937	−0.990	−0.998	−0.977
2.6	−0.400	−1.278	−1.061	−1.101	−1.107	−1.094
2.8	−0.700	−1.211	−0.080	−1.104	−1.107	−1.101
3.0	−1.000	(−1.000)	(−1.000)	(−1.000)	(−1.000)	(−1.000)

in its effect and when no good initial estimate is available, iterative methods applied to the algebraic finite-difference equations may not even converge. Although the shooting method often involves a somewhat greater computational effort, it is usually more certain of giving a solution.

6.3 Derivative Boundary Conditions

The conditions that the solution of a boundary-value problem must satisfy need not necessarily be just the value of the function. In many applied problems, some derivative of the function may be known at the boundaries of an interval. In the more general case a linear combination of the function and its derivatives is specified. The finite-difference procedure needs modification for this type of boundary conditions.* We illustrate by an example:

$$\frac{d^2y}{dt^2} = y,$$

$$y'(1) = 1.1752,$$

$$y'(3) = 10.0179.$$

---

* With the shooting method, derivative boundary conditions do not require any change in procedure, but we will need to use finite-difference approximations to know when the results match with specified derivatives at the far end.

(This problem has the same differential equation as the example of Section 6.2 but, with the values specified for the derivative at $x = 1$ and $x = 3$, it now has the analytical solution $y = \cosh x$.)

We begin just as before. We change variables to make the interval $(0, 1)$ by letting $x = 2t + 1$:

$$\frac{d^2y}{dt^2} = 4y.$$

We now replace the derivative by a central-difference approximation and write the difference equation at each point where y is unknown. With $h = 0.25$,

$$
\begin{array}{lll}
t = 0: & y_\ell - 2.25y_1 + y_2 = 0, & \\
t = 0.25: & y_1 - 2.25y_2 + y_3 = 0, & \\
t = 0.50: & y_2 - 2.25y_3 + y_4 = 0, & \textbf{(6.7)} \\
t = 0.75: & y_3 - 2.25y_4 + y_5 = 0, & \\
t = 1.00: & y_4 - 2.25y_5 + y_r = 0. &
\end{array}
$$

Two more equations are required than in Section 6.2 because y is unknown at $t = 0$ and $t = 1$ as well as at the interior points. To write these added equations, we added fictitious points, y_ℓ and y_r, points one space to the left and to the right of the interval $(0, 1)$. We assume that the domain of the differential equation can be so extended.

A difficulty appears: Equations (6.7) contain seven unknowns, and we have only five equations. The boundary conditions, however, have not yet been involved. Let us also express these as difference quotients, preferring central-difference approximations of $O(h^2)$ error as used in replacing derivatives in the original equation:

$$\frac{dy}{dx}\bigg|_{x=1} = \frac{dy}{dt}\frac{dt}{dx}\bigg|_{t=0} \doteq \left(\frac{1}{2}\right)\frac{y_2 - y_\ell}{2h} = 1.1752, \qquad y_\ell = y_2 - (4)(0.25)(1.1752),$$

$$\frac{dy}{dx}\bigg|_{x=3} = \frac{dy}{dt}\frac{dt}{dx}\bigg|_{t=1} \doteq \left(\frac{1}{2}\right)\frac{y_r - y_4}{2h} = 10.0179, \qquad y_r = y_4 + (4)(0.25)(10.0179). \qquad \textbf{(6.8)}$$

The relations of (6.8) when substituted into (6.7) reduce the number of unknowns to five, and we solve the equations in the usual way. The solution is given in Table 6.8.

Table 6.8

t	x	y	$\cosh x$	Error
0	1.0	1.5522	1.5431	−0.0091
0.25	1.5	2.3338	2.3524	0.0186
0.50	2.0	3.6989	3.7622	0.0633
0.75	2.5	5.9887	6.1323	0.1436
1.00	3.0	9.7757	10.0677	0.2920

We observe here that the errors are much greater than for the previous example, being very large at $x = 3.0$. The explanation is that our approximation for the derivative is poor at large x-values. Although the central-difference approximations are $O(h^2)$, the third derivative appearing in the error term is large in magnitude. In the previous example, knowing the function at the endpoints eliminated the errors there.

Repeating the calculations with the step size halved reduces the errors about fourfold. For example, at $t = 0.5$, we get 3.7459 (error 0.0163 versus 0.0633); and at $t = 1.0$, we get 9.9921 (error 0.0756 versus 0.2920). Halving h again gives another fourfold reduction in errors ($y = 10.0486$ at $t = 1.0$; error of 0.0191). An extrapolation based on errors of $O(h^2)$ will give nearly perfect results.

It is appropriate to summarize the finite-difference method for solving a boundary-value problem.

The Finite-Difference Method for Boundary-Value Problems

To solve a boundary-value problem, replace the differential equation with a finite-difference equation by replacing the derivatives with central-difference quotients. Subdivide the interval into a suitable number of equal subintervals, and write the difference equation at each point where the value of the function is unknown. When the boundary values involve derivatives, the domain will have to be extended beyond the interval. Utilize the derivative boundary conditions to write difference quotients that permit the elimination of the fictitious points outside the interval.

Solve the system of equations so created to obtain approximate values for the solution of the differential equation at discrete points on the interval. If the original differential equation is nonlinear, the system of equations will also be nonlinear. In such situations, the shooting method will normally be preferred.

We could, of course, use more accurate finite-difference approximations to the derivatives, not only for the boundary values, but for the equation itself. The disadvantage of doing this is that the system of equations is then not tridiagonal, and the solution is more expensive.

If the more general form of boundary condition applies, $ay + by' = c$, equations similar to (6.8) result, and after the boundary conditions have been approximated and the exterior y-values eliminated from the set of equations, the solution of the problem proceeds as before.

Equations of order higher than the second will involve approximations for the third or higher derivatives. Central-difference formulas will involve points more than h away from the point where the derivative is being approximated, and may be unsymmetrical in the case of the odd derivatives. Probably the method of undetermined coefficients is the easiest way to derive the necessary formulas (see Appendix B). Again, nontridiagonal systems result. Fortunately, most of the important physical problems are simulated by equations of order 2.

Besides the two methods presented so far in this chapter, there are others that are commonly referred to as Rayleigh–Ritz, collocation, Galerkin, and finite-element methods. These approximate the solution to the boundary-value problem, $y(x)$, by writing it as a linear combination $\Sigma c_i v_i(x), i = 1, 2, \ldots, n$, where the v's are specially chosen functions. These methods have been found to be very effective for certain kinds of problems. We will introduce you to these methods in Sections 6.4 and 6.5.

6.4   Rayleigh–Ritz, Collocation, and Galerkin Methods

The Rayleigh–Ritz Method

In addition to the previous two methods for boundary-value problems, you should know something about the Rayleigh–Ritz method. It is based on an elegant branch of mathematics, the calculus of variations. In this method we solve a boundary-value problem by approximating the solution with a finite linear combination of simple basis functions that are chosen to fulfill certain criteria, including meeting the boundary conditions.

The calculus of variations seeks to optimize (often minimize) a special class of functions called *functionals*. The usual form for the functional (in problems of one independent variable) is

$$I[y] = \int_a^b F\left(x, y, \frac{dy}{dx}\right) dx. \qquad (6.9)$$

Observe that $I[y]$ is not a function of x because x disappears when the definite integral is evaluated. The argument y of $I[y]$ is not a simple variable but a function, $y = y(x)$. The square brackets in $I[y]$ emphasize this fact. A functional can be thought of as a "function of functions." The value of the right-hand side of Eq. (6.9) will change as the function $y(x)$ is varied, but when $y(x)$ is fixed, it evaluates to a scalar quantity (a constant). We seek the $y(x)$ that minimizes $I[y]$.

Let us illustrate this concept by a very simple example where the solution is obvious in advance—find the function $y(x)$ that minimizes the distance between two points. While we know what $y(x)$ must be, let's pretend we don't. The figure suggests that we are to choose from among the set of curves $y_i(x)$ of which $y_1(x)$, $y_2(x)$, and $y_3(x)$ are representative. In this simple case, the functional is the integral of the distance along any of these curves:

$$I[y] = \int_{x_1}^{x_2} \sqrt{(dx)^2 + (dy)^2} = \int_{x_1}^{x_2} \sqrt{1 + \left(\frac{dy}{dx}\right)^2} \, dx.$$

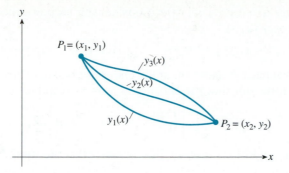

To minimize $I[y]$, just as in calculus, we set its derivative to zero. There are certain restrictions on all the curves $y_i(x)$. Obviously, each must pass through the points (x_1, y_1) and (x_2, y_2). In addition, for the optimal trajectory, the Euler–Lagrange equation must be satisfied:

$$\frac{d}{dx}\left[\frac{\partial}{\partial y'} F(x,y,y')\right] = \frac{\partial}{\partial y} F(x,y,y'). \tag{6.10}$$

Applying this to the functional for shortest distance, we have

$$F(x,y,y') = (1 + (y')^2)^{1/2},$$

$$\frac{\partial F}{\partial y} = 0,$$

$$\frac{\partial F}{\partial y'} = \frac{1}{2}(1 + (y')^2)^{-1/2}(2y'),$$

$$\frac{d}{dx}\frac{\partial F}{\partial y'} = \frac{d}{dx}\left(\frac{y'}{\sqrt{1 + (y')^2}}\right) = \frac{\partial F}{\partial y} = 0.$$

(The last comes from Eq. 6.10.)

From this, it follows that

$$\frac{y'}{\sqrt{1 + (y')^2}} = c.$$

Solving for y' gives

$$y' = \sqrt{\frac{c^2}{1 - c^2}} = \text{a constant} = b,$$

and, on integrating,

$$y = bx + a.$$

As stated, $y(x)$ must pass through P_1 and P_2; this condition is used to evaluate the constants a and b.

Let us advance to a less trivial case. Consider this second-order linear boundary-value problem over $[a, b]$:*

$$y'' + Q(x)y = F(x), \quad y(a) = y_0, \quad y(b) = y_n. \quad \textbf{(6.11)}$$

(An equation that has y = constant at the endpoints is said to be subject to Dirichlet conditions.) It turns out that the functional that corresponds to Eq. (6.11) is

$$I[u] = \int_a^b \left[\left(\frac{du}{dx} \right)^2 - Qu^2 + 2Fu \right] dx. \quad \textbf{(6.12)}$$

(If the boundary equations involve a derivative of y, the functional must be modified.)

We can transform Eq. (6.12) to Eq. (6.11) through the Euler–Lagrange conditions, so optimizing Eq. (6.12) gives the solution to Eq. (6.11). Observe carefully the benefit of operating with the functional rather than the original equation: We now have only first-order instead of second-order derivatives. This not only simplifies the mathematics but permits us to find solutions even when there are discontinuities that cause y not to have sufficiently high derivatives.

If we know the solution to our differential equation, substituting it for u in Eq. (6.12) will make $I[u]$ a minimum. If the solution isn't known, perhaps we can approximate it by some (almost) arbitrary function and see whether we can minimize the functional by a suitable choice of the parameters of the approximation. The Rayleigh–Ritz method is based on this idea. We let $u(x)$, which is the approximation to $y(x)$ (the exact solution), be a sum:

$$u(x) = c_0 v_0 + c_1 v_1 + \cdots + c_n v_n = \sum_{i=0}^{n} c_i v_i. \quad \textbf{(6.13)}$$

There are two conditions on the v's in Eq. (6.13): They must be chosen such that $u(x)$ meets the boundary conditions, and the individual v's must be linearly independent (meaning that no one v can be obtained by a linear combination of the others). We call the v's *trial functions*; the c's and v's are to be chosen to make $u(x)$ a good approximation to the true solution to Eq. (6.11).

If we have some prior knowledge of the true function, $y(x)$, we may be able to choose the v's to closely resemble $y(x)$. Most often we lack such knowledge, and the usual choice then is to use polynomials. We must find a way of getting values for the c's to force $u(x)$ to be close to $y(x)$. We will use the functional of Eq. (6.12) to do this.

If we substitute $u(x)$ as defined by Eq. (6.13) into the functional, Eq. (6.12), we get

$$I(c_0, c_1, \ldots, c_n) = \int_a^b \left[\left(\frac{d}{dx} \Sigma c_i v_i \right)^2 - Q(\Sigma c_i v_i)^2 + 2F \Sigma c_i v_i \right] dx. \quad \textbf{(6.14)}$$

We observe that I is an ordinary function of the unknown c's after this substitution, as reflected in our notation. To minimize I, we take its partial derivatives with respect to each unknown c and set to zero, resulting in a set of equations in the c's that we can solve. This will define $u(x)$ in Eq. (6.13).

---

* This equation is a prototype of many equations in applied mathematics. Equations for heat conduction, elasticity, electrostatics, and so on in a one-dimensional situation are of this form.

We now substitute the $u(x)$ of Eq. (6.13) into the functional. If we partially differentiate with respect to, say, c_i where this is one of the unknown c's, we will get

$$\frac{\partial I}{\partial c_i} = \int_a^b 2\left(\frac{du}{dx}\right)\frac{\partial}{\partial c_i}\left(\frac{du}{dx}\right)dx - \int_a^b 2Qu\left(\frac{\partial u}{\partial c_i}\right)dx + 2\int_a^b F\left(\frac{\partial u}{\partial c_i}\right)dx, \quad \textbf{(6.15)}$$

where we have broken the integral into three parts.

An example will clarify the procedure.

EXAMPLE 6.2   Solve the equation $y'' + y = 3x^2$, with boundary points $(0,0)$ and $(2, 3.5)$. (Here $Q = 1$ and $P = 3x^2$.) Use polynomial trial functions up to degree 3. If we define $u(x)$ as

$$u(x) = \frac{7x}{4} + c_2(x)(x - 2) + c_3(x^2)(x - 2), \quad \textbf{(6.16)}$$

we have linearly independent v's. The boundary conditions are met by the first term, and since the other terms are zero at the boundaries, $u(x)$ also meets the boundary conditions. [It is customary to match the boundary conditions with the initial term(s) of $u(x)$ and then make the succeeding terms equal zero at the boundaries, as we have done here.]

Examination of Eq. (6.15) shows that we need these quantities:

$$\frac{du}{dx} = \frac{7}{4} + c_2(2x - 2) + c_3(3x^2 - 4x),$$

$$\frac{\partial}{\partial c_2}\left(\frac{du}{dx}\right) = 2x - 2, \qquad \frac{\partial}{\partial c_3}\left(\frac{du}{dx}\right) = 3x^2 - 4x, \quad \textbf{(6.17)}$$

$$\frac{\partial u}{\partial c_2} = x(x - 2), \qquad \frac{\partial u}{\partial c_3} = x^2(x - 2).$$

We now substitute from Eqs. (6.17) into Eqs. (6.15). Note that we have two equations, one for the partial with respect to c_2 and the other from the partial with respect to c_3. The results from this step are:

$$\frac{\partial I}{\partial c_2}: \quad 0 = \int_0^2 2\left[\frac{7}{4} + c_2(2x - 2) + c_3(3x^2 - 4x)\right](2x - 2)\,dx$$

$$- \int_0^2 2(1)\left[\frac{7x}{4} + c_2(x^2 - 2x) + c_3(x^3 - 2x^2)\right](x^2 - 2x)\,dx$$

$$+ 2\int_0^2 (3x^2)(x^2 - 2x)\,dx, \quad \textbf{(6.18)}$$

$$\frac{\partial I}{\partial c_3}: \quad 0 = \int_0^2 2\left[\frac{7}{4} + c_2(2x - 2) + c_3(3x^2 - 4x)\right](3x^2 - 4x)\,dx$$

$$- \int_0^2 2(1)\left[\frac{7x}{4} + c_2(x^2 - 2x) + c_3(x^3 - 2x^2)\right](x^3 - 2x^2)\,dx$$

$$+ 2\int_0^2 (3x^2)(x^3 - 2x^2)\,dx. \quad \textbf{(6.19)}$$

We now carry out the integrations. Though there are quite a few of them, all are quite simple in our example. With a more complicated $Q(x)$ and $F(x)$, this might require numerical integrations. The result of this step is the pair of equations

$$\frac{16}{5}c_2 + \frac{16}{5}c_3 = \frac{74}{15},$$

$$\frac{16}{5}c_2 + \frac{128}{21}c_3 = \frac{36}{5},$$ (6.20)

which we solve to get the coefficients in our $u(x)$. On expanding, we find that

$$u(x) = \left(\frac{119}{152}\right)x^3 - \left(\frac{46}{57}\right)x^2 + \left(\frac{53}{228}\right)x.$$ (6.21)

(DERIVE was used in all of this so we could find rational fraction coefficients.)

Figure 6.2 shows that our $u(x)$ agrees well with the exact solution, which is $6\cos(x) + 3(x^2 - 2)$ over the interval $[0, 2]$. Table 6.9 compares computed values and the error of $u(x)$. The largest errors occur near the middle of the interval.

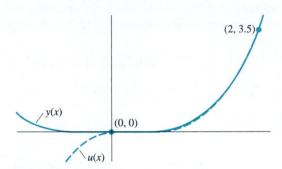

(2, 3.5)

$y(x)$

(0, 0)

$u(x)$

Figure 6.2

Table 6.9

x	$y(x)$	$u(x)$	Error	x	$y(x)$	$u(x)$	Error
0.00	0.00	0.00	0.00	1.10	0.35	0.32	0.03
0.10	0.00	0.02	−0.02	1.20	0.49	0.47	0.02
0.20	0.00	0.02	−0.02	1.30	0.67	0.66	0.02
0.30	0.00	0.02	−0.02	1.40	0.90	0.89	0.01
0.40	0.01	0.01	−0.01	1.50	1.17	1.18	−0.00
0.50	0.02	0.01	0.00	1.60	1.50	1.51	−0.01
0.60	0.03	0.02	0.01	1.70	1.90	1.91	−0.01
0.70	0.06	0.04	0.02	1.80	2.36	2.37	−0.01
0.80	0.10	0.07	0.03	1.90	2.89	2.90	−0.01
0.90	0.16	0.13	0.03	2.00	3.50	3.50	0.00
1.00	0.24	0.21	0.03				

Collocation

There are other ways to approximate $y(x)$ in Example 6.2. The *collocation method* is what is called a "residual method." We begin by defining the residual, $R(x)$, as equal to the left-hand side of Eq. (6.11) minus the right-hand side:

$$R(x) = y'' + Qy - F. \tag{6.22}$$

We approximate $y(x)$ again with $u(x)$ equal to a sum of trial functions, usually chosen as linearly independent polynomials, just as for the Rayleigh–Ritz method. We substitute $u(x)$ into $R(x)$ and attempt to make $R(x) = 0$ by a suitable choice of the coefficients in $u(x)$. Of course, normally we cannot do this everywhere in the interval $[a, b]$, so we select several points at which we make $R(x) = 0$. [The number of points where we do this must equal the number of unknown coefficients in $u(x)$.] An example will clarify the procedure.

EXAMPLE 6.3   Solve the same equation as in Example 6.2, but this time use collocation.
   The equation we are to solve is

$$y'' + y = 3x^2, \qquad y(0) = 0, \qquad y(2) = 3.5. \tag{6.23}$$

We take $u(x)$ as before to satisfy the boundary conditions:

$$u(x) = \frac{7x}{4} + c_2(x)(x - 2) + c_3(x^2)(x - 2). \tag{6.24}$$

The residual is, after substituting $u(x)$ for $y(x)$,

$$R(x) = u'' + u - 3x^2, \tag{6.25}$$

which becomes, when we differentiate u twice to get u'',

$$R(x) = c_2(2) + c_3(6x - 4) + \frac{7x}{4} + c_2(x^2 - 2x) + c_3(x^3 - 2x^2) - 3x^2. \tag{6.26}$$

Since there are two unknown constants, we can force $R(x)$ to be zero at two points in $[0, 2]$. Since we do not know which two points will be the best choices, we arbitrarily take them as $x = 0.7$ and $x = 1.3$. (These points are more or less equally spaced in the interval.) Setting $R(x) = 0$ for these choices gives a pair of equations in the c's:

From $x = 0.7$: $\qquad \dfrac{1090c_2 - 437c_3 - 245}{1000} = 0,$

From $x = 1.3$: $\qquad \dfrac{1090c_2 + 2617c_3 - 2795}{1000} = 0. \tag{6.27}$

When these are solved for the c's, we get, for $u(x)$,

$$u(x) = \left(\frac{425}{509}\right)x^3 - \left(\frac{61607}{55481}\right)x^2 + \left(\frac{140023}{221924}\right)x, \tag{6.28}$$

in which the coefficients are quite different than in Eq. (6.21). Figure 6.3 shows that this approximation is not as good as that obtained by the Rayleigh–Ritz technique. (But the amount of arithmetic is certainly less! We could improve the approximation by using more terms in $u(x)$.) Table 6.10 compares the approximation with the exact solution.

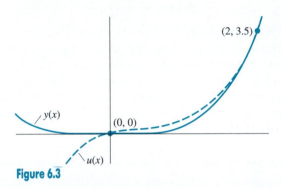

Figure 6.3

Table 6.10

x	$y(x)$	$u(x)$	Error	x	$y(x)$	$u(x)$	Error
0.00	0.00	0.00	0.00	1.10	0.35	0.32	0.03
0.10	0.00	0.05	−0.05	1.20	0.49	0.60	−0.11
0.20	0.00	0.09	−0.09	1.30	0.67	0.78	−0.10
0.30	0.00	0.11	−0.11	1.40	0.90	1.00	−0.10
0.40	0.01	0.13	−0.12	1.50	1.17	1.27	−0.09
0.50	0.02	0.14	−0.13	1.60	1.50	1.59	−0.08
0.60	0.03	0.16	−0.13	1.70	1.90	1.97	−0.07
0.70	0.06	0.18	−0.12	1.80	2.36	2.41	−0.05
0.80	0.10	0.22	−0.12	1.90	2.89	2.92	−0.03
0.90	0.16	0.28	−0.12	2.00	3.50	3.50	0.00
1.00	0.24	0.36	−0.11				

The Galerkin Method

The Galerkin method is widely used, especially in the very popular technique that we will describe in Section 6.5. It is important to know the Galerkin method because of its widespread application.

　　Like collocation, Galerkin is a "residual method" that uses the $R(x)$ of Eq. (6.22), except that now we multiply $R(x)$ by weighting functions, $W_i(x)$. The $W_i(x)$ can be chosen in many ways, but Galerkin showed that using the individual trial functions, v_i, of Eq. (6.13) is an especially good choice.

Once we have selected the v's for Eq. (6.13), we compute the unknown coefficients by setting the integral over $[a, b]$ of the weighted residual to zero:

$$\int_a^b W_i(x)\, R(x)\, dx = 0, \qquad i = 0, 1, \ldots, n, \tag{6.29}$$

where $W_i(x) = v_i$. (Observe that using Dirac delta functions for the $W_i(x)$ gives the collocation method.)

Let us use the Galerkin method on the same example as before.

EXAMPLE 6.4   Solve

$$y'' + y = 3x^2, \qquad y(0) = 0, \qquad y(2) = 3.5$$

by the Galerkin method. Use the same $u(x)$ as before:

$$u(x) = \frac{7x}{4} + c_2(x)(x - 2) + c_3(x^2)(x - 2),$$

so that $v_2 = x(x - 2)$ and $v_3 = x^2(x - 2)$.

The residual is

$$R(x) = y'' + y - 3x^2,$$

which becomes, after substituting u'' and u for y'' and y, respectively,

$$R(x) = c_2(2) + c_3(6x - 4)$$
$$+ \frac{7x}{4} + c_2(x)(x - 2) + c_3(x^2)(x - 2)$$
$$- 3x^2. \tag{6.30}$$

We now carry out two integrations (because there are two unknown c's):

Using v_2 as a W_i: $\displaystyle\int_0^2 [x(x - 2)] * R(x)\, dx = 0,$

Using v_3 as a W_i: $\displaystyle\int_0^2 [x^2(x - 2)] * R(x)\, dx = 0,$

which gives two equations in the c's:

$$-\frac{24c_2 + 24c_3 - 37}{15} = 0,$$

$$-\frac{2(84c_2 + 160c_3 - 180)}{105} = 0. \tag{6.31}$$

Although Eqs. (6.31) do not look the same as Eqs. (6.20), which resulted from the Rayleigh–Ritz procedure, the solution is identical, which gives our $u(x)$ as

$$u(x) = \left(\frac{119}{152}\right)x^3 - \left(\frac{46}{57}\right)x^2 + \left(\frac{53}{228}\right)x. \tag{6.32}$$

We get the same result as with the Rayleigh–Ritz method, but this is no coincidence; on this type of problem the Galerkin technique is identical to Rayleigh–Ritz. Forming the equations with Galerkin is much easier, however, and we never have to find the variational form—which may be difficult.

6.5   The Finite-Element Method

The disadvantages of the methods of the previous section are twofold: Finding good trial functions (the v's in Eq. 6.13) is not easy, and polynomials [the usual choice when we have no prior knowledge of the behavior of $y(x)$] may interpolate poorly. (We can think of $u(x)$ as an interpolation function between the boundary conditions that also obeys the differential equation.) This is especially true when the interval $[a, b]$ is large.

The remedy to our problem is based on the observations of Chapters 3 and 4: that even low-degree polynomials can reflect the behavior of a function if based on values that are closely spaced. Accordingly, we hope to successfully apply a technique of Section 6.4 (specifically the Galerkin method) using low-degree polynomials if we subdivide $[a, b]$ into smaller subintervals. It will turn out that our hope is fulfilled.

Applying this technique, we have what is known as the *finite-element method*. Our strategy is as follows:

1. Subdivide $[a, b]$ into n subintervals, called *elements*, that join at $x_1, x_2, \ldots, x_{n-1}$. Add to this array $x_0 = a$ and $x_n = b$. We call the x_i the *nodes* of the interval. Number the elements from 1 to n where element (i) runs from x_{i-1} to x_i. The x_i need not be evenly spaced.
2. Apply the Galerkin method to each element separately to interpolate (subject to the differential equation) between the end nodal values, $u(x_{i-1})$ and $u(x_i)$, where these u's are approximations to the $y(x_i)$'s that are the true solution to the differential equation. [These nodal values are actually the c's in our adaptation of Eq. (6.13), the equation for $u(x)$.]
3. Use a low-degree polynomial for $u(x)$. Our development will use a first-degree polynomial, although quadratics or cubics are often used. (The development for these higher-degree polynomials parallels what we will do but is more complicated.)
4. The result of applying Galerkin to element (i) is a pair of equations in which the unknowns are the nodal values at the ends of element (i), the c's. When we have done this for each element, we have equations that involve all the nodal values, which we combine to give a set of equations that we can solve for the unknown nodal values. (The process of combining the separate *element equations* is called *assembling the system*.)
5. These equations are adjusted for the boundary conditions and solved to get approximations to $y(x)$ at the nodes; we get intermediate values for $y(x)$ by linear interpolation.

We now begin the development. While it involves several steps, each step is straightforward. The differential equation that we will solve is

$$y'' + Q(x)y = F(x) \quad \text{subject to boundary conditions at } x = a \text{ and } x = b. \quad \textbf{(6.33)}$$

(We will specify the boundary conditions later.)

Step 1: Subdivide $[a, b]$ into n elements, as discussed. Focus attention on element (i) that runs between x_{i-1} and x_i. To simplify the notation, call the left node L and the right node R.

Step 2: Write $u(x)$ for element (i):

$$u(x) = c_L N_L + c_R N_R = c_L \frac{x - R}{L - R} + c_R \frac{x - L}{R - L}$$

$$= c_L \frac{x - R}{-h_i} + c_R \frac{x - L}{h_i}, \quad \textbf{(6.34)}$$

where $h_i = R - L$. In Eq. (6.34) we have used the symbol N for the trial functions, rather than the v's that we have used before, to agree with the literature on finite elements. The N's are called *shape functions* in that literature.

Recognize that the N's in Eq. (6.34) are really first-degree Lagrangian polynomials. When we use such linear interpolation, the shape functions are often called *hat functions*. (*Chapeau functions*, from the French, is another name.) The reason for this name will become apparent later on.

Figure 6.4 sketches N_L and N_R within element (i). Since the values of the N's vary (from unity to zero) as x goes from x_L to x_R, they are functions of x. Note also that the c's in Eq. (6.34) are independent of x.

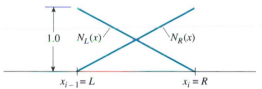

Figure 6.4 N_L and N_R within element (i)

The reason that our N's are called "hat functions" is apparent when we look at a sketch of the N's for several adjacent elements in Fig. 6.5. Observe that we combine the $N_R(x)$ and $N_L(x)$ of Fig. 6.4 that join at x_i into a quantity that we call N_i.

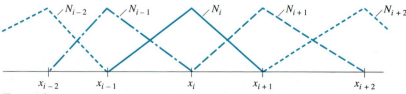

Figure 6.5

Step 3: Apply the Galerkin method to element (i). The residual is

$$R(x) = y'' + Qy - F = u'' + Qu - F, \qquad (6.35)$$

where we have substituted $u(x)$ for $y(x)$. The Galerkin method sets the integral of R weighted with each of the N's (over the length of the element) to zero:

$$\int_L^R N_L R(x)\, dx = 0,$$

$$\int_L^R N_R R(x)\, dx = 0. \qquad (6.36)$$

Now expand Eqs. (6.36):

$$\int_L^R u'' N_L\, dx + \int_L^R Qu N_L\, dx - \int_L^R F N_L\, dx = 0. \qquad (6.37)$$

$$\int_L^R u'' N_R\, dx + \int_L^R Qu N_R\, dx - \int_L^R F N_R\, dx = 0. \qquad (6.38)$$

Step 4: Transform Eqs. (6.37) and (6.38) by applying integration by parts* to the first integral. In the second integral, we will take Q out from the integrand as Q_{av}, an average value within the element. We also take F outside the third integral. When this is done, Eq. (6.37) becomes

$$-\int_L^R \left(\frac{du}{dx}\right)\left(\frac{dN_L}{dx}\right) dx + Q_{av}\int_L^R u N_L\, dx$$
$$\qquad - F_{av}\int_L^R N_L\, dx + N_L\frac{du}{dx}\bigg|_{x=R} - N_L\frac{du}{dx}\bigg|_{x=L} = 0. \qquad (6.39)$$

In the last two terms of Eq. (6.39), $N_L = 1$ at L and is zero at R, so the equation can be simplified:

$$-\int_L^R \left(\frac{du}{dx}\right)\left(\frac{dN_L}{dx}\right) dx + Q_{av}\int_L^R u N_L\, dx - F_{av}\int_L^R N_L\, dx - \frac{du}{dx}\bigg|_{x=L} = 0. \qquad (6.40)$$

Doing similarly with Eq. (6.38) gives

$$-\int_L^R \left(\frac{du}{dx}\right)\left(\frac{dN_R}{dx}\right) dx + Q_{av}\int_L^R u N_R\, dx - F_{av}\int_L^R N_R\, dx + \frac{du}{dx}\bigg|_{x=R} = 0. \qquad (6.41)$$

Step 5: Change signs in Eqs. (6.40) and (6.41); substitute from Eq. (6.34) for u,

---

*From $d(UV) = U\,dV + V\,dU$, we have $\int_a^b U\,dV = -\int_a^b V\,dU,\ UV\big|_a^b$

so $\int_L^R u'' N_i\, dx = -\int_L^R u'\,(dN_i/dx) + N_i u'\big|_L^R$

du/dx, dN_L/dx, and dN_R/dx; and carry out the integrations. We show this separately for each term in Eq. (6.40):

$$\int_L^R \left(\frac{du}{dx}\right)\left(\frac{dN_L}{dx}\right)dx = \int_L^R \left[\frac{-c_L}{h_i} + \frac{c_R}{h_i}\right]\left(\frac{-1}{h_i}\right)dx = \left(\frac{c_L}{h_i^2} - \frac{c_R}{h_i^2}\right)\int_L^R dx$$

$$= \left(\frac{1}{h_i}\right)c_L - \left(\frac{1}{h_i}\right)c_R. \tag{6.42a}$$

$$-Q_{av}\int_L^R (c_L N_L + c_R N_R)N_L\, dx = -c_L Q_{av}\int_L^R N_L^2\, dx - c_R Q_{av}\int_L^R N_R N_L\, dx$$

$$= -c_L Q_{av}\int_L^R \left(\frac{x-R}{h_i}\right)^2 dx \tag{6.42b}$$

$$- c_R Q_{av}\int_L^R \left(\frac{x-L}{h_i}\right)\left(\frac{x-R}{-h_i}\right)dx$$

$$= -\left(Q_{av}*\frac{h_i}{3}\right)c_L - \left(Q_{av}*\frac{h_i}{6}\right)c_R.$$

$$F_{av}\int_L^R N_L\, dx = F_{av}\int_L^R \frac{(x-R)}{-h_i}\, dx = F_{av}*\frac{h_i}{2}. \tag{6.42c}$$

Doing the same with Eq. (6.41) gives

$$\int_L^R \left(\frac{du}{dx}\right)\left(\frac{dN_R}{dx}\right)dx = -\left(\frac{1}{h_i}\right)c_L + \left(\frac{1}{h_i}\right)c_R. \tag{6.43a}$$

$$-Q_{av}\int_L^R (c_L N_L + c_R N_R)N_R\, dx = -\left(Q_{av}*\frac{h_i}{6}\right)c_L - \left(Q_{av}*\frac{h_i}{3}\right)c_R. \tag{6.43b}$$

$$F_{av}\int_L^R N_R\, dx = F_{av}*\frac{h_i}{2}. \tag{6.43c}$$

Step 6: Substitute the result of Step 5 (Eqs. 6.42 and 6.43) into Eqs. (6.40) and (6.41), and rearrange to give two linear equations in the unknown c_L and c_R:

$$\left(\frac{1}{h_i} - \frac{Q_{av}h_i}{3}\right)c_L + \left(\frac{-1}{h_i} - \frac{Q_{av}h_i}{6}\right)c_R = \frac{-F_{av}h_i}{2} - \frac{du}{dx}\Big|_{x=L},$$

$$\left(\frac{-1}{h_i} - \frac{Q_{av}h_i}{6}\right)c_L + \left(\frac{1}{h_i} - \frac{Q_{av}h_i}{3}\right)c_R = \frac{-F_{av}h_i}{2} + \frac{du}{dx}\Big|_{x=R}. \tag{6.44}$$

We call the pair of equations in (6.44) the *element equations*. We can do the same for each element to get n such pairs.

Step 7: Combine (assemble) all the element equations together to form a system of linear equations for the problem. We now recognize that point R in element (i) is precisely the same as point L in element $(i + 1)$. Renumber the c's as $c_0, c_1, \ldots, c_n$. Also notice that the gradient (du/dx) must be the same on either side of the join of the elements—that is, $(du/dx)_{x=R}$ in element (i) equals $(du/dx)_{x=L}$ in element $(i + 1)$. This means that these terms cancel when we do the assembling except in the first and last equations. (On rare occasions this is not true, but in that case the difference in the two gradients is a known value.)

The result of this step is this set of $n + 1$ equations (numbered from 0 to n),

$$[K]\{c\} = \{b\}, \tag{6.45}$$

where the diagonal elements of $[K]$ are

$$\left(\frac{1}{h_1} - Q_{av,1} * \frac{h_1}{3}\right) \quad \text{in row 0,}$$

$$\left(\frac{1}{h_i} - Q_{av,i} * \frac{h_i}{3}\right) + \left(\frac{1}{h_{i+1}} - Q_{av,i+1} * \frac{h_{i+1}}{3}\right) \quad \text{in rows 1 to } n - 1,$$

$$\left(\frac{1}{h_n} - Q_{av,n} * \frac{h_n}{3}\right) \quad \text{in row } n;$$

and elements above and to the left of the diagonal in rows 1 to n are

$$\left(\frac{-1}{h_i} - Q_{av,i} * \frac{h_i}{6}\right).$$

The elements of $\{c\}$ are $c_i, i = 0$ to n.

The elements of $\{b\}$ are:

$$-F_{av,1} * \frac{h_1}{2} - \left(\frac{du}{dx}\right)_{x=a} \quad \text{in row 0,}$$

$$-F_{av,i} * \frac{h_i}{2} - F_{av,i+1} * \frac{h_{i+1}}{2} \quad \text{in rows 1 to } n - 1,$$

$$-F_{av,n} * \frac{h_n}{2} + \left(\frac{du}{dx}\right)_{x=b} \quad \text{in row } n.$$

In the preceding equations, $Q_{av,i}$ and $F_{av,i}$ are values of Q and F at the midpoints of element (i).

Step 8:   Adjust the set of equations from step 6 for the boundary conditions. We will handle two cases: Case (1), a Dirichlet condition is specified—$y(a) = $ constant [and/or $y(b) = $ constant]. Case (2), a Neumann condition is specified—$dy/dx = $ constant at $x = a$ and/or at $x = b$.

(If $Q = 0$, we cannot have a Neumann condition at both ends, since the solution would be known only to within an additive constant.)

Case (1): Dirichlet condition.   In this case, c is known at the end node. Suppose this is $y(a) = A$. Then the equation in row 0 is redundant, and so we remove it from the set of equations of step 6. In the next row, we move $k_{10} * A$ to the right-hand side (subtracting this from the element computed in step 6). If the condition is $y(b) = B$, we do the same but with the last and next to last equations.

Case(2): Neumann condition.   In this case, c is not known at the end node. Suppose the condition is $dy/dx = A$ at $x = a$. We retain the equation in row 0 and substitute the given value of dy/dx into the right-hand side. If the condition is $dy/dx = B$ at $x = b$, we do the same with the last equation.

Step 9:   Solve the set of equations for the unknown c's after adjusting, in step 8, for the boundary conditions. The c's are approximations to $y(x)$ at the nodes. If intermediate values of y are needed between the nodes, we obtain them by linear interpolation.

Examples will clarify the procedure.

EXAMPLE 6.5 Solve $y'' + y = 3x^2$, $y(0) = 0$, $y(2) = 3.5$. (We solved this same equation in Section 6.4.) Subdivide into seven elements that join at $x = 0.4, 0.7, 0.9, 1.1, 1.3,$ and 1.6. Table 6.11 shows the values we need to build the system of equations.

Table 6.11

Element	L	R	Midpoint	h_i	Q_{av}	F_{av}
1	0	0.4	0.2	0.4	1	0.12
2	0.4	0.7	0.55	0.3	1	0.9075
3	0.7	0.9	0.8	0.2	1	1.92
4	0.9	1.1	1.0	0.2	1	3
5	1.1	1.3	1.2	0.2	1	4.32
6	1.3	1.6	1.45	0.3	1	6.3075
7	1.6	2.0	1.8	0.4	1	9.72

The augmented matrix of the set of equations from step 6 is

$$
\left[
\begin{array}{cccccccc|c}
2.367 & -2.567 & 0.000 & 0.000 & 0.000 & 0.000 & 0.000 & 0.000 & -0.024 \\
-2.567 & 5.600 & -3.383 & 0.000 & 0.000 & 0.000 & 0.000 & 0.000 & -0.160 \\
0.000 & -3.383 & 8.167 & -5.033 & 0.000 & 0.000 & 0.000 & 0.000 & -0.328 \\
0.000 & 0.000 & -5.033 & 9.867 & -5.033 & 0.000 & 0.000 & 0.000 & -0.492 \\
0.000 & 0.000 & 0.000 & -5.033 & 9.867 & -5.033 & 0.000 & 0.000 & -0.732 \\
0.000 & 0.000 & 0.000 & 0.000 & -5.033 & 8.167 & -3.383 & 0.000 & -1.378 \\
0.000 & 0.000 & 0.000 & 0.000 & 0.000 & -3.383 & 5.600 & -2.567 & -2.890 \\
0.000 & 0.000 & 0.000 & 0.000 & 0.000 & 0.000 & -2.567 & 2.367 & -1.944 \\
\end{array}
\right]
\qquad \textbf{(6.46)}
$$

To adjust for the boundary conditions, we eliminate the first and last equations and subtract $(0)(-2.567) = 0$ from the right-hand side of row 1 and subtract $(3.50)(-2.567) = -8.9845$ from the right-hand side of row 6 to get

$$
\left[
\begin{array}{cccccc|c}
5.600 & -3.383 & 0.000 & 0.000 & 0.000 & 0.000 & -0.160 \\
-3.383 & 8.167 & -5.033 & 0.000 & 0.000 & 0.000 & -0.328 \\
0.000 & -5.033 & 9.867 & -5.033 & 0.000 & 0.000 & -0.492 \\
0.000 & 0.000 & -5.033 & 9.867 & -5.033 & 0.000 & -0.732 \\
0.000 & 0.000 & 0.000 & -5.033 & 8.167 & -3.383 & -1.378 \\
0.000 & 0.000 & 0.000 & 0.000 & -3.383 & 5.600 & 6.094 \\
\end{array}
\right]
\qquad \textbf{(6.47)}
$$

When we solve the set of Eqs. (6.47) we get the solution tabulated in Table 6.12.

Table 6.12

x	$u(x)$	Anal.	Error
0.000	0.0000	0.0000	0
0.400	−0.0011	0.0064	7.45303E-03
0.700	0.0455	0.0591	1.352482E-02
0.900	0.1398	0.1597	1.986827E-02
1.100	0.3262	0.3516	2.532825E-02
1.300	0.6452	0.6750	2.982068E-02
1.600	1.4793	1.5048	2.551758E-02
2.000	3.5031	3.5031	1.907349E-05

Observe that we always have a tridiagonal (and symmetrical) system and that this system can be solved quickly.

▲

EXAMPLE 6.6 Solve $y'' - (x + 1)y = -e^{-x}(x^2 - x + 2)$ subject to Neumann conditions of

$$y'(2) = 0, \quad y'(4) = -0.036631.$$

Use four elements of equal lengths. Compare to the analytical solution

$$y(x) = e^{-x}(x - 1).$$

Table 6.13 gives values that we need to set up the equations.

Table 6.13

Element	L	R	Midpoint	h_i	Q_{av}	F_{av}
1	2	2.5	2.25	0.5	−3.25	−0.5072
2	2.5	3.0	2.75	0.5	−3.75	−0.4355
3	3.0	3.5	3.25	0.5	−4.25	−0.3611
4	3.5	4.0	3.75	0.5	−4.75	−0.2896

The initial matrix of equations is

$$\begin{bmatrix} 2.542 & -1.729 & 0.000 & 0.000 & 0.000 & | & 0.127 \\ -1.729 & 5.167 & -1.688 & 0.000 & 0.000 & | & 0.236 \\ 0.000 & -1.688 & 5.333 & -1.646 & 0.000 & | & 0.199 \\ 0.000 & 0.000 & -1.646 & 5.500 & -1.604 & | & 0.163 \\ 0.000 & 0.000 & 0.000 & -1.604 & 2.792 & | & 0.072 \end{bmatrix}.$$

After adjusting for boundary conditions, we get

$$
\begin{bmatrix}
2.542 & -1.729 & 0.000 & 0.000 & 0.000 & \vline & 0.127 \\
-1.729 & 5.167 & -1.688 & 0.000 & 0.000 & \vline & 0.236 \\
0.000 & -1.688 & 5.333 & -1.646 & 0.000 & \vline & 0.199 \\
0.000 & 0.000 & -1.646 & 5.500 & -1.604 & \vline & 0.163 \\
0.000 & 0.000 & 0.000 & -1.604 & 2.792 & \vline & 0.036
\end{bmatrix}.
$$

and the solution is

x	$u(x)$	Anal.	Error
2.000	0.1268	0.1353	8.526847E-03
2.500	0.1228	0.1231	3.230497E-04
3.000	0.0996	0.0996	-2.031028E-05
3.500	0.0758	0.0755	-3.280044E-04
4.000	0.0564	0.0549	-1.431525E-03

6.6 Characteristic-Value Problems

Problems in the fields of elasticity and vibration (including applications of the wave equations of modern physics) fall into a special class of boundary-value problems known as *characteristic-value problems*. (Certain problems in statistics also reduce to such problems.) We discuss only the most elementary forms of characteristic-value problems here.

Consider the homogeneous* second-order equation with homogeneous boundary conditions:

$$\frac{d^2y}{dx^2} + k^2y = 0, \qquad y(0) = 0, \qquad y(1) = 0, \qquad \textbf{(6.48)}$$

where k^2 is a parameter. We first solve this equation nonnumerically to show that there is a solution for only certain particular, or "characteristic," values of the parameter. The general solution is

$$y = a \sin kx + b \cos kx,$$

which can easily be verified by substituting into the differential equation; the solution contains the two arbitrary constants a and b, because the differential equation is second-order. The constants a and b are to be determined to make the general solution agree with the boundary conditions.

---

*Homogeneous here means that all the terms are alike in that they are functions of y or its derivatives.

At $x = 0$, $y = 0 = a \sin(0) + b \cos(0) = b$. Then b must be zero. At $x = 1$, $y = 0 = a \sin(k)$; we may have either $a = 0$ or $\sin(k) = 0$ to satisfy this condition. The former leads to $y(x) \equiv 0$, which we call the *trivial solution*. This function, y everywhere zero, will always be a solution to any homogeneous differential equation with homogeneous boundary conditions. (This trivial solution is usually of no interest.) To get a nontrivial solution, we must choose the other alternative, $\sin(k) = 0$, which is satisfied only for certain values of k, the characteristic values of the system. The solution to Eq. (6.48) then requires that

$$k = \pm n\pi, \qquad n = 1, 2, \ldots,$$

so that we have the solutions

$$y = a \sin n\pi x. \tag{6.49}$$

Note that the arbitrary constant a can have any value and still permit the function y to meet the boundary conditions, so that the solution is determined only to within a multiplicative constant. In Fig. 6.6 we sketch several of the solutions as given by Eq. (6.49).

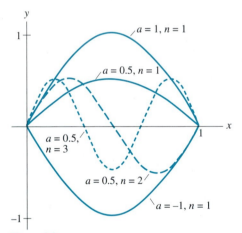

Figure 6.6

The information of interest in characteristic-value problems are these characteristic values, or *eigenvalues*, for the system. If we are dealing with a vibration problem, they give the natural frequencies of the system, which are especially important because, with external loads applied at or near these frequencies, resonance will cause an amplification of motion so that failure is likely. There is an eigenfunction, $y(x)$, corresponding to each eigenvalue, and these determine the possible shapes of the elastic curve when the system is in equilibrium. Often the smallest nonzero value of the parameter is especially important; this gives the fundamental frequency of the system.

To illustrate the use of numerical methods, we will solve Eq. (6.48) again. Our attack is the previous one of replacing the differential equation by a difference equation, and writing this at all points into which the x-interval has been subdivided and where y is unknown.* Using equal-sized intervals and replacing the derivative in (6.48) by a central-difference approximation, we have

$$\frac{y_{i+1} - 2y_i + y_{i-1}}{h^2} + k^2 y_i = 0. \tag{6.50}$$

Letting $h = 0.2$, and writing out Eq. (6.50) at each of the four interior points, we get

$$
\begin{aligned}
y_1 - (2 - 0.04k^2)y_2 + y_3 &= 0, \\
y_2 - (2 - 0.04k^2)y_3 + y_4 &= 0, \\
y_3 - (2 - 0.04k^2)y_4 + y_5 &= 0, \\
y_4 - (2 - 0.04k^2)y_5 + y_6 &= 0.
\end{aligned}
\tag{6.51}
$$

The boundary conditions give $y_1 = y_6 = 0$. Making this substitution and writing in matrix form, after multiplying all equations by -1, we get

$$
\begin{bmatrix}
2 - 0.04k^2 & -1 & 0 & 0 \\
-1 & 2 - 0.04k^2 & -1 & 0 \\
0 & -1 & 2 - 0.04k^2 & -1 \\
0 & 0 & -1 & 2 - 0.04k^2
\end{bmatrix}
\begin{bmatrix}
y_2 \\
y_3 \\
y_4 \\
y_5
\end{bmatrix}
= 0. \tag{6.52}
$$

Note that we can write this as the matrix equation $(A - \lambda I)y = 0$, where

$$
A = \begin{bmatrix}
2 & -1 & 0 & 0 \\
-1 & 2 & -1 & 0 \\
0 & -1 & 2 & -1 \\
0 & 0 & -1 & 2
\end{bmatrix}, \qquad \lambda = 0.04k^2,
$$

and I is the identity matrix of order 4. We will consider the problem in this form in the next section.

The solution to our characteristic-value problem, Eq. (6.48), reduces to solving the set of equations in (6.51) or (6.52).

Such a set of homogeneous linear equations has a nontrivial solution only if the determinant of the coefficient matrix is equal to zero. Hence

$$
\mathrm{Det}
\begin{bmatrix}
2 - 0.04k^2 & -1 & 0 & 0 \\
-1 & 2 - 0.04k^2 & -1 & 0 \\
0 & -1 & 2 - 0.04k^2 & -1 \\
0 & 0 & -1 & 2 - 0.04k^2
\end{bmatrix}
= 0.
$$

---

* There is an alternative trial-and-error approach, analogous to the shooting method, but it generally involves more work.

Expanding this determinant will give an eighth-degree polynomial in k (this is called the characteristic polynomial of matrix A); the roots of this polynomial will be approximations to the characteristic values of the system. Letting $2 - 0.04k^2 = Z$ makes the expansion simpler, and we get

$$Z^4 - 3Z^2 + 1 = 0.$$

The roots of this biquadratic are $Z^2 = (3 \pm \sqrt{5})/2$, or

$$Z = 1.618, \qquad -1.618, \qquad 0.618, \qquad -0.618.$$

We will get an estimate of the principal eigenvalues from $2 - 0.04k^2 = 1.618$, giving $k = \pm 3.09$ (compare to $\pm \pi$). The next eigenvalues are obtained from

$$2 - 0.04k^2 = 0.618, \qquad \text{giving } k = \pm 5.88 \text{ (compare to } \pm 2\pi = \pm 6.28\text{);}$$
$$2 - 0.04k^2 = -0.618, \qquad \text{giving } k = \pm 8.09 \text{ (compare to } \pm 3\pi = \pm 9.42\text{);}$$
$$2 - 0.04k^2 = -1.618, \qquad \text{giving } k = \pm 9.51 \text{ (compare to } \pm 4\pi = \pm 12.57\text{).}$$

Our estimates of the characteristic values get progressively worse. Fortunately, the smallest values of k are of principal interest in many applied problems. To improve accuracy, we will need to write our difference equation using smaller values of h. The work soon becomes inordinately long, however, and we look for a simpler method to find the eigenvalues of a matrix. Such a method is the subject of Section 6.7.

Another reason, beyond the amount of computational effort, that argues against expanding the determinant and then finding the roots of the characteristic equation is accuracy. Finding the zeros of a polynomial of high degree is a process that is often subject to large errors due to round-off. The process is very ill-conditioned.

Observe that the eigenvalues of a diagonal matrix or an upper- or lower-triangular matrix are equal to the terms on the diagonal. [Why? *Hint:* Expand $\det(A - \lambda I)$.] This is the basis for the methods of Section 6.8.

6.7 Eigenvalues by Iteration—The Power Method

We have seen that particular values of the parameter occurring in a characteristic-value problem are of special interest, and the numerical solution involves finding the eigenvalues of the coefficient matrix of the set of difference equations. To do this, we need to find the values of λ that satisfy the matrix equation

$$Ax = \lambda x \qquad \text{or} \qquad (A - \lambda I)x = 0,$$

where A is a square matrix. In effect, we look for a certain vector x that gives a scalar multiple of itself when multiplied by the matrix A. For most vectors chosen at random, this will not be true. If A is $n \times n$ and nonsingular, there are usually exactly n essentially different vectors x that, when multiplied by the matrix A, give a scalar multiple of x. In the terminology of linear algebra these special vectors are called *eigenvectors* or *characteristic vectors*. The multipliers λ, which are n in number, are termed the eigenvalues or characteristic values of the matrix A. As we have seen in the previous section, one method is to solve the equivalent matrix equation

$$(A - \lambda I)x = 0$$

by determining the values of λ that make the determinant of $(A - \lambda I)$ equal to zero. This involves expanding the determinant to give a polynomial in λ, the *characteristic equation*, whose roots must be evaluated. This tells us, of course, that there will be n eigenvalues for the system, but they will not necessarily all be different in value. In fact, it is possible that some eigenvalues will be complex-valued. Solving for the roots of the characteristic equation is laborious.* This section develops an alternative iterative procedure that involves less effort.

Consider a simple 3×3 system:

$$\begin{bmatrix} 3 & -1 & 0 \\ -2 & 4 & -3 \\ 0 & -1 & 1 \end{bmatrix} \begin{Bmatrix} x_1 \\ x_2 \\ x_3 \end{Bmatrix} = \lambda \begin{Bmatrix} x_1 \\ x_2 \\ x_3 \end{Bmatrix},$$

$$\begin{bmatrix} 3 - \lambda & -1 & 0 \\ -2 & 4 - \lambda & -3 \\ 0 & -1 & 1 - \lambda \end{bmatrix} \begin{Bmatrix} x_1 \\ x_2 \\ x_3 \end{Bmatrix} = \begin{Bmatrix} 0 \\ 0 \\ 0 \end{Bmatrix}. \tag{6.53}$$

We wish to find the values of λ and vectors x that satisfy Eq. (6.53). Let us first see what they are by repeating the method of the last section.

The characteristic equation is found by expanding the determinant:

$$\lambda^3 - 8\lambda^2 + 14\lambda - 1 = 0.$$

The roots are

$$\lambda_1 = 5.4773, \qquad \lambda_2 = 2.4481, \qquad \text{and} \qquad \lambda_3 = 0.07458.$$

We can find the eigenvectors corresponding to these from the three sets of equations:

1. $\begin{bmatrix} 3 & -1 & 0 \\ -2 & 4 & -3 \\ 0 & -1 & 1 \end{bmatrix} \begin{Bmatrix} x_1 \\ x_2 \\ x_3 \end{Bmatrix} = 5.4773 \begin{Bmatrix} x_1 \\ x_2 \\ x_3 \end{Bmatrix}$ or

$$\begin{aligned} -2.4473\, x_1 \qquad\qquad - x_2 \qquad\qquad\qquad &= 0 \\ -2\, x_1 - 1.4773\, x_2 \qquad\quad - 3\, x_3 &= 0 \\ -x_2 - 4.4773\, x_3 &= 0 \end{aligned}$$

2. $\begin{bmatrix} 3 & -1 & 0 \\ -2 & 4 & -3 \\ 0 & -1 & 1 \end{bmatrix} \begin{Bmatrix} x_1 \\ x_2 \\ x_3 \end{Bmatrix} = 2.4481 \begin{Bmatrix} x_1 \\ x_2 \\ x_3 \end{Bmatrix}$ or

$$\begin{aligned} -0.5519\, x_1 \qquad\qquad - x_2 \qquad\qquad\qquad &= 0 \\ -2\, x_1 + 1.5519\, x_2 \qquad\quad - 3\, x_3 &= 0 \\ -x_2 - 1.4487\, x_3 &= 0 \end{aligned}$$

---

* The labor involved is not only in finding the roots of the characteristic polynomial but in evaluating the coefficients in the first place.

3. $\begin{bmatrix} 3 & -1 & 0 \\ -2 & 4 & -3 \\ 0 & -1 & 1 \end{bmatrix} \begin{Bmatrix} x_1 \\ x_2 \\ x_3 \end{Bmatrix} = 0.07458 \begin{Bmatrix} x_1 \\ x_2 \\ x_3 \end{Bmatrix}$ or

$$
\begin{aligned}
2.9254\,x_1 \quad\quad\quad - x_2 \quad\quad\quad\quad &= 0 \\
-2\,x_1 + 3.9254\,x_2 \quad\quad - 3\,x_3 &= 0 \\
-x_2 + 0.9254\,x_3 &= 0
\end{aligned}
$$

The solutions to these are:

1. $x_1 = -0.4037, x_2 = 1, x_3 = -0.2233,$
2. $x_1 = 1, x_2 = 0.5519, x_3 = -0.3811,$
3. $x_1 = 0.3163, x_2 = 0.9254, x_3 = 1.$

In solving these equations, we must set one of the vectors to any arbitrary nonzero value. We have made x_2 in (1), x_1 in (2), and x_3 in (3) each equal to unity. (There are other ways to scale the vectors—a frequent choice is to make the length equal to one.) In the matrix of Eq. (6.53), notice that the sum of the eigenvalues $(5.4773 + 2.4481 + 0.0746 = 8)$ is exactly equal to the trace: $3 + 4 + 1 = 8$. This is always true for any square matrix.

We have observed that a triangular matrix has its eigenvalues on the diagonal. Since we already know how to convert a matrix to a triangular one, we might hope that this would give the eigenvalues. Unfortunately, our elimination methods change the eigenvalues (but in Section 6.8 we will present a technique of triangularization that does preserve the eigenvalues).

The Power Method

We now present an iterative technique for finding some of the eigenvalues of a matrix. This is called the *power method*. We start with an arbitrary vector, multiply it by the matrix, and then normalize. We repeat this process until the normalized product converges. Let us begin with this vector:

$$
\begin{Bmatrix} 1 \\ 1 \\ 1 \end{Bmatrix}.
$$

We get the following results. At each step, we normalize the vector by making its largest component equal to unity.

$$
\begin{vmatrix} 3 & -1 & 0 \\ -2 & 4 & -3 \\ 0 & -1 & 1 \end{vmatrix} \begin{Bmatrix} 1 \\ 1 \\ 1 \end{Bmatrix} = \begin{Bmatrix} 2 \\ -1 \\ 0 \end{Bmatrix} = 2 \begin{Bmatrix} 1 \\ -0.5 \\ 0 \end{Bmatrix}
$$

$$\begin{vmatrix} 3 & -1 & 0 \\ -2 & 4 & -3 \\ 0 & -1 & 1 \end{vmatrix} \begin{Bmatrix} 1 \\ -0.5 \\ 0 \end{Bmatrix} = \begin{Bmatrix} 3.5 \\ -4 \\ 0.5 \end{Bmatrix} = -4 \begin{Bmatrix} -0.875 \\ 1 \\ -0.125 \end{Bmatrix}$$

$$\begin{vmatrix} 3 & -1 & 0 \\ -2 & 4 & -3 \\ 0 & -1 & 1 \end{vmatrix} \begin{Bmatrix} -0.875 \\ 1 \\ -0.125 \end{Bmatrix} = \begin{Bmatrix} -3.625 \\ 6.125 \\ -1.125 \end{Bmatrix} = 6.125 \begin{Bmatrix} -0.5918 \\ 1 \\ -0.1837 \end{Bmatrix}$$

$$\begin{vmatrix} 3 & -1 & 0 \\ -2 & 4 & -3 \\ 0 & -1 & 1 \end{vmatrix} \begin{Bmatrix} -0.5918 \\ 1 \\ -0.1837 \end{Bmatrix} = \begin{Bmatrix} -2.7755 \\ 5.7347 \\ -1.1837 \end{Bmatrix} = 5.7347 \begin{Bmatrix} -0.4840 \\ 1 \\ -0.2064 \end{Bmatrix}$$

$$\vdots$$

$$\begin{vmatrix} 3 & -1 & 0 \\ -2 & 4 & -3 \\ 0 & -1 & 1 \end{vmatrix} \begin{Bmatrix} -0.4037 \\ 1 \\ -0.2233 \end{Bmatrix} = \begin{Bmatrix} -2.2111 \\ 5.4774 \\ -1.2233 \end{Bmatrix} = 5.4774 \begin{Bmatrix} -0.4037 \\ 1 \\ -0.2233 \end{Bmatrix}$$

When the change in the normalizing factor becomes negligible, its value is an estimate of the eigenvalue. Here the change was less than 0.0001 after 14 steps.

We see that the successive values of the product vector approach closer and closer to multiples of the eigenvector

$$\begin{Bmatrix} -0.4037 \\ 1 \\ -0.2233 \end{Bmatrix}.$$

The normalization factors approach the value of the largest eigenvalue.

The method is slow to converge if the magnitudes of the largest eigenvalues are nearly the same.

With more realistically sized matrices, the iteration takes many steps and requires a computer for its practical execution. The method will converge only when the eigenvalue largest in modulus is uniquely determined. In Section 6.10, we explain why this is true.

A special advantage of the power method is that the eigenvector that corresponds to the dominant eigenvalue is generated at the same time. For most methods of determining eigenvalues, a separate computation is needed to obtain the eigenvector. A disadvantage, of course, is that it gives only one eigenvalue. Sometimes the largest eigenvalue is the most important, but if it is not, we must modify the method.

The Inverse Power Method

In some problems, the most important eigenvalue and eigenvector is that where the eigenvalue is of least magnitude. We can find this one if we apply the power method to the inverse of A. This is because the inverse matrix has a set of eigenvalues that are the reciprocals of the eigenvalues of A. This is readily shown:

Given that $Ax = \lambda x$.

Multiply by A^{-1}:

$$A^{-1}Ax = A^{-1}\lambda x = \lambda A^{-1}x.$$

Thus we see that

$$x = \lambda A^{-1}x \quad \text{or} \quad A^{-1}x = \frac{1}{\lambda}x.$$

(It is inefficient to actually invert A before applying the power method to get the eigenvalue. We use the LU equivalent of A, obtainable at much less effort, and solve for $v^{(n+1)}$ from

$$LUv^{(n+1)} = Av^{(n+1)} = v^{(n)}, \quad \text{equivalent to} \quad A^{-1}v^{(n)} = v^{(n+1)}. \quad \textbf{(6.54)}$$

The solution of Eq. (6.54) when the LU is known requires the same amount of effort as a matrix multiplication.)

For the preceding matrix, the inverse is

$$\begin{bmatrix} 1 & 1 & 3 \\ 2 & 3 & 9 \\ 2 & 3 & 10 \end{bmatrix},$$

and the power method, beginning with $v_0 = (1, 1, 1)^T$, gives $\lambda = 13.4090$. The reciprocal, 0.0746, is the eigenvalue of smallest magnitude for the original matrix. The corresponding eigenvector is $(0.3163, 0.9254, 1)^T$.

Accelerating the Power Method—Shifting

The power method is apt to be slowly convergent. We can accelerate it if we know an approximate value for the eigenvalue. This is based on the result of shifting the eigenvalues. Observe that subtracting a constant from the diagonal elements of A gives a system whose eigenvalues are those of A with the same constant subtracted:

Given $Ax = \lambda x$.

Subtract $sIx = sx$ from both sides:

$$Ax - sIx = \lambda x - sx,$$
$$(A - sI)x = (\lambda - s)x.$$

This relationship can be applied in two ways. Suppose we wish to determine the value of an eigenvalue near to some number s. We shift the eigenvalues by subtracting s from the diagonal elements; there is then an eigenvalue very near to zero in the shifted matrix. We use the power method on the inverse of the shifted matrix. This is often rapidly convergent because the reciprocal of the very small value is very large, and is usually much larger than the next largest one (for the shifted inverse system). After we obtain it, we reverse the transformations to obtain the desired value for the original matrix. This process is called the *inverse power method*. For the preceding matrix, if we shift by 5.5 and use this technique, we get the dominant eigenvalue in only 5 iterations, rather than 14.

Another application of shifting is to determine the eigenvalue at the other extreme after the dominant one has been computed. Suppose a matrix has eigenvalues whose values are 8, 4, 1, and -5. After applying the power method to get the one at 8, we shift by subtracting 8. The eigenvalues of the shifted system are then 0, -4, -7, and -13. The power method will get the dominant one, at -13. When we add 8 back to reverse the effect of the shifting, we obtain -5, the eigenvalue of the original matrix at the opposite extreme from 8.

EXAMPLE 6.7   We illustrate the use of the power method and its several variations with the 3×3 matrix:

$$A = \begin{bmatrix} 4 & -1 & 1 \\ 1 & 1 & 1 \\ -2 & 0 & -6 \end{bmatrix}.$$

If we begin with the initial vector $(1, 1, 1)^T$, the regular power method gives the results shown in Table 6.14. The iterations were terminated when the normalization factor changed by less than 0.0001 from the previous value.

Table 6.14

Iteration number	Normalization factor	Resultant vector
1	-8	$(-0.5, -0.375, 1)^T$
2	-5	$(0.125, -0.025, 1)^T$
3	-6.25	$(-0.244, -0.176, 1)^T$
.	.	.
.	.	.
.	.	.
22	-5.76849	$(-0.1157, -0.1306, 1)^T$

After the dominant eigenvalue at -5.76849 was determined, the matrix was shifted by subtracting this value from the elements on the diagonal. For this shifted matrix, the power method gave, after 32 iterations, a normalization factor of 9.23759 with a vector of $(1, 0.31963, -0.21121)^T$. We then know that the original matrix had an eigenvalue at 3.4691 [from $9.23759 + (-5.76849)$]. (The corresponding eigenvector is the same as that determined by the power method for the shifted matrix.)

Applying the power method to A^{-1} (through application of the LU equivalent, of course) gave, after nine iterations, the value of 0.76989 with a vector of $(0.4121, 1, -0.1129)^T$. Now we know that the original matrix A had its smallest eigenvalue at $1/0.76989 = 1.29889$. For this 3×3 system, we have computed all its eigenvalues.

The dominant eigenvalue came with some slowness; 22 iterations were needed. If we shift by -5.77 and apply the method to the inverse, only four iterations are required to obtain a value of 672.438 and a vector of $(-0.1157, -0.1306, 1)^T$. This is the same vector as that obtained in the first application of the method. We get the dominant eigenvalue for A by adding -5.77 to the reciprocal of 672.438:

$$\frac{1}{672.438} + (-5.77) = -5.76851.$$

Many iterations have been saved.

An Algorithm for the Power Method

Given an $n \times n$ matrix A and a tolerance value, to find the dominant eigenvalue and its corresponding eigenvector:

SET $v_0 = n$-component vector with all elements equal to one.
REPEAT
    SET LSAV = largest component of v_0.
    Multiply A times v_0 to give a new vector v.
    SET L = largest component of v.
    SET $v_0 = v/L$. (Normalize)
UNTIL $|\text{LSAV} - L| <$ tolerance.

On termination, L = dominant eigenvalue and v_0 is the corresponding eigenvector.

Note: If the initial v_0 is perpendicular to (dot product is zero) the eigenvector of the dominant eigenvalue, convergence may be to a different eigenvalue. Also, if the dominant eigenvalue is very near in magnitude to a second eigenvalue, the method will be slow to converge.

An Algorithm for the Inverse Power Method

Given an $n \times n$ matrix A and a tolerance value, to find the eigenvalue nearest to a value S and its corresponding eigenvector:

SET $B = A - S*I$, where I is the $n \times n$ unit matrix.
Carry out the power method on the inverse of B.
On termination, $1/L + S$ is the eigenvalue desired and v_0 is the corresponding eigenvector.

Note 1: If the initial v_0 is perpendicular to (dot product is zero) the eigenvector of the desired eigenvalue, convergence may be to a different eigenvalue.

Note 2: If the inverse power method is applied to matrix A without shifting, the eigenvalue of A of smallest magnitude is $1/L$.

Gerschgorin's Theorems

The key to the inverse power method is finding an approximate value for the desired eigenvalue. Sometimes the physical problem itself can suggest a value. Applying the standard power method for only a few iterations frequently gives an approximate value (the many successive iterations are needed to refine the value). Gerschgorin's theorems can give good estimates if there is strong diagonal dominance. Let D_i be a circle whose center is a_{ii} and whose radius is

$$\Sigma |a_{ij}|, \quad j = 1, 2, \ldots, n \quad \text{and} \quad j \neq i.$$

Then Gerschgorin I says that every eigenvalue of A must lie in the union of those circles. Gerschgorin II says that if k of these circles do not touch the other $n - k$ circles, then exactly k eigenvalues (counting multiplicities) lie in the union of those k circles. For the previous example we have

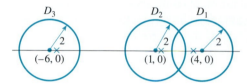

The dominant eigenvalue, -5.76851, was found to lie in D_3, and the other two are in $D_1 \cup D_2$. The $\times$'s in the figure indicate the eigenvalues.

Problems with Equal Eigenvalues

When a matrix has two eigenvalues of equal largest magnitude,* the power method as described will fail. In the case of two largest eigenvalues of opposite sign, the normalization factor does not converge but oscillates. For example, when the power method is applied to

$$A = \begin{bmatrix} -10 & -7 & 7 \\ 5 & 5 & -4 \\ -7 & -5 & 6 \end{bmatrix},$$

which has eigenvalues of 4, -4, and 1, the power method gives the results shown in Table 6.15 and the normalizations continue indefinitely to be -10 and -1.6. In such a case, the magnitude of the largest eigenvalue(s) is given by taking the square root of the product of the alternating normalization factors; in this example, -10 and -1.6 have a product, 16, whose square roots are 4 and -4.

---

*This may occur if the eigenvalues are complex; for a matrix with real coefficients, complex eigenvalues will come in complex conjugate pairs.

Table 6.15

Iteration number	Normalization factor	Resultant vector
1	−10	$(1, -0.6, 0.6)^T$
2	−1.6	$(1, 0.25, 0.25)^T$
3	−10	$(1, -0.525, 0.675)^T$
4	−1.6	$(1, 0.2031, 0.2031)^T$

6.8 Eigenvalues by the *QR* Method

When all the eigenvalues of a large matrix are desired, the power method is not efficient, nor is determination from the characteristic polynomial. The usual method for getting all the eigenvalues is the *QR* technique. It depends on similarity transformations—transformations of the matrix that preserve the eigenvalues.

Employing Similarity Transformations

For any nonsingular matrix, M, $M^{-1}AM$ has the same eigenvalues as A (but not necessarily the same eigenvectors). Proving this statement is not difficult: Suppose that $B = M^{-1}AM$. The eigenvalues of B are the roots of

$$\det(B - \lambda I) = \det(M^{-1}AM - \lambda I) = \det(M^{-1}(A - \lambda I)M)$$
$$= \det(M^{-1}) * \det(A - \lambda I) * \det(M)$$
$$= \det(A - \lambda I).$$

In this transformation, we have used the fact that the identity matrix and the scalar λ can be multiplied in any order. Since A and B have the same characteristic polynomial, they have the same eigenvalues. To get the eigenvectors of B, if x is an eigenvector of A, then, with its corresponding eigenvalue, λ, we have

$$Ax = \lambda x, \quad \text{or} \quad MBM^{-1}x = \lambda x, \quad \text{or} \quad B(M^{-1}x) = \lambda(M^{-1}x),$$

which says that the eigenvectors of B are $M^{-1}x$.

When $M^{-1}AM = B$, matrices B and A are said to be *similar*.

The *QR* method uses similarity transformations to convert matrix A to upper- (or lower-) triangular. We have already seen that a triangular matrix has its eigenvalues on the diagonal. What we want is a way to put zeros below the main diagonal. It is not difficult to find a matrix that puts a zero in any desired position—Gaussian elimination does this. (But that results in a matrix that is not similar; we must do something different.) The *QR* method employs this different technique.

Suppose that

$$
A = \begin{bmatrix} 7 & 8 & 6 & 6 \\ 1 & 6 & -1 & -2 \\ 1 & -2 & 5 & -2 \\ 3 & 4 & 3 & 4 \end{bmatrix}
$$

and we wish to change a_{42} to zero. Matrix Q^{-1} will do that:

$$
Q^{-1} = \begin{bmatrix} 1 & 0 & 0 & 0 \\ 0 & c & 0 & s \\ 0 & 0 & 1 & 0 \\ 0 & -s & 0 & c \end{bmatrix}, \quad
\begin{aligned}
d &= \sqrt{a_{22}^2 + a_{42}^2}, \\
c &= \frac{a_{22}}{d}, \\
s &= \frac{a_{42}}{d}.
\end{aligned}
$$

(Q^{-1} is called a *rotation* because $Q^{-1} = Q^T$.) Observe that if we wish to put a zero in row R and column C, the numerator of c is a_{CC} and the numerator of s is a_{RC}, while each denominator is the square root of the sum of the squares of these two elements. (The reason for this definition of c and s will become apparent later.) Multiplying $Q^{-1}A$ gives

$$
\begin{bmatrix} 7 & 8 & 6 & 6 \\ 2.496 & 7.211 & 0.832 & 0.555 \\ 1 & -2 & 5 & -2 \\ 1.941 & 0 & 3.051 & 4.438 \end{bmatrix}.
$$

The element in row 4, column 2 is zero, as expected. Notice that the eigenvalues will not be the same of those of A, because the trace is different. To get a similarity transformation, we must postmultiply by Q, the inverse of Q^{-1}. With our definition of the c- and s-values of Q^{-1}, this is easy to obtain because the transpose of Q is its inverse, so we just have to transpose Q^{-1}. Doing the postmultiply, we get

$$
\begin{bmatrix} 7 & 9.985 & 6 & 0.555 \\ 2.496 & 6.308 & 0.832 & -3.538 \\ 1 & -2.774 & 5 & 0.555 \\ 1.941 & 2.462 & 3.051 & 3.692 \end{bmatrix}.
$$

Now the trace is the same as that of A (equal to 22), and the eigenvalues are unchanged.

It may seem that the change in A is not significant, but observe that the sum of the off-diagonal elements in row 4 is smaller and, from Gerschgorin's theorem, 3.692 is closer to an eigenvalue than the original value of 4. (The actual eigenvalue for row 4 will turn out to be 1.) Also the diagonal element in row 2, 6.308, is closer to the eigenvalue of 7 than the original element, which was 6.

This suggests that we can get all the eigenvalues to appear on the diagonal if we continue to apply similarity transformations to convert all elements below the diagonal to zero. This is in fact true; applying similarity transformations systematically to make all elements below the diagonal disappear eventually gives

$$\begin{bmatrix} 10 & 1.581 & 11.068 & 3 \\ 0 & 7 & 0.998 & -0.001 \\ 0 & 0 & 4 & -3.162 \\ 0 & 0 & 0 & 1 \end{bmatrix},$$

which shows that the eigenvalues of A are 10, 7, 4, and 1. (The zeros shown here are really values less than $5 * 10^{-4}$. It took more than 100 successive transformations to accomplish this.) The QR method is an iterative process.

The Householder Transformation

The only disadvantage of the QR technique is its poor efficiency. Although the multiplications are not very costly (only two rows or columns are affected), we hope to make an improvement. There are two tricks that we can use.

First, there is a similarity transformation called a Householder transformation that puts zeros below the *subdiagonal* of A that stay zero on subsequent operations. A matrix that has all zeros below the diagonal except for elements immediately below the diagonal (we might call this "almost triangular") is called a lower *Hessenberg* matrix.

Once the $N \times N$ matrix A has been converted to Hessenberg, there are only $N - 1$ subdiagonal elements to convert to zeros, but we can do this with even fewer iterations if we shift A before each iteration (borrowing from the idea of shifting in the power method) and, of course, reverse the shift after the iteration. This is our second trick. Some authors compute a shift value from the eigenvalues of the 2×2 matrix at the lower-right corner of A, but shifting by the value of a_{NN} itself is about as effective and saves having to compute roots of a quadratic from the 2×2 matrix.

Here is the scheme we use:

1. Convert A to upper Hessenberg. (If A is symmetric, it stays symmetric, meaning it becomes tridiagonal.)
2. Shift by a_{NN}, then perform similarity transformations, as done before, for all columns from 1 to $N - 1$. Make sure the transformations give similar matrices by doing a postmultiply for each column from $N - 1$ to 1. Then reverse the shift.
3. Repeat step 2 until the element to the left of a_{NN} is essentially zero; an eigenvalue appears as the diagonal of the last row, in position a_{NN}.
4. Once an eigenvalue has been found, the last row and column of A can be ignored because they do not change. Repeat the process of steps 2 and 3 to annihilate the elements to the left of the diagonal in successive rows from $N - 1$ to 2. Once all the subdiagonal elements are essentially zero, all the eigenvalues appear on the main diagonal of A.

The reason the method is called QR is that we do these successive multiplications to get matrix R:

$$Q_{N-1} Q_{N-2} \ldots Q_1 A = R$$

Then we do

$$RQ_1 Q_2 \ldots Q_{N-1} = A^*,$$

where A^* represents the result of this stage of iteration.

It remains to be seen how a Householder transformation can be accomplished. For the same matrix A as before,

$$A = \begin{bmatrix} 7 & 8 & 6 & 6 \\ 1 & 6 & -1 & -2 \\ 1 & -2 & 5 & -2 \\ 3 & 4 & 3 & 4 \end{bmatrix},$$

we reduce in row 3, column 1, by forming $B^{-1}AB$, where

$$B_1^{-1} = \begin{bmatrix} 1 & 0 & 0 & 0 \\ 0 & 1 & 0 & 0 \\ 0 & -b_3 & 1 & 0 \\ 0 & -b_4 & 0 & 1 \end{bmatrix}, \qquad \begin{aligned} b_3 &= \frac{a_{31}}{a_{21}} = \frac{1}{1} = 1, \\[2mm] b_4 &= \frac{a_{41}}{a_{21}} = \frac{3}{1} = 3. \end{aligned}$$

$$B_1 = \begin{bmatrix} 1 & 0 & 0 & 0 \\ 0 & 1 & 0 & 0 \\ 0 & b_3 & 1 & 0 \\ 0 & b_4 & 0 & 1 \end{bmatrix}.$$

Thus we have

$$B_1^{-1}AB_1 = \begin{bmatrix} 7 & 32 & 6 & 8 \\ 1 & -1 & -1 & -2 \\ 0 & -2 & 6 & 0 \\ 0 & 22 & 6 & 10 \end{bmatrix}.$$

We continue with the second column, using

$$B_2^{-1} = \begin{bmatrix} 1 & 0 & 0 & 0 \\ 0 & 1 & 0 & 0 \\ 0 & 0 & 1 & 0 \\ 0 & 0 & -b_4 & 1 \end{bmatrix}, \qquad b_4 = \frac{a_{42}}{a_{32}} = \frac{22}{-2} = -11,$$

with the B_2 matrix the same except that b_{43} is the positive value.

When we compute $B^{-1}AB$ again, we get the Hessenberg matrix:

$$B_2^{-1}B_1^{-1}AB_1B_2 = \begin{bmatrix} 7 & 32 & -60 & 6 \\ 1 & -1 & 21 & -2 \\ 0 & -2 & 6 & 0 \\ 0 & 0 & -38 & 10 \end{bmatrix}.$$

The rule we have used is this:

> To reduce column C of matrix A, form matrix B from the identity matrix by replacing the elements in row J and column $C + 1$ of I with $a_{J,C}/a_{C+1,C}$ for $J = C + 2$ to N. Form matrix B^{-1} with the negatives of these values. Multiply $B^{-1}AB$.
>
> Perform the reductions in columns 1 through $N - 2$ of A.

There is one potential flaw in this reduction to Hessenberg. The divisor used in forming the B matrices could be zero or very small, causing overflow (or large round-off errors). We avoid this by examining the elements in each column below the diagonal to find the entry of largest absolute value, then interchanging rows and columns to put the largest element in the subdiagonal position. The interchange of columns is necessary to make sure that the diagonal elements remain the same even though they may be scrambled, thus preserving the eigenvalues. We really do $P^{-1}AP$ where P is the permutation matrix.

If the preceding reduction to Hessenberg is done with row/column interchanges, the resulting matrix is

$$\begin{bmatrix} 7 & 10.667 & 12.286 & 8 \\ 3 & 6.333 & 6.143 & 4 \\ 0 & -3.111 & 1.381 & -3.333 \\ 0 & 0 & 0.582 & 7.286 \end{bmatrix}.$$

Once the matrix has been reduced to Hessenberg form, we perform the QR algorithm until all subdiagonal elements disappear, giving the eigenvalues on the diagonal.*

The EISPACK programs and IMSL contain subroutines for finding eigenvalues and eigenvectors that are based on the QR method. In particular, these will handle the complex case. In addition, the LINPACK library uses QR in the singular-value-decomposition of a (possibly rectangular) matrix. This latter is applied to least-squares problems (Chapter 3). There are other variations on this theme. For symmetric matrices, the Jacobi method has been widely used.

---

* There are other transformation that reduce to Hessenberg that may be more efficient; see Stewart (1973). The method given here is patterned after Fox (1965).

6.9 Applications of Eigenvalues

As an illustration of the importance of eigenvalues in numerical analysis, this section discusses their application to the convergence of iterative methods for solving a set of linear equations. We also describe their use in getting the solution to systems of initial-value problems.

Convergence Rates in Iterative Solutions

In Section 2.10, we considered an example system of equations, $Ax = b$, where

$$A = \begin{bmatrix} 6 & -2 & 1 \\ 1 & 2 & -5 \\ -2 & 7 & 2 \end{bmatrix}, \qquad b = \begin{Bmatrix} 11 \\ -1 \\ 5 \end{Bmatrix}.$$

In Section 2.10, we also showed that both the Jacobi and Gauss–Seidel methods can be expressed in the form

$$x^{(n+1)} = Gx^{(n)} = b' - Bx^{(n)}. \tag{6.55}$$

To apply these methods, however, we must have the diagonal coefficients dominate, so we interchange rows 2 and 3 and get

$$A = \begin{bmatrix} 6 & -2 & 1 \\ -2 & 7 & 2 \\ 1 & 2 & -5 \end{bmatrix}, \qquad b = \begin{Bmatrix} 11 \\ 5 \\ -1 \end{Bmatrix}. \tag{6.56}$$

As shown in Chapter 2, for the Jacobi method

$$x^{(n+1)} = -D^{-1}(L + U)x^{(n)} + D^{-1}b,$$

so we have

$$B = D^{-1}(L + U) = \begin{bmatrix} 0 & -0.3333 & 0.1667 \\ -0.2857 & 0 & 0.2857 \\ -0.2000 & 0.4000 & 0 \end{bmatrix} \tag{6.57}$$

and

$$D^{-1}b = (1.8333, \quad 0.7143, \quad 0.2000)^T.$$

For the Gauss–Seidel method, we iterate with

$$x^{(n+1)} = -(L + D)^{-1}Ux^{(n)} + (L + D)^{-1}b,$$

so that

$$B = (L + D)^{-1}U = \begin{bmatrix} 0 & -0.3333 & 0.1667 \\ 0 & -0.0952 & 0.3333 \\ 0 & -0.1048 & 0.1667 \end{bmatrix} \tag{6.58}$$

and

$$(L + D)^{-1}b = (1.8333, \quad 1.2381, \quad 1.0619)^T.$$

We now will show that whether an iterative method converges is determined by the eigenvalues of its B matrix and, further, that the rate of convergence is also related to their magnitude.

The general formulation for iteration is $x^{(n+1)} = b' - Bx^{(n)}$. If an iterative method converges, then $x^{(n)} \to \bar{x}$, where the symbol $\bar{x}$ represents the solution. Hence $A\bar{x} = b$. Equation (6.58) becomes, for $x = \bar{x}$,

$$\bar{x} = b' - B\bar{x}.$$

Let $e^{(n)}$ be the error of the nth iteration:

$$e^{(n)} = x^{(n)} - \bar{x}.$$

When the method converges (and it certainly does if the matrix A is diagonally dominant), $e^{(n)} \to 0$, the zero vector, as n increases. Using Eq. (6.58), we have

$$e^{(n+1)} = -Be^{(n)} = B^2 e^{(n-1)} = \cdots = (-B)^{n+1} e^{(0)}.$$

If it should be that $B^n \to 0$, the zero matrix, obviously $e^{(n)} \to 0$. In linear algebra it is shown that any square matrix B can be written as $B = UDU^{-1}$. If the eigenvalues of B are distinct, then D is a diagonal matrix with the eigenvalues of B on its diagonal. (If some of the eigenvalues of B are repeated, then D may be triangular rather than diagonal. Our argument holds in either case.)

We have

$$B = UDU^{-1}, \qquad B^2 = UD^2U^{-1}, \qquad B^3 = UD^3U^{-1}, \qquad \ldots, \qquad B^n = UD^nU^{-1}.$$

If all the eigenvalues of B (which are the elements on the diagonal of D) have magnitudes less than 1, it is clear that D^n will approach the zero matrix. This implies that $B^n \to$ 0-matrix, $e^{(n)} \to$ 0-vector, and $x^{(n)} \to \bar{x}$. In other words, if all the eigenvalues of B, the iteration matrix, are less than 1 in magnitude, the iteration will always converge (even if the system is not diagonally dominant). It is also easy to see that the rate of convergence is determined by the largest eigenvalue in magnitude of B, since this controls how rapidly B^n approaches the zero matrix.

We now look at the eigenvalues of the B matrices in Eqs. (6.57) and (6.58). For the Jacobi matrix (6.57), we find that two are complex and one is real:

$$\lambda_1 = -0.1425 + 0.3366i, \qquad \lambda_2 = -0.1425 - 0.3366i, \qquad \lambda_3 = 0.2851.$$

The largest magnitude is $(0.1425^2 + 0.3366^2)^{1/2} = 0.3655$. Since all of its eigenvalues are less than 1 in magnitude, Jacobi converges for the example problem. (This confirms the demonstration in Section 2.10.)

For the Gauss–Seidel B matrix (Eq. 6.58), the eigenvalues are

$$\lambda_1 = 0, \qquad \lambda_2 = 0.0357 + 0.1333i, \qquad \lambda_3 = 0.0357 - 0.1333i,$$

of which the largest magnitude is 0.1380. Therefore the Gauss–Seidel also converges (again confirming the demonstration). It converges faster than Jacobi because its eigenvalue of largest magnitude is smaller. The more superficial argument about the relative rates of convergence given in Section 2.10 is now put on a sound basis.

Application to Systems of Differential Equations

In Chapter 5 we considered initial-value differential equations. The eigenvalues and eigenvectors of a matrix are important in the solution of a system of differential equations with constant coefficients. Consider the system

$$
\begin{aligned}
x' &= 10x, \\
y' &= \quad x - 3y - 7z, \\
z' &= \qquad 2y + 6z,
\end{aligned}
$$

where $x(0) = 1, y(0) = -1, z(0) = 2$. We may rewrite these equations as

$$
X'(t) = \begin{bmatrix} 10 & 0 & 0 \\ 1 & -3 & -7 \\ 0 & 2 & 6 \end{bmatrix} X(t), \qquad X(0) = \begin{bmatrix} 1 \\ -1 \\ 2 \end{bmatrix}. \tag{6.59}
$$

The general solution to this problem is

$$
X(t) = Ae^{10t} \begin{bmatrix} 1 \\ \frac{2}{33} \\ \frac{1}{33} \end{bmatrix} + Be^{4t} \begin{bmatrix} 0 \\ -1 \\ 1 \end{bmatrix} + Ce^{-t} \begin{bmatrix} 0 \\ 1 \\ -\frac{2}{7} \end{bmatrix}. \tag{6.60}
$$

That is,

$$
x(t) = Ae^{10t},
$$

$$
y(t) = \frac{2}{33} Ae^{10t} - Be^{4t} + Ce^{-t},
$$

$$
z(t) = \frac{1}{33} Ae^{10t} + Be^{4t} - \frac{2}{7} Ce^{-t},
$$

where, at $t = 0$,

$$
1 = A,
$$

$$
-1 = \frac{2}{33} A - B + C,
$$

$$
2 = \frac{1}{33} A + B - \frac{2}{7} C.
$$

It seems remarkable that the multipliers of t in the exponential are the eigenvalues and that the vectors in Eq. (6.60) are the eigenvectors of the matrix of Eq. (6.59).

6.10 Theoretical Matters

Existence and Uniqueness of Solutions to Boundary-Value Problems

The theory of boundary-value problems is involved. The existence and uniqueness of a solution requires that conditions be imposed on the problem. For example, a linear boundary-value problem of the form

$$\frac{d^2y}{dx^2} = py' + qy + r \quad \text{for } x \text{ on } [a,b], \quad y(a) = y_0, \quad y(b) = y_1,$$

where p, q, and r are functions of x only, has a unique solution if two conditions are met:

p, q, and r are continuous on $[a,b]$, and $q > 0$ on $[a,b]$.

If the problem is nonlinear, more severe conditions are needed involving the partial derivatives of the right-hand side with respect to y and y'.

A simple example will illustrate why these conditions are necessary.

EXAMPLE 6.8 Show that $y'' = y' + y + x^2, y(1) = 1, y(3) = -1$ has a unique solution. Here $p = 1, q = 1, r = x^2$; the conditions are satisfied.

The general solution to the problem is

$$e^{x/2}[c_1 e^{\sqrt{5}x/2} - c_2 e^{-\sqrt{5}x/2}] - x^2 + 2x - 4.$$

Imposing the boundary conditions allows us to solve for the constants: $c_1 = 0.038154$ and $c_2 = -7.06409$. (DERIVE was of great help here.)

On the other hand, if we change the value of q to -1 so that $y'' = y' - y + x^2$, then $p = 1, q = -1, r = x^2$, and q is not positive as the conditions require. For this equation, the general solution is

$$e^{x/2}\left[c_1 \sin\left(\frac{\sqrt{3}x}{2}\right) - c_2 \cos\left(\frac{\sqrt{3}x}{2}\right)\right] + x^2 + 2x.$$

When we impose the boundary conditions, we find that we cannot solve for the constants! [Still, there are special situations where the solution exists: If $y(0) = 0$ and $y(\pi/3) = 1$, we can solve for the constants: $c_1 = -1.6479, c_2 = 0.$]

The preceding conditions on a linear boundary-value problem must be interpreted to say that, if the conditions are met, a solution exists for any real values for y at the boundaries. When they are violated, the solution may not exist except in special cases.

A Linear Boundary-Value Problem Can Be Solved with Two Trials

It is easy to show this truth, which implies that the true solution (except for truncation and round-off errors) is a linear combination of the two trial solutions.

Suppose that $x_1(t)$ satisfies $x'' + Fx' + Gx = H$, where F, G, and H are functions of t only. Suppose also that $x_2(t)$ is a solution. Then

$$y = \frac{c_1 x_1 + c_2 x_2}{c_1 + c_2}$$

will be a solution. We show that this will always be a solution in the following manner:

Since x_1 and x_2 are solutions,

$$x_1'' + Fx_1' + Gx_1 = H,$$

and

$$x_2'' + Fx_2' + Gx_2 = H.$$

Substituting y into the differential equation, with

$$y' = \frac{c_1 x_1' + c_2 x_2'}{c_1 + c_2}, \qquad y'' = \frac{c_1 x_1'' + c_2 x_2''}{c_1 + c_2},$$

we get

$$\frac{c_1 x_1'' + c_2 x_2''}{c_1 + c_2} + F\frac{c_1 x_1' + c_2 x_2'}{c_1 + c_2} + G\frac{c_1 x_1 + c_2 x_2}{c_1 + c_2}$$

$$= \frac{c_1 x_1'' + c_1 Fx_1' + c_1 Gx_1}{c_1 + c_2} + \frac{c_2 x_2'' + c_2 Fx_2' + c_2 Gx_2}{c_1 + c_2}$$

$$= \frac{c_1 H}{c_1 + c_2} + \frac{c_2 H}{c_1 + c_2} = H.$$

Richardson Extrapolations Work on Solutions by Sets of Equations

We only need to observe that all the substitutions for the derivatives in Sections 6.2 and 6.3 have errors of $O(h^2)$. When we extrapolate the results of two computations with the value of h halved in the second set of equations, we can improve the accuracy of the solutions by extrapolation.

Be sure to observe that round-off errors are not considered in this statement. If h is very small, the accuracy of the derivative approximations is affected by round-off, since we are subtracting quantities in the computation of derivatives. For this reason, the shooting method may have an advantage. Even so, if the computed values of y increase rapidly as x goes from $x = a$ to $x = b$, shooting can also be severely affected by round-off. A remedy in this case may be to solve the problem in the reverse direction, or to solve from both ends to meet at a middle point.

Remark on the Finite-Element Method

Our treatment of methods based on variational principles is skimpy and far from complete. We presented this abbreviated version only to alert you to a very powerful technique that is of great usefulness. A better set of weighting functions than the linear "hat" functions of Section 6.5 would be cubic splines that are continuous in slope and curvature from one subinterval to the next. We used the linear weighting functions to keep the mathematics simple.

Remark on Section 6.6

Again, our treatment of characteristic-value problems does not do full justice to this very important area of applied mathematics. Still, we feel that the section gives you some appreciation for these special types of problems.

The Power Method

The utility of the power method is that it computes the eigenvalue of greatest magnitude (and its corresponding eigenvector) in a simple and straightforward manner. Its disadvantage is that convergence is slow if there is a second eigenvalue whose magnitude is near that of the largest. The truth of this is shown by the following discussion. You will also see that some starting vectors are not suitable—if they are "perpendicular" (dot product is zero) to the eigenvector corresponding to the eigenvalue being sought, the procedure fails in that what is obtained is not the expected eigen-pair but another whose vector is nearly parallel to the starting vector.

The power method works because the eigenvectors are a set of *basis vectors*. By this we mean that they are a set of linearly independent vectors that *span the space*; that is, any n-component vector can be written as a unique linear combination of them. Let $v^{(0)}$ be any vector and $x_1, x_2, \ldots, x_n$ be eigenvectors. Then

$$v^{(0)} = c_1 x_1 + c_2 x_2 + \cdots + c_n x_n.$$

If we multiply $v^{(0)}$ by the matrix A, we have (since the x_i are eigenvectors with corresponding eigenvalues λ_i),

$$v^{(1)} = Av^{(0)} = c_1 A x_1 + c_2 A x_2 + \cdots + c_n A x_n$$
$$= c_1 \lambda_1 x_1 + c_2 \lambda_2 x_2 + \cdots + c_n \lambda_n x_n. \qquad \textbf{(6.61)}$$

Upon repeated multiplication by A we get, after m times,

$$v^{(m)} = A^m v^{(0)} = c_1 \lambda_1^m x_1 + c_2 \lambda_2^m x_2 + \cdots + c_n \lambda_n^m x_n.$$

If one eigenvalue, say λ_1, is larger in magnitude than all the rest, the values of $\lambda_i^m, i \neq 1$, will be negligibly small in comparison to λ_1^m when m is large and

$$A^m v \to c_1 \lambda^m x_1, \quad \text{or} \quad A^m v \to \text{Multiple of eigenvector } x_1,$$

with the normalization factor of λ_1, provided that $c_1 \neq 0$. This is the principle behind the power method.

In applying the power method to find the dominant eigenvalue (the eigenvalue whose magnitude exceeds all others), we normalize at each step; this in effect is just scaling the c's after each multiplication, and will not change the relation. If some eigenvalue other than λ_1 were the dominant one, we would merely renumber the eigenvalues. It is easy to see why the rate of convergence depends strongly on the ratio of the two eigenvalues of largest magnitude.

Obviously, if we are able to choose the starting vector $v^{(0)}$ close to the eigenvector x_1, we will converge more rapidly; the coefficients $c_i, i \neq 1$, in Eq. (6.61) will be small in this case. We usually have no knowledge of this eigenvector, so we start with an arbitrary vector with all components equal to unity. This choice could be a bad one. If the value of c_1 in Eq. (6.61) is zero for this arbitrary choice, the process may never converge. Actually, the process often does converge in spite of this; round-off errors that are incurred as we repeatedly multiply by A produce components in the direction of x_1 so that we do converge, although slowly.

These examples will shed more light on the method.

EXAMPLE 6.9   Use the power method to find the eigen-pairs for

$$
\begin{bmatrix}
4 & -1 & 1 \\
-1 & 1 & 1 \\
1 & 1 & -4.1
\end{bmatrix}.
$$

The power method, with a starting vector of $(1, 1, 1)^T$, converges slowly to the eigenvalue of largest magnitude, -4.45274, requiring 461 iterations to meet a tolerance of 0.0001. This is because there is another eigenvalue of nearly the same magnitude, 4.35820. Using the inverse power method, we easily get the smallest eigenvalue, 0.994508, requiring only seven iterations.

The eigenvector corresponding to the largest eigenvalue is $(-0.143106, -0.209639, 1)^T$. If we start with a vector that is perpendicular $(0, 1, -.209637)^T$, we converge quickly to the next greatest eigenvalue, 4.35820, in nine iterations. This happy result is due to the fact that its corresponding eigenvector is also nearly perpendicular to that for the greatest eigenvalue. (It is $(1, -0.272154, 0.086052)^T$).

EXAMPLE 6.10   The QR method shows that the eigenvalues of the matrix

$$
\begin{bmatrix}
1 & 0 & 2 & 4 \\
2 & 2 & 4 & 3 \\
4 & 0 & 1 & 3 \\
1 & 3 & 2 & 4
\end{bmatrix}
$$

are 9.0903, 0.4587, 0.4588, and -2.0078. Can these be found with the power method?

The regular power method finds 9.0903 in only 10 iterations. Using the inverse power method, we get -2.0078 in 52 iterations. We would expect to get the one of smallest magnitude, but that does not occur because it has a very close companion. Even after shifting by 0.4587 and using the inverse method, we cannot get convergence to the smallest; instead we converge to the negative eigenvalue!

This example shows that the power method is poor when there are two nearly equal eigenvalues.

6.11   Using DERIVE

DERIVE does not have routines to solve boundary-value problems numerically, although it does provide for imposing boundary conditions on the analytical solution of a second-order differential equation. This function, contained in the file ODE2.MTH, is used after a solution to the second-order equation has been solved. Because our emphasis is on numerical methods, we will not discuss this technique.

The Shooting Method

DERIVE can help with the shooting method through the use of the TAY_ODES or RK routines, which are found in the file ODE_APPR.MTH. As an example, suppose we use DERIVE to assist in solving the linear second-order equation of Section 6.1. The equation is

$$\frac{d^2x}{dt^2} = t + \left(1 - \frac{t}{5}\right)x, \qquad x(1) = 2, \qquad x(3) = -1.$$

We first transform to a pair of first-order equations:

$$x' = y, \qquad\qquad x(1) = 2,$$
$$y' = t + \left(1 - \frac{t}{5}\right)x, \qquad y(1) = \,?$$

in which we will guess at the required value of $y(1)$ to give the required $x(3) = -1$. Let us start with the guess of $y(1) = -1.5$. We solve the system in DERIVE with

$$\text{RK}\!\left(\left[y, t + \left(1 - \frac{t}{5}\right)x\right], [t, x, y], [1, 2, -1.5], 0.2, 10\right),$$

which, when approximated, gives the display of Fig. 6.7(a). (These values are not identical to those in Section 6.1 because we are now using a more accurate solution method.)

We now repeat the solution, except that this time we will use a starting value of

−3 for y. The display shown in Fig. 6.7(b) results, whose values are again more accurate than those of the second trial of Section 6.1. We interpolate between the y_0-values of −1.5 and −3 [which gave $x(3)$-values of 4.7859 and 0.435991] to arrive at an improved starting y-value of −3.494987. When we use this value in a third trial, we find that $x(2)$ is exactly 1, the value we have hoped for.

We can do the same for a nonlinear equation except that now we will have to make more trials to arrive at the correct answer.

1	2	−1.5
1.2	1.75137	−0.988559
1.4	1.60429	−0.481395
1.6	1.55969	0.0388976
1.8	1.62175	0.587622
2	1.79754	1.17827
2.2	2.09665	1.82269
2.4	2.53088	2.53101
2.6	3.11387	3.31155
2.8	3.86074	4.17057
3	4.78759	5.11186

(a)

1	2	−3
1.2	1.44981	−2.51181
1.4	0.992070	−2.07190
1.6	0.619209	−1.65982
1.8	0.327462	−1.25803
2	0.116335	−0.851221
2.2	−0.0117883	−0.425896
2.4	−0.0519908	0.0298642
2.6	0.00288748	0.526542
2.8	0.161959	1.07317
3	0.435991	1.67726

(b)

Figure 6.7
Results for the shooting method example for DERIVE

Solution Through a Set of Equations

DERIVE can help here by solving the equations that represent the boundary-value problem. DERIVE provides no easy way to get the equations themselves; further, in the student version of DERIVE, the system is limited in size. Still, the program is an excellent alternative to using an ordinary subroutine. The same is true for the set of equations that result from applying the Rayleigh–Ritz procedure.

Eigenvalues and Eigenvectors

DERIVE can get the characteristic polynomial of a matrix through the built-in function CHARPOLY which then can be solved for the eigenvalues, or more directly through the built-in function EIGENVALUES. The later is easier to use. (Actually, this latter function just gets the roots of the characteristic polynomial.)

We now illustrate with some examples using DERIVE on eigenvalues.

EXAMPLE 6.11 What are the eigenvalues of this matrix?

$$\begin{bmatrix} 10 & 0 & 0 \\ 1 & -3 & -7 \\ 0 & 2 & 6 \end{bmatrix}$$

Here is the statement we enter into DERIVE:

EIGENVALUES ([[10,0,0],[1,-3,-7],[0,2,6]],z)

On simplifying, we get the response

$$[z = -1 \quad z = 4 \quad z = 10]$$

which are the correct values. (The last parameter of the EIGENVALUES command is the variable. If omitted, a default value of w is used.)

▲

Some aspects of this function are peculiar. For example, if we change the element in row 1, column 2, to 1, we do not get the eigenvalues directly.

EXAMPLE 6.12   What results from

EIGENVALUES([[10,1,0],[1,-3,-7],[0,2,6]])?

We get the characteristic polynomial! The display is

$$[-w^3 + 13w^2 - 25w - 46 = 0]$$

which we can solve (through approximations) to get the roots

$$w = -1.12496, \quad w = 4.06439, \quad w = 10.0605.$$

▲

The output of the polynomial appears to be related to whether the roots of the polynomial are integers, because inputting

$$\text{EIGENVALUES} \begin{bmatrix} 7 & 8 & 6 & 6 \\ 1 & 6 & -1 & -2 \\ 1 & -2 & 5 & -2 \\ 3 & 4 & 3 & 4 \end{bmatrix}$$

gives $[w = 1, w = 4, w = 7, w = 10]$. Conversely, changing one element so that the eigenvalues are not integers gives again the characteristic polynomial.

If we want the eigenvectors, the VECTOR.MTH utility file has two functions that will obtain them. EXACT_EIGENVECTOR (A,w) gives the eigenvector corresponding to the eigenvalue w, but w must be exact. The function APPROX_EIGENVECTOR (A,w) gives the eigenvector corresponding to eigenvalue w, but now w must *not* be an exact value! (This last function uses an inverse iteration method to get the eigenvector. Apparently the first solves a set of equations.) Here are the results from the preceding examples.

EXAMPLE 6.13   What results from

EXACT_EIGENVECTOR ([[10, 0, 0],[1, −3, −7],[0, 2, 6]], 4)?

We get

$$[x1=0 \quad x2=@1 \quad x3=@-1]$$

which is the correct vector corresponding to $\lambda = 4$. (The @ before a value indicates that it is a value only within a multiplicative constant.)

What results from

APPROX_EIGENVECTOR ([[10, 0, 0],[1, −3, −7],[0, 2, 6]], 3.9)?

We get

$$[-0.00589707, \quad 0.705909, \quad -0.708275]$$

in which the values are normalized to unit magnitude.

If we change the approximate value to 3.9999, we get a better result:

$$[-5.89257 \ 10^{-5}, \quad 0.707094, \quad -0.707118]$$

Chapter Summary

Can you do all of the following? You can if you really understand Chapter 6.

1. Solve boundary-value problems either using the shooting method or through a set of equations.

2. Explain why no more than two trials are necessary to solve a linear boundary-value problem by the shooting method.

3. Apply the Rayleigh–Ritz, collocation, and Galerkin's methods to get the solution to a boundary-value problem.

4. Describe to an associate the finite-element procedure for a boundary-value problem, and outline the steps used in setting up the equations that are to be solved.

5. Understand what is meant by a characteristic-value problem, and solve simple examples.

6. Find the largest and smallest eigenvalues of a matrix by the power method. You know why shifting works and why it accelerates the convergence to an eigenvalue.

7. Know what a similarity transformation is, how it can be used to convert a matrix to upper Hessenberg, and why it is advantageous in the QR method.

8. Outline the argument that relates the convergence rate of iterative solutions to a set of equations to the eigenvalues of a matrix.

9. Use a computer program (perhaps even write your own) to solve typical boundary-value problems.

Computer Programs

We present two programs to illustrate how some algorithms of this chapter can be implemented. The first is a program written in Pascal that solves a boundary-value problem by the shooting method. The second program, written in FORTRAN, finds the principal eigenvalue and its eigenvector by the power method.

Program 6.1 The Shooting Method

PROGRAM *boundary* solves a second-order boundary-value problem, as listed in Fig. 6.8. The illustrative problem contained in the program is nonlinear, so iteration is required to solve it. A driver program that makes calls to the correct sequence of procedures is at the end of the figure. The logic is:

1. Initialize a number of variables including two guesses at the initial slope.
2. Solve the equations with these guesses using the Runge–Kutta–Fehlberg method, writing out the results from each solution.
3. Interpolate from these results to find an improved value for the initial slopes.
4. Repeat this process, writing out the results each time, until a tolerance value for agreement with the specified final function value is obtained (or until a limit to the number of iterations is reached).

A message is written when the process terminates to indicate whether it was successful.

Embedded in this program is an RKF procedure that adjusts step sizes to maintain a desired level of accuracy.

```
PROGRAM boundary;
(*
                              Chapter 6

     Gerald/Wheatley, APPLIED NUMERICAL ANALYSIS(fifth edition)
                       Addison-Wesley, 1994

     ----------------------------------------------------------------
     This program solves a nonlinear second-order BOUNDARY VALUE
     problem by the SHOOTING method. It employs the Runge-Kutta-
     Fehlberg Method to solve the equations with two assumed values
     at initial time.

     ----------------------------------------------------------------
     The equation is: X'' + 3X X' + X = T; X(1) = 2; X(3) = 7.
     It is rewritten as two first order equations:

         x1' = x2                              x1(1) = 2
         x2' = t - x1 - 3x1*x2                 x2(2) = ?

     where x1 = x   and x2 = x'
```

Figure 6.8   Program 6.1

Figure 6.8 *(continued)*

```
-------------------------------------------------------------
    The program interpolates from the initial results to find a
    better estimate of the unknown value of x'(1) = x2(1).
    Iterations are continued until successive values of x(3) = 7
    within a specified tolerance. In the program, variables x1 &
    x2 are the first and second components of the vector x0.
*)
CONST
    maxN = 10;          (* Maximum number of first order equations *)
    hSTART = 0.1;       (* beginning stepsize at beginning of interval *)
    tol = 0.0001;       (* tolerance needed for RKF procedure *)
    Dtol = 0.001;       (* stopping criterion for shooting method *)
    MaxCount = 10;      (* limit on changes of stepsize in RKF *)
    tSTART = 1.0;
    tFINAL = 3.0;
    xSTART = 2.0;
    d = 7.0;            (* the desired value for x at time t = 3 *)

TYPE
    vector = Array[1..maxN] OF Real;
    matrix = Array[1..6, 1..maxN] OF Real;
VAR
    x0,          (* state vector at beginning of interval *)
    xend,        (* state vector at end of the interval    *)



    f            (* array that holds the values of the derivatives *)
        : vector;
    h,           (* adjustable stepsize *)
    t0,          (* initial time value at beginning of interval *)
    g1, g2,      (* first TWO guesses at initial slope *)
    r1, r2       (* corresponding results from the two guesses *)
        : Real;

    i, iter,
    numberOFequations,
    count,
    steps                      : Integer;

(*
    INITIALIZE VARIABLES
*)
PROCEDURE initialize_variables;
BEGIN
    h := hSTART;   iter := 1;
    t0 := tSTART;
    g1 := -1.0;   g2 := 1.0;
    numberOFequations := 2;
    x0[1] := xSTART;
    x0[2] := g1;
    steps := ROUND( (tFINAL - tSTART) /hSTART )
END;         (* initialize_variables *)
(*
    DEFINE THE TWO FIRST ORDER DIFFERENTIAL EQUATIONS
*)
PROCEDURE getDERIVS(x0 : vector; VAR f : vector; t0 : Real);
BEGIN
    f[1] := x0[2];
    f[2] := t0 - x0[1] - 3.0*x0[1]*x0[2]
END;          (* get derivs *)
(*
                MAIN PROCEDURE
*)
PROCEDURE rkf(t0, h : Real;
              VAR x0, xend : vector);
```

Figure 6.8 *(continued)*

```
VAR
   workarray   (* a two-dimensional array to store the k values *)
            : matrix;
   i,row,col,
   count       : Integer;
   tEND, error    (* tEND is t[i+1] = t[i] + hSTART *)
            : Real;
(*
   -----------------------------------------------------------------
   This function returns the Larger of two real numbers.
*)
FUNCTION TheLargerOf(a,b : Real): Real;
BEGIN
   IF a > b THEN TheLargerOf := a
   ELSE TheLargerOf := b
END;                        (* TheLargerOf *)
(*
   This procedure computes the estimated error from the new
   ki's (workarray) in getting the next state vector.
*)
PROCEDURE computeERROR;
VAR
   temp : Real;
BEGIN
   error := 0.0;
   FOR i := 1 TO   numberOFequations DO BEGIN
      temp := ABS(workarray[1,i]/360.0
               - 128.0*workarray[3,i]/4275.0
               - 2197.0*workarray[4,i]/75240.0
               + workarray[5,i]/50.0
               + 2.0*workarray[6,i]/55.0);
      error := TheLargerOf(error,temp)   END
END;                                    (* ComputeError *)
(*
   This procedure updates the state vector from
      time[i] to time[i+1].
*)
PROCEDURE ComputeNewX;
VAR
   i  : Integer;
BEGIN
   FOR i := 1 TO numberOFequations DO
      xend[i] := x0[i] + 16.0*workarray[1,i]/135.0
               + 6656.0*workarray[3,i]/12825.0
               + 28561.0*workarray[4,i]/56430.0
               - 9.0*workarray[5,i]/50.0
               + 2.0*workarray[6,i]/55.0;
END;                                    (* ComputeNewX *)

BEGIN     (* rkf *)
   tEND := t0 + hSTART;
   count := 0;
   WHILE (t0 < tEND) AND (count < MaxCount) DO
BEGIN
      count := count + 1;

(*
   Get first estimate of the delta X's
*)
   getDERIVS(x0,f,t0);
   FOR i := 1 TO numberOFequations DO      BEGIN
      workarray[1,i] := h*f[i];
      xend[i] := x0[i] + workarray[1,i]/4.0  END;
(*
   Get second estimate. The xend vector holds the x values.
*)
```

Figure 6.8 *(continued)*

```
   getDERIVS(xend,f,t0+h/4.0);
   FOR i := 1 TO numberOFequations DO                BEGIN
      workarray[2,i] := h*f[i];
      xend[i] := x0[i] + (workarray[1,i]*3.0
                        + workarray[2,i]*9.0)/32.0    END;
(*
   Repeat for third estimate
*)

   getDERIVS(xend,f, t0+3.0*h/8.0);
   FOR i := 1 TO numberOFequations DO                BEGIN
      workarray[3,i] := h*f[i];
      xend[i] := x0[i] + (workarray[1,i]*1932.0
                        - workarray[2,i]*7200
                        + workarray[3,i]*7296)/2197.0 END;
(*
   Fourth estimates, i.e. k4's
*)
   getDERIVS(xend,f, t0+12.0*h/13.0);
   FOR i := 1 TO numberOFequations DO                BEGIN
      workarray[4,i] := h*f[i];
      xend[i] := x0[i] + 439.0*workarray[1,i]/216.0
                       - 8.0*workarray[2,i]
                       + 3680.0*workarray[3,i]/513.0
                       - 845.0*workarray[4,i]/4104.0 END;
(*
   Fifth estimates
*)
   getDERIVS(xend,f,t0+h);
   FOR i := 1 TO numberOFequations DO                BEGIN
      workarray[5,i] := h*f[i];
      xend[i] := x0[i]  - 8.0*workarray[1,i]/27.0
                       + 2.0*workarray[2,i]
                       - 3544.0*workarray[3,i]/2565.0
                       + 1859.0*workarray[4,i]/4104.0
                       - 11.0*workarray[5,i]/40.0     END;
(*
   Sixth estimates
*)
   getDERIVS(xend,f,t0+h/2.0);
   FOR i := 1 TO numberOFequations do  workarray[6,i] := h*f[i];
(*
   Compute error estimate
*)

   error := 0.0;
   computeERROR;

   IF (error < tol) THEN                     BEGIN
      t0 := t0 + h;
      ComputeNewX;
      FOR i := 1 TO numberOFequations DO
         x0[i] := xend[i]            END;

   IF (error>tol) THEN  h := h/2.0;
   IF (error < h*tol/10.0) THEN  h := 2.0*h;
   IF (t0 + h > tEND) THEN h := tEND - t0
                                                      END;
                                      (* while *)
END;                                  (* rkf PROCEDURE *)
(*
   PRODUCE THE FIRST RESULT:  r1
*)
PROCEDURE GET_r1;
VAR
   i : Integer;
BEGIN
```

Figure 6.8 *(continued)*

```
            FOR i := 1 TO steps DO    BEGIN
                rkf(t0, h, x0, xend);
                t0 := t0 + hSTART      END;
        r1 := x0[1]
END;      (* GET_r1 *)
(*
    PRODUCE THE SECOND RESULT: r2
*)
PROCEDURE GET_r2;
VAR
    i : Integer;
BEGIN
    t0 := tSTART;   x0[1] := xSTART;   x0[2] := g2;
    FOR i := 1 TO steps DO        BEGIN
        rkf(t0, h, x0, xend);
        t0 := t0 + hSTART      END;
    r2 := x0[1]
END;         (* GET_r2 *)
(*
    PRODUCE A NEW GUESS VALUE BY INTERPOLATION. THEN USE
    RKF TO COMPUTE THE NEW RESULT.
*)
PROCEDURE GET_newG_and_newR;
VAR
    i : Integer;
BEGIN
    t0 := tSTART;   x0[1] := xSTART;

    x0[2] := g1 + (g2 - g1)/(r2 - r1)*(d - r1);

    g1 := g2;   g2 := x0[2];   r1 := r2;
    FOR i := 1 TO steps DO        BEGIN
        rkf(t0, h, x0, xend);
        t0 := t0 + hSTART        END;
    r2 := x0[1]
END;    (* GET_newG_and_newR *)
(*
    IF THE SEQUENCE OF R2'S DO NOT CONVERGE TO  D WE PRINT OUT
    THE MESSAGE AND THE FINAL VALUES FOR R2 AND G2
*)
PROCEDURE printERRORmessage;
BEGIN
    WriteLn;
    WriteLn(' TOLERANCE NOT MET IN 20 ITERATIONS. ',
                    'FINAL VALUES WERE: R2 = ', r2:7:4,
                    ' WITH INITIAL SLOPE OF G2 = ', g2:7:4);
    WriteLn; WriteLn
END;      (* printERRORmessage *)

(*
    PRINT THE SOLUTION, R2 AND G2, IF THERE IS CONVERGENCE
    BEFORE ITER EQUALS 20.
*)
PROCEDURE printSOLUTIONmessage;
BEGIN
    WriteLn; WriteLn;
    WriteLn(' TOLERANCE CONDITION MET IN ', iter:2,
                ' ITERATIONS',
                    ' FINAL VALUE AT END OF INTERVAL IS: ');
    WriteLn;
    WriteLn('':20, r2:7:4, ' USING AN INITIAL SLOPE VALUE ',
                    ' OF ', g2:7:4);
```

Figure 6.8   *(continued)*

```
    WriteLn; WriteLn
END;        (* printSOLUTIONmessage *)
(*
   RECOMPUTE AND WRITE OUT THE LAST SET OF COMPUTATIONS
*)
PROCEDURE printINITIALvalueSOLUTION;
VAR
   i : Integer;
BEGIN
   t0 := tSTART; x0[1] := xSTART;  x0[2] := g2;

   WriteLn;
   WriteLn(' LAST COMPUTATIONS WERE ');
   WriteLn; WriteLn;
   WriteLn('':6,'TIME','':5,'X1 VALUE', '':5, 'X2 VALUE');
   WriteLn;
   WriteLn('':7, t0:3:1, '':6, x0[1]:7:4, '':6, x0[2]:7:4);

   FOR i := 1 TO steps DO                              BEGIN
      rkf(t0, h, x0, xend);
      t0 := t0 + hSTART;
      WriteLn('':7, t0:3:1, '':6, x0[1]:7:4, '':6, x0[2]:7:4) END
END; (* printINITIALvalueSOLUTION *)

BEGIN
   initialize_VARIABLES;
   GET_r1;  GET_r2;
   WriteLn(g1:9:5, g2:9:5, r1:11:5, r2:9:5);
   iter := 0;

WHILE (iter <= 20) AND ( ABS(r2 - d) > Dtol)  DO     BEGIN
   GET_newG_and_newR;
   WriteLn(g1:9:5, g2:9:5, r1:11:5, r2:9:5);
   iter := iter + 1                                   END;

WriteLn;
IF (ABS(r2 - d) < Dtol) THEN      BEGIN
   printSOLUTIONmessage;
   printINITIALvalueSOLUTION      END
ELSE printERRORmessage;

END.


*****************************************************************


          OUTPUT FOR BNDARY.PAS

      G1         G2         R1        R2


  -1.00000   1.00000    1.86228   2.11531
   1.00000  39.60961    2.11531   5.12416
  39.60961  63.68034    5.12416   6.38546
  63.68034  75.40842    6.38546   6.92072
  75.40842  77.14556    6.92072   6.99681
  77.14556  77.21832    6.99681   6.99998
```

Figure 6.8 *(continued)*

```
      TOLERANCE CONDITION MET IN  5 ITERATIONS. FINAL VALUE AT END OF
      INTERVAL IS:


       7.0000    USING AN INITIAL SLOPE VALUE OF    77.2183


      LAST COMPUTATIONS WERE

          TIME      X1 VALUE      X2 VALUE

          1.0        2.0000       77.2183
          1.1        6.3170       23.0528
          1.2        7.2522        3.5130
          1.3        7.3754        0.2760
          1.4        7.3702       -0.2117
          1.5        7.3457       -0.2608
          1.6        7.3194       -0.2619
          1.7        7.2934       -0.2578
          1.8        7.2679       -0.2530
          1.9        7.2428       -0.2481
          2.0        7.2183       -0.2432
          2.1        7.1942       -0.2383
          2.2        7.1706       -0.2334
          2.3        7.1475       -0.2284
          2.4        7.1249       -0.2234
          2.5        7.1028       -0.2184
          2.6        7.0812       -0.2133
          2.7        7.0602       -0.2083
          2.8        7.0396       -0.2032
          2.9        7.0195       -0.1981
          3.0        7.0000       -0.1929
```

Program 6.2 The Power Method

Figure 6.9 lists the second program of Chapter 6. This FORTRAN program calls a subroutine that does the power method. The driver program sends these parameters to the subroutine:

A	The matrix whose principal eigen-pair is desired.
X	An initial vector that starts the power method.
TOL	A tolerance value to terminate iterations. This is the difference between successive normalization factors.
NLIMIT	The maximum number of iterations to be used.
NDIM	The first dimension of A (actually A is square).
XWRK	A vector used to store intermediate values.

The logic is simple: The product of A and the initial X is computed and then normalized, and the normalized vector is repeatedly multiplied with A and normalized. If the process converges, successive normalization factors become the eigenvalue and the normalized vector the eigenvector.

Successive iterates produced by the program are seen at the end of Fig. 6.9.

```
          PROGRAM PPOWER
C
C    -------------------------------------------------------
C                      CHAPTER 6
C
C    GERALD/WHEATLEY, APPLIED NUMERICAL ANALYSIS(Fifth edition)
C                     ADDISON-WESLEY, 1994
C
C    -----------------------------------------------------------
C
C    THE SUBROUTINE POWER THAT IS CALLED HERE IMPLEMENTS THE BASIC
C    ITERATIVE METHOD FOR FINDING THE LARGEST EIGENVALUE (IN
C    MAGNITUDE) AND ITS CORRESPONDING EIGENVALUE.
C
          REAL A(10,10), X(10), TOL, C
          INTEGER N, NDIM, NLIMIT
C
          DATA A/7.0, 2*1.0,3.0,6*0.0, 8.0, 6.0, -2.0, 4.0, 6*0,
        +       6.0, -1.0, 5.0, 3.0, 6*0.0, 6.0, 2*-2.0, 4.0, 6*0.0,
        +       60*0.0/
          DATA NDIM, N, NLIMIT/10, 4, 15/
          DATA X,TOL/1.0, 9*0.0, 0.001/
C
          CALL PRINTA(A, NDIM, N)
C
          CALL POWER(A,X,C,TOL,NLIMIT,N,NDIM)
C
          END
C
C    ------------------------------------------------------------
C    THIS SUBROUTINE PRINTS OUT THE INPUT N by N MATRIX A
C
C    ------------------------------------------------------------
C
          SUBROUTINE PRINTA(A, NDIM, N)
C
          INTEGER N, NDIM, I, J
          REAL A(NDIM, N)
C
          PRINT 100
          DO 10 I = 1, N
             PRINT 200, (A(I,J), J = 1,N)
10           CONTINUE
          PRINT '(//)'
100       FORMAT(//, '     INPUT MATRIX IS'/)
200       FORMAT(10F7.2)
          RETURN
          END
C    ------------------------------------------------------------
C    THIS SUBROUTINE PRINTS OUT SCALAR VALUE AND THE
C    NORMALIZED VECTOR.
C
C    ------------------------------------------------------------
          SUBROUTINE PRINTVC(X, C, NDIM, N)
C
          INTEGER N, NDIM, I
          REAL X(NDIM), C
C

          PRINT 100, (X(I), I = 1,N)
          PRINT 200, C
100       FORMAT('    THE NORMALIZED VECTOR IS:  ', 10F7.4)
200       FORMAT('    THE VALUE FOR C IS: ', F10.6/)
          RETURN
          END
```

Figure 6.9   Program 6.2

Figure 6.9   *(continued)*

```
C
C     ------------------------------------------------------------
C         SUBROUTINE POWER :
C         THIS SUBROUTINE COMPUTES THE LARGEST EIGENVALUE
C         (in MAGNITUDE)AND ITS CORRESPONDING EIGENVECTOR BY THE
C         POWER METHOD.
C     ------------------------------------------------------------
C
          SUBROUTINE POWER(A,X,C,TOL,NLIMIT,N,NDIM)
C
C         PARAMETERS ARE :
C
C         A       - AN N X N MATRIX WHOSE EIGENS ARE BEING DETERMINED
C         X       - ESTIMATE FOR THE EIGENVECTOR USED TO BEGIN
C                   ITERATIONS. IF NO APPROXIMATION TO THIS IS KNOWN,
C                   THE USUAL CHOICE IS A VECTOR WITH ALL COMPONENTS
C                   EQUAL TO UNITY. X ALSO RETURNS THE FINAL
C                   EIGENVECTOR TO THE CALLER.
C         C       - RETURNS THE VALUE OF THE EIGENVALUE
C         TOL     - TOLERANCE VALUE USED TO DETERMINE CONVERGENCE.
C                   ITERATIONS CONTINUE UNTIL SUCCESSIVE ESTIMATES OF
C                   THE EIGENVALUEARE THE SAME WITHIN TOL IN VALUE.
C         NLIMIT  - LIMIT TO THE NUMBER OF ITERATIONS IF METHOD DOES
C                            NOT CONVERGE.
C         N       - SIZE OF THE MATRIX AND THE VECTOR X
C         NDIM    - FIRST DIMENSION OF MATRIX A IN THE CALLING PROGRAM
C         XWRK    - VECTOR USED TO STORE INTERMEDIATE VALUES. MUST BE
C                   DIMENSIONED TO HOLD AT LEAST N ELEMENTS IN THE
C                   MAIN PROGRAM.
C
          INTEGER N,NDIM,NLIMIT,IROW,ITER,JCOL
          REAL A(NDIM,N),X(N),XWRK(10),C,TOL,SAVE
          LOGICAL CONVERGENT
C
C     ------------------------------------------------------------
C BEGIN THE ITERATIONS. GET THE PRODUCT OF A AND X, THEN
C NORMALIZE. THE NORMALIZATION FACTORS SHOULD CONVERGE TO THE
C EIGENVALUE ON REPEATED MULTIPLICATIONS. WE STORE THE CURRENT
C VALUE OF C FOR COMPARISON WITH THE NEXT VALUE TO TEST
C CONVERGENCE. TO  BEGIN, WE MAKE SAVE = 0.
C
          SAVE = 0.0
          DO 50 ITER = 1,NLIMIT
            DO 20 IROW = 1,N
              XWRK(IROW) = 0.0
              DO 10 JCOL = 1,N
                XWRK(IROW) = XWRK(IROW) + A(IROW,JCOL)*X(JCOL)
10          CONTINUE
20    CONTINUE
C
C
C     ------------------------------------------------------------
C
C FIND THE LARGEST ELEMENT OF THE PRODUCT VECTOR FOR NORMALIZING
C
          C = 0.0
          DO 30 IROW = 1,N
            IF ( ABS(C) .LT. ABS(XWRK(IROW)) )  C = XWRK(IROW)
30        CONTINUE
C
C     ------------------------------------------------------------
C
C NOW NORMALIZE THE PRODUCT VECTOR AND PUT INTO
C X FOR NEXT ITERATION.
C
```

Figure 6.9 *(continued)*

```
           DO 40 IROW = 1,N
               X(IROW) = XWRK(IROW) / C

40         CONTINUE
C
           CALL PRINTVC(X, C, NDIM, N)
C
C
C  SEE IF TOLERANCE IS MET.
C
           CONVERGENT = ( ABS(C - SAVE) .LE. TOL )
           SAVE = C
50         CONTINUE
C
C  ------------------------------------------------------------
C
C
C  IF WE DO NOT MEET THE TOLERANCE FOR CONVERGENCE, PRINT A
C  MESSAGE AND RETURN LAST VALUES CALCULATED.
C
       IF  ( .NOT. CONVERGENT)  THEN
           PRINT 200, NLIMIT
200        FORMAT(/' CONVERGENCE NOT REACHED IN ',I5,'ITERATIONS')
           RETURN
           ENDIF
       END


    *************************************************************

                      OUTPUT FOR PPOWER.F

                      INPUT MATRIX IS

              7.00    8.00    6.00    6.00
              1.00    6.00   -1.00   -2.00
              1.00   -2.00    5.00   -2.00
              3.00    4.00    3.00    4.00


THE NORMALIZED VECTOR IS:   1.0000   .1429   .1429   .4286
THE VALUE FOR C IS:   7.000000

THE NORMALIZED VECTOR IS:   1.0000   .0741   .0494   .4938
THE VALUE FOR C IS:   11.571428

THE NORMALIZED VECTOR IS:   1.0000   .0375   .0102   .4994
THE VALUE FOR C IS:   10.851852

THE NORMALIZED VECTOR IS:   1.0000   .0209  -.0022   .4999
THE VALUE FOR C IS:   10.358361

THE NORMALIZED VECTOR IS:   1.0000   .0126  -.0057   .5000
THE VALUE FOR C IS:   10.153433

THE NORMALIZED VECTOR IS:   1.0000   .0080  -.0051   .5000
THE VALUE FOR C IS:   10.069348

THE NORMALIZED VECTOR IS:   1.0000   .0053  -.0041   .5000
THE VALUE FOR C IS:   10.033591

THE NORMALIZED VECTOR IS:   1.0000   .0036  -.0031   .5000
THE VALUE FOR C IS:   10.017592

THE NORMALIZED VECTOR IS:   1.0000   .0025  -.0023   .5000
THE VALUE FOR C IS:   10.009959
```

Figure 6.9 *(continued)*

```
THE NORMALIZED VECTOR IS:    1.0000   .0017 -.0016   .5000
THE VALUE FOR C IS:   10.006031

THE NORMALIZED VECTOR IS:    1.0000   .0012 -.0012   .5000
THE VALUE FOR C IS:   10.003847

THE NORMALIZED VECTOR IS:    1.0000   .0008 -.0008   .5000
THE VALUE FOR C IS:   10.002542

THE NORMALIZED VECTOR IS:    1.0000   .0006 -.0006   .5000
THE VALUE FOR C IS:   10.001719

THE NORMALIZED VECTOR IS:    1.0000   .0004 -.0004   .5000
THE VALUE FOR C IS:   10.001180

THE NORMALIZED VECTOR IS:    1.0000   .0003 -.0003   .5000
THE VALUE FOR C IS:   10.000816
```

Exercises

Section 6.1

▶**1.** Solve the boundary-value problem

$$y'' = xy' + 3y = 11x, \qquad y(1) = 1.5, \qquad y(2) = 15$$

by the shooting method. Assume two values for $y'(1)$ (which is near 5). Use $h = 0.25$, and compare results with Runge–Kutta–Fehlberg and the modified Euler methods. Why are the results different? Can you get essentially the same results with the modified Euler method as with Runge–Kutta–Fehlberg? Find the analytical solution, and compare to the computed values of $y(x)$.

2. In Exercise 1, you obtained the solution by using an initial slope interpolated from the first two trials and then repeating the integration method. Show that the values of $y(x)$ can be computed without using an integration method, by interpolating from the values obtained in the trials.

▶**3.** In Exercise 1, the exact analytical value for $y'(1)$ is 5.5000. You probably found a different value with $h = 0.25$ and the modified Euler method. Explain the discrepancy. Can you match the analytical solution using the modified Euler method and a smaller value for h? If not, explain.

4. The shooting method can work backward as well as forward through the interval. Solve Exercise 1 by working backward from $x = 2$ (h is then negative). Do you get the same results?

5. Use the shooting method to solve

$$y'' + yy' = e^{x/2}, \qquad y(-1) = 2, \qquad y(1) = 3.$$

Use either the computer program of Chapter 5 or another one that you have written to solve with assumed values for the initial slope. Why is this more difficult to solve than Exercise 1?

6. Exercise 5 solves a nonlinear boundary-value problem. The solution will not be exact because there are errors in the numerical procedure, depending on the step size. If the step size is reduced, the solution should be more accurate. Vary the step size in Exercise 5 until you believe the value of $y(0)$ is correct to within $5 * 10^{-5}$. Justify your belief that you have attained this accuracy.

7. Use the shooting method to solve

$$\frac{d^3y}{dt^3} + 10\frac{dy}{dt} - 5y^3 = t^3 - ty, \qquad y(0) = 0,$$
$$y(2) = -1, \qquad y''(2) = 0.$$

Section 6.2

8. Given the boundary-value problem

$$\frac{d^2y}{d\theta^2} + \frac{y}{4} = 0, \qquad y(0) = 0, \qquad y(\pi) = 2,$$

a. normalize to the interval $[0, 1]$ by a change of variable.

b. solve the normalized problem by the finite-difference method, replacing the derivative with an

$O(h^2)$ approximation. Use $h = 0.25$ to set up the equations.

c. solve the original equation (without normalizing). Take $h = \pi/4$.

d. compare your solutions to the analytical solution, $y = 2 \sin(\theta/2)$.

e. Repeat part (b) with $h = 0.5$, and extrapolate the value at the midpoint. How much more accurate is it?

f. How small a value of h must be used to give a solution that agrees with the analytical values at each computed point to within four decimal places?

▶**9.** Solve the boundary-value problem by finite differences, taking $h = 0.2$:

$$y'' + xy' - x^2 y = 2x^3, \qquad y(0) = 1, \qquad y(1) = -1.$$

10. Solve the equation in Exercise 1 by the finite-difference method. What value of h is required to match the analytical solution to four significant figures?

11. Solve the equation in Exercise 5 with finite-difference approximations. Use $h = \frac{1}{3}$. You will need to solve a set of nonlinear equations.

12. Use finite differences to solve Exercise 7. Compare your answers when using various step sizes with those of Exercise 7.

Section 6.3

13. Repeat Exercise 8, except with these derivative boundary conditions:

$$y'(0) = 0, \qquad y'(\pi) = 1.$$

In part (d), compare to $y = -2 \cos(\theta/2)$.

14. Suppose the right boundary in Exercise 13 were $y'(2\pi) = 0$. An analytical solution is still $y = -2 \cos(\theta/2)$. Can you find this solution from a set of difference equations?

▶**15.** Solve through finite differences with four subintervals:

$$\frac{d^2 y}{dx^2} + y = 0, \qquad y'(0) + y(0) = 2,$$

$$y'\left(\frac{\pi}{2}\right) + y\left(\frac{\pi}{2}\right) = -1.$$

16. Write a computer program to set up the equations for Exercise 10, but this time for end conditions of

$$y'(0) + y(0) = A, \qquad y'\left(\frac{\pi}{2}\right) + y\left(\frac{\pi}{2}\right) = B.$$

17. The most general form of boundary conditions normally encountered for second-order equations is a linear combination of the function and its derivative at both ends of the interval. Set up equations, with $h = 0.2$, to solve

$$y'' - xy' + x^2 y = x^3, \qquad \text{subject to}$$

$$y(0) + y'(0) + y(1) + y'(1) = 4,$$

$$y(0) - y'(0) + y(1) - y'(1) = 3.$$

18. Write a computer program to set up the equations for Exercise 17 for $h = 1/n$, where n is an integer.

19. Solve the third-order boundary-value problem with $h = 0.2$:

$$y''' - y' = e^x, \quad y(0) = 0, \quad y(1) = 1, \quad y'(1) = 0.$$

Approximate the third derivative in terms of an average of third differences:

$$y_0''' = \frac{y_2 - 2y_1 + 2y_{-1} - y_{-2}}{2h^3} + O(h^2)$$

at $x = 0.4, 0.6$, and 0.8. Using the derivative condition at $x = 1$ will eliminate the assumed function value at $x = 1.2$, but we are still short one equation. Obtain this by writing the equation at $x = 0.2$ using an unsymmetrical approximation for y''':

$$y_0''' = \frac{-y_3 + 6y_2 - 12y_1 + 10y_0 - 3y_{-1}}{2h^3} + O(h^2).$$

▶**20.** Solve, through finite differences, this fourth-order problem:

$$y^{iv} - y''' + y = x^3, \qquad y(0) = 2, \qquad y'(0) = -1,$$
$$y(1) = -2, \qquad y'(1) = 1.$$

Use unsymmetrical expressions for the derivatives and take $h = 0.2$.

21. Solve the differential equation in Exercise 20, subject to the conditions

$$y(0) = y'(0) = y''(0) = 0, \qquad y(1) = 2.$$

Unsymmetrical expressions may not be required!

22. Solve the nonlinear problem

$$y'' = 2 - \frac{4y^2}{\sin^2(x)}, \qquad y'(1) = 0.9093,$$
$$y(2) = 0.8268,$$

by a set of difference equations. Compare to the analytical solution, $y = \sin^2(x)$.

Section 6.4

23. Show that the integrand of Eq. (6.12) is equivalent to Eq. (6.11) if the Euler–Lagrange condition is used. This means that Eq. (6.12) is the functional for any second-order boundary-value problem of the form

$$y'' + Q(x)y = F(x),$$

subject to Dirichlet boundary conditions

$$y(a) = A, \qquad y(b) = B,$$

where A and B are constants.

▶**24.** Use the Rayleigh–Ritz method to approximate the solution of

$$y'' = 3x + 1, \qquad y(0) = 0, \qquad y(1) = 0,$$

using a quadratic in x as the approximating function. Compare to the analytical solution by graphing the approximation and the analytical solution.

25. Repeat Exercise 24, but this time, for the approximating function, use

$$ax(x - 1) + bx^2(x - 1).$$

Show that this reproduces the analytical solution.

26. Another approximating function that meets the boundary condition of Exercise 25 is

$$ax(x - 1) + bx(x - 1)^2.$$

Use this to solve by the Rayleigh–Ritz technique.

27. Suppose that the boundary conditions in Exercise 25 are $y(0) = 1, y(1) = 3$. Modify the procedure of Exercise 25 to get a solution.

▶**28.** Solve Exercise 24 by collocation, setting the residual to zero at $x = \frac{1}{3}$ and $x = \frac{2}{3}$. Compare this solution to that from Exercise 24.

29. Repeat Exercise 28, except now use different points within $[0, 1]$ for setting the residual to zero. Are some pairs of points better than others?

30. Repeat Exercise 25, but now use collocation. Does it matter where within $[0, 1]$ you set the residual to zero?

▶**31.** Use Galerkin's technique to solve Exercise 24. Is the same solution obtained?

32. Repeat Exercise 25, but now use Galerkin.

Section 6.5

33. Suppose that $Q(x) = \sin(x)$ and $F(x) = x^2 + 2$ in Eq. (6.33). For an element that occupies $[0.33, 0.45]$,
  a. find N_L and N_R (Eq. 6.34).
  b. set up the Galerkin integrals (Eq. 6.36), writing out $R(x)$ in full.
  c. write the element equations (Eq. 6.44).
  d. compute the correct average values for Q and F.

34. Repeat Exercise 33 for the two adjacent elements. These occupy $[0.21, 0.33]$ and $[0.45, 0.71]$.

35. Assemble the three pairs of element equations of Exercises 33 and 34 to form a set of four equations with the nodal values at $x = 0.21$, $x = 0.33$, $x = 0.45$, and $x = 0.71$ as unknowns.

▶**36.** Solve by the finite-element method of Section 6.5:

$$y'' + xy = x^3 - \frac{4}{x^3}, \qquad y(1) = -1, \qquad y(2) = 3.$$

Put nodes at $x = 1.2, 1.5,$ and 1.75 as well as at the ends of $[1, 2]$. Compare your solution to the analytical solution, which is $y = x^2 - 2/x$.

37. Repeat Exercise 36 except for the end condition at $x = 1$ of $y'(1) = 4$.

Section 6.6

38. Consider the characteristic-value problem with k restricted to real values:

$$y'' - k^2 y = 0, \qquad y(0) = 0, \qquad y(1) = 0.$$

  a. Show analytically that there is no solution except the trivial solution $y = 0$.
  b. Show, by setting up a set of difference equations corresponding to the differential equation with $h = 0.2$, that there are no real values for k for which a solution to the set exists.
  c. Show, using the shooting method, that it is impossible to match $y(1) = 0$ for any real value of k [except if $y'(0) = 0$, which gives the trivial solution].

▶**39.** For the equation

$$y'' - 3y' + 2k^2 y = 0, \qquad y(0) = 0, \qquad y(1) = 0,$$

find the principal eigenvalue and compare to $|k| = 2.46166$,
  a. using $h = \frac{1}{2}$.
  b. using $h = \frac{1}{3}$.

c. using $h = \frac{1}{4}$.

d. Assuming errors are proportional to h^2, extrapolate from parts (a) and (c) to get an improved estimate.

40. Estimate the principal eigenvalue of

$$y'' + k^2 x^2 y = 0, \qquad y(0) = 0, \qquad y(1) = 0.$$

41. Using the principal eigenvalue, $k = 2.46166$, in Exercise 39, find y as a function of x over $[0, 1]$. This is the corresponding eigenfunction.

42. The second eigenvalue for Exercise 40 is 12.61105. Find the corresponding eigenfunction.

▶43. Parallel the computations of Exercise 39 to estimate the second eigenvalue. Compare to the analytical value of 4.56773.

Section 6.7

44. Find the dominant eigenvalue and the corresponding eigenvector by the power method:

a. $\begin{bmatrix} 3 & 1 \\ 2 & 9 \end{bmatrix}$
b. $\begin{bmatrix} 2 & 3 \\ 6 & 5 \end{bmatrix}$
c. $\begin{bmatrix} 2 & 3 \\ 3 & -2 \end{bmatrix}$

d. $\begin{bmatrix} 6 & 2 & 0 \\ 2 & 4 & 1 \\ 0 & 1 & -1 \end{bmatrix}$
e. $\begin{bmatrix} 1 & 2 & 3 \\ 0 & 1 & 3 \\ 2 & 2 & 1 \end{bmatrix}$

[In part (c), the two eigenvalues are equal but of opposite sign.]

45. For the two matrices

$$A = \begin{bmatrix} -5 & 2 & 1 \\ 1 & -9 & -1 \\ 2 & -1 & 7 \end{bmatrix},$$

$$B = \begin{bmatrix} -4 + 2i & -1 & -5i \\ -3 & 7 + i & -i \\ 2 & -1 & 4 - i \end{bmatrix},$$

a. put bounds on the eigenvalues using Gerschgorin's theorem.

b. can you tell from part (a) whether either of the matrices is singular?

46. Use the power method or its variations to determine all of the eigenvalues for the matrices of parts (a), (b), and (c) of Exercise 39.

47. Invert the matrices of Exercise 44, and then find the smallest eigenvalues with the power method.

▶48. Find all the eigenvalues of matrix A of Exercise 45. Get these from the characteristic polynomial. Then invert A and repeat. Show that the two sets are reciprocals and that the corresponding eigenvectors are the same.

49. Repeat Exercise 48, but now use the power method to get the dominant eigenvalue. Then shift by that amount and get another. Finally use $\text{tr}(A)$ to get the third.

▶50. Find all the eigenvalues of

$$A = \begin{bmatrix} 3 & 1 & 2 \\ 2 & 1 & 3 \\ 1 & 3 & 3 \end{bmatrix}.$$

Now find the eigenvalues of $M^{-1}AM$, where

$$M = \begin{bmatrix} 2 & 0 & 0 \\ 0 & 3 & 1 \\ 0 & 1 & 1 \end{bmatrix}.$$

Do you get the same eigenvalues? ($M^{-1}AM$ is called a similarity transformation.)

Section 6.8

51. For matrix A of Exercise 50, find three matrices each of which will convert one of the below-diagonal elements to zero.

52. Use the matrices of Exercise 51 successively to make one element below the diagonal of A equal to zero, then multiply that product by the inverse of the rotation matrix (which is easy to find because it is just its transpose). We keep the eigenvalues the same because the two multiplications are a similarity transformation.

Repeat this process until all elements below the diagonal are less than 1.0E-4. When this is done, compare the elements now on the diagonal to the eigenvalues of A obtained by iteration. (This will take many steps. You will want to write a short computer program to carry it out.)

▶53. How many repetitions of the scheme of Exercise 52

are required to reduce all below-diagonal elements to less than 0.0001 for

$$A = \begin{bmatrix} 4 & 2 & 0 \\ 1 & 3 & 1 \\ 1 & 1 & 2 \end{bmatrix}?$$

54. Use similarity transformations to reduce the matrix to upper Hessenberg. (Do no column or row interchanges.)

$$C = \begin{bmatrix} 3 & -1 & 2 & 7 \\ 1 & 2 & 0 & -1 \\ 4 & 2 & 1 & 1 \\ 2 & -1 & -2 & 2 \end{bmatrix}$$

55. Repeat Exercise 54, this time with row/column interchanges that maximize the magnitude of divisors.

56. Show that each pair of transformation matrices used in Exercise 54 are in fact inverses.

57. Repeat Exercise 52 but first convert to upper Hessenberg.

▶**58.** Repeat Exercise 53 but first convert to upper Hessenberg. How many fewer iterations are required?

Section 6.9

▶**59.** Matrix A is defined as

$$A = \begin{bmatrix} 4 & 3 & 2 \\ 2 & 3 & 4 \\ 2 & 4 & a \end{bmatrix}.$$

What is the smallest value for a in matrix A for which convergence will be obtained with
a. the Jacobi method?
b. the Gauss–Seidel method?

60. Let T be

$$T = \begin{bmatrix} L & 0 \\ x & L \end{bmatrix}.$$

Show that

$$T^n = \begin{bmatrix} L^n & 0 \\ nL^{n-1}x & L^n \end{bmatrix}.$$

From this infer that $T^n \to 0$-matrix if $|L| < 1$.

61. Let matrices S and T be

$$S = \begin{bmatrix} Z_1 & 0 & 0 \\ a & Z_2 & 0 \\ b & c & Z_3 \end{bmatrix}, \quad T = \begin{bmatrix} Z & 0 & 0 \\ a & Z & 0 \\ b & c & Z \end{bmatrix},$$

where $Z = \text{Max}\{|Z_1|, |Z_2|, |Z_3|\}$.
Write $U = L + D$, where

$$L = \begin{bmatrix} 0 & 0 & 0 \\ a & 0 & 0 \\ b & c & 0 \end{bmatrix}, \quad D = \begin{bmatrix} Z & 0 & 0 \\ 0 & Z & 0 \\ 0 & 0 & Z \end{bmatrix}.$$

a. Show that $L^3 = 0$-matrix.
b. Since $LD = DL$, show that

$$T^n = (L + D)^n$$
$$= \binom{n}{n-2} L^2 D^{n-2} + \binom{n}{n-1} LD^{n-1} + D^n,$$

from the binomial formula. From this, we see that $T^n \to 0$-matrix and also S^n, if $Z < 1$.
c. Prove this for any general triangular matrix T where $|T_{i,i}| < 1$ for $i = 1, 2, \ldots, n$.

Section 6.10

62. Which of these boundary-value problems meet the conditions for a unique solution? For those that do not, find a special case for the boundary values for which a solution does exist.
a. $y'' = (x - 1)y' + (x + 1)y + (x^2 - 1)$,
  $y(-1) = -1$, $\quad y(1) = 1$.
b. $y'' = (x + 1)y' + (x - 1)y + (x^2 + 1)$,
  $y(-1) = -1$, $\quad y(1) = 1$.
c. $y'' = y' + q*y + x$,
$$q = \begin{cases} -1, -1 < x < 0 \\ 0, x = 0 \\ 1, 0 < x < 1 \end{cases}$$
  $y(-1) = -1$, $\quad y(1) = 1$.

▶**63.** For each of the equations in Exercise 62, set up the difference equations to solve with $h = 0.5$. What is the determinant of each matrix?

64. The equation in Exercise 5 is nonlinear. Solve with three assumed values of $y'(-1)$, and then use Muller's method to get the next value.

65. Continue Exercise 64 until you converge on $y(2)$ within 0.0001. Are fewer trials needed than if you linearly interpolate from two values of $y'(-1)$?

▶**66.** Repeat Exercise 64; start with $y'(1) = -1.5$ and -3, then interpolate, and solve again. Form a quadratic with these to estimate a fourth start value, then use the four values to form a cubic for the next estimate. Compare to Exercise 65.

67. Show that the power method does not converge if there are two eigenvalues of the same largest magnitude.

Section 6.11

68. Use the TAY_ODES routine in DERIVE to obtain the results shown in Table 6.2.

69. Repeat Exercise 68, but now use the RK routine.

►70. Use DERIVE to solve the first set of equations in Section 6.2,
a. through getting the inverse matrix.
b. through soLve.
c. with ROW_REDUCE.

71. Use DERIVE or some other computer algebra system to get the eigenvalues of

$$\begin{bmatrix} 5 & 3 & 2 \\ -2 & 6 & 3 \\ 3 & 2 & 4 \end{bmatrix}.$$

72. Add 0.1 to each element of the matrix in Exercise 71, and then get its eigenvalues with DERIVE. Compare the sum of these to the sum of the previous ones.

73. Use DERIVE to get the eigenvectors for the matrix in Exercise 71
a. with the utility functions in VECTOR.MTH.
b. by setting up and solving equations.

►74. Find the characteristic polynomial of the matrix in Exercise 72 through DERIVE or some other computer algebra system.

75. Using DERIVE or some other computer algebra system, do similarity transformations on the following matrix to obtain another that has no nonzero elements. Then show that they have the same eigenvalues. Do they have the same eigenvectors?

$$\begin{bmatrix} 1 & 0 & 0 \\ 0 & 2 & 0 \\ 0 & 0 & 3 \end{bmatrix}$$

Applied Problems and Projects

76. If a cantilever beam of length L, which bends due to a uniform load of w lb/ft, is also subject to an axial force P at its free end (see Fig. 6.10), the equation of its elastic curve is

$$EI\frac{dy^2}{dx^2} = Py - \frac{wx^2}{2}.$$

For this equation, the origin O has been taken at the free end. I is the moment of inertia; here $I = bh^3/12$. At the point $x = L$, $dy/dx = 0$; at $(0, 0)$, $y = 0$. Solve this boundary-value problem by the shooting method for a 2 in. × 4 in. × 10-ft wooden beam for which $E = 12 \times 10^5$ lb/in². Find y versus x when the beam has the 4-in. dimension vertical with $w = 25$ lb/ft and a tension force of $P = 500$ lb. Also solve for the deflections if the beam is turned so that the 4-in. dimension is horizontal.

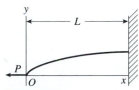

Figure 6.10

77. Solve Problem 76 by replacing the differential equation by difference equations. Compare the solutions by each method.

78. In Problem 76 y'' is used although the radius of curvature should be employed in the equation:

$$EI\frac{y''}{[1 + (y')^2]^{3/2}} = Py - \frac{wx^2}{2}.$$

If the deflection of the beam is small, the difference is negligible, but in the second part of Problem 76 at least, this is not true. Furthermore, if there is considerable bending of the beam, the horizontal distance from the origin to the wall is less than L, the original length of the beam. Solve Problem 76 taking these factors into account, and determine by how much the deflections differ from those previously calculated.

79. A cylindrical pipe has a hot fluid flowing through it. Since the pressure is very high, the walls of the pipe are thick. For such a situation, the differential equation that relates temperatures in the metal wall to radial distance is

$$r\frac{d^2u}{dr^2} + \frac{du}{dr} = 0,$$

where

$$r = \text{radial distance from the centerline,}$$
$$u = \text{temperature.}$$

Solve for the temperatures within a pipe whose inner radius is 1 cm and whose outer radius is 2 cm if the fluid is at 540°C and the temperature of the outer circumference is 20°C.

80. The pipe in Problem 79 is insulated to reduce the heat loss. The insulation used has properties such that the gradient du/dr at the outer circumference is proportional to the difference in temperatures from the outer wall to the surroundings:

$$\left.\frac{du}{dr}\right|_{r=2} = 0.083[u(2) - 20].$$

Solve Problem 79 with this boundary condition.

81. When air flows at a velocity of u_∞ past a flat plate placed horizontally to the flow, experiments show that the air velocity at the surface of the plate is zero and that a boundary layer within which the velocities are less than u_∞ builds up as flow progresses along the plate (Fig. 6.11).

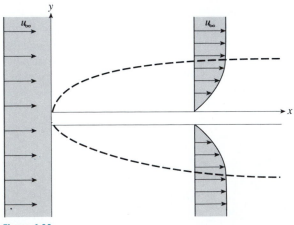

Figure 6.11

This is a typical problem of interest to aeronautical engineers. To solve for the velocities within the boundary layer, a stream function f is defined for which

$$\frac{d^3f}{dz^3} + \frac{f}{2}\frac{d^2f}{dz^2} = 0,$$

where z is a dimensionless distance involving x, y, u_∞, and the viscosity of air. This is a boundary-value problem because $f = 0$ and $f' = 0$ at $z = 0$, while $f' = 1$ at $z = \infty$. In solving this problem, one does not use $z = \infty$ as a boundary; rather, the integration is taken to a value of z large enough so that there is an insignificant change when it is increased. Solve this problem to find f and df/dz as functions of z.

82. A simple spring–mass system obeys the equation

$$\frac{d^2y}{dt^2} + \alpha^2 y = 0,$$

where the positive constant α^2 equals k/m, the ratio of the spring constant to the mass. If it is known that $y = 0$ at $t = 0$ and again at $t = 1.26$ sec (so that its period is 2.52 sec), this is a characteristic-value problem that has a solution for only certain values of α. These characteristic values are discussed in Section 6.6 for this equation. Suppose, however, that the spring is not an ideal one with constant k, but that the spring force and elongation vary with y according to the equation $k = b(1 - y^{0.1})$. If it is still true that $y = 0$ at $t = 0$ and also at $t = 1.26$, is this a characteristic-value problem requiring that the ratio k/m be at certain fixed values in order to have a nontrivial solution? If it is, find these values.

83. A shaft rotates in fixed bearings (Fig. 6.12) so that at these points $y = 0$ and $dy/dx = 0$. The governing equation is

$$EI\frac{d^4y}{dx^4} = \frac{w}{g}\omega^2 y,$$

where E is Young's modulus, I is the moment of inertia of the cross-sectional area of the shaft, y is the displacement from the horizontal, x is the distance measured from the point midway between the bearings, w is the weight of the shaft per unit length, g is the acceleration of gravity, $2L$ is the length of the shaft between the bearings, and ω is the rotational speed.

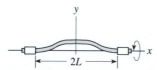

Figure 6.12

At certain critical speeds, the shaft will distort from a straight line. These critical speeds are very important to know in designing the machine of which the shaft is a part. For parameter values given here in consistent units, find the first two values of critical rotational speeds. Also find the curve $y(x)$ at these speeds.

$$E = 2 \times 10^6 \text{ lb/in}^2 \qquad g = 386.4 \text{ in./sec}^2$$
$$I = 0.46 \text{ in}^4 \qquad L = 16 \text{ in.}$$
$$w = 0.025 \text{ lb/in}$$

7

Numerical Solution of Partial-Differential Equations

Contents of This Chapter

Chapter 7 introduces you to the area of partial-differential equations, a subject of great importance in applied mathematics. These equations involve two or more independent variables that determine the behavior of the dependent variable as described by a differential equation, usually of second or higher order. They find application in the fields of heat flow, fluid flow, electrical potential distribution, electrostatics, diffusion of matter, and so on.

In this chapter, we concentrate on the so-called steady-state problem in which the dependent variable (the *potential*) does not vary with time. The type of equations that are treated fall in the class of partial-differential equations known as *elliptic equations*. The chapter discusses two numerical techniques—finite-difference methods and the finite-element method.

7.1 Equilibrium Temperatures in a Heated Slab

Presents a typical two-dimensional problem that illustrates one practical application of an elliptic partial-differential equation.

7.2 Equation for Steady-State Heat Flow

Derives the equation that governs this situation, introduces you to some terminology, and shows how partial-differential equations are classified.

7.3 Representation as a Difference Equation

Utilizes finite-difference approximations to the derivatives in the equation to change the differential equation to a difference equation. When written at specified points in the region of the problem, these difference equations create a system of simultaneous equations that can be solved to approximate the solution to the differential equation.

7.12 Theoretical Matters

Collects a number of items of theoretical importance that underlie the problems and methods of the chapter.

7.13 Using *Mathematica*

Explains the facilities in this software package that apply to solving elliptic partial-differential equations.

Chapter Summary

Lets you systematically review your knowledge of Chapter 7.

Computer Programs

Presents some illustrative examples of computer programs that solve some of the problems of the chapter.

Parallel Processing

The methods of this chapter, like those of Chapter 6, almost always involve the solving of sets of equations. The set may sometimes be very large, but it is sparse, and iterative methods are frequently used. As you learned in Chapter 2, both direct and iterative methods can be assisted by parallel processing. However, there is nothing new to know about parallelism in connection with this chapter.

7.1 Equilibrium Temperatures in a Heated Slab

Many important scientific and engineering problems fall into the field of partial-differential equations. Here is a situation that is typical, in which the temperatures are a function of the coordinates of position of the point in question.

A piece of metal is 12 in. × 3 in. × 6 ft. Three feet of the slab is kept inside a furnace, but half of the slab protrudes (see Fig. 7.1). To decrease heat losses to the air, the protruding half is covered with a 1-in. thickness of insulation. If the furnace is maintained at 950°F, we ask whether all points of the metal reach a temperature of 800°F or higher, in spite of heat loss through the insulation. Such a question might arise in heat-treating the slab when the only furnace available to heat the metal is too small to contain the whole slab. The sketch in Fig. 7.1 will explain the problem.

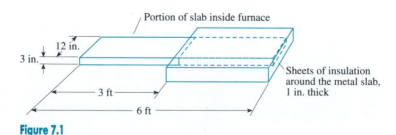

Figure 7.1

You will recognize that this is a boundary-value problem: The temperature is known for the part of the slab inside the furnace, and something is known about the temperature gradients for the insulated part. The problem differs from those discussed in the previous chapter in that more than one independent variable is involved in specifying the coordinates of points within the slab. (Specifically, these variables are the x-, y-, and z-coordinates.)

This chapter presents methods to solve this problem. These methods are generally referred to as finite-difference methods. However, since the 1950s an equally important method has been developed for solving elliptic partial-differential equations, the finite-element method. This method is very suitable for computer coding and for handling very complex figures and irregularities. The finite-element method breaks the complex region into simple subregions such as triangles and pyramids and applies solution techniques to these simpler parts (see Vichnevetsky, 1981).

There are well-known codes that use either or both of these methods. Among them are ELLPACK, which uses a variety of methods including those presented in this chapter, and TWODEPEP (IMSL), which uses the finite-element method for two-dimensional problems. However, in this chapter we consider mainly finite-difference methods; we also show how finite elements can be used.

Many physical phenomena are a function of more than one independent variable and must be represented by a partial-differential equation, by which term we mean an equation involving partial derivatives. Most scientific problems have mathematical models that are second-order equations, with the highest order of derivative being the second. Structural analysis problems often are fourth-order differential equations.

If u is a function of the two independent variables x and y, there are three second-order partial derivatives:

$$\frac{\partial^2 u}{\partial x^2}, \qquad \frac{\partial^2 u}{\partial x\, \partial y}, \qquad \frac{\partial^2 u}{\partial y^2}.$$

(We will treat only functions for which the order of differentiation is unimportant, so $\partial^2 u/\partial y\, \partial x = \partial^2 u/\partial x\, \partial y$.) Depending on the values of the coefficients of the second-derivative terms, second-order partial-differential equations are classified as elliptic, parabolic, or hyperbolic. The most important distinctions to us are the kinds of problems and the nature of boundary conditions that lead to one type of equation or another. Steady-state potential distribution problems fall in the class of elliptic equations, which we discuss first. The equilibrium temperatures attained by the metal piece protruding from the furnace is such a potential distribution problem.

The distinction as to elliptic, parabolic, or hyperbolic for second-order partial-differential equations depends on the coefficients of the second-derivative terms. We can write any such second-order equation (in two independent variables) as

$$A\frac{\partial^2 u}{\partial x^2} + B\frac{\partial^2 u}{\partial x\, \partial y} + C\frac{\partial^2 u}{\partial y^2} + D\left(x, y, u, \frac{\partial u}{\partial x}, \frac{\partial u}{\partial y}\right) = 0.$$

Depending on the value of $B^2 - 4AC$, we classify the equation as

Elliptic, if $B^2 - 4AC < 0$;

Parabolic, if $B^2 - 4AC = 0$;

Hyperbolic, if $B^2 - 4AC > 0$.

If the coefficients A, B, and C are functions of x, y, and/or u, the equation may change from one classification to another at various points in the domain.

For Laplace's and Poisson's equations, $B = 0, A = C = 1$, so these are always elliptic. We study the other types in later chapters.

In Section 7.2, Eqs. (7.2) and (7.3) will describe how u varies within the interior of a closed region. The function u is determined on the boundaries of the region by boundary conditions. The boundary conditions may specify the value of u at all points on the boundary, or some combination of the potential and the normal derivative.

7.2 Equation for Steady-State Heat Flow

We derive the relationship for temperature u as a function of the two space variables x and y for the equilibrium temperature distribution on a flat plate (Fig. 7.2). Consider the element of the plate whose surface area is $dx\,dy$. We assume that heat flows only in the x- and y-directions and not in the perpendicular direction. If the plate is very thin, or if the upper and lower surfaces are both well insulated, the physical situation will agree with our assumption. Let t be the thickness of the plate.

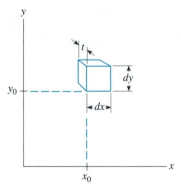

Figure 7.2

Heat flows at a rate proportional to the cross-sectional area, to the temperature gradient ($\partial u/\partial x$ or $\partial u/\partial y$), and to the thermal conductivity k, which we will assume constant at all points. The flow of heat is from high to low temperature, of course, meaning opposite to the direction of increasing gradient. We use a minus sign in the equation to account for this:

Rate of heat flow into element at $x = x_0$, in the x-direction:

$$-k(t\,dy)\frac{\partial u}{\partial x} = -(\text{conductivity})(\text{area})(\text{gradient}).$$

The gradient at $x_0 + dx$ is the gradient at x_0 plus the increment in the gradient over the distance dx:

Gradient at $x_0 + dx$:

$$\frac{\partial u}{\partial x} + \frac{\partial}{\partial x}\left(\frac{\partial u}{\partial x}\right)dx.$$

Rate of heat flow out of element at $x = x_0 + dx$:

$$-k(t\,dy)\left[\frac{\partial u}{\partial x} + \frac{\partial}{\partial x}\left(\frac{\partial u}{\partial x}\right)dx\right].$$

Net rate of heat flow into element in x-direction:

$$-k(t\,dy)\left[\frac{\partial u}{\partial x} - \left(\frac{\partial u}{\partial x} + \frac{\partial}{\partial x}\left(\frac{\partial u}{\partial x}\right)dx\right)\right] = kt\,dx\,dy\,\frac{\partial^2 u}{\partial x^2}.$$

Similarly, in the y-direction we have the following.

Net rate of heat flow into element in y-direction:

$$-k(t\,dx)\left[\frac{\partial u}{\partial y} - \left(\frac{\partial u}{\partial y} + \frac{\partial}{\partial y}\left(\frac{\partial u}{\partial y}\right)dy\right)\right] = kt\,dx\,dy\,\frac{\partial^2 u}{\partial y^2}.$$

The total heat flowing into the elemental volume by conduction is the sum of these net flows in the x- and y-directions. If heat were generated within the element, it would be added to the heat entering by conduction. The sum must equal the rate of heat lost by other mechanisms, or else there would be a buildup of heat within the element, causing its temperature to increase with time. In this chapter we will consider only the case where the temperatures do not change with time, the *steady-state* case.

If there is equilibrium as to temperature distribution (i.e., steady state), the total rate of heat flow into the element plus heat generated must be zero. Hence,

$$kt(dx)(dy)\left(\frac{\partial^2 u}{\partial x^2} + \frac{\partial^2 u}{\partial y^2}\right) + Qt(dx)(dy) = 0,$$

where Q is the rate of heat generation per unit volume and $t(dx)(dy)$ is the volume of the element. (Obviously, if there is heat generation, there must be heat flow *from* the element by conduction that just balances it.) Q will often be a function of x and y.

If $Q = 0$, we have

$$kt\,dx\,dy\left(\frac{\partial^2 u}{\partial x^2} + \frac{\partial^2 u}{\partial y^2}\right) = 0,$$

$$\frac{\partial^2 u}{\partial x^2} + \frac{\partial^2 u}{\partial y^2} = \nabla^2 u = 0.$$

(7.1)

The operator

$$\nabla^2 = \left(\frac{\partial^2}{\partial x^2} + \frac{\partial^2}{\partial y^2}\right)$$

is called the *Laplacian*, and Eq. (7.1) is called *Laplace's equation*. For three-dimensional heat-flow problems, we would have, analogously,

$$\nabla^2 u = \left(\frac{\partial^2}{\partial x^2} + \frac{\partial^2}{\partial y^2} + \frac{\partial^2}{\partial z^2}\right)u = 0.$$

(7.2)

Equations (7.1) and (7.2), derived with reference to heat flow, apply as well to steady-state diffusion problems (where u is now concentration of material) and to electrical-potential distribution (where u is electromotive force). In fact, Laplace's equation holds for the steady-state distribution of the potential of any quantity where the rate of flow is proportional to the gradient, and where the proportionality constant does not vary with position or the value of u. In our examples, we will generally use terminology corresponding to heat flow as being more closely related to the average student's everyday experience.

With Laplace's equation, we assume that no heat is being generated at points in the plate, as by electric heaters embedded in it, nor does any removal take place, as by cooling coils or other means. In the presence of such "sources" or "sinks," we would not equate the net flow into the element to zero, as in Eq. (7.1), but the net flow into the element of volume would equal the net rate of heat removal from the element, if steady state applies. Assuming this removal rate to be a function of the location of the element in the xy-plane, $f(x,y)$, we would have, with Q equal to the rate of heat generation per unit area,

$$\nabla^2 u = -\frac{1}{k}Q(x,y) = f(x,y).$$

(7.3)

This equation is called *Poisson's equation*. Our numerical methods of solving elliptic differential equations apply equally well to both Laplace's and Poisson's equations.

Analytical methods, however, find the solution of Poisson's equations or even Laplace's equations with complicated boundary conditions considerably more difficult. These two equations include most of the physical applications of elliptic partial-differential equations.

7.3 Representation as a Difference Equation

One scheme for solving all kinds of partial-differential equations is to replace the derivatives by difference quotients, converting the equation to a difference equation.* We then write the difference equation corresponding to each point at the intersections (nodes) of a gridwork that subdivides the region of interest at which the function values are unknown. Solving these equations simultaneously gives values for the function at each node that approximate the true values. We begin with the two-dimensional case.

Approximating derivatives by difference quotients was the subject matter of Chapter 4. They may also be derived by the method of undetermined coefficients (Appendix B). We rederive the few relations we need independently.

Let $h = \Delta x = $ equal spacing of gridwork in the x-direction (see Fig. 7.3). We assume that the function $f(x)$ has a continuous fourth derivative. By Taylor series,

$$f(x_n + h) = f(x_n) + f'(x_n)h + \frac{f''(x_n)}{2}h^2 + \frac{f'''(x_n)}{6}h^3 + \frac{f^{iv}(\xi_1)}{24}h^4,$$

$$x_n < \xi_1 < x_n + h,$$

$$f(x_n - h) = f(x_n) - f'(x_n)h + \frac{f''(x_n)}{2}h^2 - \frac{f'''(x_n)}{6}h^3 + \frac{f^{iv}(\xi_2)}{24}h^4,$$

$$x_n - h < \xi_2 < x_n.$$

It follows that

$$\frac{f(x_n + h) - 2f(x_n) + f(x_n - h)}{h^2} = f''(x_n) + \frac{f^{iv}(\xi)}{12}h^2, \qquad x_n - h < \xi < x_n + h.$$

A subscript notation is convenient:

$$\frac{f_{n+1} - 2f_n + f_{n-1}}{h^2} = f''_n + O(h^2). \tag{7.4}$$

In Eq. (7.4) the subscript on f indicates the x-value at which it is evaluated. The order relation $O(h^2)$ signifies that the error approaches proportionality to h^2 as $h \to 0$.

---

*This is exactly the same as the method of solving a boundary-value problem through replacing it by difference equations, as discussed in Chapter 6.

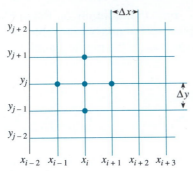

Figure 7.3

Similarly, the first derivative is approximated,

$$\frac{f(x_n + h) - f(x_n - h)}{2h} = f'(x_n) + \frac{f'''(\xi)}{6} h^2, \qquad x_n - h < \xi < x_n + h,$$

$$\frac{f_{n+1} - f_{n-1}}{2h} = f_n' + O(h^2). \tag{7.5}$$

[The first derivative could also be approximated by the forward or backward difference, but this would have an error of $O(h)$. We prefer the more accurate central-difference approximation.]

When f is a function of both x and y, we get the second partial derivative with respect to x, $\partial^2 u / \partial x^2$, by holding y constant and evaluating the function at three points where x equals x_n, $x_n + h$, and $x_n - h$. The partial derivative $\partial^2 u / \partial y^2$ is similarly computed, holding x constant. We require that fourth derivatives with respect to both variables exist.

To solve the Laplace equation on a region in the xy-plane, we subdivide the region with equispaced lines parallel to the x- and y-axes. Consider a portion of the region near (x_i, y_j). (See Fig. 7.3.) We wish to approximate

$$\nabla^2 u = \frac{\partial^2 u}{\partial x^2} + \frac{\partial^2 u}{\partial y^2} = 0.$$

Replacing the derivatives by difference quotients that approximate the derivatives at the point (x_i, y_j), we get

$$\nabla^2 u(x_i, y_j) = \frac{u(x_{i+1}, y_j) - 2u(x_i, y_j) + u(x_{i-1}, y_j)}{(\Delta x)^2}$$
$$+ \frac{u(x_i, y_{j+1}) - 2u(x_i, y_j) + u(x_i, y_{j-1})}{(\Delta y)^2} = 0.$$

It is convenient to let double subscripts on u indicate the x- and y-values:

$$\nabla^2 u_{i,j} = \frac{u_{i+1,j} - 2u_{i,j} + u_{i-1,j}}{(\Delta x)^2} + \frac{u_{i,j+1} - 2u_{i,j} + u_{i,j-1}}{(\Delta y)^2} = 0.$$

It is common to take $\Delta x = \Delta y = h$, resulting in considerable simplification, so that

$$\nabla^2 u_{i,j} = \frac{1}{h^2}[u_{i+1,j} + u_{i-1,j} + u_{i,j+1} + u_{i,j-1} - 4u_{i,j}] = 0. \qquad (7.6)$$

Note that five points are involved in the relationship of Eq. (7.6), points to the right, left, above, and below the central point (x_i, y_j). It may help to remember this if we rewrite Eq. (7.6) as

$$\nabla^2 u = \left(\frac{1}{h^2}\right)(u_R + u_L + u_A + u_B - 4u_0)$$

where the subscripts indicate right, left, above, and below the central point u_0.

It is convenient to represent the relationship pictorially, where the linear combination of u's is represented symbolically. Equation (7.6) becomes

$$\nabla^2 u_{i,j} = \frac{1}{h^2}\left\{ \begin{array}{ccc} & 1 & \\ 1 & -4 & 1 \\ & 1 & \end{array} \right\} u_{i,j} = 0. \qquad (7.7)$$

The representation of the Laplacian operator by the pictorial operator that represents the linear combination of u-values is fundamental to the finite-difference method for elliptic partial-differential equations. The approximation has $O(h^2)$ error, provided that u is sufficiently smooth. This formula is referred to as the five-point or five-point star formula.

In addition to the five-point formula just given, we can derive the nine-point formula for Laplace's equation by similar methods to get

$$\nabla^2 u_{i,j} = \frac{1}{6h^2}\left\{ \begin{array}{ccc} 1 & 4 & 1 \\ 4 & -20 & 4 \\ 1 & 4 & 1 \end{array} \right\} u_{i,j} = 0. \qquad (7.8)$$

In this case the approximation has $O(h^6)$ error, provided that u is sufficiently smooth.

7.4 Laplace's Equation on a Rectangular Region

A typical steady-state heat-flow problem is the following: A thin steel plate is a 10×20 cm rectangle. If one of the 10-cm edges is held at 100°C and the other three edges are held at 0°C, what are the steady-state temperatures at interior points? For steel, $k = 0.16 \, \text{cal/sec} \cdot \text{cm}^2 \cdot °\text{C/cm}$. We can state the problem mathematically in this way if we assume that heat flows only in the x- and y-directions:

Find $u(x, y)$ such that

$$\frac{\partial^2 u}{\partial x^2} + \frac{\partial^2 u}{\partial y^2} = 0, \tag{7.9}$$

with

$$u(x, 0) = 0,$$
$$u(x, 10) = 0,$$
$$u(0, y) = 0,$$
$$u(20, y) = 100.$$

In this statement of the problem, we imagine one corner of the plate at the origin, with boundary conditions as sketched in Fig. 7.4.

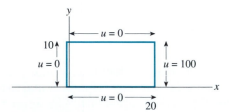

Figure 7.4

A problem with temperatures known on each boundary is said to have *Dirichlet boundary conditions*. This problem can be solved analytically, giving $u(x, y)$ as an infinite Fourier series, but we will use it to illustrate the numerical technique. We replace the differential equation by a difference equation, as described in the previous section:

$$\frac{1}{h^2} [u_{i+1,j} + u_{i-1,j} + u_{i,j+1} + u_{i,j-1} - 4u_{ij}] = 0,$$

which relates the temperature at the point (x_i, y_j) to the temperatures at four neighboring points, each the distance h away from (x_i, y_j). An approximation to the solution of Eq. (7.9) results when we select a set of such points (these are often called *nodes*) and find the solution to the set of difference equations that result. Suppose we choose $h = 5$ cm. (In normal practice we would choose a smaller mesh size.) Figure 7.5 shows the notation we use.

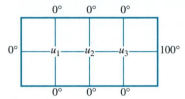

Figure 7.5

The system of equations is

$$\frac{1}{5^2}(0 + 0 + u_2 + 0 - 4u_1) = 0,$$

$$\frac{1}{5^2}(u_1 + 0 + u_3 + 0 - 4u_2) = 0, \qquad \textbf{(7.10)}$$

$$\frac{1}{5^2}(u_2 + 0 + 100 + 0 - 4u_3) = 0.$$

The pictorial operator of Section 7.3,

$$\nabla^2 u_{ij} = \frac{1}{h^2}\left\{ 1 \quad \begin{matrix} 1 \\ -4 \\ 1 \end{matrix} \quad 1 \right\} u_{ij} = 0,$$

makes the equations very easy to write. Observe that the physical properties of the material do not enter the relation* and that the constant $1/h^2 = 1/5^2$ can be divided out of Eqs. (7.10).

The solution to the set of equations is easy when there are only three of them:

$$u_1 = 1.786, \qquad u_2 = 7.143, \qquad u_3 = 26.786.$$

Of course, with such a coarse grid as that shown in Fig. 7.5, we don't expect high accuracy. Redoing the problem with a smaller value for h should improve the

---

* If the conductivity varied at different regions of the plate, this would not be true.

precision. Figure 7.6 and Eqs. (7.11) result when we make $h = 2.5$ cm. (The equations are expressed in matrix form. We use v instead of u for the temperature variable in order to distinguish it from the previous solution.)

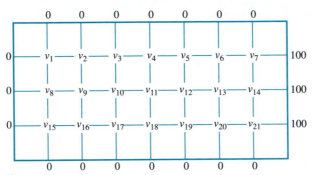

Figure 7.6

$$
\left[
\begin{array}{ccccccc|ccccccc|ccccccc}
-4 & 1 & & & & & & 1 & & & & & & & & & & & & & \\
1 & -4 & 1 & & & & & & 1 & & & & & & & & & & & & \\
 & 1 & -4 & 1 & & & & & & 1 & & & & & & & & & & & \\
 & & 1 & -4 & 1 & & & & & & 1 & & & & & & & & & & \\
 & & & 1 & -4 & 1 & & & & & & 1 & & & & & & & & & \\
 & & & & 1 & -4 & 1 & & & & & & 1 & & & & & & & & \\
 & & & & & 1 & -4 & & & & & & & 1 & & & & & & & \\
\hline
1 & & & & & & & -4 & 1 & & & & & & 1 & & & & & & \\
 & 1 & & & & & & 1 & -4 & 1 & & & & & & 1 & & & & & \\
 & & 1 & & & & & & 1 & -4 & 1 & & & & & & 1 & & & & \\
 & & & 1 & & & & & & 1 & -4 & 1 & & & & & & 1 & & & \\
 & & & & 1 & & & & & & 1 & -4 & 1 & & & & & & 1 & & \\
 & & & & & 1 & & & & & & 1 & -4 & 1 & & & & & & 1 & \\
 & & & & & & 1 & & & & & & 1 & -4 & & & & & & & 1 \\
\hline
 & & & & & & & 1 & & & & & & & -4 & 1 & & & & & \\
 & & & & & & & & 1 & & & & & & 1 & -4 & 1 & & & & \\
 & & & & & & & & & 1 & & & & & & 1 & -4 & 1 & & & \\
 & & & & & & & & & & 1 & & & & & & 1 & -4 & 1 & & \\
 & & & & & & & & & & & 1 & & & & & & 1 & -4 & 1 & \\
 & & & & & & & & & & & & 1 & & & & & & 1 & -4 & 1 \\
 & & & & & & & & & & & & & 1 & & & & & & 1 & -4
\end{array}
\right]
\left[
\begin{array}{c}
v_1 \\ v_2 \\ v_3 \\ v_4 \\ v_5 \\ v_6 \\ v_7 \\ v_8 \\ v_9 \\ v_{10} \\ v_{11} \\ v_{12} \\ v_{13} \\ v_{14} \\ v_{15} \\ v_{16} \\ v_{17} \\ v_{18} \\ v_{19} \\ v_{20} \\ v_{21}
\end{array}
\right]
=
\left[
\begin{array}{c}
0 \\ 0 \\ 0 \\ 0 \\ 0 \\ 0 \\ -100 \\ 0 \\ 0 \\ 0 \\ 0 \\ 0 \\ 0 \\ -100 \\ 0 \\ 0 \\ 0 \\ 0 \\ 0 \\ 0 \\ -100
\end{array}
\right]
\quad (7.11)
$$

The dashed lines in (7.11) show a subdivision of the matrix by rows of nodes.

We can solve Eqs. (7.11) by elimination. It turns out that the largest element always occurs on the diagonal, so pivoting isn't required. The solution vector is

$$v_1 = 0.3530, \qquad v_8 = 0.4988, \qquad v_{15} = 0.3530,$$
$$v_2 = 0.9132, \qquad v_9 = 1.2894, \qquad v_{16} = 0.9132,$$
$$v_3 = 2.0103, \qquad v_{10} = 2.8323, \qquad v_{17} = 2.0103,$$
$$v_4 = 4.2957, \qquad v_{11} = 6.0193, \qquad v_{18} = 4.2957,$$
$$v_5 = 9.1531, \qquad v_{12} = 12.6537, \qquad v_{19} = 9.1531,$$
$$v_6 = 19.6631, \qquad v_{13} = 26.2893, \qquad v_{20} = 19.6631,$$
$$v_7 = 43.2101, \qquad v_{14} = 53.1774, \qquad v_{21} = 43.2101.$$

Symmetry in the system is apparent by inspection, and we could have reduced the number of equations by taking this into account. The regular pattern of coefficients in the matrix of Eqs. (7.11) would have been lost, however.

We now are able to examine the effect of grid size h. Three points are identical in the two sets of equations. We show these results side by side in Table 7.1.

Table 7.1

Point (Fig. 7.5)	From Eq. (7.10) with $h = 5$ cm		From Eq. (7.11) with $h = 2.5$ cm		From Eq. (7.12) with $h = 2.5$ cm		Analytical value
	Value	Error	Value	Error	Value	Error	
u_1	1.786	−0.692	1.289	−0.195	1.0938	+0.0005	1.0943
u_2	7.143	−1.655	6.019	−0.531	5.4853	+0.0032	5.4885
u_3	26.786	−0.692	26.289	−0.195	26.0938	+0.0006	26.0944

The answers with $h = 2.5$ cm show significant improvement, as expected. In fact, since the error in approximating the derivatives by central-difference quotients is of $O(h^2)$, we might anticipate that the errors in the solution through sets of difference equations would also vary as h^2; this is roughly true. When we have such knowledge of the effect of h, an extrapolation can be made. When this is done, temperatures within 3% of the analytical values are obtained.

Another way to reduce the error is to use the nine-point formula for the Laplacian, Eq. (7.8). Doing so gives the matrix equation of Eq. (7.12); here we assumed values for the corner points equal to the average of adjacent boundary points. The solution to this gives much better accuracy. Table 7.1 shows three of these values.

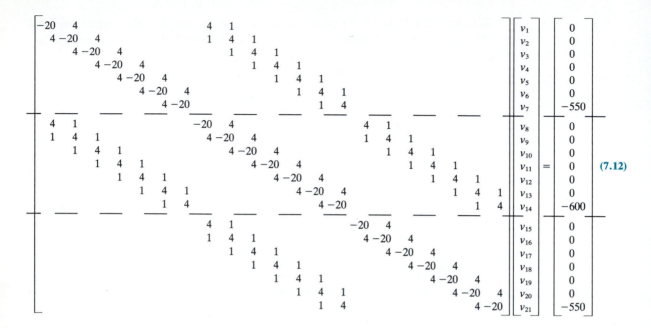

$$(7.12)$$

In principle, then, we can use the numerical method to solve the steady-state heat-flow equation to any desired accuracy; all we need to do is to continue to make h smaller. In practice, however, this procedure runs into severe difficulties. It is very apparent that the number of equations increases inordinately fast. With $h = 1.25$, we would have 105 discrete interior points; with $h = 0.625$, we have 465, and so on. Storing a matrix with 105 rows and 105 columns would require 105^2 or about 11,000 words (44,000 bytes) of computer memory; with 465 equations we would need 216,000 words (865,000 bytes). If there is not that much memory available, the use of overlays from disk storage would be extremely time-consuming. There must be a better way, you are thinking, and there is. Along with memory requirements, we worry about execution times.

The remedy to the problem of overlarge matrices is to take advantage of the sparseness exhibited by Eqs. (7.11). That system of 21 equations has nonzero terms in only 85 out of 441 elements, or 19%. The larger matrices that result when h is smaller have an even smaller proportion of nonzero terms (4% and 1% with 105 and 465 equations, respectively). Unfortunately, the systems are not tridiagonal. If they were, we could work with only the nonzero terms, as we found in Chapter 2. With banded matrices such as we have here, none of the zero elements outside the bands change, but the elements between the side bands do fill up with nonzeros. The preferred method for large numbers of points in our mesh is *iteration*, which we discuss next.

7.5 Iterative Methods for Laplace's Equation

When the steady-state heat-flow equation in two dimensions,

$$\nabla^2 u = \frac{\partial^2 u}{\partial x^2} + \frac{\partial^2 u}{\partial y^2} = 0,$$

is approximated by the pictorial relation

$$\nabla^2 u = \frac{1}{h^2} \left\{ 1 \quad \begin{matrix} 1 \\ -4 \\ 1 \end{matrix} \quad 1 \right\} u_{ij} = 0,$$

at a set of grid points (x_i, y_i), a set of simultaneous linear equations results, as Eqs. (7.10), (7.11), and (7.12) illustrate. Proper ordering of the equations gives a *diagonally dominant* system. (Diagonal dominance means that the magnitude of the coefficient on the diagonal is not less than the sum of the magnitudes of the other coefficients in the same row.) As we saw in Chapter 2, such systems can be solved by Gauss–Seidel iteration.

Liebmann's Method

When applied to Laplace's equation, the iterative technique is called *Liebmann's method*. We illustrate this procedure with the same problem as in the last section. The difference equations, when $h = 5$ cm, are

$$\begin{aligned} -4u_1 + u_2 \quad\quad &= 0, \\ u_1 - 4u_2 + u_2 &= 0, \\ u_2 - 4u_3 &= -100. \end{aligned} \tag{7.13}$$

The first step in applying Liebmann's method is to rearrange these equations into a new form by solving each equation for the variable on the diagonal:

$$\begin{aligned} u_1 &= \frac{u_2}{4}, \\ u_2 &= \frac{u_1 + u_3}{4}, \\ u_3 &= \frac{u_2 + 100}{4}. \end{aligned} \tag{7.14}$$

Observe that the pictorial operator can give this rearrangement directly:

$$\left\{ 1 \quad \begin{matrix} 1 \\ -4 \\ 1 \end{matrix} \quad 1 \right\} u_{ij} = 0 \rightarrow u_{ij} = \frac{u_{i+1,j} + u_{i-1,j} + u_{i,j+1} + u_{i,j-1}}{4}. \tag{7.15}$$

We solve Eqs. (7.13) by beginning with some initial approximation $\bar{u}$ to u, the solution vector, substituting values from this approximation into the right-hand sides of Eqs. (7.14) and generating new components of $\bar{u}$ that converge to the solution. (Convergence is usually faster if each new component of $\bar{u}$ is used as soon as it is available. The former procedure is called the Jacobi method, and the latter is called Gauss–Seidel.)

The better the beginning approximation, the more rapid the convergence, but, because the system is diagonally dominant, even a poor initial vector will eventually converge. Inspection of the problem underlying the equations lets us guess at reasonably good starting values: we know that u_1 will be small, that u_2 will be about four times as large as u_1, and that u_3 will be a little more than one fourth of 100. Suppose we begin with $u_3 = 30, u_2 = 7.5, u_1 = 2$. Performing the iterations gives the values shown in Table 7.2.

Table 7.2 Successive iterates for example problem

Point	Initial value	\multicolumn{6}{c}{Values at iteration number:}					
		1	2	3	4	5	6
1	2.0	1.875	1.992	1.812	1.789	1.786	1.786
2	7.5	7.969	7.246	7.156	7.144	7.143	7.143
3	30.0	26.992	26.812	26.789	26.786	26.786	26.786

The iterations converge as anticipated and to the same values as in the previous section when we solved the equations by elimination. We terminate the iterations when all components of $\bar{u}$ converge to constant values. For only three equations, iteration is surely more computational effort than elimination, but for large systems iteration has important advantages.

The most obvious advantage is a greatly reduced demand for computer memory space. In our example problem, if h is 1.25 cm, there are 105 interior points at grid intersections. To store the full 105×105 matrix, we need 44,000 bytes, not even taking into account the right-hand sides and the solution vector. With Liebmann's method, we need to store only the 105 values of u_{ij} plus 44 boundary values, requiring about 600 bytes. The difference in storage requirements is even more striking when $h = 0.625$ cm. The 865,000 bytes for the full matrix is reduced to a very moderate 2300 bytes.

With large systems there is also a very significant reduction in execution time as well, although the magnitude depends on how good a starting vector we can supply, and very strongly on the precision we require. It is customary to terminate the iterations when the largest change in any component of $\bar{u}$ meets a supplied tolerance value. This tolerance should be chosen with consideration for the importance of precision in the final answers (many engineering applications do not need extreme accuracy) and only after determining whether the parameters of the problem (boundary conditions, dimensions, material properties, and so on) are themselves known with great accuracy.

Since the speed of convergence depends on the initial value of $\bar{u}$, it is frequently worthwhile to spend some effort in obtaining a good initial estimate. Some guidelines can be helpful. It is obvious, since the temperature at each point is the *average* of temperatures at surrounding points, that no interior point can have a temperature *greater* than the hottest boundary point, or a temperature *lower* than the coldest. Sometimes the simple expedient of setting all interior points equal to the mean of all the boundary temperatures is used to start iterations. Interior points near cold edges have low temperatures and those near hot edges have higher temperatures. This may let us make good guesses from which to begin iterations. A more refined technique solves the problem with a coarse mesh, getting a quick solution; these values then are used as a basis to fill in initial values at intermediate points.

When our sample problem was solved with $h = 2.5$ (21 interior points), employing the average of boundary values as the starting temperatures, these results were obtained after 21 iterations. (The iterations were stopped when the largest change in any $\bar{u}$ component was less than 0.001.)

$$
\begin{aligned}
&u_1 = 0.354, &u_8 = 0.500, &\quad u_{15} = 0.354, \\
&u_2 = 0.914, &u_9 = 1.291, &\quad u_{16} = 0.914, \\
&u_3 = 2.012, &u_{10} = 2.834, &\quad u_{17} = 2.011, \\
&u_4 = 4.297, &u_{11} = 6.021, &\quad u_{18} = 4.296, \\
&u_5 = 9.154, &u_{12} = 12.655, &\quad u_{19} = 9.154, \\
&u_6 = 19.664, &u_{13} = 26.290, &\quad u_{20} = 19.663, \\
&u_7 = 43.210, &u_{14} = 53.178, &\quad u_{21} = 43.210.
\end{aligned}
$$

Comparison with the values in Section 7.4 shows that the same results are obtained as when the equations are solved by elimination.

Liebmann's method essentially takes care of the problem of memory requirement for the steady-state heat-flow problem (though three-dimensional problems can still be demanding). The chief drawback in Liebmann's method is slow convergence, which is especially acute when there are a larger number of points, because then each iteration is lengthy and, in addition, more iterations are required to reach a given tolerance.

Accelerating Convergence—S.O.R.

The relaxation method of Southwell, as discussed in Section 2.11, would be a way of attaining faster convergence in the iterative method. (In fact, Southwell developed his method to solve potential problems.) As pointed out in that section, relaxation is not adapted to computer solution of sets of equations. Based on Southwell's technique, the use of an overrelaxation factor can give significantly faster convergence, however. Since we handle each equation in a standard and repetitive order, this method is called *successive overrelaxation*, frequently abbreviated as S.O.R.

To show how successive overrelaxation can be applied to Laplace's equation, we begin with Eq. (7.15), adding superscripts to show that a new value is computed from previous iterates,

$$u_{ij}^{(k+1)} = \frac{u_{i+1,j}^{(k)} + u_{i-1,j}^{(k+1)} + u_{i,j+1}^{(k)} + u_{i,j-1}^{(k+1)}}{4}.$$

We now both add and subtract $u_{ij}^{(k)}$ on the right-hand side, getting

$$u_{ij}^{(k+1)} = u_{ij}^{(k)} + \left[\frac{u_{i+1,j}^{(k)} + u_{i-1,j}^{(k+1)} + u_{i,j+1}^{(k)} + u_{i,j-1}^{(k+1)} - 4u_{ij}^{(k)}}{4} \right]. \tag{7.16}$$

[The numerator term will be zero when final values, after convergence, are used. The term in brackets is what Southwell called the "residual," which he "relaxed" to zero by changes in the temperature at (x_i, y_j).]

We can consider the bracketed term in Eq. (7.16) to be an adjustment to the old value $u_{ij}^{(k)}$, to give the new and improved value $u_{ij}^{(k+1)}$. If, instead of adding just the bracketed term, we add a larger value (thus "overrelaxing"), we get faster convergence. We modify Eq. (7.16) by including an overrelaxation factor ω to get the new iterating relation

$$u_{ij}^{(k+1)} = u_{ij}^{(k)} + \omega \left[\frac{u_{i+1,j}^{(k)} + u_{i-1,j}^{(k+1)} + u_{i,j+1}^{(k)} + u_{i,j-1}^{(k+1)} - 4u_{ij}^{(k)}}{4} \right]. \tag{7.17}$$

Maximum acceleration is obtained for some optimum value of ω. This optimum value will always lie between 1.0 and 2.0 for Laplace's equation.

Table 7.3 shows some results that demonstrate how S.O.R. can speed up the convergence in our example problem for $h = 2.5$ (21 interior points) and for $h = 1.25$ (105 interior points.)

Table 7.3 Effect of overrelaxation factor on speed of convergence

$h = 2.5$ cm (21 interior points)		$h = 1.25$ cm (105 interior points)	
ω	Number of iterations	ω	Number of iterations
1.00	20	1.00	70
1.10	15	1.10	58
1.20	13	1.20	46
1.30	**12**	1.30	35
1.40	15	1.40	29
1.50	18	**1.50**	**26**
1.60	23	1.60	28
		1.70	36
ω_{opt} at about 1.3		ω_{opt} at about 1.5	

Note: The example problem using S.O.R. with iterations was continued until maximum change in any component of $\bar{u}$ was less than 0.001.

The optimum value for ω is not always predictable in advance. There are methods of using the results of the first few iterations to find a value of ω that is near optimum.* We do not discuss these here. For a rectangular region having constant boundary conditions (Dirichlet conditions), a reasonable estimate of the optimum ω can be determined as the smaller root of the quadratic equation

$$\left(\cos\frac{\pi}{p} + \cos\frac{\pi}{q}\right)^2 \omega^2 - 16\omega + 16 = 0, \tag{7.18}$$

where p and q are the number of mesh divisions on each side of the rectangular region. Solving the above equation for ω gives

$$\omega_{\text{opt}} = \frac{4}{2 + \sqrt{4 - c^2}},$$

with

$$c = \left(\cos\frac{\pi}{p} + \cos\frac{\pi}{q}\right).$$

For our example problem, Eq. (7.18) predicts $\omega_{\text{opt}} = 1.267$ for $h = 2.5$ and $\omega_{\text{opt}} = 1.532$ for $h = 1.25$. We see from Table 7.3 that there is good agreement with actual results.

To show the range of values for ω_{opt} that are customary in steady-state problems, in Table 7.4 we show solutions to Eq. (7.18) for several values of p and q where $p = q$ (a square region).

Table 7.4

Value of $p = q$	ω_{opt}
2	1.000
3	1.072
5	1.260
10	1.528
20	1.729
100	1.939
∞	2.000

This section has used a somewhat intuitive approach to S.O.R. and values for ω, the overrelaxation factor. Actually this is an eigenvalue problem, as you will see in Section 7.12.

---

* Hageman and Young (1981) present details.

7.6 The Poisson Equation

The methods of the previous section are readily applied to Poisson's equation. We illustrate with an analysis of torsion in a rectangular bar subject to twisting. The torsion function ϕ satisfies the Poisson equation:

$$\nabla^2\phi + 2 = 0, \qquad \phi = 0 \text{ on boundary.}^* \tag{7.19}$$

The tangential stresses are proportional to the partial derivatives of ϕ for a twisted prismatic bar of constant cross section. Let us find ϕ over the cross section of a rectangular bar 6×8 in. in size.

Subdivide the cross section into 2-in. squares, so that there are six interior points, as in Fig. 7.7. In terms of difference quotients, Eq. (7.19) becomes

$$\frac{1}{h^2}\left\{1 \quad \begin{matrix} 1 \\ -4 \\ 1 \end{matrix} \quad 1\right\}\phi_{i,j} + 2 = 0,$$

or

$$\frac{1}{4}\left\{1 \quad \begin{matrix} 1 \\ -4 \\ 1 \end{matrix} \quad 1\right\}\phi_{i,j} + 2 = 0. \tag{7.20}$$

(The function ϕ is dimensional; with our choice of h, ϕ will have square inches as units.)

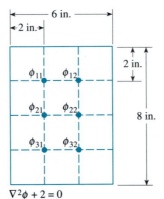

$\nabla^2\phi + 2 = 0$
Figure 7.7

---

* The second term in Eq. (7.19)—here equal to 2—actually should be multiplied by the angular twist per unit length and by the modulus of rigidity for the material.

The set of equations, when (7.20) is applied at each interior point, is

$$0 + 0 + \phi_{12} + \phi_{21} - 4\phi_{11} + 8 = 0,$$
$$\phi_{11} + 0 + 0 + \phi_{22} - 4\phi_{12} + 8 = 0,$$
$$0 + \phi_{11} + \phi_{22} + \phi_{31} - 4\phi_{21} + 8 = 0,$$
$$\phi_{21} + \phi_{12} + 0 + \phi_{32} - 4\phi_{22} + 8 = 0,$$
$$0 + \phi_{21} + \phi_{32} + 0 - 4\phi_{31} + 8 = 0,$$
$$\phi_{31} + \phi_{22} + 0 + 0 - 4\phi_{32} + 8 = 0.$$

(7.21)

Symmetry considerations show that $\phi_{11} = \phi_{12} = \phi_{31} = \phi_{32}$ and $\phi_{21} = \phi_{22}$, so only two unknowns are left in (7.21) after substitutions:

$$\phi_{21} - 3\phi_{11} + 8 = 0,$$
$$2\phi_{11} - 3\phi_{21} + 8 = 0.$$

Obviously, these equations would be solved by elimination; $\phi_{11} = 4.56, \phi_{21} = 5.72$.

Now let us solve the problem with a 1-in.-square mesh using iteration. To use Liebmann's method, we need initial estimates of ϕ. As shown in Fig. 7.8(a), we estimate them, using our previous results as a guide. We converge in 25 iterations (tolerance = 0.001) to the values shown in Fig. 7.8(b) employing Eq. (7.22):*

$$\phi_{ij} = \frac{1}{4}(\phi_{i+1,j} + \phi_{i-1,j} + \phi_{i,j+1} + \phi_{i,j-1} + 8).$$

(7.22)

1.4	2.6	3.0	2.6	1.4
2.5	**4.6**	5.0	**4.6**	2.5
3.0	5.2	5.5	5.2	3.0
3.5	**5.7**	6.0	**5.7**	3.5
3.0	5.2	5.5	5.2	3.0
2.5	**4.6**	5.0	**4.6**	2.5
1.4	2.6	3.0	2.6	1.4

2.042	3.047	3.353	3.047	2.043
3.123	4.794	5.319	4.794	3.123
3.657	5.686	6.335	5.686	3.657
3.181	5.960	6.647	5.960	3.818
3.657	5.686	6.335	5.686	3.657
3.124	4.794	5.319	4.794	3.124
2.043	3.048	3.354	3.048	2.043

(a) Initial values of ϕ for iteration   (b) Final value for ϕ by iterative methods

Figure 7.8
Torsion in a rectangular bar; mesh size of 1 in.

As we have seen before, applying an overrelaxation factor should speed up the convergence. Accordingly, we rewrite Eq. (7.22) with an overrelaxation factor:

$$\phi_{ij}^{(k+1)} = \phi_{ij}^{(k)} + \frac{\omega}{4}(\phi_{i+1,j}^{(k)} + \phi_{i-1,j}^{(k+1)} + \phi_{i,j+1}^{(k)} + \phi_{i,j-1}^{(k+1)} - 4\phi_{ij}^{(k)} + 8).$$

(7.23)

---

*Equation (7.22) differs from Eq. (7.21), the Liebmann method for Laplace's equation, only in that the $f(x, y)$ term of Poisson's equation must be included.

The optimum overrelaxation factor calculated from Eq. (7.18) is 1.383. Using this in Eq. (7.23), we converge in 13 iterations to the same set of values as before and as tabulated in Fig. 7.8(b). Using S.O.R., rather than the standard Liebmann method, cuts down the number of iterations by nearly 50%.

The nine-point formula for the Poisson equation,

$$\nabla^2\phi + f = 0,$$

becomes

$$\frac{1}{6h^2}\begin{Bmatrix} 1 & 4 & 1 \\ 4 & -20 & 4 \\ 1 & 4 & 1 \end{Bmatrix}\phi_{i,j} + f_{i,j} = 0,$$

provided that f is a function of x and y only. The truncation error in this case is still $O(h^6)$. Equation (7.19) becomes

$$\frac{1}{24}\begin{Bmatrix} 1 & 4 & 1 \\ 4 & -20 & 4 \\ 1 & 4 & 1 \end{Bmatrix}\phi_{i,j} + 2 = 0,$$

and the set of equations becomes

$$
\begin{aligned}
-20\phi_{11} + 4\phi_{12} + 4\phi_{21} + \phi_{22} \qquad\qquad\qquad + 48 &= 0, \\
4\phi_{11} - 20\phi_{12} + \phi_{21} + 4\phi_{22} \qquad\qquad\qquad + 48 &= 0, \\
4\phi_{11} + \phi_{12} - 20\phi_{21} + 4\phi_{22} + 4\phi_{31} + \phi_{32} \qquad &= 0, \\
\phi_{11} + 4\phi_{12} + 4\phi_{21} - 20\phi_{22} + \phi_{31} + 4\phi_{32} + 48 &= 0, \\
4\phi_{21} + \phi_{22} - 20\phi_{31} + 4\phi_{32} + 48 &= 0, \\
\phi_{21} + 4\phi_{22} + 4\phi_{31} - 20\phi_{32} + 48 &= 0.
\end{aligned}
$$

7.7 Derivative Boundary Conditions

In the previous examples, we solved for the steady-state temperatures at interior points of a rectangular plate, with the boundary temperatures being specified. In many problems, instead of knowing the boundary temperatures, we know the temperature gradient in the direction normal to the boundary, as for example when heat is being lost from the surface by radiation and conduction.

EXAMPLE 7.1   Consider a rectangular plate within which heat is being uniformly generated at each point at a rate of Q cal/cm³ · sec. The plate is steel and is 4 × 8 cm, 1 cm thick. $k = 0.16$ cal/sec · cm² · °C/cm. For this situation, Poisson's equation holds in the form

$$\nabla^2 u = -\frac{Q}{k} = \frac{1}{h^2} \left\{ \begin{matrix} & 1 & \\ 1 & -4 & 1 \\ & 1 & \end{matrix} \right\} u_{i,j}. \qquad (7.24)$$

Suppose that $Q = 5$ for our example. The top and bottom faces are perfectly insulated so that no heat is lost, while the upper and lower edges lose heat such that

$$\frac{\partial u}{\partial y} = -15°\text{C/cm}$$

in the outward direction, and the right and left edges are held at a constant temperature $u = 20°$C. We will find the steady-state temperatures at points in the plate.

In Fig. 7.9 we sketch the plate, with a gridwork to give 21 interior points with $h = 1$ cm. In this problem, the upper and lower edge temperatures are also unknown, increasing to 35 the total number of points at which u is to be determined. Because of symmetry, the equations can be written in terms of 12 quantities, as indicated on the diagram.

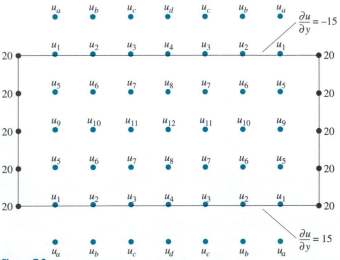

Figure 7.9

We now write Eq. (7.24) at each unknown point, which is our general procedure in elliptic partial-differential equations, except that at the upper and lower edges we cannot form the five-point combination—there are not enough points to form the star. We get around this by the device of extending our network to a row of exterior points. We utilize these fictitious exterior points to include in the set of equations those whose central point is on the upper or lower edge:

$$
\begin{aligned}
(20 + u_a + u_2 + u_5 - 4u_1) &= -5/0.16, \\
(u_1 + u_b + u_3 + u_6 - 4u_2) &= -5/0.16, \\
(u_2 + u_c + u_4 + u_7 - 4u_3) &= -5/0.16, \\
(u_3 + u_d + u_3 + u_8 - 4u_4) &= -5/0.16, \\
(20 + u_1 + u_6 + u_9 - 4u_5) &= -5/0.16, \\
(u_5 + u_2 + u_7 + u_{10} - 4u_6) &= -5/0.16, \\
(u_6 + u_3 + u_8 + u_{11} - 4u_7) &= -5/0.16, \\
(u_7 + u_4 + u_7 + u_{12} - 4u_8) &= -5/0.16, \\
(20 + u_5 + u_{10} + u_5 - 4u_9) &= -5/0.16, \\
(u_9 + u_6 + u_{11} + u_6 - 4u_{10}) &= -5/0.16, \\
(u_{10} + u_7 + u_{12} + u_7 - 4u_{11}) &= -5/0.16, \\
(u_{11} + u_8 + u_{11} + u_8 - 4u_{12}) &= -5/0.16.
\end{aligned}
\tag{7.25}
$$

It would appear that we have not helped ourselves by the fictitious points outside the plate. We have made it possible to write twelve equations, but four new unknowns have been introduced. However, we have not yet utilized the gradient conditions. Write the derivative conditions as central-difference quotients, as discussed in Chapter 4:

$$
\begin{aligned}
-\left(\frac{\partial u}{\partial y}\right)_1 &= \frac{u_5 - u_a}{2(1)} = 15, & u_a &= u_5 - 30, \\
-\left(\frac{\partial u}{\partial y}\right)_2 &= \frac{u_6 - u_b}{2(1)} = 15, & u_b &= u_6 - 30, \\
-\left(\frac{\partial u}{\partial y}\right)_3 &= \frac{u_7 - u_c}{2(1)} = 15, & u_c &= u_7 - 30, \\
-\left(\frac{\partial u}{\partial y}\right)_4 &= \frac{u_8 - u_d}{2(1)} = 15, & u_d &= u_8 - 30.
\end{aligned}
\tag{7.26}
$$

We write these approximations for the gradient condition, choosing the proper order of points so that the outward normal is negative, since heat is flowing outwardly. Using a central-difference approximation of error $O(h^2)$ makes these compatible with our other difference-quotient approximations. Each of the difference quotients allows us to write the fictitious temperature values in terms of points within the rectangle.

We solve the set of equations in (7.25) after eliminating the fictitious points by using Eq. (7.26). Elimination, iteration (Liebmann's method), or relaxation may be used. The solution of the equations is left as an exercise.

▲

We can apply the same method to the more general boundary condition,

$$au - b\frac{\partial u}{\partial x} = c,$$

where a, b, and c are constants, in an obvious application of the relationships of Eq. (7.26).

7.8 Irregular Regions and Nonrectangular Grids

When the boundary of the region is not such that the network can be drawn to have the boundary coincide with the nodes of the mesh, we must proceed differently at points near the boundary. Consider the general case of a group of five points whose spacing is nonuniform, arranged in an unequal-armed star. We represent each distance by $\theta_i h$, where θ_i is the fraction of the standard spacing h that the particular distance represents (Fig. 7.10). Along the line from u_1 to u_0 to u_3, we may approximate the first derivatives:

$$\left(\frac{\partial u}{\partial x}\right)_{1-0} \doteq \frac{u_0 - u_1}{\theta_1 h} ; \qquad \left(\frac{\partial u}{\partial x}\right)_{0-3} \doteq \frac{u_3 - u_0}{\theta_3 h} .$$

Since

$$\frac{\partial^2 u}{\partial x^2} = \frac{\partial}{\partial x}\left(\frac{\partial u}{\partial x}\right),$$

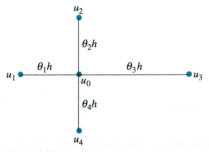

Figure 7.10

we have

$$\frac{\partial^2 u}{\partial x^2} \doteq \frac{(u_3 - u_0)/(\theta_3 h) - (u_0 - u_1)/(\theta_1 h)}{\frac{1}{2}(\theta_1 + \theta_3)h} = \frac{2}{h^2}\left[\frac{u_1 - u_0}{\theta_1(\theta_1 + \theta_3)} + \frac{u_3 - u_0}{\theta_3(\theta_1 + \theta_3)}\right]. \quad \textbf{(7.27)}$$

Similarly,

$$\frac{\partial^2 u}{\partial y^2} \doteq \frac{2}{h^2}\left[\frac{u_2 - u_0}{\theta_2(\theta_2 + \theta_4)} + \frac{u_4 - u_0}{\theta_4(\theta_2 + \theta_4)}\right]. \quad \textbf{(7.28)}$$

The expressions in Eqs. (7.27) and (7.28) have errors $O(h)$, which introduce larger errors in the computations than for points that are arranged in an equal-armed star.

Combining, we get

$$\nabla^2 u = \frac{\partial^2 u}{\partial x^2} + \frac{\partial^2 u}{\partial y^2}$$

$$= \frac{2}{h^2}\left[\frac{u_1}{\theta_1(\theta_1 + \theta_3)} + \frac{u_2}{\theta_2(\theta_2 + \theta_4)} + \frac{u_3}{\theta_3(\theta_1 + \theta_3)} + \frac{u_4}{\theta_4(\theta_2 + \theta_4)}\right.$$

$$\left. -\left(\frac{1}{\theta_1\theta_3} + \frac{1}{\theta_2\theta_4}\right)u_0\right]. \quad \textbf{(7.29)}$$

Instead of our standard operator, we use the operator of Eq. (7.29) for points adjacent to boundary points when the boundary points do not coincide with the mesh. If the boundary conditions involve normal derivatives, great complications arise, especially for curved boundaries. The finite-element method of Section 7.11 offers less difficulty.

EXAMPLE 7.2 Let us illustrate this procedure with an example. A semicircular plate of radius a has the temperature at the base (the straight side) held at $0°$ while the circumference is held at $c°$. We desire the steady-state temperatures. The analytical solution to this problem is given by the infinite series

$$u(r, \theta) = \frac{4c}{\pi}\sum_{n=1}^{\infty}\frac{1}{2n - 1}\left(\frac{r}{a}\right)^{2n-1}\sin(2n - 1)\theta, \quad \textbf{(7.30)}$$

where (r, θ) are the polar coordinates of a point on the plate. (For points near the circumference, several hundred terms are needed to compute temperatures with any degree of accuracy.) There is right-to-left symmetry in the temperatures.

The finite-difference method superimposes a gridwork on the plate. Suppose we take these values for the parameters of the problem: $a = 1, c = 100$. With $h = 0.2$, the diagram of Fig. 7.11 results; there are 17 unknowns after we utilize the left-to-right symmetry. For points u_2, u_3, u_{12}, and u_{17}, the grid does not coincide with the boundary. It is easy to find by analytic geometry that the short arms at u_2 and u_{17} have a length of $0.8990h$ and those at u_3 and u_{12} have a length of $0.5826h$. Applying Eq. (7.29), we get these pictorial operators:

$$\text{At point 2:} \qquad \frac{1}{(0.2)^2} \left\{ \begin{matrix} & 1.1715 & \\ 1 & -4.2247 & 1 \\ & 1.0532 & \end{matrix} \right\} u_2 = 0;$$

$$\text{At point 3:} \qquad \frac{1}{(0.2)^2} \left\{ \begin{matrix} & 2.1691 & \\ 1 & -5.4328 & 1 \\ & 1.2637 & \end{matrix} \right\} u_3 = 0;$$

$$\text{At point 12:} \qquad \frac{1}{(0.2)^2} \left\{ \begin{matrix} & 1 & \\ 1.2637 & -5.4328 & 2.1691 \\ & 1 & \end{matrix} \right\} u_{12} = 0;$$

$$\text{At point 17:} \qquad \frac{1}{(0.2)^2} \left\{ \begin{matrix} & 1 & \\ 1.0532 & -4.2247 & 1.1715 \\ & 1 & \end{matrix} \right\} u_{17} = 0.$$

(When computing these operators, it is useful to observe that the central number is the negative of the sum of the other four values. This helps check for accuracy.)

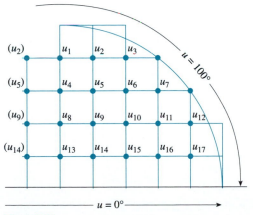

Figure 7.11

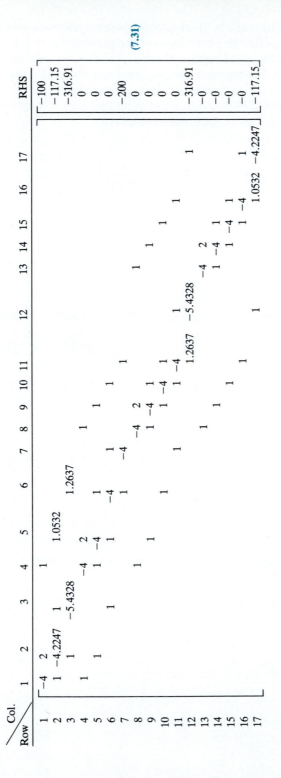

(7.31)

At all the other points, the standard operator applies. The matrix for the set of equations is given in Eq. (7.31).

Observe how sparse matrix (7.31) is. For a larger number of equations, the density of nonzero values will be even less, since no row ever has more than five nonzero values.

When the equations of (7.31) are solved, the values in Table 7.5 result. The analytical results given by Eq. (7.30) are included for comparison, using values of r and θ to coincide with the grid points. In general, the agreement is good. One might expect greater errors at points where the less accurate unequal-arm operators were used. This is not the case for this example, however.

Table 7.5   Comparison of results for Example 7.2 by finite differences, and analytical solution

x	y	Point number	u, finite differnces	u, series solution, Eq. (7.30)
0	0.8	1	86.053	85.906
0.2	0.8	2	87.548	87.417
0.4	0.8	3	92.124	92.094
0	0.6	4	69.116	68.807
0.2	0.6	5	70.773	70.482
0.4	0.6	6	75.994	75.772
0.6	0.6	7	85.471	85.405
0	0.4	8	48.864	48.448
0.2	0.4	9	50.436	50.000
0.4	0.4	10	55.606	55.151
0.6	0.4	11	65.891	65.593
0.8	0.4	12	84.189	84.195
0	0.2	13	25.466	25.133
0.2	0.2	14	26.501	26.109
0.4	0.2	15	30.102	29.527
0.6	0.2	16	38.300	37.436
0.8	0.2	17	57.206	57.006

If our mesh of points is chosen very fine (as we would do to get high accuracy), there is an even simpler way to handle irregular boundaries. One uses the closest mesh point as the boundary point, thus in effect perturbing the actual region to one that coincides with the network, and then the standard operator of Eq. (7.24) applies everywhere. This introduces some error, of course, but often its effect is no worse than the $O(h)$ operator of (7.29). It is also usually easier to program a computer solution using this perturbed-region technique.

Sometimes the region, while not fit by a rectangular mesh, can be fit with nodes in a different arrangement.

Circular Regions—Polar Coordinates

For circular regions, one may derive a finite-difference approximation to the Laplacian in polar coordinates. Consider the group of points that are the nodes of a polar-coordinate network (Fig. 7.12):

$$\nabla^2 u = \frac{\partial^2 u}{\partial r^2} + \frac{1}{r}\frac{\partial u}{\partial r} + \frac{1}{r^2}\frac{\partial^2 u}{\partial \theta^2}$$

$$= \frac{u_3 - 2u_0 + u_1}{(\Delta r)^2} + \frac{1}{r_0}\frac{u_3 - u_1}{2\Delta r} + \frac{1}{r_0^2}\frac{u_2 - 2u_0 + u_4}{(\Delta \theta)^2}.$$

No "standard" operator can be written for this relation, so finding the coefficients and solving the set of equations by iteration when the problem is in polar coordinates is awkward, but it helps to group the various coefficients of each term.

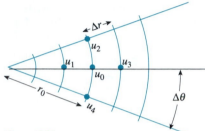

Figure 7.12

Calling u_0 the central point, u_1 the point nearer the center, u_3 the point farther from the center, and u_2 and u_4 the points on either side at the same radial distance from the center, as shown in Fig. 7.12, we get, for Laplace's equation

$$\nabla^2 u = \frac{1}{(\Delta r)^2}\left[\left(1 - \frac{\Delta r}{2r_0}\right)u_1 + \left(1 + \frac{\Delta r}{2r_0}\right)u_3 + \left(\frac{\Delta r}{r_0\,\Delta \theta}\right)^2 u_2\right.$$
$$\left. + \left(\frac{\Delta r}{r_0\,\Delta \theta}\right)^2 u_4 - \left(2 + 2\left(\frac{\Delta r}{r_0\,\Delta \theta}\right)^2\right)u_0\right] = 0. \tag{7.32}$$

EXAMPLE 7.3 Let us illustrate the Laplacian in polar coordinates by again solving the example presented at the beginning of this section. We computed the steady-state temperatures on a semicircular plate of radius equal to one, with its base held at $0°$ and its circumference at $100°$. Now we superimpose a polar-coordinate system; with $\Delta r = 0.2$ and $\Delta \theta = \pi/8$, Fig. 7.13 results. There are 16 u-values to be determined, after we allow for left-to-right symmetry.

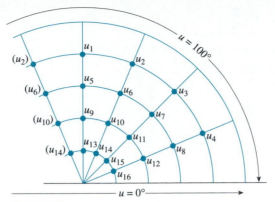

Figure 7.13

When we compute the coefficients for Eq. (7.32) applied to each of the 16 points, the matrix representation of matrix (7.33) results. Table 7.6 compares the results by finite differences with those computed from the infinite-series solution, Eq. (7.30).

The agreement in Table 7.6 is about as good as that observed in Table 7.5 with a rectangular grid. However, one point, no. 4, stands out in Table 7.6 as having an unusually large error. Large errors often occur in finite-difference solutions near a discontinuity in the boundary conditions, and this could possibly be the explanation. On the other hand, this phenomenon is absent in Table 7.5, possibly due to a cancellation of errors.

Table 7.6 Results for Example 7.3 by finite differences in polar coordinates, and analytical solution

r	θ	Point number	u, finite differences	u, series solution Eq. (7.30)
0.8	$\pi/2$	1	85.661	85.906
0.8	$3\pi/8$	2	84.425	84.792
0.8	$\pi/4$	3	79.459	80.394
0.8	$\pi/8$	4	63.679	66.186
0.6	$\pi/2$	5	68.371	68.808
0.6	$3\pi/8$	6	66.126	66.673
0.6	$\pi/4$	7	58.058	58.862
0.6	$\pi/8$	8	39.165	39.624
0.4	$\pi/2$	9	48.048	48.448
0.4	$3\pi/8$	10	45.543	45.938
0.4	$\pi/4$	11	37.457	37.730
0.4	$\pi/8$	12	22.374	22.252
0.2	$\pi/2$	13	25.003	25.133
0.2	$3\pi/8$	14	23.301	23.393
0.2	$\pi/4$	15	18.245	18.241
0.2	$\pi/8$	16	10.147	10.066

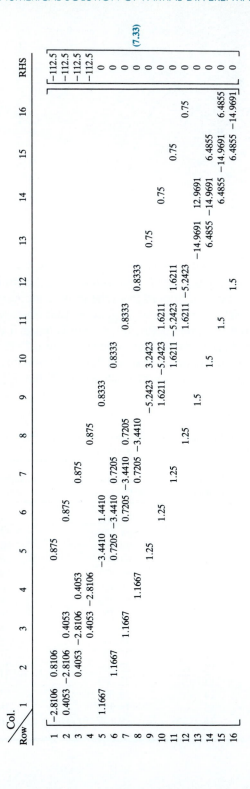

(7.33)

We observe another sparse matrix in Eq. (7.33), this time with a band structure. The structure will depend on the ordering of the points, however.

▲

Cylindrical and Spherical Coordinates

Some three-dimensional problems can be more easily expressed in terms of cylindrical coordinates. In that coordinate system, the Laplacian is

$$\nabla^2 u = \frac{\partial^2 u}{\partial r^2} + \frac{1}{r}\frac{\partial u}{\partial r} + \frac{1}{r^2}\frac{\partial^2 u}{\partial \theta^2} + \frac{\partial^2 u}{\partial z^2},$$

in agreement with that for polar coordinates except for the addition of the z-derivative. Thus we need only to add a central-difference approximation for the z-derivative to the expression in Eq. (7.32). This involves points a distance of Δz away from node 0. (These nodes are above and below the plane of the paper in Fig. 7.12.)

Spherical coordinates are also frequently used in three-dimensional problems. In this coordinate system, where a node is defined by values of r, θ, and ϕ, the Laplacian is defined as

$$\nabla^2 u = \frac{\partial^2 u}{\partial r^2} + \frac{2}{r}\frac{\partial u}{\partial r} + \frac{1}{r^2}\frac{\partial^2 u}{\partial \theta^2} + \frac{\cot(\theta)}{r^2}\frac{\partial u}{\partial \theta} + \frac{1}{r^2 \sin^2(\theta)}\frac{\partial^2 u}{\partial \phi^2}.$$

Occasionally, a two-dimensional problem can be represented numerically by nodes in an equispaced triangular network. A pictorial operator can be derived for this by using formulas for rotation of axes. We omit a detailed development, giving only the final result for an approximation to the Laplacian:

$$\nabla^2 u = \frac{2}{3}\left(\frac{\partial^2 u}{\partial a^2} + \frac{\partial^2 u}{\partial b^2} + \frac{\partial^2 u}{\partial c^2}\right),$$

where a, b, and c represent three axes, each 60° apart. A pictorial operator is

$$\nabla^2 u = \frac{2}{3h^2}\left\{ \begin{matrix} & 1 & & 1 & \\ 1 & & -6 & & 1 \\ & 1 & & 1 & \end{matrix} \right\}.$$

Notice that, for Laplace's equation, the potential is the arithmetic average of the potentials at its six equidistant neighbors.

7.9 The Laplacian Operator in Three Dimensions

If we have a region whose sides are parallel to the rectangular coordinate axes, writing a difference equation to approximate the three-dimensional Laplacian is straightforward. We use triple subscripts to indicate the spatial position of points, and take the mesh distance the same in each direction:

$$\nabla^2 u = \frac{\partial^2 u}{\partial x^2} + \frac{\partial^2 u}{\partial y^2} + \frac{\partial^2 u}{\partial z^2}$$

$$= \frac{u_{i+1,j,k} - 2u_{i,j,k} + u_{i-1,j,k}}{h^2}$$

$$+ \frac{u_{i,j+1,k} - 2u_{i,j,k} + u_{i,j-1,k}}{h^2} + \frac{u_{i,j,k+1} - 2u_{i,j,k} + u_{i,j,k-1}}{h^2}$$

$$\nabla^2 u = \frac{1}{h^2} \left\{ 1 \underset{1}{\overset{1}{\underset{\diagdown}{\diagup}}} -6 \overset{1}{\diagup} 1 \right\} u_{i,j,k}.$$

We see again, for Laplace's equation $\nabla^2 u = 0$, that the potential u is the arithmetic average of its six nearest neighboring values. The set of equations for this case is more extensive and more tedious to solve, but in principle the methods are unchanged. In hand calculations using iterative methods, keeping track of the successive values of u is awkward. An isometric projection of the points is generally recommended over the use of superimposed sheets of paper. In a computer, triple subscripting solves the problem, but large storage requirements are imposed by problems encountered in practice.

In three-dimensional problems, it is easy to exceed the memory space of even large computer systems. For example, if the volume under consideration has 100 points on a side, a total of $100^3 = 1,000,000$ points are involved. If we were to represent the coefficients of the one million equations in a square matrix, no computer system could hold all the values in real memory. Virtual memory systems could perhaps accommodate that many values, but the speed of computation would be excruciatingly slow.

Of course, there are at most seven nonzero coefficients in any of these one million equations, but even 8 million numerical values (the right-hand sides are needed too) will almost always exceed the memory capacity. Obviously, the problem must be cut down in size. However, even only 30 points on each side of a cubical volume results in 27,000 equations, with about 216,000 nonzero numerical values. While such a number of coefficients can be stored in large real memories or in virtual memory systems, lots of computer time is involved in iterating through the large set of equations. Unfortunately, many more iterations are required for convergence where the number of equations is great. It is little wonder that three-dimensional problems, while perfectly tractable in principle, are terribly expensive to solve. A solution is presented in the next section.

7.10 Matrix Patterns, Sparseness, and the A.D.I. Method

All of the examples in this chapter illustrate the fact that the coefficient matrix is sparse when an elliptic partial-differential equation is solved by the finite-difference method. Especially in a three-dimensional case, the number of nonzero coefficients is a small fraction of the total.

The relative sparseness increases as the number of equations increases. In the two-dimensional example of Section 7.4, the 21×21 coefficient matrix of Eq. (7.12) has 81% of its positions filled with zeros. If symmetry were taken into account, there would have been a 14×14 coefficient matrix with 71% zeros. Decreasing the value of h so as to have 105 points within the region gives rise to a coefficient matrix with 96% zeros. For a $30 \times 30 \times 30$ three-dimensional case, there would be 90,176 nonzero coefficients, but this is only 0.012% of the 729×10^6 values in the coefficient matrix!

In Chapter 2 it was shown that iterative methods are usually preferred for sparse matrices unless they have a tridiagonal structure. The reason is that elimination does not preserve the sparseness (unless the matrix is tridiagonal), so that one cannot work only with nonzero terms, as illustrated by Fig. 7.14. The original matrix in part (a) is shown in partially triangularized form in part (b). The shaded portion of the matrix, which originally had 50% zeros, has lost all but one of these.

Frequently the coefficient matrix has a band structure, as illustrated by Eqs. (7.31) and (7.33). Equation (7.33) illustrates a special regularity for the nonzero elements; the ordering of the points must be carefully done to attain this. A band structure is worth working for, however, since elimination does not introduce nonzero terms outside of the limits defined by the original bands. Zeros in the gaps between the parallel lines are not preserved, though, so the "tightest" possible bandedness is preferred. Sometimes it is possible to order the points so that a pentadiagonal matrix results.

$$
\begin{bmatrix}
3 & 1 & 0 & 0 & 1 & 0 & 0 & 2 \\
1 & 2 & 1 & 0 & 0 & 1 & 0 & 0 \\
0 & 2 & 3 & 1 & 0 & 0 & 1 & 0 \\
1 & 0 & 0 & 2 & 0 & 0 & 1 & 1 \\
0 & 1 & 1 & 3 & 0 & 0 & 1 & 0 \\
0 & 2 & 0 & 1 & 0 & 2 & 0 & 3 \\
2 & 0 & 0 & 3 & 0 & 1 & 1 & 0 \\
0 & 1 & 0 & 1 & 0 & 0 & 2 & 3
\end{bmatrix}
\qquad
\begin{bmatrix}
3 & 1 & 0 & 0 & 1 & 0 & 0 & 2 \\
0 & 1.677 & 1 & 0 & -0.333 & 1 & 0 & -0.666 \\
0 & 0 & 1.8 & 1 & 0.4 & -1.2 & 1 & 0.8 \\
0 & 0 & 0.6 & 2 & -0.4 & 0.2 & 1 & 0.2 \\
0 & 0 & 0.4 & 2.4 & 0.2 & -0.6 & 1 & 0.4 \\
0 & 0 & -1.2 & 1 & 0.4 & 0.8 & 0 & 3.8 \\
0 & 0 & 0.4 & 3 & -0.8 & 1.4 & 1 & -1.6 \\
0 & 0 & -0.6 & 1 & 0.2 & -0.6 & 2 & 3.5
\end{bmatrix}
$$

(a)   (b)

Figure 7.14

The best of the band structures is tridiagonal, with corresponding economy of storage and speed of solution, as discussed in Chapter 2. A method for the steady-state heat equation, called the *alternating-direction-implicit* (A.D.I.) method, results in tridiagonal matrices and is of growing popularity. It was initially developed for unsteady-state problems, which are the subject of Chapter 8.

A.D.I. is particularly useful in three-dimensional problems, but the method is more easily explained in two dimensions. When we use A.D.I. in two dimensions, we write Laplace's equation as

$$\nabla^2 u = \frac{u_L - 2u_0 + u_R}{(\Delta x)^2} + \frac{u_A - 2u_0 + u_B}{(\Delta y)^2} = 0, \tag{7.34}$$

where the subscripts L, R, A, and B indicate nodes left, right, above, and below the central node 0. If $\Delta x = \Delta y$, we can rearrange to the iterative form

$$u_L^{(k+1)} - 2u_0^{(k+1)} + u_R^{(k+1)} = -u_A^{(k)} + 2u_0^{(k)} - u_B^{(k)}. \tag{7.35}$$

In using Eq. (7.35), we proceed through the nodes by rows, solving a set of equations (they are tridiagonal) that consider the values at nodes above and below as fixed quantities that are put into the right-hand sides of the equations.

After the rowwise traverse, we then do a similar set of computations but traverse the nodes columnwise:

$$u_A^{(k+2)} - 2u_0^{(k+2)} + u_B^{(k+2)} = -u_L^{(k+1)} + 2u_0^{(k+1)} - u_R^{(k+1)}. \tag{7.36}$$

This removes the bias that would be present if we used only Eq. (7.35).

The name *alternating-direction-implicit* comes from the fact that we alternate the direction after each traverse. It is implicit, because we do not get u_0-values directly but only through solving a set of equations.

As in other iterative methods, we can accelerate convergence. We introduce an acceleration factor, ρ, by rewriting Eqs. (7.35) and (7.36):

$$u_0^{(k+1)} = u_0^{(k)} + \rho(u_A^{(k)} - 2u_0^{(k)} + u_B^{(k)}) + \rho(u_L^{(k+1)} - 2u_0^{(k+1)} + u_R^{(k+1)}),$$
$$u_0^{(k+2)} = u_0^{(k+1)} + \rho(u_L^{(k+1)} - 2u_0^{(k+1)} + u_R^{(k+1)}) + \rho(u_A^{(k+2)} - 2u_0^{(k+2)} + u_B^{(k+2)}).$$

Rearranging further to give the tridiagonal systems, we get

$$-u_L^{(k+1)} + \left(\frac{1}{\rho} + 2\right)u_0^{(k+1)} - u_R^{(k+1)} = u_A^{(k)} + \left(\frac{1}{\rho} - 2\right)u_0^{(k)} + u_B^{(k)} \tag{7.37}$$

and

$$-u_A^{(k+2)} + \left(\frac{1}{\rho} + 2\right)u_0^{(k+2)} - u_B^{(k+2)} = u_L^{(k+1)} + \left(\frac{1}{\rho} - 2\right)u_0^{(k+1)} + u_R^{(k+1)}. \tag{7.38}$$

EXAMPLE 7.4   A rectangular plate is 6 in. × 8 in. The top edge (an 8-in. edge) is held at 100°, the right edge at 50°, and the other two edges at 0°. Use A.D.I. to find the steady-state temperatures at nodes spaced 1 in. apart within the plate.

There are $(5)(7) = 35$ interior nodes, so we have 35 equations in each set. Program 7.3, at the end of this chapter, solves this problem. With $\rho = 1.5$, convergence to a tolerance value (the maximum change in any node-value) of 0.001 is achieved in 23 iterations. The starting values for interior nodes are the average of the boundary node-values. Table 7.7 shows the results. Figure 7.15 shows the maximum change in the computed values after 23 iterations when ρ is changed.

Table 7.7   Steady-state temperatures at interior nodes

48.523	66.829	74.670	78.204	79.342	77.985	71.465
27.262	44.123	53.645	58.804	61.179	61.133	57.873
16.404	28.755	36.984	42.190	45.436	47.496	48.895
9.599	17.509	23.345	27.536	30.879	34.519	40.210
4.484	8.336	11.352	13.728	16.025	19.492	27.426

For this particular example, the number of nodes is small enough that the standard S.O.R. method can be used. This method is slightly more efficient because the same interior temperatures are found after only 15 iterations with an overrelaxation factor of 1.4.

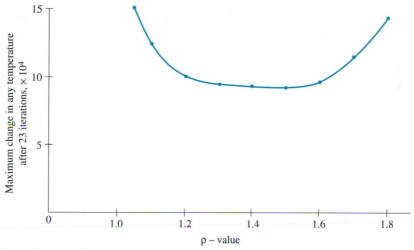

Figure 7.15   Effect of varying ρ

For elliptic problems in three dimensions, the A.D.I. method has great advantages. The directions alternate cyclically through the three coordinates, with new values of u being used in only one direction for each step, keeping the coefficient matrices tridiagonal.

7.11   The Finite-Element Method

The finite-difference method, thus far the major topic in this chapter, poses great difficulties when applied to regions that are irregular. Although Section 7.8 provided some ways to cope with this situation, they are awkward to use and often sacrifice accuracy. Further, it is not easy to locate nodes more closely in subregions where improved accuracy is needed. Analytical methods also are not well adapted to regions of unusual shapes.

The finite-element method has no such problems. As we saw in Chapter 6 for one-dimensional boundary-value problems, nodes can be placed wherever the problem solver desires with the finite-element method. This is also true for two- and three-dimensional regions. They can be placed along any boundary so as to approximate it closely. It is the method of choice for solving elliptic partial-differential equations on regions of arbitrary shape.

While setting up the equations that solve partial-differential equations is no easy task, computer programs are available that do so. It is important to understand how this method works, although this text cannot give everything that today's scientists and engineers might want to know. Our treatment will give a basic knowledge.

The introduction to finite elements in Chapter 6 is important background for what we shall do here. Recall that two ways of applying variational methods to subdivisions of the region of interest were presented: Rayleigh–Ritz and Galerkin. The first of these minimized the functional for the problem by setting partial derivatives to zero; the second set integrals of a weighted residual to zero. The two methods are equivalent for most problems, and both can be used for elliptic equations. We choose the former, in part to provide variety from the presentation in Chapter 6.

The elliptic equation that we will solve in this section is

$$u_{xx} + u_{yy} + Q(x,y)u = F(x,y) \tag{7.39}$$

on region R that is bounded by curve L, with boundary conditions

$$u(x,y) = u_0 \text{ on } L_1, \qquad \frac{\partial u}{\partial n} = \alpha u + \beta \quad \text{on } L_2,$$

where $\partial u/\partial n$ is the outward normal gradient.

Observe that we have Dirichlet conditions on some parts of the boundary and mixed boundary conditions on other parts. For our notation, we will use $u(x,y)$ as the exact solution to Eq. (7.39) and $v(x,y)$ as our approximation to $u(x,y)$. Although the finite-element method is most often used when the region is three dimensional, we will simplify the development by doing it in only two dimensions.

Here is our plan of attack:

Step 1. Find the functional that corresponds to the partial-differential equation. This is well known for a large class of problems.

Step 2. Subdivide the region into subregions (elements). While many kinds of elements can be used, our treatment will consider only triangular elements. The elements must span the entire region and approximate the boundary relatively closely. Every node (the vertices of our triangular elements) and every side of the triangles must be common with adjacent elements except for sides on the boundaries.

Step 3. Write an interpolating relation that gives values for the dependent variable within an element based on the values at the nodes (the vertices of the triangles). We will use linear interpolation from the three nodal values for the element. We will write the interpolation function as the sum of three terms; each term involves a quantity c_i, the value of $v(x, y)$ at a node.

Step 4. Substitute the interpolating relation into the functional, and set the partial derivatives of the functional with respect to each c to zero. This gives three equations, with the c's as unknowns for each element.

Step 5. Combine together (assemble) the element equations of step 4 to get a set of system equations. Adjust these for the boundary conditions of the problem, then solve. This will give the values for the unknown nodal values, the c's, that are approximations to $u(x, y)$ at the nodes. We can get approximations to $u(x, y)$ at intermediate points in the region by using the interpolating relations.

We will discuss each step in turn.

Step 1. Find the Functional

For Eq. (7.39) the functional is well known:

$$I[u] = \iint\limits_{Region} \left[\left(\frac{\partial u}{\partial x} \right)^2 + \left(\frac{\partial u}{\partial y} \right)^2 - Qu^2 + 2Fu \right] dx\, dy - \int_{L_2} [\alpha u^2 + 2\beta u]\, d_L. \quad \textbf{(7.40)}$$

It is possible to develop Eq. (7.40) using the Galerkin technique. Workers in the field of structural analysis usually derive it from the principle of virtual work. We will take it as a given.

Step 2. Subdivide the Region

As stipulated, we will use triangular elements, which will be defined by our choice of nodes. The placement of nodes is, in part, an art. In general, we place nodes close together in subregions where the solution is expected to vary rapidly. It is advantageous to make the sides run in the direction of the largest gradient. Along the curved parts of the boundary, nodes should be placed so that a side of the triangle closely approximates the boundary.

Some of these recommendations depend on knowing the nature of the solution in advance. Often, however, a better placement for the nodes can be accomplished after some preliminary computations or after preliminary trials using the finite-element method with nodes placed arbitrarily.

The chore of defining the nodes' coordinates is facilitated by computer programs that allow the user to place nodes with a pointing device on a graphical display of the region. These programs even permit rotating three-dimensional regions or looking at cross sections. Once the nodes have been located, the program connects them to create the elements.

Computer routines are available that can divide any given planar region into triangles automatically, but they usually do not have the expertise of an experienced engineer.

Step 3. Write the Interpolating Relations

This part of the development is longer than the previous one. As stated, we will use a linear relation. Figure 7.16(a) is a sketch of typical element (i) whose nodes are numbered r, s, and t in counterclockwise direction. The nodal values are c_r, c_s, and c_t.

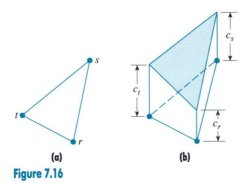

(a)   **(b)**

Figure 7.16

Within typical element (i), we write

$$v(x,y) = N_r c_r + N_s c_s + N_t c_t = \sum_{j=r,s,t} N_j c_j$$

$$= (N_r \quad N_s \quad N_t)\begin{Bmatrix} c_r \\ c_s \\ c_t \end{Bmatrix} = (N)\{c\}, \tag{7.41}$$

where the N's (called *shape functions*) will be defined so that $v(x,y)$ at an interior point is a linear interpolation from the nodal values, the c's. We have shown in Eq. (7.41) that $v(x,y)$ can be expressed as the product of vectors (N) and $\{c\}$. (We use parentheses to enclose a row vector and curly brackets to enclose a column vector.) Vector and matrix notation will be useful in this section. We will indicate matrix M by $[M]$.

Figure 7.16(b) suggests that $v(x,y)$ lies on the plane above the element that passes through the nodal values. Equation (7.41) does not define $v(x,y)$ outside of element (i); there will be similar expressions for the other elements, but their N's and c's will differ.

A sketch of the entire region would not show $v(x,y)$ as a plane. Instead, it would be a surface composed of planar facets, each in a plane above an element. $v(x,y)$ for the entire region is continuous, but $v'(x,y)$ is not. (This is one of the flaws in our choice of element. Some other element definitions do not have this flaw.)

Another name for the N's of Eq. (7.41) is *pyramid function*. The reason for this name is illustrated in Fig. 7.17, where N_s of Fig. 7.16 is drawn. Its height at node s is unity and zero at the other nodes. It looks like an unsymmetrical pyramid whose base is the element with its apex directly above node s. The other two N's are similar. It is obvious that the N's are functions of x and y and that the c's are independent of x and y. We now develop expressions for the N's.

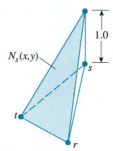

Figure 7.17

Since $v(x,y)$ varies linearly with position within the element, an alternative way to write the linear relation is

$$v(x,y) = a_1 + a_2 x + a_3 y = (1 \quad x \quad y)\{a\}, \tag{7.42}$$

which must agree with the nodal values when $(x,y) = (x_j, y_j), j = r, s, t$. Hence

$$v \text{ at } r: \quad c_r = a_1 + a_2 x_r + a_3 y_r,$$
$$v \text{ at } s: \quad c_s = a_1 + a_2 x_s + a_3 y_s,$$
$$v \text{ at } t: \quad c_t = a_1 + a_2 x_t + a_3 y_t.$$

This is a system of equations

$$[M]\{a\} = \{c\}, \quad \text{(Curly brackets show a column vector)} \tag{7.43}$$

where

$$[M] = \begin{bmatrix} 1 & x_r & y_r \\ 1 & x_s & y_s \\ 1 & x_t & y_t \end{bmatrix}, \quad \{a\} = \begin{Bmatrix} a_1 \\ a_2 \\ a_3 \end{Bmatrix}, \quad \{c\} = \begin{Bmatrix} c_r \\ c_s \\ c_t \end{Bmatrix}.$$

Solving for $\{a\}$:

$$\{a\} = [M^{-1}]\{c\}.$$

The inverse of M is not difficult to find:

$$[M^{-1}] = \frac{1}{2(\text{Area})}\begin{bmatrix} (x_s y_t - x_t y_s) & (x_t y_r - x_r y_t) & (x_r y_s - x_s y_r) \\ (y_s - y_t) & (y_t - y_r) & (y_r - y_s) \\ (x_t - x_s) & (x_r - x_t) & (x_s - x_r) \end{bmatrix}, \qquad \textbf{(7.44)}$$

with $2(\text{Area}) = \det(M)$. The value of the determinant is the sum of the elements in row 1 of Eq. (7.44) within the brackets. Area is the area of the triangular element.* You should verify that $[M^{-1}][M] = [I]$ to ensure that Eq. (7.44) truly gives the inverse matrix.

To apply the interpolating function to the minimization of the quadratic functional, Eq. (7.40), we prefer to write $v(x, y)$ in terms of the shape functions of Eq. (7.41). This task is easy. We have, from Eqs. (7.42) and (7.43),

$$v(x, y) = a_1 + a_2 x + a_3 y = (1 \quad x \quad y)\{a\}$$
$$= (1 \quad x \quad y)[M^{-1}]\{c\}.$$

But, in terms of N (from Eq. 7.41),

$$v(x, y) = (N)\{c\}. \qquad \textbf{(7.45)}$$

Comparing the two expressions, we have

$$(N) = (1 \quad x \quad y)[M^{-1}], \qquad \textbf{(7.46)}$$

where M^{-1} is given by Eq. (7.44). Observe carefully that Eq. (7.46) says that each N is a linear function of x and y of the form

$$N_j = A_j + B_j x + C_j y, \qquad j = r, s, t \qquad \textbf{(7.47)}$$

and that the coefficients are in column j of $[M^{-1}]$.

We have found the expressions for the N's. Before we go on, we digress to show an example that will clarify this step.

EXAMPLE 7.5   For the triangular element shown in Fig. 7.18 with nodes r, s, and t in counterclockwise order, find $\{a\}$, $\{N\}$, and $v(0.8, 0.4)$.

Node	x	y	c
r	0	0	100
s	2	0	200
t	0	1	300

---

*That Area $= \frac{1}{2} \det(M)$ is shown in most books on vectors where the cross product is explained.

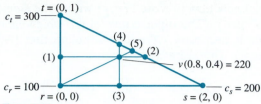

Figure 7.18

Before we do any computations, we can find $v(0.8, 0.4)$ by inspection. (See Fig. 7.18.) Point (1) is at $(0, 0.4)$, so v there is 180 by linear interpolation between nodes r and t. Similarly, v at point (2) is 240. The point $(0.8, 0.4)$ is $\frac{2}{3}$ of the distance from points (1) and (2), so $v(0.8, 0.4) = 180 + \frac{2}{3}(240 - 180) = 220$. We get the same result by interpolating between points (3) and (4), and between node r and (5).

To get $\{a\}$ we first compute $[M^{-1}]$:

$$[M] = \begin{bmatrix} 1 & 0 & 0 \\ 1 & 2 & 0 \\ 1 & 0 & 1 \end{bmatrix}, \qquad [M^{-1}] = \begin{bmatrix} 1 & 0 & 0 \\ -0.5 & 0.5 & 0 \\ -1 & 0 & 1 \end{bmatrix}.$$

Then we compute

$$\{a\} = [M^{-1}]\{c\} = \begin{bmatrix} 1 & 0 & 0 \\ -0.5 & 0.5 & 0 \\ -1 & 0 & 1 \end{bmatrix} \begin{Bmatrix} 100 \\ 200 \\ 300 \end{Bmatrix} = \begin{Bmatrix} 100 \\ 50 \\ 200 \end{Bmatrix},$$

giving $v(x, y) = 100 + 50x + 200y$. (You should confirm that this gives the correct values at each of the nodes.) If we substitute $x = 0.8, y = 0.4$, we get $v = 220$, as we should.

From Eq. (7.46),

$$(N) = (1 \quad x \quad y)[M^{-1}] = (1 - 0.5x - y, \quad 0.5x, \quad y).$$

In other words, we have

$$N_r = 1 - 0.5x - y,$$
$$N_s = 0.5x,$$
$$N_t = y.$$

(You should confirm that these also have the proper values at each of the nodes.) It is important to notice that the coefficients of the N's (the A_i, B_i, and C_i of Eq. 7.47) can be read directly from the columns of $[M^{-1}]$.

In what follows we will need the partial derivatives of the N's with respect to x and to y. From $N_j = A_j + B_j x + C_j y$, we see that these are constants that can be read from rows 2 and 3 of $[M^{-1}]$ in column j.

At this point we know how to write $v(x,y)$ within the single triangular element (i) as $v(x,y) = (N^{(i)})\{c^{(i)}\}$. (The superscripts (i) tell which element is being considered whenever this is necessary.) We now stipulate that $(N^{(i)}) \equiv (0)$ everywhere outside of element (i). Therefore we can write

$$v(x,y) = \sum_{i\,=\,\text{all elements}} (N^{(i)})\{c^{(i)}\}.$$

This is a mathematical statement of the previous observation that $v(x,y)$ is a surface composed of joined planar facets.

We are now ready for step 4 of our plan. This too is lengthy but each portion is easy.

Step 4. Substitute $v(x)$ into the Functional and Minimize

We continue to work with typical element (i) whose nodes are r, s, and t. Repeating Eq. (7.45), our $v(x)$ is

$$v(x) = (N)\{c\} = N_r c_r + N_s c_s + N_t c_t,$$

where the N's are given by Eqs. (7.46) and (7.44). Recall that each N is $A_j + B_j x + C_j y$ with the coefficients given by the elements in column j of $[M^{-1}]$.

Our objective is to develop a set of three equations for element (i), which is, in matrix form,

$$[K]\{c\} = \{b\},$$

and which is a prototype of similar equations for all other elements.

When we substitute $v(x)$ for element (i) into the functional of Eq. (7.40), we get

$$I(c_r, c_s, c_t) = \iint\limits_{\text{element } (i)} \left[\left(\frac{\partial v}{\partial x} \right)^2 + \left(\frac{\partial v}{\partial y} \right)^2 - Qv^2 + 2Fv \right] dx\,dy$$

$$- \oint_{L_2} [\alpha v^2 + 2\beta v]\,dL. \tag{7.48}$$

(I is now an ordinary function of the c's. The integral is only over the area of element (i) because the N's that define $v(x)$ in element (i) are zero outside of (i). The last term appears only if element (i) has a side on the boundary. Actually, we will postpone handling this last term for now and handle it as an adjustment to the equations after they have been developed.)

We minimize I by setting the three partials (with respect to each of the three c's) to zero. We now develop expressions for these partials. Consider first $\partial I/\partial c_r$:

For the first term in the integrand:

$$\frac{\partial}{\partial c_r} \left(\frac{\partial v}{\partial x} \right)^2 = 2 \left(\frac{\partial v}{\partial x} \right) \left(\frac{\partial}{\partial c_r} \left(\frac{\partial v}{\partial x} \right) \right).$$

But

$$\frac{\partial v}{\partial x} = \left(\frac{\partial N_r}{\partial x}\right)c_r + \left(\frac{\partial N_s}{\partial x}\right)c_t + \left(\frac{\partial N_t}{\partial x}\right)c_t$$

$$= B_r c_r + B_s c_s + B_t c_t,$$

by virtue of Eq. (7.47). (The B's come from row 2 of $[M^{-1}]$.)

Also, $\partial/\partial c_r(\partial v/\partial x) = B_r$ because c_s and c_t are independent of c_r. Hence

$$\frac{\partial}{\partial c_r}\left[\left(\frac{\partial v}{\partial x}\right)^2\right] = 2(B_r^2 + B_r B_s + B_r B_t).$$

The result for the second term is similar:

$$\frac{\partial}{\partial c_r}\left[\left(\frac{\partial v}{\partial y}\right)^2\right] = 2(C_r^2 + C_r C_s + C_r C_t),$$

where the C's come from row 3 of $[M^{-1}]$.

We next consider the Q term. Q is independent of c_r, so

$$\frac{\partial}{\partial c_r}(-Qv^2) = -Q\left[2v\left(\frac{\partial v}{\partial c_r}\right)\right]$$

$$-2Q(N_r c_r + N_s c_s + N_t c_t)(N_r).$$

Finally we work with the F term. F is independent of c_r, so

$$\frac{\partial}{\partial c_r}(2Fv) = 2F\left(\frac{\partial v}{\partial c_r}\right) = 2F(N_r).$$

Putting all this together, we have

$$\frac{\partial I}{\partial c_r} = 0 = \iint\limits_{(i)} 2[B_r^2 c_r + B_r B_s c_s + B_r B_t c_t]\,dx\,dy$$

$$+ \iint\limits_{(i)} 2[C_r^2 c_r + C_r C_s c_s + C_r C_t c_t]\,dx\,dy$$

$$- \iint\limits_{(i)} 2Q[N_r^2 c_r + N_r N_s c_s + N_r N_t c_t]\,dx\,dy$$

$$+ \iint\limits_{(i)} 2FN_r\,dx\,dy. \tag{7.49}$$

Equation (7.49) really is a formulation with the c's unknown:

$$K_{rr}c_r + K_{rs}c_s + K_{rt}c_t = b_r, \tag{7.50}$$

where

$$K_{rr} = \iint\limits_{(i)} 2B_r^2 \, dx \, dy + \iint\limits_{(i)} 2C_r^2 \, dx \, dy - \iint\limits_{(i)} 2QN_r^2 \, dx \, dy,$$

$$K_{rs} = \iint\limits_{(i)} 2B_r B_s \, dx \, dy + \iint\limits_{(i)} 2C_r C_s \, dx \, dy - \iint\limits_{(i)} 2QN_r N_s \, dx \, dy,$$

$$K_{rt} = \iint\limits_{(i)} 2B_r B_t \, dx \, dy + \iint\limits_{(i)} 2C_r C_t \, dx \, dy - \iint\limits_{(i)} 2QN_r N_t \, dx \, dy,$$

$$b_r = -\iint\limits_{(i)} 2FN_r \, dx \, dy.$$

(Remember, we postpone handling the last part of Eq. (7.48).)

Now we recognize that the B's and C's of Eq. (7.50) are constants, so we can bring them out from under the integral sign. If we use average values for Q and F within the element, we can also bring these out as their average values. (The best average value to use is the value of Q and F at the centroid of the triangular element.)

This means that we have to evaluate these five integrals:

$$I_1: \quad \iint\limits_{(i)} dx \, dy$$

$$I_2: \quad \iint\limits_{(i)} N_r^2 \, dx \, dy \qquad I_3: \quad \iint\limits_{(i)} N_r N_s \, dx \, dy \qquad I_4: \quad \iint\limits_{(i)} N_r N_t \, dx \, dy$$

$$I_5: \quad \iint\limits_{(i)} N_r \, dx \, dy.$$

The first of these is easy: I_1 = Area of the element, which we already know from having computed $[M^{-1}]$. The other integrals are laborious to compute directly, but there is a useful formula for the integral of the product of powers of linear functions over a triangle:

$$\iint\limits_{(\text{triangle})} N_r^\ell N_s^m N_t^n \, dx \, dy = \frac{2\ell! \, m! \, n!}{(\ell + m + n + 2)!} (\text{Area}).$$

Using this with the proper values for the exponents, ℓ, m, and n, gives:

$$I_2 = \frac{(\text{Area})}{6},$$

$$I_3 = \frac{(\text{Area})}{12},$$

$$I_4 = \frac{(\text{Area})}{12},$$

$$I_5 = \frac{(\text{Area})}{3}.$$

The terms in Eq. (7.50) are then

$$K_{rr} c_r + K_{rs} c_s + K_{rt} c_t = b_r, \tag{7.51}$$

where

$$K_{rr} = 2(\text{Area})\left(B_r^2 + C_r^2 - \frac{Q_{av}}{6} \right),$$

$$K_{rs} = 2(\text{Area})\left(B_r B_s + C_r C_s - \frac{Q_{av}}{12} \right),$$

$$K_{rt} = 2(\text{Area})\left(B_r B_t + C_r C_t - \frac{Q_{av}}{12} \right),$$

$$b_r = -2(\text{Area})\left(\frac{F_{av}}{3} \right).$$

If we do the same with $\partial I / \partial c_s$ and $\partial I / \partial c_t$, we get two more equations in the c's for element (i). All together we have three equations, which we call the *element equations*. We simplify these equations somewhat by omitting the common factor of 2 for each of them to get

$$[K]\begin{Bmatrix} c_r \\ c_s \\ c_t \end{Bmatrix} = \begin{Bmatrix} b_r \\ b_s \\ b_t \end{Bmatrix},$$

where

(diagonals) $\quad K_{jj} = \text{Area}\left[B_j^2 + C_j^2 - \frac{Q_{av}}{6} \right], \quad j = r, s, t;$

(off-diagonals) $\quad K_{jk} = \text{Area}\left[B_j B_k + C_j C_k - \frac{Q_{av}}{12} \right], \quad \begin{cases} j \neq k, \\ j = r, s, t, \\ k = r, s, t; \end{cases}$

(rhs) $\quad b_j = -\text{Area}\left[\frac{F_{av}}{3} \right], \quad j = r, s, t.$

Observe that $[K]$ is symmetrical: $K_{ij} = K_{ji}$.

Here is an example to clarify the formation of the element equations.

EXAMPLE 7.6 Find the element equations for the element of Example 7.5 if $Q(x, y) = (xy)/2$ and $F(x, y) = x + y$.

The nodes are $(x, y) = (0, 0), (2, 0),$ and $(0, 1)$. We had, for $[M^{-1}]$,

$$[M^{-1}] = \begin{bmatrix} 1 & 0 & 0 \\ -0.5 & 0.5 & 0 \\ -1 & 0 & 1 \end{bmatrix}.$$

Area = 1. Centroid is at $x = (0 + 2 + 0)/3 = \frac{2}{3}, y = (0 + 0 + 1)/3 = \frac{1}{3}$. $Q_{av} = (\frac{2}{3})(\frac{1}{3})/2 = \frac{1}{9}$. $F_{av} = \frac{2}{3} + \frac{1}{3} = 1$.

Using Eq. (7.51), we find that the element equations are

$$
\begin{bmatrix}
1.2315 & -0.2592 & -1.0092 \\
-0.2592 & 0.2315 & -0.0093 \\
-1.0092 & -0.0093 & 0.9815
\end{bmatrix}
\begin{Bmatrix} c_r \\ c_s \\ c_t \end{Bmatrix}
=
\begin{Bmatrix} -0.3333 \\ -0.3333 \\ -0.3333 \end{Bmatrix}.
$$

▲

We are now ready for step 5 of the plan.

Step 5. Assemble the Equations, Adjust for Boundary Conditions, Solve

There are three separate operations in step 5: (i) assemble the equations, (ii) adjust for boundary conditions, and (iii) solve the equations.

(i) Do the Assembly.   As we have seen, there are three equations for every element. However, some or all nodes of element (i) are shared with other elements; the c-value for a shared node then appears in the equations of all elements that share the node. Combining all of the element equations will create a global system coefficient matrix with as many rows and columns as there are nodes in the system. We combine (assemble) the system matrix in the following way.

Suppose there are n nodes in the system. Number the nodes in order, from 1 to n. Associate the number of each node with the row and column of every element matrix where the c for that node appears on the diagonal. Also associate the node numbers with the rows and columns of the system matrix in the same way.

We get the entry in row (i) and column (j) of the system matrix by adding the values from row (i) of every element matrix that has row (i), then adding these in the columns where the column-node numbers match. We also add the b_i's from these rows to get the b_i of the system matrix. An example will clarify this operation.

EXAMPLE 7.7   Suppose there are five nodes that define three elements, as shown in Fig. 7.19 with the element matrices of Eqs. (7.52) below. Construct the system matrix without adjusting for boundary conditions.

$$
\text{Element [1]} \quad
\begin{matrix} (1)\rightarrow \\ (2)\rightarrow \\ (4)\rightarrow \end{matrix}
\begin{bmatrix}
K_{11} & K_{12} & K_{13} \\
K_{21} & K_{22} & K_{23} \\
K_{31} & K_{32} & K_{33}
\end{bmatrix}
\begin{Bmatrix} c_1 \\ c_2 \\ c_4 \end{Bmatrix}
=
\begin{Bmatrix} b_1 \\ b_2 \\ b_3 \end{Bmatrix}
\qquad \textbf{(7.52a)}
$$

$$
\begin{matrix} \uparrow & \uparrow & \uparrow \\ (1) & (2) & (4) \end{matrix}
$$

$$
\text{Element [2]} \quad
\begin{matrix} (2)\rightarrow \\ (3)\rightarrow \\ (4)\rightarrow \end{matrix}
\begin{vmatrix}
K_{11} & K_{12} & K_{13} \\
K_{21} & K_{22} & K_{23} \\
K_{31} & K_{32} & K_{33}
\end{vmatrix}
\begin{Bmatrix} c_2 \\ c_3 \\ c_4 \end{Bmatrix}
=
\begin{Bmatrix} b_1 \\ b_2 \\ b_3 \end{Bmatrix}
\qquad \textbf{(7.52b)}
$$

$$
\begin{matrix} \uparrow & \uparrow & \uparrow \\ (2) & (3) & (4) \end{matrix}
$$

$$\text{Element [3]} \quad \begin{matrix} (4) \rightarrow \\ (5) \rightarrow \\ (1) \rightarrow \end{matrix} \begin{vmatrix} K_{11} & K_{12} & K_{13} \\ K_{21} & K_{22} & K_{23} \\ K_{31} & K_{32} & K_{33} \end{vmatrix} \begin{Bmatrix} c_4 \\ c_5 \\ c_1 \end{Bmatrix} = \begin{Bmatrix} b_1 \\ b_2 \\ b_3 \end{Bmatrix} \qquad \textbf{(7.52c)}$$

$$\begin{matrix} \uparrow & \uparrow & \uparrow \\ (4) & (5) & (1) \end{matrix}$$

[The rows and columns of Eqs. (7.52) could have been in a different order, although we always go counterclockwise around the element in selecting the nodes.]

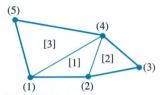

Figure 7.19

We construct the system matrix as follows, where the superscripts indicate the element number that provides the value:

For row 1	**For row 2**
col 1: $K_{11}^{[1]} + K_{33}^{[3]}$	col 1: $K_{21}^{[1]}$
col 2: $K_{12}^{[1]}$	col 2: $K_{22}^{[1]} + K_{11}^{[2]}$
col 3: 0	col 3: $K_{12}^{[2]}$
col 4: $K_{13}^{[1]} + K_{31}^{[3]}$	col 4: $K_{23}^{[1]} + K_{13}^{[2]}$
col 5: $K_{32}^{[3]}$	col 5: 0
b: $b_1^{[1]} + b_3^{[3]}$	b: $b_2^{[1]} + b_1^{[2]}$

and so forth.

[A zero appears in column 3 of row 1 because node (3) is not in any element that includes node (1). A zero appears in column 5 of row 2 because node (5) is not in any element that includes node (2).]

Once the system matrix has been assembled, we make the adjustments for boundary conditions.

(ii) Adjust for Boundary Conditions. There are two types of boundary conditions: non-Dirichlet conditions on some parts of the boundary (L_2) and Dirichlet conditions on other parts (L_1). We will always select nodes such that only one of the two types of conditions pertains to any side of the element. Hence there will always be a node at the point where the two types join. These two types of boundary conditions require two separate adjustments. We prefer to apply the adjustment for a boundary condition that involves the outward normal derivative to the system equations first and then do the adjustment for Dirichlet conditions.

Adjusting for Non-Dirichlet Conditions. Non-Dirichlet conditions (those that involve the outward normal derivative) are associated, not with the nodes, but with sides of the triangular elements, sides that correspond to part of L_2 of Eq. (7.40). Consider an element that has a non-Dirichlet condition on one side that lies between nodes r and s. The effect of the boundary condition on the equations comes from differentiating the last term of Eq. (7.48) with respect to the c's. But if we take α and β out from the integrand as average values, we see from Eq. (7.48) that they are of the same form as the Q and F terms except they are line integrals rather than area integrals. That similarity lets us immediately write the result of the differentiation with respect to c_r as

$$-2\alpha_{av} \oint (N_r^2 c_r + N_r N_s c_s)\, dL - 2\beta_{av} \oint N_r\, dL, \qquad (7.53)$$

where the line integrals are along the side between nodes r and s. It is important to note that we have not included c_t because node t is not on the side we are considering. (The average values of α and β should be taken at the midpoint of the side.) When we integrate, we have

$$-2\alpha_{av} L\left(\frac{c_r}{3} + \frac{c_s}{6}\right) - 2\beta_{av}\frac{L}{2}. \qquad (7.54)$$

(It is easy to evaluate the integrals when we remember that the N's are linear from 1 to 0 between the two nodes.)

Precisely the same relations result when we differentiate with respect to node c_s, except the roles of r and s are interchanged in Eqs. (7.53) and (7.54). The net result would be to add $2\beta L/2$ to the right-hand sides of the rows for c_r and c_t. We also would subtract the multipliers of c_r and c_s in Eq. (7.54) from the coefficients for c_r and c_s in row r. The similar equations from the partials with respect to c_s provide subtractions from the coefficients in row s.

Recall, however, that we canceled a 2 factor when we constructed the element equations and so we must do so here. We make this adjustment to the element equations for every element that has a derivative condition on a side.

Adjusting for Dirichlet Conditions. For every node that appears on the boundary where there is a Dirichlet condition, the u-value is specified. We insert this known value in place of the c of that node in every equation where it appears and transpose to the right-hand side. (Actually, if the node number is m, all entries in column m of the matrix are multiplied by the value and subtracted from the right-hand side of the corresponding row.) We also remove the row corresponding to the number of the known node from the set of equations. (The column for this node has already been "removed" by being transferred to the right-hand side.)

Removing the rows for those nodes with a Dirichlet condition is simplified in a computer program if these rows are at the top or the bottom of the matrix. There are other ways to handle Dirichlet conditions that avoid having to remove the rows.

This completes our construction of the system equations.

(iii) Getting the Solution.   We solve the system in the usual way, perhaps preferring an iterative procedure if the system is large.

An example, intentionally simple, follows.

EXAMPLE 7.8 The region shown in Fig. 7.20 has four nodes. It is divided into just two elements. The values for u are specified at nodes (3) and (4), and the outward normal gradient is specified on three sides as indicated. The equation we are to solve is

$$u_{xx} + u_{yy} - \left(\frac{y}{10}\right)u = \frac{x}{4} + y - 12.$$

Find the solution by the finite element method.

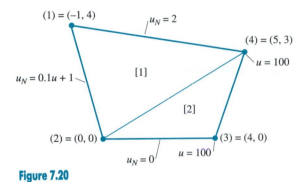

Figure 7.20

The element-matrix inverses are:

M inverse for element 1 area is 11.5			M inverse for element 2 area is 6		
1.000	0.000	0.000	1.000	0.000	0.000
−0.043	0.174	−0.130	−0.250	0.250	0.000
−0.261	0.043	0.217	0.083	−0.417	0.333
(2)	(4)	(1)	(2)	(3)	(4)

$Q_{av} = -0.233 \quad F_{av} = -9.333$ $\qquad$ $Q_{av} = -0.1 \quad F_{av} = -10.25$

From these we get these element equations:

Element equations for element 1

2→	1.2516	0.0062	−0.3633	35.7778
4→	0.0062	0.8168	0.0714	35.7778
1→	−0.3633	0.0714	1.1864	35.7778
	(2)	(4)	(1)	

Element equations for element 2

$$
\begin{array}{c}
2\rightarrow \\
3\rightarrow \\
4\rightarrow
\end{array}
\left[
\begin{array}{cccc}
0.5167 & -0.5333 & 0.2167 & 20.5000 \\
-0.5333 & 1.5167 & -0.7833 & 20.5000 \\
0.2167 & -0.7833 & 0.7667 & 20.5000
\end{array}
\right]
$$
$$
\begin{array}{cccc}
(2) & (3) & (4)
\end{array}
$$

These equations assemble to give this unadjusted system matrix:

$$
\left[
\begin{array}{cccc|c}
1.186 & -0.363 & 0.000 & 0.071 & 35.778 \\
-0.363 & 1.768 & -0.533 & 0.223 & 56.278 \\
0.000 & -0.533 & 1.517 & -0.783 & 20.500 \\
0.071 & 0.223 & -0.783 & 1.583 & 56.278
\end{array}
\right].
$$

We need the lengths of sides 4–1, 1–2, and 2–3. They are:

$$\text{Side 4–1:}\quad 6.083, \qquad \text{Side 1–2:}\quad 4.123, \qquad \text{Side 2–3:}\quad 4.$$

We now adjust for the derivative conditions to get the modified system:

$$
\left[
\begin{array}{cccc|c}
1.049 & -0.432 & 0.000 & 0.003 & 43.922 \\
-0.432 & 1.631 & -0.533 & 0.154 & 64.422 \\
0.000 & -0.533 & 1.517 & -0.783 & 20.500 \\
0.003 & 0.154 & -0.783 & 1.446 & 64.422
\end{array}
\right].
$$

The second adjustment is for the known values at nodes (3) and (4), giving

$$
\left[
\begin{array}{cc|c}
1.049 & -0.432 & 43.650 \\
-0.432 & 1.631 & 102.339
\end{array}
\right],
$$

which we solve to get these estimates of u_1 and u_2:

$$u_1 = 75.71, \qquad u_2 = 82.80.$$

7.12 Theoretical Matters

Laplace's and Poisson's Equations

The theory associated with partial-differential equations is rich. Some of the most important points with respect to Laplace's and Poisson's equations are these.

(1) If the boundary conditions for a Laplace problem are everywhere $u = f(s)$, where s represents locations on the boundary, the problem is said to have Dirichlet conditions. Such a problem has a unique solution that is determined by the boundary

conditions. Further, the maximum–minimum principle applies: No point on the interior of the region has a u-value greater than the maximum or less than the minimum of the boundary points.

The truth of this principle is easy to see from visualizing heat flow in a plate. Any interior point that exceeds the maximum boundary temperature will lose heat until it is no hotter than the maximum boundary point. Likewise, any interior point that is colder than the minimum temperature on the boundary will receive heat, since the boundary points are held at the given temperatures.

(2) A superposition principle holds for Laplace's equation. This says that if u_1 and u_2 are both solutions to the equation

$$u_{xx} + u_{yy} = 0,$$

then the weighted sum $c_1 u_1 + c_2 u_2$ is also a solution. The truth of this principle is obvious when we substitute the sum into the equation. The usual analytical methods for solving Laplace's equation depend on this principle. We first determine that there is an infinite set of solutions to the equation (ignoring the boundary conditions). We then attempt to match the conditions on the boundaries with a weighted sum of these solutions. Getting this match often leads to a Fourier series, which also has a rich body of theory.

(3) If the boundary conditions are that the normal derivative of u is specified at all points on the boundary, the problem is said to have Neumann conditions (the outward normal gradient is called the *flux*). The solution to a Neumann problem is not unique in the same way as for a Dirichlet problem, for if u is a solution, so is $u + c$. Again, one can see the truth of this principle by visualizing steady-state heat flow. Heat is lost or gained from the surroundings in proportion to the flux and the difference in temperatures between the boundary and the surroundings. If we change the surrounding temperature and maintain the flux, the body must change in temperature to maintain the same temperature difference.

(4) The superposition principle does not apply to Poisson's equation, but a subtraction principle does. This says that if u_1 and u_2 are solutions to $u_{xx} + u_{yy} = f(x, y)$, then $u_2 - u_1$ solves $u_{xx} + u_{yy} = 0$. As a consequence, the solution to Poisson's equation can be found by adding a particular solution of the problem to the general solution of the associated Laplace equation. That is, if u_P satisfies

$$u_{xx} + u_{yy} = f(x, y),$$

and if u_L is the general solution to

$$u_{xx} + u_{yy} = 0,$$

then $u_P + u_L$ is the general solution to the Poisson problem. We normally solve Poisson problems analytically by this technique.

Two references for further information on these points are Pinsky (1991) and Andrews (1986).

Eigenvalues and S.O.R.

The S.O.R. method used in Sections 7.5 and 7.6 is actually a special case of the relaxation method of Section 2.11. In fact, Eqs. (7.17) and (7.23) are just special cases of Eq. (2.51), which we repeat here:

$$x_i^{(k+1)} = x_i^{(k)} + \frac{\omega}{a_{ii}}\left(b_i - \sum_{j=1}^{i-1} a_{ij} x_j^{(k+1)} - \sum_{j=i}^{n} a_{ij} x_j^{(k)} \right),$$

$$i = 1, 2, \ldots, n. \tag{7.55}$$

Since this equation reduces to Gauss–Seidel for $\omega = 1$, we will see in this section that the purpose of introducing ω is to reduce the eigenvalues of the iteration matrix in order to accelerate convergence (as explained in Section 6.9).

Let us review the argument of Section 6.9 about speed of convergence. We prefer to work with matrices rather than the individual equations. We start with $Ax = b$, where A is $n \times n$. If we rewrite A as $L + D + U$, then

$$Ax = (L + D + U)x = b, \quad \text{or}$$

$$(L + D)x = b - Ux.$$

We can put this into an iterative form:

$$(L + D)x^{(k+1)} = b - Ux^{(k)},$$

$$Dx^{(k+1)} = b - Lx^{(k+1)} - Ux^{(k)}, \tag{7.56}$$

so that

$$x^{(k+1)} = -(L + D)^{-1}Ux^{(k)} + (L + D)^{-1}b, \tag{7.57}$$

giving the Gauss–Seidel equation, Eq. (2.46). In using this, we begin with an initial approximation $x^{(0)}$. We see that Eq. (7.57) can be written as

$$x^{(k+1)} = b' - Bx^{(k)}, \quad b' = (L + D)^{-1}b, \quad B = (L + D)^{-1}U.$$

Suppose that $\bar{x}$ is the solution to $Ax = b$. Then $\bar{x}$ certainly satisfies

$$\bar{x} = b' - B\bar{x}.$$

Subtracting this from Eq. (7.57), we get

$$x^{(k+1)} - \bar{x} = e^{(k+1)} = -B(x^{(k)} - \bar{x}) = -Be^{(k)},$$

where $e^{(i)} = x^{(i)} - \bar{x}$. This means that

$$e^{(k+1)} = -Be^{(k)} = B^2 e^{(k-1)} = \cdots = (-B)^{k+1}e^{(0)}$$

with $e^{(0)} = x^{(0)} - \bar{x}$, the error in the initial approximation to the solution. Now, assuming that we can write $e^{(0)}$ in terms of the eigenvectors of B, where $Bv_i = \lambda_i v_i$,

$$e^{(0)} = c_1 v_1 + c_2 v_2 + \cdots + c_n v_n.$$

Now

$$B^k e^{(0)} = c_1 B^k v_1 + c_2 B^k v_2 + \cdots + c_n B^k v_n$$
$$= c_1 \lambda_1^k v_1 + c_2 \lambda_2^k v_2 + \cdots + c_n \lambda_n^k v_n,$$

and we see that, if all the eigenvalues of B are less than 1 in magnitude, the iteration will converge and that the rate of convergence is faster when the eigenvalue of greatest magnitude is small. We can speed up the convergence if we can find a way to reduce the eigenvalue of maximum magnitude of B.

Consider now a simple example that shows how the relaxation factor can reduce the eigenvalues of the iteration matrix and hence accelerate convergence. Let $Ax = b$ be

$$\begin{bmatrix} 2 & 1 \\ 1 & 3 \end{bmatrix} x = \begin{bmatrix} 6 \\ -2 \end{bmatrix} \qquad \text{(solution is } \{4 \quad -2\}\text{).}^*$$

First, let us relate Eq. (7.55) (single-equation form for overrelaxation) to Eq. (7.57) (the matrix form for iteration). Let $A = L + D + U$ and write

$$\omega(b - Ax) = 0 = \omega(b - (L + D + U)x). \tag{7.58}$$

(Actually, $b - (L + D + U)x$ is the residual in Southwell's relaxation method of Section 2.11.) Obviously, we can write

$$Dx = Dx.$$

Add (7.58) to both sides:

$$Dx + 0 = Dx - \omega Lx - \omega Dx - \omega Ux + \omega b.$$

Rearrange to give the iterative form:

$$(D + \omega L)x^{(k+1)} = [(1 - \omega)D - \omega U]x^{(k)} + \omega b, \qquad \text{or}$$
$$x^{(k+1)} = (D + \omega L)^{-1}[(1 - \omega)D - \omega U]x^{(k)} + \omega(D + \omega L)^{-1}b. \tag{7.59}$$

This is the matrix formulation of Eq. (7.55) for overrelaxation.

Let us apply this to the simple problem, $Ax = b$, where

$$A = \begin{bmatrix} 2 & 1 \\ 1 & 3 \end{bmatrix}, \quad b = \begin{bmatrix} 6 \\ -2 \end{bmatrix}, \quad D = \begin{bmatrix} 2 & 0 \\ 0 & 3 \end{bmatrix}, \quad L = \begin{bmatrix} 0 & 0 \\ 1 & 0 \end{bmatrix}, \quad U = \begin{bmatrix} 0 & 1 \\ 0 & 0 \end{bmatrix}.$$

---

* In this section, braces indicate a column vector.

In the form of Eq. (7.59), this is

$$\underbrace{\begin{bmatrix} 2 & 0 \\ \omega & 3 \end{bmatrix}}_{(D\,+\,\omega L)} x^{(k+1)} = \underbrace{\begin{bmatrix} 2(1-\omega) & -\omega \\ 0 & 3(1-\omega) \end{bmatrix}}_{(D\,-\,\omega D\,-\,\omega U)} x^{(k)} + \underbrace{\omega \begin{bmatrix} 6 \\ -2 \end{bmatrix}}_{\omega b}.$$

The inverse of $(D + \omega L)$ is not difficult to compute. Doing so and multiplying the matrices gives

$$x^{(k+1)} = \begin{bmatrix} 1-\omega & -\omega/2 \\ \omega(\omega-1)/3 & (\omega^2/6 - \omega + 1) \end{bmatrix} x^{(k)} + \begin{bmatrix} 3\omega \\ -\omega^2 - 2\omega/3 \end{bmatrix}. \qquad \textbf{(7.60)}$$

(You should verify that, for $\omega = 1$, this reduces to the Gauss–Seidel matrix.) In fact, this is

$$\begin{bmatrix} 0 & -\frac{1}{2} \\ 0 & \frac{1}{6} \end{bmatrix} \qquad \text{whose eigenvalues are } \{0 \quad \tfrac{1}{6}\}.$$

We wish to determine the optimum value ω_{opt} of the overrelaxation factor in the matrix of Eq. (7.60) that reduces the largest magnitude eigenvalue of the matrix as much as possible. To do this, recall four properties of matrices:

1. $\det(AB) = \det(A)\det(B)$.
2. $\det(A) = $ product of its eigenvalues.
3. $\text{trace}(A) = $ sum of diagonal elements $= $ sum of its eigenvalues.
4. The eigenvectors of a matrix A form a basis for R^n if the matrix can be diagonalized. We used this earlier.

Applying these to the matrix of Eq. (7.60), we have

$$\lambda_1 \lambda_2 = (\omega - 1)^2,$$

where $0 < \omega < 2$. This follows because we want all the eigenvalues to be less than 1 in absolute value. Moreover, since we want the largest eigenvalue as small as possible, we take

$$\lambda_1 = \lambda_2 = \omega - 1.$$

By the trace property, it follows that

$$\lambda_1 + \lambda_2 = \frac{\omega^2}{6} - 2\omega + 2 = 2(\omega - 1).$$

This quadratic equation has one solution, $\omega_{opt} = 1.0455488$, which falls in the range from 0 to 2.

To summarize what we have done: We wanted to solve $Ax = b$ by iteration. If we had used Gauss–Seidel, our iteration matrix would have been

$$\begin{bmatrix} 0 & -\frac{1}{2} \\ 0 & \frac{1}{6} \end{bmatrix} \quad \text{with eigenvalues 0 and } \tfrac{1}{6}.$$

Starting with $x^{(0)} = \{0 \ \ 0\}$, succeeding iterates are $\{3 \ \ -1.6675\}$, $\{3.8333 \ \ -1.9444\}$, $\{3.9722 \ \ -1.9907\}, \rightarrow \{4 \ \ -2\}$.

However, by using Eq. (7.60) with $\omega = 1.0455488$, the iteration matrix becomes

$$\begin{bmatrix} -0.0455 & -0.5228 \\ 0.0159 & 0.1366 \end{bmatrix} \quad \text{with eigenvalues 0.0455 and 0.0455.}$$

Starting again with $\{0 \ \ 0\}$, we get for $x^{(1)}$, $\{3.1366 \ \ -1.7902\}$, for $x^{(2)}$, $\{3.9296 \ \ -1.9850\}$, then $\{3.9953 \ \ -1.9991\}$. Reducing the largest-magnitude eigenvalue from $\frac{1}{6}$ to 0.0455 gives faster convergence. If the reduction in magnitude had been greater, a more significant speeding up would have been observed.

You will do another example as an exercise. For larger matrices, a different approach to finding the optimal ω_{opt} would be necessary. Our purpose here is only to interrelate eigenvalues of the relaxation matrices to the convergence that is obtained. We discussed one of these other methods for eigenvalues in Chapter 6, the power method. We also described the *QR* method in Chapter 6; this is based on similarity transformations.

7.13 Using *Mathematica*

Mathematica does not have facilities to directly solve partial-differential equations, but it can solve ordinary differential equations and boundary-value problems either analytically or numerically. Both DERIVE and MAPLE have similar capabilities.

All of these computer algebra systems can get the solution to the sets of equations that are involved in the methods of Chapter 7.

In *Mathematica*, the solution to a system of equations is obtained by issuing a command of this form:

```
Solve [{2 x - 3 y == 3, x + 4 y = 7},{x,y}]
```

from which we get the solution

$$\{\{x \rightarrow 3, \ y \rightarrow 1\}\}.$$

(If there are many equations in the set, it is easier to define them in matrix form.) *Mathematica* can even get the solution to some nonlinear equations.

In getting the analytical solution to an elliptic partial-differential equation, the most widely used technique is to assume that the solution $u(x,y)$ can be expressed as the product of two functions, $X(x)$ and $Y(y)$, where each of these is a function of the single variable. Then, depending on the boundary conditions, it is shown that each of these can be solved by an expression that includes one or more of the forms

$$\sin(kx), \quad \cos(kx), \quad \sinh(kx), \quad \text{or} \quad \cosh(kx)$$

and the analogues for y. The parameter k in the above is related to the eigenvalues of the system. A solution to the original problem that both satisfies the equation and meets the boundary conditions is then expressed as an infinite sum of these terms, of which the following is typical:

$$u(x,y) = \sum_{n=1}^{\infty} A_n \sin(k_1 x) * \cosh(k_2 y),$$

and the values of A_n are found through evaluating an integral. Such a series is called a Fourier series (or a generalized Fourier series). *Mathematica* can construct such Fourier series through the command

```
FourierTrigSeries[ ]
```

that is contained in one of the "standard packages" that is supplied with the program. (MAPLE and DERIVE can do the same.) See also Section 10.4.

Chapter Summary

If you fully understand Chapter 7, you should be able to do the following.

1. Explain how to analyze heat flow to derive the partial-differential equation that governs this and similar potential flow situations.

2. Tell in which class a given differential equation falls.

3. Set up the system of equations that results from replacing derivatives with finite-difference approximations and solve by either elimination or iteration. You should be able to handle a variety of boundary conditions.

4. List the advantages of iteration over elimination for solving large systems, and explain why overrelaxation speeds up convergence.

5. Show how the finite-difference method can be adapted to nonrectangular regions.

6. Discuss the problems encountered in three-dimensional situations.

7. Use the A.D.I. procedure to solve problems, and tell why it is advantageous.

8. Solve a small problem by the finite-element method, and compare it to the finite-difference method.

9. Utilize the computer programs of this chapter to get the numerical solution to typical examples.

Computer Programs

We present three programs in this chapter. The first is not a "program" in the normal sense but shows how a spreadsheet can perform the iterations to solve Laplace's equation on a rectangle. The other two programs are FORTRAN implementations for (1) solving Poisson's equation using iteration with overrelaxation and (2) solving the steady-state heat-flow problem through the A.D.I. method. Both programs work with a rectangular region.

Program 7.1 Spreadsheet Iteration

Figure 7.21(a) shows an initial layout of u-values in a rectangular pattern. (Most of these entries can be done with a copy operation.) We see that the boundaries are everywhere zero except along the right-hand side.

The computations of Liebmann's method (Eq. 7.15) will be performed if we enter in cell $B2$ the formula

$$= (A2 + B1 + B3 + C2)/4$$

and replicate it into the block of cells $B2, \ldots, H4$. Then we request recomputations until convergence is obtained, getting the results shown in Fig. 7.21(b). Figure 7.21(c) lists the spreadsheet entries.

The reader may want to modify the spreadsheet program so that it computes a Poisson equation or solves the equation on a nonrectangular region (but one whose boundaries are approximated by mesh points).

```
      Initial Layout of u-values

        A       B       C       D       E       F       H       I
  1   0.00    0.00    0.00    0.00    0.00    0.00    0.00   50.00
  2   0.00    0.00    0.00    0.00    0.00    0.00    0.00  100.00
  3   0.00    0.00    0.00    0.00    0.00    0.00    0.00  100.00
  4   0.00    0.00    0.00    0.00    0.00    0.00    0.00  100.00
  5   0.00    0.00    0.00    0.00    0.00    0.00    0.00   50.00
```

Figure 7.21(a) Program 7.1: Initial layout of u-values

```
      Solution using Liebmann's method with spreadsheet

        A       B       C       D       E       F       G       H      I
  1   0.00    0.00    0.00    0.00    0.00    0.00    0.00    0.00   50.00
  2   0.00    0.35    0.91    2.01    4.30    9.15   19.66   43.21  100.00
  3   0.00    0.50    1.29    2.83    6.02   12.65   26.29   53.18  100.00
  4   0.00    0.35    0.91    2.01    4.30    9.15   19.66   43.21  100.00
  5   0.00    0.00    0.00    0.00    0.00    0.00    0.00    0.00   50.00
```

Figure 7.21(b) Program 7.1: Results of computation

```
                    Solution of Poisson equation using a spreadsheet.

                    A1  VALUE =0
                    B1  VALUE =0
                    C1  VALUE =0
                    D1  VALUE =0
                    E1  VALUE =0
                    F1  VALUE =0
                    G1  VALUE =0
                    H1  VALUE =0
                    I1  VALUE =50
                    A2  VALUE =0
                    B2  FORMULA = (A2+B1+C2+B3)/4
                    C2  FORMULA = (B2+C1+D2+C3)/4
                    D2  FORMULA = (C2+D1+E2+D3)/4
                    E2  FORMULA = (D2+E1+F2+E3)/4
                    F2  FORMULA = (E2+F1+G2+F3)/4
                    G2  FORMULA = (F2+G1+H2+G3)/4
                    H2  FORMULA = (G2+H1+I2+H3)/4
                    I2  VALUE =100
                    A3  VALUE =0
                    B3  FORMULA = (A3+B2+C3+B4)/4
                    C3  FORMULA = (B3+C2+D3+C4)/4
                    D3  FORMULA = (C3+D2+E3+D4)/4
                    E3  FORMULA = (D3+E2+F3+E4)/4
                    F3  FORMULA = (E3+F2+G3+F4)/4
                    G3  FORMULA = (F3+G2+H3+G4)/4
                    H3  FORMULA = (G3+H2+I3+H4)/4
                    I3  VALUE =100
                    A4  VALUE =0
                    B4  FORMULA = (A4+B3+C4+B5)/4
                    C4  FORMULA = (B4+C3+D4+C5)/4
                    D4  FORMULA = (C4+D3+E4+D5)/4
                    E4  FORMULA = (D4+E3+F4+E5)/4
                    F4  FORMULA = (E4+F3+G4+F5)/4
                    G4  FORMULA = (F4+G3+H4+G5)/4
                    H4  FORMULA = (G4+H3+I4+H5)/4
                    I4  VALUE =100
                    A5  VALUE =0
                    B5  VALUE =0
                    C5  VALUE =0
                    D5  VALUE =0
                    E5  VALUE =0
                    F5  VALUE =0
                    G5  VALUE =0
                    H5  VALUE =0
                    I5  VALUE =50
```

Figure 7.21(c) Program 7.1: Entries in the spreadsheet

Program 7.2 The Overrelaxation Method

Figure 7.22 is a listing of PROGRAM POISSON, a FORTRAN program that solves the Poisson equation by iteration with overrelaxation employed to speed convergence. (See Eq. (7.23).) The equation that is solved is

$$u_{xx} + u_{yy} = f(x, y),$$

although this particular example has $f(x, y) = 0$ (so we really are solving Laplace's equation!).

The program uses a DATA statement to define the size of the region (in terms of numbers of interior nodes), a tolerance value to stop iterating (TOL), a value for the overrelaxation factor (W), and the distance between nodes (H). This is the sequence of operations:

1. The boundary values are put into matrix U with DO loops.
2. The average of these boundary values is computed as the initial approximations to the interior u-values.
3. After printing a heading, the interior u-values are computed by iterating with the computational equation until the change in any u-value is less than the tolerance value. The final values for the u's are then displayed along with the count for the number of iterations.
4. The overrelaxation factor is then incremented and the process repeated. This continues until $W = 1.8$, and we obtain an idea of the optimal value for W.

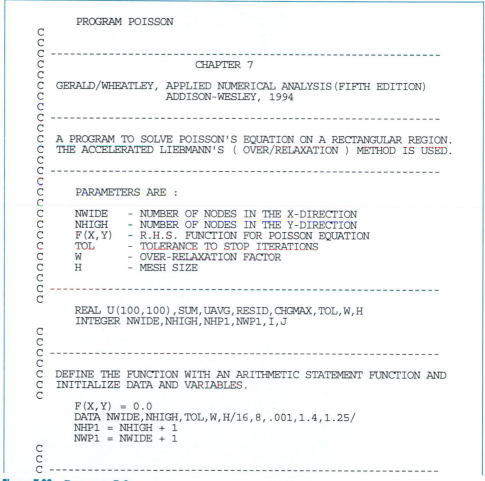

```
        PROGRAM POISSON
C
C
C  --------------------------------------------------------------
C
C                          CHAPTER 7
C
C  GERALD/WHEATLEY, APPLIED NUMERICAL ANALYSIS(FIFTH EDITION)
C                    ADDISON-WESLEY, 1994
C
C  --------------------------------------------------------------
C
C  A PROGRAM TO SOLVE POISSON'S EQUATION ON A RECTANGULAR REGION.
C  THE ACCELERATED LIEBMANN'S ( OVER/RELAXATION ) METHOD IS USED.
C
C  --------------------------------------------------------------
C
C     PARAMETERS ARE :
C
C     NWIDE     - NUMBER OF NODES IN THE X-DIRECTION
C     NHIGH     - NUMBER OF NODES IN THE Y-DIRECTION
C     F(X,Y)    - R.H.S. FUNCTION FOR POISSON EQUATION
C     TOL       - TOLERANCE TO STOP ITERATIONS
C     W         - OVER-RELAXATION FACTOR
C     H         - MESH SIZE
C
C  --------------------------------------------------------------
C
        REAL U(100,100),SUM,UAVG,RESID,CHGMAX,TOL,W,H
        INTEGER NWIDE,NHIGH,NHP1,NWP1,I,J
C
C
C  --------------------------------------------------------------
C
C  DEFINE THE FUNCTION WITH AN ARITHMETIC STATEMENT FUNCTION AND
C  INITIALIZE DATA AND VARIABLES.
C
        F(X,Y) = 0.0
        DATA NWIDE,NHIGH,TOL,W,H/16,8,.001,1.4,1.25/
        NHP1 = NHIGH + 1
        NWP1 = NWIDE + 1
C
C
C  --------------------------------------------------------------
```

Figure 7.22 Program 7.2

Figure 7.22 *(continued)*

```
C
C   USE DO LOOPS TO ESTABLISH BOUNDARY VALUES
C

        DO 1 I = 1,NHP1
          U(I,1) = 0.0
          U(I,NWP1) = 100.0
1       CONTINUE
        DO 2 I = 2,NWIDE
          U(1,I) = 0.0
          U(NHP1,I) = 0.0
2       CONTINUE
C
C
C   ----------------------------------------------------------------
C
C   COMPUTE MEAN VALUE OF THE BOUNDARY VALUES TO USE FOR
C   INITIAL VALUE OF INTERIOR POINTS
C
5       SUM = 0.0
        DO 10 I = 1,NHP1
          SUM = SUM + U(I,1) + U(I,NWP1)
10      CONTINUE
        DO 20 I = 2,NWIDE
          SUM = SUM + U(1,I) + U(NHP1,I)
20      CONTINUE
        UAVG = SUM / FLOAT(2*NWP1 + 2*(NHIGH - 1))
        X = 0.0
        Y = 0.0
        DO 30 I = 2,NHIGH
          DO 30 J = 2,NWIDE
            U(I,J) = UAVG + H*H*F(X,Y)
30      CONTINUE
C
C   ----------------------------------------------------------------
C
C   PRINT A HEADING AND BEGIN THE ITERATIONS. LIMIT TO 100
C   ITERATIONS OR UNTIL THE MAX CHANGE IN U VALUES IS < TOL.
C
        PRINT 199, W
        DO 50 KNT = 1,100
          CHGMAX = 0.0
          DO 40 I = 2,NHIGH
            Y = (I-1) * H
            DO 35 J = 2,NWIDE
              X = (J-1) * H
              RESID = W/4.0*( U(I+1,J) + U(I-1,J) + U(I,J+1) +
     +                U(I,J-1) - 4.0*U(I,J) + H*H*F(X,Y) )
              IF ( CHGMAX .LT. ABS(RESID) ) CHGMAX = ABS(RESID)
              U(I,J) = U(I,J) + RESID
35          CONTINUE
40        CONTINUE
          IF ( CHGMAX .LT. TOL ) GO TO 55
50      CONTINUE
55      PRINT 200, KNT,CHGMAX
        DO 45 I = 1,NHP1
          PRINT 201, ( U(I,J), J = 1,NWP1 )
45      CONTINUE
        W = W + 0.1
        IF ( W .LT. 1.8 ) GO TO 5
C
C
C   ----------------------------------------------------------------
```

Figure 7.22 *(continued)*

```
C
199      FORMAT(///
+             ' ITERATIONS WITH OVERRELAXATION FACTOR OF ',F5.2)
200      FORMAT(/' AFTER ITERATION NO. ',I3,' MAX CHANGE IN U = ',
+             F8.4,' U MATRIX IS '/)
201      FORMAT(1X,9F8.2)
         STOP
         END


***********************************************************************

                        PARTIAL OUTPUT FOR POISSON.F

   ITERATIONS WITH OVER-RELAXATION FACTOR OF  1.40

   AFTER ITERATION NO.   29 MAX CHANGE IN U =     .0008 U MATRIX IS

      .00      .00      .00      .00      .00      .00      .00      .00      .00
      .00      .00      .00      .00      .00      .00      .00   100.00
      .00      .08      .17      .28      .44      .66      .99     1.46     2.16
     3.20     4.75     7.10    10.73    16.64    27.10    48.34   100.00
      .00      .14      .31      .52      .81     1.22     1.82     2.70     3.99
     5.90     8.72    12.91    19.19    28.72    43.41    66.25   100.00
      .00      .19      .40      .68     1.05     1.59     2.38     3.52     5.20
     7.68    11.31    16.63    24.40    35.64    51.58    73.24   100.00
      .00      .20      .43      .73     1.14     1.73     2.57     3.81     5.63
     8.30    12.20    17.90    26.14    37.87    54.04    75.13   100.00
      .00      .19      .40      .68     1.06     1.60     2.38     3.52     5.21
     7.68    11.31    16.63    24.40    35.64    51.58    73.24   100.00
      .00      .14      .31      .52      .81     1.22     1.82     2.70     3.99
     5.90     8.72    12.91    19.19    28.72    43.41    66.25   100.00
      .00      .08      .17      .28      .44      .66      .99     1.46     2.16
     3.20     4.76     7.10    10.73    16.64    27.10    48.34   100.00
      .00      .00      .00      .00      .00      .00      .00      .00     00
      .00      .00      .00      .00      .00      .00      .00   100.00

   ITERATIONS WITH OVER-RELAXATION FACTOR OF  1.50

   AFTER ITERATION NO.   26 MAX CHANGE IN U =     .0008 U MATRIX IS

      .00      .00      .00      .00      .00      .00      .00      .00      .00
      .00      .00      .00      .00      .00      .00      .00   100.00
      .00      .08      .17      .28      .44      .66      .99     1.46     2.16
     3.20     4.76     7.10    10.73    16.64    27.10    48.34   100.00
      .00      .14      .31      .52      .81     1.22     1.82     2.70     3.99
     5.90     8.72    12.91    19.19    28.72    43.41    66.25   100.00
      .00      .19      .40      .68     1.06     1.60     2.38     3.52     5.21
     7.68    11.31    16.63    24.40    35.64    51.59    73.24   100.00
      .00      .20      .43      .73     1.14     1.73     2.58     3.81     5.63
     8.30    12.20    17.90    26.14    37.87    54.04    75.13   100.00
      .00      .19      .40      .68     1.06     1.60     2.38     3.53     5.21
     7.68    11.31    16.63    24.40    35.64    51.59    73.24   100.00
      .00      .14      .31      .52      .81     1.22     1.82     2.70     3.99
     5.90     8.72    12.91    19.19    28.72    43.41    66.25   100.00
      .00      .08      .17      .28      .44      .66      .99     1.46     2.17
     3.20     4.76     7.10    10.73    16.64    27.10    48.34   100.00
      .00      .00      .00      .00      .00      .00      .00      .00      .00
      .00      .00      .00      .00      .00      .00      .00   100.00

                                  .
                                  .
                                  .
```

Program 7.3 The A.D.I. Method

Our third program, PROGRAM ADIELL, carries out the A.D.I. method for Laplace's equation on a rectangular region. The procedures described in Section 7.10 are followed closely.

The program, given in Fig. 7.23, begins with a series of DATA statements that define initial values for the elements of the matrices and the acceleration factor. These matrix-element values are then overwritten to set up the matrices for solving Eqs. (7.37) and (7.38). In this embodiment, U is used for the nodal values during the odd traverse and V for those during the even traverse. Then the boundary u-values are put into some vectors to be used for constructing the right-hand sides of the equations.

After these preliminaries, the LU equivalents of the coefficient matrices for the two sets of equations are computed. A long loop then begins (DO 190...) that repeats the solution of the two sets of equations ITMAX times (here ITMAX = 30 and it is assumed that the u-values will have converged by then). The right-hand sides for each set of equations have to be recomputed for every iteration, of course.

The resulting u-values are printed after every second pass through the equations, since only every other set is more accurate.

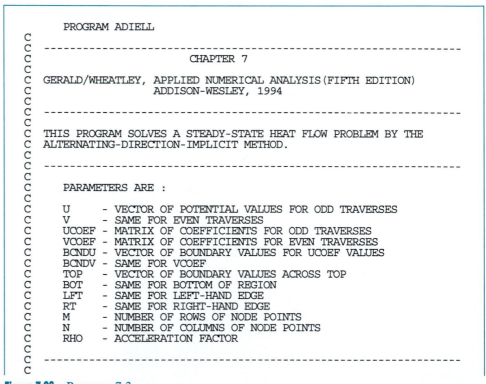

```
      PROGRAM ADIELL
C
C     -----------------------------------------------------------------
C
C                          CHAPTER 7
C
C     GERALD/WHEATLEY, APPLIED NUMERICAL ANALYSIS(FIFTH EDITION)
C                      ADDISON-WESLEY, 1994
C
C     -----------------------------------------------------------------
C
C     THIS PROGRAM SOLVES A STEADY-STATE HEAT FLOW PROBLEM BY THE
C     ALTERNATING-DIRECTION-IMPLICIT METHOD.
C
C     -----------------------------------------------------------------
C
C
C        PARAMETERS ARE :
C
C        U      - VECTOR OF POTENTIAL VALUES FOR ODD TRAVERSES
C        V      - SAME FOR EVEN TRAVERSES
C        UCOEF  - MATRIX OF COEFFICIENTS FOR ODD TRAVERSES
C        VCOEF  - MATRIX OF COEFFICIENTS FOR EVEN TRAVERSES
C        BCNDU  - VECTOR OF BOUNDARY VALUES FOR UCOEF VALUES
C        BCNDV  - SAME FOR VCOEF
C        TOP    - VECTOR OF BOUNDARY VALUES ACROSS TOP
C        BOT    - SAME FOR BOTTOM OF REGION
C        LFT    - SAME FOR LEFT-HAND EDGE
C        RT     - SAME FOR RIGHT-HAND EDGE
C        M      - NUMBER OF ROWS OF NODE POINTS
C        N      - NUMBER OF COLUMNS OF NODE POINTS
C        RHO    - ACCELERATION FACTOR
C
C     -----------------------------------------------------------------
C
```

Figure 7.23 Program 7.3

Figure 7.23   *(continued)*

```
C    SET UP THE VECTORS AND MATRICES
C
         REAL UCOEF(500,3),VCOEF(500,3),BCNDU(500),BCNDV(500)
         REAL TOP(100),BOT(100),LFT(100),RT(100),U(500),V(500)
         INTEGER M,N,MSIZE,ML1,MSL1,NL1,I,J,K,ITMAX,KNT,L,JROW
C
C    -----------------------------------------------------------------
C
C    INITIALIZE SOME VALUES WITH DATA STATEMENTS
C
         DATA UCOEF,VCOEF / 3000 * -1.0 /
         DATA M,N,RHO / 7,15,1.0 /
         DATA V / 500 * 0.0 /
         DATA RT,LFT,TOP,BOT / 100 * 100.0, 300 * 0.0 /
         DATA BCNDU,BCNDV / 1000 * 0.0 /
C
C    -----------------------------------------------------------------
C
C    ESTABLISH THE COEFFICIENT MATRICES BY OVER-WRITING ON THE
C    DIAGONAL AND CERTAIN OFF DIAGONAL TERMS.
C
         MSIZE = M * N
         DO 10 I = 1,MSIZE
            UCOEF(I,2) = 1.0/RHO + 2.0
            VCOEF(I,2) = 1.0/RHO + 2.0
10       CONTINUE
         ML1 = M - 1
         NL1 = N - 1
         MSL1 = MSIZE - 1
         DO 20 I = N,MSL1,N
            UCOEF(I,3) = 0.0
            UCOEF(I+1,1) = 0.0
20       CONTINUE
         DO 30 I = M,MSL1,M
            VCOEF(I,3) = 0.0
            VCOEF(I+1,1) = 0.0
30       CONTINUE
C
C    -----------------------------------------------------------------
C
C    NOW GET VALUES INTO THE BCOND VECTORS
C
         DO 40 I = 1,N
            BCNDU(I) = TOP(I)
            J = MSIZE - N + I
            BCNDU(J) = BOT(I)
40       CONTINUE
         DO 45 I = 1,M
            J = (I-1)*N + 1
            BCNDU(J) = BCNDU(J) + LFT(I)
            J = I * N
            BCNDU(J) = BCNDU(J) + RT(I)
45       CONTINUE
         DO 50 I = 1,M
            BCNDV(I) = LFT(I)
            J = MSIZE - M + I
            BCNDV(J) = RT(I)
50       CONTINUE
         DO 55 I = 1,N
            J = (I-1)*M + 1
            BCNDV(J) = BCNDV(J) + TOP(I)
            J = I * M
            BCNDV(J) = BCNDV(J) + BOT(I)
55       CONTINUE
C
```

Figure 7.23 *(continued)*

```
C       --------------------------------------------------------------
C
C  NOW WE GET THE LU DECOMPOSITIONS OF THE TWO COEFFICIENT MATRICES.
C  THESE ARE VERY EASY TO COMPUTE. WE STORE THEM BACK IN THE
C  ORIGINAL VECTOR SPACE.
C
        DO 60 I = 2,MSIZE
          UCOEF(I-1,3) = UCOEF(I-1,3)/UCOEF(I-1,2)
          UCOEF(I,2) = UCOEF(I,2) - UCOEF(I,1)*UCOEF(I-1,3)
          VCOEF(I-1,3) = VCOEF(I-1,3)/VCOEF(I-1,2)
          VCOEF(I,2) = VCOEF(I,2) - VCOEF(I,1)*VCOEF(I-1,3)
60      CONTINUE
C
C       --------------------------------------------------------------
C
C  NOW WE BEGIN THE ITERATIONS, LIMIT THEM TO ITMAX IN NUMBER.
C
        ITMAX = 30
        DO 190 KNT = 2,ITMAX,2
C
C       --------------------------------------------------------------
C
C  COMPUTE R.H.S. FOR THE U EQUATIONS AND STORE THESE IN THE U
C  VECTOR. FIRST DO THE TOP AND BOTTOM SETS OF TERMS.
C
        DO 65 I = 1,N
          J = (I-1)*M + 1
          U(I) = (1.0/RHO-2.0)*V(J) + V(J+1) + BCNDU(I)
          K = MSIZE - N + I
          J = I * M
          U(K) = V(J-1) + (1.0/RHO-2.0)*V(J) + BCNDU(K)
65      CONTINUE
C
C       --------------------------------------------------------------
C
C  NOW DO THE INTERMEDIATE ONES.
C
        DO 75 I = 2,ML1
          DO 70 J = 1,N
            K = (I-1)*N + J
            L = I + (J-1)*M
            U(K) = V(L-1) + (1.0/RHO-2.0)*V(L) + V(L+1) + BCNDU(K)
70        CONTINUE
75      CONTINUE
C
C       --------------------------------------------------------------
C
C  NOW GET THE SOLUTION FOR THE HORIZONTAL TRAVERSE. FIRST Y=L(-1)*B
C
        U(1) = U(1) / UCOEF(1,2)
        DO 80 I = 2,MSIZE
          U(I) = ( U(I) - UCOEF(I,1)*U(I-1) ) / UCOEF(I,2)
80      CONTINUE
C
C       --------------------------------------------------------------
C
C  READY NOW TO GET X = U(-1)*Y.
C
        DO 90 JROW = MSL1,1,-1
          U(JROW) = U(JROW) - UCOEF(JROW,3)*U(JROW+1)
90      CONTINUE
C
C       --------------------------------------------------------------
C
C  WE DO THE SAME FOR THE VERTICAL TRAVERSE - U AND V EXCHANGE
C  ROLES. COMPUTE R.H.S. FOR THE V EQUATIONS, STORE IN V. DO
C  THE TOP AND BOTTOM SETS.
```

Figure 7.23 *(continued)*

```
C
          DO 95 I = 1,M
            J = (I-1)*N +1
            V(I) = (1.0/RHO-2.0)*U(J) + U(J+1) + BCNDV(I)
            K = MSIZE - M + I
            J = I * N
            V(K) = U(J-1) + (1.0/RHO-2.0)*U(J) + BCNDV(K)
95        CONTINUE
C
C   -------------------------------------------------------------------
C
C   DO THE INTERMEDIATE ROWS
C
          DO 105 I = 2,NL1
            DO 100 J = 1,M
              K = (I-1)*M + J
              L = I + (J-1)*N
              V(K) = U(L-1) + (1.0/RHO-2.0)*U(L) + U(L+1) + BCNDV(K)
100         CONTINUE
105       CONTINUE
C
C   -------------------------------------------------------------------
C
C   GET THE SOLUTION - FIRST Y = L(-1)*B
C
          V(1) = V(1) / VCOEF(1,2)
          DO 110 I = 2,MSIZE
            V(I) = ( V(I) - VCOEF(I,1)*V(I-1) ) / VCOEF(I,2)
110       CONTINUE
C
C   -------------------------------------------------------------------
C
C   THEN X = U(-1) * Y
C
          DO 120 JROW = MSL1,1,-1
            V(JROW) = V(JROW) - VCOEF(JROW,3)*V(JROW+1)
120       CONTINUE
C
C   -------------------------------------------------------------------
C
C   PRINT OUT THE LATEST RESULT.
C
          PRINT 202, KNT, (( V(I),I=J,MSIZE,M ), J=1,M)
202       FORMAT(///1X,' AFTER ITERATION NUMBER',
     +                  I3/(1X,10F7.3/1X,5F7.3))) )
C
C   END OF THE ITERATION LOOP
C
190       CONTINUE
          STOP
          END


     ***************************************************************
                     PARTIAL OUTPUT FOR ADIELL.F

       AFTER ITERATION NUMBER    2

    .000    .000    .000    .001    .003    .008    .021    .056    .146    .383
   1.003   2.627   6.877  18.004  47.136
    .000    .000    .001    .002    .004    .011    .029    .077    .202    .529
   1.384   3.623   9.486  24.834  65.015
    .000    .000    .001    .002    .005    .012    .032    .085    .222    .581
   1.522   3.986  10.434  27.317  71.517
    .000    .000    .001    .002    .005    .013    .033    .087    .227    .595
   1.557   4.076  10.671  27.938  73.142
    .000    .000    .001    .002    .005    .012    .032    .085    .222    .581
```

Figure 7.23 *(continued)*

```
1.522   3.986 10.434 27.317 71.517
 .000    .000   .001   .002   .004    .011   .029   .077   .202   .529
1.384   3.623  9.486 24.834 65.015
 .000    .000   .000   .001   .003    .008   .021   .056   .146   .383
1.003   2.627  6.877 18.004 47.136


    AFTER ITERATION NUMBER    4

 .000    .001   .003   .006   .015    .036   .087   .206   .483  1.114
2.524   5.583 11.930 24.205 44.964
 .001    .002   .005   .012   .030    .072   .170   .397   .918  2.085
4.625   9.922 20.255 38.047 60.386
 .001    .003   .007   .016   .038    .091   .216   .504  1.159  2.617
5.760  12.222 24.516 44.594 65.464
 .001    .003   .007   .017   .041    .097   .230   .536  1.232  2.779
6.104  12.914 25.781 46.476 66.647
 .001    .003   .007   .016   .038    .091   .216   .504  1.159  2.617
5.760  12.222 24.516 44.594 65.464
 .001    .002   .005   .012   .030    .072   .170   .397   .918  2.085
4.625   9.922 20.255 38.047 60.386
 .000    .001   .003   .006   .015    .036   .087   .206   .483  1.114
2.524   5.583 11.930 24.205 44.964


                                          .
                                          .
                                          .


    AFTER ITERATION NUMBER   28

 .072    .156   .266   .419   .640    .962  1.436  2.137  3.176  4.728
7.074  10.711 16.620 27.085 48.330
 .134    .289   .492   .775  1.181   1.776  2.651  3.940  5.846  8.667
12.864 19.152 28.690 43.391 66.236
 .174    .378   .643  1.012  1.543   2.319  3.460  5.138  7.609 11.241
16.570 24.351 35.603 51.557 73.226
 .189    .409   .696  1.095  1.670   2.510  3.743  5.557  8.223 12.132
17.837 26.088 37.824 54.012 75.115
 .174    .378   .643  1.012  1.543   2.319  3.460  5.138  7.609 11.241
16.570 24.351 35.603 51.557 73.226
 .134    .289   .492   .775  1.181   1.776  2.651  3.940  5.846  8.667

12.864 19.152 28.690 43.391 66.236
 .072    .156   .266   .419   .640    .962  1.436  2.137  3.176  4.728
7.074  10.711 16.620 27.085 48.330


    AFTER ITERATION NUMBER   30

 .074    .159   .271   .425   .647    .970  1.445  2.146  3.185  4.737
7.082  10.718 16.626 27.089 48.332
 .136    .295   .500   .785  1.194   1.791  2.667  3.957  5.862  8.684
12.879 19.165 28.700 43.398 66.240
 .178    .385   .653  1.026  1.560   2.339  3.481  5.160  7.631 11.263
16.590 24.368 35.617 51.566 73.231
 .193    .417   .707  1.110  1.688   2.531  3.766  5.581  8.247 12.155
17.858 26.106 37.839 54.022 75.120
 .178    .385   .653  1.026  1.560   2.339  3.481  5.160  7.631 11.263
16.590 24.368 35.617 51.566 73.231
 .136    .295   .500   .785  1.194   1.791  2.667  3.957  5.862  8.684
12.879 19.165 28.700 43.398 66.240
 .074    .159   .271   .425   .647    .970  1.445  2.146  3.185  4.737
7.082  10.718 16.626 27.089 48.332
```

Exercises

Section 7.2

1. In Section 7.2, the equation for net heat flow into the element that has dimension t, dx, and dy is derived. The governing relation for rate of flow is

$$-kA\left(\frac{\partial u}{\partial x}\right),$$

where u = temperature and t = time. In that derivation, k and A were taken as constant. Suppose this is not true, but instead that k varies with position. Derive the equation for net heat flow into the element for variable k. Is Laplace's equation obtained again?

▶2. Repeat Exercise 1 but, in addition to having k be $k(x, y)$, allow the thickness to also vary with position.

3. Repeat Exercise 1, but now for k varying with temperature, such as $k = a + bu + cu^2$.

Section 7.3

4. The mixed second derivative $\partial^2 u/(\partial x\,\partial y)$ can be considered as

$$\frac{\partial}{\partial x}\left(\frac{\partial u}{\partial y}\right) = \frac{\partial^2 u}{\partial x\,\partial y} = \frac{\partial}{\partial y}\left(\frac{\partial u}{\partial x}\right).$$

If the nodes are spaced apart a distance h in both the x- and y-directions, show that this derivative can be represented by the pictorial operator

$$\frac{1}{4h^2}\begin{Bmatrix}-1 & 1 \\ 1 & -1\end{Bmatrix} + O(h^2).$$

▶5. If d^2u/dx^2 is represented as this fourth-order central-difference formula

$$\frac{d^2 u}{dx^2} = \frac{-u_{i+2} + 16u_{i+1} - 30u_i + 16u_{i-1} - u_{i-2}}{12h^2},$$

find the fourth-order operator for the Laplacian. (This requires the function to have a continuous sixth derivative.)

6. It is sometimes necessary to approximate a differential equation that contains a fourth derivative (bending of plates is an example). Find a central-difference approximation for this case. (You may want to use the method of undetermined coefficients that is discussed in Appendix B or the Taylor series method of Section 7.3.)

7. Suppose a differential equation contains a third par-

tial derivative. Repeat Exercise 6 for this case. Can you find an expression that uses function values symmetrical about the point of application?

▶8. A rectangular plate of constant thickness has heat flow only in the x- and y-directions (k is constant). If the top and bottom edges are perfectly insulated and the left edge is at $100°$ and the right edge at $200°$, it is obvious that there is no heat flow except in the x-direction and that temperatures vary linearly with x and are constant along vertical lines.
  a. Show that such a temperature distribution satisfies Eq. (7.7) and also Eq. (7.8).
  b. Show that this temperature distribution also satisfies the relation derived in Exercise 5. What about nodes adjacent to the edges?

9. Derive the nine-point approximation of the Laplacian in Eq. (7.8).

Section 7.4

10. Set up equations analogous to Eqs. (7.10) for the example problem of Section 7.4, but use a grid spacing of 2 cm. Solve the equations by elimination.

11. Solve for the steady-state temperatures in a rectangular plate, 12 in. × 15 in., if one 15-in. edge is held at $100°$ and the other 15-in. edge is held at $50°$; both 12-in. edges are held at $20°$. The material is aluminum. Space the nodes 3 in. apart in both directions. Consider heat to flow only laterally. Sketch the approximate location of the $65°$ isothermal curve.

▶12. Solve for the steady-state temperatures in the plate of Fig. 7.24 when the edge temperatures are as shown. The plate is 10 cm × 8 cm, and the nodal spacing is 2 cm.

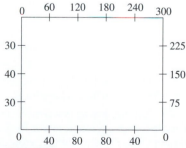

Figure 7.24

13. Repeat Exercise 10, but this time use the nine-point formula of Eq. (7.8). Assume that temperatures at $(20, 10)$ and $(20, 0)$ are 50°.

14. Repeat Exercise 12, but use the nine-point formula.

▶15. Suppose the differential equation for the plate is

$$3u_{xx} + 2u_{yy} = 0.$$

What will Eq. (7.6) and the pictorial operator of Eq. (7.7) look like?

16. The region on which we solve Laplace's equation does not have to be rectangular. We can apply the methods of Section 7.4 to any region where the nodes fall on the boundary. Solve for the steady-state temperatures at the eight interior points of Fig. 7.25.

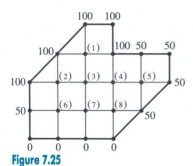

Figure 7.25

Section 7.5

▶17. Repeat Exercise 10, but this time use Liebmann's method.

18. Solve Exercise 12 by Liebmann's method with all elements of the initial u-vector equal to zero. Then repeat with all elements equal to 300°, the upper bound to the steady-state temperatures. Repeat again with the initial values all equal to the arithmetic average of the boundary temperatures. Compare the number of iterations needed to reach a given tolerance for convergence in each case. What is the effect of the tolerance value that is used?

19. Solve Exercise 11 by Liebmann's method. Before you begin, make some quick hand computations to find good values for the starting values for u.

▶20. Repeat Exercise 10, employing successive over-relaxation. Vary the overrelaxation factor to determine the optimum value experimentally. How does this compare to that predicted by Eq. (7.18)?

21. Repeat Exercise 18, but now use overrelaxation with the factor given by Eq. (7.18).

22. Repeat Exercise 19 but use overrelaxation with the factor given by Eq. (7.18).

Section 7.6

23. Find the torsion function ϕ for a 2 in. × 2 in. square bar.
    a. Subdivide the region into nine equal squares, so that there are four interior nodes. Because of symmetry, all of the nodes will have equal ϕ-values.
    b. Repeat but subdivide into 36 equal squares with 25 interior nodes. Use the results of part (a) to get starting values for iteration.

24. Solve

$$\nabla^2 u = 2 + x^2 + y^2$$

over a hollow square bar, 5 in. in outside dimension and with walls 2 in. thick (so that the inner square hole is 1 in. on a side). The origin for x and y is the center of the object. On the inner and outer surfaces, $u = 0$.

▶25. Solve for the torsion in a prismatic bar of cross section equal to the region of Exercise 16. The interior points are 0.5 cm apart.

26. Solve

$$\nabla^2 u = f(x, y)$$

with $f(x, y) = xy$. Use $h = \frac{1}{3}$. The values of u on the boundary are everywhere zero. The region is a square with corners at $(0, 0)$ and $(2, 2)$.

27. Repeat Exercise 26, but with $f(x, y) = (1 - x)(1 - y)$.

▶28. Suppose the function defined over the rectangular plate of Fig. 7.7 satisfied the equation

$$3\phi_{xx} + \phi_{yy} + 2 = 0, \qquad \phi = 0 \text{ on boundary.}$$

What would the pictorial diagram and set of equations (7.20) and (7.21) look like? (Use the five-point approximation for the Laplacian.)

29. Repeat part (b) of Exercise 23, but now use over-relaxation. Find the optimum overrelaxation factor experimentally. Does this match the value from Eq. (7.18)?

30. Repeat Exercise 24, but this time use overrelaxation. Vary the overrelaxation factor until you find the optimum. Compare to the value from Eq. (7.18).

Section 7.7

31. Solve the set of equations in Eqs. (7.25), using Eq. (7.26) to eliminate the fictitious exterior points.
  a. Use elimination.
  b. Use Liebmann's method.
  c. Use S.O.R.

▶**32.** Solve a modification of Example 7.1 in Section 7.7: The 8-cm edges are held at 20°, and along the 4-cm edges an outward gradient of −15°C/cm is imposed.

▶**33.** Solve another modification of Example 7.1 of Section 7.7: Along all the edges an outward gradient of −15° C/cm is imposed. Is it possible to get a unique solution?

34. Solve for the steady-state temperatures in the region of Exercise 16, except now the plate is insulated along the edge at each point marked zero. All other edge temperatures are held at the values as marked.

35. Consider a region that is obtained from that of Exercise 16 by reflecting across the edge marked with zeros. Show that the steady-state potentials are the same as for Exercise 16.

Section 7.8

36. Construct the pictorial operator for Eq. (7.29) if the theta values are $\theta_1 = 0.25$, $\theta_2 = 0.37$, $\theta_3 = 1.00$, and $\theta_4 = 0.55$ (as defined in Fig. 7.10).

▶**37.** A coaxial cable has a circular outer conductor 10 cm in diameter, and a concentric square inner conductor that is 3 cm on a side. The outer conductor is at zero volts, and the inner conductor at 100 volts. Find the potential at points in the space between. (Because of symmetry, only one octant needs to be computed.) Use a grid of 1-cm squares.

38. If a hollow shaft has an outer square cross section that is 10 in. on a side and a concentric circular inner cross section that is 3 in. in diameter, what is the value of the torsion function at the nodes of a 1-in. grid?

39. Use the Laplacian in polar coordinates to set up the equations that solve

$$\nabla^2 u = x^2 y$$

on a semicircular region whose radius is 4. Take $\Delta r = 1$ and $\Delta \theta = \pi/6$. The boundary conditions are $u = 10$ on the straight edge and $u = 0$ on the curved boundary.

40. Solve Exercise 39, but use a square mesh, $h = 1$. Compare results for nodes where $x = 0$.

Section 7.9

41. A cube is 4 cm along each edge. Two opposite faces are held at 100°, the other four faces at 0°. Find internal temperatures at the nodes of a 1-cm network.

▶**42.** Repeat Exercise 41, but now two adjacent 0° faces have been insulated so that no heat flows.

43. Repeat Exercise 42, except now the insulated faces are imperfectly insulated so that $u_N = 2 - 0.1u$. (There is heat leakage through the insulation proportional to the difference in surface temperature and 20°.)

Section 7.10

44. Solve Exercise 12 by the A.D.I. method, using $\rho = 1.0$. Begin with the initial u-vector equal to the arithmetic average of the boundary values. Compare the number of iterations needed to those required with Liebmann's method (Exercise 18) and S.O.R. using the optimum overrelaxation factor (Exercise 21).

45. Solve the sets of equations in (7.25) and (7.26) by A.D.I. using $\rho = 1.0$. Compare the number of iterations with the number needed by S.O.R. as in Exercise 31c.

▶**46.** Repeat Exercise 23, but use the A.D.I. method. Vary the value of ρ to find the optimum values.

47. Set up the equations to apply A.D.I. to the problem in Exercise 42.

48. Set up the equations to apply A.D.I. to the problem in Exercise 43.

Section 7.11

49. Confirm that Eq. (7.44) is in fact the inverse of matrix M in Eq. (7.43).

50. Find M^{-1}, a, N, and $u(x,y)$ for these triangular elements:
  a. Nodes: $(1.2, 3.1), (-0.2, 4), (-2, -3)$; u-values at these nodes: 5, 20, 7; point where u is to be determined: $(-1, 0)$
  b. Nodes: $(20, 40), (50, 10), (5, 10)$; u-values at these nodes: 12.5, 6.2, 10.1; point where u is to be determined: $(20, 20)$
  ▶**c.** Nodes: $(12.1, \ 11.3), \ (8.6, \ 9.3), \ (13.2, \ 9.3)$;

u-values at these nodes: $121, 215, 67$; point where u is to be determined: $(10.6, 9.6)$

51. Confirm that the sum of the entries in the first row of M^{-1} is equal to twice the area for each of the elements in Exercise 50.

▶**52.** Find the element equations for the element in part (c) of Exercise 50 if $Q = x^2 y$ and $F = -x/y$ (these refer to Eq. 7.39). There are no derivative conditions on any of the element boundaries.

53. Solve Exercise 12 by finite elements. Locate four internal nodes at $(4, 2)$, $(8, 4)$, $(2, 4)$ and $(6, 6)$ relative to the lower-left corner, and connect these to nodes at the four corners of the region. Compare answers from this solution to the values obtained through finite differences with a grid spacing of 2 cm.

54. Solve Example 7.1 by finite elements. Place nodes at the points marked 10, 11, and 4 and at the corners of the region; then draw triangular elements.

Section 7.12

In Exercises 55 and 56, the region is a square with four evenly spaced interior nodes. The boundaries are everywhere zero except $u = 10$ on the right side.

55. Demonstrate the maximum–minimum principle for Laplace's equation by solving the problem with $h = 2$,
a. starting with $u = 20$ at the interior nodes.
b. starting with $u = -10$ at the interior nodes.

56. The superposition principle works somewhat differently for numerical methods. Show that solutions of $u_{xx} + u_{yy} = 0$ and $u_{xx} + u_{yy} = f(x, y)$ differ by a value that depends on $f(x, y)$ and h.
a. Solve analytically.

b. Demonstrate by comparing the solutions of both equations for several values of h and several definitions of $f(x, y)$.

▶**57.** For each matrix, find the optimum overrelaxation factor by the technique of Section 7.12, then compare to the value predicted by Eq. (7.18) or experimentally.

a. $\begin{bmatrix} -4 & 1 \\ 1 & -4 \end{bmatrix}$

b. $\begin{bmatrix} 4 & 3 \\ 3 & 4 \end{bmatrix}$

c. $\begin{bmatrix} -4 & 1 & 0 \\ 1 & -4 & 1 \\ 0 & 1 & -4 \end{bmatrix}$

58. The system of equations given here (as an augmented matrix) can be solved by either Gauss–Jacobi or Gauss–Seidel. Both can be speeded by applying overrelaxation. Make trials with varying values of the factor to find the optimum value. (In this case you will probably find this to be less than unity, meaning it is underrelaxed.)

$$\begin{bmatrix} 8 & 1 & -1 & | & 8 \\ 1 & -7 & 2 & | & -4 \\ 2 & 1 & 9 & | & 12 \end{bmatrix}$$

59. What are the eigenvalues of the iteration matrices of Exercise 58 when the optimum relaxation factor is applied?

60. Use the technique of Section 7.12 to determine the optimum relaxation factor in Exercise 58.

Applied Problems and Projects

61. Write and run a program for Poisson's equation. Use it to solve

$$\nabla^2 u = xy(x - 2)(y - 2),$$

on the region $0 \le x \le 2, 0 \le y \le 2$, with $u = 0$ on all boundaries except for $y = 0$, where $u = 1.0$.

62. Modify the program of Exercise 61 so that it permits boundary conditions of the form

$$au + b\frac{\partial u}{\partial x} = c \quad \text{and} \quad a'u + b'\frac{\partial u}{\partial y} = c'.$$

Test it with the equation of Exercise 61, modified in that the boundary where $y = 0$ has $\partial u / \partial y = 1$.

63. Repeat Exercise 61, but now the program uses the A.D.I. method. Provide for a Poisson-equation problem as well as derivative boundary conditions.

64. A classic problem in elliptic partial-differential equations is to solve $\nabla^2 u = 0$ on a region defined by $0 \le x \le \pi, 0 \le y \le \infty$, with boundary condition of $u = 0$ at $x = 0$, at $x = \pi$, and at $y = \infty$. The boundary at $y = 0$ is held at $u = F(x)$. This can be quite readily solved by the method of separation of variables, to give the series solution

$$u = \sum_{n=1}^{\infty} B_n e^{-ny} \sin nx,$$

with

$$B_n = 2 \int_0^{\pi} F(x) \sin nx \, dx.$$

Solve this equation numerically for various definitions of $F(x)$. (You will need to redefine the region so that $0 \le y \le M$, where M is large enough that changes in u with y at $y = M$ are negligible.) Compare your results to the series solution. You might try

$$F(x) = 100; \qquad F(x) = 100 \sin x; \qquad F(x) = 400x(1 - x); \qquad F(x) = 100(1 - |2x - 1|).$$

65. The equation

$$2 \frac{\partial^2 u}{\partial x^2} + \frac{\partial^2 u}{\partial y^2} - \frac{\partial u}{\partial x} = 2$$

is an elliptic equation. Solve it on the unit square, subject to $u = 0$ on the boundaries. Approximate the first derivative by a central-difference approximation. Investigate the effect of size of Δx on the results, to determine at what size reducing it does not have further effect.

66. Solve the steady-state heat-flow problem that was discussed at the beginning of the chapter. Assume that the metal is aluminum and that the insulation is mineral wool. Look up the thermal properties in handbooks.

Solve it first as a three-dimensional problem. Then, considering that the plate is not very thick relative to its length and that aluminum is a good conductor so that there is little temperature variation across the thickness, solve it as a two-dimensional problem. Do you reach the same conclusion about the feasibility of the scheme?

Consider carefully how you will handle the problem of dissimilar materials—metal and insulation. Can you approximate the effect of the insulation by some proper value of the gradient at the surface of the aluminum?

8

Parabolic Partial-Differential Equations

Contents of This Chapter

Chapter 7 discussed the solution of partial-differential equations that were time-independent. Such steady-state problems were described by elliptic equations. Unsteady-state problems in which the function is dependent on time are of great importance. We study such parabolic equations in this chapter.

The method of finite-difference equations will be further developed here and in Chapter 9, since it is very important in solving both parabolic and hyperbolic partial-differential equations. Available commercial software includes PDECOL, distributed by IMSL, that solves systems of parabolic or hyperbolic partial-differential equations as well as boundary-value problems. Finite-element programs can also solve these systems and are especially valuable when the geometry is not simple. Most treatments of finite elements for time-dependent problems combine methods that use finite differences.

8.1 Equation for Time-Dependent Heat Flow

Develops the partial-differential equations for unsteady-state heat flow, wherein the temperatures in an object vary with time. This leads to an equation of the parabolic type.

8.2 The Explicit Method

Discusses one of the simplest ways to represent the partial-differential equation as a difference equation. This method divides both space and time into discrete uniform intervals and replaces the derivatives by finite-difference approximations. In this method, a forward difference replaces the time derivative and a central difference replaces the space derivative. Unfortunately, this simple approach places limitations on the step sizes to avoid instability. The solution marches forward in time from the initial temperature distribution, which is a prescribed initial condition.

8.3 The Crank–Nicolson Method

Overcomes the limitations that the explicit method imposes on the size of the time steps and improves accuracy at the expense of having to solve a set of equations at each time step. Fortunately, the system is tridiagonal for a one-space-dimension problem. The Crank–Nicolson method is unconditionally stable—any time step size can be used.

8.4 A Generalization—The Theta Method

Rederives the finite-difference equation using a parameter, theta (hence the name for this method). This gives a more rational basis for discussing stability. A suitable choice of the parameter value results in both the explicit and the Crank–Nicolson methods as well as other unconditionally stable methods.

8.5 Derivative Boundary Conditions

Adds nothing new to problems—they are handled exactly as when solving elliptic partial-differential equations. We extend the region artificially to write difference approximations for the derivatives on the boundaries.

8.6 Parabolic Equations in Two or Three Dimensions

Extends the finite-difference methods to regions in space. As expected, we discover that the number of equations to be solved becomes very large and, unfortunately, they are no longer tridiagonal. A modification of the A.D.I. method helps to keep the number of simultaneous equations manageable.

8.7 Finite Elements for Heat Flow

Demonstrates that this newer method readily applies, although it requires many simple but tedious steps to set up the equations. Once the steps are at hand, however, the solution is obtained in the usual way. Actually, finite elements are applied only to the space variables—the time derivative is replaced by a finite difference, so the technique combines finite elements with finite differences and, consequently, stability considerations are present. Our treatment is less than complete due to space constraints.

8.8 Theoretical Matters

Discusses in some detail the questions of stability and convergence of our numerical methods.

Chapter Summary

Helps you to review the important topics of the chapter.

Computer Programs

Provides some computer programs to solve parabolic equations.

Parallel Processing
Chapter 8 is similar to Chapter 7 in respect to parallel processing: A set of equations must be solved, repetitively in the present case. Since these sets can be very large, there is a good possibility for speeding up. The discussion of Chapter 2 applies here.

8.1   Equation for Time-Dependent Heat Flow

When temperatures in an object change with time, we have a problem of unsteady-state flow of heat, a physical situation that can be represented by a parabolic partial-differential equation. The simplest situation is for flow of heat in one direction. Imagine a rod that is uniform in cross section and insulated around its perimeter so that heat flows only longitudinally. Consider a differential portion of the rod, dx in length with cross-sectional area A (see Fig. 8.1).

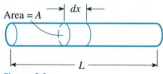

Figure 8.1

We let u represent the temperature at any point in the rod, whose distance from the left end is x. Heat is flowing from left to right under the influence of the temperature gradient $\partial u/\partial x$. Make a balance of the rate of heat flow into and out of the element. Use k for thermal conductivity, cal/sec/[(cm^2)(°C/cm)], which we assume is constant.

Rate of flow of heat in: $-kA\dfrac{\partial u}{\partial x}$;

Rate of flow of heat out: $-kA\left(\dfrac{\partial u}{\partial x} + \dfrac{\partial}{\partial x}\left(\dfrac{\partial u}{\partial x}\right)dx\right)$.

The difference between the rate of flow in and the rate of flow out is the rate at which heat is being stored in the element. If c is the heat capacity, cal/[(g)(°C)], and ρ is the density, g/cm^3, we have, with t for time,

$$-kA\frac{\partial u}{\partial x} - \left(-kA\left(\frac{\partial u}{\partial x} + \frac{\partial}{\partial x}\left(\frac{\partial u}{\partial x}\right)dx\right)\right) = c\rho(A\,dx)\frac{\partial u}{\partial t}.$$

Simplifying, we have

$$k\frac{\partial^2 u}{\partial x^2} = c\rho\frac{\partial u}{\partial t}.$$

This is the basic mathematical model for unsteady-state flow. We have derived it for heat flow, but it applies equally to diffusion of material, flow of fluids (under conditions of laminar flow), flow of electricity in cables (the telegraph equations), and so on.

The differential equation is classed as parabolic because, comparing to the standard form of an equation in two independent variables, x and t,

$$A\frac{\partial^2 u}{\partial x^2} + B\frac{\partial^2 u}{\partial x\,\partial t} + C\frac{\partial^2 u}{\partial t^2} + D\left(x,t,u,\frac{\partial u}{\partial x},\frac{\partial u}{\partial t}\right) = 0,$$

we find that $B^2 - 4AC = 0$.

In two or three space dimensions, analogous equations apply:

$$k\left(\frac{\partial^2 u}{\partial x^2} + \frac{\partial^2 u}{\partial y^2}\right) = c\rho\frac{\partial u}{\partial t},$$

$$k\left(\frac{\partial^2 u}{\partial x^2} + \frac{\partial^2 u}{\partial y^2} + \frac{\partial^2 u}{\partial z^2}\right) = c\rho\frac{\partial u}{\partial t}.$$

The function that we call the solution to the problem not only must obey the preceding differential equation, but also must satisfy an initial condition and a set of boundary conditions. For the one-dimensional heat-flow problem that we first consider, the initial condition will be the initial temperatures at all points along the rod,

$$u(x,t)|_{t=0} = u(x,0) = f(x).$$

The boundary conditions will describe the temperature at each end of the rod as functions of time. Our first examples will consider the case where these temperatures are held constant:

$$u(0,t) = c_1,$$
$$u(L,t) = c_2.$$

More general (and more practical) boundary conditions will involve not only the temperature but the temperature gradients, and these may vary with time:

$$A_1 u(0,t) + B_1 \frac{\partial u(0,t)}{\partial x} = F_1(t),$$

$$A_2 u(L,t) + B_2 \frac{\partial u(L,t)}{\partial x} = F_2(t).$$

In the general case, the thermal properties may vary along the rod and there may be heat generation of Q cal/sec/cm^2. In this case the equation becomes

$$\frac{\partial}{\partial x}\left(k(x)\frac{\partial u}{\partial x}\right) + Q(x) = c(x)\rho(x)\frac{\partial u}{\partial t}.$$

For unsteady-state heat flow in two or three dimensions, these properties may vary with each coordinate:

$$\frac{\partial}{\partial x}\left(k(x,y,z)\frac{\partial u}{\partial x}\right) + \frac{\partial}{\partial y}\left(k(x,y,z)\frac{\partial u}{\partial y}\right) + \frac{\partial}{\partial z}\left(k(x,y,z)\frac{\partial u}{\partial z}\right)$$

$$+ Q(x,y,z) = c(x,y,z)\rho(x,y,z)\frac{\partial u}{\partial t}.$$

We will stay with the simpler situations.

8.2 The Explicit Method

Our approach to solving parabolic partial-differential equations by a numerical method is to replace the partial derivatives by finite-difference approximations. For the one-dimensional heat-flow equation,

$$\frac{\partial u}{\partial t} = \frac{k}{c\rho}\frac{\partial^2 u}{\partial x^2}, \tag{8.1}$$

we can use the relations

$$\left.\frac{\partial u}{\partial t}\right|_{\substack{x=x_i \\ t=t_j}} = \frac{u_i^{j+1} - u_i^j}{\Delta t} + O(\Delta t) \tag{8.2}$$

and

$$\left.\frac{\partial^2 u}{\partial x^2}\right|_{\substack{x=x_i \\ t=t_j}} = \frac{u_{i+1}^j - 2u_i^j + u_{i-1}^j}{(\Delta x)^2} + O(\Delta x)^2. \tag{8.3}$$

We use subscripts to denote position and superscripts for time. Note that the error terms are of different orders, since a forward difference is used in Eq. (8.2). This introduces some special limitations, but it does simplify the procedure.

Substituting Eqs. (8.2) and (8.3) into (8.1) and solving for u_i^{j+1} gives the equation for the forward-difference method:

$$u_i^{j+1} = \frac{k \, \Delta t}{c\rho(\Delta x)^2} (u_{i+1}^j + u_{i-1}^j) + \left(1 - \frac{2k \, \Delta t}{c\rho(\Delta x)^2}\right) u_i^j. \qquad (8.4)$$

We have solved for u_i^{j+1} in terms of the temperatures at time t_j in Eq. (8.4) in view of the normally known conditions for a parabolic partial-differential equation. We subdivide the length into uniform subintervals and apply our finite-difference approximation to Eq. (8.1) at each point where u is not known. Equation (8.4) then gives the values of u at each interior point at $t = t_1$, since the values at $t = t_0$ are given by the initial conditions. It can then be used to get values at t_2 using the values at t_1 as initial conditions, so we can step the solution forward in time. At the endpoints, the boundary conditions will determine u.

The relative size of the time and distance steps, Δt and Δx, affects Eq. (8.4). If the ratio of $\Delta t/(\Delta x)^2$ is chosen so that $k \, \Delta t/c\rho(\Delta x)^2 = \frac{1}{2}$, the equation is simplified in that the last term vanishes and we have

$$u_i^{j+1} = \frac{1}{2} (u_{i+1}^j + u_{i-1}^j). \qquad (8.5)$$

We will let $k \, \Delta t/c\rho(\Delta x)^2 = r$. As we will see, the value of r is critical.

If the value of $k \, \Delta t/c\rho(\Delta x)^2$ is chosen as less than one-half, there will be improved accuracy* (limited, of course, by the errors dependent on the size of Δx). If the value is chosen greater than one-half, which would reduce the number of calculations required to advance the solution through a given interval of time, the phenomenon of *instability* sets in. (We will show later in this chapter why this ratio affects stability and convergence.)

We illustrate the method with a simple example, varying the value of $r = k \, \Delta t/c\rho(\Delta x)^2$ to demonstrate its effect.

EXAMPLE 8.1 A large flat steel plate is 2 cm thick. If the initial temperatures (°C) within the plate are given, as a function of the distance from one face, by the equations

$$u = 100x \qquad \text{for } 0 \leq z \leq 1,$$
$$u = 100(2 - x) \qquad \text{for } 1 \leq x \leq 2,$$

find the temperatures as a function of x and t if both faces are maintained at 0°C.

Since the plate is large, we can neglect lateral flow of heat relative to the flow perpendicular to the faces and hence use Eq. (8.1) for heat flow in one direction.

---

* It can be shown that choosing the ratio equal to $\frac{1}{6}$ is especially advantageous in minimizing the truncation error.

For steel, $k = 0.13$ cal/sec $\cdot$ cm $\cdot$ °C, c $= 0.11$ cal/g $\cdot$ °C, and $\rho = 7.8$ g/cm³. To use Eq. (8.5) as an approximation to the physical problem, we subdivide the total thickness into an integral number of spaces. Let us use $\Delta x = 0.25$, giving eight subdivisions. If we wish to use Eq. (8.5), we then fix Δt by the relation

$$\frac{k\,\Delta t}{c\rho(\Delta x)^2} = \frac{1}{2}, \qquad \Delta t = \frac{(0.11)(7.8)(0.25)^2}{(2)(0.13)} = 0.206 \text{ sec.}$$

In our case, the boundary conditions are

$$u(0,t) = 0, \qquad u(2,t) = 0.$$

The initial condition is

$$u(x,0) = 100x \qquad \text{for } 0 \le x \le 1,$$
$$u(x,0) = 100(2 - x) \qquad \text{for } 1 \le x \le 2.$$

Our calculations are conveniently recorded in a table, as in Table 8.1, where each row of figures is at a particular time. We begin by filling in the initial conditions along the first row, at $t = 0.0$. The simple algorithm of Eq. (8.5) tells us that, at each interior point, the temperature at any point at the end of a time step is just the arithmetic average of the temperatures at the adjacent points at the beginning of that time step. The end temperatures are given by the boundary conditions. Because the temperatures are symmetrical on either side of the center line, we compute only for $x \le 1.0$. The temperature at $x = 1.25$ is the same as at $x = 0.75$.

Table 8.1

| | Calculated temperatures at | | | | | |
t	$x = 0$	$x = 0.25$	$x = 0.50$	$x = 0.75$	$x = 1.0$	$x = 1.25$
0.0	0	25.0	50.0	75.0	100.0	75.0
0.206	0	25.0	50.0	75.0	75.0	75.0
0.412	0	25.0	50.0	62.5	75.0	62.5
0.619	0	25.0	43.75	62.5	62.5	62.5
0.825	0	21.88	43.75	53.12	62.5	53.12
1.031	0	21.88	37.5	53.12	53.12	53.12
1.238	0	18.75	37.5	45.31	53.12	45.31
1.444	0	18.75	32.03	45.31	45.31	45.31
1.650	0	16.16	32.03	38.67	45.31	38.67
1.856	0	16.16	27.34	38.67	38.67	38.67
2.062	0	13.67	27.34	33.01	38.67	33.01

In Fig. 8.2 we compare some of the numerical results with the analytical solution, which is

$$u = 800 \sum_{n=0}^{\infty} \frac{1}{\pi^2(2n + 1)^2} \cos \frac{(2n + 1)\pi(x - 1)}{2} e^{-0.3738(2n+1)^2 t}.$$

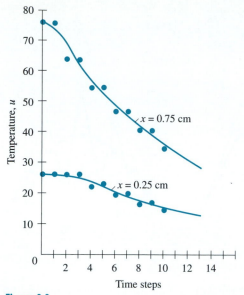

Figure 8.2

Note that the numerical values are close to and oscillate about the curves that are drawn to represent the analytical solution. In general the errors of Table 8.1 are less than 4%. If the size of Δx were less, the errors would be smaller. Unfortunately, this easy method is not always so accurate.

When the value for $k\,\Delta t/c\rho(\Delta x)^2$ in Eq. (8.4) is other than 0.5, the resulting equation is slightly more complicated, but the related computer program is not significantly slower. The major effect is the size of the time steps; reducing the value requires more successive calculations to reach a given time after the start of heat flow. Increasing the ratio to a value greater than 0.5 would decrease the number of successive calculations and hence reduce the computer time needed to solve a given problem. An extremely important phenomenon occurs, however, when the ratio is >0.5; as illustrated in Fig. 8.3, where temperature profiles are shown at two different times, $t = 0.99$ sec and $t = 1.98$ sec, very inaccurate results may occur. The curves show the solution from the infinite series. The open circles, with $r = 0.6$, show extreme oscillation: The calculated values are very imprecise, and completely impossible behavior is represented. The data for $r = 0.4$ are smooth and almost exactly match the series solution.

The reason for this great difference in behavior is that the choice of the ratio $k\,\Delta t/c\rho(\Delta x)^2 = 0.6$ introduces instability in which errors grow at an accelerating rate as time increases. (At values of $t > 2$, negative values of u are computed, a patently impossible situation.) As indicated in Fig. 8.3, when this ratio is taken at 0.4, the results are excellent.

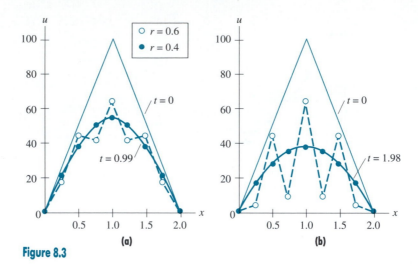

Figure 8.3

In Fig. 8.2 with $k \, \Delta t/c\rho(\Delta x)^2 = 0.5$, the limiting value to avoid instability was used. Even here, the oscillation of points around the curves indicates its borderline value, and the accuracy is hardly acceptable. The phenomena of stability and convergence, more fully discussed in a later section of this chapter, set a maximum value of 0.5 to the ratio $k \, \Delta t/c\rho(\Delta x)^2$. When derivative boundary conditions are involved, the ratio must be less than 0.5. Discontinuities in the initial conditions—even a discontinuity in the gradient, $\partial u/\partial x$, at $t = 0$—cause the accuracy to be poor at the maximum value. The next example will demonstrate this.

We will now solve a problem in diffusion, which is governed by the same mathematical equation as is heat conduction in a solid.

EXAMPLE 8.2   A hollow tube 20 cm long is initially filled with air containing 2% of ethyl alcohol vapors. At the bottom of the tube is a pool of alcohol that evaporates into the stagnant gas above. (Heat transfers to the alcohol from the surroundings to maintain a constant temperature of 30°C, at which temperature the vapor pressure is 0.1 atm.) At the upper end of the tube, the alcohol vapors dissipate to the outside air, so the concentration is essentially zero. Considering only the effects of molecular diffusion, determine the concentration of alcohol as a function of time and the distance x measured from the top of the tube.

Molecular diffusion follows the law

$$\frac{\partial c}{\partial t} = D \frac{\partial^2 c}{\partial x^2},$$

where D is the diffusion coefficient, with units of cm^2/sec in the cgs system. (This is the same as for the ratio $k/c\rho$, which is often termed *thermal diffusivity*.) For ethyl alcohol, $D = 0.119$ cm^2/sec at 30°C, and the vapor pressure is such that 10 % alcohol by volume in air is present at the surface.

Subdivide the length of the tube into five intervals, so $\Delta x = 4$ cm. Using the maximum value permitted for Δt yields

$$D\frac{\Delta t}{(\Delta x)^2} = \frac{1}{2} = 0.119\frac{\Delta t}{(4)^2}, \qquad \Delta t = 67.2 \text{ sec.}$$

Our initial condition is $c(x, 0) = 2.0$. The boundary conditions are

$$c(0, t) = 0.0, \qquad c(20, t) = 10.0.$$

The concentrations are measured by the percent of alcohol vapor in the air.

The computations are shown in Table 8.2. Again, using Eq. (8.5), each interior value of c is given by the arithmetic average of concentrations on either side in the line above. A little reflection is required to determine the proper concentrations to be used for $x = 0$ and $x = 20$ at $t = 0$. While they are initially at 2%, we assume they are changed instantaneously to 0% and 10% because the effective concentrations acting during the first time interval are not the initial values but the changed values. We have accordingly rewritten these values as shown in the first row of Table 8.2.

Table 8.2 Diffusion of alcohol vapors in a tube—solution of the equation

$$u_i^{j+1} = \frac{1}{2}(u_{i-1}^j + u_{i+1}^j)$$

Time, sec	Concentration of alcohol at					
	$x = 0$	$x = 4$	$x = 8$	$x = 12$	$x = 16$	$x = 20$
0	0.0	2.0	2.0	2.0	2.0	10.0
67.2	0.0	1.00	2.00	2.00	6.00	10.0
134.4	0.0	1.00	1.50	4.00	6.00	10.0
201.6	0.0	0.75	2.50	3.75	7.00	10.0
268.8	0.0	1.25	2.25	4.75	6.875	10.0
336.0	0.0	1.125	3.00	4.562	7.375	10.0
403.2	0.0	1.500	2.844	5.188	7.281	10.0
470.4	0.0	1.422	3.344	5.062	7.594	10.0
537.6	0.0	1.672	3.242	5.469	7.531	10.0
604.8	0.0	1.621	3.570	5.386	7.734	10.0
Steady state	0.0	2.0	4.0	6.0	8.0	10.0

As time passes, the concentration will become a linear function of x, from 10% at $x = 20$ to 0% at $x = 0$. The calculated values approach the steady-state concentrations as time passes. However, as Fig. 8.4 shows, the successive calculated values oscillate about the values calculated from the analytical solution, given by the infinite series shown in Fig. 8.4.

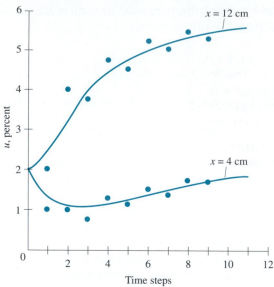

Figure 8.4 Analytical solution: $c(x, t) = \dfrac{x}{2} + \displaystyle\sum_{n=1}^{\infty} \dfrac{1}{n\pi}(4 + 16(-1)^n)\sin\left(\dfrac{n\pi x}{20}\right)e^{-0.119n^2\pi^2 t/20^2}$

While the algorithm of Eq. (8.5) is simple to use, the results leave something to be desired in the way of accuracy. The oscillatory nature of the results, when a smooth trend of the concentration is expected, is also of concern. We can improve accuracy by smaller steps, but if we decrease Δx, the time steps must also decrease, because the ratio $D\,\Delta t/(\Delta x)^2$ cannot exceed $\frac{1}{2}$. Cutting Δx in half will require making Δt one-fourth of its previous value, giving a total of eight times as many calculations. With the mixed order of error terms, it is not obvious what reduction of error this would give, but they should be reduced about fourfold.

We can reduce Δt without change in Δx, however, because the method is stable for any value of the ratio less than $\frac{1}{2}$. Suppose we make $D\,\Delta t/(\Delta x)^2 = \frac{1}{4}$. The basic difference equation, (8.4), now becomes

$$u_i^{j+1} = \frac{1}{4}(u_{i+1}^j + u_{i-1}^j) + \frac{1}{2}u_i^j. \tag{8.6}$$

Table 8.3 summarizes the use of Eq. (8.6) on the data of Example 8.2. As shown more clearly by Fig. 8.5, this last calculation causes the calculated concentrations to follow smooth curves, and the error is reduced considerably. The poorest accuracy is near the beginning. The later values are close to the curve.

Table 8.3   Diffusion of alcohol vapors in a tube:

$$u_i^{j+1} = \frac{1}{4}\left(u_{i-1}^j + u_{i+1}^j\right) + \frac{1}{2}u_i^j$$

Time, sec	Concentration of alcohol at					
	$x = 0$	$x = 4$	$x = 8$	$x = 12$	$x = 16$	$x = 20$
0.0	0.0	2.0	2.0	2.0	2.0	10.0
33.6	0.0	1.50	2.00	2.00	4.00	10.0
67.2	0.0	1.25	1.875	2.50	5.00	10.0
100.8	0.0	1.094	1.875	2.969	5.625	10.0
134.4	0.0	1.015	1.953	3.360	6.055	10.0
168.0	0.0	0.996	2.070	3.682	6.368	10.0
201.6	0.0	1.015	2.204	3.950	6.604	10.0
235.2	0.0	1.058	2.343	4.177	6.790	10.0
268.8	0.0	1.115	2.480	4.372	6.939	10.0
302.4	0.0	1.177	2.612	4.541	7.062	10.0
336.0	0.0	1.241	2.736	4.689	7.166	10.0
369.6	0.0	1.304	2.850	4.820	7.255	10.0
403.2	0.0	1.364	2.956	4.936	7.332	10.0
436.8	0.0	1.421	3.053	5.040	7.400	10.0
470.4	0.0	1.474	3.142	5.133	7.460	10.0
504.0	0.0	1.522	3.223	5.217	7.513	10.0
537.6	0.0	1.567	3.296	5.292	7.561	10.0
Steady state	0.0	2.0	4.0	6.0	8.0	10.0

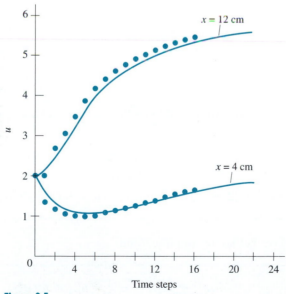

Figure 8.5

This initial poor accuracy is a result of the discontinuities in the boundary conditions. The abrupt change of concentrations at the ends of the tube makes the explicit method inaccurate. In a situation with continuity of the initial values of the potential plus continuity of its derivatives and the absence of discontinuities in the boundary conditions, one would get better accuracy throughout.

The method presented in this section is called the *explicit method* because each new value of the potential can be immediately calculated from quantities that are already known. It is simple and economical to calculate but has a severely limited upper value for the ratio $k\,\Delta t/c\rho(\Delta x)^2$. We can remove this limitation by going to an *implicit method*.

8.3 The Crank–Nicolson Method

When the difference equation, Eq. (8.4), was derived, we noted that a mixed order of error was involved, because a forward difference was used to replace the time derivative while a central difference was used for the distance derivative. However, the difference quotient, $(u_i^{j+1} - u_i^j)/\Delta t$, can be considered a central difference if we take it as corresponding to the midpoint of the time interval. Suppose we do consider $(u_i^{j+1} - u_i^j)/\Delta t$ as a central-difference approximation to $\partial u/\partial t$, and equate it to a central-difference quotient for the second derivative with distance, also corresponding to the midpoint in time, by averaging difference quotients at the beginning and end of the time step. Then

$$\frac{\partial u}{\partial t} = \frac{k}{c\rho}\frac{\partial^2 u}{\partial x^2}$$

is approximated by

$$\left(\frac{u_i^{j+1} - u_i^j}{\Delta t}\right) = \frac{1}{2}\left(\frac{k}{c\rho}\right)\left(\frac{u_{i+1}^j - 2u_i^j + u_{i-1}^j}{(\Delta x)^2} + \frac{u_{i+1}^{j+1} - 2\,u_i^{j+1} + u_{i-1}^{j+1}}{(\Delta x)^2}\right).$$

Central difference	Central difference	Central difference
at $t_{j+1/2}$	at t_j	at t_{j+1}

When this is rearranged we get the Crank–Nicolson formula, where

$$r = \frac{k\,\Delta t}{c\rho(\Delta x)^2},$$

$$-ru_{i-1}^{j+1} + (2 + 2r)u_i^{j+1} - ru_{i+1}^{j+1} = ru_{i-1}^j + (2 - 2r)u_i^j + ru_{i+1}^j. \qquad \textbf{(8.7)}$$

Letting $r = 1$, we get some simplification:

$$-u_{i-1}^{j+1} + 4u_i^{j+1} - u_{i+1}^{j+1} = u_{i-1}^j + u_{i+1}^j.$$

As we will discuss in the next section, one advantage of the Crank–Nicolson method is that it is stable for any value of r, although small values are more accurate.

Equation (8.7) is the usual formula for using the Crank–Nicolson method. Note that the new temperature u_i^{j+1} is not given directly in terms of known temperatures one time step earlier, but is a function of unknown temperatures at adjacent positions as well. It is therefore termed an *implicit method* in contrast to the explicit method of the previous section. With implicit methods, the values of u at $t = t_1$ are not just a function of values at $t = t_0$, but also involve the other u values at the same time step. This requires us to solve a set of simultaneous equations at each time step.

EXAMPLE 8.3

We illustrate the method by solving the problem of diffusion of alcohol vapors by this implicit method, the same as that previously attacked by the explicit method. Again, let us take $\Delta x = 4$ cm and $r = 1$. Restating the problem, we have

$$\frac{\partial^2 u}{\partial x^2} = \frac{1}{D} \frac{\partial u}{\partial t}, \qquad u(x,0) = 2.0, \qquad \begin{cases} u(0,t) = 0, \\ u(20,t) = 10.0, \end{cases}$$

$$D = 0.119 \text{ cm}^2/\text{sec}.$$

If $D\,\Delta t/(\Delta x)^2 = r = 1$ and $\Delta x = 4$ cm, then $\Delta t = 134.4$ sec. For $t = 134.4$, at the end of the first time step, we write the equations at each point whose concentration is unknown. The effective concentrations at the two ends, 0% and 10%, respectively, are again used:

$$
\begin{aligned}
-0.0 + 4u_1 - u_2 &= 0.0 + 2.0, \\
-u_1 + 4u_2 - u_3 &= 2.0 + 2.0, \\
-u_2 + 4u_3 - u_4 &= 2.0 + 2.0, \\
-u_3 + 4u_4 - 10.0 &= 2.0 + 10.0.
\end{aligned}
$$

This method requires more work because, at each time step, we must solve a similar set of equations. Fortunately, the system is tridiagonal. Since we will need to repeatedly solve the system with the same coefficient matrix (we transpose the constant values to the right side in the first and last equations), it is desirable to use an LU method. We then solve by back-substitutions, using the elements of L and U:

$$
L = \begin{bmatrix} 4 & 0 & 0 & 0 \\ -1 & 3.75 & 0 & 0 \\ 0 & -1 & 3.733 & 0 \\ 0 & 0 & -1 & 3.732 \end{bmatrix}, \qquad
U = \begin{bmatrix} 1 & -0.25 & 0 & 0 \\ 0 & 1 & -0.267 & 0 \\ 0 & 0 & 1 & -0.268 \\ 0 & 0 & 0 & 1 \end{bmatrix}.
$$

Remember that, to solve $Ax = b = LUx$, we first solve $Ly = b$ (which requires no reduction of L, because it is already triangular) and then we solve $Ux = y$ by back-substitution. Observe that when A is tridiagonal, its zeros are preserved in L

and U, and, in fact we need to compute only $2(n - 1)$ new values to get the LU equivalent of the $n \times n$ tridiagonal matrix A. Not only is the computational effort minimized, but we can also minimize computer memory space by storing only the three nonzero values in each row. In fact, as discussed in Chapter 2, we can store the elements of both L and U in place of the original values in A.

The results of calculations by the Crank–Nicolson method are listed in Table 8.4. For the first line of calculations the b-vector has components 2.0, 4.0, 4.0, and 22.0, and the calculated values are the components of $A^{-1}b$. In computing the second line, the components of b are 2.019, 4.052, 8.011, and 23.072. Succeeding lines use the proper sums of values from the line above to determine the b-vector.

When the calculated values are compared to the analytical results, it is observed that the errors of this method, in this example, are about the same as those made in the explicit method with $D\,\Delta t/(\Delta x)^2 = \frac{1}{4}$. We require only one-fourth as many time steps, however, to reach 537.6 sec. With smaller values of Δx, this freedom of choice in the value of $D\,\Delta t/(\Delta x)^2$ is especially valuable.

Table 8.4 Diffusion of alcohol vapors in a tube—Crank–Nicolson method

Time, sec	Calculated concentration values at					
	$x = 0$	$x = 4$	$x = 8$	$x = 12$	$x = 16$	$x = 20$
0.0	0.0	2.0	2.0	2.0	2.0	10.0
134.4	0.0	1.005	2.019	3.072	6.268	10.0
268.8	0.0	1.103	2.391	4.386	6.864	10.0
403.2	0.0	1.328	2.922	4.873	7.315	10.0
537.6	0.0	1.543	3.250	5.255	7.686	10.0

Time, sec	Analytical values					
	$x = 0$	$x = 4$	$x = 8$	$x = 12$	$x = 16$	$x = 20$
134.4	0.0	1.078	1.956	3.190	5.826	10.0
268.8	0.0	1.108	2.430	4.271	6.350	10.0
403.2	0.0	1.339	2.904	4.872	7.286	10.0
537.6	0.0	1.542	3.254	5.247	7.531	10.0

If we use $\Delta x = 2$ and $r = 1$ with the Crank–Nicolson method so that the value of Δt is the same as for the explicit method, errors are reduced almost tenfold.

It is almost always preferable to reduce Δx rather than Δt, so methods that are unconditionally stable are always the better choice.

8.4 A Generalization—The Theta Method

In the Crank–Nicolson method, unconditional stability is achieved by interpreting the finite-difference approximation for $\partial u/\partial t$ to apply at the midpoint of the time interval. We can make a more general statement if we interpret the approximation to apply at a point within the time interval that is the fraction θ from t_j. We then will approximate the space derivative at time $t_j + \theta \Delta t$ with a weighted average of approximations at the start and end of the time interval. So we approximate the differential equation

$$\frac{\partial u}{\partial t} = \frac{k}{c\rho} \frac{\partial^2 u}{\partial x^2}$$

by

$$\frac{u_i^{j+1} - u_i^j}{\Delta t} = \frac{k}{c\rho} \left[(1 - \theta) \frac{u_{i-1}^j - 2u_i^j + u_{i+1}^j}{(\Delta x)^2} + (\theta) \frac{u_{i-1}^{j+1} - 2u_i^{j+1} + u_{i+1}^{j+1}}{(\Delta x)^2} \right].$$

Observe that using 0.5 for θ gives the Crank–Nicolson method and, with $\theta = 0$, we get the explicit method. The other extreme is to take θ equal to 1. This gives a method that has been called the *implicit method*. As the name indicates, it is implicit as is Crank–Nicolson, and it also is unconditionally stable, as we will see later on.

The implicit method is frequently used because of its stability. The analogue of Eq. (8.7) is

$$r = \frac{k \, \Delta t}{c\rho(\Delta x)^2} \, ,$$
$$-ru_{i-1}^{j+1} + (1 + 2r)u_i^{j+1} - ru_{i+1}^{j+1} = u_i^j.$$

For any value of θ, the typical equation is

$$-r\theta u_{i-1}^{j+1} + (1 + 2r\theta)u_i^{j+1} - r\theta u_{i+1}^{j+1}$$
$$= r(1 - \theta)u_{i-1}^j + [1 - 2r(1 - \theta)]u_i^j + r(1 - \theta)u_{i+1}^j. \quad \textbf{(8.8)}$$

EXAMPLE 8.4   We illustrate the implicit method by solving the previous example for the diffusion of alcohol vapors. Using $r = 1$ and $\Delta x = 4$ cm, Δt is again 134.4 sec. The equations for u after the first time step, at $t = 134.4$ sec, are

$$
\begin{aligned}
-0.0 + 3u_1 - u_2 \phantom{{}- u_3} &= 2.0, \\
-u_1 + 3u_2 - u_3 \phantom{{}- u_4} &= 2.0, \\
-u_2 + 3u_3 - u_4 &= 2.0, \\
-u_3 + 3u_4 - 10.0 &= 2.0,
\end{aligned}
$$

(where θ was taken equal to 1).

Here we have, for L and U:

$$L = \begin{bmatrix} 3 & 0 & 0 & 0 \\ -1 & 2.667 & 0 & 0 \\ 0 & -1 & 2.625 & 0 \\ 0 & 0 & -1 & 2.619 \end{bmatrix}, \quad U = \begin{bmatrix} 1 & -0.333 & 0 & 0 \\ 0 & 1 & -0.375 & 0 \\ 0 & 0 & 1 & -0.381 \\ 0 & 0 & 0 & 1 \end{bmatrix}$$

The initial b-vector is

$$b = \{2 \quad 2 \quad 2 \quad 12\}.$$

Solving the system $Au = b$, we get, at the end of the first time step,

$$u = \{1.3818 \quad 2.1455 \quad 3.0545 \quad 5.0182\}.$$

Table 8.5 compares the results from the explicit, Crank–Nicolson, and implicit methods after the first time step.

Table 8.5 Comparison of results for alcohol diffusion

Method	Computed concentrations at $t = 134.4$ sec						Average error
	$x = 0$	$x = 4$	$x = 8$	$x = 12$	$x = 16$	$x = 20$	
Explicit	0.0	1.015	1.953	3.360	6.055	10.0	0.116
Crank–Nicolson	0.0	1.005	2.019	3.072	6.268	10.0	0.174
Implicit	0.0	1.382	2.146	3.055	5.018	10.0	0.359
Analytical	0.0	1.078	1.956	3.190	5.826	10.0	

The results in Table 8.5 show that the explicit method resulted in the least error, but this method used a value of $r = 0.25$ (Table 8.3) whereas the others used $r = 1.0$. If the comparison is made with results from $r = 0.5$ for the explicit method, where the difference in Δt is not so great, the explicit method comes out least accurate.

This comparison of methods, though quite superficial, indicates that there may be an optimum value for θ in Eq. (8.8). Burnett (1987) suggests that $\theta = \frac{2}{3}$ is near optimal, but he also points out that an argument can be made to use $\theta = 0.878$.

8.5 Derivative Boundary Conditions

In heat-conduction problems, the most usual situation at the endpoints is not that they are held at a constant temperature, but that heat is lost by conduction or radiation at a rate proportional to some power of the temperature difference between the surface of the body and its surroundings. This leads to a relationship involving the space derivative of temperature at the surface. In the analytical solution of heat-flow problems through Fourier series, this adds considerable complication in determining the coefficients, but our numerical technique requires only minor modifications.

The rate of heat loss from the surface of a solid is generally expressed as

$$\text{Rate of heat loss} = hA(u - u_0), \qquad \textbf{(8.9)}$$

where A is the surface area, u is the surface temperature, u_0 is the temperature of the surrounding medium, and h is a coefficient of heat transfer; h is increased by motion of the surrounding medium. To facilitate heat flow, the surrounding medium is often caused to flow rapidly by mechanical means (as in some salt evaporators) or by proper design (as in vertical-tube heat exchangers in distillation columns). This situation is called *forced convection*.

In other situations, h is also a function of the surface temperature, as in *natural convection*, in which the motion of surrounding fluid is caused by thermal currents. Another important situation is heat loss by radiation, in which the rate of heat loss is proportional to the difference between the fourth powers of the temperatures of the surface and the surrounding surfaces to which heat is being radiated. In both these situations, the rate of heat loss is proportional to some power of the surface temperature. This gives rise to nonlinear equations that are not as readily solved. One usually prefers to force the mathematical model to a linear form even though this is not exactly true. One way to approximate this is to use Eq. (8.9), absorbing the nonlinear aspects into the coefficient h, and to change h appropriately through the progress of the calculations so that it takes on a reasonably correct average value. Our examples will treat only the simpler situations where Eq. (8.9) applies directly.

As we will see, heat loss from the surface by conduction leads to a derivative boundary condition. Our procedure will be to replace the derivatives in both the differential equation and in the boundary conditions by difference quotients. We illustrate with a simple example.

EXAMPLE 8.5   An aluminum cube is 4 in. × 4 in. × 4 in. ($k = 0.52, c = 0.226, \rho = 2.70$ in cgs units). All but one face is perfectly insulated, and the cube is initially at 1000°F. Heat is lost from the uninsulated face to a fluid flowing past it according to the equation

$$\text{Rate of heat loss (in Btu per sec)} = hA(u - u_0),$$

where $h = 0.15$ Btu/sec $\cdot$ ft^2 $\cdot$ °F,
A = surface area in ft^2,
u = surface temperature, °F,
u_0 = temperature of fluid, °F.

If u_0, the temperature of fluid flowing past the aluminum cube, is constant at 70°F, find the temperatures inside the cube as a function of time. Although we could work the problem in cgs units, we elect to use English units, making suitable changes in k, c, and ρ.

Because of the insulation on the lateral faces, again the only direction in which heat will flow is perpendicular to the uninsulated face, and the equation again is

$$\frac{\partial u}{\partial t} = \frac{k}{c\rho} \frac{\partial^2 u}{\partial x^2}.$$

The initial condition, with x representing distance from the uninsulated face, is

$$u(x, 0) = 1000.$$

For boundary conditions, at the open surface we have

$$-kA\frac{\partial u}{\partial x}\bigg|_{x=0} = -hA(u - 70),$$

because the rate at which heat leaves the surface must be equal to the rate at which heat flows to the surface. The negative sign on the left is required because heat flows in a direction opposite to a positive gradient; on the right it occurs because heat is being lost in the direction of negative x. At the other side of the cube, the rate of heat flow is zero because of insulation:

$$\frac{\partial u}{\partial x}\bigg|_{x=4} = 0.$$

We plan to use the explicit method with the following ratio:*

$$r = \frac{k\,\Delta t}{c\rho(\Delta x)^2} = \frac{1}{4}.$$

Suppose we let $\Delta x = 1$ in. To calculate Δt, we need k, c, and ρ expressed in the units of inches, pounds, seconds, and °F. We first convert units:

$$k = \left(0.52\,\frac{\text{cal}}{\text{sec}\cdot\text{cm}\cdot°\text{C}}\right)\left(\frac{1\ \text{Btu}}{252\ \text{cal}}\right)\left(\frac{2.54\ \text{cm}}{1\ \text{in.}}\right)\left(\frac{1°\text{C}}{1.8°\text{F}}\right)$$

$$= 0.00291\,\frac{\text{Btu}}{\text{sec}\cdot\text{in.}\cdot°\text{F}};$$

$$c = \left(0.226\,\frac{\text{cal}}{\text{g}\cdot°\text{C}}\right)\left(\frac{1\ \text{Btu}}{252\ \text{cal}}\right)\left(\frac{454\ \text{g}}{1\ \text{lb}}\right)\left(\frac{1°\text{C}}{1.8°\text{F}}\right)$$

$$= 0.226\,\frac{\text{Btu}}{\text{lb}\cdot°\text{F}};$$

$$\rho = \left(2.70\,\frac{\text{g}}{\text{cm}^3}\right)\left(\frac{1\ \text{lb}}{454\ \text{g}}\right)\left(\frac{2.54\ \text{cm}}{1\ \text{in.}}\right)^3$$

$$= 0.0975\,\frac{\text{lb}}{\text{in.}^3};$$

$$\Delta t = \frac{rc\rho(\Delta x)^2}{k} = \frac{(0.25)(0.226)(0.0975)(1)^2}{(0.00291)} = 1.89\ \text{sec.}$$

---

* The ratio must be smaller than $\frac{1}{2}$ to give stability with the derivative end conditions.

For our choice of r value (0.25), our differential equation becomes

$$u_i^{j+1} = \frac{1}{4}(u_{i+1}^j + u_{i-1}^j) + \frac{1}{2}u_i^j, \qquad (8.10)$$

which is to be applied at every point where u is unknown. In this example, this includes the points at $x = 0$ and $x = 4$ as well as the interior points. To enable us to write Eq. (8.10), we extend the domain of u one step on either side of the boundary. Let x_R be a fictitious point to the right of $x = 4$, and let x_L be a fictitious point to the left of $x = 0$. If x_1 signifies $x = 0$, and if x_5 signifies $x = 4$, we have the relations, from Eq. (8.10):

$$u_1^{j+1} = \frac{1}{4}(u_2^j + u_L^j) + \frac{1}{2}u_1^j,$$

$$u_5^{j+1} = \frac{1}{4}(u_R^j + u_4^j) + \frac{1}{2}u_5^j.$$

Now we use the boundary conditions to eliminate the fictitious points, writing them as central-difference quotients:

$$-k\frac{\partial u}{\partial x}\Big|_{x=0} \doteq -(0.00291)\left(\frac{u_2^j - u_L^j}{2(1)}\right) = -\frac{0.15}{144}(u_1^j - 70);$$

$$-k\frac{\partial u}{\partial x}\Big|_{x=4} \doteq -(0.00291)\left(\frac{u_R^j - u_4^j}{2(1)}\right) = 0.$$

The 144 factor in the first equation changes h to a basis of in.^{2}. Solving for u_L and u_R, we have

$$u_L = u_2 - 0.716u_1 + 50.1,$$

$$u_R = u_4,$$

and the set of equations that give the temperatures becomes

$$u_1^{j+1} = 0.32u_1^j + \frac{1}{2}u_2^j + 12.525,$$

$$u_2^{j+1} = \frac{1}{4}(u_1^j + u_3^j) + \frac{1}{2}u_2^j,$$

$$u_3^{j+1} = \frac{1}{4}(u_2^j + u_4^j) + \frac{1}{2}u_3^j, \qquad (8.11)$$

$$u_4^{j+1} = \frac{1}{4}(u_3^j + u_5^j) + \frac{1}{2}u_4^j,$$

$$u_5^{j+1} = \frac{1}{2}u_4^j + \frac{1}{2}u_5^j.$$

A similar treatment of the boundary conditions using the Crank–Nicolson method with $r = k \, \Delta t / c\rho(\Delta x)^2 = 1$ leads to the set of simultaneous equations:

$$
\begin{aligned}
4.716u_1^{j+1} - 2u_2^{j+1} \qquad\qquad\qquad &= -0.716u_1^j + 2u_2^j + 100.2, \\
-u_1^{j+1} + 4u_2^{j+1} - u_3^{j+1} \qquad\qquad &= \qquad u_1^j \qquad + u_3^j, \\
-u_2^{j+1} + 4u_3^{j+1} - u_4^{j+1} \qquad &= \qquad\qquad u_2^j \qquad + u_4^j, \\
-u_3^{j+1} + 4u_4^{j+1} - u_5^{j+1} &= \qquad\qquad\qquad u_3^j \qquad + u_5^j, \\
-2u_4^{j+1} + 4u_5^{j+1} &= \qquad\qquad\qquad\qquad 2u_4^j.
\end{aligned}
\qquad \textbf{(8.12)}
$$

For the set of equations in (8.12), Δt will be 7.56 sec. We advance the solution one time step at a time by repeatedly solving either (8.11) or (8.12), using the proper values of u_i^j.

8.6   Parabolic Equations in Two or Three Dimensions

Rectangular Regions

In principle, we can readily extend the preceding methods to higher space dimensions, especially when the region is rectangular. The heat-flow equation in two dimensions is

$$
\frac{\partial u}{\partial t} = \frac{k}{c\rho}\left(\frac{\partial^2 u}{\partial x^2} + \frac{\partial^2 u}{\partial y^2} \right).
$$

Taking $\Delta x = \Delta y = \Delta$, and letting $r = k \, \Delta t / c\rho\Delta^2$, we find that the explicit scheme becomes

$$
u_{i,j}^{k+1} - u_{i,j}^k = r(u_{i+1,j}^k - 2u_{i,j}^k + u_{i-1,j}^k + u_{i,j+1}^k - 2u_{i,j}^k + u_{i,j-1}^k)
$$

or

$$
u_{i,j}^{k+1} = r(u_{i+1,j}^k + u_{i-1,j}^k + u_{i,j+1}^k + u_{i,j-1}^k) + (1 - 4r)u_{i,j}^k.
$$

In this scheme, the maximum value permissible for r in the simple case of constant end conditions is $\frac{1}{4}$. (Note that this corresponds again to the numerical value that gives a particularly simple formula.) In the more general case with $\Delta x \neq \Delta y$, the criterion is

$$\frac{k\,\Delta t}{c\rho[(\Delta x)^2 + (\Delta y)^2]} \leq \frac{1}{8}.$$

The analogous equation in three dimensions, with equal grid spacing each way, has the coefficient $(1 - 6r)$, and $r \leq \frac{1}{6}$ is required for convergence and stability.

The difficulty with the use of the explicit scheme is that the restrictions on Δt require inordinately many rows of calculations. One then looks for a method in which Δt can be made larger without loss of stability. In one dimension, the Crank–Nicolson method was such a method. In the two-dimensional case, using averages of central-difference approximations to give $\partial^2 u/\partial x^2$ and $\partial^2 u/\partial y^2$ at the midvalue of time, we get

$$u_{i,j}^{k+1} - u_{i,j}^{k} = \frac{r}{2}\,[u_{i+1,j}^{k+1} - 2u_{i,j}^{k+1} + u_{i-1,j}^{k+1} + u_{i+1,j}^{k} - 2u_{i,j}^{k} + u_{i-1,j}^{k}$$
$$+ u_{i,j+1}^{k+1} - 2u_{i,j}^{k+1} + u_{i,j-1}^{k+1} + u_{i,j+1}^{k} - 2u_{i,j}^{k} + u_{i,j-1}^{k}].$$

The problem now is that a set of $(M)(N)$ simultaneous equations must be solved at each time step, where M is the number of unknown values in the x-direction and N in the y-direction. Furthermore, the coefficient matrix is no longer tridiagonal, so the solution to each set of equations is slower and memory space to store the elements of the matrix may be exorbitant.

The advantage of a tridiagonal matrix is retained in the alternating-direction-implicit scheme (A.D.I.) proposed by Peaceman and Rachford (1955). It is widely used in modern computer programs for the solution of parabolic partial-differential equations. In this method, we approximate $\nabla^2 u$ by adding a central-difference approximation to $\partial^2 u/\partial x^2$ written at the beginning of the interval to a similar expression for $\partial^2 u/\partial y^2$ written at the end. We will use subscripts L, R, A, and B to indicate nodes at the left, right, above, and below the central point u_0. We then have

$$u_0^{j+1} - u_0^{j} = r[u_L^{j} - 2u_0^{j} + u_R^{j}] + r[u_A^{j+1} - 2u_0^{j+1} + u_B^{j+1}], \qquad \textbf{(8.13)}$$

where $r = k\,\Delta t/c\rho\Delta^2$ and $\Delta = \Delta x = \Delta y$. The obvious bias in this formula is balanced by reversing the order of the second derivative approximations in the next time span:

$$u_0^{j+2} - u_0^{j+1} = r[u_L^{j+2} - 2u_0^{j+2} + u_R^{j+2}] + r[u_A^{j+1} - 2u_0^{j+1} + u_B^{j+1}]. \qquad \textbf{(8.14)}$$

Observe that in using Eq. (8.13), we make a vertical traverse through the nodes, computing new values for each column of nodes. Similarly, in using Eq. (8.14), we make a horizontal traverse, computing new values row by row. In effect, we consider u_L and u_R as fixed when we do a vertical traverse; we consider u_A and u_B as fixed for horizontal traverses.

EXAMPLE 8.6   A square plate of steel is 8 in. wide and 6 in. high. Initially all points on the plate are at 50°. The edges are suddenly brought to the temperatures shown in Fig. 8.6 and held at these temperatures. Trace the history of temperatures at nodes spaced 2 in. apart using the A.D.I. method, assuming that heat flows only in the x- and y-directions.

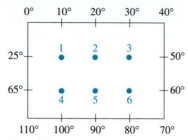

Figure 8.6

Figure 8.6 shows a numbering system for the internal nodes, all of which start at 50°, as well as the temperatures at boundary nodes.

Using Eq. (8.13), the typical equation for a vertical traverse is

$$-ru_A + (1 + 2r)u_0 - ru_B = ru_L + (1 - 2r)u_0 + ru_R.$$

If we use this equation and the numbering system of Fig. 8.6 to set up the equations for a vertical traverse, we do not get the tridiagonal system that we desire, but we do if the nodes are renumbered as shown in Fig. 8.7. To keep track of the different numbering systems, we will use v for temperatures when a vertical traverse is made (numbered as in Fig. 8.7) and u when a horizontal traverse is made (numbered as in Fig. 8.6).

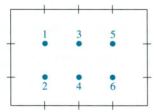

Figure 8.7

This is the set of equations for a vertical traverse:

$$\begin{bmatrix} (1+2r) & -r & & & & \\ -r & (1+2r) & & & & \\ & & (1+2r) & -r & & \\ & & -r & (1+2r) & & \\ & & & & (1+2r) & -r \\ & & & & -r & (1+2r) \end{bmatrix} \begin{Bmatrix} v_1 \\ v_2 \\ v_3 \\ v_4 \\ v_5 \\ v_6 \end{Bmatrix} = \begin{Bmatrix} r(25) + (1-2r)u_1 + ru_2 \ \ + r(10) \\ r(65) + (1-2r)u_4 + ru_5 \ \ + r(100) \\ ru_1 \ \ + (1-2r)u_2 + ru_3 \ \ + r(20) \\ ru_4 \ \ + (1-2r)u_5 + ru_6 \ \ + r(90) \\ ru_2 \ \ + (1-2r)u_3 + r(50) + r(30) \\ ru_5 \ \ + (1-2r)u_6 + r(60) + r(80) \end{Bmatrix}.$$

When we apply Eq. (8.14) to get a set of equations for a horizontal traverse, we get (the dashed lines show they break into subsets)

$$
\begin{bmatrix}
(1+2r) & -r \\
-r & (1+2r) & -r \\
& -r & (1+2r) \\
\hline
& & & (1+2r) & -r \\
& & & -r & (1+2r) & -r \\
& & & & -r & (1+2r)
\end{bmatrix}
\begin{Bmatrix} u_1 \\ u_2 \\ u_3 \\ \hline u_4 \\ u_5 \\ u_6 \end{Bmatrix}
=
\begin{Bmatrix}
r(10) + (1-2r)v_1 + rv_2 & + r(25) \\
r(20) + (1-2r)v_3 + rv_4 \\
r(30) + (1-2r)v_5 + rv_6 & + r(50) \\
rv_1 & + (1-2r)v_2 + r(100) + r(65) \\
rv_3 & + (1-2r)v_4 + r(90) \\
rv_5 & + (1-2r)v_6 + r(80) + r(60)
\end{Bmatrix}
$$

A value must be specified for r. Small r's give better accuracy but smaller Δt's, so more time steps are required to compute the history. If we take $r = 1$, Δt is 26.4 sec.

The first vertical traverse gives results for $t = 26.4$ sec. We get the first set of v's from

$$
\begin{bmatrix}
3 & -1 \\
-1 & 3 \\
& & 3 & -1 \\
& & -1 & 3 \\
& & & & 3 & -1 \\
& & & & -1 & 3
\end{bmatrix}
\begin{Bmatrix} v_1 \\ v_2 \\ v_3 \\ v_4 \\ v_5 \\ v_6 \end{Bmatrix}
=
\begin{Bmatrix}
25 - (1)50 + 50 + 10 \\
65 - (1)50 + 50 + 100 \\
50 - (1)50 + 50 + 20 \\
50 - (1)50 + 50 + 40 \\
50 - (1)50 + 50 + 30 \\
50 - (1)50 + 60 + 80
\end{Bmatrix}
=
\begin{Bmatrix}
35 \\ 165 \\ 70 \\ 140 \\ 80 \\ 140
\end{Bmatrix}.
$$

Solving, we get these values:

$$\{33.75 \quad 66.25 \quad 43.75 \quad 61.25 \quad 47.5 \quad 62.5\}.$$

These values are used to build the right-hand sides for the next computations, a horizontal traverse, getting these equations for $t = 52.8$ sec:

$$
\begin{bmatrix}
3 & -1 \\
-1 & 3 & -1 \\
& -1 & 3 \\
& & & 3 & -1 \\
& & & -1 & 3 & -1 \\
& & & & -1 & 3
\end{bmatrix}
\begin{Bmatrix} u_1 \\ u_2 \\ u_3 \\ u_4 \\ u_5 \\ u_6 \end{Bmatrix}
=
\begin{Bmatrix}
10 & - (1)33.75 + 66.25 + 25 \\
20 & - (1)43.75 + 61.25 \\
30 & - (1)47.5 + 62.5 + 50 \\
33.75 - (1)66.25 + 100 & + 65 \\
43.75 - (1)61.25 + 90 \\
47.5 - (1)62.5 + 80 & + 60
\end{Bmatrix}
=
\begin{Bmatrix}
67.5 \\ 37.5 \\ 95 \\ 132.5 \\ 72.5 \\ 125
\end{Bmatrix},
$$

which have the solution (a set of u's)

$$\{35.595 \quad 39.286 \quad 44.762 \quad 66.786 \quad 67.857 \quad 64.286\}.$$

We continue by alternating between vertical and horizontal traverses to get the results shown in Table 8.6. This also shows the steady-state temperatures that are reached after a long time. The steady-state temperatures could have been computed by the methods of Chapter 7. We observe that the A.D.I. algorithm for steady-state temperatures of Chapter 7 is essentially identical to what we have seen here.

Table 8.6   Results for Example 8.6 using the A.D.I. method

AT START, TEMPS ARE

0.0000	10.0000	20.0000	30.0000	40.0000
25.0000	50.0000	50.0000	50.0000	50.0000
65.0000	50.0000	50.0000	50.0000	60.0000
110.0000	100.0000	90.0000	80.0000	70.0000

AFTER ITERATION 1 TIME = 26.4 – VALUES ARE

0.0000	10.0000	20.0000	30.0000	40.0000
25.0000	33.7500	43.7500	47.5000	50.0000
65.0000	66.2500	61.2500	62.5000	60.0000
110.0000	100.0000	90.0000	80.0000	70.0000

AFTER ITERATION 2 TIME = 52.8 – VALUES ARE

0.0000	10.0000	20.0000	30.0000	40.0000
25.0000	35.5952	39.2857	44.7619	50.0000
65.0000	66.7857	67.8571	64.2857	60.0000
110.0000	100.0000	90.0000	80.0000	70.0000

AFTER ITERATION 3 TIME = 79.2 – VALUES ARE

0.0000	10.0000	20.0000	30.0000	40.0000
25.0000	35.2679	42.0536	45.8929	50.0000
65.0000	67.1131	65.0893	63.1548	60.0000
110.0000	100.0000	90.0000	80.0000	70.0000

AFTER ITERATION 4 TIME = 105.6 – VALUES ARE

0.0000	10.0000	20.0000	30.0000	40.0000
25.0000	36.2443	41.8878	46.3832	50.0000
65.0000	66.1366	65.2551	62.6644	60.0000
110.0000	100.0000	90.0000	80.0000	70.0000

STEADY–STATE TEMPERATURES:

0.0000	10.0000	20.0000	30.0000	40.0000
25.0000	35.8427	41.8323	46.1760	50.0000
65.0000	66.5383	65.3106	62.8716	60.0000
110.0000	100.0000	90.0000	80.0000	70.0000

The compensation of errors produced by this alternation of direction gives a scheme that is convergent and stable for all values of r, although accuracy requires that r not be too large. The three-dimensional analog alternates three ways, returning to each of the three formulas after every third step. (Unfortunately the three-dimensional case is not stable for all fixed values of $r > 0$. A variant due to Douglas (1962) is unconditionally stable, however.) When the nodes are renumbered, in each case tridiagonal coefficient matrices result.

Note that the equations can be broken up into two independent subsets, corresponding to the nodes in each column or row. (See the first set of equations of Example 8.6.) This is always true in the A.D.I. method; each row gives a set independent of the equations from the other rows. For columns, the same thing occurs. For very large problems, this is important, because it permits the ready overlay of main memory in solving the independent sets. Observe also that each subset can be solved at the same time by parallel processors.

Nonrectangular Regions

When the region in which the heat-flow equation is to be satisfied is not rectangular, we can perturb the boundary to make it agree with a square mesh, or we can interpolate from boundary points to estimate u at adjacent mesh points, as discussed in Chapter 7 for elliptic equations.

The frequency with which circular or spherical regions occur makes it worthwhile to mention the heat equation in polar and spherical coordinates. The basic equation

$$\frac{\partial u}{\partial t} = \frac{k}{c\rho} \nabla^2 u$$

becomes, in polar coordinates (r, θ),

$$\frac{\partial u}{\partial t} = \frac{k}{c\rho} \left(\frac{\partial^2 u}{\partial r^2} + \frac{1}{r} \frac{\partial u}{\partial r} + \frac{1}{r^2} \frac{\partial^2 u}{\partial \theta^2} \right),$$

and in spherical coordinates (r, ϕ, θ),

$$\frac{\partial u}{\partial t} = \frac{k}{c\rho} \left(\frac{\partial^2 u}{\partial r^2} + \frac{2}{r} \frac{\partial u}{\partial r} + \frac{1}{r^2} \frac{\partial^2 u}{\partial \theta^2} + \frac{\cot \phi}{r} \frac{\partial u}{\partial \theta} + \frac{1}{r^2 \sin^2 \theta} \frac{\partial^2 u}{\partial \phi^2} \right).$$

Using finite-difference approximations to convert these to difference equations is straightforward except at the origin where $r = 0$. For this point, consider $\nabla^2 u$ in rectangular coordinates, so that, in two dimensions,

$$\partial^2 u = \frac{u_{i+1,j} + u_{i-1,j} + u_{i,j+1} + u_{i,j-1} - 4u_{i,j}}{(\Delta r)^2}, \qquad u_{ij} \text{ at } r = 0.$$

This is exactly the same as the expression for the Laplacian in Chapter 7, Eq. (7.6). This expression for $\nabla^2 u$ is obviously independent of the orientation of the axes. We get the best value by using the average value of all points that are a distance Δr from the origin, so that for $r = 0$,

$$\nabla^2 u = \frac{4(u_{av} - u_0)}{(\Delta r)^2} \qquad \text{at } r = 0.$$

The corresponding relation for spherical coordinates is

$$\nabla^2 u = \frac{6(u_{av} - u_0)}{(\Delta r)^2} \qquad \text{at } r = 0.$$

8.7 Finite Elements for Heat Flow

As we saw in Chapters 6 and 7, the finite-element method is often preferred for solving boundary-value problems. It is also the preferred method for solving the heat equation when the region of interest is not regular. You should know something about this application of finite elements, but space constraints preclude a full treatment.

Consider the heat-flow equation in two-space dimension with heat generation given by $F(x, y)$:

$$\frac{\partial u}{\partial t} = \frac{k}{c\rho}\left(\frac{\partial^2 u}{\partial x^2} + \frac{\partial^2 u}{\partial y^2}\right) + F(x, y), \tag{8.15}$$

which is subject to initial conditions at $t = 0$ and boundary conditions that may be Dirichlet or may involve the outward normal gradient. Although this is really a three-dimensional problem (in x, y, and t), it is customary to approximate the time derivative with a finite difference and apply finite elements only to the spatial region. Doing so, we can rewrite Eq. (8.15) as

$$\frac{u^{m+1} - u^m}{\Delta t} = \frac{k}{c\rho}\left(\frac{\partial^2 u}{\partial x^2} + \frac{\partial^2 u}{\partial y^2}\right) + F(x, y), \tag{8.16}$$

where we have used a forward difference as in the explicit method. (We might prefer Crank–Nicolson or the implicit method, but we will keep things simple.)

To apply finite elements to the region, we do exactly as described in Chapter 7—cover the region with joined elements, write element equations for the right-hand side of Eq. (8.15), assemble these, adjust for boundary conditions, and solve. But we must also consider the time variable. We do so by considering Eq. (8.16) to apply at a fixed point in time, t_m. Since we know the values of u everywhere within the region at $t = t_0$, we surely know the initial nodal values. We then can solve Eq. (8.16) for the u-values at $t = t_0 + \Delta t$, where the size of Δt is chosen small enough to ensure stability.

We will use the Galerkin procedure to derive the element equations to provide some variety from Chapter 7. In this procedure you will remember that we integrate the residual weighted with each of the shape functions and set them to zero. (The integrations are done over the element area.) If we stay with linear triangular elements, there are three shape functions, N_r, N_s, and N_t, where the subscripts denote the three vertices (nodes) of the element taken in counterclockwise order.

The residual for Eq. (8.15) is

$$\text{Residual} = u_t - \alpha(u_{xx} + u_{yy}) - F, \tag{8.17}$$

where we have used the subscript notation for derivatives and have abbreviated $k/c\rho$ with α.

As stated, we will use linear triangular elements; within each element we approximate u with

$$u(x,y) \approx v(x,y) = N_r c_r + N_s c_s + N_t c_t. \tag{8.18}$$

This means that Galerkin integrals are

$$\iint N_j[v_t - \alpha(v_{xx} + v_{yy}) + F]\, dx\, dy = 0, \qquad j = r, s, t. \tag{8.19}$$

If we apply integration by parts (as we did in Chapter 6) to the second derivatives of Eq. (8.19), we can reduce the order of these derivatives. Doing so and replacing v from Eq. (8.18) gives a set of three equations for each element, which we write in matrix form:

$$[C]\left\{\frac{\partial c}{\partial t}\right\} + [K]\{c\} = \{b\}. \tag{8.20}$$

The components of $\{c\}$ are the nodal temperatures of the element, of course; those of $\{\partial c/\partial t\}$ are the time derivatives. The components of the matrices of Eq. (8.20) are

$$C_{ij} = \iint N_i\, dx\, dy, \qquad i = r, s, t,$$

$$K_{ij} = \alpha_{av} \iint \left[\left(\frac{\partial N_i}{\partial x}\right)\left(\frac{\partial N_j}{\partial x}\right) + \left(\frac{\partial N_i}{\partial y}\right)\left(\frac{N_j}{\partial y}\right)\right] dx\, dy, \qquad i, j = r, s, t, \tag{8.21}$$

$$b_i = \iint F N_i\, dx\, dy + \oint u_N N_i\, dL, \qquad i = r, s, t.$$

In Eqs. (8.21), the line integral in the b's is present only along a side of an element on the boundary of the region where the outward normal gradient u_N is specified as a boundary condition.

From the development of Chapter 7, we know how to evaluate all of the integrals of Eqs. (8.21) when the elements are triangles. [See, for example, Burnett (1987) for the evaluations for other types of elements.]

As stated, we will use a finite-difference approximation for $\partial c/\partial t$. If this is a forward difference as suggested, we get the explicit formula

$$\frac{1}{\Delta t}[C]\{c^{m+1}\} = \frac{1}{\Delta t}[C]\{c^m\} - [K]\{c^m\} + \{b\}, \tag{8.22}$$

where all the c's on the right are nodal temperatures at $t = t_m$ and the nodal temperatures on the left in $\{c^{m+1}\}$ are at $t = t_{m+1}$.

We can put Eq. (8.22) into a more familiar iterating form by multiplying through by $\Delta t [C]^{-1}$:

$$\{c^{m+1}\} = \{c^m\} - \Delta t [C]^{-1} [K] \{c^m\} + \Delta t [C]^{-1} \{b\}. \tag{8.23}$$

[We can make Eq. (8.23) more compact by combining the multipliers of $\{c^m\}$.]

In principle, we have solved the heat-flow problem by finite elements. We construct the equations for every element from Eq. (8.23) and assemble them to get the global matrix, then adjust for boundary conditions just as in Chapter 7. This gives a set of equations in the unknown nodal values that we use to step forward in time from the initial point. With the explicit method illustrated here, each time step is just a matrix multiplication of the current nodal temperatures (and a vector addition) to get the next set of values. If we had used an implicit method such as Crank–Nicolson, we would have had to solve a set of equations at each step, but, unfortunately, they are not tridiagonal. We might hope for some equivalent to the A.D.I. method, but A.D.I. requires that the nodes be uniformly spaced. The conclusion is that the finite-element method in two- or three-space dimension is a problem that is expensive to solve. In one dimension, however, the system is tridiagonal, so that situation is not bad.

8.8   Theoretical Matters

While there is much to say about the theory behind parabolic equations, we will discuss only stability and convergence of our numerical procedures. We already know that, to ensure stability and convergence in the explicit method, the ratio $r = k \, \Delta t / c \rho (\Delta x)^2$ must be 0.5 or less. The Crank–Nicolson and the implicit methods, however, have no such limitations.

By *convergence*, we mean that the results of the method approach the analytical values as Δt and Δx both approach zero. By *stability*, we mean that errors made at one stage of the calculations do not cause increasingly large errors as the computations are continued, but rather will eventually damp out.

We will first discuss convergence, limiting ourselves to the simple case of the unsteady-state heat-flow equation in one dimension:*

$$\frac{\partial U}{\partial t} = \frac{k}{c\rho} \frac{\partial^2 U}{\partial x^2}. \tag{8.24}$$

Let us use the symbol U to represent the exact solution to Eq. (8.24), and u to represent the numerical solution. At the moment we assume that u is free of round-off errors, so the only difference between U and u is the error made by

---

* We could have treated the simpler equation $\partial U / \partial T = \partial^2 U / \partial X^2$ without loss of generality, since with the change of variables—$X = \sqrt{c\rho}\, x$, $T = kt$—the two equations are seen to be identical.

replacing Eq. (8.24) by the difference equation. Let $e_i^j = U_i^j - u_i^j$, at the point $x = x_i, t = t_j$. By the explicit method, Eq. (8.24) becomes

$$u_i^{j+1} = r(u_{i+1}^j + u_{i-1}^j) + (1 - 2r)u_i^j, \qquad (8.25)$$

where $r = k\,\Delta t/c\rho(\Delta x)^2$. Substituting $u = U - e$ into Eq. (8.25), we get

$$e_i^{j+1} = r(e_{i+1}^j + e_{i-1}^j) + (1 - 2r)e_i^j - r(U_{i+1}^j + U_{i-1}^j) - (1 - 2r)U_i^j + U_i^{j+1}. \qquad (8.26)$$

By using Taylor-series expansions, we have

$$U_{i+1}^j = U_i^j + \left(\frac{\partial U}{\partial x}\right)_{i,j}\Delta x + \frac{(\Delta x)^2}{2}\frac{\partial^2 U(\xi_1, t_j)}{\partial x^2}, \qquad x_i < \xi_1 < x_{i+1},$$

$$U_{i-1}^j = U_i^j - \left(\frac{\partial U}{\partial x}\right)_{i,j}\Delta x + \frac{(\Delta x)^2}{2}\frac{\partial^2 U(\xi_2, t_j)}{\partial x^2}, \qquad x_{i-1} < \xi_2 < x_i,$$

$$U_i^{j+1} = U_i^j + \Delta t\frac{\partial U(x_i, \eta)}{\partial t}, \qquad t_j < \eta < t_{j+1}.$$

Substituting these into Eq. (8.26) and simplifying, remembering that $r(\Delta x)^2 = k\,\Delta t/c\rho$, we get

$$e_i^{j+1} = r(e_{i+1}^j + e_{i-1}^j) + (1 - 2r)e_i^j + \Delta t\left[\frac{\partial U(x_i, \eta)}{\partial t} - \frac{k}{c\rho}\frac{\partial^2 U(\xi, t_j)}{\partial x^2}\right],$$

$$t_j \le \eta \le t_{j+1}, \qquad x_{i-1} \le \xi \le x_{i+1}. \qquad (8.27)$$

Let E^j be the magnitude of the maximum error in the row of calculations for $t = t_j$, and let $M > 0$ be an upper bound for the magnitude of the expression in brackets in Eq. (8.27). If $r \le \frac{1}{2}$, all the coefficients in Eq. (8.27) are positive (or zero) and we may write the inequality

$$|e_i^{j+1}| \le 2rE^j + (1 - 2r)E^j + M\,\Delta t = E^j + M\,\Delta t.$$

This is true for all the e_i^{j+1} at $t = t_{j+1}$, so

$$E^{j+1} \le E^j + M\,\Delta t.$$

Since this is true at each time step,

$$E^{j+1} \le E^j + M\,\Delta t \le E^{j-1} + 2M\,\Delta t \le \cdots \le E^0 + (j + 1)M\,\Delta t = E^0 + Mt_{j+1}$$
$$= Mt_{j+1},$$

because E^0, the errors at $t = 0$, are zero, since U is given by the initial conditions.

Now, as $\Delta x \to 0$, $\Delta t \to 0$ if $k\,\Delta t/c\rho(\Delta x)^2 \le \frac{1}{2}$, and $M \to 0$, because, as both Δx and Δt get smaller,

$$\left[\frac{\partial U(x_i, \eta)}{\partial t} - \frac{k}{c\rho}\frac{\partial^2 U(\xi, t_j)}{\partial x^2}\right] \to \left(\frac{\partial U}{\partial t} - \frac{k}{c\rho}\frac{\partial^2 U}{\partial x^2}\right)_{i,j} = 0.$$

This last is by virtue of Eq. (8.24), of course. Consequently, we have shown that the explicit method is convergent for $r \leq \frac{1}{2}$, because the errors approach zero as Δt and Δx are made smaller.

For the solution to the heat-flow equation by the Crank–Nicolson method, the analysis of convergence may be made by similar methods. The treatment is more complicated, but it can be shown that each E^{j+1} is no greater than a finite multiple of E^j plus a term that vanishes as both Δx and Δt become small, and this is independent of r. Hence, since the initial errors are zero, the finite-difference solution approaches the analytical solution as $\Delta t \to 0$ and $\Delta x \to 0$, requiring only that r stay finite. This is also true for the θ method whenever $0.5 \leq \theta \leq 1$.

Let us begin our discussion of stability with a numerical example. Since the heat-flow equation is linear, if two solutions are known, their sum is also a solution. We are interested in what happens to errors made in one line of the computations as the calculations are continued, and because of the additivity feature, the effect of a succession of errors is just the sum of the effects of the individual errors. We follow, then, a single error,* which most likely occurred due to round-off. If this single error does not grow in magnitude, we will call the method *stable*, since then the cumulative effect of all errors affects the later calculations no more than a linear combination of the previous errors would. (Since round-off errors are both positive and negative, we can expect some cancellation.)

Table 8.7 illustrates the principle. We have calculated for the simple case where the boundary conditions are fixed, so that the errors at the endpoints are zero. We assume that a single error of size e occurs at $t = t_1$ and $x = x_2$. The explicit method, $k \, \Delta t / c \rho (\Delta x)^2 = \frac{1}{2}$, was used. The original error quite obviously dies out. As an exercise, it is left to the student to show that with $r > 0.5$, errors have an increasingly large effect on later computations. Table 8.8 shows that errors damp out for the Crank–Nicolson method with $r = 1$ even more rapidly than in the explicit method with $r = 0.5$. The errors with the implicit method also die out with $r = 1$, more rapidly than with the explicit method but less rapidly than with Crank–Nicolson.

Table 8.7 Propagation of errors—explicit method

t	Endpoint x_1	x_2	x_3	x_4	Endpoint x_5
t_0	0	0	0	0	0
t_1	0	e	0	0	0
t_2	0	0	$0.50e$	0	0
t_3	0	$0.25e$	0	$0.25e$	0
t_4	0	0	$0.25e$	0	0
t_5	0	$0.125e$	0	$0.125e$	0
t_6	0	0	$0.125e$	0	0
t_7	0	$0.062e$	0	$0.062e$	0
t_8	0	0	$0.062e$	0	0

---

*A computation made assuming that each of the interior points has an error equal to e at $t = t_1$ demonstrates the effect more rapidly.

Table 8.8 Propagation of errors—Crank–Nicolson method

t	x_1	x_2	x_3	x_4	x_5
t_0	0	0	0	0	0
t_1	0	e	0	0	0
t_2	0	$0.071e$	$0.286e$	$0.071e$	0
t_3	0	$0.092e$	$0.082e$	$0.092e$	0
t_4	0	$0.036e$	$0.064e$	$0.036e$	0
t_5	0	$0.024e$	$0.030e$	$0.024e$	0
t_6	0	$0.012e$	$0.018e$	$0.012e$	0
t_7	0	$0.007e$	$0.009e$	$0.007e$	0
t_8	0	$0.004e$	$0.005e$	$0.004e$	0

A More Analytical Argument

To discuss stability in a more analytical sense, we need some material from linear algebra. In Chapter 6 we discussed eigenvalues and eigenvectors of a matrix. We recall that for the matrix A and vector x, if

$$Ax = \lambda x,$$

then the scalar λ is an eigenvalue of A and x is the corresponding eigenvector. If the N eigenvalues of the $N \times N$ matrix A are all different, then the corresponding N eigenvectors are linearly independent, and any N-component vector can be written uniquely in terms of them.

Consider the unsteady-state heat-flow problem with fixed boundary conditions. Suppose we subdivide into $N + 1$ subintervals so there are N unknown values of the temperature being calculated at each time step. Think of these N values as the components of a vector. Our algorithm for the explicit method (Eq. 8.4) can be written as the matrix equation*

$$\begin{bmatrix} u_1^{j+1} \\ u_2^{j+1} \\ \vdots \\ u_N^{j+1} \end{bmatrix} = \begin{bmatrix} (1 - 2r) & r & & \\ r & (1 - 2r) & r & \\ & & \ddots & \\ & & r & (1 - 2r) \end{bmatrix} \begin{bmatrix} u_1^{j} \\ u_2^{j} \\ \vdots \\ u_N^{j} \end{bmatrix}, \qquad \textbf{(8.28)}$$

or

$$u^{j+1} = Au^j,$$

---

* A change of variable is required to give boundary conditions of $u = 0$ at each end. This can always be done for fixed end conditions.

where A represents the coefficient matrix and u^j and u^{j+1} are the vectors whose N components are the successive calculated values of temperature. The components of u^0 are the initial values from which we begin our solution. The successive rows of our calculations are

$$u^1 = Au^0,$$
$$u^2 = Au^1 = A^2u^0,$$
$$\vdots$$
$$u^j = Au^{j-1} = A^2u^{j-2} = \cdots = A^ju^0.$$

(The superscripts on the A are here exponents; on the vectors they indicate time.)

Suppose that errors are introduced into u^0, so that it becomes $\bar{u}^0$. We will follow the effects of this error through the calculations. The successive lines of calculation are now

$$\bar{u}^j = A\bar{u}^{j-1} = \cdots = A^j\bar{u}^0.$$

Let us define the vector e^j as $u^j - \bar{u}^j$ so the e^j represents the errors in u^j caused by the errors in $\bar{u}^0$. We have

$$e^j = u^j - \bar{u}^j = A^ju^0 - A^j\bar{u}^0 = A^je^0. \tag{8.29}$$

This shows that errors are propagated by using the same algorithm as that by which the temperatures are calculated, as was implicitly assumed earlier in this section.

Now the N eigenvalues of A are distinct (see below) so that its N eigenvectors $x_1, x_2, \ldots, x_N$ are independent, and

$$Ax_1 = \lambda_1 x_1,$$
$$Ax_2 = \lambda_2 x_2,$$
$$\vdots$$
$$Ax_N = \lambda_N x_N.$$

We now write the error vector e^0 as a linear combination of the x_i:

$$e^0 = c_1 x_1 + c_2 x_2 + \cdots + c_N x_N,$$

where the c's are constants. Then e^1 is, in terms of the x_i,

$$e^1 = Ae^0 = \sum_{i=1}^{N} Ac_i x_i = \sum_{i=1}^{N} c_i Ax_i = \sum_{i=1}^{N} c_i \lambda_i x_i,$$

and for e^2,

$$e^2 = Ae^1 = \sum_{i=1}^{N} Ac_i \lambda_i x_i = \sum_{i=1}^{N} c_i \lambda_i^2 x_i.$$

(Again, the superscripts on vectors indicate time; on λ they are exponents.) After j steps, Eq. (8.29) can be written

$$e^j = \sum_{i=1}^{N} c_i \lambda_i^j x_i.$$

If the magnitudes of all of the eigenvalues are less than or equal to unity, errors will not grow as the computations proceed; that is, the computational scheme is stable. This then is the analytical condition for stability: that the largest eigenvalue of the coefficient matrix for the algorithm be one or less in magnitude.

The eigenvalues of matrix A (Eq. 8.28) can be shown to be

$$1 - 4r \sin^2 \frac{n\pi}{2(N + 1)}, \qquad n = 1, 2, \ldots, N$$

(note that they are all distinct). We will have stability for the explicit scheme if

$$-1 \leq 1 - 4r \sin^2 \frac{n\pi}{2(N + 1)} \leq 1.$$

The limiting value of r is given by

$$-1 \leq 1 - 4r \sin^2 \frac{n\pi}{2(N + 1)}$$

$$r \leq \frac{\dfrac{1}{2}}{\sin^2 \left(\dfrac{n\pi}{2(N + 1)} \right)}.$$

Hence, if $r \leq \frac{1}{2}$, the explicit scheme is stable.

The Crank–Nicolson scheme, in matrix form, is

$$\begin{bmatrix} (2 + 2r) & -r & & \\ -r & (2 + 2r) & -r & \\ & \ddots & \ddots & \\ & & -r & (2 + 2r) \end{bmatrix} \begin{bmatrix} u_1^{j+1} \\ u_2^{j+1} \\ \vdots \\ u_N^{j+1} \end{bmatrix} = \begin{bmatrix} (2 - 2r) & r & & \\ r & (2 - 2r) & r & \\ & \ddots & \ddots & \\ & & r & (2 - 2r) \end{bmatrix} \begin{bmatrix} u_1^j \\ u_2^j \\ \vdots \\ u_N^j \end{bmatrix},$$

or

$$A u^{j+1} = B u^j.$$

We can write

$$u^{j+1} = (A^{-1}B) u^j, \tag{8.30}$$

so that stability is given by the magnitudes of the eigenvalues of $A^{-1}B$. These are

$$\frac{2 - 4r \sin^2 \left(\dfrac{n\pi}{2(N - 1)} \right)}{2 + 4r \sin^2 \left(\dfrac{n\pi}{2(N - 1)} \right)}, \qquad n = 1, 2, \ldots, N.$$

Clearly, all the eigenvalues are no greater than one in magnitude for any positive value of r. A similar argument shows that the implicit method is also unconditionally stable.

With derivative boundary conditions, a similar analysis shows that the Crank–Nicolson method is stable for any positive value of r. For the explicit scheme, $r = \frac{1}{2}$ leads to instability with a finite surface coefficient. Smith (1978) shows that the limitation on r for stability is

$$r \leq \frac{1}{2 + P \, \Delta x},$$

where P is the ratio of surface coefficient to conductivity, h/k.

Chapter Summary

Test your understanding of this chapter by asking yourself if you can

1.   Solve a parabolic partial-differential equation in one-space dimension by both the explicit and implicit (Crank–Nicolson) methods with boundary conditions that may involve derivatives.

2.   Explain the differences between explicit and implicit methods, giving the advantages of each.

3.   Show that the theta method represents a generalization of all of the finite-difference methods of this chapter.

4.   Discuss how finite-difference methods can be applied to parabolic equations in two and three space dimensions, describing the problems that are involved.

5.   Set up the equations for the A.D.I. method in a two-dimensional problem, and employ a computer program for this on a rectangular region. You can outline how the method might be applied to an irregular region.

6.   Explain how finite elements can be applied to parabolic equations and solve simpler situations with this method. You should be able to tell when this method reduces to the same equations as with a finite-difference method.

7.   Outline the arguments that show whether a method is stable and convergent.

8.   Use, modify, and critique computer programs that solve parabolic equations.

Computer Programs

We show how two methods of this chapter can be implemented: the theta method through a program in QuickBASIC version 4.5 and the A.D.I. method applied to a rectangular region through a program written in FORTRAN.

Program 8.1 The Theta Method

Figure 8.8 lists the BASIC program. It handles unsteady-state heat flow in one-dimensional space. CONST and DATA statements provide values for the thermal properties of the bar and its length, its end temperatures (derivative end conditions are not provided for), the number of uniform subintervals, and the initial temperatures at the interior nodes. This information allows for the dimensioning of a number of arrays.

The user inputs values for θ and r (from which the program can determine the size of the time steps). Recall that we can get the explicit, Crank–Nicolson, or implicit methods by a suitable choice for theta.

The program prepares for solving the problem by computing Δx and Δt as well as N, the number of interior nodes. It reads in the initial nodal temperatures and sets up the A matrix for the set of equations to be solved. The solution is obtained by forming the LU equivalent of A.

After printing the initial temperatures, the program enters a repetitive loop in which temperatures are computed for five time steps. The temperatures are obtained by forming the right-hand side vector and completing the solution, which is then printed. The user can get another set of five computations by pressing a key. (This loop must be broken by a "break" command.) The output is for $o = 1$, $r = 1$.

```
' THETA.BAS
'
' ***************************************************************
'
'                          Chapter 8
'   Gerald/Wheatley, APPIED NUMERICAL ANALYSIS (fifth edition)
'                      Addison-Wesley, 1994
'
' ***************************************************************
'
'   This uses the theta method for 1-D heat equation
'     with ends at fixed values.
'
'   Use CONST for thermal properties and length, boundary values
'   Get number of intervals and initial u-values from DATA
'   Get values for theta and r = kappa*Dt/(c*rho*Dx^2) from user
'
' ***************************************************************
'
'   Varibles used:
'         kappa, c, rho   Thermal properties
'         L               Length
'         UL              u-value at left end
'         UR              u-value at right end
'         th              Theta, fraction of step where derivatives are
'                             evaluated. If 0, explicit method; if 0.5,
'                             Crank-Nicolson; if 1, implicit method.
'         Dx, Dt          Delta x, step size for t
'         r               Ratio of kappa*Dt/(c*rho*Dx^2)
'         Nint            Number of uniform intervals in region
'         N               Number of interior nodes = Nint -1
'         u               Vector of u-values at nodes
```

Figure 8.8   Program 8.1

Figure 8.8 *(continued)*

```
'       A              Matrix of coefficients of equations
'       b              Vector of right hand sides
'       LL, UU         Matrices for LU decompostion
'       y              Intermediate vector used in solving system
'
'
' **************************************************************

CONST kappa = .119, c = 1, rho = 1, L = 20
CONST UL = 0: UR = 10
DATA 5
DATA 2,2,2,2

' Initalize: get values, do DIM, set up u's

CLS

READ Nint
DIM u(Nint), A(Nint, 3), b(Nint), y(Nint)
DIM LL(Nint, 2), UU(Nint)

INPUT "Values for theta, r ", th, r

Dx = L / Nint
Dt = c * rho * Dx ^ 2 / kappa * r
N = Nint - 1        ' Number of interior nodes

FOR i = 1 TO N: READ u(i): NEXT i
u(0) = UL: u(Nint) = UR

' Set up A matrix
FOR i = 1 TO N
  A(i, 1) = -r * th
  A(i, 2) = 1 + 2 * r * th
  A(i, 3) = -r * th
NEXT i

' Get LU equivalent of A
LL(1, 2) = A(1, 2): LL(2, 1) = A(2, 1): UU(1) = A(1, 3) / LL(1, 2)
FOR i = 2 TO N - 1
  LL(i, 2) = A(i, 2) - LL(i, 1) * UU(i - 1)
  LL(i + 1, 1) = A(i + 1, 1)
  UU(i) = A(i, 3) / LL(i, 2)
NEXT i

LL(N, 2) = A(N, 2) - LL(N, 1) * UU(N - 1)

' Print initial u's, then begin loop
PRINT "At start, u's are"
FOR i = 0 TO Nint
  PRINT USING "###.### "; u(i);
NEXT i: PRINT

' Loop: get rhs, solve, step t, print u's -- in sets of 5

t = Dt

DO

  PRINT "Press a key to get 5 more"

  DO                    ' Wait for keypress
    more$ = INKEY$
  LOOP UNTIL more$ <> ""
```

Figure 8.8 *(continued)*

```
      FOR j = 1 TO 5      ' Compute five sets of values

'  Set up rhs
      rth = r * (1 - th)
      FOR i = 1 TO N
        b(i) = rth * u(i - 1) + (1 - 2 * rth) * u(i) + rth * u(i + 1)
      NEXT i
      b(1) = b(1) - A(1, 1) * UL
      b(Nint - 1) = b(N) - A(N, 3) * UR

'  Solve the system using LL and UU

'  Get intermediate vector, y = LL*u
        y(1) = b(1) / LL(1, 2)
        FOR i = 2 TO N
          y(i) = (b(i) - LL(i, 1) * y(i - 1)) / LL(i, 2)
        NEXT i

'  Solve for u from U*u = y
        u(N) = y(N)
        FOR i = N - 1 TO 1 STEP -1
          u(i) = y(i) - UU(i) * u(i + 1)
        NEXT i

        PRINT "At t = "; t; "u's are"
        FOR i = 0 TO Nint
          PRINT USING "###.### "; u(i);
        NEXT i: PRINT
        t = t + Dt

    NEXT j

LOOP    ' End of DO (forever)


****************************************************

          OUTPUT FOR THETA.BAS


At start, u's are
  0.000   2.000   2.000   2.000   2.000  10.000
At t =  134.4538 u's are
  0.000   1.382   2.145   3.055   5.018  10.000
At t =  268.9076 u's are
  0.000   1.279   2.456   3.944   6.321  10.000
At t =  403.3613 u's are
  0.000   1.358   2.793   4.567   6.962  10.000
At t =  537.8151 u's are
  0.000   1.482   3.089   4.991   7.318  10.000
At t =  672.2689 u's are
  0.000   1.602   3.325   5.283   7.534  10.000
At t =  806.7227 u's are
  0.000   1.702   3.505   5.487   7.674  10.000
At t =  941.1765 u's are
  0.000   1.780   3.639   5.631   7.768  10.000
At t =  1075.63 u's are
  0.000   1.839   3.738   5.734   7.834  10.000
At t =  1210.084 u's are
  0.000   1.883   3.810   5.808   7.881  10.000
At t =  1344.538 u's are
  0.000   1.915   3.862   5.861   7.914  10.000
```

Program 8.2 The A.D.I. Method

Our second program, listed in Fig. 8.9 and written in FORTRAN, uses the A.D.I. method for unsteady-state heat flow in a rectangle (so it solves a two-dimensional problem). It resembles the A.D.I. program of Chapter 7 quite closely, so the comments for that program are pertinent here. The major difference is that each "iteration" of A.D.I. computes values for the next time step, continuing until the value of TMAX is reached. Another obvious difference is that the size of the time steps is determined from the size of Δx, r, and the thermal properties.

Because this program has logic so closely related to the A.D.I. program of Chapter 7 and to its copious comments, it is unnecessary to discuss it any further here.

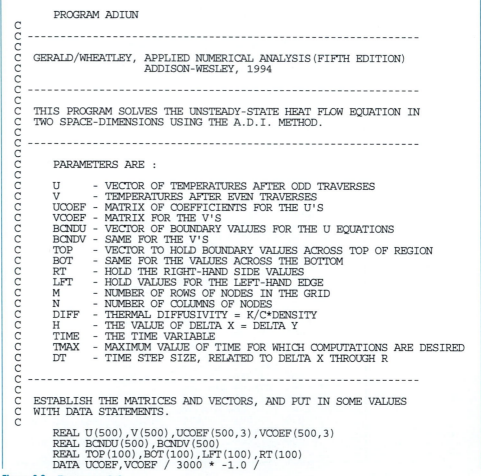

```
      PROGRAM ADIUN
C
C  ------------------------------------------------------------
C
C  GERALD/WHEATLEY, APPLIED NUMERICAL ANALYSIS (FIFTH EDITION)
C                   ADDISON-WESLEY, 1994
C
C  ------------------------------------------------------------
C
C  THIS PROGRAM SOLVES THE UNSTEADY-STATE HEAT FLOW EQUATION IN
C  TWO SPACE-DIMENSIONS USING THE A.D.I. METHOD.
C
C  ------------------------------------------------------------
C
C     PARAMETERS ARE :
C
C     U     - VECTOR OF TEMPERATURES AFTER ODD TRAVERSES
C     V     - TEMPERATURES AFTER EVEN TRAVERSES
C     UCOEF - MATRIX OF COEFFICIENTS FOR THE U'S
C     VCOEF - MATRIX FOR THE V'S
C     BCNDU - VECTOR OF BOUNDARY VALUES FOR THE U EQUATIONS
C     BCNDV - SAME FOR THE V'S
C     TOP   - VECTOR TO HOLD BOUNDARY VALUES ACROSS TOP OF REGION
C     BOT   - SAME FOR THE VALUES ACROSS THE BOTTOM
C     RT    - HOLD THE RIGHT-HAND SIDE VALUES
C     LFT   - HOLD VALUES FOR THE LEFT-HAND EDGE
C     M     - NUMBER OF ROWS OF NODES IN THE GRID
C     N     - NUMBER OF COLUMNS OF NODES
C     DIFF  - THERMAL DIFFUSIVITY = K/C*DENSITY
C     H     - THE VALUE OF DELTA X = DELTA Y
C     TIME  - THE TIME VARIABLE
C     TMAX  - MAXIMUM VALUE OF TIME FOR WHICH COMPUTATIONS ARE DESIRED
C     DT    - TIME STEP SIZE, RELATED TO DELTA X THROUGH R
C
C  ------------------------------------------------------------
C
C  ESTABLISH THE MATRICES AND VECTORS, AND PUT IN SOME VALUES
C  WITH DATA STATEMENTS.
C
      REAL U(500),V(500),UCOEF(500,3),VCOEF(500,3)
      REAL BCNDU(500),BCNDV(500)
      REAL TOP(100),BOT(100),LFT(100),RT(100)
      DATA UCOEF,VCOEF / 3000 * -1.0 /
```

Figure 8.9 Program 8.2

Figure 8.9   *(continued)*

```
          DATA BCNDU,BCNDV / 1000 * 0.0 /
          DATA M,N,R / 7,15,0.5 /
          DATA H,DIFF / 0.125,0.152 /OVER
          DATA TMAX / 2.0 /
          DATA RT,LFT,TOP,BOT / 100 * 100.0, 300 * 0.0 /
C
C    -----------------------------------------------------------------
C
C    FOR THE TEST CASE, WE BEGIN WITH TEMPERATURES EVERYWHERE EQUAL
C    TO ZERO. ESTABLISH THESE BY A DATA STATEMENT.
C
          DATA V / 500 * 0.0 /
C
C    -----------------------------------------------------------------
C
C    SET UP THE COEFFICIENT MATRICES BY WRITING OVER THE
C    DIAGONAL AND CERTAIN OFF DIAGONAL TERMS.
C
          MSIZE = M * N
          DO 10 I = 1,MSIZE
             UCOEF(I,2) = 1.0/R + 2.0
             VCOEF(I,2) = 1.0/R + 2.0
10        CONTINUE
          ML1 = M - 1
          NL1 = N - 1
          MSL1 = MSIZE - 1
C
          DO 20 I = N,MSL1,N
             UCOEF(I,3) = 0.0
             UCOEF(I+1,1) = 0.0
20        CONTINUE
        DO 30 I = M,MSL1,M
             VCOEF(I,3) =0.0
             VCOEF(I+1,1) = 0.0
30        CONTINUE
C
C    -----------------------------------------------------------------
C
C    NOW GET VALUES INTO THE BCND VECTORS
C
          DO 40 I = 1,N
             BCNDU(I) = TOP(I)
             J = MSIZE - N + I
             BCNDU(J) = BOT(I)
40        CONTINUE
          DO 45 I =1,M
             J = (I-1)*N + 1
             BCNDU(J) = BCNDU(J) + LFT(I)
             J = I * N
             BCNDU(J) = BCNDU(J) + RT(I)
45        CONTINUE
          DO 50 I = 1,M
             BCNDV(I) = LFT(I)
             J = MSIZE - M + I
             BCNDV(J) = RT(I)
50        CONTINUE

          DO 55 I = 1,N
             J = (I-1)*M + 1
             BCNDV(J) = BCNDV(J) + TOP(I)
             J = I * M
             BCNDV(J) = BCNDV(J) + BOT(I)
55        CONTINUE
C
C
C    -----------------------------------------------------------------
```

Figure 8.9 *(continued)*

```
C
C   WE NOW GET THE LU DECOMPOSITIONS OF UCOEF AND VCOEF
C
      DO 60 I = 2,MSIZE
         UCOEF(I-1,3) = UCOEF(I-1,3)/UCOEF(I-1,2)
         UCOEF(I,2) = UCOEF(I,2) - UCOEF(I,1)*UCOEF(I-1,3)
         VCOEF(I-1,3) = VCOEF(I-1,3)/VCOEF(I-1,2)
         VCOEF(I,2) = VCOEF(I,2) - VCOEF(I,1)*VCOEF(I-1,3)
60    CONTINUE
C
C
C
C   ------------------------------------------------------------
C
C   NOW WE DO THE ITERATIONS UNTIL TIME EQUALS TMAX.
C
      TIME = 0.0
      DT = R / DIFF*H*H
62    IF ( TIME .GT. TMAX ) STOP
C
C   ------------------------------------------------------------
C
C   COMPUTE THE R.H.S. FOR THE U EQUATIONS AND STORE IN THE U
C   VECTOR. FIRST DO THE TOP AND BOTTOM ROWS.
C
      DO 65 I = 1,N
         J = (I-1)*M + 1
         U(I) = ( 1.0/R - 2.0 )*V(J) + V(J+1) + BCNDU(I)
         K = MSIZE - N + I
         J = I * M
         U(K) = V(J-1) + ( 1.0/R - 2.0 )*V(J) + BCNDU(K)
65    CONTINUE
C
C   ------------------------------------------------------------
C
C   NOW FOR THE OTHER ONES
C
      DO 75 I = 2,ML1
         DO 70 J = 1,N
            K = (I-1)*N + J
            L = I + (J-1)*M
            U(K) = V(L-1) + ( 1.0/R - 2.0 )*V(L) + V(L+1) + BCNDU(K)
70       CONTINUE
75    CONTINUE
C
C   ------------------------------------------------------------
C
C   NOW GET THE SOLUTION FOR THE ODD TRAVERSE
C
      U(1) = U(1) / UCOEF(1,2)
      DO 80 I = 2,MSIZE
         U(I) = ( U(I) - UCOEF(I,1)*U(I-1)) / UCOEF(I,2)
80    CONTINUE
      DO 90 I = 1,MSL1
         JROW = MSIZE - I
         U(JROW) = U(JROW) - UCOEF(JROW,3)*U(JROW+1)
90    CONTINUE
C
C   ------------------------------------------------------------
C
C   COMPUTE THE R.H.S. FOR THE EVEN TRAVERSE, STORE IN V. DO THE
C   TOP AND BOTTOM ONES.
C
```

Figure 8.9   *(continued)*

```
          DO 95 I = 1,M
            J = (I-1)*N + 1
            V(I) = ( 1.0/R - 2.0 )*U(J) + U(J+1) + BCNDV(I)
            K = MSIZE - M + I
            J = I * N
            V(K) = U(J-1) + ( 1.0/R - 2.0 )*U(J) + BCNDV(K)
 95       CONTINUE
C
C
C  ----------------------------------------------------------
C
C  NOW THE REST OF THE ROWS
C
          DO 105 I = 2,NL1
            DO 100 J = 1,M
              K = (I-1)*M + J
              L = I + (J-1)*N
              V(K) = U(L-1) + (1.0/R-2.0)*U(L) + U(L+1) + BCNDV(K)
100       CONTINUE
105       CONTINUE
C
C
C  ----------------------------------------------------------
C
C  GET THE SOLUTION FOR THE EVEN TRAVERSE
C
          V(1) = V(1) / VCOEF(1,2)
          DO 110 I = 2,MSIZE
            V(I) = ( V(I) - VCOEF(I,1)*V(I-1)) / VCOEF(I,2)
110       CONTINUE
          DO 120 I = 1,MSL1
            JROW = MSIZE - I
            V(JROW) = V(JROW) - VCOEF(JROW,3)*V(JROW+1)
120       CONTINUE
          TIME = TIME + 2.0*DT
C
C  ----------------------------------------------------------
C
C  PRINT OUT THE LAST RESULT.
C
          PRINT 202, TIME, ( V(I), I = 1,MSIZE )
202       FORMAT(//1X,' WHEN T = ',F6.3,' V VALUES ARE' /
     +           (1X,10F7.1/1X, 5F7.1))
          GO TO 62
          END


****************************************************************

                PARTIAL OUTPUT FOR ADIUN.F

   WHEN T =   .103 V VALUES ARE

   .0     .0     .0     .0     .0     .0     .0     .0     .0     .0
   .0     .0     .0     .0     .0
   .0     .0     .0     .0     .0     .0     .0     .0     .0     .0
   .0     .0     .0     .0     .0
   .0     .0     .0     .0     .0     .0     .0     .0     .0     .0
   .0     .0     .0     .0     .0
   .0     .0     .0     .0     .0     .0     .0     .0     .0     .0
   .0     .0     .0     .0     .0
   .0     .0     .0     .1     .1     .1     .1     .1     .1     .1
   .2     .3     .3     .3     .3
   .3     .2     .8    1.0    1.0    1.0    1.0    1.0     .8    2.8
  3.6    3.8    3.8    3.8    3.6
  2.8   10.5   13.3   14.1   14.2   14.1   13.3   10.5   39.2   49.7
 52.5   53.0   52.5   49.7   39.2
```

Figure 8.9 *(continued)*

WHEN T =   .206 V VALUES ARE

.0	.0	.0	.0	.0	.0	.0	.0	.0	0
.0	.0	.0	.0	.0					
.0	.0	.0	.0	.0	.0	.0	.0	.0	.0
.0	.0	.0	.0	.0					
.0	.0	.0	.0	.0	.0	.0	.0	.0	.0
.0	.0	.0	.0	.0					
.0	.0	.0	.0	.0	.0	.0	.0	.0	.0
.0	.1	.1	.1	.2					
.1	.1	.1	.2	.4	.5	.5	.5	.4	.2
.8	1.3	1.5	1.5	1.5					
1.3	.8	2.3	3.8	4.4	4.5	4.4	3.8	2.3	6.8
10.8	12.3	12.7	12.3	10.8					
6.8	18.4	27.6	31.1	31.9	31.1	27.6	18.4	42.7	56.0
60.0	60.8	60.0	56.0	42.7					

.
.
.

WHEN T =   2.056 V VALUES ARE

.1	.1	.1	.1	.1	.1	.1	.1	.2	.3
.3	.3	.2	.1	.2					
.4	.5	.6	.5	.4	.2	.4	.7	.9	.9
.9	.7	.4	.6	1.0					
1.4	1.5	1.4	1.0	.6	.9	1.6	2.1	2.3	2.1
1.6	.9	1.3	2.5	3.2					
3.5	3.2	2.5	1.3	2.0	3.7	4.9	5.3	4.9	3.7
2.0	3.1	5.6	7.3	7.9					
7.3	5.6	3.1	4.6	8.5	11.0	11.8	11.0	8.5	4.6
7.0	12.7	16.3	17.6	16.3					
12.7	7.0	10.6	19.0	24.1	25.8	24.1	19.0	10.6	16.5
28.6	35.4	37.6	35.4	28.6					
16.5	27.0	43.3	51.4	53.9	51.4	43.3	27.0	48.3	66.2
73.2	75.0	73.2	66.2	48.3					

Exercises

Section 8.1

1. The parameters of the basic equation for unsteady-state heat flow are dimensional. If it is desired to measure u in °F and x in feet, how must the units of k, c, and ρ be chosen in

$$\frac{\partial^2 u}{\partial x^2} = \frac{c\rho}{k}\frac{\partial u}{\partial t}?$$

2. Suppose that k, c, and ρ vary with x instead of being constant, as assumed in Section 8.1. Rederive the equation for unsteady-state heat flow in one dimension.

▶**3.** Suppose that the rod sketched in Fig. 8.1 is tapered, with the diameter varying linearly from 2 in. at the left end to 1.25 in. at the right end; the rod is 14 in. long and is made of steel. If 200 BTU/hr of heat flows from left to right (the flow is the same at each x-value along the rod—steady state), what are the values of the gradient at

a. the left end?

b. the right end?

c. $x = 3$ in.?

Section 8.2

▶**4.** Solve for the temperatures at $t = 2.06$ sec in the 2-cm thick steel slab of Example 8.1 if the initial temperatures are given by

$$u(x,0) = 100 \sin\left(\frac{\pi x}{2}\right).$$

Use the explicit method with $\Delta x = 0.25$ cm. Compare to the analytical solution: $100e^{-0.3738t} \sin(\pi x/2)$.

5. Solve for the temperatures in a cylindrical copper rod that is 8 in. long and whose curved outer surface is insulated so that heat flows only in one direction. The initial temperature is linear from 0°C at one end to 100°C at the other, when suddenly the hot end is brought to 0°C and the cold end is brought to 100°C. Use $\Delta x = 1$ in. and an appropriate value of Δt so that $k \, \Delta t/c\rho(\Delta x)^2 = \frac{1}{2}$. Look up values for k, c, and ρ in a handbook. Carry out the solution for 10 time steps.

6. Repeat Exercise 5 but with $\Delta x = 0.5$ in., and compare the temperature at points 1 in., 3 in., and 6 in. from the cold end with those of the previous exercise. You will need to compute more time steps to match the 10 steps done previously.

   You will find it instructive to graph the temperatures for both sets of computations.

7. Repeat Exercise 5 but with $\Delta x = 1$ in. and Δt such that $k \, \Delta t/c\rho(\Delta x)^2 = \frac{1}{4}$. Compare results with Exercises 5 and 6.

▶8. Repeat the diffusion example of Example 8.2 but with the following parameters, and compare to the analytical solution given in Fig. 8.4. Continue the computations until $t = 268.8$ sec.
   a. With $\Delta x = 4$ cm and $D \, \Delta t/(\Delta x)^2 = 0.125$
   b. With $\Delta x = 2$ cm and $D \, \Delta t/(\Delta x)^2 = 0.5$
   c. With $\Delta x = 1$ cm and $D \, \Delta t/(\Delta x)^2 = 0.5$

   If you combine these with the results shown in Section 8.2, you now have data that illustrate the separate effects of Δx and r on the accuracy. Considering the amount of computations required, which is the more effective way to improve accuracy?

Section 8.3

9. Solve Exercise 4 by the Crank–Nicolson method with $r = 1$.

10. Solve Exercise 5 by the Crank–Nicolson method with $r = 1$. Compare your result with those from Exercises 5, 6, and 7.

11. Repeat Exercise 9, but this time vary the value of r between 0.5 and 2.0. How does r affect accuracy? (Compare to the analytical solution given in Exercise 4.)

▶12. The methods of Sections 8.2 and 8.3 can be readily applied to more complicated situations. For exam-

ple, if heat is being generated at points along the bar at a rate that is a function of x, $f(x)$, the unsteady-state heat equation becomes

$$k \frac{\partial^2 u}{\partial x^2} - c\rho \frac{\partial u}{\partial t} = f(x)$$

Solve this when $f(x) = x(x - 1)$ cal/cm³, subject to conditions

$$u(0, t) = 0, \qquad u(1, t) = 0, \qquad u(x, 0) = 0.$$

Take $\Delta x = 0.2, k = 0.37, c\rho = 0.433$ in cgs units. (These are the properties of magnesium.) Use the Crank–Nicolson method, $r = 1$, and solve for five time steps.

13. Compare the computational effort between Crank–Nicolson and the explicit method, with a Δx one-fourth as large in the latter (where the examples of Sections 8.2 and 8.3 suggest they have similar accuracy).

Section 8.4

14. Solve Exercise 4 by the implicit method, $r = 1$.

15. Solve Exercise 5 by the implicit method, $r = 1$. Compare the results with those of Exercises 5, 6, and 10. You may want to make graphs at several time periods.

▶16. Solve Exercise 12 by the implicit method, and compare results to those of that exercise. If you vary r, can you get closer agreement?

▶17. Solve Exercise 4 by the implicit method, but this time vary r to determine how the errors are related to that factor. Use the average of the errors at all nodes as the measure of accuracy. You can vary r by changing either Δx or Δt. Which of these has the greater effect on the errors?

18. Repeat Exercise 17 but use Crank–Nicolson. Is the relation between r, Δx, and Δt and the errors the same?

19. Some authors suggest that the best value for theta is $\frac{2}{3}$ or perhaps 0.878. Experiment with the value of theta in Exercise 14 to see whether you can reach a decision about this matter.

Section 8.5

20. Use the set of equations (8.11) to get the numerical solution to Example 8.5 through eight time steps.

21. Solve the set of equations (8.12) to find another solution to Example 8.5—the Crank–Nicolson solution. Compare the results with those of Exercise 20.

22. Set up the equations to solve Example 8.5 by the implicit method with $r = 1$. Solve up to a value of t equal to that reached in Exercise 20. Compare the results with those of Exercises 20 and 21.

▶**23.** Solve Example 8.5 but with changed boundary conditions: Two opposite faces lose heat, one at a rate given by the relation $0.15A(u - 70)$, the other at a rate with a different heat transfer coefficient, $0.25A(u - 70)$. (In these, u is the surface temperature and A is the area; 70 is the temperature of the surroundings.) Use $\Delta x = 1$ in., and employ the explicit method with $k\,\Delta t/c\rho(\Delta x)^2 = \frac{1}{4}$.

▶**24.** Repeat Exercise 23, but now use Crank–Nicolson. Compare the results with those from the explicit method.

25. Repeat Exercise 23, but now use the implicit method. Compare the results with those from the explicit method and from Crank–Nicolson.

Section 8.6

▶**26.** A rectangular plate 2 in. × 3 in. is initially at 50°. At $t = 0$, one 2-in. edge is suddenly raised to 100°, and one 3-in. edge is suddenly cooled to 0°. The temperature on these two edges is held constant at these temperatures. The other two edges are perfectly insulated. Use a 1-in. grid to subdivide the plate and write the A.D.I. equations for each of the six nodes where unknown temperatures are involved. Use $r = 2$, and solve the equations for four time steps.

27. Suppose the cube used in Example 8.5 (4 in. × 4 in. × 4 in.) had heat flowing in all four directions, such as by having three adjacent faces lose heat by conduction to the flowing liquid. Set up the equations to solve for the temperature at any point. Do this in terms of the surrounding temperatures one Δt previously, using the explicit method and with the same spacing between nodes in all three directions. How many time steps are needed to reach $t = 15.12$ sec? How many equations are involved at each time step?

28. Repeat Exercise 27 for the Crank–Nicolson method, $r = 1$.

29. Repeat Exercise 27 for the implicit method, $r = 1$.

▶**30.** Repeat Exercise 27 for the A.D.I. method, $r = 1$.

Section 8.7

▶**31.** For a triangular element that has nodes at points $(1.2, 3.2)$, $(4.3, 2.7)$, and $(2.4, 4.1)$, find the components of each matrix in the element equations (Eqs. 8.20 and 8.21) if the material is aluminum.

32. Consider heat flow in one-space dimension; it will be governed by Eq. (8.1). For this situation, repeat the development that leads to the analog of Eq. (8.21).

▶**33.** Use the relations that you derived in Exercise 32 to solve Exercise 6 by finite elements. Place the nodes exactly as those used in the finite-difference solution. Are the resulting equations the same?

34. Use finite elements to solve Exercise 26. Place interior nodes at three arbitrarily chosen points (but do not make them symmetrical). Create triangular elements with these nodes and the four corner points. Then set up the element equations, assemble, and solve for four time steps. Use the resulting nodal temperatures to estimate the same set of temperatures that were computed in Exercise 26, and compare the two different methods of solving the problem.

35. Solve Example 8.2 by finite elements, placing nodes closer together near the ends of the tube where concentrations change most rapidly. Does this give better agreement with the analytical solution (shown in Fig. 8.4) than that of the example?

36. Redo Eq. (8.23), but this time for the theta method.

Section 8.8

37. Demonstrate by performing calculations similar to those in Table 8.7 that the explicit method is unstable for $r = 0.6$.

38. Demonstrate by performing calculations similar to those in Table 8.7 that the explicit method is stable for $r = 0.25$ but that the errors damp out less rapidly.

39. Suppose that the end conditions are not $u =$ a constant as in Table 8.7 but rather $u_x = 0$. Demonstrate by performing calculations similar to those in Table 8.7 that the explicit method is still stable for $r = 0.5$ but that the errors damp out much more slowly. Observe that the errors at a later stage become a linear combination of earlier errors.

▶**40.** Demonstrate by performing calculations similar to those in Table 8.7 that the Crank–Nicolson method is stable even if $r = 10$. You will need to solve a system of equations in this exercise.

41. Repeat Exercise 40 but for the implicit method.

▶**42.** Compute the largest eigenvalue of the coefficient matrix in Eq. (8.28) for $r = 0.5$, then for $r = 0.6$. Do you find that the statements in the text relative to eigenvalues are confirmed?

43. Repeat Exercise 42 but for Eq. (8.30) using $A^{-1}B$. Use $r = 1.0$ and $r = 2.0$.

44. Starting with the matrix form of the implicit method, show that for $A^{-1}B$ none of the eigenvalues exceed 1 in magnitude.

Applied Problems and Projects

45. When steel is forged, billets are heated in a furnace until the metal is of the proper temperature, between 2000°F and 2300°F. It can then be formed by the forging press into rough shapes that are later given their final finishing operations. To produce a certain machine part, a billet of size 4 in. × 4 in. × 20 in. is heated in a furnace whose temperature is maintained at 2350°F. You have been requested to estimate how long it will take all parts of the billet to reach a temperature above 2000°F. Heat transfers to the surface of the billet at a very high rate, principally through radiation. It has been suggested that you can solve the problem by assuming that the surface temperature becomes 2250°F instantaneously and remains at that temperature. Using this assumption, find the required heating time.

Since the steel piece is relatively long compared to its width and thickness, it may not introduce significant error to calculate as if it were infinitely long. This will simplify the problem, permitting a two-dimensional treatment rather than three-dimensional. Such a calculation should also give a more conservative estimate of heating time. Compare the estimates from two- and three-dimensional approaches.

46. After you have calculated the answers to Problem 45, your results have been challenged on the basis of assuming constant surface temperature of the steel. Radiation of heat flows according to the equation

$$q = E\sigma(u_F^4 - u_S^4) \text{ Btu/hr} \cdot \text{ft}^2,$$

where E = emissivity (use 0.80), σ is the Stefan–Boltzmann constant (0.171×10^{-8} Btu/hr·ft²·°F⁴), u_F and u_S are the furnace and surface absolute temperatures, respectively (°F + 460°).

The heat radiating to the surface must also flow into the interior of the billet by conduction, so

$$q = -k\frac{\partial u}{\partial x},$$

where k is the thermal conductivity of steel (use 26.2 Btu/hr·ft²·°F/ft) and $(\partial u/\partial x)$ is the temperature gradient at the surface in a direction normal to the surface. Solve the problem with this boundary condition, and compare your solution to that of Problem 45. (Observe that this is now a nonlinear problem. Think carefully how your solution can cope with it.)

47. Shipment of liquefied natural gas by refrigerated tankers to industrial nations may become an important means of supplying the world's energy needs. It must be stored at the receiving port, however. [A. R. Duffy and his co-workers (1967) discuss the storage of liquefied natural gas in underground tanks.] A commercial design, based on experimental verification of its feasibility, contemplated a prestressed concrete tank 270 ft in diameter and 61 ft deep, holding some 600,000 bbl of liquefied gas at −258°F. Convection currents in the liquid were shown to keep the temperature uniform at this value, the boiling point of the liquid.

Important considerations of the design are the rate of heat gained from the surroundings (causing evaporation of the liquid gas) and variation of temperatures in the earth below the tank (relating to the safety of the tank, which could be affected by possible settling or frost-heaving.)

The tank itself is to be made of concrete 6 in. thick, covered with 8 in. of insulation (on the liquid side). (A sealing barrier keeps the insulation free of liquid; otherwise its insulating capacity would be impaired.) The experimental tests showed that there is a very small temperature drop through the concrete: 12°F. This observed 12°F temperature difference seems reasonable in light of the relatively high thermal conductivity of concrete. We expect then that most of the temperature drop occurs in the insulation or in the earth below the tank.

Since the commercial-design tank is very large, if we are interested in ground temperatures near the center of the tank (where penetration of cold will be a maximum), it should be satisfactory to consider heat flowing in only one dimension, in a direction directly downward from the base of the tank. Making this simplifying assumption, compute how long it will take for the temperature to decrease to 32°F (freezing point of water) at a point 8 ft away from the tank wall. The necessary thermal data are

	Insulation	Concrete	Earth
Thermal conductivity (Btu/hr · ft · °F)	0.013	0.90	2.6
Density (lb/ft³)	2.0	150	132
Specific heat (Btu/lb · °F)	0.195	0.200	0.200

Assume the following initial conditions: temperature of liquid, −258°F; temperature of insulation, −258°F to 72°F (inner surface to outer); temperature of concrete, 72°F to 60°F; temperature of earth, 60°F.

Hyperbolic Partial-Differential Equations

Contents of This Chapter

This chapter describes numerical methods for solving the third type of partial-differential equation, hyperbolic equations. These equations model the vibration of structures (buildings, beams, machines) and are the basis for the fundamental equations of atomic physics. We solve some typical problems by two methods, but we admit that our treatment is not thorough.

9.7 Theoretical Matters

Demonstrates that the finite-difference solution is stable when we restrict the values for Δt and Δx.

Chapter Summary

Reviews the chapter in our accustomed style.

Computer Program

Presents a program that solves the one-dimensional wave equation.

Parallel Processing

Except for the method of characteristics (Section 9.4), the methods of this chapter involve sets of equations for which the remarks of Chapters 7 and 8 are pertinent. The somewhat complicated method of characteristics requires the solution of initial-value problems. As was observed in the opening to Chapter 5, parallel processing is of minor help with these problems.

9.1 The Wave Equation

The *wave equation* is the simplest example of a hyperbolic partial-differential equation. It models a vibrating string that is fixed at its ends. Imagine an elastic string stretched between two fixed endpoints and set to vibrating by plucking it with a finger.

We make a number of simplifying assumptions: the string is perfectly elastic, and we neglect gravitational forces, so that the only force is the tension force in the direction of the string; the string is uniform in density and thickness and has a weight per unit length of w lb/ft; the lateral displacement of the string is so small that the tension can be considered to be a constant value of T lb; and the slope of the string is hence small enough that $\sin \alpha \doteq \tan \alpha$, where α is the angle of inclination. Figure 9.1 illustrates the problem with the lateral displacement greatly exaggerated.

We take $x = 0$ at the left end. L is the total length of the string. Focus attention on the element of length dx between points A and B. The uniform tension T acts at each end. We are interested in how y, the lateral displacement, varies with time t and with distance x along the string.

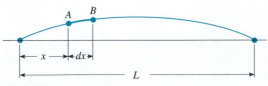

Figure 9.1

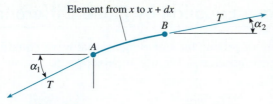

Figure 9.2

The forces acting in the y-direction are the vertical components of the two tensions (see Fig. 9.2). We take the upward direction as positive:

Upward force at left end: $\quad -T \sin \alpha_1 \doteq -T \tan \alpha_1 = -T \left(\dfrac{\partial y}{\partial x} \right)_A$;

Upward force at right end: $\quad +T \sin \alpha_2 \doteq +T \tan \alpha_2 = T \left(\dfrac{\partial y}{\partial x} \right)_B$

$$= T \left(\left(\frac{\partial y}{\partial x} \right)_A + \frac{\partial}{\partial x} \left(\frac{\partial y}{\partial x} \right) dx \right);$$

Net force: $T \dfrac{\partial^2 y}{\partial x^2} dx$.

Partials are used to express the slope because y is a function of both t (time) and x (horizontal distance). We use Newton's law, and equate the force to mass times acceleration in the vertical direction. Our simplifying assumptions permit us to use $w\, dx$ as the weight:

$$\left(\frac{w\, dx}{g} \right) \frac{\partial^2 y}{\partial t^2} = T \frac{\partial^2 y}{\partial x^2} dx, \qquad \frac{\partial^2 y}{\partial t^2} = \frac{Tg}{w} \frac{\partial^2 y}{\partial x^2}. \qquad (9.1)$$

This is the wave equation in one dimension. The conditions imposed on the solution are the end conditions $[ay + b(\partial y/\partial x) = f(t)]$ at $x = 0$ and $x = L$, and the initial conditions at $t = 0$. Initial conditions specifying both $y = f(x)$ and the velocity $\partial y/\partial t = g(x)$ are usual for this problem.

Comparing Eq. (9.1) to the general form of a second-order partial-differential equation,

$$A \frac{\partial^2 y}{\partial t^2} + B \frac{\partial^2 y}{\partial t\, \partial x} + C \frac{\partial^2 y}{\partial x^2} + D \left(t, x, y, \frac{\partial y}{\partial t}, \frac{\partial y}{\partial x} \right) = 0,$$

shows $B^2 - 4AC > 0$, since $B = 0$, $A = 1$, and $C = -Tg/w$. We see that Eq. (9.1) falls in the class of hyperbolic equations.

9.2 Solving the Wave Equation by Finite Differences

We attack the problem of solving the one-dimensional wave equation by replacing derivatives by difference quotients. We use superscripts to denote time, and subscripts for position:

$$\frac{\partial^2 y}{\partial t^2} = \frac{Tg}{w} \frac{\partial^2 y}{\partial x^2},$$

$$\frac{y_i^{j+1} - 2y_i^j + y_i^{j-1}}{(\Delta t)^2} = \frac{Tg}{w}\left(\frac{y_{i+1}^j - 2y_i^j + y_{i-1}^j}{(\Delta x)^2}\right).$$

Solving for the displacement at the end of the current interval, at $t = t_{j+1}$, we get

$$y_i^{j+1} = \frac{Tg(\Delta t)^2}{w(\Delta x)^2}(y_{i+1}^j + y_{i-1}^j) - y_i^{j-1} + 2\left(1 - \frac{Tg(\Delta t)^2}{w(\Delta x)^2}\right)y_i^j.$$

Note that selecting the ratio $Tg(\Delta t)^2/w(\Delta x)^2 = 1$ gives some simplification, though y_i^{j+1} is still a function of conditions at both t_j and t_{j-1}:

$$y_i^{j+1} = y_{i+1}^j + y_{i-1}^j - y_i^{j-1}, \qquad \Delta t = \frac{\Delta x}{\sqrt{Tg/w}} \qquad \text{(9.2)}$$

Equation (9.2) is a common way to solve the one-dimensional wave equation numerically.* We are, of course, interested in the validity of choosing the ratio at unity. Our finite-difference replacements to the derivatives do not have mixed error terms here, as they did in the explicit method for the heat equation; but, as we will later show, after discussing the method of characteristics, if the ratio is greater than one, we cannot be sure of convergence. Stability also sets a limit of unity to the ratio. It is surprising to find that, if one uses a value of $w(\Delta t)^2/Tg(\Delta x)^2$ of less than one, the results are less accurate, while with the ratio equal to one we can get exact analytical answers!†

There is still a problem in applying Eq. (9.2). We know y at $t = t_0 = 0$ from the initial condition, but to compute y at $t = \Delta t = t_1$, we need values at $t = t_{-1}$. We may at first be bothered by a need to know displacements *before* the start of the problem; but if we imagine the function $y = y(x,t)$ to be extended backward in time, the term t_{-1} makes good sense. Since we will ordinarily get periodic functions for y versus t at a given point, we can consider zero time as an arbitrary point at which we know the displacement and velocity, within the duration of an ongoing process.

One commonly used way to get values for the fictitious point at $t = t_{-1}$ is by

---

* Finite elements are also possible; we discuss this techinque briefly in a later section.
† What this means is that, with $w(\Delta t)^2/Tg(\Delta x)^2$ less than one, errors will decrease as $\Delta x \rightarrow 0$; that is, the method is stable. However, with finite values for Δx, the computed values are not exact unless the ratio is equal to one. Our discussion of characteristics later in this chapter will clarify this unusual situation.

employing the specification for the initial velocity $\partial y/\partial t = g(x)$ at $t = 0$. Using a central-difference approximation, we have

$$\frac{\partial y}{\partial t}(x_i, 0) \doteq \frac{y_i^1 - y_i^{-1}}{2\Delta t} = g(x_i),$$

$$y_i^{-1} = y_i^1 - 2g(x_i)\Delta t \qquad \text{at } t = 0 \text{ only.} \qquad \textbf{(9.3)}$$

Equation (9.3) is valid only at $t = 0$; substituting into Eq. (9.2) gives us, for $t = t_1$,

$$y_i^1 = \frac{1}{2}(y_{i+1}^0 + y_{i-1}^0) + g(x_i)\Delta t. \qquad \textbf{(9.4)}$$

After computing the first line with Eq. (9.4), we use Eq. (9.2) thereafter.* If the boundary conditions involve derivatives, we rewrite them in terms of difference quotients to incorporate them in our difference equation.

EXAMPLE 9.1

A banjo string is 80 cm long and weighs 1.0 g. It is stretched with a tension of 40,000 g. At a point 20 cm from one end it is pulled 0.6 cm from the equilibrium position and then released. Find the displacement of points along the string as a function of time. How long does it take for one complete cycle of motion? From this, compute the frequency with which it vibrates.

Let $\Delta x = 10$ cm. In Table 9.1 we show the results of using Eqs. (9.4) and (9.2). Because the string is just released from its initially displaced position, the initial

Table 9.1 Solution to banjo string example

t	0	10	20	30	40	50	60	70	80
0	0.0	0.3	0.6	0.5	0.4	0.3	0.2	0.1	0.0
Δt	0.0	0.3	0.4	0.5	0.4	0.3	0.2	0.1	0.0
$2\Delta t$	0.0	0.1	0.2	0.3	0.4	0.3	0.2	0.1	0.0
$3\Delta t$	0.0	−0.1	0.0	0.1	0.2	0.3	0.2	0.1	0.0
$4\Delta t$	0.0	−0.1	−0.2	−0.1	0.0	0.1	0.2	0.1	0.0
$5\Delta t$	0.0	−0.1	−0.2	−0.3	−0.2	−0.1	0.0	0.1	0.0
$6\Delta t$	0.0	−0.1	−0.2	−0.3	−0.4	−0.3	−0.2	−0.1	0.0
$7\Delta t$	0.0	−0.1	−0.2	−0.3	−0.4	−0.5	−0.4	−0.3	0.0
$8\Delta t$	0.0	−0.1	−0.2	−0.3	−0.4	−0.5	−0.6	−0.3	0.0
$9\Delta t$	0.0	−0.1	−0.2	−0.3	−0.4	−0.5	−0.4	−0.3	0.0
$10\Delta t$	0.0	−0.1	−0.2	−0.3	−0.4	−0.3	−0.2	−0.1	0.0
$11\Delta t$	0.0	−0.1	−0.2	−0.3	−0.2	−0.1	0.0	0.1	0.0
$12\Delta t$	0.0	−0.1	−0.2	−0.1	0.0	0.1	0.2	0.1	0.0
$13\Delta t$	0.0	−0.1	0.0	0.1	0.2	0.3	0.2	0.1	0.0
$14\Delta t$	0.0	0.1	0.2	0.3	0.4	0.3	0.2	0.1	0.0
$15\Delta t$	0.0	0.3	0.4	0.5	0.4	0.3	0.2	0.1	0.0
$16\Delta t$	0.0	0.3	0.6	0.5	0.4	0.3	0.2	0.1	0.0

---

* We will later discuss a more accurate way to begin the solution. Equation (9.4) is satisfactory when the initial velocity is constant.

velocity at all points is zero, and Eq. (9.4) becomes simply

$$y_i^1 = \frac{1}{2}(y_{i+1}^0 + y_{i-1}^0).$$

The initial conditions imply that y is linear in x from $y = 0$ at $x = 0$ to $y = 0.6$ at $x = 20$ and also linear to $x = 80$. The size of time steps is given by

$$\frac{Tg(\Delta t)^2}{w(\Delta x)^2} = 1 = \frac{(40,000)(980)(\Delta t)^2}{(1.0/80)(10)^2}, \qquad \Delta t = 0.000179 \text{ sec.}$$

After we have completed calculations for eight time steps, we observe that the y-values are reproducing the original steps but with negative signs and the end of the string reversed; that is, half a cycle has been completed in eight steps. For a complete cycle, 16 Δt's are needed. The frequency of vibration is then

$$f = \frac{1}{(16)(0.00179)} = 350 \text{ cycles/sec.}$$

(This value is exactly the same as that given by the standard formula,

$$f = \frac{1}{2L}\sqrt{\frac{Tg}{w}}.\Big)$$

▲

The solution to our example as given in Table 9.1 is exactly the analytical solution, as we will now demonstrate. We will compare our finite-difference equation (Eq. 9.2) with the d'Alembert solution to the wave equation in the next section.

9.3 Comparison to the d'Alembert Solution

The method of d'Alembert lets us find the analytical solution to the wave equation in one dimension. The one-dimensional wave equation, with $\sqrt{Tg/w} = c$, is

$$\frac{\partial^2 y}{\partial t^2} = c^2 \frac{\partial^2 y}{\partial x^2}. \qquad (9.5)$$

By direct substitution, it is readily seen that, for any arbitrary functions F and G, Eq. (9.5) is solved by

$$y(x,t) = F(x + ct) + G(x - ct). \qquad (9.6)$$

The demonstration is easy, since

$$\frac{\partial y}{\partial t} = F'\frac{\partial(x + ct)}{\partial t} + G'\frac{\partial(x - ct)}{\partial t} = cF' - cG',$$

$$\frac{\partial^2 y}{\partial t^2} = c^2 F'' - c(-c)G'' = c^2 F'' + c^2 G''; \qquad (9.7)$$

$$\frac{\partial y}{\partial x} = F' \frac{\partial(x + ct)}{\partial x} + G' \frac{\partial(x - ct)}{\partial x} = F' + G',$$

$$\frac{\partial^2 y}{\partial x^2} = F'' + G''. \tag{9.8}$$

Equation (9.5) results immediately from Eqs. (9.7) and (9.8).

The solution to the problem is found, then, if we can find a pair of functions of the form of (9.6) whose sum matches the initial conditions and the boundary conditions for the problem. If the initial conditions are

$$y(x, 0) = f(x),$$

$$\frac{\partial y}{\partial t}(x, 0) = g(x),$$

it is again readily seen that $y(x, t)$ is given by

$$y(x, t) = \frac{1}{2}[f(x + ct) + f(x - ct)] + \frac{1}{2c} \int_{x - ct}^{x + ct} g(v)\, dv. \tag{9.9}$$

Note that we can rewrite Eq. (9.9) in the form of Eq. (9.6). In Eq. (9.9) we have changed to the dummy variable v under the integral sign. The demonstration parallels that preceding when we recall how to differentiate an integral:

$$\frac{\partial}{\partial t}\left[\frac{1}{2c} \int_{x - ct}^{x + ct} g(v)\, dv\right] = \frac{1}{2c}[cg(x + ct) - (-c)g(x - ct)],$$

$$\frac{\partial^2}{\partial t^2}\left[\frac{1}{2c} \int_{x - ct}^{x + ct} g(v)\, dv\right] = \frac{\partial}{\partial t}\left[\frac{1}{2}g(x + ct) + \frac{1}{2}g(x - ct)\right] = \frac{1}{2}cg' - \frac{1}{2}cg' = 0,$$

$$\frac{\partial}{\partial x}\left[\frac{1}{2c} \int_{x - ct}^{x + ct} g(v)\, dv\right] = \frac{1}{2c}[g(x + ct) - g(x - ct)],$$

$$\frac{\partial^2}{\partial x^2}\left[\frac{1}{2c} \int_{x - ct}^{x + ct} g(v)\, dv\right] = \frac{\partial}{\partial x}\left[\frac{1}{2c}g(x + ct) - \frac{1}{2c}g(x - ct)\right] = \frac{1}{2c}g' - \frac{1}{2c}g' = 0.$$

Equation (9.9) gives the value of y at any time t provided that y is known at points ct to the right and left of the point at $t = 0$, and in terms of the integral of the initial velocity between the lateral points. Hence it is useful to find the displacement of points in the interior portions of a vibrating string.

We are now in a position to verify that our numerical procedure, given by Eq. (9.2), gives the exact solution (except for round-off), provided that two lines of correct y-values are known. The algorithm we use for the numerical procedure is

$$y_i^{j+1} = y_{i-1}^j + y_{i+1}^j - y_i^{j-1}. \tag{9.10}$$

We now show that the solution to this difference equation is a solution to the differential equation. Consider the function

$$y_i^j = F(x_i + ct_j) + G(x_i - ct_j). \tag{9.11}$$

Because we use

$$\frac{Tg(\Delta t)^2}{w(\Delta x)^2} = 1 = \frac{c^2(\Delta t)^2}{(\Delta x)^2},$$

then

$$\Delta x = c\,\Delta t.$$

Write x_i and t_j in terms of starting values, x_0, and $t_0 = 0$:

$$x_i = x_0 + i\,\Delta x,$$
$$ct_j = c(t_0 + j\,\Delta t) = cj\,\Delta t = j\,\Delta x.$$

Substituting into Eq. (9.11), we get

$$y_i^j = F(x_0 + i\,\Delta x + j\,\Delta x) + G(x_0 + i\,\Delta x - j\,\Delta x)$$
$$= F[x_0 + (i + j)\Delta x] + G[x_0 + (i - j)\Delta x].$$

Use this relation to rewrite the right-hand side of the difference equation, Eq. (9.10):

$$
\begin{aligned}
y_{i-1}^j + y_{i+1}^j - y_i^{j-1} = {}& F[x_0 + (i - 1 + j)\Delta x] \\
& + G[x_0 + (i - 1 - j)\Delta x] \\
& + F[x_0 + (i + 1 + j)\Delta x] \\
& + G[x_0 + (i + 1 - j)\Delta x] \\
& - F[x_0 + (i + j - 1)\Delta x] \\
& - G[x_0 + (i - j + 1)\Delta x] \\
= {}& F[x_0 + (i + j + 1)\Delta x] \\
& + G[x_0 + (i - j - 1)\Delta x] = y_i^{j+1}. \qquad \textbf{(9.12)}
\end{aligned}
$$

Equation (9.12) shows that, if the previous two lines of the numerical solution are of the form of Eq. (9.11) (and hence must be exact solutions of the wave equation in view of Eq. (9.6)), it follows that the values on the next line are exact solutions also, because they also are of the form of Eq. (9.11).

For our simple algorithm to give the analytical solution, it is then necessary only that two lines of the computation be correct. The first line is correct because it is given by the initial conditions. While Eq. (9.4) is frequently recommended for giving the second line in terms of the initial velocity, it is sometimes inaccurate because it assumes that the initial velocity and the average velocity from t_{-1} to t_1 are the same, and this may not be true.*

Equation (9.9) offers a more exact way to get the second line of y-values. If we employ that relationship at $t = \Delta t$, we have

---

* The relationship of Eq. (9.4) is correct whenever $g(x) = $ constant. It is therefore correct for the case of $g(x) = 0$.

$$y(x_i, \Delta t) = y_i^1 = \frac{1}{2}[f(x_i + \Delta x) + f(x_i - \Delta x)] + \frac{1}{2c}\int_{x_i - \Delta x}^{x_i + \Delta x} g(v)\, dv$$

$$= \frac{1}{2}[y_{i+1}^0 + y_{i-1}^0] + \frac{1}{2c}\int_{x_{i-1}}^{x_{i+1}} g(v)\, dv. \qquad (9.13)$$

Equation (9.13) differs from Eq. (9.4) only in the last term. We now see why Example 9.1 gave correct values in spite of using Eq. (9.4). In the example, $g(x)$ was everywhere zero, so the form of the last term was unimportant. In the next example, we illustrate the difference between the two different ways to begin the solution.

EXAMPLE 9.2   A string whose length is 9 units is initially in its equilibrium position. It is set into motion by striking it so that its initial velocity is given by $\partial y/\partial t = 3\sin(\pi x/L)$. Take $\Delta x = 1$ unit and assume that $c = \sqrt{Tg/w} = 2$. When the ratio of $c^2(\Delta t)^2/(\Delta x)^2$ is unity, $\Delta t = 0.5$ time unit.

If the ends are fixed, find the displacements at $t = 0.5$ time unit later. The length is subdivided into nine intervals because $\Delta x = 1$.

The displacements we require are after one time step. Table 9.2 summarizes the computations. We first compute them using Eq. (9.4). The values disagree by several percent from the analytical values, which are given by

$$y(x, t) = \frac{1}{2c}\int_{x-ct}^{x+ct} 3\sin\frac{\pi v}{L}\, dv = \frac{3L}{2c\pi}\left\{\cos\left(\frac{\pi x}{L} - \frac{\pi ct}{L}\right) - \cos\left(\frac{\pi x}{L} + \frac{\pi ct}{L}\right)\right\}.$$

Table 9.2   Comparison of ways to begin the solution of the wave equation in one dimension:

$$\frac{\partial^2 y}{\partial t^2} = c^2\frac{\partial^2 y}{\partial x^2} \qquad \text{over } x = 0 \text{ to } x = 9, \text{ with } c = 2;$$

$$y(x, 0) = 0, \qquad \frac{\partial y}{\partial t}(x, 0) = 3\sin\left(\frac{\pi x}{9}\right)$$

	Value of displacements at $t = 0.5$			
Using Eq. (9.4), $\Delta x = 1$:				
x	1	2	3	4
y	0.5130	0.9642	1.2990	1.4772
Using Eq. (9.4), $\Delta x = 0.5$:				
x	1	2	3	4
y	0.5052	0.9495	1.2793	1.4548
Using Eq. (9.13), $\Delta x = 1$, Simpson's-rule integration:				
x	1	2	3	4
y	0.5027	0.9448	1.2729	1.4475
Analytical solution:				
x	1	2	3	4
y	0.50267	0.94472	1.27282	1.44740

When Δx is cut in half, Eq. (9.4) gives improved values; it now requires two time steps to reach $t = 0.5$, of course.

Using Eq. (9.13), and evaluating the integral with Simpson's rule, we obtain results with $\Delta x = 1$ that are accurate to within one in the fourth decimal place.

9.4 Method of Characteristics

The properties of the solution to the wave equation are further elucidated by considering the "characteristic curves" of the equation. This will also permit us to apply numerical methods to more general hyperbolic equations.

Consider the second-order partial-differential equation in two variables x and t:

$$au_{xx} + bu_{xt} + cu_{tt} + e = 0. \tag{9.14}$$

Here we have used the subscript notation to represent partial derivatives. The coefficients a, b, c, and e may be functions of x, t, u_x, u_t, and u, so the equation is very general.* We take $u_{xt} = u_{tx}$. To facilitate manipulations, let

$$p = \frac{\partial u}{\partial x} = u_x, \qquad q = \frac{\partial u}{\partial t} = u_t.$$

Write out the differentials of p and q:

$$dp = \frac{\partial p}{\partial x}\,dx + \frac{\partial p}{\partial t}\,dt = u_{xx}\,dx + u_{xt}\,dt,$$

$$dq = \frac{\partial q}{\partial x}\,dx + \frac{\partial q}{\partial t}\,dt = u_{xt}\,dx + u_{tt}\,dt.$$

Solving these last equations for u_{xx} and u_{tt}, respectively, we have

$$u_{xx} = \frac{dp - u_{xt}\,dt}{dx} = \frac{dp}{dx} - u_{xt}\frac{dt}{dx},$$

$$u_{tt} = \frac{dq - u_{xt}\,dx}{dt} = \frac{dq}{dt} - u_{xt}\frac{dx}{dt}.$$

Substituting in Eq. (9.14) and rearranging, we obtain

$$-au_{xt}\frac{dt}{dx} + bu_{xt} - cu_{xt}\frac{dx}{dt} + a\frac{dp}{dx} + c\frac{dq}{dt} + e = 0.$$

---

* When the coefficients are independent of the function u or its derivatives, it is linear. If they are functions of u, u_x, or u_t (but not u_{xx} or u_{tt}), it is called quasilinear.

Now multiplying by $-dt/dx$, we get

$$u_{xt}\left[a\left(\frac{dt}{dx}\right)^2 - b\left(\frac{dt}{dx}\right) + c\right] - \left[a\frac{dp}{dx}\frac{dt}{dx} + c\frac{dq}{dx} + e\frac{dt}{dx}\right] = 0.$$

Suppose, in the xt-plane, we define curves such that the first bracketed expression is zero. On such curves, the original differential equation is equivalent to setting the second bracketed expression equal to zero. That is, if

$$am^2 - bm + c = 0, \tag{9.15}$$

where $m = dt/dx$, then the solution to the original equation (Eq. 9.14) can be found by solving

$$am\,dp + c\,dq + e\,dt = 0. \tag{9.16}$$

We have elected to write Eq. (9.16) in the form of differentials. It will be seen that we reduce the original problem, which is a second-order partial-differential equation, to solving a pair of first-order equations of the form of Eq. (9.16).

The curves whose slope m is given by Eq. (9.15) are called the *characteristics* of the differential equation. Since the equation is a quadratic, it may have one, two, or no real solutions, depending on the value of $b^2 - 4ac$. The value of this discriminant is the usual basis for classifying partial-differential equations. If

$$b^2 - 4ac < 0,$$

the equation is called *elliptic*, and there are no (real) characteristics. If $b^2 - 4ac = 0$, there is a single characteristic at any point, and the equation is termed *parabolic*. When $b^2 - 4ac > 0$, at every point there will be a pair of characteristic curves whose slopes are given by the two distinct, real roots of Eq. (9.15). Such equations are called *hyperbolic*, and our present discussion considers only this type.

Along the characteristics, the solution has special and desirable properties. For example, discontinuities in the initial conditions are propagated along them. On the characteristic curves, the numerical solution can be developed in the general case also.

We will now outline a method of solving equations of the form of Eq. (9.14) by numerical integration along the characteristics.* We visualize the initial conditions

---

* We discuss only the solution of hyperbolic differential equations by the method of characteristics, but the technique can also be applied to parabolic equations.

as specifying the function u on some curve in the tx-plane,* as well as its normal derivative. Consider two points, P and Q on this initial curve (Fig. 9.3). When Eq. (9.14) is hyperbolic, there are two characteristic curves through each point. The rightmost curve through P intersects the leftmost curve through Q, and these curves are such that their slopes are given by the appropriate roots of Eq. (9.15). Call m_+ the values of the slope on curve PR, and m_- the values on curve QR.

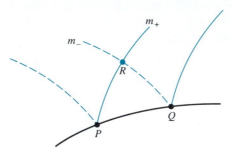

Figure 9.3

Since these curves are characteristics, the solution to the problem can be found by solving Eq. (9.16) along them.

Our procedure will be to first find point R (perhaps only as a first approximation if a, b, or c involves the unknown function u). This is done by solving the equation

$$\Delta t = \left(\frac{dt}{dx}\right)_{\text{av}} \Delta x = m_{\text{av}} \Delta x, \tag{9.17}$$

applied over the arcs PR and QR simultaneously. When dt/dx is a function of x and/or t, it may be possible to integrate Eq. (9.17) analytically. When dt/dx varies with u, we will use a procedure resembling the Euler predictor–corrector method, by predicting with m_{av} taken as equal to m_+ at P or m_- at Q to start the solution. We correct by using the arithmetic average of m at the endpoints of each arc as soon as the value of m at R can be evaluated.

We then integrate Eq. (9.16) in the form

$$a_{\text{av}} m_{\text{av}} \Delta p + c_{\text{av}} \Delta q + e_{\text{av}} \Delta t = 0,$$

starting first from P and then from Q, using the appropriate values of m for each.

---

* This curve must not itself be one of the characteristics, or advancing the solution would be impossible.

This will estimate p and q at point R. Finally we evaluate the function u at R from

$$du = \frac{\partial u}{\partial x} dx + \frac{\partial u}{\partial t} dt,$$

used in the form

$$\Delta u = p_{av} \Delta x + q_{av} \Delta t.$$

The equation for Δu can be applied to either the change along

$$P \rightarrow R \quad \text{or} \quad Q \rightarrow R;$$

when these do not give the same result, we will *average* the two values. It may be necessary to iterate this procedure to get improved values at R. Average values are used for all varying quantities. The calculations are repeated for a second point S that is the intersection of characteristics through Q and another initial point W, and then continued in a like manner to evaluate u throughout the region in the xt-plane as desired. We illustrate with three examples, first with dt/dx equal to a constant, then with dt/dx varying with x, and finally a more complex example with dt/dx varying with u.

For the simple wave equation,

$$u_{tt} = c^2 u_{xx},$$

the slopes of the characteristics are

$$m = \pm \frac{1}{c},$$

and the characteristics are the lines

$$t = \pm \frac{1}{c}(x - x_i).$$

Consider the curves from points P and Q, taken on the line for $t = 0$ (Fig. 9.4). The network of points used in the explicit finite-difference method of Section 9.2 are seen to be the intersections of characteristics through pairs of points spaced $2\Delta x$ apart.

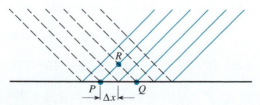

Figure 9.4

The finite-difference method, with

$$\frac{Tg(\Delta x)^2}{w(\Delta t)^2} = 1,$$

will be found to be the equivalent of integration along the characteristics, lending further support to the likelihood that it will give exact answers.

EXAMPLE 9.3   Solve

$$\frac{\partial^2 u}{\partial t^2} = 2\frac{\partial^2 u}{\partial x^2} - 4,$$

with initial conditions

$$u = 12x \qquad \text{for} \quad 0 \le x \le 0.25,$$
$$u = 4 - 4x \qquad \text{for} \quad 0.25 \le x \le 1.0,$$
$$\frac{\partial u}{\partial t} = 0 \qquad \text{for} \quad 0 \le x \le 1.0;$$

boundary conditions are $u = 0$ at $x = 0$ and at $x = 1.0$.
  Putting the equation into the standard form,

$$a\frac{\partial^2 u}{\partial x^2} + b\frac{\partial^2 u}{\partial x\,\partial t} + c\frac{\partial^2 u}{\partial t^2} + e = 0,$$

gives $a = -2$, $b = 0$, $c = 1$, and $e = 4$. (The equation is linear since a, b, c, and e are independent of u, u_x, and u_t.)
  The slopes of the characteristics are the roots of

$$-2m^2 + 1 = 0,$$
$$m = \pm\frac{\sqrt{2}}{2},$$

so the characteristic curves are straight lines in the xt-plane, as shown in Fig. 9.5. Consider points P, Q, and R—$(0.25, 0)$, $(0.75, 0)$, and $(0.5, 0.1768)$—and solve Eq. (9.16), which is

$$am_{av}\,\Delta p + c\,\Delta q + e\,\Delta t = 0.$$

Along $P \to R$: $-2\left(\dfrac{\sqrt{2}}{2}\right)\Delta p + \Delta q + 4\Delta t = 0,$

  $-\sqrt{2}(p_R - p_P) + (q_R - q_P) + 4(0.1768) = 0;$

Along $Q \to R$: $-2\left(\dfrac{-\sqrt{2}}{2}\right)\Delta p + \Delta q + 4\Delta t = 0,$

  $\sqrt{2}(p_R - p_Q) + (q_R - q_Q) + 4(0.1768) = 0.$

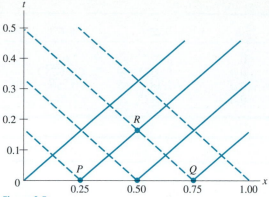

Figure 9.5

Using

$$p_P = \left(\frac{\partial u}{\partial x}\right)_P = -4,^* \qquad q_P = \left(\frac{\partial u}{\partial t}\right)_P = 0, \qquad p_Q = \left(\frac{\partial u}{\partial x}\right)_Q = -4,$$

$$q_Q = \left(\frac{\partial u}{\partial t}\right)_Q = 0,$$

we find $p_R = -4$, $q_R = -\sqrt{2}/2$ by solving the equations $P \rightarrow R$ and $Q \rightarrow R$ simultaneously.

Now we evaluate u at point R through its change along $P \rightarrow R$:

$$\Delta u = p_{av}\,\Delta x + q_{av}\,\Delta t = -4(0.25) + \left(\frac{0 - \sqrt{2}/2}{2}\right)(0.1768)$$

$$= -1.0625;$$

$$u_R = 3 + (-1.0625) = 1.9375.$$

(If we compute through evaluating Δu along $Q \rightarrow R$, we get the same result.)

For this simple problem, the finite-difference method is much simpler, and we expect it to give the same results. Following the procedure of Section 9.2,[†] we compute with $\Delta x = 0.25$, $\Delta t = \Delta x / \sqrt{c} = 0.1768$, and obtain Table 9.3. The circled value agrees exactly with that calculated by the method of characteristics.

---

[*] The gradient has a discontinuity at $x = 0.25$. The value of $\partial u / \partial x$ for points to the right of P applies for the region PRQ.

[†] The algorithm is $u_i^{j+1} = (u_{i+1}^j + u_{i-1}^j) - u_i^{j-1} - 4(\Delta t)^2$ with $\Delta t = \Delta x / \sqrt{2}$. For the first time step, $u_i^1 = \frac{1}{2}(u_{i+1}^0 + u_{i-1}^0) - \frac{1}{2}(4)(\Delta t)^2$.

Table 9.3

x	0	0.25	0.5	0.75	1.0
$u(t = 0)$	0.0	3.0	2.0	1.0	0.0
$u(t = 0.1768)$	0.0	0.9375	1.9375	0.9375	0.0
$u(t = 0.3535)$	0.0	−1.1875	−0.2500	0.8125	0.0
$u(t = 0.5303)$	0.0	−1.3125	−2.4375	−1.3125	0.0

EXAMPLE 9.4   Solve

$$\frac{\partial^2 u}{\partial t^2} = (1 + 2x)\frac{\partial^2 u}{\partial x^2}$$

over $(0, 1)$ with fixed boundaries and the initial conditions

$$u(x, 0) = 0, \qquad \frac{\partial u}{\partial t}(x, 0) = x(1 - x).$$

For this problem, $a = -(1 + 2x), b = 0, c = 1, e = 0$. Then $am^2 + bm + c = 0$ gives

$$m = \pm\sqrt{\frac{1}{(1 + 2x)}}.$$

The characteristic curves are found by solving the differential equations $dt/dx = \sqrt{1/(1 + 2x)}$ and $dt/dx = -\sqrt{1/(1 + 2x)}$. Integrating* from the initial point x_0 and t_0, we have

$$t = t_0 + \sqrt{1 + 2x} - \sqrt{1 + 2x_0} \quad \text{from } m_+,$$
$$t = t_0 - \sqrt{1 + 2x} + \sqrt{1 + 2x_0} \quad \text{from } m_-.$$

Figure 9.6 shows several of the characteristic curves. We select two points on the initial curve for $t = 0$, at $P = (0.25, 0)$ and $Q = (0.75, 0)$, whose characteristics intersect at point R. Solving for the intersection, we find $R = (0.4841, 0.1782)$.

We now solve Eq. (9.6) to obtain $p = \partial u/\partial x$ and $q = \partial u/\partial t$ at R:

At point P: $x = 0.25, \quad t = 0, \quad u = 0, \quad p = \left(\dfrac{\partial u}{\partial x}\right)_P = 0,$

$$q = \left(\frac{\partial u}{\partial t}\right)_P = x - x^2 = 0.1875,$$

$$m = \sqrt{\frac{1}{(1 + 2x)}} = 0.8165,$$

$$a = -(1 + 2x) = -1.5, \quad b = 0, \quad c = 1, \quad e = 0.$$

---

* In this example, the integration methods of calculus are easy to use. We could use a numerical method if they were not.

At point Q: $x = 0.75$, $t = 0$, $u = 0$, $p = 0$, $q = x - x^2 = 0.1875$,

$$m = -\sqrt{\frac{1}{(1 + 2x)}} = -0.6325,$$

$$a = -2.5, \quad b = 0, \quad c = 1, \quad e = 0.$$

At point R: $x = 0.4841$, $t = 0.1783$,

$$m_+ = \sqrt{\frac{1}{(1 + 2x)}} = 0.7128,$$

$$m_- = -\sqrt{\frac{1}{(1 + 2x)}} = -0.7128,$$

$$a = -1.9682, \quad b = 0, \quad c = 1, \quad e = 0.$$

Equation (9.16) becomes, when we use average values for a and m,

$P \rightarrow R$: $-1.7341(0.7646)(p_R - 0) + (1)(q_R - 0.1875) = 0$;

$Q \rightarrow R$: $-2.2341(-0.6726)(p_R - 0) + (1)(q_R - 0.1875) = 0$.

Solving simultaneously, we get $p_R = 0$, $q_R = 0.1875$.
 We calculate the change in u along the characteristics:

$P \rightarrow R$: $\Delta u = 0(0.2341) + 0.1875(0.1783) = 0.0334$,

$Q \rightarrow R$: $\Delta u = 0(-0.2659) + 0.1875(0.1783) = 0.0334$,

$u_R = 0 + 0.0334 = 0.0334$.

Figure 9.6 gives the results at several other intersections of characteristics. Students should verify these results to be sure they understand the method of characteristics.

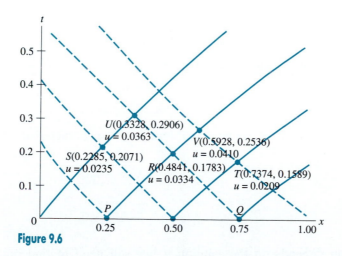

Figure 9.6

EXAMPLE 9.5   Solve the quasilinear equation, with conditions as shown, by numerical integration along the characteristics. (This might be a vibrating string with tension related to the displacement u and subject to an external lateral force.)

$$\frac{\partial^2 u}{\partial x^2} - u\frac{\partial^2 u}{\partial t^2} + (1 - x^2) = 0, \qquad u(x,0) = x(1 - x), \qquad u_t(x,0) = 0,$$

$$u(0,t) = 0, \qquad u(1,t) = 0. \tag{9.18}$$

We will advance the solution beyond the start from P, at $x = 0.2, t = 0$, and Q, at $x = 0.4, t = 0$, to one new point R. Comparing Eq. (9.18) to the standard form,

$$au_{xx} + bu_{xt} + cu_{tt} + e = 0,$$

we have $a = 1, b = 0, c = -u, e = 1 - x^2$. We first compute u, p, and q at points P and Q,

$$u = x(1 - x)$$

(from the initial conditions), so

$$u_P = 0.2(1 - 0.2) = 0.16,$$
$$u_Q = 0.4(1 - 0.4) = 0.24.$$

Also,

$$p = \frac{\partial u}{\partial x} = 1 - 2x$$

(by differentiating the initial conditions), so

$$p_P = 1 - 2(0.2) = 0.6,$$
$$p_Q = 1 - 2(0.4) = 0.2;$$

and

$$q = \frac{\partial u}{\partial t} = 0$$

(from the initial conditions), so

$$q_P = 0,$$
$$q_Q = 0.$$

To locate point R, we need the slope m of the characteristic. Using $am^2 - bm + c = 0$, we get

$$m = \frac{b \pm \sqrt{b^2 - 4ac}}{2a},$$

$$m = \frac{\pm\sqrt{4u}}{2} = \pm\sqrt{u}.$$

Since m depends on the solution u, we will need to find point R through the predictor–corrector approach. In the first trial, use the initial values over the whole arc; that is, take $m_+ = +m_P$ and $m_- = -m_Q$:

$$m_+ = \sqrt{u_P} = \sqrt{0.16} = 0.4,$$
$$m_- = -\sqrt{u_Q} = -\sqrt{0.24} = -0.490.$$

We now estimate the coordinates of R by solving simultaneously

$$t_R = m_+(x_R - x_P) = 0.4(x_R - 0.2),$$
$$t_R = m_-(x_R - x_Q) = -0.490(x_R - 0.4).$$

These give

$$x_R = 0.310, \qquad t_R = 0.044.$$

We write Eq. (9.16) along each characteristic, again using the initial values of m, since m at R is still unknown:

$$am\,\Delta p + c\,\Delta q + e\,\Delta t = 0,$$

$$(1)(0.4)(p_R - 0.6) + (-0.16)(q_R - 0) + \left(1 - \frac{0.04 + 0.096}{2}\right)(0.044) = 0,$$

$$(1)(-0.490)(p_R - 0.2) + (-0.24)(q_R - 0) + \left(1 - \frac{0.16 + 0.096}{2}\right)(0.044) = 0.$$

In these equations we used the arithmetic average of x^2 in the last terms. Solving simultaneously, we get

$$p_R = 0.399, \qquad q_R = -0.246.$$

As a first approximation for u at R, then,

$$\Delta u = p\,\Delta x + q\,\Delta t,$$

$$u_R - 0.16 = \frac{0.6 + 0.399}{2}(0.310 - 0.2) + \frac{0 - 0.246}{2}(0.044 - 0),$$

$$u_R = 0.2095.$$

The last computation was along PR, using average values of p and q. We could have alternatively proceeded along QR. If this is done,

$$u_R - 0.24 = \frac{0.2 + 0.399}{2}(0.310 - 0.4) + \frac{0 - 0.246}{2}(0.044 - 0),$$

$$u_R = 0.2076.$$

The two values should be close to each other. Let us use the average value, 0.2086, as our initial estimate of u_R. We now repeat the work. In getting the coordinates of R, we now use average values of the slopes,

$$t_R = \frac{0.4 + \sqrt{0.2086}}{2}(x_R - 0.2),$$

$$t_R = \frac{-0.490 - \sqrt{0.2086}}{2}(x_R - 0.4),$$

$$x_R = 0.305, \qquad t_R = 0.045;$$

$$(1)\left(\frac{0.4 + \sqrt{0.2086}}{2}\right)(p_R - 0.6) - \left(\frac{0.16 + 0.2086}{2}\right)(q_R - 0)$$

$$+ \left(1 - \frac{0.04 + 0.0930}{2}\right)(0.045) = 0,$$

$$(1)\left(\frac{-0.490 - \sqrt{0.2086}}{2}\right)(p_R - 0.2) - \left(\frac{0.24 + 0.2086}{2}\right)(q_R - 0)$$

$$+ \left(1 - \frac{0.16 + 0.0930}{2}\right)(0.045) = 0,$$

$$p_R = 0.398, \qquad q_R = -0.242;$$

$$u_R = 0.16 + \frac{0.6 + 0.398}{2}(0.305 - 0.2) + \frac{0 - 0.242}{2}(0.045 - 0),$$

$$u_R = 0.2071 \qquad \text{(along } PR);$$

$$u_R = 0.24 + \frac{0.2 + 0.398}{2}(0.305 - 0.4) + \frac{0 - 0.242}{2}(0.045 - 0),$$

$$u_R = 0.2063 \qquad \text{(along } QR).$$

The average value is 0.2067.

Another round of calculations gives $u_R = 0.2066$, which checks the previous value sufficiently. This method is, of course, very tedious by hand. ▲

9.5 The Wave Equation in Two Space Dimensions

The finite-difference method can be applied to hyperbolic partial-differential equations in two or more space dimensions. A typical problem is the vibrating membrane. Consider a thin flexible membrane stretched over a rectangular frame and set to vibrating. A development analogous to that for the vibrating string gives

$$\frac{\partial^2 u}{\partial t^2} = \frac{Tg}{w}\left(\frac{\partial^2 u}{\partial x^2} + \frac{\partial^2 u}{\partial y^2}\right),$$

in which u is the displacement, t is the time, x and y are the space coordinates, T is the uniform tension per unit length, g is the acceleration of gravity, and w is the weight per unit area. For simplification, let $Tg/w = \alpha^2$. Replacing each derivative by its central-difference approximation, and using $h = \Delta x = \Delta y$, gives (we recognize the Laplacian on the right-hand side)

$$\frac{u_{i,j}^{k+1} - 2u_{i,j}^k + u_{i,j}^{k-1}}{(\Delta t)^2} = \alpha^2 \frac{u_{i+1,j}^k + u_{i-1,j}^k + u_{i,j+1}^k + u_{i,j-1}^k - 4u_{i,j}^k}{h^2}. \tag{9.19}$$

Solving for the displacement at time t_{k+1}, we obtain

$$u_{i,j}^{k+1} = \frac{\alpha^2(\Delta t)^2}{h^2}\left\{1 \quad \begin{matrix} 1 \\ 0 \\ 1 \end{matrix} \quad 1\right\}u_{i,j}^k - u_{i,j}^{k-1} + \left(2 - 4\frac{\alpha^2(\Delta t)^2}{h^2}\right)u_{i,j}^k. \tag{9.20}$$

In Eqs. (9.19) and (9.20), we use superscripts to denote the time. If we let $\alpha^2(\Delta t)^2/h^2 = \frac{1}{2}$, the last term vanishes and we get

$$u_{i,j}^{k+1} = \frac{1}{2}\left\{1 \quad \begin{matrix} 1 \\ 0 \\ 1 \end{matrix} \quad 1\right\}u_{i,j}^k - u_{i,j}^{k-1}. \tag{9.21}$$

For the first time step, we get displacements from Eq. (9.22), which is obtained by approximating $\partial u/\partial t$ at $t = 0$ by a central-difference approximation involving $u_{i,j}^1$ and $u_{i,j}^{-1}$.

$$u_{i,j}^1 = \frac{1}{4}\left\{1 \quad \begin{matrix} 1 \\ 0 \\ 1 \end{matrix} \quad 1\right\}u_{i,j}^0 + (\Delta t)g(x_i, y_j). \tag{9.22}$$

In Eq. (9.22), $g(x, y)$ is the initial velocity.

It should not surprise us to learn that this ratio $\alpha^2(\Delta t)^2/h^2 = \frac{1}{2}$ is the maximum value for stability, in view of our previous experience with explicit methods. However, in contrast with the wave equation in one space dimension, we do not get exact answers from the numerical procedure of Eq. (9.21), and we further observe that we must use smaller time steps in relation to the size of the space interval. Therefore we advance in time more slowly. However, the numerical method is straightforward, as the following example will show.

EXAMPLE 9.6   A membrane for which $\alpha^2 = Tg/w = 3$ is stretched over a square frame that occupies the region $0 \le x \le 2, 0 \le y \le 2$, in the xy-plane. It is given an initial displacement described by

$$u = x(2 - x)y(2 - y),$$

and has an initial velocity of zero. Find how the displacement varies with time.

We divide the region with $h = \Delta x = \Delta y = \frac{1}{2}$, obtaining nine interior nodes. Initial displacements are calculated from the initial conditions: $u^0(x, y) = x(2 - x)y(2 - y)$; Δt is taken at its maximum value for stability, $h/(\sqrt{2}\alpha) = 0.2041$. The values at the end of one time step are given by

$$u_{i,j}^1 = \frac{1}{4}\left\{1 \quad \begin{matrix} 1 \\ 0 \\ 1 \end{matrix} \quad 1\right\}u_{i,j}^0$$

because $g(x, y)$ in Eq. (9.22) is everywhere zero. For succeeding time steps, Eq. (9.21) is used. Table 9.4 gives the results of our calculations. Also shown in Table 9.4 (in parentheses) are analytical values, computed from the double infinite series:

$$u(x, y, t) = \sum_{m=1}^{\infty} \sum_{n=1}^{\infty} B_{mn} \sin\frac{m\pi x}{a} \sin\frac{n\pi y}{b} \cos\left(\alpha\pi t\sqrt{\frac{m^2}{a^2} - \frac{n^2}{b^2}}\right),$$

$$B_{mn} = \frac{16a^2 b^2 A}{\pi^6 m^3 n^3}(1 - \cos m\pi)(1 - \cos n\pi),$$

which gives the displacement of a membrane fastened to a rectangular framework, $0 \le x \le a, 0 \le y \le b$, with initial displacements of $Ax(a - x)y(b - y)$.

We observe that the finite-difference results do not agree exactly with the analytical calculations. The finite-difference values are symmetrical with respect to position and repeat themselves with a regular frequency. The very regularity of the values itself indicates that the finite-difference computations are in error, since they predict that the membrane could emit a musical note. We know from experience that a drum does not give a musical tone when struck; therefore the vibrations do not have a cycle pattern of constant frequency, as exhibited by our numerical results.

Decreasing the ratio of $\alpha^2(\Delta t)^2/h^2$ and using Eq. (9.20) gives little or no improvement in the average accuracy; to approach closely to the analytical results, $h = \Delta x = \Delta y$ must be made smaller. When this is done, Δt will need to decrease in proportion, requiring many time steps and leading to many repetitions of the

Table 9.4   Displacements of a vibrating membrane—finite-difference method: $\Delta t = h/(\sqrt{2}\alpha)$

				Grid location				
t	$(0.5, 0.5)$	$(1.0, 0.5)$	$(1.5, 0.5)$	$(0.5, 1.0)$	$(1.0, 1.0)$	$(1.5, 1.0)$	$(0.5, 1.5)$	$(1.0, 1.5)$	$(1.5, 1.5)$
0	0.5625 (0.5625)	0.750 (0.750)	0.5625	0.750	1.000 (1.000)	0.750	0.5625	0.750	0.5625
0.204	0.375 (0.380)	0.531 (0.536)	0.375	0.531	0.750 (0.755)	0.531	0.375	0.531	0.375
0.408	−0.031 (−0.044)	0.000 (−0.009)	−0.031	0.000	0.062 (0.083)	0.000	−0.031	0.000	−0.031
0.612	−0.375 (−0.352)	−0.531 (−0.539)	−0.375	−0.531	−0.750 (−0.813)	−0.531	−0.375	−0.531	−0.375
0.816	−0.500 (−0.502)	−0.750 (−0.746)	−0.500	−0.750	−1.125 (−1.114)	−0.750	−0.500	−0.750	−0.500
1.021	−0.375 (−0.407)	−0.531 (−0.535)	−0.375	−0.531	−0.750 (−0.691)	−0.531	−0.375	−0.531	−0.375
1.225	−0.031 (−0.015)	0.000 (0.008)	−0.031	0.000	0.062 (0.030)	0.000	−0.031	0.000	−0.031
1.429	0.375 (0.410)	0.531 (0.534)	0.375	0.531	0.750 (0.688)	0.531	0.375	0.531	0.375

Note: Analytical values are in parentheses.

algorithm and extravagant use of computer time. One remedy is the use of implicit methods, which allow the use of larger ratios of $\alpha^2(\Delta t)^2/h^2$. However, with many nodes in the xy-grid, this requires large, sparse matrices similar to the Crank–Nicolson method for parabolic equations in two space dimensions. A.D.I. methods have been used for hyperbolic equations—tridiagonal systems result. We do not discuss these methods.

9.6 Finite Elements and the Wave Equation

We will only outline how finite elements are applied to the wave equation, because this topic is too complex for full coverage here. Just as for the heat equation, finite elements are used for the space region and finite differences for time derivatives. We will develop only the vibrating string case (one dimension); two or three space dimensions are handled analogously but are harder to follow.

The equation that is usually solved is a more general case of the simple wave equation we have been discussing. In engineering applications, damping forces that serve to decrease the amplitude of the vibrations are important, and external forces that excite the system are usually involved. We therefore use, for a one-dimensional case, this equation for the displacement of points on the vibrating string, $y(x,t)$:

$$\frac{\partial}{\partial x}\left[T(x)\frac{\partial y}{\partial x}\right] - h(x)\frac{\partial y}{\partial t} + F(x,t) = \frac{w(x)}{g}\frac{\partial^2 y}{\partial t^2}. \tag{9.23}$$

Here, T represents the tension, which is allowed to vary with x; h represents a damping coefficient that opposes motion in proportion to the velocity; F is the external force; and w/g is the mass density. There are boundary conditions (at $x = a$ and $x = b$) as well as initial conditions that specify initial displacements and velocities.

The approach is essentially identical to that used for unsteady-state heat flow: Apply finite elements to x and finite differences to the time derivatives. We will use linear one-dimensional elements, so we subdivide $[a, b]$ into portions (elements) that join at points that we call nodes. Within each element, we approximate $y(x,t)$ with $v(x,t)$,

$$y(x,t) \approx v(x,t) = N_L c_L + N_R c_R, \tag{9.24}$$

where c_L and c_R are the approximations to the displacements at the nodes at the left and right ends of a typical linear element. The N's are shape functions (in this one-dimensional case, we have called them "hat functions").

By using the Galerkin procedure, we can get this integral equation, which we will eventually transform into the element equations:

$$\int_L^R N_i\left(\frac{w}{g}\right)y_{tt}\,dx + \int_L^R N_i'(T)y_x\,dx + \int_L^R N_i(h)y_t\,dx$$
$$= \int_L^R N_i(F)\,dx + N_i(T)y_x\bigg]_{x=R} - N_i(T)y_x\bigg]_{x=L}, \qquad i = L, R \tag{9.25}$$

In Eq. (9.25) we have used subscript notation for the partial derivatives of y with respect to t and x and primes to represent the derivatives of the N's with respect to x (since the N's are functions of x only).

We now use Eq. (9.24) to find substitutions for y and its derivatives:

$$y(x,t) \approx N_L c_L + N_R c_R,$$

$$\frac{\partial y}{\partial x} \approx N_L' c_L + N_R' c_R,$$

$$\frac{\partial y}{\partial t} \approx N_L \dot{c}_L + N_R \dot{c}_R, \tag{9.26}$$

$$\frac{\partial^2 y}{\partial t^2} \approx N_L \ddot{c}_L + N_R \ddot{c}_R.$$

Here we employ the dot notation for time derivatives. (The c's vary with time, of course, but the N's do not.)

We now substitute from Eqs. (9.26) into Eq. (9.25) to get a pair of equations for each element (we write them in matrix form):

$$[M]\{\ddot{c}\} + [C]\{\dot{c}\} + [K]\{c\} = \{b\},$$

$$\left.\begin{aligned}
M_{ij} &= \int_L^R N_i \left(\frac{w}{g}\right) N_j \, dx, \\[4pt]
C_{ij} &= \int_L^R N_i(h) N_j \, dx, \\[4pt]
K_{ij} &= \int_L^R N_i'(T) N_j' \, dx,
\end{aligned}\right\} \quad i, j = L, R, \tag{9.27}$$

$$b_i = \int_L^R N_i(F) \, dx + N_i(T)\frac{\partial y}{\partial x}\bigg]_{x=R} - N_i(T)\frac{\partial y}{\partial x}\bigg]_{x=L}, \quad i = L, R.$$

We will replace the time derivatives with finite differences, selecting central differences because they worked so well in the finite-difference solution to the simple wave equation. Thus we get

$$[M]\frac{\{c^{m+1} - 2c^m + c^{m-1}\}}{(\Delta t)^2} + [C]\frac{\{c^{m+1} - c^{m-1}\}}{2\Delta t} + [K]\{c^m\} = \{b^m\}. \tag{9.28}$$

Now we solve Eq. (9.28) for $\{c^{m+1}\}$:

$$\left(\frac{1}{(\Delta t)^2}[M] + \frac{1}{2\Delta t}[C]\right)\{c^{m+1}\}$$

$$= \left(\frac{2}{(\Delta t)^2}[M] - [K]\right)\{c^m\} - \left(\frac{1}{(\Delta t)^2}[M] - \frac{1}{2\Delta t}[C]\right)\{c^{m-1}\} + \{b^m\}. \tag{9.29}$$

Notice that we need two previous sets of displacements to advance to the new time, t_{m+1}. We faced this identical problem when we solved the simple wave equation with finite differences, and we solve it in the same way. We use the initial velocities (given as one of the initial conditions) to get $\{c^{-1}\}$ to start the solution:

$$\{c^{-1}\} = \{c^1\} - 2\Delta t\{g(x)\}, \tag{9.30}$$

where $\{g(x)\}$ is the vector of initial velocities. (In view of our earlier work, we expect improved results if we use a weighted average of the g-values if the $g(x)$'s are not constants.)

We have not specifically developed the formulas for the components of the matrices and vector of Eqs. (9.27), but they are identical to those we derived when we applied finite elements to boundary-value problems in Chapter 6, because we will take out w, h, T, and F as average values within the elements. So we just copy from Section 6.5:

$$M_{11} = M_{22} = \left(\frac{w}{g}\right)\frac{\Delta}{3}, \qquad M_{12} = M_{21} = \left(\frac{w}{g}\right)\frac{\Delta}{6},$$

$$C_{11} = C_{22} = h\frac{\Delta}{3}, \qquad C_{12} = C_{21} = h\frac{\Delta}{6},$$

$$K_{11} = K_{22} = \frac{T}{\Delta}, \qquad K_{12} = K_{21} = \frac{T}{\Delta}, \qquad \text{(9.31)}$$

$$b_1 = F\frac{\Delta}{2} - \left[T\frac{\partial y}{\partial x}\right]_{x=L} \qquad b_2 = F\frac{\Delta}{2} + \left[T\frac{\partial y}{\partial x}\right]_{x=R}.$$

In this set, Δ represents the length of the element.

We now have everything we need to construct the element equations. Except for the end elements (and then only if the boundary conditions involve the gradient), the gradient terms in Eqs. (9.31) cancel between adjacent elements. Assembly in this case is very simple because there are always two elements that share each node (except at the ends).

What advantage is there to finite elements over finite differences? The major one is that we can use nodes that are unevenly spaced without having to modify the procedure. The advantage becomes really significant in two- and three-dimensional situations, but the other side of the coin is that solving the equations for each time step is not inexpensive.

9.7 Theoretical Matters

In this chapter we will demonstrate only that our finite-difference method is stable when applied to the wave equation in one dimension. Since we will ordinarily solve the one-dimensional wave equation numerically only with

$$\frac{Tg(\Delta t)^2}{w(\Delta x)^2} = 1,$$

it is sufficient to demonstrate stability for that scheme. We assume that a set of errorless computations have been made when a single error of size 1 occurs; we trace the effects of the single error. If the error does not have increasingly great effect on subsequent calculations, we call the method *stable*.

This simple procedure is adequate because, since the problem is linear, the principle of superposition lets us total the effects of all errors and lets us add these errors to the true solution to obtain the actual results. Table 9.5 demonstrates the principle, assuming that the displacement of the endpoints is specified so that these are always free of error.

As the arrows indicate, the wave equation propagates disturbances in opposite directions, with reflections occurring at fixed ends, with a reversal of sign on reflection. Stability is demonstrated because the original error does not grow in size.

Table 9.5   Propagation of single error in numerical solution to wave equation

Initially error-free values	0.0	0.0	0.0	0.0	0.0	0.0	0.0
	0.0	0.0	0.0	0.0	0.0	0.0	0.0
Error made here ⟶	0.0	0.0	1.0	0.0	0.0	0.0	0.0
	0.0	1.0	0.0	1.0	0.0	0.0	0.0
	0.0	0.0	1.0	0.0	1.0	0.0	0.0
	0.0	0.0	0.0	1.0	0.0	1.0	0.0
	0.0	0.0	0.0	0.0	1.0	0.0	0.0
	0.0	0.0	0.0	0.0	0.0	0.0	0.0
	0.0	0.0	0.0	0.0	−1.0	0.0	0.0
	0.0	0.0	0.0	−1.0	0.0	−1.0	0.0
	0.0	0.0	−1.0	0.0	−1.0	0.0	0.0
	0.0	−1.0	0.0	−1.0	0.0	0.0	0.0
	0.0	0.0	−1.0	0.0	0.0	0.0	0.0
	0.0	0.0	0.0	0.0	0.0	0.0	0.0
	0.0	0.0	1.0	0.0	0.0	0.0	0.0
	0.0	1.0	0.0	1.0	0.0	0.0	0.0
	0.0	0.0	1.0	0.0	1.0	0.0	0.0

Chapter Summary

You can test your understanding of Chapter 9 by seeing whether you can

1. Determine whether a partial-differential equation falls in the hyperbolic class.

2. Set up equations to find the displacements of a vibrating string after one time step, and, from these, determine the displacements after n time steps.

3. Explain the d'Alembert method for the wave equation and how this shows that the finite-difference method can give exact answers.

4. Outline the argument that demonstrates stability for the finite-difference procedure.

5. Explain what characteristic curves are, find them for a typical problem, and compute a few points on the characteristics.

6. Apply the finite-difference method to a simple hyperbolic partial-differential equation with two space dimensions.

7. Outline how finite elements are applied to the one-dimensional wave equation.

Computer Program

This chapter illustrates just one program. This FORTRAN program, listed in Fig. 9.7, solves the vibrating string problem using the procedure of Eq. (9.13) to begin the solution and Eq. (9.2) to compute displacements after the first time step. As we have seen, if the integral in Eq. (9.13) is evaluated accurately, this technique gives the analytical solution. In the program, the integration is done by Simpson's $\frac{1}{3}$ rule with 20 panels in each subinterval.

It is easy to follow the program when the algorithm is so simple. It begins with DATA statements that give values to

1. The starting values for X and time (T).
2. N, the number of uniform subintervals.
3. TDM = tension/mass = c^2.
4. XLEN = length of the string.
5. TLAST, how far in time to compute.
6. PI, the value of pi.

Initial displacements are computed from a function F(X) and put into an array. They are written to begin a table of displacements versus time. The displacements at the end of the first step are then computed with the integrations of initial velocities that are given by function G(X). These are added to the table.

A DO loop then begins, computing for the successive time steps, and the table of displacements is added to for each of these.

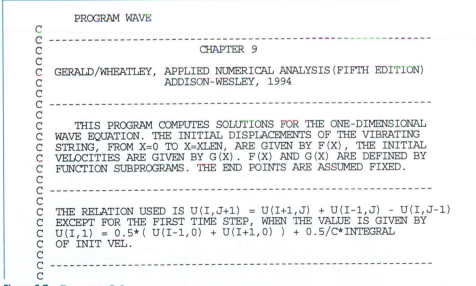

```
      PROGRAM WAVE
C
C ------------------------------------------------------------
C
C                         CHAPTER 9
C
C   GERALD/WHEATLEY, APPLIED NUMERICAL ANALYSIS (FIFTH EDITION)
C                     ADDISON-WESLEY, 1994
C
C ------------------------------------------------------------
C
C      THIS PROGRAM COMPUTES SOLUTIONS FOR THE ONE-DIMENSIONAL
C   WAVE EQUATION. THE INITIAL DISPLACEMENTS OF THE VIBRATING
C   STRING, FROM X=0 TO X=XLEN, ARE GIVEN BY F(X), THE INITIAL
C   VELOCITIES ARE GIVEN BY G(X). F(X) AND G(X) ARE DEFINED BY
C   FUNCTION SUBPROGRAMS. THE END POINTS ARE ASSUMED FIXED.
C
C ------------------------------------------------------------
C
C   THE RELATION USED IS U(I,J+1) = U(I+1,J) + U(I-1,J) - U(I,J-1)
C   EXCEPT FOR THE FIRST TIME STEP, WHEN THE VALUE IS GIVEN BY
C   U(I,1) = 0.5*( U(I-1,0) + U(I+1,0) ) + 0.5/C*INTEGRAL
C   OF INIT VEL.
C
C ------------------------------------------------------------
C
```

Figure 9.7 Program 9.1

Figure 9.7 *(continued)*

```
C        PARAMETERS ARE :
C
C        X       - DISTANCE ALONG THE STRING
C        DX      - INCREMENTS OF DISTANCE
C        XLEN    - TOTAL LENGTH OF STRING
C        N       - NUMBER OF SUBDIVISIONS
C        T       - TIME
C        TLAST   - FINAL VALUE OF TIME FOR WHICH SOLUTION IS DESIRED
C        F(X)    - INITIAL DISPLACEMENTS
C        G(X)    - INITIAL VELOCITIES
C        TDM     - VALUE OF TENSION / MASS = C SQUARED
C        U       - DISPLACEMENTS AT EVEN-TIME INTERVALS
C        V       - DISPLACEMENTS AT ODD-TIME INTERVALS
C
C   -----------------------------------------------------------------
C
         REAL U(100),V(100),X,DX,XLEN,T,TLAST,F,G,TDM,SUBDX,XSUB,PI
         INTEGER N,NP1,I,J
         COMMON XLEN,PI
C
C   -----------------------------------------------------------------
C
C  DEFINE SOME DATA STATEMENTS FOR INITIAL VALUES
C
         DATA X,N,T/0.0,9,0.0/
         DATA TDM,TLAST/4.0,4.5/
         PI = 4.0*ATAN(1.0)
         XLEN = 9.0
C
C   -----------------------------------------------------------------
C
C  GET INITIAL DISPLACEMENTS
C
         NP1 = N + 1
         DX = XLEN / FLOAT(N)
         U(1) = 0.0
         U(NP1) = 0.0
         V(1) = 0.0
         V(NP1) = 0.0
         DO 10 I = 2,N
           X = X + DX
           U(I) = F(X)
10       CONTINUE
C
C   -----------------------------------------------------------------
C
C  WRITE OUT THE INITIAL DISPLACEMENTS
C
         PRINT 200, ( U(I), I = 1,NP1/2)
C
C   ------------------------------------------------------------------
C
C  NOW GET DISPLACEMENTS AFTER FIRST TIME STEP. USE SIMPSON'S RULE
C  TO PERFORM INTEGRATIONS, USING 20 INTERVALS BETWEEN X(N-1) AND
C  X(N+1). WE DO THIS FOR EACH INTERMEDIATE X-VALUE.
C
         SUBDX = DX / 10.0
         XSUB = DX
         DO 30 I = 2,N
           SUM = 0.0
           XSUB = XSUB - DX
           DO 20 J = 1,19,2
             SUM = SUM + G(XSUB) + 4.0*G(XSUB+SUBDX) +
     +             G(XSUB+2.0*SUBDX)
             XSUB = XSUB + 2.0*SUBDX
20         CONTINUE
```

Figure 9.7   *(continued)*

```
              V(I) = 0.5*( U(I-1) + U(I+1) ) +
     +                  0.5/SQRT(TDM)*SUBDX/3.0*SUM
30    CONTINUE
C
C   ----------------------------------------------------------
C
C   WRITE OUT DISPLACEMENTS AFTER FIRST TIME STEP
C
      T = DX / SQRT(TDM)
      PRINT 201, T,( V(I), I = 1,NP1/2)
C
C   ----------------------------------------------------------
C
C   NOW COMPUTE UNTIL TLAST IS REACHED
C
35    IF ( T .GE. TLAST ) STOP
      DO 40 I = 2,N
        U(I) = V(I-1) + V(I+1) -U(I)
40    CONTINUE
      T = T + DX/SQRT(TDM)
      PRINT 201, T,( U(I), I = 1,NP1/2)
      DO 50 I = 2,N
        V(I) = U(I-1) + U(I+1) - V(I)
50    CONTINUE
      T = T + DX/SQRT(TDM)
      PRINT 201, T,( V(I), I = 1,NP1/2)
      GO TO 35
C
C   ----------------------------------------------------------
C
200   FORMAT(//' SOLUTION TO VIBRATING STRING PROBLEM ',///,
     +        ' INITIAL DISPLACEMENTS ARE '//(1X,11F9.4) )
201   FORMAT(/' AT T = ',F5.2/(1X,11F9.4) )
      END
C
C   ----------------------------------------------------------
C
C      DISPLACEMENTS AT T = 0
C
C   ----------------------------------------------------------
      REAL FUNCTION F(X)
C
      REAL X
          F = 0.0
      RETURN
      END
C
C   ----------------------------------------------------------
C
C      VELOCITY AT T = 0
C
C   ----------------------------------------------------------
C
      REAL FUNCTION G(X)
C
      REAL X
      COMMON XLEN,PI
      G = 3.0*SIN(PI*X/XLEN)
      RETURN
      END


      ************************************************************
```

Figure 9.7 *(continued)*

```
              OUTPUT FOR WAVE.F

    SOLUTION TO VIBRATING STRING PROBLEM

    INITIAL DISPLACEMENTS ARE
         .0000    .0000    .0000    .0000    .0000

    AT T =  .50
         .0000    .5027    .9447   1.2728   1.4474
    AT T = 1.00
         .0000    .9447   1.7755   2.3921   2.7202
    AT T = 1.50
         .0000   1.2728   2.3921   3.2229   3.6649
    AT T = 2.00
         .0000   1.4474   2.7202   3.6649   4.1676
    AT T = 2.50
         .0000   1.4474   2.7202   3.6649   4.1676
    AT T = 3.00
         .0000   1.2728   2.3921   3.2229   3.6649
    AT T = 3.50
         .0000    .9447   1.7755   2.3921   2.7202
    AT T = 4.00
         .0000    .5027    .9447   1.2728   1.4474
    AT T = 4.50
         .0000    .0000    .0000    .0000    .0000
```

Exercises

1. If the banjo string of Example 9.1 is tightened or if it is shortened (as by holding it against one of the frets with a finger), the frequency of vibration is raised and the pitch of the sound is higher. What would the frequency be if the tension is made 42,500 g and the effective length is 65 cm? Determine this by finding the number of time steps for the original displacement to be repeated. Compare this to

$$f = \left(\frac{1}{2L}\right)\sqrt{\frac{Tg}{w}} .$$

2. A vibrating string has $Tg/w = 4$ cm^2/sec^2 and is 48 cm long. Divide the length into subintervals so that $\Delta x = L/8$. Find the displacement for $T = 0$ to $T = L$ if both ends are fixed and the initial conditions are

   a. $y = x(x - L)/L^2, y_t = 0$. ($y_t$ is the velocity.)

b. the string is displaced $+2$ units at $L/4$ and -1 unit at $5L/8$, $y_t = 0$.

c. $y = 0$, $y_t = x(L - x)/L^2$. (Use Eq. 9.4.)

d. the string is displaced 1 unit at $L/2$, $y_t = -y$.

e. Compare part (a) to the analytical solution,

$$y = \frac{8}{\pi^3} \sum_{n=1}^{\infty} \frac{1}{(2n-1)^3} \sin\left[(2n-1)\frac{\pi x}{L}\right] \cos\left[(4n-2)\frac{\pi t}{L}\right].$$

▶**3.** The function u satisfies the equation

$$u_{xx} = u_{tt},$$

with boundary conditions of $u = 0$ at $x = 0$ and $u = 0$ at $x = 1$, and with initial conditions

$$u = \sin(\pi x), \qquad u_t = 0, \qquad \text{for } 0 \le x \le 1.$$

Solve by the finite-difference method and show that the results are the same as the analytical solution,

$$u(x, t) = \sin(\pi x) \cos(\pi t).$$

4. The ends of the vibrating string do not have to be fixed. Solve the equation $u_{xx} = u_{tt}$ with $y(x, 0) = 0$, $y_t(x, 0) = 0$ for $0 \le x \le 1$, and end conditions of

$$y(0, t) = 0, \quad y(1, t) = \sin\left(\frac{\pi t}{4}\right), \quad y_x(1, t) = 0.$$

Section 9.3

5. Why can we not use Eq. (9.9) to solve Example 9.1?

▶**6.** Equation (9.4) is inaccurate when the initial velocity is not zero or a constant, so the solutions to parts (c) and (d) of Exercise 2 are not exact. Solve again, but use Eq. (9.13) to begin the solution. (Use Simpson's $\frac{1}{3}$ rule as in Example 9.2.) How different are these from the original computations?

7. Repeat Exercise 6, but now use more points between x_{i-1} and x_{i+1} in Simpson's rule. Does this make much difference?

8. A string that weighs w lb/ft is tightly stretched between $x = 0$ and $x = L$ and is initially at rest. Each point is given an initial velocity of

$$y_t(x, 0) = v_0 \sin^3\left(\frac{\pi x}{L}\right).$$

The analytical solution is

$$y(x, t) = \frac{v_0 L}{12 a \pi}\left(9 \sin\frac{\pi x}{L} \sin\frac{a \pi t}{L} - \sin\frac{3\pi x}{L} \sin\frac{3 a \pi t}{L}\right),$$

where $a = \sqrt{Tg/w}$, T being the tension and g the acceleration due to gravity. When $L = 3$ ft, $w = $

0.02 lb/ft, and $T = 5$ lb, with $v_0 = 1$ ft/sec, the analytical formula predicts $y = 0.081$ in. at the midpoint when $t = 0.01$ sec. Solve the problem numerically to confirm this. Does your solution agree with the analytical solution at other values of x and t?

Section 9.4

9. For the partial-differential equation given by

$$a u_{xx} + b u_{xt} + c u_{tt} + e = 0,$$

sketch the characteristic curves through the point $x = 0.5$, $t = 0$ when:

a. $a = 1$, $b = 2$, $c = 1$.

b. $a = 1$, $b = 2$, $c = -3$.

c. $a = 1$, $b = 2$, $c = 3$.

d. $a = 1$, $b = 2$, $c = -1$.

e. $a = t^2$, $b = xt$, $c = -2x^2$.

10. Verify the values given at points S, T, U, and V in Fig. 9.6 for Example 9.4 by the method of characteristics.

11. Can the equation

$$u_{tt} = (1 + 2x) u_{xx}$$

can be solved through finite-difference approximations for the derivatives? If it can, use that technique to solve it over the interval $[0, 1]$ if

$$u(x, 0) = 0, \qquad u_t(x, 0) = x(1 - x).$$

Compare answers with those of Example 9.4.

▶**12.** Given the equation

$$u_{xx} + u_{xt} - u_{tt} = 1,$$

subject to initial conditions of

$$u(x, 0) = 0, \qquad u_t(x, 0) = x(1 - x).$$

Use the method of characteristics to find the solution at the intersection of characteristics through $(0.4, 0)$, $(0.5, 0)$, and $(0.6, 0)$.

▶**13.** Use the method of characteristics to find the solution at several points in that part of the xt-plane bounded by $x = 0.4$ and $x = 0.6$, for

$$u_{xx} + xt u_{xt} - u_{tt} = 0$$

with initial and boundary conditions of

$$u(x, 0) = 2x, \qquad u_t(x, 0) = 0;$$
$$u(0, t) = 0, \qquad u(1, t) = 2.$$

14. Continue the solution of Example 9.5 by finding the solution at the intersections of characteristics that start at $(0.2, 0)$, $(0.3, 0)$, and $(0.7, 0)$ with those that start at $(0.4, 0)$, $(0.5, 0)$ and $(0.6, 0)$. This shows how the solution can be advanced in time.

Section 9.5

15. Solve the vibrating membrane of Example 9.6 but with initial conditions of

$$u(x, y) = 0 \quad \text{at } t = 0,$$
$$u_t(x, y) = x^2(2 - x)y^2(2 - y) \quad \text{at } t = 0.$$

16. Solve the vibrating membrane of Example 9.6 but with

$$u(x, y) = x^2(2 - x)y^2(2 - y) \quad \text{at } t = 0,$$
$$u_t(x, y) = 0 \quad \text{at } t = 0.$$

▶17. A membrane is stretched over a frame that occupies the region in the xy-plane bounded by

$$x = 0, \quad x = 3, \quad y = 0, \quad y = 2.$$

At $t = 0$, the point on the membrane at $(1, 1)$ is lifted 1 unit above the xy-plane and then released. If $T = 6$ lb/in. and $w = 0.55$ lb/in.^{2}, find the displacement of the point $(2, 1)$ as a function of time.

18. How do the vibrations of Exercise 17 change if $w = 0.055$, the other parameters being unchanged?

19. The frame holding the membrane of Exercise 17 is distorted by lifting the corner at $(3, 2)$ 1 unit above the xy-plane. (The members of the frame elongate

so that the corner moves vertically.) The membrane is set to vibrating in the same way as in Exercise 17. Follow the vibrations through time. [Assume that the rest positions of points on the membrane lie on the two planes defined by the adjacent edges that meet at $(0, 0)$ and at $(3, 2)$.]

Section 9.6

20. Set up the finite-element equations (Eqs. 9.29, 9.31) for advancing the solution to part (a) of Exercise 2.

21. Set up the finite-element equations (Eqs. 9.29, 9.31) for starting the solution to part (a) of Exercise 2. (Use Eq. 9.30.)

▶22. If we were to solve part (c) of Exercise 2 by finite elements, would there be an advantage to using shorter elements near the middle of the string, where the displacements depart more from linearity?

Section 9.7

23. Paraphrase Table 9.5 to investigate the stability of the solution when the equivalent of Eq. (9.2) is used but when the ratio $Tg(\Delta t)^2/w(\Delta x)^2$ is not 1 but is
   a. equal to 0.5.
   b. equal to 2.0.

▶24. Redo Table 9.5 [with the ratio $Tg(\Delta t)^2/w(\Delta x)^2 = 1$] but for end conditions given in Exercise 4. Is stability still indicated?

Applied Problems and Projects

25. A vibrating string, with a damping force opposing its motion that is proportional to the velocity, follows the equation

$$\frac{\partial^2 y}{\partial t^2} = \frac{Tg}{w} \frac{\partial^2 y}{\partial x^2} - B \frac{\partial y}{\partial t},$$

where B is the magnitude of the damping force. Solve the problem if the length of the string is 5 ft with $T = 24$ lb, $w = 0.1$ lb/ft, and $B = 2.0$. Initial conditions are

$$y(x)|_{t=0} = \frac{x}{3}, \quad 0 \le x < 3,$$

$$y(x)|_{t=0} = \frac{5}{2} - \frac{x}{2}, \quad 3 \le x \le 5,$$

$$\frac{\partial y}{\partial t}\bigg|_{t=0} = x(x - 5).$$

Compute a few points of the solution by both difference equations and the method of characteristics. Compare the effort involved in the two methods.

26. A horizontal elastic rod is initially undeformed and is at rest. One end, at $x = 0$, is fixed, and the other end, at $x = L$ (when $t = 0$), is pulled with a steady force of F lb/ft². It can be shown that the displacements $y(x,t)$ of points originally at the point x are given by

$$\frac{\partial^2 y}{\partial t^2} = a^2 \frac{\partial^2 y}{\partial x^2}, \qquad y(0, t) = 0, \qquad \left.\frac{\partial y}{\partial t}\right|_{x=L} = \frac{F}{E},$$

$$y(x, 0) = 0, \qquad \left.\frac{\partial y}{\partial t}\right|_{t=0} = 0,$$

where $a^2 = Eg/\rho$; E = Young's modulus (lb/ft²); g = acceleration of gravity; ρ = density (lb/ft³). Find y versus t for the midpoint of a 2-ft-long piece of rubber for which $E = 1.8 \times 10^6$ and $\rho = 70$ if $F/E = 0.7$.

27. A circular membrane, when set to vibrating, obeys the equation (in polar coordinates)

$$\frac{1}{r} \frac{\partial}{\partial r}\left(r \frac{\partial u}{\partial r}\right) + \frac{1}{r^2} \frac{\partial^2 u}{\partial \theta^2} = \frac{w}{Tg} \frac{\partial^2 y}{\partial t^2}.$$

A 3-ft-diameter kettledrum is started to vibrating by depressing the center $\frac{1}{2}$ in. If $w = 0.072$ lb/ft² and $T = 80$ lb/ft, find how the displacements at 6 in. and 12 in. from the center vary with time. The problem can be solved in polar coordinates, or it can be solved in rectangular coordinates using the method of Section 7.7 to approximate $\nabla^2 u$ near the boundaries.

28. A flexible chain hangs freely, as shown in Fig. 9.8. For small disturbances from its equilibrium position (hanging vertically), the equation of motion is

$$x\frac{\partial^2 y}{\partial x^2} + \frac{\partial y}{\partial x} = \frac{1}{g}\frac{\partial^2 y}{\partial t^2}.$$

In this equation, x is the distance from the end of the chain, y is the displacement from the equilibrium position, t is the time, and g is the acceleration of gravity. A 10-ft-long chain is originally hanging freely. It is set into motion by striking it sharply at its midpoint, imparting a velocity there of 1 ft/sec. Find how the chain moves as a result of the blow. If you find you need additional information at $t = 0$, make reasonable assumptions.

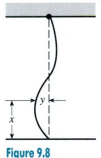

Figure 9.8

29. Write a computer program that solves hyperbolic partial-differential equations by the method of characteristics, given the values of u and $\partial u/\partial t$ along a curve that is not one of the characteristics. Your program should be a subroutine that accepts the coordinates of two points on the initial-condition curve and computes the value of u and $\partial u/\partial t$ at the intersection of the characteristic curves through the two points.

10

Approximation of Functions

Contents of This Chapter

In this chapter we will consider two ways to approximate the values of a known function. Both are of great importance. We will first discuss how to represent a function by a polynomial or a ratio of polynomials in the most efficient way. "Efficient way" refers to obtaining values, at any point in an interval, with the smallest error for a given number of arithmetic operations. This topic is essential when developing computer procedures that compute sines, cosines, logarithms, and other nonalgebraic functions. We limit our search of approximating functions to polynomials and ratios of these because such computation is the only kind a computer can do. (Modern math-coprocessor chips can do more.)

The second topic of the chapter is the technique of approximating a function with a series containing only sines and cosines. Such series are called Fourier series. We consider how the coefficients can be determined analytically and then numerically. This leads us to the consideration of the fast Fourier transform (FFT).

10.1 Chebyshev Polynomials

Develops the theory of a class of orthogonal polynomials that are the basis for fitting nonalgebraic functions with polynomials of maximum efficiency.

10.2 Economized Power Series

Shows how the Chebyshev polynomials can be used to create polynomial approximations that are significantly more efficient than a Maclaurin series.

10.3 Approximation with Rational Functions

Applies the previous techniques to generate the coefficients of rational functions (ratios of polynomials) that are even more efficient. The section also discusses continued fractions, which offer a means of evaluating a rational function with fewer operations.

10.4   Fourier Series

Discusses these approximating series composed of sine and cosine terms. The classical method of determining the coefficients involves many integrations. However, for almost all functions, a Fourier series can approximate a function throughout an interval quite closely. Fourier series are important in many areas, particularly in the analytical solution of partial-differential equations.

10.5   Getting Fourier Coefficients Numerically and the Fast Fourier Transform

Shows that numerical integration can be used to find the coefficients of a Fourier series but at the expense of much computational effort. The fast Fourier transform reduces this effort tremendously.

10.6   Theoretical Matters

Presents a small part of the large theory associated with this chapter's topics.

10.7   Using DERIVE

Describes those elements of the DERIVE symbolic algebra program that can help in the computations of this chapter.

Chapter Summary

Reviews the important topics of this chapter.

Computer Programs

Illustrates how some of the procedures in the chapter can be implemented on a computer.

Parallel Processing   The numerical procedures of this chapter offer many possibilities for applying parallel processing, but all have been discussed before:

Sections 10.2 and 10.3 evaluate polynomials.

Section 10.3 solves systems of equations.

Section 10.4 evaluates a sum.

Section 10.5 performs numerical integrations and does matrix operations.

10.1   Chebyshev Polynomials

We turn now to the problem of representing a function with a minimum error. This is a central problem in the software development of digital computers because it is more economical to compute the values of the common functions using an efficient approximation than to store a table of values and employ interpolation techniques. Since digital computers are essentially only arithmetic devices, the most elaborate function they can compute is a rational function, a ratio of polynomials. We will hence restrict our discussion to representation of functions by polynomials or rational functions.

One way to approximate a function by a polynomial is to use a truncated Taylor series. This is not the best way, in most cases. To study better ways, we first need to introduce the Chebyshev polynomials.

The familiar Taylor-series expansion represents the function with very small error near the point of the expansion, but the error increases rapidly (proportional to a power) as we employ it at points farther away. In a digital computer, we have no control over where in an interval the approximation will be used, so the Taylor series is not usually appropriate. We would prefer to trade some of its excessive precision at the center of the interval to reduce the errors at the ends.

We can do this while still expressing functions as polynomials by the use of Chebyshev polynomials. The first few of these are[*]

$$
\begin{aligned}
T_0(x) &= 1, \\
T_1(x) &= x, \\
T_2(x) &= 2x^2 - 1, \\
T_3(x) &= 4x^3 - 3x, \\
T_4(x) &= 8x^4 - 8x^2 + 1, \\
T_5(x) &= 16x^5 - 20x^3 + 5x, \\
T_6(x) &= 32x^6 - 48x^4 + 18x^2 - 1, \\
T_7(x) &= 64x^7 - 112x^5 + 56x^3 - 7x, \\
T_8(x) &= 128x^8 - 256x^6 + 160x^4 - 32x^2 + 1, \\
T_9(x) &= 256x^9 - 576x^7 + 432x^5 - 120x^3 + 9x, \\
T_{10}(x) &= 512x^{10} - 1280x^8 + 1120x^6 - 400x^4 + 50x^2 - 1.
\end{aligned}
\tag{10.1}
$$

The members of this series of polynomials can be generated from the two-term recursion formula

$$
T_{n+1}(x) = 2xT_n(x) - T_{n-1}(x), \qquad T_0(x) = 1, \qquad T_1(x) = x. \tag{10.2}
$$

Note that the coefficient of x^n in $T_n(x)$ is always 2^{n-1}. In Fig. 10.1 we plot the first four polynomials of Eq. (10.1).

---

[*] The commonly accepted symbol $T(x)$ comes from the older spelling, Tschebycheff.

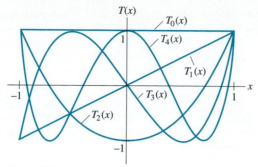

Figure 10.1

These polynomials have some unusual properties. They form an orthogonal set, in that

$$\int_{-1}^{1} \frac{T_n(x)T_m(x)}{\sqrt{1-x^2}} \, dx = \begin{cases} 0, & n \neq m, \\ \pi, & n = m = 0, \\ \dfrac{\pi}{2}, & n = m \neq 0. \end{cases} \tag{10.3}$$

The orthogonality of these functions will not be of immediate concern to us.

The Chebyshev polynomials are also terms of a Fourier series,* since

$$T_n(x) = \cos n\theta, \tag{10.4}$$

where $\theta = \arccos x$. Observe that $\cos 0 = 1, \cos \theta = \cos(\arccos x) = x$.

To demonstrate the equivalence of Eq. (10.4) to Eqs. (10.1) and (10.2), we recall some trigonometric identities, such as

$$\cos 2\theta = 2 \cos^2 \theta - 1,$$
$$T_2(x) = 2x^2 - 1;$$

$$\cos 3\theta = 4 \cos^3 \theta - 3 \cos \theta,$$
$$T_3(x) = 4x^3 - 3x;$$

$$\cos(n + 1)\theta + \cos(n - 1)\theta = 2 \cos \theta \cos n\theta,$$
$$T_{n+1}(x) + T_{n-1}(x) = 2x T_n(x).$$

Because of the relation $T_n(x) = \cos n\theta$, it is apparent that the Chebyshev polynomials have a succession of maximums and minimums of alternating signs, each of magnitude one. Further, since $|\cos n\theta| = 1$ for $n\theta = 0, \pi, 2\pi, \ldots$, and since θ varies from 0 to π as x varies from 1 to -1, $T_n(x)$ assumes its maximum magnitude of unity $n + 1$ times on the interval $[-1, 1]$.

---

* We discuss Fourier series later in this chapter.

Most important for our present application of these polynomials is the fact that, of all polynomials of degree n where the coefficient of x^n is unity, the polynomial

$$\frac{1}{2^{n-1}} T_n(x)$$

has a smaller upper bound to its magnitude in the interval $[-1, 1]$ than any other. We prove this in a later section. Because the maximum magnitude of $T_n(x)$ is one, the upper bound referred to is $1/2^{n-1}$. This is important because we will be able to write power-series representations of functions whose maximum errors are given in terms of this upper bound.

10.2 Economized Power Series

We are now ready to use Chebyshev polynomials to "economize" a power series. Consider the Maclaurin series for e^x:

$$e^x = 1 + x + \frac{x^2}{2} + \frac{x^3}{6} + \frac{x^4}{24} + \frac{x^5}{120} + \frac{x^6}{720} + \cdots .$$

If we would like to use a truncated series to approximate e^x on the interval $[0, 1]$ with precision of 0.001, we will have to retain terms through that in x^6, since the error after the term in x^5 will be more than $1/720$. Suppose we subtract

$$\left(\frac{1}{720}\right)\left(\frac{T_6}{32}\right)$$

from the truncated series. We note from Eq. (10.1) that this will exactly cancel the x^6 term and at the same time make adjustments in other coefficients of the Maclaurin series. Since the maximum value of T_6 on the interval $[0, 1]$ is unity, this will change the sum of the truncated series by only

$$\frac{1}{720} \cdot \frac{1}{32} < 0.00005,$$

which is small with respect to our required precision of 0.001. Performing the calculations, we have

$$e^x \doteq 1 + x + \frac{x^2}{2} + \frac{x^3}{6} + \frac{x^4}{24} + \frac{x^5}{120} + \frac{x^6}{720}$$

$$- \frac{1}{720}\left(\frac{1}{32}\right)(32x^6 - 48x^4 + 18x^2 - 1), \qquad \textbf{(10.5)}$$

$$e^x \doteq 1.000043 + x + 0.499219x^2 + \frac{x^3}{6} + 0.043750x^4 + \frac{x^5}{120} .$$

This gives a fifth-degree polynomial that approximates e^x on [0, 1] almost as well as the sixth-degree one derived from the Maclaurin series. (The actual maximum error of the fifth-degree expression is 0.000270; for the sixth-degree expression it is 0.000226.) We hence have "economized" the power series in that we get nearly the same precision with fewer terms.

By subtracting $\frac{1}{120}(T_5/16)$ we can economize further, getting a fourth-degree polynomial that is almost as good as the economized fifth-degree one. It is left as an exercise to do this and to show that the maximum error is now 0.000781, so that we have found a fourth-degree power series that meets an error criterion that requires us to use two additional terms of the original Maclaurin series. Because of the relative ease with which they can be developed, such economized power series are frequently used for approximations to functions and are much more efficient than power series of the same degree obtained by merely truncating a Taylor or Maclaurin series. Table 10.1 compares the errors of these power series.

Table 10.1 Comparison of errors of economized power series and a Maclaurin series for e^x

x	e^x	Maclaurin, sixth-degree	Economized, fifth-degree	Economized, fourth-degree	Maclaurin, fourth-degree
0	1.00000	1.00000	1.00004	1.00004	1.00000
0.2	1.22140	1.22140	1.22142	1.22098	1.22140
0.4	1.49182	1.49182	1.49179	1.49133	1.49173
0.6	1.82212	1.82211	1.82208	1.82212	1.82140
0.8	2.22554	2.22549	2.22553	2.22605	2.22240
1.0	2.71828	2.71806	2.71801	2.71749	2.70833
Maximum error		0.00023	0.00027	0.00078	0.00995

The maximum error in the economized fifth-degree polynomial is only slightly greater than the sixth-degree Maclaurin series. The economized fourth-degree polynomial incurs a maximum error about three and one-half times as much, but still within the 0.001 limit that was initially imposed, and will require significantly reduced computational effort. In addition, there is a proportionately reduced memory-space requirement to store the constants of the polynomial. In contrast, a fourth-degree Maclaurin series has an error nearly ten times greater than the 0.001 tolerance, and its error is over twelve times that of the fourth-degree economized form.

Chebyshev Series

By rearranging the Chebyshev polynomials, we can express powers of x in terms of them:

$$1 = T_0,$$

$$x = T_1,$$

$$x^2 = \frac{1}{2}(T_0 + T_2),$$

$$x^3 = \frac{1}{4}(3T_1 + T_3),$$

$$x^4 = \frac{1}{8}(3T_0 + 4T_2 + T_4),$$

$$x^5 = \frac{1}{16}(10T_1 + 5T_3 + T_5),$$ **(10.6)**

$$x^6 = \frac{1}{32}(10T_0 + 15T_2 + 6T_4 + T_6),$$

$$x^7 = \frac{1}{64}(35T_1 + 21T_3 + 7T_5 + T_7),$$

$$x^8 = \frac{1}{128}(35T_0 + 56T_2 + 28T_4 + 8T_6 + T_8),$$

$$x^9 = \frac{1}{256}(126T_1 + 84T_3 + 36T_5 + 9T_7 + T_9).$$

By substituting these identities into an infinite Taylor series and collecting terms in $T_i(x)$, we create a Chebyshev series. For example, we can get the first four terms of a Chebyshev series by starting with the Maclaurin expansion for e^x. Such a series converges more rapidly than does a Taylor series on $[-1, 1]$:

$$e^x = 1 + x + \frac{x^2}{2} + \frac{x^3}{6} + \frac{x^4}{24} + \cdots.$$

Replacing terms by Eq. (10.6), but omitting polynomials beyond $T_3(x)$, since we want only four terms,* we have

$$e^x = T_0 + T_1 + \frac{1}{4}(T_0 + T_2) + \frac{1}{24}(3T_1 + T_3) + \frac{1}{192}(3T_0 + 4T_2 + \cdots)$$

$$+ \frac{1}{1920}(10T_1 + 5T_3 + \cdots) + \frac{1}{23,040}(10T_0 + 15T_2 + \cdots) + \cdots$$

$$= 1.2661T_0 + 1.1302T_1 + 0.2715T_2 + 0.0443T_3 + \cdots.$$

To compare the Chebyshev expansion with the Maclaurin series, we convert back to powers of x, using Eq. (10.1):

---

*The number of terms that are employed determines the accuracy of the computed values, of course.

$$e^x = 1.2661 + 1.1302(x) + 0.2715(2x^2 - 1) + 0.0443(4x^3 - 3x) + \cdots .$$
$$e^x = 0.9946 + 0.9973x + 0.5430x^2 + 0.1772x^3 + \cdots . \qquad \textbf{(10.7)}$$

Table 10.2 and Fig. 10.2 compare the error of the Chebyshev expansion, Eq. (10.7), with the Maclaurin series, using terms through x^3 in each case. The figure shows how the Chebyshev expansion attains a smaller maximum error by permitting the error at the origin to increase. The errors can be considered to be distributed more or less uniformly throughout the interval. In contrast to this, the Maclaurin expansion, which gives very small errors near the origin, allows the error to bunch up at the ends of the interval.

If the function is to be expressed directly as an expansion in Chebyshev polyno-

Table 10.2 Comparison of Chebyshev series for e^x with Maclaurin series:
$$e^x = 0.9946 + 0.9973x + 0.5430x^2 + 0.1772x^3;$$
$$e^x = 1 + x + 0.5x^2 + 0.1667x^3$$

x	e^x	Chebyshev	Error	Maclaurin	Error
−1.0	0.3679	0.3631	0.0048	0.3333	0.0346
−0.8	0.4493	0.4536	−0.0042	0.4346	0.0147
−0.6	0.5488	0.5534	−0.0046	0.5440	0.0048
−0.4	0.6703	0.6712	−0.0009	0.6693	0.0010
−0.2	0.8187	0.8154	0.0033	0.8187	0.0001
0	1.0000	0.9946	0.0054	1.0000	0.0000
0.2	1.2214	1.2172	0.0042	1.2213	0.0001
0.4	1.4918	1.4917	0.0001	1.4907	0.0012
0.6	1.8221	1.8267	−0.0046	1.8160	0.0061
0.8	2.2255	2.2307	−0.0051	2.2054	0.0202
1.0	2.7183	2.7121	0.0062	2.6667	0.0516

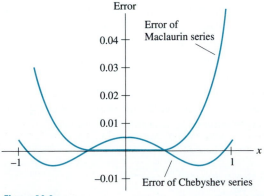

Figure 10.2

mials, the coefficients can be obtained by integration. Based on the orthogonality property, the coefficients are computed from*

$$a_i = \frac{2}{\pi} \int_{-1}^{1} \frac{f(x) T_i(x)}{\sqrt{1 - x^2}} \, dx,$$

and the series is expressed as

$$f(x) = \frac{a_0}{2} + \sum_{i=1}^{\infty} a_i T_i(x).$$

A change of variable will be required if the desired interval is other than $(-1, 1)$. In some cases, the definite integral that defines the coefficients can be profitably evaluated by numerical methods as described in Chapter 4.

Since the coefficients of the terms of a Chebyshev expansion usually decrease even more rapidly than the terms of a Maclaurin expansion, we can get an estimate of the magnitude of the error from the next nonzero term after those that were retained. For the truncated Chebyshev series given by Eq. (10.7), the $T_4(x)$ term would be

$$\frac{1}{192} (T_4) + \frac{1}{23,040} (6T_4) + \cdots = 0.00525 T_4.$$

Since the maximum value of $T_4(x)$ on $(-1, 1)$ is 1.0, we estimate the maximum errors of Eq. (10.7) to be 0.00525. The maximum error in Table 10.2 is 0.0062. This good agreement is caused by the very rapid decrease in coefficients in this example.

The computational economy to be gained by economizing a Maclaurin series, or by using a Chebyshev series, is even more dramatic when the Maclaurin series is slowly convergent. The previous example for $f(x) = e^x$ is a case in which the Maclaurin series converges rapidly. The power of the methods of this section is better demonstrated in the following example.

EXAMPLE 10.1   A Maclaurin series for $(1 + x)^{-1}$ is

$$(1 + x)^{-1} = 1 - x + x^2 - x^3 + x^4 - \cdots \qquad (-1 < x < 1).$$

Table 10.3 compares the accuracy of truncated Maclaurin series with the economized series derived from them.

In Table 10.3, we see that the error of the Maclaurin series is small for $x = 0.2$, and this also would be true for other values near $x = 0$, while the economized polynomial has less accuracy. At $x = 0.8$, the situation is reversed, however. Economized polynomials of degrees 8 and 6, derived from truncated Maclaurin series of degrees 10 and 8, actually have smaller errors than their precursors. Further economization, giving polynomials of degrees 6 and 4, have lesser or only slightly greater errors than their precursors, at significant savings of computational effort and with smaller storage requirements in a computer's memory for the coefficients of the polynomials.

---

*The integration is not easy because the integrand is infinite at the endpoints.

Table 10.3 Comparison of Maclaurin and economized series for $(1 + x)^{-1}$

Maclaurin			Economized					
Degree	Value	Error	Degree*	Value	Error	Degree*	Value	Error
				$x = 0.2$				
2	0.840000	0.006667						
4	0.833600	0.000267	2	0.758600	-0.0747333			
6	0.833344	$11 * 10^{-6}$	4	0.764594	-0.068739			
8	0.833334	$1 * 10^{-6}$	6	0.803646	-0.029687			
10	0.833333	0	8	0.822786	-0.010547			
				$x = 0.8$				
2	0.840000	0.284445						
4	0.737600	0.182045	2	0.812600	0.257045			
6	0.672064	0.116509	4	0.678314	0.122759			
8	0.630121	0.074566	6	0.628558	0.073003	4	0.658246	0.102691
10	0.603277	0.047722	8	0.602106	0.046551	6	0.598199	0.042644

*Economized series were derived from Maclaurin series of corresponding degree.

10.3 Approximation with Rational Functions

We have seen that expansion of a function in terms of Chebyshev polynomials gives a power-series expansion that is much more efficient on the interval $(-1, 1)$ than the Maclaurin expansion, in that it has a smaller maximum error with a given number of terms. These are not the best approximations for use in most digital computers, however. In this application, we measure efficiency by the computer time required to evaluate the function, plus some consideration of storage requirements for the constants. Since the arithmetic operations of a computer can directly evaluate only polynomials, we limit our discussion of more efficient approximations to rational functions, which are the ratios of two polynomials.

Our discussion of methods of finding efficient rational approximations will be elementary and introductory only. Obtaining truly best approximations is a difficult subject. In its present stage of development it is as much art as science, and requires successive approximations from a "suitably close" initial approximation. Our study will serve to introduce the student to some of the ideas and procedures used. The topic is of great importance, however, since the saving of just 1 msec of time in the generation of a frequently used elementary function may save many dollars' worth of machine time each year.

Padé Approximations

We start with a discussion of Padé approximations. Suppose we wish to represent a function as the quotient of two polynomials:

$$f(x) \doteq R_N(x) = \frac{a_0 + a_1 x + a_2 x^2 + \cdots + a_n x^n}{1 + b_1 x + b_2 x^2 + \cdots + b_m x^m}, \qquad N = n + m.$$

The constant term in the denominator can be taken as unity without loss of generality, since we can always convert to this form by dividing numerator and denominator by b_0. The constant b_0 will generally not be zero, for, in that case, the fraction would be undefined at $x = 0$. The most useful of the Padé approximations are those with the degree of the numerator equal to, or one more than, the degree of the denominator. Note that the number of constants in $R_N(x)$ is $N + 1 = n + m + 1$.

The Padé approximations are related to Maclaurin expansions in that the coefficients are determined in a similar fashion to make $f(x)$ and $R_N(x)$ agree at $x = 0$ and also to make the first N derivatives agree at $x = 0$.*

We begin with the Maclaurin series for $f(x)$ (we use only terms through x^N) and write

$$
\begin{aligned}
&f(x) - R_N(x) \\
&\doteq (c_0 + c_1 x + c_2 x^2 + \cdots + c_N x^N) - \frac{a_0 + a_1 x + \cdots + a_n x^n}{1 + b_1 x + \cdots + b_m x^m} \\
&= \frac{(c_0 + c_1 x + \cdots + c_N x^N)(1 + b_1 x + \cdots + b_m x^m) - (a_0 + a_1 x + \cdots + a_n x^n)}{1 + b_1 x + \cdots + b_m x^m}. \quad \textbf{(10.8)}
\end{aligned}
$$

The coefficients c_i are $f^{(i)}(0)/(i!)$ of the Maclaurin expansion. Now if $f(x) = R_N(x)$ at $x = 0$, the numerator of Eq. (10.8) must have no constant term. Hence

$$c_0 - a_0 = 0.$$

For the first N derivatives of $f(x)$ and $R_N(x)$ to be equal at $x = 0$, the coefficients of the powers of x up to and including x^N in the numerator must all be zero also. This gives N additional equations for the a's and b's. The first n of these involve a's, the rest only b's and c's:

$$
\begin{aligned}
b_1 c_0 + c_1 - a_1 &= 0, \\
b_2 c_0 + b_1 c_1 + c_2 - a_2 &= 0, \\
b_3 c_0 + b_2 c_1 + b_1 c_2 + c_3 - a_3 &= 0, \\
&\;\;\vdots \\
b_m c_{n-m} + b_{m-1} c_{n-m+1} + \cdots + c_n - a_n &= 0, \\
b_m c_{n-m+1} + b_{m-1} c_{n-m+2} + \cdots + c_{n+1} &= 0, \\
b_m c_{n-m+2} + b_{m-1} c_{n-m+3} + \cdots + c_{n+2} &= 0, \\
&\;\;\vdots \\
b_m c_{N-m} + b_{m-1} c_{N-m+1} + \cdots + c_N &= 0.
\end{aligned}
\qquad \textbf{(10.9)}
$$

---

* A similar development can be derived for the expansion about a nonzero value of x, but the manipulations are not as easy. By a change of variable we can always make the region of interest contain the origin.

Note that, in each equation, the sum of the subscripts on the factors of each product is the same, and is equal to the exponent of the x-term in the numerator. The $N + 1$ equations of Eqs. (10.8) and (10.9) give the required coefficients of the Padé approximation. We illustrate by an example.

EXAMPLE 10.2   Find $\arctan x \doteq R_9(x)$. Use in the numerator a polynomial of degree five.
The Maclaurin series through x^9 is

$$\arctan x \doteq x - \frac{1}{3}x^3 + \frac{1}{5}x^5 - \frac{1}{7}x^7 + \frac{1}{9}x^9. \qquad (10.10)$$

We form, analogously to Eq. (10.8),

$$\frac{f(x) - R_9(x)}{} $$

$$= \frac{\left(x - \frac{1}{3}x^3 + \frac{1}{5}x^5 - \frac{1}{7}x^7 + \frac{1}{9}x^9\right)(1 + b_1 x + b_2 x^2 + b_3 x^3 + b_4 x^4) - (a_0 + a_1 x + \cdots + a_5 x^5)}{(1 + b_1 x + b_2 x^2 + b_3 x^3 + b_4 x^4)}.$$

$$(10.11)$$

Making coefficients through that of x^9 in the numerator equal to zero, we get

$$a_0 = 0,$$
$$a_1 = 1,$$
$$a_2 = b_1,$$
$$a_3 = -\frac{1}{3} + b_2,$$
$$a_4 = -\frac{1}{3}b_1 + b_3,$$
$$a_5 = \frac{1}{5} - \frac{1}{3}b_2 + b_4,$$
$$\frac{1}{5}b_1 - \frac{1}{3}b_3 = 0,$$
$$-\frac{1}{7} + \frac{1}{5}b_2 - \frac{1}{3}b_4 = 0,$$
$$-\frac{1}{7}b_1 + \frac{1}{5}b_3 = 0,$$
$$\frac{1}{9} - \frac{1}{7}b_2 + \frac{1}{5}b_4 = 0.$$

Solving first the last four equations for the b's, and then getting the a's, we have

$$b_1 = 0, \qquad b_2 = \frac{10}{9}, \qquad b_3 = 0, \qquad b_4 = \frac{5}{21},$$

$$a_0 = 0, \qquad a_1 = 1, \qquad a_2 = 0, \qquad a_3 = \frac{7}{9}, \qquad a_4 = 0, \qquad a_5 = \frac{64}{945}.$$

A rational function that approximates $\arctan x$ is then

$$\arctan x \doteq \frac{x + \dfrac{7}{9}x^3 + \dfrac{64}{945}x^5}{1 + \dfrac{10}{9}x^2 + \dfrac{5}{21}x^4}. \qquad (10.12)$$

In Table 10.4 we compare the errors for Padé approximation (Eq. 10.12) to the Maclaurin series expansion (Eq. 10.10). Enough terms are available in the Maclaurin series to give five-decimal precision at $x = 0.2$ and 0.4, but at $x = 1$ (the limit for convergence of the series) the error is sizable. Even though we used no more information in establishing it, the Padé formula is surprisingly accurate, having an error only 1/275 as large at $x = 1$. It is then particularly astonishing to realize that the Padé approximation is still not the best one of its form, for it violates the minimax principle. If the extreme precision near $x = 0$ is relaxed, we can make the maximum error smaller in the interval.

Table 10.4 Comparison of Padé approximation to Maclaurin series for $\arctan x$

x	True value	Padé (Eq. 10.12)	Error	Maclaurin (Eq. 10.10)	Error
0.2	0.19740	0.19740	0.00000	0.19740	0.00000
0.4	0.38051	0.38051	0.00000	0.38051	0.00000
0.6	0.54042	0.54042	0.00000	0.54067	−0.00025
0.8	0.67474	0.67477	−0.00003	0.67982	−0.00508
1.0	0.78540	0.78558	−0.00018	0.83492	−0.04952

Figure 10.3 shows how closely the Padé approximation matches $\arctan(x)$, especially on $[-1, 1]$.

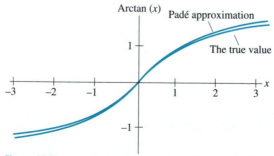

Figure 10.3

The error of a Padé approximation can often be roughly estimated by computing the next nonzero term in the numerator of Eq. (10.12). For Example 10.2, the coefficient of x^{10} is zero, and the next term is

$$\left(-\frac{1}{7}b_4 + \frac{1}{9}b_2 - \frac{1}{11}\right)x^{11} = \left[-\frac{1}{7}\left(\frac{5}{21}\right) + \frac{1}{9}\left(\frac{10}{9}\right) - \frac{1}{11}\right]x^{11}$$
$$= -0.0014x^{11}.$$

Dividing by the denominator, we have

$$\text{Error} \doteq \frac{-0.0014x^{11}}{1 + 1.1111x^2 + 0.2381x^4}.$$

At $x = 1$ this estimate gives -0.00060, which is about three times too large, but still of the correct order of magnitude. It is not unusual that such estimates are rough; analogous estimates of error by using the next term in a Maclaurin series behave similarly. The validity of the rule of thumb that "next term approximates the error" is poor when the coefficients do not decrease rapidly.

The preference for Padé approximations with the degree of the numerator the same as or one less than the degree of the denominator rests on the empirical fact that the errors are usually more for these. But there are even more efficient rational functions.

Continued Fractions

Before we discuss such better approximations in the form of rational functions, remarks on the amount of effort required for the computation using Eq. (10.12) are in order. If we implement the equation in a computer as it stands, we would, of course, use the constants in decimal form, and we would evaluate the polynomials in nested form:

$$\text{Numerator} = [(0.0677x^2 + 0.7778)x^2 + 1]x,$$
$$\text{Denominator} = (0.2381x^2 + 1.1111)x^2 + 1.$$

Since additions and subtractions are generally much faster than multiplications or divisions, we generally neglect them in a count of operations. We have then three multiplications for the numerator, two for the denominator, plus one to get x^2, and one division, for a total of seven operations. The Maclaurin series is evaluated with six multiplications, using the nested form. If division and multiplication consume about the same time, there is about a standoff in effort, but greater precision for Eq. (10.12).*

---

* On many computers, division is slower than multiplication. This will modify the conclusion reached here.

Since small differences in effort accumulate for a frequently used function, it is of interest to see whether we can further decrease the number of operations to evaluate Eq. (10.12). By means of a succession of divisions we can re-express it in continued-fraction form:*

$$\frac{0.0677x^5 + 0.7778x^3 + x}{0.2381x^4 + 1.1111x^2 + 1} = \frac{0.2844x^5 + 3.2667x^3 + 4.2x}{x^4 + 4.6667x^2 + 4.2}$$

$$= \frac{0.2844x(x^4 + 11.4846x^2 + 14.7659)}{x^4 + 4.6667x^2 + 4.2}$$

$$= \frac{0.2844x}{(x^4 + 4.6667x^2 + 4.2)/(x^4 + 11.4846x^2 + 14.7659)}$$

$$= \frac{0.2844x}{1 - (6.8179x^2 + 10.5659)/(x^4 + 11.4846x^2 + 14.7659)}$$

$$= \frac{0.2844x}{1 - 6.8179(x^2 + 1.5497)/(x^4 + 11.4846x^2 + 14.7659)}$$

$$= \frac{0.2844x}{1 - 6.8179/[(x^4 + 11.4846x^2 + 14.7659)/(x^2 + 1.5497)]}$$

$$= \frac{0.2844x}{1 - 6.8179/[x^2 + 9.9348 - 0.6304/(x^2 + 1.5497)]}.$$

In this last form, we see that three divisions and two multiplications are needed (one multiplication by x and one to get x^2), for a total of five operations. We have saved two steps. In most cases there is an even greater advantage to the continued-fraction form; in this example the missing powers of x favored the evaluation as polynomials.

Better Rational Function Approximations

One can get somewhat improved rational-function approximations by starting with the Chebyshev expansion and operating analogously to the method for Padé approximations. We illustrate with an approximation for e^x. The Chebyshev series was derived in Section 10.2, Eq. (10.7):

$$e^x = 1.2661T_0 + 1.1302T_1 + 0.2715T_2 + 0.0443T_3.$$

Using this approximation, we form the difference

$$f(x) - \frac{P_n(x)}{Q_m(x)}$$

$$= \frac{(1.2661 + 1.302T_1 + 0.2715T_2 + 0.0443T_3)(1 + b_1 T_1) - (a_0 + a_1 T_1 + a_2 T_2)}{1 + b_1 T_1}.$$

---

* Acton (1970) is an excellent reference.

Here we have chosen the numerator as a second-degree Chebyshev polynomial and the denominator as first degree. We again make the first $N = n + m$ powers of x in the numerator vanish. Expanding the numerator, we get

$$\text{Numerator} = 1.2661 + 1.1302T_1 + 0.2715T_2 + 0.0443T_3 + 1.2661b_1 T_1$$
$$+ 1.1302b_1 T_1^2 + 0.2715b_1 T_1 T_2 + 0.0443b_1 T_1 T_3 - a_0$$
$$- a_1 T_1 - a_2 T_2.$$

Before we can equate coefficients to zero, we need to resolve the products of Chebyshev polynomials that occur. Recalling that $T_n(x) = \cos n\theta$, we can use the trigonometric identity

$$\cos n\theta \cos m\theta = \frac{1}{2}[\cos(n + m)\theta + \cos(n - m)\theta],$$

$$T_n(x)T_m(x) = \frac{1}{2}[T_{n+m}(x) + T_{|n-m|}(x)].$$

The absolute value of the difference $n - m$ occurs because $\cos(z) = \cos(-z)$. Using this relation we can write the equations

$$a_0 = 1.2661 + \frac{1.1302}{2}b_1,$$

$$a_1 = 1.1302 + \left(\frac{0.2715}{2} + 1.2661\right)b_1,$$

$$a_2 = 0.2715 + \left(\frac{1.1302}{2} + \frac{0.0443}{2}\right)b_1,$$

$$0 = 0.0443 + \frac{0.2715}{2}b_1.$$

Solving, we get $b_1 = -0.3263$, $a_0 = 1.0817$, $a_1 = 0.6727$, $a_2 = 0.0799$, and

$$e^x \doteq \frac{1.0817 + 0.6727T_1 + 0.0799T_2}{1 - 0.3263T_1},$$

$$e^x \doteq \frac{1.0018 + 0.6727x + 0.1598x^2}{1 - 0.3263x}.$$

(10.13)

The last expression results when the Chebyshev polynomials are written in terms of powers of x. In Table 10.5 the error of this rational approximation is compared to the Chebyshev expansion. We see that the maximum error is reduced by 22%. Note that we do not, nevertheless, yet have a "best approximation." The error should reach equal maximums at five points in the interval—instead the error is large near $x = 1$ and too small elsewhere.

The basis for this last statement is the minimax theorem. Based on a theorem due to Chebyshev, we may state a principle whereby we may determine whether the approximation represented by a given polynomial or rational function is optimum, in the sense that it gives the least maximum error of any rational function of the same degree of numerator and denominator on a given interval. An expression is "minimax" if and only if there are at least $N + 2$ maxima in the deviations, and these are all equal in magnitude and of alternating sign, on the interval of approximation. (Here N is the sum of the degrees of numerator and denominator of the rational function.) In the discussion that follows, we shall be referring to the magnitude of the errors and will not be concerned about their sign.

Table 10.5 Comparison of rational approximations (Eq. 10.13) with Chebyshev series for e^x

x	e^x	Chebyshev	Error	Rational function	Error
−1.0	0.3679	0.3631	0.0048	0.3686	−0.0007
−0.8	0.4493	0.4536	−0.0042	0.4488	0.0006
−0.6	0.5488	0.5534	−0.0046	0.5484	0.0005
−0.4	0.6703	0.6712	−0.0009	0.6707	−0.0004
−0.2	0.8187	0.8154	0.0033	0.8201	−0.0014
0	1.0000	0.9946	0.0054	1.0018	−0.0018
0.2	1.2214	1.2172	0.0042	1.2225	−0.0011
0.4	1.4918	1.4917	0.0001	1.4911	0.0008
0.6	1.8221	1.8267	−0.0046	1.8191	0.0030
0.8	2.2255	2.2307	−0.0051	2.2224	0.0032
1.0	2.7183	2.7121	0.0062	2.7227	−0.0044

A second important consequence of this principle is that we can put bounds on the error of the minimax expression from the range of the errors of a function that is not minimax. Suppose we have an expression like Eq. (10.13). It has five maxima of alternating sign on the interval $[-1, 1]$, as shown by Table 10.5. This is the correct number, but we know the function isn't minimax because the maxima aren't equal in magnitude. While some of the maxima are not given precisely by the table, we can see that the smallest is 0.0007 (at $x = -1$) and the largest appears to be 0.0044 (at $x = 1$). From this range of maximum error values, we can bound the maximum error of the minimax rational function [of degree $(2, 1)$]: The minimax expression will have a maximum error on $[-1, 1]$ no less than 0.0007 and no greater than 0.0044.

Similarly, the truncated Chebyshev series of degree 3 whose errors are tabulated in Table 10.2 is not truly minimax; it has five alternating maxima to its errors but they are not quite equal in magnitude. From an examination of Table 10.2 we can say, however, that the minimax polynomial of degree 3 will have a maximum error bounded by 0.0046 and 0.0062. (A tighter bound might result from a more careful computation of the errors of Eq. 10.7.) Such a prediction about the amount of improvement that will be provided by a minimax expression can help us decide whether the additional effort to find it is worthwhile.

To obtain the optimum rational function that approximates the function with equal-magnitude errors distributed through the interval is beyond the scope of this text. The approach that is used is to improve an initial estimate of a function, such as Eq. (10.13), by successive trials, often modifying the constants on the basis of experience until eventually one has a satisfactory formula. Systematic methods of determining the constants in such minimax rational approximations have also been determined. They are iteration methods beginning from an initial "sufficiently good" approximation. They are expensive to compute because the iterations involve solving a set of nonlinear equations. Ralston (1965) describes one such method. Prenter (1975) discusses the approximation of functions of several variables.

An Algorithm to Construct a Padé Approximation, $R_n(x)$

Given a power series $P_N(x)$ (a Maclaurin series if the expansion is about $x = 0$) that includes powers of x to x^N:

$$\text{SET } R_N(x) = \sum_{j=0}^{n} a_j x^j \bigg/ \left(1 + \sum_{i=1}^{m} b_i x^i\right), \ (n = m \text{ or } n = m + 1, N = n + m).$$

Subtract $R_N(x)$ from $P_N(x)$ to form a fraction.
Multiply out the numerator and combine like powers of x.
Set the coefficient of each power of x to zero to give a set of equations in the a's and b's.
Solve the equations for the coefficients of $R_n(x)$.

Note: An improved rational function approximation is obtained if the initial power series is a Chebyshev series.

10.4 Fourier Series

Polynomials are not the only functions that can be used to approximate known functions. Another means for representing known functions are approximations that use sines and cosines, called *Fourier series* after the French mathematician who first proposed, in the early 1800s, that "any function can be represented by an infinite sum of sine and cosine terms."

Fourier used these terms in his studies of heat conduction. His belief that any function can be represented in the form of a sum of sine and cosine terms with the proper coefficients, possibly with an infinite number of terms, was disputed by other mathematicians because he did not adequately develop the theory. Actually, the belief is false, for there are functions (mostly esoteric) that do not have a representation as a Fourier series. However, most functions can be so represented; we will discuss the conditions for this in another section.

Representing a function as a trigonometric series is important in solving some partial-differential equations analytically. In this section we will see how to determine the coefficients of a Fourier series.

Since a Fourier series is a sum of sine and/or cosine terms, it will obviously always be a periodic function.

Any function, $f(x)$, is periodic of period P if it has the same value for any two x-values that differ by P, or

$$f(x) = f(x + P) = f(x + 2P) = \cdots = f(x - P) = f(x - 2P) = \cdots.$$

Figure 10.4 shows such a periodic function. Additional occurrences are shown as dashed on the plot. Observe that the period can be started at any point on the x-axis. $\text{Sin}(x)$ and $\cos(x)$ are periodic of period 2π; $\sin(2x)$ and $\cos(2x)$ are periodic of period π; $\sin(nx)$ and $\cos(nx)$ are periodic of period $2\pi/n$.

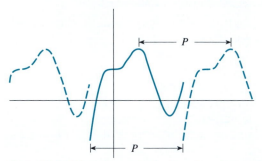

Figure 10.4 Plot of a periodic function of period $= P$

We now discuss the issue of how to find the A's and B's in a Fourier series of the form

$$f(x) \approx \frac{A_0}{2} + \sum_{n=1}^{\infty} [A_n \cos(nx) + B_n \sin(nx)]. \qquad \textbf{(10.14)}$$

(We read the symbol "$\approx$" in Eq. (10.14) as "is represented by.") The determination of the coefficients of a Fourier series [when a given function, $f(x)$, can be so represented] is based on the *property of orthogonality* for sines and cosines. For integer values of n, m:

$$\int_{-\pi}^{\pi} \sin(nx)\, dx = 0; \qquad \textbf{(10.15)}$$

$$\int_{-\pi}^{\pi} \cos(nx)\, dx = \begin{cases} 0, & n \neq 0, \\ 2\pi, & n = 0; \end{cases} \qquad \textbf{(10.16)}$$

$$\int_{-\pi}^{\pi} \sin(nx)\cos(mx)\, dx = 0; \qquad \textbf{(10.17)}$$

$$\int_{-\pi}^{\pi} \sin(nx)\sin(mx)\, dx = \begin{cases} 0, & n \neq m, \\ \pi, & n = m; \end{cases} \qquad \textbf{(10.18)}$$

$$\int_{-\pi}^{\pi} \cos(nx)\cos(mx)\, dx = \begin{cases} 0, & n \neq m, \\ \pi, & n = m. \end{cases} \qquad \textbf{(10.19)}$$

Although the term *orthogonal* should not be interpreted geometrically, it is related to the same term used for orthogonal (perpendicular) vectors whose dot product is zero. Many functions, besides sines and cosines, are orthogonal, such as the Chebyshev polynomials that were discussed previously.

To begin, we assume that $f(x)$ is periodic of period 2π and can be represented as in Eq. (10.14). We find the values of A_n and B_n in Eq. (10.14) in the following way.

1. Multiply both sides of Eq. (10.14) by $\cos(0x) = 1$, and integrate term by term between the limits of $-\pi$ and π. (We assume that this is a proper operation; you will find that it works.)

$$\int_{-\pi}^{\pi} f(x)\,dx = \int_{-\pi}^{\pi} \frac{A_0}{2}\,dx + \sum_{n=1}^{\infty} \int_{-\pi}^{\pi} A_n \cos(nx)\,dx + \sum_{n=1}^{\infty} \int_{-\pi}^{\pi} B_n \sin(nx)\,dx \quad \textbf{(10.20)}$$

Because of Eqs. (10.15) and (10.16), every term on the right vanishes except the first, giving

$$\int_{-\pi}^{\pi} f(x)\,dx = \frac{A_0}{2}(2\pi), \quad \text{or} \quad A_0 = \frac{1}{\pi} \int_{-\pi}^{\pi} f(x)\,dx. \quad \textbf{(10.21)}$$

Hence A_0 is found and it is equal to the twice the average value of $f(x)$ over one period.

2. Multiply both sides of Eq. (10.14) by $\cos(mx)$, where m is any positive integer, and integrate:

$$\int_{-\pi}^{\pi} \cos(mx) f(x)\,dx = \int_{-\pi}^{\pi} \frac{A_0}{2} \cos(mx)\,dx + \sum_{n=1}^{\infty} \int_{-\pi}^{\pi} A_n \cos(mx) \cos(nx)\,dx$$

$$+ \sum_{n=1}^{\infty} \int_{-\pi}^{\pi} B_n \cos(mx) \sin(nx)\,dx. \quad \textbf{(10.22)}$$

Because of Eqs. (10.16), (10.17), and (10.19), the only nonzero term on the right is when $n = m$ in the first summation, so we get a formula for the A's:

$$A_n = \frac{1}{\pi} \int_{-\pi}^{\pi} f(x) \cos(nx)\,dx, \quad n = 1, 2, 3, \ldots. \quad \textbf{(10.23)}$$

3. Multiply both sides of Eq. (10.14) by $\sin(mx)$, where m is any positive integer, and integrate:

$$\int_{-\pi}^{\pi} \sin(mx) f(x)\,dx = \int_{-\pi}^{\pi} \frac{A_0}{2} \sin(mx)\,dx + \sum_{n=1}^{\infty} \int_{-\pi}^{\pi} A_n \sin(mx) \cos(nx)\,dx$$

$$+ \sum_{n=1}^{\infty} \int_{-\pi}^{\pi} B_n \sin(mx) \sin(nx)\,dx. \quad \textbf{(10.24)}$$

Because of Eqs. (10.15), (10.17), and (10.18), the only nonzero term on the right is when $n = m$ in the second summation, so we get a formula for the B's:

$$B_n = \frac{1}{\pi} \int_{-\pi}^{\pi} f(x) \sin(nx)\,dx, \quad n = 1, 2, 3, \ldots. \quad \textbf{(10.25)}$$

By comparing Eqs. (10.21) and (10.23), you now see why Eq. (10.14) had $A_0/2$ as its first term. That makes the formula for all of the A's the same:

$$A_n = \frac{1}{\pi} \int_{-\pi}^{\pi} f(x) \cos(nx)\,dx, \quad n = 0, 1, 2, \ldots. \quad \textbf{(10.26)}$$

Fourier Series for Periods Other Than 2π

What if the period of $f(x)$ is not 2π? No problem—we just make a change of variable. If $f(x)$ is periodic of period P, the function can be considered to have one period between $-P/2$ and $P/2$. The functions $\sin(2\pi x/P)$ and $\cos(2\pi x/P)$ are periodic between $-P/2$ and $P/2$. (When $x = -P/2$, the angle becomes $-\pi$; when $x = P/2$, it is π.) We can repeat the preceding developments for sums of $\cos(2n\pi x/P)$ and $\sin(2n\pi x/P)$, or rescale the preceding results. In any event, the formulas become, for $f(x)$ periodic of period P:

$$A_n = \frac{2}{P} \int_{-P/2}^{P/2} f(x) \cos\left(\frac{n\pi x}{P/2}\right) dx, \qquad n = 0, 1, 2, \ldots, \qquad \textbf{(10.27)}$$

$$B_n = \frac{2}{P} \int_{-P/2}^{P/2} f(x) \sin\left(\frac{n\pi x}{P/2}\right) dx, \qquad n = 1, 2, 3, \ldots. \qquad \textbf{(10.28)}$$

Since a function that is periodic with period P between $-P/2$ and $P/2$ is also periodic with period P between A and $A + P$, the limits of integration in Eqs. (10.27) and (10.28) can be from 0 to P.

EXAMPLE 10.3   Let $f(x) = x$ be periodic between $-\pi$ and π. (See Fig. 10.5.) Find the A's and B's of its Fourier expansion.

For A_0:

$$A_0 = \frac{1}{\pi} \int_{-\pi}^{\pi} f(x)\, dx = \frac{1}{\pi} \int_{-\pi}^{\pi} x\, dx = \frac{x^2}{2\pi}\Bigg]_{-\pi}^{\pi} = 0. \qquad \textbf{(10.29)}$$

For the other A's:

$$A_n = \frac{1}{\pi} \int_{-\pi}^{\pi} x \cos(nx)\, dx = \frac{1}{\pi}\left(\frac{\cos(nx)}{n^2} + \frac{x\,\sin(nx)}{n}\right)\Bigg]_{-\pi}^{\pi} = 0. \qquad \textbf{(10.30)}$$

For the B's:

$$B_n = \frac{1}{\pi} \int_{-\pi}^{\pi} x \sin(nx)\, dx = \frac{1}{\pi}\left(\frac{\sin(nx)}{n^2} - \frac{x\,\cos(nx)}{n}\right)\Bigg]_{-\pi}^{\pi}$$

$$= \frac{2(-1)^{n+1}}{n}, \qquad n = 1, 2, 3, \ldots. \qquad \textbf{(10.31)}$$

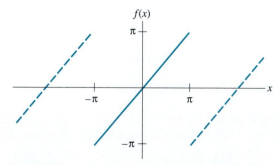

Figure 10.5   Plot of $f(x) = x$, periodic of period 2π

We then have

$$x \approx 2 \sum_{n=1}^{\infty} \frac{(-1)^{n+1}}{n} \sin(nx), \qquad -\pi < x < \pi. \qquad \textbf{(10.32)}$$

Figure 10.6 shows how the series approximates to the function when only two, four, or eight terms are used.

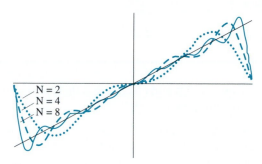

Figure 10.6 Plot of Eq. (10.32) for N = 2, 4, 8

EXAMPLE 10.4 Find the Fourier coefficients for $f(x) = |x|$ on $-\pi$ to π:

$$A_0 = \frac{1}{\pi} \int_{-\pi}^{0} (-x)\,dx + \frac{1}{\pi} \int_{0}^{\pi} x\,dx = \frac{2}{\pi} \int_{0}^{\pi} x\,dx = \pi; \qquad \textbf{(10.33)}$$

$$\begin{aligned} A_n &= \frac{1}{\pi} \int_{-\pi}^{0} (-x)\cos(nx)\,dx + \frac{1}{\pi} \int_{0}^{\pi} x\cos(nx)\,dx \\ &= \frac{2}{\pi} \left(\frac{\cos(nx)}{n^2} + \frac{x\sin(nx)}{n} \right) \Big]_{0}^{\pi} \\ &= \begin{cases} 0, & n = 2, 4, 6, \ldots, \\ \dfrac{-4}{n^2\pi}, & n = 1, 3, 5, \ldots; \end{cases} \end{aligned} \qquad \textbf{(10.34)}$$

$$B_n = \frac{1}{\pi} \int_{-\pi}^{0} (-x)\sin(nx)\,dx + \frac{1}{\pi} \int_{0}^{\pi} x\sin(nx)\,dx = 0. \qquad \textbf{(10.35)}$$

Since the definite integrals in Eq. (10.34) are nonzero only for odd values of n, it simplifies to change the index of the summation. The Fourier series is then

$$|x| \approx \frac{\pi}{2} - \frac{4}{\pi} \sum_{n=1}^{\infty} \frac{\cos((2n-1)x)}{(2n-1)^2}. \qquad \textbf{(10.36)}$$

Figure 10.7 shows how the series approximates the function when two, four, or eight terms are used.

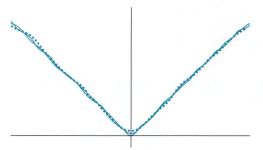

Figure 10.7 Plot of Eq. (10.36) for N = 2, 4, 8

When you compare Eqs. (10.32) and (10.36) and their plots in Figs. 10.6 and 10.7, you will notice several differences:

1. The first contains only sine terms, the second only cosines.
2. Equation (10.32) gives a value at both endpoints that is the average of the end values for $f(x)$, where there is a discontinuity.
3. Equation (10.36) gives a closer approximation when only a few terms are used.

Example 10.5 will further examine these points.

EXAMPLE 10.5 Find the Fourier coefficients for $f(x) = x(2 - x) = 2x - x^2$ over the interval $(-2, 2)$ if it is periodic of period 4. Equations (10.27) and (10.28) apply.

$$A_0 = \frac{2}{4} \int_{-2}^{2} (2x - x^2)\, dx = \frac{-8}{3} \tag{10.37}$$

$$A_n = \frac{2}{4} \int_{-2}^{2} (2x - x^2) \cos\left(\frac{n\pi x}{2}\right) dx = \frac{16(-1)^{n+1}}{n^2\pi^2}, \quad n = 1, 2, 3, \ldots \tag{10.38}$$

$$B_n = \frac{2}{4} \int_{-2}^{2} (2x - x^2) \sin\left(\frac{n\pi x}{2}\right) dx = \frac{8(-1)^{n+1}}{n\pi}, \quad n = 1, 2, 3, \ldots \tag{10.39}$$

$$(2 - x) \approx \frac{-4}{3} + \frac{16}{\pi^2} \sum_{n=1}^{\infty} \frac{(-1)^{n+1}}{n^2} \cos\left(\frac{n\pi x}{2}\right) + \frac{8}{\pi} \sum_{n=1}^{\infty} \frac{(-1)^{n+1}}{n} \sin\left(\frac{n\pi x}{2}\right) \tag{10.40}$$

You will notice that both sine and cosine terms occur in the Fourier series and that the discontinuity at the endpoints shows itself in forcing the Fourier series to reach the average value. It should also be clear that the series is the sum of separate series for $2x$ and $-x^2$. Figure 10.8 shows how the series of Eq. (10.40) approximates to the function when 40 terms are used. It is obvious that many more terms are needed to reduce the error to negligible proportions because of the extreme oscillation near the discontinuities, called the *Gibbs phenomenon*. The conclusion is that a Fourier series often involves a lot of computation as well as awkward integrations to give the formula for the coefficients.

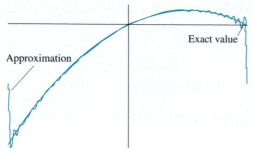

Figure 10.8   Plot of Eq. (10.40) for N = 40

Fourier Series for Nonperiodic Functions and Half-Range Expansions

The development until now has been for a periodic function. What if $f(x)$ is not periodic? Can we approximate it by a trigonometric series? We assume that we are interested in approximating the function only over a limited interval and we do not care whether the approximation holds outside of that interval. This situation frequently occurs when we want to solve partial-differential equations analytically.

Suppose we have a function defined for all x-values, but we are only interested in representing it over $(0, L)$.* Figure 10.9 is typical. Because we will ignore the behavior of the function outside of $(0, L)$, we can redefine the behavior outside that interval as we wish. Figures 10.10 and 10.11 show two possible redefinitions.

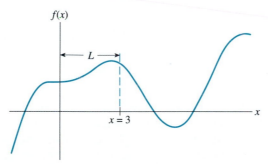

Figure 10.9   A function, $f(x)$, of interest on [0, 3]

In the first redefinition, we have reflected the portion of $f(x)$ about the y-axis and have extended it as a periodic function of period $2L$. This creates an *even* periodic

---

* If the range of interest is $[a, b]$, a simple change of variable can make this $[0, L]$.

function. If we reflect it about the origin and extend it periodically, we create an *odd* periodic function of period $2L$. More formally, we define even and odd functions through these relations:

$$f(x) \text{ is even if } f(-x) = f(x), \tag{10.41}$$

$$f(x) \text{ is odd if } f(-x) = -f(x). \tag{10.42}$$

It is easy to see that $\cos(Cx)$ is an even function and that $\sin(Cx)$ is an odd function for any real value of C.

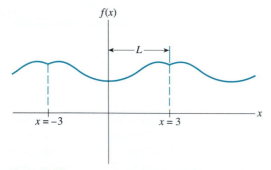

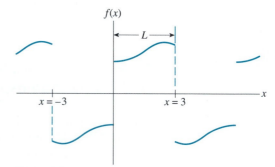

Figure 10.10   Plot of a function reflected about the y-axis

Figure 10.11   Plot of a function reflected about the origin

There are two important relationships for integrals of even and odd functions. (If you think of the integrals in a geometric interpretation, these relationships are obvious.)

$$\text{If } f(x) \text{ is even, } \int_{-L}^{L} f(x)\,dx = 2\int_{0}^{L} f(x)\,dx. \tag{10.43}$$

$$\text{If } f(x) \text{ is odd, } \int_{-L}^{L} f(x)\,dx = 0. \tag{10.44}$$

It is also easy to show that the product of two even functions is even, that the product of two odd functions is even, and that the product of an even and an odd function is odd. This means that, if $f(x)$ is even, $f(x)\cos(nx)$ is even and $f(x)\sin(nx)$ is odd. Further, if $f(x)$ is odd, $f(x)\cos(nx)$ is odd and $f(x)\sin(nx)$ is even. Because of Eq. (10.41), the Fourier series expansion of an even function will contain only cosine terms (all the B-coefficients are zero). Also, if $f(x)$ is odd, its Fourier expansion will contain only sine terms (all the A-coefficients are zero). These facts are important when we develop the "half-range" expansion of a function.

Therefore, if we want to represent $f(x)$ between 0 and L as a Fourier series and are interested only in approximating it on the interval $(0, L)$, we can redefine f within the interval $(-L, L)$ in two importantly different ways: (1) We can redefine the portion from $-L$ to 0 by reflecting about the y-axis. We then generate an even function. (2) We can reflect the portion between 0 and L about the origin to generate an odd function. Figures 10.10 and 10.11 showed these two possibilities.

Thus two different Fourier series expansions of $f(x)$ on $(0, L)$ are possible, one that has only cosine terms or one that has only sine terms. We get the A's for the even extension of $f(x)$ on $(0, L)$ from

$$A_n = \frac{2}{L} \int_0^L f(x) \cos\left(\frac{n\pi x}{L}\right) dx, \qquad n = 0, 1, 2, \ldots. \qquad \textbf{(10.45)}$$

We get the B's for the odd extension of $f(x)$ on $(0, L)$ from

$$B_n = \frac{2}{L} \int_0^L f(x) \sin\left(\frac{n\pi x}{L}\right) dx, \qquad n = 1, 2, 3, \ldots. \qquad \textbf{(10.46)}$$

EXAMPLE 10.6 Find the Fourier cosine series expansion of $f(x)$, given that

$$f(x) = \begin{cases} 0, & 0 < x < 1, \\ 1, & 1 < x < 2. \end{cases} \qquad \textbf{(10.47)}$$

Figure 10.12 shows the even extension of the function.

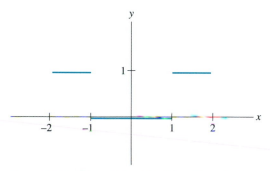

Figure 10.12 Plot of Eq. (10.47) reflected about the y-axis

Since we are dealing with an even function on $(-2, 2)$, we know that the Fourier series will have only cosine terms. We get the A's with

$$A_0 = \frac{2}{2} \int_1^2 (1)\, dx = 1;$$

$$A_n = \frac{2}{2} \int_1^2 (1) \cos\left(\frac{n\pi x}{2}\right) dx = \begin{cases} 0, & n \text{ even}, \\ \dfrac{2(-1)^{(n+1)/2}}{n\pi}, & n \text{ odd}. \end{cases} \qquad \textbf{(10.48)}$$

Then the Fourier Cosine series is

$$f(x) \approx \frac{1}{2} + \frac{2}{\pi} \sum_{n=1}^{\infty} \frac{(-1)^n \cos((2n-1)(\pi x/2))}{(2n-1)}. \qquad \textbf{(10.49)}$$

EXAMPLE 10.7 Find the Fourier sine series expansion for the same function as in Example 10.6. Figure 10.13 shows the odd extension of the function.

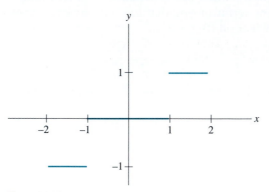

Figure 10.13 Plot of Eq. (10.47) reflected about the origin

We know that all of the A-coefficients will be zero, so we need to compute only the B's:

$$
\begin{aligned}
B_n &= \frac{2}{2}\int_1^2 (1)\sin\left(\frac{n\pi x}{2}\right)dx \\
&= \frac{2}{n\pi}\left[-\cos(n\pi) + \cos\left(\frac{n\pi}{2}\right)\right], \qquad n = 1,2,3,\dots.
\end{aligned}
\tag{10.50}
$$

The term in brackets gives the sequence $1, -2, 1, 0, 1, -2, 1, 0, \dots$. Because this sequence is awkward to reduce (except by use of the mod function), we simply write

$$
f(x) = \frac{2}{\pi}\sum_{n=1}^{\infty}\frac{[\cos(n\pi/2) - \cos(n\pi)]}{n}\sin\left(\frac{n\pi x}{2}\right).
\tag{10.51}
$$

Summary of Formulas for Computation of Fourier Coefficients

A function that is periodic of period P and meets certain criteria (see below) can be represented by Eq. (10.52):

$$
f(x) = \frac{A_0}{2} + \sum_{n=1}^{\infty} A_n \cos\left(\frac{n\pi x}{P/2}\right) + \sum_{n=1}^{\infty} B_n \sin\left(\frac{n\pi x}{P/2}\right).
\tag{10.52}
$$

The coefficients can be computed with

$$
A_n = \frac{2}{P}\int_{-P/2}^{P/2} f(x)\cos\left(\frac{n\pi x}{P/2}\right)dx, \qquad n = 0,1,2,\dots,
\tag{10.53}
$$

$$
B_n = \frac{2}{P}\int_{-P/2}^{P/2} f(x)\sin\left(\frac{n\pi x}{P/2}\right)dx, \qquad n = 1,2,3,\dots.
\tag{10.54}
$$

(The limits of the integrals can be from A to $A + P$.)

If $f(x)$ is an even function, only the A's will be nonzero. Similarly, if $f(x)$ is odd, only the B's will be nonzero. If $f(x)$ is neither even nor odd, its Fourier series will contain both cosine and sine terms.

Even if $f(x)$ is not periodic, it can be represented on just the interval $(0, L)$ by redefining the function over $(-L, 0)$ by reflecting $f(x)$ about the y-axis or, alternatively, about the origin. The first creates an even function, the second an odd function. The Fourier series of the redefined function will actually represent a periodic function of period $2L$ that is defined for $(-L, L)$.

When L is the half-period, the Fourier series of an even function contains only cosine terms and is called a Fourier cosine series. The A's can be computed by

$$A_n = \frac{2}{L} \int_0^L f(x) \cos\left(\frac{n\pi x}{L}\right) dx, \qquad n = 0, 1, 2, \ldots. \qquad \textbf{(10.55)}$$

The Fourier series of an odd function contains only sine terms and is called a Fourier sine series. The B's can be computed by

$$B_n = \frac{2}{L} \int_0^L f(x) \sin\left(\frac{n\pi x}{L}\right) dx, \qquad n = 1, 2, 3, \ldots. \qquad \textbf{(10.56)}$$

If $f(x)$ (or its redefined extension) has a finite discontinuity, the Fourier series will converge to the average of the two limiting values at the discontinuity. The Fourier series converges slowly at a point of discontinuity and exhibits more pronounced oscillations (the Gibbs phenomenon) near that point. If $f(x)$ (or the redefined function) has a discontinuity in its first derivative, convergence will be slower at that point.

10.5   Getting Fourier Coefficients Numerically and the Fast Fourier Transform

Since the coefficients of a Fourier series are obtained by integration, we can compute them numerically rather than through formal integration. Example 10.8 shows how accurate the computation is, first by the trapezoidal rule, then by Simpson's $\frac{1}{3}$ rule.

EXAMPLE 10.8   Evaluate the coefficients for the half-range expansions for the function $f(x) = x$ on $[0, 2]$ by numerical integration.

For the even extension (Fourier cosine series), we get the A's from

$$A_n = \left(\frac{2}{2}\right) \int_0^2 x \cos\left(\frac{n\pi x}{2}\right) dx, \qquad n = 0, 1, 2, \ldots.$$

For the odd extension (Fourier sine series), we get the B's from

$$B_n = \left(\frac{2}{2}\right) \int_0^2 x \sin\left(\frac{n\pi x}{2}\right) dx, \qquad n = 1, 2, 3, \ldots.$$

Table 10.6   Comparison of numerical integration with analytical results: 20 subdivisions of $[0, 2]$

	Trapezoidal rule		Simpson's rule		Analytical integration	
n	A_n	B_n	A_n	B_n	A_n	B_n
0	2		2		2	
1	−0.81224	1.27062	−0.81056	1.27324	−0.81057	1.27323
2	0	−0.63138	0	−0.63665	0	−0.63662
3	−0.09175	0.41653	−0.08999	0.42453	−0.09006	0.42441
4	0	−0.30777	0	−0.31860	0	−0.31831
5	−0.03414	0.24142	−0.03219	0.25523	−0.03242	0.25465

Table 10.6 shows the results from numerical integrations with 20 subdivisions of the interval $[0, 2]$ using both the trapezoidal rule and Simpson's rule and compares the results with analytical integrations. Table 10.7 does the same but for 200 intervals.

Table 10.7   Comparison of numerical integration with analytical results: 200 subdivisions of $[0, 2]$

	Trapezoidal rule		Simpson's rule		Analytical integration	
n	A_n	B_n	A_n	B_n	A_n	B_n
0	2		2		2	
1	−0.81059	1.27321	−0.81057	1.27324	−0.81057	1.27323
2	0	−0.63657	0	−0.63662	0	−0.63662
3	−0.09008	0.42433	−0.09006	0.42441	−0.09006	0.42441
4	0	−0.31821	0	−0.31831	0	−0.31831
5	−0.03244	0.25452	−0.03242	0.25465	−0.03242	0.25465

It is clear (as expected) that we can use numerical integrations. The accuracy with Simpson's rule is good—essentially perfect agreement is seen in Table 10.7. The errors do increase as n increases (particularly apparent in Table 10.6).

The implication of Example 10.8 is that we can compute Fourier coefficients on a table of data, say from measurements for a periodic phenomenon. Doing so performs what is termed a *discrete Fourier transform*. Other names for this procedure are *harmonic analysis* and *finite Fourier transform*. The data, which originally are most often functions of time, can now be interpreted in terms of the angles that appear in a Fourier series, more often expressed as the frequencies. It is a transform because we change a function of t (time) to an equivalent function of frequencies.

Why should we want to so transform a set of experimental data? Because knowing which frequencies of a Fourier series are most significant (have the largest coefficients) gives information on the fundamental frequencies of the system. This knowledge is important because an applied periodic external force that includes components of the same frequency as one of these fundamental frequencies causes extremely large disturbances. (Such a periodic force may come from vibrations from rotating machinery, from wind, or from earthquakes.) We normally want to avoid such extreme responses for fear that the system will be damaged. Here is an example.

EXAMPLE 10.9   An experiment (actually, these are contrived data) showed the displacements given in Table 10.8 when the system was caused to vibrate in its natural modes. The values represent a periodic function on the interval for t of $[2, 10]$ because they repeat themselves after $t = 10$.

Table 10.8   Measurements of displacements versus time

t	Displacement	t	Displacement
2.000	3.804	6.250	3.746
2.250	6.503	6.500	5.115
2.500	7.496	6.750	4.156
2.750	6.094	7.000	1.593
3.000	3.003	7.250	-0.941
3.250	-0.105	7.500	-1.821
3.500	-1.598	7.750	-0.329
3.750	-0.721	8.000	2.799
4.000	1.806	8.250	5.907
4.250	4.350	8.500	7.338
4.500	5.255	8.750	6.380
4.750	3.878	9.000	3.709
5.000	0.893	9.250	0.992
5.250	-2.048	9.500	-0.116
5.500	-3.280	9.750	1.047
5.750	-2.088	10.000	3.802
6.000	0.807		

We will use trapezoidal integration to find the Fourier series coefficients for the data. Doing so gives these values for the A's and B's:

n	A	B
0	4.6015	
1	1.5004	-0.5006
2	-0.0009	0.0016
3	-0.0017	0.0016
4	0.0008	4.0011
5	-0.0017	0.0000
6	-0.0009	0.0022
7	-0.0005	-0.0023
8	-0.0008	0.0009

This shows that only A_0, A_1, B_1, and B_4 are important. There would be no amplification of motion from forces that do not include the frequencies corresponding to these.

(Table 10.8 was constructed from

$$f(t) = 2.3 + 1.5 \cos(t) - 0.5 \sin(t) + 4 \sin(4t),$$

plus a small random variation whose values ranged from -0.01 to $+0.01$. It is the random variations that cause nonzero values for the insignificant A's and B's.)

The Fast Fourier Transform

If we need to do a finite Fourier transform on lots of data, the effort used in carrying out the computations is exorbitant. In the preceding examples, where we reevaluated cosines and sines numerous times, we should have recognized that many of these values are the same. When we evaluate the integrals for a finite Fourier transform, we compute sines and cosines for angles around the origin, as indicated in Fig. 10.14.

When we need to find $\cos(nx)$ and $\sin(nx)$, we move around the circle; when $n = 1$, we use each value in turn. For other values of n, we use every nth value, but it is easy to see that these repeat previous values. The *fast Fourier transform* (often written as *FFT*) takes advantage of this fact to avoid the recomputations.

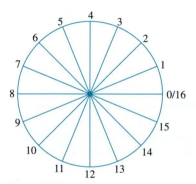

Figure 10.14 Angles used in computing for 16 points

In developing the FFT algorithm, the preferred method is to use an alternative form of the Fourier series. Instead of

$$f(x) \approx \frac{A_0}{2} + \sum_{n=1}^{\infty} [A_n \cos(nx) + B_n \sin(nx)], \qquad (\text{period} = 2\pi), \quad \textbf{(10.57)}$$

we will use an equivalent form in terms of complex exponentials. Utilizing Euler's identity (using i as $\sqrt{-1}$),

$$e^{ijx} = \cos(jx) + i \sin(jx),$$

we can write Eq. (10.57) as

$$f(x) = \sum_{j=0}^{\infty} (c_j e^{ijx} + c_{-j} e^{-ijx})$$

$$= 2c_0 + \sum_{j=1}^{\infty} [(c_j + c_{-j}) \cos(jx) + i(c_j - c_{-j}) \sin(jx)] \qquad \textbf{(10.58)}$$

$$= \sum_{j=-\infty}^{\infty} c_j e^{ijx}.$$

We can match up the A's and B's of Eq. (10.57) to the c's of (10.58):

$$a_j = c_j + c_{-j}, \qquad b_j = i(c_j - c_{-j}),$$

$$c_j = \frac{a_j - ib_j}{2}, \qquad c_{-j} = \frac{a_j + ib_j}{2}. \qquad \textbf{(10.59)}$$

For $f(x)$ real, it is easy to show that $c_0 = \bar{c}_0$ and $c_j = \bar{c}_{-j}$, where the bars represent complex conjugates.

For integers j and k, it is true that

$$\int_0^{2\pi} (e^{ikx})(e^{ijx}) \, dx = \int_0^{2\pi} e^{i(k+j)x} \, dx = \begin{cases} 0 & \text{for } k \neq -j, \\ 2\pi & \text{for } k = -j. \end{cases}$$

(You can verify the first of these through Euler's identity.) This allows us to evaluate the c's of Eq. (10.58) by the following.

For each fixed k, we get

$$f(x) e^{-ikx} = \sum_{j=-\infty}^{\infty} c_k e^{i(j-k)x},$$

$$\int_0^{2\pi} f(x) e^{-ikx} \, dx = 2\pi c_k, \qquad \text{or} \qquad \textbf{(10.60)}$$

$$c_k = \frac{1}{2\pi} \int_0^{2\pi} f(x) e^{-ikx} \, dx, \qquad k = 0, \pm 1, \pm 2, \ldots.$$

EXAMPLE 10.10 (You should verify each of these.)

1. Let $f(x) = x$; then

$$c_k = \frac{1}{2\pi} \int_0^{2\pi} x e^{-ikx} \, dx = -\frac{i}{k}, \qquad k \neq 0.$$

2. Let $f(x) = x(2\pi - x)$; then

$$c_k = \frac{1}{2\pi} \int_0^{2\pi} x(2\pi - x) e^{-ikx} \, dx = \frac{2}{k^2}, \qquad k \neq 0.$$

3. Let $f(x) = \cos(x)$; then

$$c_k = \frac{1}{2\pi} \int_0^{2\pi} \cos(x) e^{-ikx} \, dx = \begin{cases} \dfrac{1}{2} & \text{for } k = 1 \text{ or } -1, \\ 0 & \text{for all other } k. \end{cases}$$

Note that for Eq. (10.57) this makes $a_1 = 1$ and all the other a_j's = 0. Thus, for a given $f(x)$ that satisfies continuity conditions, we have

$$c_j = \frac{1}{2\pi} \int_0^{2\pi} f(x) e^{-ijx} \, dx, \qquad j = 0, \pm 1, \pm 2, \ldots .$$

The magnitudes of the Fourier series coefficients $|c_j|$ are the *power spectrum of f*; these show the frequencies that are represented in $f(x)$. If we know $f(x)$ in the time domain, we can identify f by computing the c_j's. In getting the Fourier series, we have transformed from the time domain to the frequency domain, an important aspect of wave analysis.

Suppose we have N values for $f(x)$ on the interval $[0, 2\pi]$ at equispaced points, $x_k = 2\pi k/N, k = 0, 1, \ldots, N - 1$. Since $f(x)$ is periodic, $f_N = f_0$, $f_{N+1} = f_1$, and so on. Instead of formal analytical integration, we would use a numerical integration method to get the coefficients. Even if $f(x)$ is known at all points in $[0, 2\pi]$, we might prefer to use numerical integration. This would use only certain values of $f(x)$, often those evaluated at uniform intervals. It is also often true that we do not know $f(x)$ everywhere, because we have sampled a continuous signal. In that case, however, it is better to use the discrete Fourier transform, which can be defined as

$$X(n) = \sum_{k=0}^{N-1} x_0(k) e^{-i2\pi nk/N}, \qquad n = 0, 1, 2, \ldots, N - 1. \qquad \textbf{(10.61)}$$

In Eq. (10.61), we have changed notation to conform more closely to the literature on FFT. $X(n)$ corresponds to the coefficients of N frequency terms, and the $x_0(k)$ are the N values of the signal samples in the time domain. You can think of n as indexing the X-terms and k as indexing the x_0-terms. Equation (10.61) corresponds to a set of N linear equations that we can solve for the unknown $X(n)$. Since the unknowns appear on the left-hand side of (10.61), this requires only the multiplication of an N-component vector by an $N \times N$ matrix.

It will simplify the notation if we let $W = e^{-i2\pi/N}$, making the right-hand-side terms of Eq. (10.61) become $x_0(k)W^{nk}$. To develop the FFT algorithm, suppose that $N = 4$. We write the four equations for this case:

$$X(0) = W^0 x_0(0) + W^0 x_0(1) + W^0 x_0(2) + W^0 x_0(3),$$
$$X(1) = W^0 x_0(0) + W^1 x_0(1) + W^2 x_0(2) + W^3 x_0(3),$$
$$X(2) = W^0 x_0(0) + W^2 x_0(1) + W^4 x_0(2) + W^6 x_0(3),$$
$$X(3) = W^0 x_0(0) + W^3 x_0(1) + W^6 x_0(2) + W^9 x_0(3).$$

In matrix form:

$$\begin{bmatrix} X(0) \\ X(1) \\ X(2) \\ X(3) \end{bmatrix} = \begin{bmatrix} W^0 & W^0 & W^0 & W^0 \\ W^0 & W^1 & W^2 & W^3 \\ W^0 & W^2 & W^4 & W^6 \\ W^0 & W^3 & W^6 & W^9 \end{bmatrix} x_0. \qquad \textbf{(10.62)}$$

In solving the set of N equations in the form of Eq. (10.62), we will have to make N^2 complex multiplications plus $N(N-1)$ complex additions. The value of the FFT is that the number of such operations is greatly reduced. While there are several variations on the algorithm, we will concentrate on the Cooley–Tukey formulation. The matrix of Eq. (10.62) can be factored to give an equivalent form for the set of equations. At the same time we will use the fact that $W^0 = 1$ and $W^k = W^{k \bmod(N)}$:

$$\begin{bmatrix} X(0) \\ X(2) \\ X(1) \\ X(3) \end{bmatrix} = \begin{bmatrix} 1 & W^0 & 0 & 0 \\ 1 & W^2 & 0 & 0 \\ 0 & 0 & 1 & W^1 \\ 0 & 0 & 1 & W^3 \end{bmatrix} \begin{bmatrix} 1 & 0 & W^0 & 0 \\ 0 & 1 & 0 & W^0 \\ 1 & 0 & W^2 & 0 \\ 0 & 1 & 0 & W^2 \end{bmatrix} x_0. \tag{10.63}$$

You should verify that the factored form (Eq. 10.63) is exactly equivalent to Eq. (10.62) by multiplying out. Note carefully that the elements of the X-vector are scrambled. (The development can be done formally and more generally by representing n and k as binary values, but it will suffice to show the basis for the FFT algorithm by expanding on this simple $N = 4$ case.)

By using the factored form, we now get the values of $X(n)$ by two steps (stages), in each of which we multiply a matrix times a vector. In the first stage, we transform x_0 into x_1 by multiplying the right matrix of Eq. (10.63) and x_0. In the second stage, we multiply the left matrix and x_1, getting x_2. We get X by unscrambling the components of x_2. By doing the operation in stages, the number of complex multiplications is reduced to $N(\log_2 N)$. For $N = 4$, this is a reduction by one-half, but for large N it is very significant; if $N = 1024$, there are 10 stages and the reduction in complex multiplies is a hundredfold!

It is convenient to represent the sequence of multiplications of the factored form (Eq. 10.63 or its equivalent for larger N) by flow diagrams. Figure 10.15 is for $N = 4$ and Fig. 10.16 is for $N = 16$. Each column holds values of x_{ST}, where the subscript tells which stage is being computed; ST ranges from 1 to 2 for $N = 4$ and from 1 to 4 for $N = 16$. [The number of stages, for N a power of 2, is $\log_2(N)$.] In each stage, we get x-values of the next stage from those of the present stage. Every new x-value is the sum of the two x-values from the previous stage that connect to it, with one of these multiplied by a power of W. The diagram tells which x_{ST} terms are combined

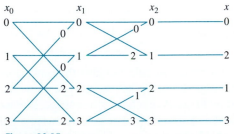

Figure 10.15

to give an x_{ST+1} term, and the numbers shown within the lines are the powers of W that are used. For example, looking at Fig. 10.16, we see that

$$x_2(6) = x_1(2) + W^8 x_1(6),$$
$$x_3(13) = x_2(13) + W^6 x_2(15),$$
$$x_4(9) = x_3(8) + W^9 x_3(9),$$

and so on.

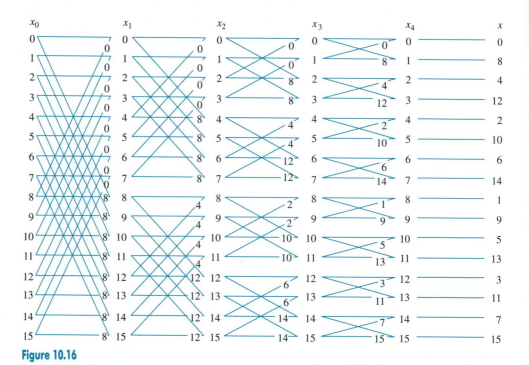

Figure 10.16

The last columns in Figs. 10.15 and 10.16 indicate how the final x-values are unscrambled to give the X-values. This relationship can be found by expressing the index k of x in the last stage as a binary number and reversing the bits; this gives n in $X(n)$. For example, in Fig. 10.16, we see that $x_4(3) = X(12)$ and $x_4(11) = X(13)$. From the bit-reversing rule, we get

$$3 = 0011_2 \rightarrow 1100_2 = 12, \qquad 11 = 1011_2 \rightarrow 1101_2 = 13.$$

Observe also that the bit-reversing rule can give the powers of W that are involved in computing the next stage. For the last stage, the powers are identical to the numbers obtained by bit reversal. At each previous stage, however, only the first half of the powers are employed, but each power is used twice as often. It is of interest to see how we can generate these values. Computer languages that facilitate bit manipulations make this an easy job, but there is a good alternative. Observe how the powers in Fig. 10.15 differ from those in Fig. 10.16 and how they progress from stage to stage. The following table pinpoints this.

Stage	$N = 4$				$N = 16$															
1:	0	0	2	2	0	0	0	0	0	0	0	0	8	8	8	8	8	8	8	8
2:	0	2	1	3	0	0	0	0	8	8	8	8	4	4	4	4	12	12	12	12
3:					0	0	8	8	4	4	12	12	2	2	10	10	6	6	14	14
4:					0	8	4	12	2	10	6	14	1	9	5	13	3	11	7	15

Can you see what a similar table for $N = 2$ would look like? Its single row would be 0   1. Now we see that the row of powers for the last stage can be divided into two halves with the numbers in the second half always 1 greater than the corresponding entry in the first half. The row above is the left half of the current row with each value repeated. This observation leads to the following algorithm.

Algorithm to Generate Powers of W in FFT

For N a power of 2, let $Q = \log_2(N)$.

Initialize an array P of length N to all zeros.
Let ST = 1.
Repeat
   Double the values of $P(K)$ for $K = 1 .. 2^{ST-1}$,
   Let each $P(K + 2^{ST-1}) = P(K) + 1$ for $K = 0 .. 2^{ST-1} - 1$,
   Increment ST,
Until ST > Q.

The successive new values for powers of W are now in array P.

EXAMPLE 10.11 Use the algorithm to generate the powers of W for $N = 8$:
$$Q = \log_2(8) = 3.$$

K:	0	1	2	3	4	5	6	7
Initial P array:	0	0	0	0	0	0	0	0
ST = 1, doubled:	0	0	0	0	0	0	0	0
add 1:	0	1	0	0	0	0	0	0
ST = 2, doubled:	0	2	0	0	0	0	0	0
add 1:	0	2	1	3	0	0	0	0
ST = 3, doubled:	0	4	2	6	0	0	0	0
add 1:	0	4	2	6	1	5	3	7

The last row of values corresponds to the bits of 000 to 111 after reversal.

Our discussion has assumed that N is a power of 2; for this case the economy of the FFT is a maximum. When N is not a power of 2 but can be factored, there are adaptations of the general idea that reduce the number of operations, but they are more than $N \log_2(N)$. See Brigham (1974) for a discussion of this as well as a fuller treatment of the theory behind FFT.

More recently there has been interest in another transform, called the discrete Hartley transform. A discussion of this transform would parallel our discussion of the Fourier transform. Moreover, it has been shown that this transform can be converted into a fast Hartley transform (FHT) that reduces to $N \log_2(N)$ computations. For a full coverage of the FHT, one should consult Bracewell (1986). The advantages of the FHT over the Fourier transform are its faster and easier computation. Moreover, it is easy to compute the FFT from the Hartley transform. However, the main power of the FHT is that all the computations are done in real arithmetic, so that we can use a language like Pascal that does not have a complex data type. An interesting and easy introduction into the FHT is found in O'Neill (1988).

An Algorithm to Perform a Fast Fourier Transform

Given n data points, (x_i, f_i), $i = 0, \ldots, n - 1$ (with n a power of 2) and x on $[0, 2\pi]$:

SET $yr_i = f_i, i = 0, \ldots, n - 1$.
SET $yi_i = 0, i = 0, \ldots, n - 1$.
SET $c_i = \cos(2i\pi/n), i = 0, \ldots, n - 1$. (These are trigonometric
SET $s_i = \sin(2i\pi/n), i = 0, \ldots, n - 1$. values that are used.)
SET nst $= \log(n)/\log(2)$. (nst = number of stages)
SET $p_i = 0, i = 0, \ldots, n - 1$. (Use the previous algorithm
DO FOR st = 1 TO nst: to get
   SET $p_i = 2p_i, i = 1, \ldots, 2^{st-1}$. "bit reversal"
   SET $p_{i+2} = p_i + 1, i = 0, \ldots, 2^{st-1}$. values)
ENDDO (FOR st).
SET stage = 1. (These values
SET nsets = 1. are for the
SET del $= n/2$. first stage—
SET $k = 0$. k indexes y-value being computed.)
REPEAT
  DO FOR set = 1 TO nsets:
    DO FOR $i = 0$ TO $n/\text{nsets} - 1$:
      SET $j = i$ MOD del $+ (\text{set} - 1) * \text{del} * 2$. (Indexes old y-values)
      SET $\ell = p_{\text{INT}(k/\text{del})}$. (Indexes c_i, s_i values)
      SET $yyr_k = yr_j + c_\ell * yr_{j+\text{del}} - s_\ell * yi_{j+\text{del}}$.
      SET $yyi_k = yi_j + c_\ell * yi_{j+\text{del}} - s_\ell * yr_{j+\text{del}}$.
      SET $k = k + 1$.
    ENDDO (FOR i).

ENDDO (FOR set).
SET $yr_i = yyr_i, i = 0 .. n - 1$.   (Reset
SET $yi_i = yyi_i, i = 0 .. n - 1$.   values
SET stage = stage + 1.   for
SET nsets = nsets $* 2$.   next
SET del = del/2.   stage.)
SET $k = 0$.
UNTIL stage > nst.

When terminated, the A's and B's of the Fourier series are contained in the yr and yi arrays. These must be divided by $n/2$ and should be unscrambled using the p-array values as indices.

Note: the algorithm can be used when the f_i are complex numbers; set the imaginary parts into yi_i in this case.)

EXAMPLE 10.12 Use the FFT algorithm to obtain the finite Fourier series coefficients for the same data as in Table 10.8.

A computer program that implements the algorithm gave these results:

n	A_n	B_n
0	4.6017	
1	1.4993	−0.4994
2	0.0017	−0.0010
3	0.0003	−0.0005
4	0.0015	3.9990
5	0.0019	0.0009
6	−0.0004	−0.0009
7	−0.0003	−0.0019
8	0.0017	−0.0008
9	−0.0023	0.0019
10	−0.0024	−0.0011
11	0.0003	0.0020
12	0.0008	−0.0033
13	−0.0004	0.0011
14	0.0025	0.0003
15	−0.0005	0.0013
16	−0.0010	

The results are essentially the same as those of Example 10.9, which were computed by the trapezoidal rule.

Observe that we compute exactly as many A's and B's as there are data points. This is not only reasonable (we cannot "manufacture" information) but is in accord with *information theory*. (Some mention of this theory is made in the next section.)

10.6 Theoretical Matters

There is much theory on functional approximation and trigonometric interpolations. (A Fourier series can be thought of as an interpolation formula.) We only scratch the surface, however.

Error Bounds for Chebyshev Polynomials

We have asserted that, of all polynomials of degree n whose highest power of x has a coefficient of one, $T_n(x)/2^{n-1}$ has the smallest error bounds on $[-1, 1]$. The proof is by contradiction.

Let $P_n(x)$ be a polynomial whose leading term* is x^n and suppose that its maximum magnitude on $[-1, 1]$ is less than that of $T_n(x)/2^{n-1}$. Write

$$\frac{T_n(x)}{2^{n-1}} - P_n(x) = P_{n-1}(x),$$

where $P_{n-1}(x)$ is a polynomial of degree $n - 1$ or less, since the x^n terms cancel. The polynomial $T_n(x)$ has $n + 1$ extremes (counting endpoints), each of magnitude one, so $T_n(x)/2^{n-1}$ has $n + 1$ extremes each of magnitude $1/2^{n-1}$, and these successive extremes alternate in sign. By our supposition about $P_n(x)$, at each of these maximums or minimums, the magnitude of $P_n(x)$ is less than $1/2^{n-1}$; hence $P_{n-1}(x)$ must change its sign at least for every extreme of $T_n(x)$, which is then at least $n + 1$ times. $P_{n-1}(x)$ hence crosses the axis at least n times and would have n zeros. But this is impossible if $P_{n-1}(x)$ is only of degree $n - 1$, unless it is identically zero. The premise must then be false and $P_n(x)$ has a larger magnitude than the polynomial we are testing or, alternatively, $P_n(x)$ is exactly the same polynomial.

Orthogonal Polynomials

The Chebyshev polynomials are only one of many that are orthogonal. We discussed the Legendre polynomials in Chapter 4 in connection with Gaussian quadrature. They are orthogonal because

$$\int_{-1}^{1} L_n L_m \, dx = \begin{cases} 0, & n \neq m, \\ >0, & n = m. \end{cases}$$

---

* We restrict the polynomials to those whose leading term is x^n so that all are scaled alike.

Other orthogonal polynomials include the Hermite, Laguerre, Jacobi, Weber, and Gegenbauer. (Observe that these are named for famous mathematicians. Each has special applications.)

A better definition of orthogonality is to say that a set of functions Q_i are orthogonal with respect to the weight function $W(x)$ if

$$\int_{-1}^{1} Q_n(x)\, W(x)\, Q_m(x)\, dx = \begin{cases} 0, & n \neq m, \\ >0, & n = m. \end{cases}$$

If the value is unity when $n = m$, the set is said to be *orthonormal*.

Theory of Fourier Series

We will only summarize the important theorems concerning Fourier series. Proofs can be found in Conte and de Boor (1980), Fike (1968), Brigham (1974), Ramirez (1985), and Ralston (1965). In the following theorems, $f(x)$ refers to the periodic function being represented or to the periodic extension given by a redefinition. It is essential that $f(x)$ be integrable if we are to compute the coefficients of a Fourier series by the formulas given previously.

1. $f(x)$ is said to be piecewise continuous on $(0, L)$ if it is continuous on $(0, L)$ except for a finite number of finite discontinuities. If $f(x)$ and/or $f'(x)$ is piecewise continuous on $(0, L)$, $f(x)$ is said to be piecewise smooth. An infinite series is said to converge pointwise to $f(x)$ if the sum of n terms of the series converges to $f(x)$ at the point in $(0, L)$ as $n \to \infty$. An infinite series is said to converge uniformly if it converges pointwise to $f(x)$ at all points in $(0, L)$.
2. If $f(x)$ is continuous and piecewise smooth, its Fourier series converges uniformly to $f(x)$. If $f(x)$ is piecewise smooth, the series converges pointwise to $f(x)$ at all points where $f(x)$ is continuous and converges to the average value where $f(x)$ has a finite discontinuity.
3. If $f(x)$ is piecewise continuous, its Fourier series can be integrated term by term to yield a series that converges pointwise to the integral of $f(x)$.
4. If $f(x)$ is continuous and $f'(x)$ is piecewise smooth, then the Fourier series of $f(x)$ can be differentiated term by term to give a series that converges pointwise to $f'(x)$ wherever $f''(x)$ exists.

The theory of Fourier series is a major topic in mathematics. Most mathematical texts that cover Fourier series at least outline proofs of the preceding theorems.

Information Theory—The Sampling Theorem

In performing a discrete Fourier transform, we work with samples of some function of t, $f(t)$. We normally have data taken at evenly spaced intervals of time. If the interval between samples is D sec, its reciprocal, $1/D$, is called the *sampling rate* (the number of samples per second).

Corresponding to the sampling interval, D, is a critical frequency, called the *Nyquist critical frequency*, f_c, where

$$f_c = \frac{1}{2}D.$$

The reason this is a critical frequency is seen from the following argument. Suppose we sample a sine wave whose frequency is f_c and get a value corresponding to its positive peak amplitude. The next sample will be at the negative peak, the next beyond that at the positive peak, and so on—that is, critical sampling is at a rate of two samples per cycle. We can construct the magnitude of the sine wave from these two samples. If the frequency is less than f_c, we will have more than two samples per cycle and again we can construct the wave correctly. On the other hand, if the frequency is greater than f_c, we have fewer than two samples per cycle and we have inadequate information to determine $f(t)$.

The significance of this theorem is that if the phenomenon described by $f(t)$ has no frequencies greater than f_c, then $f(t)$ is completely determined from samples at the rate $1/D$. Unfortunately, this also means that if there are frequencies in $f(t)$ greater than f_c, all these frequencies are spuriously crowded into the range $[0, f_c]$, causing a distortion of the power spectrum. This distortion is called *aliasing*.

All of this is very clear if we think of the results of an FFT on the samples. If we have N samples of the phenomenon, we certainly cannot determine more than a total of exactly N of the Fourier coefficients, the A's and B's. The last of these will be $A_{n/2}$ (assuming an even number of samples). We see that this corresponds to the Nyquist frequency.

10.7 Using DERIVE

DERIVE has several facilities that correlate with the topics of this chapter.

Chebyshev Polynomials

The professional version of DERIVE has a utility file, ORTH_POL.MTH, that contains derived functions for several orthogonal polynomials, including Chebyshev polynomials. Though this file is not included in the Student Edition, we can compute the Chebyshev polynomials by using the recursive ability of the program. If we Author this statement:

```
TCH(x,n) := IF(n = 0,1,IF(n = 1,x,2x*TCH(x,n - 1) - TCH(x,n - 2)))
```

we have defined the various T_n through the recursion formula, Eq. (10.2). (The IF statement in DERIVE has the form

```
IF(condition, true-value, false-value)
```

and gives the true-value if the condition is true or the false-value if not true. Here we have nested IF's.)

If we test the formula by entering TCH(x,4) and Simplify, we get

$$8x^4 - 8x^2 + 1$$

which is precisely the same as that shown in Section 10.1.

Taylor Series

The TAYLOR function is a choice in the Calculus submenu. If we enter the parameters for $f(x) = e^x$ centered about $x = 0$ up to the fifth degree, the statement looks like this:

```
TAYLOR (e^x, x, 0, 5)
```

After Simplifying, we get

$$\frac{x^5}{120} + \frac{x^4}{24} + \frac{x^3}{6} + \frac{x^2}{2} + x + 1$$

which is exactly correct.

Economized Power Series

An economized power series can be obtained by subtracting a multiple of a Chebyshev polynomial so as to cancel the last term of a truncated Maclaurin series. If we have first defined the function for Chebyshev polynomials (as already described), we can easily do this in DERIVE for $f(x) = e^x$ with

```
TAYLOR (e^x,x,0,6) − TCH(x,6)/6!/2⁵
```

On Simplifying, we then get exactly Eq. (10.5). As described in Section 10.2, we can continue to "kill" the last terms to further economize.

Padé Approximations

The professional edition of DERIVE contains PADE(f(x),x,x_0,n,d) in utility file APPROX.MTH, a function that constructs the ratio of polynomials of degree n in the numerator and degree d in the denominator for $f(x)$ centered about x_0. (n must be equal to d or to $d - 1$.)

We can apply this to $f(x) = \arctan(x)$ with

```
PADE(ATAN(x),x,0,5,5)
```

and get the same result as Eq. (10.12).

Fourier Series

DERIVE gets the Fourier series coefficients quite readily. The utility file INT_APPS.MTH contains the function

$$\mathtt{FOURIER(f,x,a,b,n)}$$

which Simplifies to give n terms of the Fourier series of f considered to be periodic between a and b. For example,

```
FOURIER(x(x - 2),x,-2,2,4)
```

gives the equivalent of Eq. (10.40) with angles up to ($4\pi x/2$).

Of course the plotting capabilities of DERIVE can produce graphs such as Fig. 10.8.

Chapter Summary

Here's what you are able to do if you understand the topics of the chapter:

1. Explain how to get a Chebyshev polynomial of degree n, what is meant by a set of orthogonal functions, and what is the most important property of the Chebyshev polynomials.

2. Use Chebyshev polynomials to economize a power series, and explain why such approximations are preferred in computers.

3. Obtain the coefficients of a Padé approximation, and describe how one can get a further improvement in efficiency. You can explain the minimax principle.

4. Tell how the coefficients of a trigonometric series can be evaluated, and outline the FFT method for fitting to a set of equispaced data.

5. Use computer programs like those of this chapter.

Computer Programs

We conclude this chapter and this book with two programs, both written in Quick-BASIC. The first program computes the coefficients of a Padé approximation starting with the coefficients of the Maclaurin series. The second does the fast Fourier transform to compute the coefficients of a finite Fourier series.

Program 10.1 Padé Approximation

Figure 10.17 lists the program for a Padé approximation. As written, the program computes the Maclaurin coefficients, but it could be changed to read them from a file or from DATA statements. The program begins by declaring some variables that define the parameters and then dimensions several arrays. The Maclaurin coefficients are computed, and then printed for user review.

Next, the program sets up the equations that relate the a's and b's of the Padé rational expression to the a's, the Maclaurin coefficients. The latter are solved first to get the b's; then the a's are obtained and the coefficients printed. Finally, a table of the approximation from the Padé formula is displayed.

Output from the program is given at the end of Fig. 10.17.

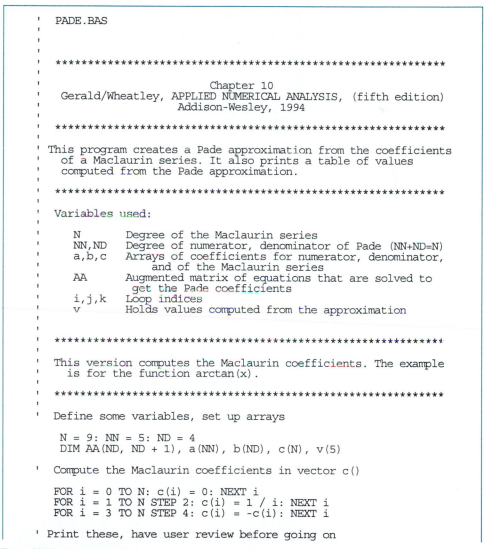

```
'   PADE.BAS
'
'
'
'   ************************************************************
'                           Chapter 10
'   Gerald/Wheatley, APPLIED NUMERICAL ANALYSIS, (fifth edition)
'                       Addison-Wesley, 1994
'
'   ************************************************************
'  This program creates a Pade approximation from the coefficients
'    of a Maclaurin series. It also prints a table of values
'    computed from the Pade approximation.
'
'   ************************************************************
'
'  Variables used:
'
'    N          Degree of the Maclaurin series
'    NN,ND      Degree of numerator, denominator of Pade (NN+ND=N)
'    a,b,c      Arrays of coefficients for numerator, denominator,
'                  and of the Maclaurin series
'    AA         Augmented matrix of equations that are solved to
'                  get the Pade coefficients
'    i,j,k      Loop indices
'    v          Holds values computed from the approximation
'
'
'   ************************************************************
'
'  This version computes the Maclaurin coefficients. The example
'    is for the function arctan(x).
'
'   ************************************************************
'
'  Define some variables, set up arrays
'
     N = 9: NN = 5: ND = 4
     DIM AA(ND, ND + 1), a(NN), b(ND), c(N), v(5)
'
'  Compute the Maclaurin coefficients in vector c()
'
     FOR i = 0 TO N: c(i) = 0: NEXT i
     FOR i = 1 TO N STEP 2: c(i) = 1 / i: NEXT i
     FOR i = 3 TO N STEP 4: c(i) = -c(i): NEXT i
'
'  Print these, have user review before going on
```

Figure 10.17 Program 10.1

Figure 10.17 *(continued)*

```
CLS
PRINT "Here are the coefficients of the Maclaurin series"
PRINT
PRINT "power of x    Coefficient"
FOR i = 0 TO N
  PRINT "   "; i; : PRINT USING "           ##.####"; c(i)
NEXT i
PRINT

INPUT "Press RETURN to continue ", a$

' Set up the equations for the B's (denominator coefficients)

FOR i = 1 TO ND: AA(i, i) = c(NN): NEXT i

FOR i = 1 TO ND - 1
  FOR j = i + 1 TO ND
    AA(i, j) = c(NN - j + i)
  NEXT j
NEXT i
FOR i = 2 TO ND
  FOR j = 1 TO i + 1
    AA(i, j) = c(NN + i - j)
  NEXT j
NEXT i

'  Now do right hand side

  FOR i = 1 TO ND: AA(i, ND + 1) = -c(NN + i): NEXT i

' Now solve for the B's. First reduce.

  FOR i = 1 TO ND - 1
    FOR j = i + 1 TO ND
      RATIO = AA(j, i) / AA(i, i)
        FOR k = i + 1 TO ND + 1
          AA(j, k) = AA(j, k) - RATIO * AA(i, k)
        NEXT k
      NEXT j
    NEXT i

' Now back substitute

  b(ND) = AA(ND, ND + 1) / AA(ND, ND)
  FOR j = ND - 1 TO 1 STEP -1
  SUM = 0
    FOR k = ND TO j + 1 STEP -1
      SUM = SUM + AA(j, k) * b(k)
    NEXT k
  b(j) = (AA(j, ND + 1) - SUM) / AA(j, j)
  NEXT j
  b(0) = 1

' Now get the a's
  FOR i = 1 TO NN
  SUM = 0
    FOR j = 1 TO i
      IF j > ND THEN EXIT FOR
      SUM = SUM + b(j) * c(i - j)
    NEXT j
    a(i) = SUM + c(i)
  NEXT i
  a(0) = c(0)
```

Figure 10.17 *(continued)*

```
' Print the coefficients of the Pade approximation

  PRINT "The coefficients of the Pade approximation are"
  PRINT
  PRINT " exp on x    coef in numerator"
  PRINT
  FOR i = 0 TO NN
    PRINT "    "; i; : PRINT USING "           ##.#### "; a(i)
  NEXT i

  PRINT
  PRINT

  PRINT " exp on x    coef in denominator"
  PRINT
  FOR i = 0 TO ND
    PRINT "    "; i; : PRINT USING "           ##.#### "; b(i)
  NEXT i
  PRINT
  PRINT

' Now do a table of approximations using Pade

  INPUT "Press RETURN to get table of computed values", a$
  PRINT
  PRINT "Here are values computed from the Pade approx"
  PRINT
  PRINT "    x", "   Value"
  PRINT

  FOR x = 0 TO 1.001 STEP .2
    v = a(0)
    FOR i = 1 TO NN
      v = v + a(i) * x ^ i
    NEXT i
  NUM = v
    v = 1
  FOR i = 1 TO ND
      v = v + b(i) * x ^ i
    NEXT i
    v = NUM / v
    PRINT x, v
  NEXT x

  *************************************************************
                  OUTPUT FOR PADE.BAS

Here are the coefficients of the Maclaurin series

power of x    Coefficient
    0            0.0000
    1            1.0000
    2            0.0000
    3           -0.3333
    4            0.0000
    5            0.2000
    6            0.0000
    7           -0.1429
    8            0.0000
    9            0.1111

Press RETURN to continue
```

Figure 10.17 *(continued)*

```
The coefficients of the Pade approximation are

exp on x      coef in numerator

   0               0.0000
   1               1.0000
   2               0.0000
   3               0.7778
   4               0.0000
   5               0.0677


exp on x      coef in denominator

   0               0.0000
   1               0.0000
   2               1.1111
   3               0.0000
   4               0.2381



Press RETURN to get table of computed values

Here are values computed from the Pade approx

     x                Value

   0                0
   .2               .1973955
   .4               .3805064
   .6               .5404217
   .8               .6747709
   1                .7855856
```

Program 10.2 Fast Fourier Transform

Our second program, titled FFT.BAS and reproduced in Fig. 10.18, does a fast Fourier transform on a set of data read from a file. (The number of data values, N, must be a power of 2.) After defining some constants and dimensioning arrays, the data are read in from the file and displayed for user review.

The program avoids complex arithmetic by computing through sine and cosine values, all at N evenly spaced points around the unit circle (see Fig. 10.14). As shown in Fig. 10.16, the final computed values are scrambled, with their indices ordered as the powers of W in the computations analogous to Eq. (10.63). (We don't actually use them as powers of W, since complex arithmetic is not used.) This ordering is obtained through the algorithm of Section 10.5.

The actual FFT computations then begin. They are not easy to follow, but they do exactly as shown in Fig. 10.16 and in the FFT algorithm. The calculations are done by stages, each of which handles subsets of the data.

Program output is shown at the end of Fig. 10.18. These correspond to data where

$$f(x) = 1.5 \cos(2\pi x) - 0.5 \sin(\pi x) + 4 \sin(4\pi x),$$

and it is seen that precisely the correct a's and b's are obtained.

Although the program does not show it, the input data can represent a complex valued function. How the file data are read in would have to be changed to handle such, of course.

```
'   FFT.BAS
'
'
'   **************************************************************
'
'                          Chapter 10
'     Gerald/Wheatley, APPLIED NUMERICAL ANALYSIS, (fifth edition)
'
'                      Addison-Wesley, 1994
'
'   **************************************************************
'
'  This program does FFT to get coefficients of the finite Fourier
'     series. This version obtains the data pairs from a file.
'
'   **************************************************************
'
'   Variables used:
'     N         Number of points (must be a power of 2)
'     NST       Number of stages in the computation
'     PI        Value of pi
'     pw        Array of powers of W, also used to unscramble results
'     c,s       Arrays of cosine, sine values
'     y         Array of data as read in
'     yr,yi     Arrays of real, imaginary parts of data
'     yyr,yyi   Arrays to preserve y-values for next stage
'     st        Stage number (loop control)
'     NSETS     Number of subsets in a stage
'     set       Set number in a stage
'     del       Distance between y-values being used
'     a,b       Fourier coefficients
'     i,j,k,l   Loop indices
'
'   **************************************************************
'   Define some constants and dimension arrays

n = 32
PI = 3.1415926#
DIM y(n), pw(n), c(n), s(n), yr(n), yi(n), yyr(n), yyi(n)

'   Read in data from a file, put into yr(), set yi() = 0

OPEN "I", #1, "A:FFT1.DTA"
FOR i = 0 TO n - 1
  INPUT #1, y(i)
NEXT i

FOR i = 0 TO n - 1
  yr(i) = y(i)
  yi(i) = 0
NEXT i
```

Figure 10.18   Program 10.2

Figure 10.18 *(continued)*

```
' Print input data, four values per line

CLS
PRINT "This is the data as read in"
PRINT
FOR i = 1 TO (n / 4)
  FOR j = 1 TO 4
    PRINT USING "   ###.####   "; yr((i - 1) * 4 + j - 1);
  NEXT j
  PRINT
NEXT i
INPUT "Press RETURN to continue", a$


' Compute cosine, sine values

FOR i = 0 TO n - 1
  c(i) = COS(i * 2 * PI / n)
  s(i) = SIN(i * 2 * PI / n)
NEXT i
' Set up powers of W. This is also data sequence in last stage.

NST = LOG(n) / LOG(2)
FOR i = 0 TO n - 1
  pw(i) = 0
NEXT i
FOR st = 1 TO NST
  FOR i = 1 TO 2 ^ (st - 1)
    pw(i) = 2 * pw(i)
  NEXT i
  FOR i = 0 TO 2 ^ (st - 1) - 1
    pw(i + 2 ^ (st - 1)) = pw(i) + 1
  NEXT i
NEXT st


' Do the FFT by stages. In this k indicates y-value being computed,
'    j and j + del indicate old y-values that are used,
'    l is index to cosine and sine values

del = n / 2
NSETS = 1
FOR st = 1 TO NST
  k = 0
  FOR set = 1 TO NSETS
    FOR i = 0 TO n / NSETS - 1
      j = i MOD del + (set - 1) * del * 2
      l = pw(INT(k / del))
      yyr(k) = yr(j) + c(l) * yr(j + del) - s(l) * yi(j + del)
      yyi(k) = yi(j) + c(l) * yi(j + del) + s(l) * yr(j + del)
      k = k + 1
    NEXT i
  NEXT set

' Reset the y-arrays for next stage, also redo del, NSETS, and k

  FOR i = 0 TO n - 1
    yr(i) = yyr(i)
    yi(i) = yyi(i)
  NEXT i
  del = del / 2
  NSETS = NSETS * 2
NEXT st


' FFT is finished. Print the results, unscrambling with pw(i)
```

Figure 10.18 *(continued)*

```
PRINT " k       a = c(real)  b = c(imag)"
FOR i = 0 TO n / 2
  PRINT USING "## "; i;
  j = pw(i)
  PRINT USING "       ###.###      "; yr(j) / n * 2;
  IF (i <> 0) OR (i <> n / 2) THEN
    PRINT USING " ###.### "; yi(j) / n * 2;
  END IF
  PRINT
NEXT i
```

```
******************************************************************
                      OUTPUT FOR FFT.BAS

This is the data as read in

      1.5000        4.1170        4.8690        3.1250
     -0.3540       -3.8180       -5.5230       -4.7050
     -2.0000        0.9520        2.4770        1.8390
     -0.3540       -2.5320       -3.1310       -1.5400
      1.5000        4.3120        5.2520        3.6800
      0.3540       -2.9870       -4.5990       -3.7240
     -1.0000        1.9330        3.4010        2.6700
      0.3540       -1.9770       -2.7480       -1.3450

Press RETURN to continue

 k        a = c(real)  b = c(imag)

 0          -0.000        0.000
 1           0.000       -0.500
 2           1.500        0.000
 3          -0.000        0.000
 4           0.000        4.000
 5          -0.000       -0.000
 6           0.000       -0.000
 7          -0.000        0.000
 8           0.000       -0.000
 9          -0.000       -0.000
10          -0.000        0.000
11          -0.000       -0.000
12           0.000        0.000
13          -0.000        0.000
14          -0.000        0.000
15          -0.000        0.000
16          -0.000        0.000
```

Exercises

Section 10.1

1. Compute $T_{11}(x)$ and $T_{12}(x)$.

2. Show that Eq. (10.3) is satisfied for these values of (m, n): $(0, 1), (1, 1), (1, 2)$. (Don't do this numerically!)

3. Extend the graphs of $T_1(x)$, $T_2(x)$, $T_3(x)$ to the interval $[-2, 2]$. Observe that the maximum magnitude

for the Chebyshev polynomials is not equal to one outside of $[-1, 1]$.

▶**4.** Graph $T_5(x)$ for $x = -1$ to $x = 1$. Locate the zeros more accurately than could be read from the graph by any method of Chapter 1.

5. Expand $\cos(6x)$ in terms of $\cos(x)$, and show that this is equivalent to $T_6(x)$. (You may have to use a formula for the sum of two angles to do the expansion.)

6. Reduce the fifth-degree polynomial of Eq. (10.5) to a fourth-degree polynomial, and confirm that its maximum error is 0.000791.

7. Graph the error of the economized fourth-degree polynomial of Exercise 6 over $[-1, 1]$, and compare it to the graph of the errors of the fourth-degree Maclaurin series for e^x.

▶**8.** The function $\arctan(x)$ can be represented by this power series:

$$\arctan(x) = x - \frac{x^3}{3} + \frac{x^5}{5} - \frac{x^7}{7} + \frac{x^9}{9} - \cdots.$$

Economize this three times to give a third-degree polynomial. Graph the errors, and compare this graph to one of the errors of a ninth-degree expansion.

9. Find the first few terms of the Chebyshev series for $\sin(x)$ by rewriting the Maclaurin series in terms of the $T(x)$'s and collecting terms. Then express this series as a power series. Compare the errors of both series after truncating to give third-degree polynomials.

10. A series expansion for $(1 + x/5)^{1/2}$ is

$$1 + \frac{x}{10} - \frac{x^2}{200} + \frac{x^3}{2000}$$

$$- \frac{x^4}{16,000} + \frac{7x^5}{800,000} - \cdots.$$

Convert this to a Chebyshev series, including terms to T^2. What is the maximum error of the truncated Chebyshev series? Compare this to the error of the power series when it is truncated to second degree.

▶**11.** To get the smaller error of a Chebyshev series or an economized power series requires that the approximation be for the interval $[-1, 1]$. Show what change of variable will change $f(x)$ on $[a, b]$ to $f(y)$ on $[-1, 1]$.

▶**12.** Find Padé approximations for the following functions, with numerators and denominators each of degree three:
a. $\sin^2(x)$,   **b.** $\cos(x^3)$,
c. e^x.

13. Compare the errors on $[-1, 1]$ for the Padé approximations of Exercise 12 with the errors of the corresponding Maclaurin series that has terms up to x^6.

14. Express the following rational fractions in continued-fraction form. In each part, compare the number of multiplication and division operations with that resulting from evaluating the polynomials by Horner's method (in nested form).
a. $\dfrac{x^2 - 2x + 2}{x^2 + 2x - 2}$

b. $\dfrac{2x^3 + x^2 + x + 3}{x^2 - x - 4}$

c. $\dfrac{2x^4 + 45x^3 + 381x^2 + 1353x + 1511}{x^3 + 21x^2 + 157x + 409}$

▶**15.** Express each of the Padé approximation of Exercise 12 as continued fractions.

16. Estimate the errors of each of the Padé approximations in Exercise 12 by computing the coefficient of the next nonzero term of the numerator. Compare these to the actual errors at $x = 1$ and at $x = -1$.

17. A Chebyshev series for $\cos(\pi x/4)$ is

$$0.851632 - 0.146437T_2 + 0.00192145T_4$$
$$- 9.965 * 10^{-6}T_6.$$

Use this series to develop a Padé-like rational function by the method of Section 10.3, where the function is $R_{4,2}$.

18. Fike (1968) gives this example of a rational fraction approximation to $\Gamma(1 + x)$ on $[0, 1]$:

$$R_{3,4}(x) =$$
$$\frac{0.999999 + 0.601781x + 0.186145x^2 + 0.0687440x^3}{1 + 1.17899x - 0.122321x^2 - 0.0260996x^3 + 0.060992x^4}.$$

Is this a minimax approximation? If not, what are the bounds of the errors of the $R_{3,4}$ minimax approximation?

19. Which of these functions are periodic? What is the period if it is periodic?
a. $\sin(2x) + 2\cos(x)$
b. $e^{-10x}\cos(x)$
c. $\sin^3(x)$
d. e^{2ix}, where $i = \sqrt{-1}$

20. Example 10.4 plots $f(x) = |x|$ between $x = -\pi$ and $x = \pi$. $f(x)$ is also periodic. Extend the plot of Fig. 10.7 for the interval $[-10, 10]$.

21. Find the Fourier coefficients for $f(x) = x^3$ if $f(x)$ is periodic and one period extends from $x = 0$ to $x = 2$.

▶22. Find the Fourier coefficients for $g(x) = x^2 - 1$ if it is periodic and one period extends from $x = 0$ to $x = 2$.

23. Show that the Fourier series of
$$x^3 + x^2 - 1$$
is just the sum of those for $f(x)$ (Exercise 21) and $g(x)$ (Exercise 22).

24. Is the Fourier series for $[f(x)] * [g(x)]$ (Exercises 21 and 22) the product of the two series?

25. Suppose we are interested in $f(x) = e^{-x} \sin(2x - 1)$ only in the interval $[0, 2]$. Sketch the half-range extensions that give
   a. an even function.
   b. an odd function.

26. Repeat Exercise 25, but for the range of interest $[-1, 3]$.

▶27. Find the Fourier coefficients for the periodic functions of parts (a) and (b) of Exercise 25.

28. Repeat Exercise 27, this time for the function of Exercise 26.

Section 10.5

29. Use trapezoidal rule integration with 20 panels to get the first four coefficients of Exercises 21 and 22. Then repeat Exercise 23, this time with trapezoidal rule integrations.

▶30. Repeat Exercise 29, but use Simpson's $\frac{1}{3}$ rule with 20 panels. Compare your answers to those of Exercise 29.

31. Repeat Exercise 29, this time with more than 20 panels. At what point, if any, do the results match those of Exercise 30?

32. Use Simpson's $\frac{1}{3}$ rule to obtain the A's and B's of Example 10.9. (You may not be able to generate the values in Table 10.8 with a computer program. If that is true, you will need to input the tabulated values.) Are the values the same as in the example? Why should there be a difference?

33. Verify that Eqs. (10.63) and (10.62) are really identical.

34. Make a flow diagram (like Fig. 10.16) for $n = 8$.

▶35. Generate the powers of W for use in the FFT with the algorithm of Section 10.5 for $n = 16$. Do these agree with Fig. 10.16?

36. Repeat Exercise 35, but use the bit-reversing rule.

37. Write a computer program to implement the algorithm for FFT in some language other than BASIC.

38. Use your program of Exercise 37 to get the A's and B's of Exercises 21 and 22.

Section 10.6

39. The Legendre polynomials (Section 4.9) resemble the Chebyshev polynomial in that they have the same number of zeros with $[-1, 1]$ and the same number of maxima and minima. Compare the graphs of several of the two polynomials of the same degree. Why are the Legendre polynomials less suited for economizing a power series?

40. Try to verify Eq. (10.3) directly by numerical integration. Use $(n, m) = (0, 0), (0, 1), (1, 1), (1, 2)$.

41. Verify Eq. (10.3) after making the substitution from Eq. (10.4) for x. Do this both analytically and numerically.

▶42. Let $f(t) = 2 \cos(t) + \sin(5t)$; this is periodic of period 2π. How many samples in the interval $[-\pi, \pi]$ are needed to determine the Fourier series coefficients correctly?

43. What is the Nyquist critical frequency for the $f(t)$ of Exercise 42?

▶44. Do an FFT on eight evenly spaced samples of the $f(t)$ of Exercise 42. Are the correct Fourier coefficients obtained? Do these samples have to be within $\{-\pi, \pi]$?

Section 10.7

45. Use DERIVE to find T_5, T_6, T_7, T_8.

▶46. Write a DERIVE statement to compute Legendre polynomials (see Section 4.9), and use it to get a display of all of these polynomials through degree 6.

47. Use DERIVE to get the economized polynomials of degree 5 from the Maclaurin series of degree 6 for
   a. $f(x) = e^x \sin(\pi x^2)$.
   b. $f(x) = \cos^2(\pi x)/e^x$.
   c. $f(x) = |x|, -2 < x < 2$, and periodic of period 4.

48. Reduce each of the results of Exercise 47 to degree 4.

49. Compare the errors of the economized series of Exercise 47 with the original Maclaurin series of degree 6 on $[-1, 1]$.

50. Use DERIVE to get six terms of the Fourier series of each function defined in Exercise 47.

51. Use DERIVE to compare the graphs of each series of Exercise 50 with the plot of the original function. Do the graphs between $x = -5$ and $x = 5$.

Applied Problems and Projects

52. In Section 3 of this chapter, the Padé rational functions were developed to approximate $f(x)$ on the interval $[-1, 1]$. If we want to approximate $f(x)$ on a different interval, say, $[a, b]$, a simple linear transformation can change the interval to $[-1, 1]$. What if we want to approximate a function on an interval with one or both endpoints infinite? Devise a transformation for such cases.

53. Investigate, for some computer system available to you, how some or all of the following transcendental functions are approximated in FORTRAN. Classify these into Taylor series formulas, Chebyshev polynomials, rational functions, or other types. Which of these are minimax?
   a. $\sin(x)$
   b. $\cos(x)$
   c. $\tan(x)$
   d. $\operatorname{atan}(x)$
   e. $\exp(x)$
   f. $\ln(x)$

54. Repeat Exercise 53 for other computer languages: BASIC, Pascal, and C.

55. As illustrated by Fig. 10.8, the sum of n terms of the Fourier series for a function that has a jump discontinuity has larger oscillations near the discontinuity—the Gibbs phenomenon. For $f(x)$ equal to a square wave, investigate whether the departure of the sum of the series from $f(x)$ for the last "hump" in the curve (the "ear"), decreases when n is increased. What can you conclude about the size of the ear?

56. One way to eliminate the Gibbs phenomenon (see Exercise 55) is to abolish the jump discontinuity by subtracting a linear function from $f(x)$. Suppose the linear function is $L(x)$. For $f(x)$ equal to square wave, find the $L(x)$ for which $f(x) - L(x)$ has no jump discontinuities. Compare the accuracy of the sum of 10 terms of the Fourier series for $f(x)$ with that for the sum of 10 terms of the Fourier series for $g(x) = f(x) - L(x)$.

57. Another way to ameliorate the problem of the Gibbs phenomenon is to use the so-called Lanczos's factors. Search the literature to find out more about this method. Apply it to obtain an improved approximation to a square wave.

58. Section 10.5 described how to perform the fast Fourier transform (FFT) on a set of evenly spaced data pairs when the number of pairs is a power of 2. While it is less efficient, the FFT can be employed when the number of pairs, n, is not a power of 2. Do a literature search to see how the algorithm of Section 10.5 must be modified for such a case. Make a critical comparison of the reduction in computational effort for the FFTs compared to doing finite Fourier transforms as in the first example of Section 10.5, for values of n ranging from 10 to 1024.

59. Extend your work in Exercise 58 by writing a computer program that does a FFT for n not equal to a power of 2.

APPENDIX A

Some Basic Information from Calculus

Since a number of results and theorems from the calculus are frequently used in the text, we collect here a number of these items for ready reference, and to refresh the student's memory.

Open and Closed Intervals

We use for the open interval $a < x < b$, the notation (a, b), and for the closed interval $a \leq x \leq b$, the notation $[a, b]$.

Continuous Functions

If a real-valued function is defined on the interval (a, b), it is said to be continuous at a point x_0 in that interval if for every $\varepsilon > 0$ there exists a positive nonzero number δ such that $|f(x) - f(x_0)| < \varepsilon$ whenever $|x - x_0| < \delta$ and $a < x < b$. In simple terms, we can meet any criterion of matching the value of $f(x_0)$ (the criterion is the quantity ε) by choosing x near enough to x_0, without having to make x equal to x_0, when the function is continuous.

If a function is continuous for all x-values in an interval, it is said to be continuous on the interval. A function that is continuous on a closed interval $[a, b]$ will assume a maximum value and a minimum value at points in the interval (perhaps the endpoints). It will also assume any value between the maximum and the minimum at some point in the interval.

Similar statements can be made about a function of two or more variables. We then refer to a domain in the space of the several variables instead of to an interval.

Sums of Values of Continuous Functions

When x is in $[a, b]$, the value of a continuous function $f(x)$ must be no greater than the maximum and no less than the minimum value of $f(x)$ on $[a, b]$. The sum of n such values must be bounded by $(n)(m)$ and $(n)(M)$, where m and M are the minimum and maximum values. Consequently, the sum is n times some intermediate value of the function. Hence

$$\sum_{i=1}^{n} f(\xi_i) = nf(\xi) \quad \text{if } a \leqslant \xi_i \leqslant b, \quad i = 1, 2, \ldots, n, \quad a \leqslant \xi \leqslant b.$$

Similarly, it is obvious that

$$c_1 f(\xi_1) + c_2 f(\xi_2) = (c_1 + c_2) f(\xi), \quad \xi_1, \xi_2, \xi \text{ in } [a, b],$$

for the continuous function f when c_1 and c_2 are both equal to or greater than one. If the coefficients are positive fractions, dividing by the smaller gives

$$c_1 f(\xi_1) + c_2 f(\xi_2) = c_1 \left[f(\xi_1) + \frac{c_2}{c_1} f(\xi_2) \right] = c_1 \left(1 + \frac{c_2}{c_1} \right) f(\xi) = (c_1 + c_2) f(\xi),$$

so the rule holds for fractions as well. If c_1 and c_2 are of unlike sign, this rule does not hold unless the values of $f(\xi_1)$ and $f(\xi_2)$ are narrowly restricted.

Mean-Value Theorem for Derivatives

When $f(x)$ is continuous on the closed interval $[a, b]$, then at some point ξ in the interior of the interval

$$f'(\xi) = \frac{f(b) - f(a)}{b - a}, \quad a < \xi < b,$$

provided, of course, that $f'(x)$ exists at all interior points. Geometrically this means that the curve has at one or more interior points a tangent parallel to the secant line connecting the ends of the curve (Fig. A.1).

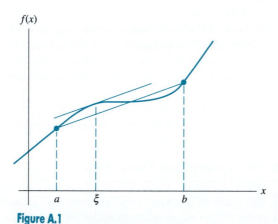

Figure A.1

Mean-Value Theorems for Integrals

If $f(x)$ is continuous and integrable on $[a, b]$, then

$$\int_a^b f(x)\, dx = (b - a) f(\xi), \qquad a < \xi < b.$$

This says, in effect, that the value of the integral is an average value of the function times the length of the interval. Since the average value lies between the maximum and minimum values, there is some point ξ at which $f(x)$ assumes this average value.

If $f(x)$ and $g(x)$ are continuous and integrable on $[a, b]$, and if $g(x)$ does not change sign on $[a, b]$, then

$$\int_a^b f(x)g(x)\, dx = f(\xi) \int_a^b g(x)\, dx, \qquad a < \xi < b.$$

Note that the previous statement is a special case $[g(x) = 1]$ of this last theorem, which is called the *second theorem of the mean for integrals*.

Taylor Series

If a function $f(x)$ can be represented by a power series on the interval $(-a, a)$, then the function has derivatives of all orders on that interval and the power series is

$$f(x) = f(0) + f'(0)x + \frac{f''(0)}{2!}x^2 + \frac{f'''(0)}{3!}x^3 + \cdots.$$

The preceding power-series expansion of $f(x)$ about the origin is called a *Maclaurin series*. Note that if the series exists, it is unique and any method of developing the coefficients gives this same series.

If the expansion is about the point $x = a$, we have the Taylor series

$$f(x) = f(a) + f'(a)(x - a) + \frac{f''(a)}{2!}(x - a)^2 + \frac{f'''(a)}{3!}(x - a)^3 + \cdots.$$

We frequently represent a function by a polynomial approximation, which we can regard as a truncated Taylor series. Usually we cannot represent a function exactly by this means, so we are interested in the error. Taylor's formula with a remainder gives us the error term. The remainder term is usually derived in elementary calculus texts in the form of an integral:

$$f(x) = f(a) + f'(a)(x - a) + \frac{f''(a)}{2!}(x - a)^2 + \cdots + \frac{f^{(n)}(a)}{n!}(x - a)^n$$

$$+ \int_a^x \frac{(x - t)^n}{n!} f^{(n+1)}(t)\, dt.$$

Since $(x - t)$ does not change sign as t varies from a to x, the second theorem of the mean allows us to write the remainder term as

$$\text{Remainder of Taylor series} = \frac{(x - a)^{n+1}}{(n + 1)!} f^{(n+1)}(\xi), \qquad \xi \text{ in } [a, x].$$

The derivative form is the more useful for our purposes. It is occasionally useful to express a Taylor series in a notation that shows how the function behaves at a distance h from a fixed point a. If we call $x = a + h$ in the preceding series, so that $x - a = h$, we get

$$f(a + h) = f(a) + f'(a)h + \frac{f''(a)}{2!} h^2 + \cdots + \frac{f^{(n)}(a)}{n!} h^n + \frac{f^{(n+1)}(\xi)}{(n + 1)!} h^{n+1}.$$

Taylor Series for Functions of Two Variables

For a function of two variables, $f(x, y)$, the rate of change of the function can be due to changes in either x or y. The derivatives of f can be expressed in terms of the partial derivatives. For the expansion in the neighborhood of the point (a, b),

$$f(x, y) = f(a, b) + f_x(a, b)(x - a) + f_y(a, b)(y - b)$$

$$+ \frac{1}{2!} [f_{xx}(a, b)(x - a)^2 + 2f_{xy}(a, b)(x - a)(y - b) + f_{yy}(a, b)(y - b)^2]$$

$$+ \cdots.$$

Descartes' Rule of Signs

Let $p(x)$ be a polynomial with real coefficients and consider the equation $p(x) = 0$. Then

1. The number of positive real roots is equal to the number of variations in the signs of the coefficients of $p(x)$ or is less than that number by an even integer.
2. The number of negative real roots is determined the same way, but for $p(-x)$. Here also the number of negative roots is equal to the number of variations in the signs of the coefficients of $p(-x)$ or is less than that number by an even integer.

For example, the polynomial $p(x) = 3x^5 - 2x^4 + 7x^2 - 12 = 0$ will have 3 or 1 positive and 2 or 0 negative real roots.

APPENDIX B

Deriving Formulas by the Method of Undetermined Coefficients

In Chapter 4 we developed formulas for integration and differentiation of functions by replacing the actual function with a polynomial that agrees at a number of points (a so-called *interpolating polynomial*), and then integrating or differentiating the polynomial. These formulas are valuable if one wishes to write computer programs for integration and differentiation, but the most important use is for solving differential equations numerically. Because computers calculate their functional values rather than interpolate in a table, there is less interest today in interpolating polynomials than in earlier times. Hence, there is reason to present an alternative method of deriving the formulas for derivatives and integrals that are needed to solve differential equations.

We will call the method that we employ in this appendix the *method of undetermined coefficients*. Basically, we impose certain conditions on a formula of desired form and use these conditions to determine values of the unknown coefficients in the formula. Hamming (1973) presents the method in considerable detail.

B.1 Derivative Formulas by the Method of Undetermined Coefficients

Since the derivative of a function is the rate of change of the function relative to changes in the independent variable, we should expect that formulas for the derivative would involve differences between function values in the neighborhood of the point where we wish to evaluate the derivative. It is, in fact, possible to approximate the derivative as a linear combination of such function values. While one can argue that formulas of greatest accuracy will use function values very near to the point in question, the formulas important in practice impose the restriction that only function values at equally spaced x-values are to be used.

For example, we can write a formula for the first derivative in terms of $n + 1$ equispaced points:

$$f'(x_0) = c_0 f(x_0) + c_1 f(x_1) + c_2 f(x_2) + \cdots + c_n f(x_n),$$

$$x_{i+1} - x_i = h = \text{constant} \tag{B.1}$$

The more terms we employ, the greater accuracy we shall expect, since more information about the function is being fed in. We will evaluate the coefficients in the equation by requiring that the formula be exact whenever the function is a polynomial of degree n or less.* (We will find, throughout this appendix, that the method of undetermined coefficients uses this criterion to develop formulas. It has validity because any function that is continuous on an interval can be approximated to any specified precision by a polynomial of sufficiently high degree. Using polynomials to replace the function also greatly simplifies the work, in contrast to replacing with other functions.)

Let us illustrate the method of undetermined coefficients with a simple case. We shall simplify the notation by defining

$$f_i = f(x_i).$$

If we write the derivative in terms of only two function values, we would have

$$f_0' = c_0 f_0 + c_1 f_1. \tag{B.2}$$

We require that the formula be exact if $f(x)$ is a polynomial of degree 1 or less. (The maximum degree is always one less than the number of undetermined constants.) Hence, since the formula is to be exact if $f(x)$ is any first-degree polynomial, it must be exact if either $f(x) = x$ or $f(x) = 1$, for these definitions of $f(x)$ are just two special cases of the general first-degree function $ax + b$. We write Eq. (B.2) for both these cases:

$$f(x) = x, \quad f'(x) = 1, \quad 1 = c_0(x_0) + c_1(x_0 + h),$$

$$f(x) = 1, \quad f'(x) = 0, \quad 0 = c_0(1) + c_1(1). \tag{B.3}$$

We solve Eqs. (B.3) simultaneously to get $c_0 = -1/h, c_1 = 1/h$. Consequently,

$$f'(x_0) \doteq \frac{f_1 - f_0}{h}. \tag{B.4}$$

The dot over the equal sign in Eq. (B.4) reminds us that the formula is only an approximation. It is exact if $f(x)$ is a polynomial of degree 1, but not exact if a polynomial of higher degree, or some transcendental function.

Similarly, a three-term formula can be derived by replacing the function with x^2 and x and 1 in

$$f_0' = c_0 f_0 + c_1 f_1 + c_2 f_2.$$

---

* Intuitively, it seems reasonable that we can satisfy this criterion, for a polynomial of degree n is determined uniquely by its $n + 1$ coefficients and our formula contains $n + 1$ constants.

The set of equations to solve is

$$2x_0 = c_0(x_0)^2 + c_1(x_0 + h)^2 + c_2(x_0 + 2h)^2,$$
$$1 = c_0(x_0) + c_1(x_0 + h) + c_2(x_0 + 2h), \tag{B.5}$$
$$0 = c_0(1) + c_1(1) + c_2(1).$$

The arithmetic is simplified by letting $x_0 = 0$. That this is valid is readily seen. Imagine the graph of $f(x)$ versus x. The derivative we desire is the slope of the curve at the point where $x = x_0$, which obviously is unchanged by a translation of the axes. Taking $x_0 = 0$ is the equivalent of a translation of axes so that the origin corresponds to x_0. Equations (B.5) become, with this change,

$$0 = c_0(0) + c_1(h)^2 + c_2(2h)^2,$$
$$1 = c_0(0) + c_1(h) + c_2(2h),$$
$$0 = c_0(1) + c_1(1) + c_2(1).$$

Solving, we get $c_0 = -3/2h$, $c_1 = 2/h$, $c_2 = -1/2h$; so a three-term formula for the derivative is

$$f'_0 \doteq \frac{-3f_0 + 4f_1 - f_2}{2h}. \tag{B.6}$$

We could extend these formulas to include more and more terms, but after a little reflection we can conclude that the original form, Eq. (B.1), is not the best to use. It utilizes only functional values to one side of the point in question, while it would be better to utilize information from both sides of x. After all, the limit in the definition of the derivative is two-sided, and further, we utilize closer and hence more pertinent information by going to both the left and the right.

We expect an improvement over Eq. (B.6) if we begin with

$$f'_0 = c_{-1}f_{-1} + c_0f_0 + c_1f_1, \tag{B.7}$$

where $f_{-1} = f(x_0 - h)$. Adopting the simplification of letting $x_0 = 0$ as before, and taking x^2, x, and 1 for $f(x)$, we have to solve

$$0 = c_{-1}(-h)^2 + c_0(0) + c_1(h)^2,$$
$$1 = c_{-1}(-h) + c_0(0) + c_1(h), \tag{B.8}$$
$$0 = c_{-1}(1) + c_0(1) + c_1(1).$$

Completing the algebra, we get $c_{-1} = -1/2h$, $c_0 = 0$, $c_1 = 1/2h$.

The following equation is particularly important:

$$f'_0 \doteq \frac{f_1 - f_{-1}}{2h}. \tag{B.9}$$

In the next section we will compare its accuracy with Eqs. (B.4) and (B.6). Because the point where the derivative is evaluated is centered among the function values whose differences appear in the formula, it is called a *central-difference approxima-tion*. Higher-order central-difference approximations to the first derivative, utilizing five or seven or more points, can be derived by this same procedure.

We now apply the method of undetermined coefficients to higher derivatives. We will discuss only the central-difference approximations since they are the more widely used. In terms of values both to the right and left of x_0.

$$f_0'' = c_{-1}f_{-1} + c_0 f_0 + c_1 f_1.$$

Taking $f(x) = x^2$, x, and 1, we get the relations ($x_0 = 0$),

$$2 = c_{-1}(-h)^2 + c_0(0) + c_1(h)^2,$$
$$0 = c_{-1}(-h) + c_0(0) + c_1(h),$$
$$0 = c_{-1}(1) + c_0(1) + c_1(1).$$

The resulting formula is

$$f_0'' \doteq \frac{f_{-1} - 2f_0 + f_1}{h^2} \qquad \text{(B.10)}$$

Equation (B.10), like Eq. (B.9), is particularly useful.

We are not confined to using only functional values in the method of unde-termined coefficients.* Suppose, for example, we have values of the first derivative as well as functional values. A formula for the second derivative might then be written as

$$f_0'' = c_0 f_0 + c_1 f_1 + c_3 f_0'.$$

As before, we take x^2, x, and 1 for $f(x)$, with $x_0 = 0$:

$$2 = c_0(0)^2 + c_1(h)^2 + c_3(0),$$
$$0 = c_0(0) + c_1(h) + c_3(1),$$
$$0 = c_0(1) + c_1(1) + c_3(0).$$

Solving, we obtain

$$f_0'' = 2\frac{f_1 - f_0 - hf_0'}{h^2}.$$

In the same way, we can derive formulas for the third and fourth derivatives. Derivatives beyond these do not often appear in applied problems. We must use a minimum of four and five terms, however, since only polynomials of degrees 3 and 4 have nonzero third or fourth derivatives. For the third-degree formula, complete

---

* We must be sure that the set of values is sufficient to determine a polynomial uniquely, however. For example, f_{-1}, f_1, and f_0' will not give a formula for f_0''.

symmetry with four points is impossible; five-term formulas for both derivatives are therefore given. We present the results only:

$$f_0''' = \frac{-f_{-2} + 2f_{-1} - 2f_1 + f_2}{2h^3} \tag{B.11}$$

$$f_0^{iv} = \frac{f_{-2} - 4f_{-1} + 6f_0 - 4f_1 + f_2}{h^4}. \tag{B.12}$$

B.2 Error Terms for Derivative Formulas

In the previous section, we used the method of undetermined coefficients to derive several formulas for the first derivative of a function, utilizing function values at equispaced x-values. These were, using the notation that $f_i = f(x_i)$,

$$f_0' \doteq \frac{f_1 - f_0}{h}, \tag{B.13}$$

$$f_0' \doteq \frac{-3f_0 + 4f_1 - f_2}{2h}, \tag{B.14}$$

$$f_0' \doteq \frac{f_1 - f_{-1}}{2h}. \tag{B.15}$$

While each of these formulas is not exact, as suggested by the $\doteq$ symbol, we argued heuristically that the error should decrease in each succeeding one. We now wish to develop expressions for the errors.

Begin with the Taylor-series expansion of $f(x_1) = f(x_0 + h)$ in terms of $x_1 - x_0 = h$:

$$f(x_1) = f(x_0) + hf'(x_0) + \frac{1}{2}h^2f''(x_0) + \frac{1}{6}h^3f'''(x_0) + \cdots.$$

Changing to our subscript notation, and truncating after the term in h, with the usual error term, we have

$$f_1 = f_0 + hf_0' + \frac{1}{2}h^2f''(\xi), \qquad x_0 < \xi < x_0 + h. \tag{B.16}$$

Solving for f_0', we have

$$f_0' = \frac{f_1 - f_0}{h} - \frac{1}{2}hf''(\xi), \qquad x_0 < \xi < x_0 + h, \tag{B.17}$$

so that the error term is $-\frac{1}{2}hf''(\xi)$ for the derivative formula, Eq. (B.13).

In Eq. (B.17), the error term involves the second derivative of the function evaluated at a place that is known only within a certain interval. In many of the applications for numerical differentiation, not only is the point of evaluation uncertain, but the function $f(x)$ is also unknown. If we do not know $f(x)$, we can hardly expect to know its derivatives. We do know, however, that the error involves the first power of $h = x_{i+1} - x_i$, and, in fact, the only way we can change the error is to change h. Making h smaller will decrease h, and in the limit as $h \to 0$ the error will go to zero. Further, as $h \to 0$, $x_1 \to x_0$, and the value of ξ is squeezed into a smaller and smaller interval. In other words, $f''(\xi)$ approaches a fixed value, specifically $f''(x_0)$, as h goes to zero.

The special importance of h in the error term is denoted by a special notation in numerical analysis, the *order relation*. We say the error of Eq. (B.17) is "of order h" and write

$$\text{Error} = O(h), \quad \text{when} \quad \lim_{h \to 0} (\text{error}) = ch,$$

where c is a fixed value not equal to zero.

To develop the error term for Eq. (B.14), we proceed similarly, except that we write expansions for both f_1 and f_2. It is also necessary to carry terms through h^3 because the second derivatives cancel, resulting in an error term involving the third derivative:

$$f_1 = f_0 + hf_0' + \frac{1}{2}h^2 f_0'' + \frac{1}{6}h^3 f'''(\xi_1), \qquad x_0 < \xi_1 < x_0 + h;$$

$$f_2 = f_0 + 2hf_0' + \frac{1}{2}(2h)^2 f_0'' + \frac{1}{6}(2h)^3 f'''(\xi_2), \qquad x_0 < \xi_2 < x_0 + 2h.$$

Note that the values of ξ_1 and ξ_2 may not be identical. If we multiply the first equation by 4, the second by -1, and add $-3f_0$ to their sum, we get

$$-3f_0 + 4f_1 - f_2 = (-3 + 4 - 1)f_0 + 2hf_0' + (2 - 2)f_0'' + \frac{4h^3}{6}[f'''(\xi_1) - 2f'''(\xi_2)].$$

Solving for f_0', we get

$$f_0' = \frac{-3f_0 + 4f_1 - f_2}{2h} + \frac{1}{3}h^2[2f'''(\xi_2) - f'''(\xi_1)],$$

$$x_0 < \xi_1 < x_0 + h, \qquad x_0 < \xi_2 < x_0 + 2h. \qquad \text{(B.18)}$$

The last term of Eq. (B.18) is the error term. As $h \to 0$, the two values of ξ approach the same value. Consequently, the error term approaches $\frac{1}{3}h^2 f'''(x_0)$. We then conclude that the error of Eq. (B.14) is $O(h^2)$.

For the error of Eq. (B.15), we proceeded similarly. It is not hard to show that

$$f_0' = \frac{f_1 - f_{-1}}{2h} - \frac{1}{6}h^2 f'''(\xi), \qquad x_0 - h < \xi < x_0 + h,$$

$$= \frac{f_1 - f_{-1}}{2h} + O(h^2). \qquad \text{(B.19)}$$

The error terms of both Eqs. (B.18) and (B.19) are $O(h^2)$, but the coefficient in Eq. (B.19) is only half the magnitude of the coefficient in Eq. (B.18). The progressive increase in accuracy we anticipated is confirmed.

By similar arguments, we can show that the formulas for third and fourth derivatives, Eqs. (B.11) and (B.12), both have errors $O(h^2)$.

B.3 Integration Formulas by the Method of Undetermined Coefficients

A numerical integration formula will estimate the value of the integral of the function by a formula involving the values of the function at a number of points in or near the interval of integration. Figure B.1 illustrates the general situation. If we desire to evaluate $\int_a^b f(x)\,dx$, it is obvious that if we could find some average value of the function on the interval $[a, b]$, the integral would be:

$$\int_a^b f(x)\,dx = (b - a)f_{av}.$$

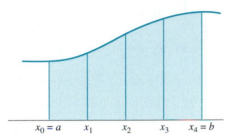

$x_0 = a \quad x_1 \quad x_2 \quad x_3 \quad x_4 = b$

Figure B.1

It seems reasonable to assume that f_{av} could be approximated by a linear combination of function values within or near the interval.* We therefore write, similarly to our procedure for derivatives,

$$\int_a^b f(x)\,dx = c_0 f(x_0) + c_1 f(x_1) + \cdots + c_n f(x_n),$$

where the coefficients c_i are to be determined. The points x_i at which the function is to be evaluated can also be left undetermined, but for the formulas which we need to derive, we will again impose the restriction that they are equally spaced with the value of $x_{i+1} - x_i = \Delta x = h = $ constant. We will impose the further restriction that the boundaries of the interval of integration, a and b, coincide with two of the x_i-values.

We start with a simple case. Suppose we wished to express the integral in terms of just two functional values, specifically $f(a)$ and $f(b)$. Our formula takes the form

$$\int_a^b f(x)\,dx = c_0 f(a) + c_1 f(b). \tag{B.20}$$

---

*One could generalize the concept by extending this to include derivatives of the function as well, but we stay with the simpler case.

Obviously, the values of c_0 and c_1 depend on $f(x)$, and probably on the values of a and b as well. The method of undetermined coefficients assumes that $f(x)$ can be approximated by a polynomial over the interval $[a, b]$ and determines c_0 and c_1 so that Eq. (B.20) is exact for all polynomials of a certain maximum degree or less.

Equation (B.20) has only two arbitrary constants, so we would not expect that it could be made exact for polynomials of degree higher than the first, since more than two parameters appear in second- or higher-degree polynomials. We therefore force Eq. (B.20) to be exact when $f(x)$ is replaced by either a first-degree polynomial or one of zero degree (a constant function). If it is to be exact for all first-degree polynomials, it must be exact for the very simple function $f(x) = x$. If it is to be exact when the function is any constant, it must hold if $f(x) = 1$. We express these two relations:

$$\int_a^b x\, dx = c_0(a) + c_1(b),$$

$$\int_a^b (1)\, dx = c_1(1) + c_1(1). \tag{B.21}$$

The integrals of Eqs. (B.21) are easily evaluated, so we have, as conditions that the constants must satisfy,

$$\int_a^b x\, dx = \frac{x^2}{2}\Big|_a^b = \frac{b^2}{2} - \frac{a^2}{2} = c_0 a + c_1 b,$$

$$\int_a^b dx = x\Big|_a^b = b - a = c_0 + c_1.$$

Solving these equations gives $c_0 = (b - a)/2$, $c_1 = (b - a)/2$.

Consequently,

$$\int_a^b f(x) \doteq \frac{b - a}{2}[f(a) + f(b)]. \tag{B.22}$$

This formula is the *trapezoidal rule,* and was derived in Chapter 4 as Eq. (4.29) through an entirely different approach. We have put a dot over the equality sign in Eq. (B.22) to remind ourselves that the relation is only approximately true, for the function $f(x)$ cannot ordinarily be replaced without error by a first-degree polynomial. It is intuitively obvious that unless the interval $[a, b]$ is very small, the error will be considerable.

We can reduce the error in the above formula by using a higher-degree polynomial to replace $f(x)$, but we would then need to take a linear combination of more than two function values. In fact, we will have to preserve a balance between the number of undetermined coefficients in the formula and the number of parameters in the polynomial. The number of coefficients hence must be one greater than the degree of the polynomial.

We now look at the next case, a three-term formula corresponding to replacing $f(x)$ by a quadratic. For three terms, using $x_0 = a$ and $x_2 = b$, with x_1 at the midpoint,

$$\int_a^b f(x)\, dx = c_0 f(a) + c_1 f\left(\frac{a + b}{2}\right) + c_2 f(b).$$

The formula must be exact if $f(x) = x^2$, or $f(x) = x$, or $f(x) = 1$, so

$$\int_a^b x^2\, dx = \frac{b^3}{3} - \frac{a^3}{3} = c_0 a^2 + c_1 \left(\frac{a+b}{2}\right)^2 + c_2 b^2,$$

$$\int_a^b x\, dx = \frac{b^2}{2} - \frac{a^2}{2} = c_0 a + c_1 \left(\frac{a+b}{2}\right) + c_2 b,$$

$$\int_a^b dx = b - a = c_0 + c_1 + c_2.$$

The solution is $c_0 = c_2 = \frac{1}{6}(b - a)$, $c_1 = \frac{4}{6}(b - a)$. Hence

$$\int_a^b f(x)\, dx \doteq \frac{b-a}{6}\left(f(a) + 4f\left(\frac{a+b}{2}\right) + f(b)\right). \qquad \textbf{(B.23)}$$

A more common form of Eq. (B.23) is found by writing $b - a = 2h$ (the interval from a to b is subdivided into two panels) and substituting x_0 for a, x_2 for b, and x_1 for the midpoint. We then have

$$\int_{x_0}^{x_0+2h} f(x)\, dx \doteq \frac{h}{3}(f_0 + 4f_1 + f_2). \qquad \textbf{(B.24)}$$

Formula (B.24) is Simpson's $\frac{1}{3}$ rule, a particularly popular formula. Again, it is not exact—as indicated by the dot over the equality sign.

The application of Eqs. (B.22) and (B.24) over an extended interval is straightforward. It is intuitively apparent that the error in these formulas will be large if the interval is not small. (We discuss these errors quantitatively in Section B.5.) To apply these to a large interval of integration and still maintain control over the error, we break the interval into a large number of small subintervals and add together the formulas applied to the subintervals. When this is done we get the *extended trapezoidal rule*:

$$\int_a^b f(x)\, dx \doteq \frac{b-a}{2n}[f(x_0) + 2f(x_1) + 2f(x_2) + \cdots + 2f(x_{n-1}) + f(x_n)]; \qquad \textbf{(B.25)}$$

and the extended Simpson's $\frac{1}{3}$ rule:

$$\int_a^b f(x)\, dx \doteq \frac{b-a}{3n}[f(x_0) + 4f(x_1) + 2f(x_2) + 4f(x_3) + \cdots$$
$$+ 2f(x_{2n-2}) + 4f(x_{2n-1}) + f(x_{2n})]. \qquad \textbf{(B.26)}$$

These forms of the trapezoidal rule (Eq. (B.25)) and of Simpson's $\frac{1}{3}$ rule (Eq. (B.26)) are widely used in computer programs for integration. By applying them with n increasing, the error can be made arbitrarily small.* For Simpson's rule, observe that the number of subintervals must be even.

---

* Except for round-off error effects, which eventually will dominate, since they are not decreased by small subdivision of the interval and, in fact, may increase as the number of computations increases.

B.4 Integration Formulas Using Points Outside the Interval

In studying numerical methods to solve differential equations we shall have need for some specific integral formulas that involve function values computed at points outside the interval of integration. Figures B.2, B.3, and B.4 sketch three special cases of importance. In Fig. B.2 the curve that passes through the four points whose abscissas are x_{n-3}, x_{n-2}, x_{n-1}, and x_n is extrapolated to x_{n+1}, and we desire the integral only over the extrapolated interval. Figure B.3 presents the case where the curve passes through four points, and the integral is taken only over the last panel. In Fig. B.4 we consider a case where a curve that fits at three points is extrapolated in both directions, and integration is taken over four panels.

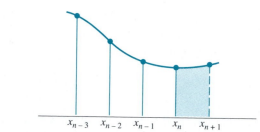

Figure B.2

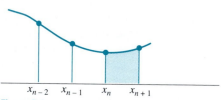

Figure B.3

Figure B.4

In the derivations of this section it will be convenient to adopt the notation that $f_i = f(x_i)$, so that subscripts on the function indicate the x-value at which it is evaluated.

For the first case (Fig. B.2) we desire a formula of the form

$$\int_{x_n}^{x_{n+1}} f(x)\,dx = c_0 f_{n-3} + c_1 f_{n-2} + c_2 f_{n-1} + c_3 f_n.$$

With four constants, we can make the formula exact when $f(x)$ is any polynomial of degree 3 or less. Accordingly, we replace $f(x)$ successively by x^3, x^2, x, and 1 to evaluate the coefficients.

It is apparent that the formula must be independent of the actual x-values. To simplify the equations, let us shift the origin to the point $x = x_n$; our integral is then taken over the interval from 0 to h, where $h = x_{n+1} - x_n$:

$$\int_0^h f(x)\, dx = c_0 f(-3h) + c_1 f(-2h) + c_2 f(-h) + c_3 f(0).$$

Carrying out the computations by replacing $f(x)$ with the particular polynomials, we have

$$\frac{h^4}{4} = c_0(-3h)^3 + c_1(-2h)^3 + c_2(-h)^3 + c_3(0),$$

$$\frac{h^3}{3} = c_0(-3h)^2 + c_1(-2h)^2 + c_2(-h)^2 + c_3(0),$$

$$\frac{h^2}{2} = c_0(-3h) + c_1(-2h) + c_2(-h) + c_3(0), \tag{B.27}$$

$$h = c_0(1) + c_1(1) + c_2(1) + c_3(1).$$

After completion of the algebra we get

$$\int_{x_n}^{x_{n+1}} f(x)\, dx \doteq \frac{h}{24}(-9f_{n-3} + 37f_{n-2} - 59f_{n-1} + 55f_n). \tag{B.28}$$

For the second case, illustrated by Fig. B.3, we again translate the origin to x_n to simplify. The set of equations analogous to Eqs. (B.27) is

$$\frac{h^4}{4} = c_0(-2h)^3 + c_1(-h)^3 + c_2(0) + c_3(h)^3,$$

$$\frac{h^3}{3} = c_0(-2h)^2 + c_1(-h)^2 + c_2(0) + c_3(h)^2,$$

$$\frac{h^2}{2} = c_0(-2h) + c_1(-h) + c_2(0) + c_3(h),$$

$$h = c_0(1) + c_1(1) + c_2(1) + c_3(1).$$

These give values of the constants in the integration formula:

$$\int_{x_n}^{x_{n+1}} f(x)\, dx \doteq \frac{h}{24}(f_{n-2} - 5f_{n-1} + 19f_n + 9f_{n+1}). \tag{B.29}$$

The third case, Fig. B.4, where extrapolation of a quadratic in both directions is involved, leads to the equation

$$\int_{x_{n-3}}^{x_{n+1}} f(x)\, dx \doteq \frac{4h}{3}(2f_{n-2} - f_{n-1} + 2f_n). \tag{B.30}$$

The particular methods for solving differential equations that use the formulas derived in this section are known as *Adams' method* and *Milne's method*.

B.5 Error of Integration Formulas

In each of the formulas that we derived so far, we used the symbol $\doteq$ to remind us that the formulas are not exact unless $f(x)$ is in fact a polynomial of a certain degree or less. It is important to derive expressions for the error.

We first determine the error term for the trapezoidal rule over one panel, Eq. (B.22). Begin with a Taylor expansion:

$$F(x_1) = F(x_0) + F'(x_0)h + \frac{1}{2}F''(x_0)h^2 + \frac{1}{6}F'''(\xi_1)h^3,$$

$$x_0 < \xi_1 < x_1 = x_0 + h. \tag{B.31}$$

Let us define $F(x) = \int_a^x f(t)\,dt$ so that $F'(x) = f(x)$, $F''(x) = f'(x)$, $F'''(x) = f''(x)$. Equation (B.31) becomes, with this definition of $F(x)$,

$$\int_a^{x_1} f(x)\,dx = \int_a^{x_0} f(x)\,dx + f(x_0)h + \frac{1}{2}f'(x_0)h^2 + \frac{1}{6}f''(\xi_1)h^3.$$

On rearranging, and using subscript notation, we get

$$\int_a^{x_1} f(x)\,dx = \int_a^{x_0} f(x)\,dx = \int_{x_0}^{x_1} f(x)\,dx = f_0 h + \frac{1}{2}f_0'h^2 + \frac{1}{6}f''(\xi_1)h^3. \tag{B.32}$$

We now replace the term f_0' by the forward-difference approximation with error term from Section B.2:

$$f_0' = \frac{f_1 - f_0}{h} - \frac{1}{2}hf''(\xi_2), \qquad x_0 < \xi_2 < x_1.$$

Equation (B.32) becomes

$$\int_{x_0}^{x_1} f(x)\,dx = f_0 h + \frac{1}{2}h(f_1 - f_0) - \frac{1}{4}h^3 f''(\xi_2) + \frac{1}{6}h^3 f''(\xi_1).$$

We can combine the last two terms into $-\frac{1}{12}h^3 f''(\xi)$, where ξ also lies in $[x_0, x_1]$. Hence,

$$\int_{x_0}^{x_1} f(x)\,dx = \frac{h}{2}(f_0 + f_1) - \frac{1}{12}h^3 f''(\xi).$$

The error of the trapezoidal rule over one panel is then $O(h^3)$. This is called the *local error*. Since we normally use a succession of applications of this rule over the interval $[a, b]$ to evaluate $\int_a^b f(x)\,dx$, by subdividing into $n = (b - a)/h$ panels, we need to determine the so-called global error for such intervals of integration. The global error will be the sum of the local errors in each of the n panels:

$$\text{Global error} = -\frac{1}{12}h^3 f''(\xi_1) - \frac{1}{12}h^3 f''(\xi_2) - \cdots - \frac{1}{12}h^3 f''(\xi_n)$$

$$= -\frac{1}{12}h^3[f''(\xi_1) + f''(\xi_2) + \cdots + f''(\xi_n)].$$

Here the subscripts on ξ indicate the panel in whose interval the value lies. If $f''(x)$ is continuous throughout the interval of integration,

$$\sum_{i=1}^{n} f''(\xi_i) = nf''(\xi), \qquad a < \xi < b.$$

Using $n = (b - a)/h$, we get

$$\text{Global error} = -\frac{1}{12}h^3\left(\frac{b - a}{h}\right)f''(\xi) = -\frac{b - a}{12}h^2f''(\xi) = O(h^2).$$

Therefore the global error is $O(h^2)$, even though the local error is $O(h^3)$.

Similar treatment shows that the local error for Simpson's $\frac{1}{3}$ rule is

$$-\frac{1}{90}h^5f^{iv}(\xi) = O(h^5),$$

and the global error is

$$-\frac{1}{180}(b - a)h^4f^{iv}(\xi) = O(h^4).$$

We can also show that the local error of each of the integration formulas in Section B.4 is $O(h^5)$.

APPENDIX C

Software Libraries

Besides the programs and subroutines presented in this book, many other excellent programs and subroutines are available, most of them for desktop computers and workstations.

DERIVE is one of the computer algebra systems used in this book. In addition to the PC version, it offers a ROM-card version for the HP 95LX Palmtop computer.
Source: Soft Warehouse, Inc., 3660 Waialae Avenue, Suite 304, Honolulu, HI 96816-3236.

IBM's ESSL (Engineering and Scientific Subroutine Library) Version 2 consists of more than 400 subroutines callable from C and FORTRAN. The package includes topics in numerical quadrature, interpolation, random number generation, FFTs, linear systems, and eigenvalue problems. This package is available for workstations.
Source: (914)385-9829.

IMSL (International Mathematical and Statistical Library) is a library of over 900 useful FORTRAN subroutines that cover both mathematical and statistical problems. There is a comparable C/Base/Library of numerical functions as well. Moreover, this library can be installed on a variety of machines, including the CRAY Y-MP.
Source: IMSL, Inc. Houston, TX (713)242-6776.

LAPACK is a library of FORTRAN 77 subroutines for solving problems in numerical linear algebra, including eigenvalue problems, and singular value problems. This library supersedes LINPACK and EISPACK.
Source: Individual routines available through *netlib*. Currently, the e-mail addresses are **netlib@ornl.gov** and **netlib@research.att.com.** For just general information send mail to one of these two addresses with this message: "send index from lapack." The complete package, consisting of about 600,000 lines of FORTRAN code, can be obtained on magnetic media from NAG (see entry for NAG).

MACSYMA is a software package that has been around for a long time. Some of its features include simplification of trig functions, differential-equation packages, and color graphics. It is now available for the PC.
Source: MACSYMA, Inc., 20 Academy Street, Arlington, MA 02174.

MAPLE V is a very powerful software package that performs both symbolic and numeric computations. It provides both two-dimensional and three-dimensional color graphing capabilities. This software runs on almost all computers including PC's, Macintoshes, workstations, and mainframes.

Source: Waterloo Maple Software, 160 Columbia Street West, Dept. 507, Waterloo, Ontario, Canada N2L 3L3. e-mail: **wmsi@daisy.waterloo.edu.**

MATHEMATICA is one of the best-known software packages for computing a large variety of mathematical problems. Moreover, it has excellent two-dimensional and three-dimensional graphing capabilities and does both symbolic and numeric computations. This system is available on a very large variety of computers.

Source: Wolfram Research, Inc., P.O. Box 6059, Champaign, IL 61821-9902.

MATLAB covers several topics in mathematics, including linear algebra, digital signal processing, graphing, and numerical methods. It is available both for the MS-DOS machines as well as the Macintosh computers.

Source: The Math Works Inc., Cochituate Place, 24 Prime Park Way, Natick, MA 01760. e-mail: **info@mathworks.com.**

NAG (Numerical Algorithms Group) has provided a large library of FORTRAN subprograms. In addition, this group has now come out with a symbolic solver software package, called AXIOM. This package also produces two-dimensional and three-dimensional graphics.

Source: Numerical Algorithms Group, Inc. 1400 Opus Place, Suite 200, Downers Grove, IL 60515-5702.

NETLIB, mentioned with regard to the LAPACK software, is a tremendous resource for numerical libraries. Rather than deal with a long list, one could send an e-mail message containing the line "help" to one of the Internet addresses: **netlib@research.att.com, netlib@ornl.gov, netlib@nec.no** for information on how to use netlib and for an overview of the mathematical software.

NUMERICAL RECIPES is a well-known set of 300-plus numerical routines for a variety of programming languages, including C, FORTRAN, Pascal, and BASIC. In addition to the books, the source code for each version is available on diskette.

Source: Cambridge University Press, 40 West 20th Street, New York, NY 10011-4211.

References

Acton, F. S. (1970). *Numerical Methods That Work*. New York: Harper and Row.

Allaire, Paul E. (1985). *Basics of the Finite Element Method*. Dubuque, Iowa: Brown.

Allen, D. N. (1954). *Relation Methods*. New York: McGraw-Hill.

Anderson, E., Z. Gai, C. Bishof, J. Demmel, J. Dongarra, et al. (1992). *LAPACK Users' Guide*. Philadelphia: SIAM.

Andrews, G., and R. Olsson (1993). *The SR Programming Language*. Redwood City, Calif.: Benjamin Cummings.

Andrews, Larry C. (1985). *Elementary Partial Differential Equations with Boundary Value Problems*. Philadelphia: Saunders College Publishing.

Arney, David C. (1987). *The Student Edition of DERIVE, Manual*. Reading, Mass.: Addison-Wesley—Benjamin/Cummings.

Atkinson, K. E. (1978). *An Introduction to Numerical Analysis*. New York: Wiley.

Bartels, Richard, J. Beatty, and B. Barsky (1987). *An Introduction to Splines for Use in Computer Graphics and Geometric Modeling*. Los Altos, Calif.: Morgan Kaufmann.

Birkhoff, Garrett, Richard Varga, and David Young (1962). Alternating direction implicit methods. *Advances in Computers* 3:187–273.

Bracewell, Ronald N. (1986). *The Hartley Transform*. New York: Oxford University Press.

Brigham, E. Oron (1974). *The Fast Fourier Transform*. Englewood Cliffs, N.J.: Prentice-Hall.

Burnett, David S. (1987). *Finite Element Analysis: From Concepts to Applications*. Reading, Mass.: Addison-Wesley.

Campbell, Leon, and Laizi Jacchia (1941). *The Story of Variable Stars*. Philadelphia: Blakiston.

Carnahan, Brice (1964). *Radiation Induced Cracking of Pentanes and Dimethylbutanes*. Ph.D. dissertation, University of Michigan.

Carnahan, Brice, et al. (1969). *Applied Numerical Methods*. New York: Wiley.

Carslaw, H. S., and J. C. Jaeger (1959). *Conduction of Heat in Solids*. 2nd ed. London: Oxford University Press.

Chandy, K. M., and S. Taylor (1992). *An Introduction to Parallel Programming*. Boston: Jones and Bartlett.

Char, B., K. Geddes, G. Gonnet, B. Leong, M. Monagan, and S. Watt (1992). *First Leaves: A Tutorial Introduction to Maple V*. New York: Springer-Verlag.

Condon, Edward, and Hugh Odishaw, eds. (1967). *Handbook of Physics*. New York: McGraw-Hill.

Conte, S. D., and C. de Boor (1980). *Elementary Numerical Analysis*. 3rd ed. New York: McGraw-Hill.

Cooley, J. W., and J. W. Tukey (1965). An algorithm for the machine calculations of complex Fourier series. *Mathematics of Computation* 19:297–301.

Corliss, G., and Y. F. Chang (1982). Solving ordinary differential equations using Taylor series. *ACM Transactions on Mathematical Software* 8:114–144.

Crow, Frank (1987). Origins of a teapot. *IEEE Computer Graphics and Applications* 7(1):8–19.

Davis, Alan J. (1980). *The Finite Element Method*. Oxford: Clarendon Press.

Davis, Phillip J., and Philip Rabinowitz (1967). *Numerical Integration*. Waltham, Mass.: Blaisdell.

de Boor, C. (1978). *A Practical Guide to Splines*. New York: Springer-Verlag.

De Santis, R., F. Gironi, and L. Marelli (1976). Vector-liquid equilibrium from a hard-sphere equation of state. *Industrial and Engineering Chemistry Fundamentals* 15(3):183–189.

Dongarra, J., I. Duff, D. Sorensen, and H. van der Vorst (1991). *Solving Linear Systems on Vector and Shared Memory Computers*. Philadelphia: SIAM.

Dongarra, J. J., J. R. Bunch, C. B. Moler, and G. W. Stewart (1979). *LINPACK User's Guide*. Philadelphia: SIAM.

Douglas, J. (1962). Alternating direction methods for three space variables. *Numerical Mathematics* 4:41–63.

Duffy, A. R., J. E. Sorenson, and R. E. Mesloh (1967). Heat transfer characteristics of belowground LNG storage. *Chemical Engineering Progress* 63(6):55–61.

Etter, D. M. (1993). *Quattro Pro—A Software Tool for Engineers and Scientists*. Redwood City, Calif.: Benjamin/Cummings.

Fike, C. T. (1968). *Computer Evaluation of Mathematical Functions*. Englewood Cliffs, N.J.: Prentice-Hall.

Forsythe, G. E., M. A. Malcolm, and C. B. Moler (1977). *Computer Methods for Mathematical Computation*. Englewood Cliffs, N.J.: Prentice-Hall.

Forsythe, G. E., and C. B. Moler (1967). *Computer Solution of Linear Algebraic Systems*. Englewood Cliffs, N.J.: Prentice-Hall.

Fox, L. (1965). *An Introduction to Numerical Linear Analysis*. New York: Oxford University Press.

Gear, C. W. (1967). The numerical integration of ordinary differential equations. *Mathematics of Computation* 21:146–156.

Gear, C. W. (1971). *Numerical Initial Value Problems in Ordinary Differential Equations*. Englewood Cliffs, N.J.: Prentice-Hall.

Hageman, L. A., and D. M. Young (1981). *Applied Iterative Methods*. New York: Academic Press.

Hamming, R. W. (1973). *Numerical Methods for Scientists and Engineers*. 2nd ed. New York: McGraw-Hill.

Hamming, R. W. (1971). *Introduction to Applied Numerical Analysis*. New York: McGraw-Hill.

Harrington, Steven (1987). *Computer Graphics: A Programming Approach*. New York: McGraw-Hill.

Henrici, P. H. (1964). *Elements of Numerical Analysis*. New York: Wiley.

Hornbeck, R. W. (1975). *Numerical Methods*. New York: Quantum.

Householder, A. S. (1970). *The Numerical Treatment of a Single Nonlinear Equation*. New York: McGraw-Hill.

IEEE Standard for Binary Floating-Point Arithmetic (1985). Institute of Electrical and Electronics Engineers, Inc., New York.

JaJa, J. (1992). *An Introduction to Parallel Algorithms*. Reading, Mass.: Addison-Wesley.

Jones, B. (1982). A note on the T transformation. *Nonlinear Analysis, Theory, Methods and Applications* 6:303–305.

Kahaner, D., C. Moler, S. Nash (1989). *Numerical Methods and Software*. Englewood Cliffs, N.J.: Prentice-Hall.

Lee, Peter, and Geoffrey Duffy (1976). Relationships between velocity profiles and drag reduction in turbulent fiber suspension flow. *Journal of the American Institute of Chemical Engineering* 22(4):750–753.

Love, Carl H. (1966). *Abscissas and Weights for Gaussian Quadrature*. National Bureau of Standards, Monograph 98.

Muller, D. E. (1956). A method of solving algebraic equations using an automatic computer. *Math Tables and Other Aids to Computation* 10:208–215.

O'Neill, Mark A. (1988). Faster Than Fast Fourier. *BYTE* 13(4):293–300.

Orvis, William J. (1987). *1-2-3 for Scientists and Engineers*. San Francisco: Sybex.

Peaceman, D. W., and H. H. Rachford (1955). The numerical solution of parabolic and elliptic differential equations. *Journal of the Society for Industrial and Applied Mathematics* 3:28–41.

Penrod, E. B., and K. V. Prasanna (1962). Design of a flat-plate collector for a solar earth heat pump. *Solar Energy* 6(1)9–22.

Pinsky, Mark A. (1991). *Partial Differential Equations and Boundary Value Problems with Applications*. 2nd ed. New York: McGraw-Hill.

Pizer, Stephen M. (1975). *Numerical Computing and Mathematical Analysis*. Chicago: Science Research Associates.

Pokorny, C., and C. Gerald (1989). *Computer Graphics: The Principles Behind the Art and Science*. Irvine, Calif.: Franklin, Beedle, and Associates.

Prenter, P. M. (1975). *Splines and Variational Methods*. New York: Wiley.

Press, W., B. Flannery, S. Teudolsky, and W. Vetterling (1992). *Numerical Recipes in FORTRAN: The Art of Scientific Computing*. 2nd ed. New York: Cambridge University Press.

Press, W., B. Flannery, S. Teudolsky, and W. Vetterling (1992). *Numerical Recipes in C: The Art of Scientific Computing*. 2nd ed. New York: Cambridge Univeristy Press.

Press, W., B. Flannery, S. Teudolsky, and W. Vetterling (1989). *Numerical Recipes in Pascal: The Art of Scientific Computing*. Rev. 1st ed. New York: Cambridge University Press.

Ralston, Anthony (1965). *A First Course in Numerical Analysis*. New York: McGraw-Hill.

Ramirez, Robert W. (1985). *The FFT, Fundamentals and Concepts*. Englewood Cliffs, N.J.: Prentice-Hall.

Rice, John R. (1983). *Numerical Methods, Software, and Analysis*. New York: McGraw-Hill.

Richtmyer, R. D. (1957). *Difference Methods for Initial Value Problems*. New York: Wiley Interscience.

Sedgwick, R. (1992). *Algorithms in C++*. Reading, Mass.: Addison-Wesley. [Other versions available: *in C, in Pascal.*]

Shampine, L., and R. Allen (1973). *Numerical Computing*. Philadelphia: Saunders.

Smith, G. D. (1978). *Numerical Solution of Partial Differential Equations*. 2nd ed. London: Oxford University Press.

Stallings, William (1990). *Computer Organization and Architecture*. New York: Macmillan.

Stewart, G. W. (1973). *Introduction to Matrix Computations*. New York: Academic Press.

Stoer, J., and R. Bulirsch (1980). *Introduction to Numerical Analysis*. New York: Springer-Verlag.

Stubbs, D. and N. Webre (1987). *Data Structures with Abstract Data Types*. Monterey, Calif.: Brooks/Cole.

Traub, J. F. (1964). *Iterative Methods for the Solution of Equations*. Englewood Cliffs, N.J.: Prentice-Hall.

Varga, Richard (1959). *p*-Cyclic matrices: A generalization of the Young–Frankel successive overrelaxation scheme. *Pacific Journal of Mathematics* 9:617–628.

Vichnevetsky, R. (1981). *Computer Methods for Partial Differential Equations. Vol. 1: Elliptic Equations and the Finite Element Method.* Englewood Cliffs, N.J.: Prentice-Hall.

Waser, S., and M. J. Flynn (1982). *Introduction to Arithmetic for Digital Systems Designers.* New York: Holt, Rinehart and Winston.

Wilkinson, J. H. (1963). *Rounding Errors in Algebraic Processes.* Englewood Cliffs, N.J.: Prentice-Hall.

Wilkinson, J. H. (1965). *The Algebraic Eigenvalue Problem.* London: Oxford University Press.

Wolfram, Stephen (1988). *Mathematica: A System for Doing Mathematics by Computer.* Reading, Mass.: Addison-Wesley.

Answers to Selected Exercises

CHAPTER 0   **4.** This begins as a trigonometric probem. If we let H = depth of well, D = diameter of well, W = width of ladder, V = width of side rails, L = length of ladder, and A = angle of inclination from the vertical, we can write

$$\tan(A) = \frac{2\left(\dfrac{D}{2}\right)^2 - \left(\dfrac{W}{2}\right)^2 - V * \cos(A)}{H - V * \sin(A)}$$

and

$$L = \frac{H - V * \sin(A)}{\cos(A)}.$$

With the given values for H, D, W, and V, DERIVE can solve the first to give $A = 0.37965$ radian, and from this we get, from the second equation, $L = 181.557$ in.

7. Endless loop if tolerance = 1E-8. Stop with "BREAK."

9. b. $x = 1.114653$ in 20 iterations with tol = lE-6.

13. Total numbers = 180,001; largest positive number = 0.9999E5; most negative number = -0.9999E5; smallest positive number = 0.1000E-4; smallest negative number = -0.1000E-4.

15.   a. 0.123E2 (chop), 0.123E2 (round)
b. -0.307E-1 (chop), -0.307E-1 (round)
c. 0.122E2 (chop), 0.123E2 (round)
d. -0.288E3 (chop), -0.289E3 (round)
e. 0.130E3 (chop), 0.130E3 (round)
f. -0.156E5 (chop), -0.156E5 (round)
g. 0.123E-6 (chop), 0.123E-6 (round)

20. Exact = -1.297387
Chopped 3 digits = -1.31, relative error = -0.00972
Round 3 digits = -1.30, relative error = -0.00201

25. The series will converge because with a very large value for N, $1/N$ = zero. Looping is impractical because of the time required. Compute through largest positive number. ($N = 3.402823$E38 in QuickBASIC, single precision. Other systems may have a smaller limit.)

29. The answer depends on the spacing between the roots. If evenly spaced, we obtain the middle one, of course. Closely spaced roots act like a multiple root. If $f(x) > 0$ beyond $x = b$, the root found tends to be a larger root than the middle one.

33. Parallel processing is applicable when iterate $n + 1$ does not require that interate n be available.

CHAPTER 1

3. There are two intersections, at $x = 0.05247$ and at $x = 4.50524$.

8. Using DERIVE, intersections are at about $x = -0.792, 0.555$. These results obtained:

From secant: $x = -0.801938$, 4 iterations
$x = 0.554598$, 3 iterations

From regula falsi: $x = -0.801937$, 17 iterations
$x = 0.554958$, 14 iterations

13. $F(x) = x^2 - N = 0$, $F'(x) = 2x$, so

$$x_1 = x_0 - \frac{x_0^2 - N}{2x_0} = \frac{x_0^2 + N}{2x_0} = \frac{1}{2}\left(x_0 + \frac{N}{x_0}\right).$$

18. From $x_0 = 0.9$ or 1.1, tol = 1E-6, converge in 26 iterations. From $x_0 = 1.9$ or 2.1, tol = 1E-6, converge in 4 iterations [to $f(x) = 0$].

20. From graph, $f(x)$ has minimum at about $x = 1.75$. This value is confirmed by secant method: $f'(1.75) = 0$.

24. For each example of Exercise 4, Muller's method does get the root nearest zero. In some cases this is not true, such as $(x + 0.3)(x - 0.2)(x - 0.3)$, or when there is a root very near to -0.5 or 0.5 in addition to a smaller root.

26. The method will attempt to get the square root of a negative number. Try different starting values.

30. Converges to 0.6180 in 11 iterations. If accelerated, it takes 6 iterations (plus extrapolations).

32. Starting from $x = 1.2$:
(1) $x = ((5 + 3x - 3x^2)/2)^{1/3}$, 16 iterations,
(2) $x = ((5 + 3x - 2x^3)/3)^{1/2}$, 81 iterations,
(3) $x = ((5 + 3x)/(2x + 3))^{1/2}$, 3 iterations.

35. $x = -4.5615528$, $x = -0.43844719$; factor is $2x^2 - 3x + 7$; roots: $0.75 \pm 1.713913i$

38.

Change in coeff	Max change in magnitude of any root
$2.00 \rightarrow 2.02$	0.73%
$7.00 \rightarrow 7.07$	0.56%
$4.00 \rightarrow 4.04$	0.71%
$29.00 \rightarrow 29.29$	0.80%
$14.00 \rightarrow 14.14$	0.89%

42. $P(x) = (x^2 - 1.5x + 4.3)(x^2 - 4.2x + 16)$; roots: $0.75 \pm 1.9333i$, $2.1 \pm 3.4040i$

44. $P(x) = (x^2 - 5x + 2)(2x^2 - 3x + 7)$; roots: -0.438445, -4.56155, $0.75 \pm 1.7139i$

48. a. After convergence, q's approximate the roots: $x = 1.8012$, -1.2462, 0.4450.
b. Two real roots: $x = -0.6474$ and -3.5526; quadratic factor: $x^2 - 2.1x + 3.1$

52. If there are complex roots, an initial estimate near their modulus asks for square root of negative number. If there are multiple roots, convergence is slow.

56. Bairstow gives these factors:

$$x^2 - 2x + 1, \quad x^2 - 5.9999x + 8.9997, \quad x - 3.0001,$$

whose roots are 1, 1; 3, 2.9999; 3.0001. The results depend on the starting value and the tolerance value.

59. $P(x)/P'(x) = (5x^2 - 18x + 21)/(5x - 9)^2$. Starting with $x_0 = 1.1$, successive errors are 0.100000, 0.00826, and 0.00005, and each of these is about the square of the previous one. Starting with $x_0 = 3.3$, errors are 0.30000, 0.02243, and 0.00017; these actually decrease faster than $O(\text{error}^2)$.

63. a. Starting with $x_0 = 2.05$, converges to root at $x = 0$ but there are wide swings at first. Near convergence, errors are -0.71630, -0.15299, and -0.00232. Last two show quadratic convergence.

b. Starting with $x_0 = 2.00$, converges to root at 9.41757. Last errors are 0.03679, 0.00130 showing about quadratic convergence.

66. c. Convergence is very, very slow. There is linear convergence; each error is 2/3 times the previous value.

69. A DERIVE plot shows a double root at zero and next at about $x = 4.49$. Using soLve in Approximate mode starting from [4, 5] gives root = 4.49340.

CHAPTER 2   **1.** b. $\begin{bmatrix} 1 & 0 & -3 & 0 \\ -3 & 0 & 0 & -4 \\ -2 & 5 & 3 & 9 \end{bmatrix}$, $\begin{bmatrix} 37 \\ 5 \\ 19 \end{bmatrix}$, $\begin{bmatrix} -29 \\ -3 \\ -20 \end{bmatrix}$

c. -35; $\begin{bmatrix} -1 & 2 & -3 & 2 \\ 3 & -6 & 9 & -6 \\ -8 & 16 & -24 & 16 \\ 2 & -4 & 6 & -4 \end{bmatrix}$

2. a. $BA = \begin{bmatrix} -18 & 7 & 9 \\ -15 & -8 & -1 \\ 8 & 11 & 26 \end{bmatrix}$

3. d. $\begin{bmatrix} 0 & 0 & 0 \\ 3 & 0 & 0 \\ 2 & 0 & 0 \end{bmatrix} + \begin{bmatrix} 1 & 0 & 0 \\ 0 & 1 & 0 \\ 0 & 0 & 1 \end{bmatrix} + \begin{bmatrix} 0 & -2 & 2 \\ 0 & 0 & 1 \\ 0 & 0 & 0 \end{bmatrix}$

4. b. $p_B(x) = (8 - x)(x + 2)(x - 1)$ implies the eigenvalues are $\{-2, 1, 8\}$.

6. $\begin{bmatrix} 3 & -6 & 2 \\ -4 & 1 & -1 \\ 1 & -3 & 7 \end{bmatrix} \begin{bmatrix} x \\ y \\ z \end{bmatrix} = \begin{bmatrix} 15 \\ -2 \\ 22 \end{bmatrix}$

7. b. $x_3 = 2$; $x_2 = (3 + 6)/3 = 3$; $x_1 = (7 - 4 + 3)/2 = 3$

11. By using the elementary row operations, we can reduce the augmented matrix to

$$\begin{bmatrix} 3 & 2 & -1 & -4 & 10 \\ 1 & -1 & 3 & -1 & -4 \\ 2 & 1 & -3 & 0 & 16 \\ 0 & -1 & 8 & -5 & 3 \end{bmatrix} \rightarrow \begin{bmatrix} 3 & 2 & -1 & -4 & 10 \\ 0 & 5 & -10 & -1 & 22 \\ 0 & 0 & -45 & 39 & 162 \\ 0 & 0 & 0 & 0 & 435 \end{bmatrix}.$$

The last line would imply that $0 = 435$!

13. Let $R_1 = (3\ \ 2\ \ -1\ \ -4), \ldots, R_4 = (0\ \ -1\ \ 8\ \ -5)$ represent the four rows of the coefficient matrix of Exercise 11. Then we have the relationship $R_1 + R_2 - 2R_3 = R_4$.

15. a. $\begin{bmatrix} 1 & 1 & -2 & 3 \\ 4 & -2 & 1 & 5 \\ 3 & -1 & 3 & 8 \end{bmatrix} \rightarrow \cdots \rightarrow \begin{bmatrix} 4 & -2 & 1 & 5 \\ \frac{1}{4} & \frac{3}{2} & -\frac{9}{4} & \frac{7}{4} \\ \frac{3}{4} & \frac{1}{3} & 3 & \frac{11}{3} \end{bmatrix}$

Using back-substitution, we get

$$ x_3 = \frac{11}{9}; \qquad x_2 = \left(\frac{7}{4} + \frac{11}{4}\right) * \frac{2}{3} = 3; \qquad x_1 = \frac{22}{9} $$

b. $\det(A) = (-1) * 4 * (3/2) * 3 = -18$

c. $\begin{bmatrix} 1 & 0 & 0 \\ \frac{1}{4} & 1 & 0 \\ \frac{3}{4} & \frac{1}{3} & 1 \end{bmatrix} \begin{bmatrix} 4 & -2 & 1 \\ 0 & \frac{3}{2} & -\frac{9}{4} \\ 0 & 0 & 3 \end{bmatrix} = \begin{bmatrix} 4 & -2 & 1 \\ 1 & 1 & -2 \\ 3 & -1 & 3 \end{bmatrix}$

16. a. $x^T = (1.30, -1.35, -0.275)$
    b. $x^T = (1.45, -1.59, -0.276)$
    c. Calculated right-hand sides are

$$ (0.02, 1.02, -0.21) \qquad \text{and} \qquad (0.04, 1.03, -0.54). $$

19. c. In general, the number of multiplications/divisions for the Gauss–Jordan method is $O(n^3/2)$ versus $O(n^3/3)$ for Gaussian elimination.

20. $Az = b$ implies $(B + Ci)(x + yi) = (Bx - Cy) + (By + Cx)i = p + qi$.

a. $\begin{bmatrix} B & -C \\ C & B \end{bmatrix} \begin{bmatrix} x \\ y \end{bmatrix} = \begin{bmatrix} p \\ q \end{bmatrix}$

b. $2n^2 + 2n$ versus $4n^2 + 2n$

21. b. $z_1 = 1.3529 - 0.4118i, \ z_2 = 0.3529 + 0.5882i$

22. $A = \begin{bmatrix} 1 & 0 & 0 \\ 4 & -6 & 0 \\ 3 & -4 & 3 \end{bmatrix} \begin{bmatrix} 1 & 1 & -2 \\ 0 & 1 & -\frac{3}{2} \\ 0 & 0 & 1 \end{bmatrix}$

First solving for $Ly = b$, we get: $y = (3, 7/6, 11/9)$; then solving $Ux = (3, 7/6, 11/9)^T$ we get the answer: $(22/9, 3, 11/9)$.

29. b. $7c_1 - 12c_2 - 3c_3 = 0$

30. b. Solving for the first three equations gives us the unique solution: $(3/2, -1/2, -3/2)$. However, substituting these values into the fourth equation does not produce the correct result. This set of four equations does not have a solution.

31. a. $\det(H)$ is very small (about 1.65×10^{-7}); one also cannot avoid a very small divisor in solving.
    b. $x^T = (1.11, 0.228, 1.95, 0.797)$
    c. $x^T = (0.988, 1.42, -0.428, 2.10)$
    d. $x^T = (1.0000, 0.9995, 1.0017, 0.9990)$

37.   b. $H^{-1} = \begin{bmatrix} 16 & -120 & 240 & -140 \\ -120 & 1200 & -2700 & 1680 \\ 240 & -2700 & 6480 & -4200 \\ -140 & 1680 & -4200 & 2800 \end{bmatrix}$

38.   Gauss Elimination: 25 mult/div; 11 add/subtracts. Gauss-Jordan: 29 mult/div; 15 add/subtracts. The system here is too small to illustrate the true difference between the methods.

39.   d. $||B||_1 = 12; ||B||_\infty = 11; ||B||_2 = 8.275$. This last follows from the fact that the eigenvalues of B are $\{8.275, 0.725, -2\}$.

40.   25/12, which is the sum of the elements of the first row

42.   a. $x^T = (1592.6, -631.911, -493.62)$

44. $x^T = (119.5, -47.14, -36.84)$. This is further evidence of ill-condition, in that small changes in the coefficients make large changes in the solution vector.

48. $\dfrac{1}{\text{C.N.}} \dfrac{||r||_2}{||b||_2} = 1.82 \times 10^{-5}$,   C.N. $\dfrac{||r||_2}{||b||_2} = 55{,}642$

$\dfrac{||e||_2}{||\bar{x}||_2} = 25.5$, $\dfrac{||e||_2}{||x||_2} = 1.04$

The relation is verified with either $||\bar{x}||$ or $||x||$.

51.   Using 3-digit arithmetic to compute e:

$$e = \{-0.00272, 0.00126, 0.00103\}, \qquad x = \{0.153, 0.144, 0.166\}.$$

$(x + e) = \{0.15029, 0.14526, -0.16497\}$; compare with $x = \{0.15094, 0.14525, -0.16592\}$. The improvement is remarkable even in one iteration.

54.   After six iterations, the Jacobi vector has diverged to $x^T = (3.16, 3.16)$, while with Gauss–Seidel, $x^T = (-154.5, 467.6)$. Gauss–Seidel is diverging much more rapidly.

55.   a. Using the Jacobi method, we converge to the solution vector $(-2, 1, -3)$ in 10 iterations.
b. With the Gauss–Seidel method there is convergence in 5 iterations.

56. $x^T = (1, -1, 2)$

59. $(0.72595, 0.50295)$ in 5 iterations from $(1, 1)$;
$(-1.6701, 0.34513)$ in 6 iterations from $(-2, 1)$

61. $(1.64304, -2.34978), (-2.07929, -3.16173)$

64.   b. Multiply $P_{1,4} * A$ where $P_{1,4}$ is the identity matrix with the first and fourth rows interchanged.
c. Multiply $A * P_{1,2}$ where $P_{1,2}$ is similarly defined.
d. $P_{2,4} * A * P_{2,4}$

CHAPTER 3   **2.** $P_3(x) = -0.99167x^3 + 2.85x^2 + 0.49167x - 2.35$

6.   From table: $P_2(3) = 7.8546$, $P_3(3) = 7.8514$, $P_4(3) = 7.8503$

15.   Best points: For $f(0.15)$: $x = 0.1, 0.0, 0.5$ or -0.2
For $f(-0.1)$: $x = -0.2, 0.0, 0.1$
For $f(1.2)$: $x = 0.7, 0.5, 0.1$

18.   Interpolate $= 1.22183$, error $= -0.00043$. Bounds on error are -0.00033 and -0.00045, which bracket the error. (If more accurate values for e^x had been used, the error would be -0.00039, so round-off has entered.)

22. Third difference =

$$0.3365 - 3(0.3001) + 3(0.2624) - 0.2231 = 0.0003$$

26. $P_2(0.203) = 0.78156$, next term increment $= 0.00007$
$P_2(0.203) = 0.78163$, next term increment $= -0.00001$
$P_2(0.203) = 0.78162$, next term increment $= -0.00000$

30. Fourth degree because all fourth differences are constant.

34. The system is

$$\begin{bmatrix} 1.220 & 0.490 & 0 & 0 & -1.326 \\ 0.490 & 1.240 & 0.130 & 0 & -1.715 \\ 0 & 0.130 & 0.620 & 0.180 & -0.829 \\ 0 & 0 & 0.180 & 2.440 & -1.865 \end{bmatrix}.$$

The S-values are $0, -0.678, -1.018, -0.972, -0.696, 0$.

39. Plot of cubic spline, interpolating polynomial, and original curve. (Evenly spaced points are not the best choice.)

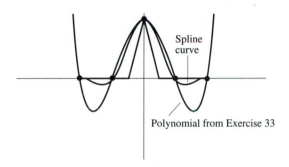

43. $[u^4, u^3, u^2, u, 1]\begin{bmatrix} 1 & -4 & 6 & -4 & 1 \\ -4 & 12 & -12 & 4 & 0 \\ 6 & -12 & 6 & 0 & 0 \\ -4 & 4 & 0 & 0 & 0 \\ 1 & 0 & 0 & 0 & 0 \end{bmatrix}\begin{bmatrix} p_0 \\ p_1 \\ p_2 \\ p_3 \\ p_4 \end{bmatrix}$

45. $dx/du = -(3/6)(1-u)^2 x_{i-1} + (1/6)(6u^2 - 6u)x_i + (1/6)(-9u^2 + 6u + 3)x_{i+1} + (3/6)(u^2)x_{i+2}$
At $u = 0$, $dx/du = -(3/6)x_{i-1} + 0 + (3/6)x_{i+1} + 0 = (3/6)(x_{i+1} - x_{i-1})$.
Similarly for dy/du, so

$$dy/dx = \frac{y_{i+1} - y_{i-1}}{x_{i+1} - x_{i-1}},$$

which is the slope between points adjacent to p_i. A similar relation holds at the right end.

52. $f(1.6, 0.33) = 1.8328$

54. Interpolating by rows: $f(1.1, 0.71) = 0.70726$, $f(3.0, 0.71) = 5.2614$, $f(3.7, 0.71) = 8.0028$, $f(5.2, 0.71) = 15.8077$. From these, $f(3.32, 0.71) = 6.4435$.

59. $y = 2.908x + 2.02533$

61. $z = 2.8530x - 1.9145y + 1.0399$

67. $\ln(S) = \ln(1.2019) + 0.009603T$, or $S = 1.2019 * e^{0.009603T}$.

73.   Fitting polynomials of increasing degree, we find the variances are degree 2, 533.3; degree 3, 85.47; degree 4, 86.73; degree 5, 64.84. This indicates that the optimum degree is 3 and, for this,
$$f(x) = 0.1223x^3 - 3.614x^2 + 24.11x - 6.392.$$

78.   Values for $P(0.1)$ from polynomials:

Degree	$P(0.1)$	Error
2	0.99	−0.39
3	0	0.60
4	0.9504	−0.3504
5	0.2256	0.3744

Cannot compare to error bounds because $f'(x)$ is discontinuous and hence requirements for Eq. (3.34) are not met.

81.   From $E^{-1} = 1 - \Delta$, $E^s = (1 - \Delta)^{-s} = 1/(1 - \Delta)^s$ and
$$f_s = E^s f_0 = [1/(1 - \Delta)^{-s}]f_0$$
$$= f_0 + s\,\Delta f_0 + \binom{s+1}{2}\Delta f_0 + \binom{s+2}{3}\Delta f_0 + \cdots$$

85.   1.8407

87. $\ln(s) = 0.183932 + 0.0096028T$

CHAPTER 4   **3.**   Derivatives values:

x	Computed	Exact
0.21	2.0904	2.0681
0.22	1.9754	1.9741
0.23	1.8973	1.8882
0.24	1.8191	1.8096
0.25	1.7409	1.7372
0.26	1.6627	1.6704
0.27	1.6230	1.6085

8.   a. $f'(0.72)$ from $P_3(x) = 1.2506$ (exact $= 1.25060$)
b. $f'(1.33)$ from $P_2(x) = 1.0790$ (exact $= 1.07949$)
c. $f'(0.50)$ from $P_4(x) = 1.2925$ (exact $= 1.29253$)

11. $f'(0.5) = 1.2905$, exact $= 1.29253$, actual error is 0.0020. Compare to bounds: 0.0021, 0.0017.

18.

No. terms	Value	Actual error	Estimated error
1	−3.988	0.446	0.508
2	−3.480	−0.062	−0.073
3	−3.552	0.010	0.016

(Exact value $= -3.54169$)

22. From Taylor series:

$$f'(x_0) = \frac{f_1 - f_0}{h} - \frac{h}{2} f''(\xi)$$

$$f'(x_0) = \frac{f_0 - f_{-1}}{h} + \frac{h}{2} f''(\xi)$$

27. $f'(0.32) = 0.15728$ after 2 extrapolations

29. The order of error is $O(h)$, so Eq. (4.21) applies with $n = 1$:

$$\text{Extrap value} = \text{More accurate} + \frac{1}{2-1} * (\text{More accurate} - \text{less})$$

On successive extrapolations, power of h in error term increases by 1, not 2.

34. (We used undetermined coefficients.)

$$\text{Integral from } x_0 \text{ to } x_0 + 4h = \left(\frac{h}{45}\right)(14f_0 + 64f_1 + 24f_2 + 64f_3 + 14f_4) + O(h^7)$$

$$\text{Integral from } x_0 \text{ to } x_0 + 5h = \left(\frac{h}{288}\right)(95f_0 + 375f_1 + 250f_2 + 250f_3 + 375f_4 + 95f_5) + O(h^7)$$

37. Analytical value is 1.7669731. Trapezoidal rule results:

h	No. panels	Value	Error	Error/h^2
0.1	8	1.7684	-0.0014	-0.140
0.2	4	1.7728	-0.0058	-0.145
0.4	2	1.7904	-0.0234	-0.146

41. Estimate of integral is 0.69315; $\ln(2) = 0.693147$

46. With 8 panels ($h = 0.125$), integral is 1.718284, analytical is 1.7182820.

51. Let $P_3(x) = a + bx + cx^2 + dx^3$. By change of variable, integration can be from $-h$ to h with midpoint at $x = 0$. The quadratic that fits at three evenly spaced points is $a + (b + dh^2)x + cx^2$. The integral of each is the same: $2ah + 2ch^3/3$.

55. We want the value of the integral to be $af_0 + bf_1$ where a and b are to be determined. By change of variable, the limits can be from $x = 0$ to $x = h$. Letting $f(x) = 1$ and then $f(x) = x$, we need to solve:

$$\begin{cases} h = a + b, \\ \dfrac{h^2}{2} = bh, \end{cases}$$

which gives $a = h/2$, $b = h/2$.

58. Three-term formula gives 1.71281, which is accurate to 6 significant figures.

60. Correct value is -0.700943. Even five terms in the Gaussian formula is not enough. Simpson's $\frac{1}{3}$ rule attains 5 digits accurate with 400 intervals. Extrapolating from Simpson's $\frac{1}{3}$ rule to 7 levels also attains this (requiring 128 intervals).

65. Analytical value $= 0.2499161$. Adaptive integration with $\text{TOL} = 0.02$ gives 0.2496683, which differs from this by 0.099%.

69. a. With x-values vertical:

$$\frac{\Delta x}{3} \frac{\Delta y}{2} \begin{Bmatrix} 1 & 2 & 2 & 1 \\ 4 & 8 & 8 & 4 \\ 2 & 4 & 4 & 2 \\ 4 & 8 & 8 & 4 \\ 1 & 2 & 2 & 1 \end{Bmatrix}$$

d. For part (a), even number in x-direction.
For part (b), even number in both directions.
For part (c), divisible by 3 in both directions.

74. Analytical value is 2/3.

x	y	Integral	Error	Error/h^2
0.5	0.5	0.75	−0.0833	−0.3333
0.25	0.5	0.7185	−0.05208	−0.3333*
0.5	0.25	0.7185	−0.05208	−0.3333*
0.25	0.25	0.6875	−0.0208	−0.3333
0.125	0.125	0.6719	−0.0052	−0.3333

* Used the average of the squares of the h-values.

78. a. For three-term formulas, let $x = t/2 + 1/2$, so x-limits of $[0, 1]$ become t-limits of $[-1, 1]$ and $dx = (1/2)dt$. Gauss-points for x are then 0.1127, 0.5, 0.8873. Hold x fixed at these values and integrate with respect to y.
    With x fixed as before, we substitute for y in terms of s: $y = (\sqrt{1-x^2})/2 * s + (\sqrt{1-x^2})/2$ so y-limits of from zero to curve become $[-1, 1]$ for s. This substitution defines three Gauss-points for y at each fixed value of x. We then evaluate $f(x, y)$ at nine Gauss-points in the region, multiply by the weights, add, and then multiply by 1/2 (the factor that changes dx to dt). The result is $0.915944/2 = 0.45797$; compare to analytical value of 0.45494.

83. Computed values follow. The errors of $f''(x)$ are shown.

x	$f'(x)$	$f''(x)$	Error in $f''(x)$
1.0	−0.1111	0.0843	−0.0102
1.5	−0.0808	0.0368	0.0098
2.0	−0.0635	0.0326	−0.0014
2.5	−0.0491	0.0250	−0.0030
3.0	−0.0400	0.0115	0.0045

84. Analytical value is 1.30176. From the cubic spline, the value is 1.299188, error = 0.00257. By Simpson's $\frac{1}{3}$ rule (4 panels), value is 1.30160, error = 0.00016.

92. The function used was $f(x) = e^x$. For the local error, two panels were used, starting at $x = 1$:

Limits	h	Integral	Error	Error/h^5
[1, 2]	0.5	4.672349	−0.001575	−0.0504
[1, 1.5]	0.25	1.763445	−0.000037	−0.0379
[1, 1.25]	0.125	0.7720621	−0.000001	−0.0317

which shows that errors are about proportional to h^5.

With $f(x) = e^x$ again, integrals between 1 and 2 with varying h:

h	Integral	Error	Error/h^4
0.5	4.672349	-0.001575	-0.0252
0.25	4.670875	-0.0001007	-0.0258
0.125	4.670781	-0.0000067	-0.0276
0.0625	4.670775	-0.0000007	-0.0478

Round-off has affected the last result.

100. Results from *Mathematica*:

$$(\text{OPERATOR})^3 =$$

$$\left(\frac{1}{h^3}\right)\left[\Delta^3 - \left(\frac{3}{2}\right)\Delta^4 + \left(\frac{7}{4}\right)\Delta^5 - \left(\frac{15}{8}\right)\Delta^6 + \left(\frac{29}{15}\right)\Delta^7 - \cdots\right]$$

$$(\text{OPERATOR})^4 =$$

$$\left(\frac{1}{h^4}\right)\left[\Delta^4 - 2\Delta^5 + \left(\frac{17}{6}\right)\Delta^6 - \left(\frac{7}{2}\right)\Delta^7 + \left(\frac{967}{240}\right)\Delta^8 - \cdots\right]$$

CHAPTER 5

1. a. $y(x) = 1 + x + (3/2)x^2 + (5/6)x^3 + (7/12)x^4 + (17/60)x^5$;
$y(0.1) = 1.11589$, $y(0.5) = 2.0245$

3. $y(0.2) = 1.2015$, $y(0.4) = 1.4128$, $y(0.6) = 1.6471$

5. With $h = 0.01$, $y(0.1) = 1.11418$, error $= 0.00171$. To reduce the error to 0.00005 (34-fold), we must reduce h 34-fold, or to about 0.00029.

7. $y(0.1) = 1.11587$ (four decimals are correct). The simple Euler method will require about 340 steps and 340 function evaluations for similar accuracy, compared to four steps and eight function evaluations with the modified Euler method.

9.
x:	0.1	0.2	0.3	0.4	0.5
y:	2.2150	2.4630	2.7473	3.0715	3.4394

13.
t:	0.2	0.4	0.6
y:	2.0933	2.1755	2.2493

18. The exact answer is $y(x) = -5e^{-x} + 2x^2 - 4x + 4$.

x	y (Runge–Kutta–Fehlberg)	Analytical
0.00	-1.000000	-1.000000
0.10	-0.904187	-0.904187
0.20	-0.813654	-0.813654
0.30	-0.724091	-0.724091
$\vdots$	$\vdots$	$\vdots$
1.80	2.453506	2.453506
1.90	2.872158	2.872157
2.00	3.323325	3.323324

20.   a. $y(0.5) = -0.28326$, error $= -0.00077$
b. $y(0.5) = -0.28387$, error $= -0.00016$
c. $y(0.5) = -0.28396$, error $= -0.00007$

24. $y(4)$ predicted $= 4.2229$, $y(4)$ corrected $= 4.1149$. From these, the error in y_c is estimated as 0.0037; equivalently y_c has three correct digits. (The actual error is 0.0082.) Obviously, the starting values must have at least three digits of accuracy also.

26.

x:	0.8	1.0	1.2	1.6	2.0
y:	2.3163	3.3780	2.4350	2.5380	2.6294

Error estimate: 0.0003   <0.00005   0   −0.00005   −0.00002

After $x = 1.2$, h was increased to 0.4.

29.   By Runge–Kutta:

x:	0	0.2	0.4	0.6
y:	0	0.0004	0.0064	0.0325

By Adams–Moulton:

x:	0.8	0.9	1.0	1.1	1.2	1.25	1.3	1.35	1.4
y:	0.1035	0.1669	0.2574	0.3836	0.5581	0.6688	0.7994	0.9541	1.1387

The step size was halved after $x = 0.8$ and $x = 1.2$ to provide four-decimal accuracy.

34.   For part (a), corrected values:

x:	1.0	1.1	1.2	1.3
y_c:	2.000000	1.818746	1.670577	1.549859
hy':	−0.200000	−0.163690	−0.133646	−0.108490
$h^2y''/2$:	0.020000	0.016232	0.013456	0.011356
$h^3y'''/6$:	−0.001333	−0.001282	−0.001044	−0.000815
Analytical:	2.000000	1.818746	1.670540	1.549833

37.   a. Corrected values:

x:	1.0	1.1	1.2	1.3
y_c:	2.000000	1.818667	1.669947	1.548795
hy':	−0.200000	−0.163680	−0.133596	−0.108416
$h^2y''/2$:	0.020000	0.016240	0.013483	0.011364
$h^3y'''/6$:	−0.001333	−0.001280	−0.001039	−0.000817

b. Corrected values:

x:	1.0	1.1	1.2	1.3
y_c:	2.000000	1.818833	1.670876	1.550462
hy':	−0.200000	−0.163695	−0.133670	−0.108532
$h^2y''/2$:	0.020000	0.016166	0.013127	0.010937
$h^3y'''/6$:	−0.001333	−0.001306	−0.0011259	−0.000945

39.   a. $f_y = \sin(x)$, so $h_{max} = (29/9)/1 = 2.67$.
b. With $h = 0.267$, D cannot exceed 10×10^{-N} for N-decimal-place accuracy.
c. If $D = 14.2 \times 10^{-N}$, h cannot exceed $(1/14.2)h_{max}$.

45.

x	y	f	$1 + hf_y$	$h^2y''/2$	Est'd error	Actual error
1.0	1.000	1.000	1.200	0.015	0	0
1.1	1.100	1.331	1.242	0.022	0.019	0.017
1.2	1.233	1.825	1.296	0.035	0.052	0.049
1.3	1.416	2.605	1.368	0.058	0.115	0.111
1.4	1.676	3.933	1.437	0.087	0.234	0.247
1.5	2.069	6.424	1.560	0.173	0.481	0.597
1.6	2.712	11.766	1.782	0.403	1.091	1.834

48. $y' = z$, $y(0) = 0$;

$$z' = \frac{M(x)}{EI}(1 - z^2)^{3/2}, \ z(0) = 0$$

50.

t:	0	0.2	0.4	0.6
y:	1	0.982	0.933	0.864
x:	0	0.022	0.093	0.200

53.

x	y	y'
0	1	−1
0.1	0.8950	−1.0995
0.2	0.7802	−1.1956
0.3	0.6561	−1.2847
0.4	0.5236	−1.3629
0.5	0.3840	−1.4263
0.6	0.2389	−1.4715

56.

t	x	Analytical
0	0	0
0.1	0.0715	0.0717
0.2	0.1983	0.1999
0.3	0.2004	0.2026
0.4	−0.0237	−0.0233
0.5	−0.3747	−0.3784
0.6	−0.5912	−0.5977
0.7	−0.4377	−0.4419
0.8	0.0898	0.0924

61. a.
```
de := diff(y(x),x) = x*x + x*y(x);
G := dsolve( {de, y(1) = 2}, y(x) );
S := series(rhs(G), x = 1, 5);
P := convert('', polynom);
subs(x = 3, P);
evalf('');
```

(This produces the answer: 66.00000000. Since this answer is not close to the exact solution, 196.29, we call for a longer series by replacing the third line with:

```
S := series( rhs(G), x = 1, 10);
```

This produces the answer: 168.9925926. Thus the Taylor series is not suitable for such a large step.)

b. ```
de := diff(x(t),t) = sin(t);
G := dsolve({de, x(0) = 1}, x(t), series);
P := convert(rhs(G), polynom);
subs(t = 2, P);
evalf('');
```

(This produces the answer: 2.33333333. We can get closer to the exact answer of 2.416146837 by going out with more terms on the series as in part (a). In this case, the longer series produces a better result.)

c. ```
de := diff(y(x),x) + x + y(x) = 2;
G := dsolve( {de, y(x0) = y0}, y(x) );
S := series( rhs(G), x = x0, 5);
P := convert('', polynom);
subs(x = x0 + 2, P );
evalf('');
```

(This produces the answer: .33333333 $y0 - .66666666\ x0$)

65. ```
with (plots);
de := diff(y(x),x,x) + 2*diff(y(x),x) = x*x + y(x);
G := dsolve({de, y(0) = 1, D(y)(0) = -1}, y(x));
plot(rhs(G), x = 0..2);
```

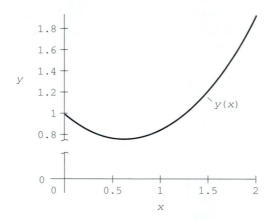

**66.** a. Satisfies the Lipschitz condition, since we have

$$|f(x, y_1) - f(x, y_2)| \le |y_1 + y_2|\,|y_1 - y_2| < 2|y_1 - y_2|.$$

b. Does not satisfy the Lipschitz condition, since

$$|f(x, y_1) - f(x, y_2)| = x^2/|y_1 y_2|\,|y_1 - y_2| \text{ is unbounded at } y = 0.$$

**68.** Examine $f(x, y) = |x|$ on the unit square. In general, consider the integral of a bounded function with a finite number of discontinuities.

**69.** Parts (b) and (c) are stable, whereas parts (a) and (d) are unstable.
```

CHAPTER 6

1. Required value of $y'(1)$ to give $y(2) = 15.0000$: with modified Euler, 5.48408; with 4th order Runge–Kutta, 5.49872; with Runge–Kutta–Fehlberg, 5.50012. Analytical solution has $y'(1) = 5.50000$.

3. Modified Euler can match analytical to 6 significant figures if $h = 0.01$. (Runge–Kutta–Fehlberg matches if $h = 0.125$.)

9.

x:	0.0	0.2	0.4	0.6	0.8	1.0
y:	1	0.5211	0.0624	−0.3556	−0.7149	−1
(By R–K:	1	0.5209	0.0621	−0.3560	−0.7152	−1)

15.

x:	0	$\pi/8$	$\pi/4$	$3\pi/8$	$\pi/2$
y:	1.509	1.586	1.417	1.031	0.485
(By R–K:	1.509	1.580	1.410	1.025	0.485)

20.

x:	0.0	0.2	0.4	0.6	0.8	1.0
y:	2	1.3222	0.1968	−0.9547	−1.7812	−2

24. Letting $u(x) = cx(x - 1)$, Rayleigh–Ritz integral gives

$$\frac{2c}{3} + 0 = -2\left(\frac{5}{12}\right), \ c = \frac{5}{4},$$

so $u(x) = (5/4)x(x - 1); y(x)$ anal. $= x^3/2 + x^2/2 - x$.

x:	0	0.2	0.4	0.6	0.8	1
$u(x)$:	0	−0.200	−0.300	−0.300	−0.200	0
$y(x)$:	0	−0.176	−0.288	−0.312	−0.242	0

28. $R(x) = y'' - 3x - 1$. Let $u(x) = cx(x - 1)$, so $u'' = 2c$. Setting $R(x) = 0$ at $x = 0.5$ gives $2c - 3(0.5) - 1 = 0$, $c = 5/4$. Answer identical to Exercise 24.

31. Galerkin integral is

$$\int_0^1 [x(x - 1)][2c - 3x - 1] \ dx = 0,$$

which gives $(5 - 4c)/12 = 0$, $c = 5/4$. Answer identical to Exercises 24 and 28.

36. In the following $u(x)$ is the approximation, $y(x)$ is exact.

x:	1.0	1.2	1.5	1.75	2.0
$u(x)$:	−1	−0.2263	0.9176	1.9205	3
$y(x)$:	−1	−0.2267	0.9167	1.9196	3

39. Exact = 2.46166
a. 2.000 ($h = 1/2$)
b. 2.25895 ($h = 1/3$)
c. 2.34774 ($h = 1/4$)
d. 2.4636 (extrapolated)

43. Cannot get a second eigenvalue with $h = 1/2$. From $h = 1/3$, 3.59125; from $h = 1/4$, 4.0000; from $h = 1/5$, 4.19885.

48. Eigenvalues and eigenvectors:

$$-4.6242; \ [0.95724, \ 0.25143, \ -0.14308]$$
$$7.2024; \ [0.37006, \ 0.92577, \ 0.077473]$$
$$-9.5782; \ [0.41719, \ -0.90275, \ -0.10479]$$

Eigenvalues of A^{-1}: $-0.10440, 0.13884, -0.21625$, which are the reciprocals of the preceding. The eigenvectors are the same.

50. Eigenvalues of A: $-1.05124, 1.62941, 6.42182$. $M^{-1}AM$ has the same eigenvalues, the matrix is

$$\frac{1}{4}\begin{bmatrix} 14 & 10 & 6 \\ -7 & -17 & -7 \\ 27 & 65 & 31 \end{bmatrix}.$$

53. After 62 rotations, matrix is

$$\begin{bmatrix} 5.3028 & 0.6412 & -0.8928 \\ 0.0000 & 2.0000 & 0.8899 \\ 0.0000 & 0.0000 & 1.6974 \end{bmatrix}.$$

58. Upper Hessenberg matrix:

$$\begin{bmatrix} 4.0000 & -1.4142 & -1.4142 \\ -1.4142 & 3.5000 & 0.5000 \\ 0.0000 & 0.5000 & 1.5000 \end{bmatrix}.$$

It then takes only 5 iterations to get the upper-triangular result (which matches Exercise 53).

59. a. 19.331
 b. 6.001

63. a. $Ax = b$ is: $(\det(A) = -6.5820)$

$$\begin{bmatrix} -2.125 & 1.375 & 0.000 & -0.4375 \\ 0.750 & -2.250 & 1.250 & -0.250 \\ 0.000 & 0.875 & -2.375 & -1.3125 \end{bmatrix}.$$

b. $Ax = b$ is: $(\det(A) = -3.0352)$

$$\begin{bmatrix} -1.625 & 0.875 & 0.000 & 1.4375 \\ 0.750 & -1.750 & 0.750 & 0.250 \\ 0.000 & 0.875 & -1.875 & -0.3125 \end{bmatrix}.$$

c. $Ax = b$ is: $(\det(A) = -4.125)$

$$\begin{bmatrix} -2.250 & 0.750 & 0.000 & 1.125 \\ 1.250 & -2.000 & 0.750 & 0.0000 \\ 0.000 & 1.259 & -1.750 & 0.625 \end{bmatrix}.$$

66. Using $h = 0.2$:

$y'(-1)$	$y(1) = 3.000$?
-1.5	1.9574
-3.0	0.8245
-0.1183	2.6114
1.6147	3.2364
0.9523	3.0145
0.9108	3.0000

70. All three parts give identical results: $[0.552013, -0.424370, -0.964409]$.

74. $-w^3 + 15.3w^2 - 69.9w + 100.2$ (roots are $8.39608, 3.45195 \pm 0.134597i$)

CHAPTER 7 **2.** Net flow into the element:

$$dx\,dy\,[kt(u_{xx} + u_{yy}) + t(k_x u_x + k_y u_y) + k(t_x u_x + t_y u_y)].$$

5.
$$\left\{ \begin{matrix} & & -1 & & \\ & & 16 & & \\ -1 & 16 & -60 & 16 & -1 \\ & & 16 & & \\ & & -1 & & \end{matrix} \right\} \frac{u_{ij}}{12h^2} = 0$$

8. The gradient is $100/L$ where L is the width of the plate. Let $h = L/n$, the nodal spacing with points at $x_i = i * h$, $(i = 0, \ldots, n)$, measured from left end. Then $u_i = 100 + i * h(L/100)$; $u_{i-1} + u_{i+1} = 2u_i$, $u_{i-2} + u_{i+2} = 2u_i$.

a.
$$\left\{ \begin{matrix} & u_i & \\ u_{i-1} & -4u_i & u_{i+1} \\ & u_i & \end{matrix} \right\} /h^2 = (2u_i - 2u_i)/h^2 = 0$$

$$\left\{ \begin{matrix} u_{i-1} & 4u_i & u_{i+1} \\ 4u_{i-1} & -20u_i & 4u_{i+1} \\ u_{i-1} & 4u_i & u_{i+1} \end{matrix} \right\} /(6h^2) = (12u_i - 12u_i)/(6h^2) = 0$$

b.
$$\left\{ \begin{matrix} & & -u_i & & \\ & & 16u_i & & \\ -u_{i-2} & 16u_{i-1} & -60u_i & 16u_{i+1} & -u_{i+2} \\ & & 16u_i & & \\ & & -u_i & & \end{matrix} \right\} /(12h^2) = (-2u_i + 32u_i - 30u_i)/(12h^2) = 0$$

For points adjacent to boundary, assume outside fictitious points with same gradient as inside.

12. Interior temperatures:

64.21	105.20	146.65	186.41
61.63	89.94	114.99	134.00
52.39	77.93	89.38	85.59

15. $3u_{xx} + 2u_{yy} = 3(u_L - 2u_0 + u_R)/h^2 + 2(u_A - 2u_0 + u_B)/h^2 = \left\{ \begin{matrix} & 2 & \\ 3 & -10 & 3 \\ & 2 & \end{matrix} \right\} \dfrac{u_{ij}}{h^2} = 0$

17. Temperatures are symmetrical. With tol = 0.00001, interior temperatures above centerline are:

0.212	0.507	0.996	1.867	3.460	6.413	11.981	22.856	45.668
0.344	0.821	1.610	3.013	3.561	10.210	18.655	33.775	59.814

20.

$w =$	1.2	1.3	1.35	1.36	1.37	1.38	1.4	1.5
iterations	36	27	22	21	20	21	21	27

(Predicted value for $w = 1.356$, from above, $w_{opt} = 1.37$.)

25. Points laid out as Fig. 7.25:

	0.239		
0.314	0.455	0.372	0.218
0.302	0.393	0.316	

28. $\left\{\begin{matrix} & 1 & \\ 3 & -8 & 3 \\ & 1 & \end{matrix}\right\} \dfrac{u_{ij}}{h^2} + 2 = 0$

Augmented matrix, using symmetry):

$$\begin{bmatrix} -5 & 1 & -8 \\ 2 & -5 & -8 \end{bmatrix}$$

32. There is two-way symmetry. Values in upper-right quadrant:

64.3	63.6	61.0	55.2	43.7
78.9	77.8	73.9	64.7	43.4

33. There is no unique solution: if $u(x,y)$ is a solution, then $u(x,y)+$ any constant also is a solution.

37. Values in first octant (others are symmetrical):

	15.23	
53.34	30.46	10.43
76.23	42.84	18.04
80.22	46.65	20.64

42. Assume that top and bottom are at 100, left and front sides are insulated, right and back sides are at 0. There is symmetry above and below centerline. Results after 105 iterations with tol = 0.001. Top two layers of interior nodes:

53.977	52.963	48.890	37.057
75.035	73.326	66.704	48.890
83.274	81.200	73.326	52.963
85.458	83.274	75.035	53.977
42.905	41.584	36.619	24.559
66.306	64.056	55.793	36.619
76.752	74.003	64.056	41.584
79.654	76.752	66.306	42.905

46. Optimum factor is 1.72 but converges to tol < 0.001 in nine iterations for any factor between 1.6 and 2.0. There is two-way symmetry. Values in upper-right quadrant:

0.342	0.312	0.211
0.521	0.472	0.312
0.577	0.521	0.342

50. c. $M^{-1} = \begin{bmatrix} -4.650 & 3.982 & 1.668 \\ 0.000 & -0.217 & 0.217 \\ 0.500 & -0.120 & -0.380 \end{bmatrix}$

$a = [405.16, -32.174, 9.3044]$

$N = \begin{bmatrix} -4.650 & & + 0.500y \\ 3.982 - 0.217x & - 0.120y \\ 1.688 + 0.217x & - 0.380y \end{bmatrix}$

$u(10.6, 9.6) = 153.444$

52. The augmented matrix:

$$\begin{bmatrix} -974.54 & -488.12 & -488.72 & 1.7385 \\ -488.12 & -975.41 & -487.83 & 1.7385 \\ -488.72 & -487.83 & -974.81 & 1.7385 \end{bmatrix}$$

57. a. $w_{opt} = 1.01612$
Eigenvalues, $w = 1$: 0, 0.0625
Eigenvalues at w_{opt}: 0.01524, 0.01524 (over 4-fold reduction)
Equation (7.18) with $p = 2$, $q = 3$ gives 1.01613
b. $w_{opt} = 1.20377$
Eigenvalues, $w = 1$: 0, 0.5625
Eigenvalues at w_{opt}: both are 0.20378
Equation (7.18) doesn't apply. Experimentally w_{opt} is about 1.21.
c. The method is difficult to apply. Equation (7.18) with $p = 2$, $q = 4$ gives $w_{opt} = 1.03337$.
Eigenvalues, $w = 1$: 0, 0.125, 0.125.
Eigenvalues with $w = 1.03337$ all have magnitude 0.0486, a reduction by a factor of 3.75.
Equation (7.18) is confirmed.

CHAPTER 8

3. (Using $k = 2.156$ Btu/(hr $*$ in. $*$ °F))
a. -29.53 °F/in.
b. -75.59 °F/in.
c. -34.91 °F/in.

4.

x:	0	0.25	0.50	0.75	1.00	1.25
Computed:	0	17.34	32.04	41.86	45.31	41.86
Analytical:	0	17.70	32.71	42.74	46.26	42.74

(Temperatures are symmetrical about $x = 1$.)

8. Results at $t = 268.8$ sec:

$x =$	4	8	12	16	Steps	Points	Relative effort
(a)	1.132	2.473	4.317	6.880	16	4	64
(b)	1.115	2.481	4.372	6.939	16	9	144
(c)	1.110	2.443	4.297	6.874	64	19	1216
Anal.	1.108	2.430	4.272	6.850			

As expected, (c) is most accurate. For most x-values, (a) is better than (b) and with much less total effort. Reducing r is the preferred way to improve accuracy.

12. The equation as written actually represents heat loss from the bar when $f(x)$ is positive. Using $\Delta t = 0.0468$:

t, sec	$x =$	0	0.2	0.4	0.6	0.8	0
0.0936	$u =$	0	-0.023	-0.037	-0.037	-0.023	0
0.234	$u =$	0	-0.037	-0.059	-0.059	-0.037	0
0.468	$u =$	0	-0.042	-0.068	-0.068	-0.042	0
Steady state:		0	-0.043	-0.069	-0.069	-0.043	0

16. With $r = 1$:

t, sec	$x =$	0	0.2	0.4	0.6	0.8	0
0.0936	$u =$	0	-0.021	-0.033	-0.033	-0.021	0
0.234	$u =$	0	-0.035	-0.055	-0.055	-0.035	0
0.468	$u =$	0	-0.042	-0.066	-0.066	-0.042	0
Steady state:		0	-0.043	-0.069	-0.043	-0.069	0

At $t = 0.0936$ sec, results match Exercise 12 if $r = 0.0625$.

17. All errors are the same fraction of the initial u's. As r is changed, they vary linearly.

23. (Take $h = 0.15$ on the left face, $h = 0.25$ on the right face.) Time step $= 1.893$ sec. After 8 steps:

x:	0	1	2	3	4
u:	614.8	764.6	797.1	701.8	490.3

24. (Take $h = 0.15$ on the left face, $h = 0.25$ on the right face.) Time step $= 1.893$ sec. After 8 steps:

x:	0	1	2	3	4
u:	617.9	768.4	802.1	709.1	496.3

26. Six points are required. Let the right-hand face be a 2 in face held at 100° and the bottom face be a 3 in face held at zero. Number the points from left to right beginning at the top row, giving these u's and v's:

$$
\begin{array}{ccc}
u_1(v_1) & u_2(v_3) & u_3(v_5) \\
u_4(v_2) & u_5(v_4) & u_6(v_6)
\end{array}
$$

The coefficient matrix for the u's is

$$
\begin{bmatrix}
(1/r + 2) & -2 & 0 & 0 & 0 & 0 \\
-1 & (1/r + 2) & -1 & 0 & 0 & 0 \\
0 & -1 & (1/r + 2) & 0 & 0 & 0 \\
0 & 0 & 0 & (1/r + 2) & -2 & 0 \\
0 & 0 & 0 & -1 & (1/r + 2) & -1 \\
0 & 0 & 0 & 0 & -1 & (1/r + 2)
\end{bmatrix}
$$

with these right-hand sides:

$$
\begin{bmatrix}
(1/r - 2)v_1 + 2v_2 \\
(1/r - 2)v_3 + 2v_4 \\
100 + (1/r - 2)v_5 + 2v_6 \\
(1/r - 2)v_2 + v_1 \\
(1/r - 2)v_4 + v_3 \\
100 + (1/r - 2)v_6 + v_5
\end{bmatrix} .
$$

The coefficient matrix for the v's is

$$
\begin{bmatrix}
(1/r + 2) & -2 & 0 & 0 & 0 & 0 \\
-1 & (1/r + 2) & 0 & 0 & 0 & 0 \\
0 & 0 & (1/r + 2) & -2 & 0 & 0 \\
0 & 0 & -1 & (1/r + 2) & 0 & 0 \\
0 & 0 & 0 & 0 & (1/r + 2) & -2 \\
0 & 0 & 0 & 0 & -1 & (1/r + 2)
\end{bmatrix}
$$

with these right-hand sides:

$$\begin{bmatrix} (1/r - 2)u_1 + 2u_2 \\ (1/r - 2)u_4 + 2u_5 \\ u_1 + (1/r - 2)u_2 + u_3 \\ u_4 + (1/r - 2)u_5 + u_6 \\ u_2 + (1/r - 2)u_3 + 100 \\ u_5 + (1/r - 2)u_6 + 100 \end{bmatrix}.$$

With $r = 2$, after four steps, the temperatures are

$$\begin{matrix} 24.70 & 31.49 & 54.90 \\ 18.39 & 24.08 & 45.07 \end{matrix}$$

30. There are 125 equations to solve at each time step, but they can be decomposed into 25 subsets of five tridiagonal equations each. If $r = 1$, only two time steps are needed to reach $t = 15.12$.

31. $c_{ij} = \begin{cases} 0.2825, & i = j \\ 0.1412, & i \ne j \end{cases}$

$$[K] = \frac{k}{c\rho} \begin{bmatrix} 0.484 & 0.089 & -0.573 \\ 0.089 & 0.196 & -0.285 \\ -0.573 & -0.285 & 0.857 \end{bmatrix}$$

$b_i = F_{av} * 0.565$

33. The equations are the same.

40. A single error of magnitude 1 becomes less than 0.0005 after 56 time steps. The line of the table for t_7 is

$$\begin{matrix} 0 & 0.199 & -0.137 & -0.002 & 0. \end{matrix}$$

42. For $N = 4$: $r = 0.5$, 0.8090; $r = 0.6$, -1.1708
For $N = 5$: $r = 0.5$, 0.8660; $r = 0.6$, -1.2392

CHAPTER 9

3. The computed values agree exactly with the analytical solution. There is symmetry about $x = 0.5$. Some values:

Time steps	0	1/8	2/8	3/8	4/8	5/8
1	0	0.354	0.653	0.854	0.924	0.854
2	0	0.271	0.500	0.633	0.707	0.633
8	0	−0.383	−0.707	−0.924	−1.000	−0.924
9	0	−0.354	−0.653	−0.854	−0.924	−0.854

(Eight time steps is one-half the period)

6. Part (c): For both, the full period is 16 time steps = 24 sec. Displacements at the center point:

t, sec:	0	3	6	9	12	15	18
Eq. (9.13):	0	0.688	1.281	1.688	1.828	1.688	1.281
Eq. (9.4):	0	0.750	1.406	1.875	2.063	1.875	1.406

12. The characteristics are straight lines. Intersections are at (0.472, 0.0447), (0.572, 0.0447), (0.545, 0.0894). At these points, $u = 0.0109, 0.0109, 0.0215$.

13.

Initial points	Intersection	Value of u
(0.2, 0) (0.4, 0)	(0.3007, 0.1001)	0.6015
(0.4, 0) (0.6, 0)	(0.5011, 0.1001)	1.0023
(0.6, 0) (0.8, 0)	(0.7016, 0.1001)	1.4023
(0.2, 0) (0.6, 0)	(0.4038, 0.2075)	0.8075
(0.4, 0) (0.8, 0)	(0.5115, 0.3018)	1.2229

The solution agrees with the analytical solution, which is $u(x, t) = 2x$.

17. Time steps are 0.00544 sec. There appears to be no repetitive pattern. Some displacements at (2, 1):

Time steps:	0	1	2	4	6
u:	0.5	0.375	0.125	-0.266	-0.711
Steps:	8	10	14	18	
u:	0.341	0.793	-0.936	0.695	

22. No. If the initial velocity is handled properly, the result will match the analytical regardless of the nodal spacing.

24. The table is identical because the error is snuffed out at the border when that value is specified.

CHAPTER 10 **4.** Zeros at 0, ±0.58778525, ±0.95105651

8. The ninth-degree Maclaurin series is very accurate near $x = 0$, but the error increases rapidly to 0.04952 at $x = \pm1$. The error of P_3 has four maxima (counting the endpoints) and has a maximum error of 0.0349 at $x = \pm1$.

11. $y = (2x - b - a)/(b - a)$

12. a. Cannot find $P_{3,3}$.
$$P_{4,2} = \frac{15x^2 - 3x^4}{15 + 2x^2}, \text{ maximum error} = 0.00219.$$
b. Cannot get Padé for $N = 6$.
c. $R_{33} = \dfrac{1 + x/2 + x^2/10 + x^3/120}{1 - x/2 + x^2/10 - x^3/120}$, maximum error = 2.81E$-$5

15. For part (c):
$$-1 + \cfrac{24}{-x + 12 - \cfrac{50}{x + 10/x}}$$

22. $\dfrac{1}{3} + \dfrac{4}{\pi^2} \displaystyle\sum_{n=1}^{\infty} \dfrac{\cos(n\pi x)}{n^2} - \dfrac{4}{\pi} \displaystyle\sum_{n=1}^{\infty} \dfrac{\sin(n\pi x)}{n}$

27. a. $f(x) = A_n\cos(n\pi x/2)$, $n = 0, 1, 2, \ldots$

n:	0	1	2	3	4
A_n:	0.9760	-0.250735	-0.306422	-0.0944593	-0.0629498

b. $f(x) = B_n\sin(n\pi x/2)$, $n = 1, 2, 3, \ldots$

n:	1	2	3	4
B_n:	0.199033	-0.159817	-0.173484	-0.133881

30. Values of A's:

n:	0	1	2	3	4
Trapezoidal rule:	2.0050	1.2259	0.3142	0.1456	0.0868
Simpson's:	1.8833	1.2157	0.3031	0.1330	0.07155
Analytical:	2.0000	1.2158	0.3040	0.1352	0.07599

Values of B's:

n:	0	1	2	3
Trapezoidal rule:	−2.1385	−1.1827	−0.7708	−0.5446
Simpson's:	−2.1595	−1.2259	−0.8384	−0.6409
Analytical:	−2.1595	−1.2249	−0.8345	−0.6306

35. **After stage**

1	0	1	0	0	0	0	0	0	0	0	0	0	0	0	0	0
2	0	2	1	3	0	0	0	0	0	0	0	0	0	0	0	0
3	0	4	2	6	1	5	3	7	0	0	0	0	0	0	0	0
4	0	8	4	12	2	10	6	14	1	9	5	13	3	11	7	15

42. In principle, only 12 samples, but with the FFT as written, 16 are required.

44. Using eight samples gives $2\cos(t) - \sin(3t)$, which matches $f(x)$ at eight evenly spaced points but does not represent $f(x)$.

46. $L(x, n) := \text{IF}(n = 0, 1, \text{IF}(n = 1, x, (2n - 1)/n\,xL(x, n - 1) - (n - 1)/n\,L(x, n - 2)))$

Index